Environmental and Pollution Science

Environmental and Pollution Science

THIRD EDITION

Mark L. Brusseau, Ph.D.
Professor of Subsurface Hydrology and
Environmental Chemistry, The University of
Arizona, Tucson, AZ, United States

Ian L. Pepper, Ph.D.
Professor of Environmental Microbiology, The
University of Arizona, Tucson, AZ, United States

Charles P. Gerba, Ph.D.
Professor of Environmental Microbiology, The
University of Arizona, Tucson, AZ, United States

ACADEMIC PRESS

An imprint of Elsevier

Academic Press is an imprint of Elsevier
125 London Wall, London EC2Y 5AS, United Kingdom
525 B Street, Suite 1650, San Diego, CA 92101, United States
50 Hampshire Street, 5th Floor, Cambridge, MA 02139, United States
The Boulevard, Langford Lane, Kidlington, Oxford OX5 1GB, United Kingdom

Notices
Knowledge and best practice in this field are constantly changing. As new research and experience broaden our understanding, changes in research methods, professional practices, or medical treatment may become necessary.

Practitioners and researchers must always rely on their own experience and knowledge in evaluating and using any information, methods, compounds, or experiments described herein. In using such information or methods they should be mindful of their own safety and the safety of others, including parties for whom they have a professional responsibility.

To the fullest extent of the law, neither the Publisher nor the authors, contributors, or editors, assume any liability for any injury and/or damage to persons or property as a matter of products liability, negligence or otherwise, or from any use or operation of any methods, products, instructions, or ideas contained in the material herein.

Library of Congress Cataloging-in-Publication Data
A catalog record for this book is available from the Library of Congress

British Library Cataloguing-in-Publication Data
A catalogue record for this book is available from the British Library

ISBN 978-0-12-814719-1

For information on all Academic Press publications
visit our website at https://www.elsevier.com/

Working together
to grow libraries in
developing countries

www.elsevier.com • www.bookaid.org

Publisher: Candice Janco
Acquisition Editor: Laura S. Kelleher
Editorial Project Manager: Lindsay Lawrence
Production Project Manager: Omer Mukthar
Cover Designer: Mark Rogers

Typeset by SPi Global, India

Printed in the United States of America
Last digit is the print number: 9 8 7 6 5 4 3 2 1

Dedication

"I dedicate this to my Dad, Howard Brusseau"

Mark L. Brusseau

"My first viable memory is crawling in the dirt as an infant, in between rows of vegetables in my father's garden. Since then I have always been involved with earth, so I dedicate this book to soil."

Ian L. Pepper

"To Peggy, Peter, and Phillip for putting up with me all these years."

Charles P. Gerba

Contents

Part I
Fundamental Concepts
1. The Extent of Global Pollution

M.L. Brusseau, I.L. Pepper and C.P. Gerba

2. Physical-Chemical Characteristics of Soils and the Subsurface

I.L Pepper and M.L. Brusseau

3. Physical-Chemical Characteristics of Water

M.L. Brusseau, D.B. Walker and K. Fitzsimmons

9. Biological Processes Affecting Contaminants Transport and Fate

R.M. Maier

Part II
Environmental Pollution
10. The Role of Environmental Monitoring in Pollution Science

J.F. Artiola and M.L. Brusseau

11. Physical Contaminants

J. Walworth and I.L. Pepper

12. Chemical Contaminants

M.L. Brusseau and J.F. Artiola

13. Microbial Contaminants

C.P. Gerba and I.L. Pepper

14. Soil and Land Pollution

*J.F. Artiola, J.L. Walworth, S.A. Musil
and M.A. Crimmins*

15. Subsurface Pollution

M.L. Brusseau

16. Surface Water Pollution

*D.B. Walker, D.J. Baumgartner, C.P. Gerba
and K. Fitzsimmons*

17. Atmospheric Pollution

M.L. Brusseau, A.D. Matthias, A.C. Comrie and S.A. Musil

18. Urban and Household Pollution

J.F. Artiola, K.A. Reynolds and M.L. Brusseau

Part III
Remediation, Restoration, Treatment, and Reuse
19. Soil and Groundwater Remediation

M.L. Brusseau

26. Environmental Impacts on Human Health and Well-Being

M.L. Brusseau, M. Ramirez-Andreotta, I.L. Pepper and J. Maximillian

27. Medical Geology and the Soil Health-Human Health Nexus

M.L. Brusseau and I.L. Pepper

28. Environmental Toxicology

C.P. Gerba

Contributing Authors

All currently or formerly with The University of Arizona, Tucson, AZ, United States

Janick Artiola, Professor of Soil, Water and Environmental Science

Donald J. Baumgartner, Former Professor of Soil, Water and Environmental Science

Hinrich Bohn, Former Professor of Soil, Water and Environmental Science

Jon Chorover, Professor of Soil, Water and Environmental Science

Andrew Comrie, Professor, School of Geography & Development

Michael Crimmins, Professor of Soil, Water and Environmental Science

Kevin Fitzsimmons, Professor of Soil, Water and Environmental Science

Edward Glenn, Former Professor of Soil, Water and Environmental Science

Raina Maier, Professor of Environmental Microbiology

Allan Matthias, Former Professor of Soil, Water and Environmental Science

Jacqueline Maximillian, Assitant Professor of Practice of Soil, Water and Environmental Science

Sheri Musil, Systems Administrator in Soil, Water and Environmental Science

Monica Ramirez-Andreotta, Assitant Professor of Soil, Water and Environmental Science

Kelly Reynolds, Professor of Environmental Health Sciences

David Walker, Research Scientist in Soil, Water and Environmental Science

James Walworth, Professor of Soil, Water and Environmental Science

Author Bio

Dr Mark Brusseau's work is focused on developing a fundamental understanding of the physical, chemical, and biological factors and processes influencing the transport and fate of contaminants in the subsurface environment. At the university he also teaches courses on Contaminant Transport in Porous Media as well as Soil and Groundwater Remediation. He has published over 250 works and is a fellow of the American Geophysical Union and the Soil Science Society of America.

Dr. Ian Pepper is currently Professor at the University of Arizona. He is also Director of the University of Arizona, Environmental Research Laboratory (ERL) and the NSF Water and Environmental Technology (WET) Center. Dr. Pepper is an environmental microbiologist specializing in the molecular ecology of the environment. His research has focused on the fate and transport of pathogens in air, water, soils, and wastes. His expertise has been recognized by membership on six National Academy of Science Committees and former memberships on an EPA FIFRA Science and Advisory Panel. Dr. Pepper is a Fellow of the American Association for the Advancement of Science, American Academy of Microbiology, the Soil Science Society of America, and the American Society of Agronomy. He is also a Board Certified Environmental Scientist within the American Academy of Environmental Engineers and Scientists. He is the author or coauthor of six textbooks, 40 book chapters, and over 180 peer-review journal articles.

Dr. Charles P. Gerba is a Professor at the University of Arizona. He conducts research on the transmission of pathogens through the environment. His recent research encompasses the transmission of pathogens by water, food, and fomites; fate of pathogens in land applied wastes; development of new disinfectants; domestic microbiology and microbial risk assessment. He has been an author on more than 500 articles including several books in environmental microbiology and pollution science. He is a fellow of the American Academy of Microbiology and the American Association for the Advancement of Science. In 1998 he received the A. P. Black Award from the American Water Works Association for outstanding contributions to water science and in 1996 he received the McKee medal from the Water Environment Federation for outstanding contributions to groundwater protection. He received the 1999 Award of Excellence in Environmental Health from National Association of County and City Health Officials.

Preface

This textbook focuses on (i) the continuum of the environment, namely, the lithosphere, hydrosphere, atmosphere, biosphere, and technosphere epitomized by the term "Environmental Science"; (ii) science-based aspects concerning pollution of and disturbances to the environment; and (iii) the human dimensions to pollution. The textbook is designed to provide the scientific knowledge necessary to understand complex pollution issues; to support sound decision-making; and to develop effective, sustainable solutions for managing, mitigating, and preventing environmental pollution.

In general, pollution can be defined as the accumulation and adverse interaction of contaminants within the environment. Pollution is ubiquitous and can occur on or within land, oceans, or the atmosphere. Contaminants can consist of chemical compounds/elements, biological entities, particulate matter, or energy, and they may be of natural or anthropogenic origin. In addition to pollution, environmental systems can be disturbed by other means, such as persistent drought or habitat destruction. Pollution and other disturbances of the environment can have numerous negative consequences for human health and well-being. Given the complexities of environmental systems and of the interactions between humans and the environment, multidisciplinary approaches are needed to address environmental pollution issues. This text provides a rigorous science-based integration of the physical, chemical, and biological properties and processes that influence the environment and that also affect the transport and fate of contaminants in the environment. It also incorporates the human dimensions, which are critical to a full understanding of pollution issues and the development of successful solutions.

Part I of the text covers the fundamental concepts of environmental and pollution science. Within that section, the properties of environmental systems are presented as well as the various processes that affect the transport and fate of pollutants within the environment. Part II covers the various types and sources of pollution. Part III is focused on the treatment and reuse of wastes, and the reclamation, remediation, and restoration of polluted and otherwise disturbed environments. Part IV highlights the human dimensions of pollution.

This text is designed for a science-based junior/senior-level undergraduate course or introductory graduate course.

Students that will benefit from this text originate from a variety of science backgrounds, including environmental science, microbiology, hydrology, earth science, geography, social and political sciences, and environmental engineering. The text will also serve as an introductory text for those with nonscience backgrounds.

This third edition represents a continued evolution of "Pollution Science," which was published in 1996, and the second edition of "Environmental and Pollution Science," published in 2006. New and significantly revised material introduced in this third edition includes the following. Chapter 6, Ecosystems and Ecosystem Services, is a new chapter that presents the principles and concepts of ecology and the services that ecosystems provide. Chapter 20, Ecosystem Restoration and Land Reclamation, is greatly revised and expanded to cover the primary methods used for reclamation and restoration of terrestrial and aquatic systems. Chapter 25, Pollution and Environmental Perturbations in the Global System, is updated and expanded to focus attention on the interconnectedness of human activities and global-scale effects. Chapter 26, Environmental Human Health, is a new chapter focused on the impacts of pollution and other disturbances to human health and well-being. Chapter 27, Medical Geology and the Soil Health:Human Health Nexus, is a new chapter that illustrates how the natural environment can impact human health in both positive and negative ways. Chapter 31, Environmental Justice, is a new chapter highlighting how pollution and its many impacts are distributed unevenly among the human population, with the most vulnerable generally being impacted the greatest. Finally, Chapter 32, Sustainable Development and Other Solutions to Pollution and Global Change, is a new chapter focused on the various methods we have available to reduce, mitigate, and manage the impacts of human activities on the environment.

Ultimately, we hope this text will illustrate that the integration of physical, chemical, earth, and biological sciences with engineering, health, and social sciences can be successfully accomplished, thereby inspiring readers to implement inter-, multi-, and transdisciplinary solutions for past, present, and future environmental pollution issues.

Mark L. Brusseau,
Ian L. Pepper and Charles P. Gerba

Part I

Fundamental Concepts

Chapter 1

The Extent of Global Pollution

M.L. Brusseau, I.L. Pepper and C.P. Gerba

Pollution is ubiquitous, and can even cause beautiful sunsets. *Photo courtesy Ian Pepper.*

1.1 THE SCIENCE OF POLLUTION

Pollution is ubiquitous and takes many forms and shapes. For example, the beautiful sunsets that we may see in the evening are often due to the interaction of light and atmospheric contaminants, as illustrated in the photo above.

Pollution can be defined as the accumulation and adverse effects of contaminants or other constituents on human health and welfare, and/or the environment. But in order to truly understand pollution, we must define the identity and nature of potential contaminants. Contaminants can result from waste materials produced from the activity of living organisms, especially humans. However, contamination can also occur from natural processes such as arsenic dissolution from bedrock into groundwater or air pollution from smoke that results from natural fires.

The major categories of pollutants and their predominant routes of human exposure are illustrated in Fig. 1.1. Clearly, many of the agents identified in Fig. 1.1 are generated directly through human activities such as manufacturing, mining, or agriculture. However, pollution is also produced as an indirect result of human activity. For example, fossil fuel burning increases atmospheric carbon dioxide levels and increases global climate change. Global climate change, in turn, can create or exacerbate existing pollution.

Many contaminants are chemicals, either naturally occurring or synthetic (human made). Chemical contaminants can occur in the environment in the solid, liquid, or gaseous state. Some common chemical contaminants that find their way into the environment, with the potential to adversely affect human health and well-being, are presented in Table 1.1. Physical contaminants also exist, such as dust particles, heat, and noise. There are also biological contaminants, such as pathogenic microorganisms. Some examples of microbial pathogens and associated diseases are given in Table 1.2.

In this textbook, we will discuss the many different sources and types of pollution in a science-based context, hence the name: *Environmental and Pollution Science* (Information Box 1.1). Pollution in the environment, and the impacts on human health and environmental quality, is a very complex and challenging issue. The many diverse scientific disciplines needed to study pollution are shown in Fig. 1.2. It is the holistic integration of these diverse and complex entities that presents the means to understanding both Environmental and Pollution Science, and ultimately solving pollution issues.

Environmental and Pollution Science. https://doi.org/10.1016/B978-0-12-814719-1.00001-X

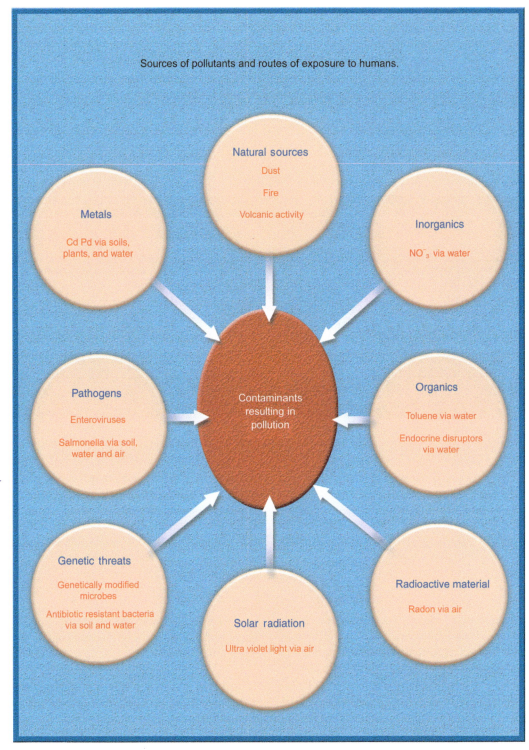

FIG. 1.1 Major sources of pollutants. *(From* Environmental and Pollution Science © 2006, Academic Press, San Diego, CA.)*

1.2 ENVIRONMENTAL INTERACTIONS

We know that pollution can have negative impacts on environmental quality, producing for example, contaminated air or water. But, the environment itself plays a key role in the fate of pollutants, and thus mediates their impacts on the environment (and ultimately on humans). The environment consists of land, water, the atmosphere, and the biosphere. All sources of pollution are initially released into one of these phases of the environment. The environment thus acts as a receptacle into which all pollutants are placed. As

TABLE 1.1 Common Organic and Inorganic Contaminants Found in the Environment

Chemical Class	Frequency of Occurrence
Gasoline, fuel oil	Very frequent
Polycyclic aromatic hydrocarbons	Common
Creosote	Infrequent
Alcohols, ketones, esters	Common
Ethers	Common
Chlorinated organics	Very frequent
Polybrominated diphenyl ethers (PBDEs)	
Polychlorinated biphenyls (PCBs)	Infrequent
Nitroaromatics (TNT)	Common
Metals (Cd, Cr, Cu, Hg, Ni, Pb, Zn)	Common
Nitrate	Common

From *Environmental Microbiology* © 2000, Academic Press, San Diego, CA.

TABLE 1.2 Microbes That Have had a Significant Impact on Human Health

Agent	Mode of Transmission	Disease/Symptoms
Rotavirus	Waterborne	Diarrhea
Legionella	Waterborne	Legionnaire's disease
Escherichia coli O157:H7	Foodborne Waterborne	Enterohemorrhagic fever, kidney failure
Hepatitis E virus	Waterborne	Hepatitis
Cryptosporidium	Waterborne	Diarrhea
	Foodborne	
Calicivirus	Waterborne	Diarrhea
	Foodborne	
Helicobacter pylori	Foodborne	Stomach ulcers
	Waterborne	
Cyclospora	Foodborne	Diarrhea
	Waterborne	

pollutants interact with the environment, they undergo physical and chemical changes, and are ultimately incorporated into the environment. The pollutants obey the conservation of mass law: matter cannot be destroyed; it is merely converted from one form to another. Thus, taken together, the manner in which pollutants are added to the environment, the rate at which these pollutants are added, and the subsequent changes that occur within the environment determine the ultimate impact of the pollutant on the environment, and concomitantly on humans and other organisms.

INFORMATION BOX 1.1

Environmental and Pollution Science is the study of the physical, chemical, and biological processes fundamental to the transport, fate, and mitigation of contaminants that arise from human activities as well as natural processes.

FIG. 1.2 The disciplines of pollution science. *(From* Environmental and Pollution Science © *2006, Academic Press, San Diego, CA.)*

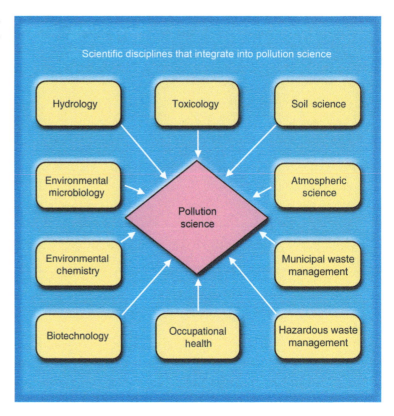

1.3 POLLUTION AND POPULATION PRESSURES

To understand the relationship between population and pollution, let us examine a typical curve for the growth of a pure culture of bacteria in a liquid medium (Fig. 1.3). Early on, the bacteria growing in the medium do not increase significantly in number, due to low initial population densities. Under these conditions, the organisms have minimal impact on each other or their environment. This initial low-growth phase is known as the *lag period*. Next, the number of organisms increases exponentially for a finite period of time. This phase of growth is known as the *exponential phase* or *log phase*. After this exponential phase of growth, a *stationary phase* occurs, during which the total number of organisms remains constant as new organisms are constantly being produced while other organisms are dying. Finally, we observe the *death phase*, in which the total number of organisms decreases. We know that bacteria reproduce by binary fission, so it is easy to see how a doubling of bacteria occurs during exponential growth. But what causes the stationary and death phases of growth? Two mechanisms prevent the number of organisms from increasing ad infinitum: first, the organisms begin to run out of nutrients; and second, waste products build up within the growth medium and become toxic to the organisms.

An analogous situation exists for humans. Initially, in prehistoric times, population densities were low and population numbers did not increase significantly or rapidly (Fig. 1.4.). During this time resources were plentiful; thus the environment could easily accommodate the numbers of humans present and the amount of wastes they produced. Later, populations began to increase very rapidly. Although not exponential, this phase of growth was comparable to the log phase of microbial growth. During this period then, large amounts of resources were used, and wastes were produced in ever-greater quantities. This period of growth is still under way. However, we seem to be approaching a

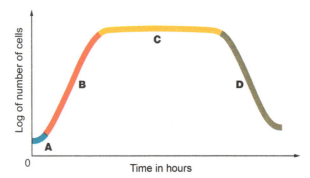

FIG. 1.3 Typical growth curve for a pure culture of bacteria. A = lag period, B = exponential phase, C = stationary phase, D = death phase. *(From* Pollution Science © *1996, Academic Press, San Diego, CA.)*

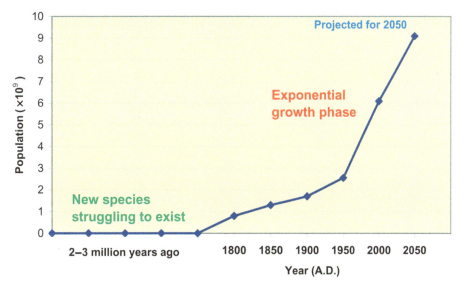

FIG. 1.4 World population increases from the inception of the human species. *(From Population Reference Bureau, Inc., 1990. Adapted from* Pollution Science © *1996, Academic Press, San Diego, CA.)*

period in which lack of resources or buildup of wastes (i.e., pollution) will limit continued growth—hence the great interest in recycling materials as well as in controlling, managing, and cleaning up waste materials—what is termed sustainable development. To do this, we must have a clear understanding of the biotic and abiotic characteristics of the environment and the hydrobiogeochemical processes occurring therein.

Currently, the world population is 7.6 billion and increasing rapidly. This population pressure has caused intense industrial, resource extraction, and agricultural activities that produce hazardous contaminants in their own right. In addition, increased populations result in the production of greater amounts of wastes. Finally, note that as the world population increases, people tend to relocate from sparsely populated rural areas to more congested urban centers or "mega-cities." Typically, urbanized areas consume more natural resources and produce more waste per capita than rural areas. The continued trend toward mega-cities will intensify this problem.

1.4 THE HUMAN DIMENSIONS TO POLLUTION

Poor management of waste materials, overuse of resources, and uncontrolled development all have negative impacts on the environment. These impacts reverberate throughout the environment, leading to impaired soil, water, and air quality; damaged habitats; and stress to wildlife populations. These impacts, in turn, have a myriad of direct and indirect effects on humans. The presence of pollutants may cause illness and even death for some members of an exposed population. This is a direct consequence of pollution. The presence of pollution in a particular environmental medium reduces the quality of that medium, which makes it less available and usable for humans, possibly leading to scarcity of that resource. This in turn may affect human well-being, which is an indirect consequence. Similarly, overuse of a particular resource can lead to scarcity, and hence indirect effects on human health and well-being.

For example, consider people living in a small town that uses groundwater as their sole potable water source. If that groundwater is unknowingly contaminated with a carcinogenic compound, some of the population may develop cancer after a long period of groundwater use. Once the contaminant is identified, its presence may prevent further use of that water source if it is too difficult or costly to treat the contamination. This contamination then reduces the availability of potable water, which may lead to water scarcity, depending upon the availability of alternate water sources. Let us say that the town switches to a local surface water reservoir for their potable water. If they use the water at a rate greater than it is replenished, they will eventually arrive at a condition where their only other source of potable water is depleted, possibly affecting the very viability of that town.

From the above we see that human activities that create pollution and resource-scarcity issues are posing direct and indirect threats to their own health and well-being. Thus it is in our own best interests to reduce our impacts on the environment. The impacts of humans on the environment, the resultant effects on humans of these impacts, and how humans respond to these issues constitute what we may call the *human dimensions to pollution.*

1.5 ORGANIZATION OF TEXTBOOK AND NEW MATERIAL

The focus of the text will be to identify the basic scientific processes that control the transport and fate of pollutants in the environment. We will also define the potential for adverse effects to human health and well-being. Finally, we will present and discuss various approaches for mitigating the effects of pollution and cleaning up contaminated environments.

Part I of the text covers the fundamental concepts of pollution science. Within that section, the properties of environmental systems are presented as well as the various processes that affect the transport and fate of pollutants within the environment. Part II covers the various types and sources of pollution. Part III is focused on the treatment and reuse of wastes, and the reclamation, remediation, and restoration of polluted and otherwise disturbed environments. Part IV highlights the human dimensions of pollution.

New and greatly revised material introduced in this third edition includes the following. Chapter 6, Ecosystems and Ecosystem Services, is a new chapter that presents the principles and concepts of ecology and the services that ecosystems provide. Chapter 20, Ecosystem Restoration and Land Reclamation, is greatly revised and expanded to cover the primary methods used for reclamation and restoration of terrestrial and aquatic systems. Chapter 25, Pollution and Environmental Perturbations in the Global System, is greatly revised and expanded to focus attention on the interconnectedness of human activities and global-scale effects. Chapter 26, Environmental Human Health, is a new chapter focused on the impacts of pollution and other disturbances to human health and well-being. Chapter 27, Medical Geology and the Soil Health:Human Health Nexus, is a new chapter that illustrates how the natural environment can impact human health in both positive and negative ways. Chapter 31, Environmental Justice, is a new chapter highlighting how pollution and its many impacts are distributed unevenly among the human population, with the most vulnerable generally being impacted the greatest. Finally, Chapter 32, Sustainable Development and Other Solutions to Pollution and Global Change, is a new chapter focused on the various methods we have available to reduce, mitigate, and manage the impacts of human activities on the environment.

We hope that the revised and expanded third edition enhances the contributions of the text to improved understanding of environmental systems and the development of innovative solutions to pollution issues in the 21st century.

Chapter 2

Physical-Chemical Characteristics of Soils and the Subsurface

I.L Pepper and M.L. Brusseau

Cover Art: Photograph of top soil. *Source: (https://en.wikipedia.org/wiki/Topsoil)*

2.1 SOIL AND SUBSURFACE ENVIRONMENTS

The human environment is located at the earth's surface and is heavily dependent on the soil/water/atmosphere continuum. Ultimately, this continuum moderates all of our activities, and the physical, chemical, and biological properties of each component are interactive. Soil is the thin veneer of material that covers much of the earth's surface. This fragile part of the earth's skin is frequently less than a meter thick, yet is absolutely vital for human life. It has a rich texture and fragrance and teems with plants, insects, and microorganisms. Young and Crawford (2004) described it as "the most complicated biomaterial on the planet." The complexity of soil is driven by two components: the abiotic soil architecture; and biotic diversity, which is driven and supported by large amounts of energy from the sun through photosynthesis. Bacterial diversity in soil ranges from 2000 to 8.3 million per gram of soil

depending on the measurement methods used (Roesch et al., 2007). This diversity results in bacterial populations of 10^8–10^{10} per gram of soil (Pepper et al., 2015). Integrated together these components result in amazing physical, chemical, and biological heterogeneity among soils globally. The geological zone between the land surface and groundwater consists of unsaturated material and is known as the *vadose zone*. A subset of the vadose zone is the near-surface *soil* environment, which is in direct contact with both surface water and the atmosphere. Since pollutants are often disposed of into surface soils, the transport of these contaminants into both the atmosphere and groundwater is influenced by the properties of soil and the vadose zone. In addition, since plants are grown in surface soils, the potential for uptake of contaminants such as heavy metals by plants is also controlled by soil properties.

Soil is an intricate, yet durable entity that directly and indirectly influences our quality of life (Pepper, 2013). Colloquially known as dirt, soil is taken for granted by most people, and yet it is essential to our daily existence. It is responsible for plant growth, for the cycling of all nutrients through microbial transformations, for maintaining the oxygen/carbon dioxide balance of the atmosphere; it is also the ultimate site of disposal for most waste products. Soil is a complex mixture of weathered rock particles, organic residues, water, and billions of living organisms. It can be as thin as six inches or it may be hundreds of feet thick. Generally, most soils are 2–3 ft in depth. Because soils are derived from unique sources of parent material under specific environmental conditions, no two soils are exactly alike. Hence, there are literally thousands of different kinds of soils just within the United States. These soils have different properties that influence the way soils are used optimally.

Soil is the weathered end-product of the action of climate and living organisms on soil parent material with a particular topography over time. We refer to these factors

Environmental and Pollution Science. https://doi.org/10.1016/B978-0-12-814719-1.00002-1

2.1.1   Vadose Zone

The *vadose zone* is defined as the subsurface unsaturated oligotrophic environment that lies between the surface soil and the saturated zone. The vadose zone contains mostly unweathered parent material and has a very low organic carbon content (generally <0.1%). Thus the availability of carbon and micronutrients is very limited compared with that of surface soils. The thickness of the vadose zone varies considerably. When the saturated zone is shallow or near the surface, the unsaturated zone is narrow or sometimes even nonexistent, as in a wetland area. In contrast, there are many arid or semiarid areas of the world where the unsaturated zone can be hundreds of meters thick. These unsaturated regions, especially deep unsaturated regions, may receive little or no moisture recharge from the surface, and normally have limited microbial activity because of low nutrient and/or moisture status. However, these regions are receiving more attention from a biogeochemical and site-remediation perspective because pollutants that are present from surface contamination must pass through the vadose zone before they can reach groundwater (Fig. 2.2).

There are several parameters of soil that vitally affect the transport and fate of environmental pollutants. We will now discuss these parameters while providing an overview of soil as a natural body as it affects pollution.

2.2   SOLID PHASE

2.2.1   Soil Profiles

The process of soil formation generates different horizontal layers, or *soil horizons*, that are characteristic of that particular soil. It is the number, nature, and extent of these horizons that give a particular soil its unique character. A typical soil profile is illustrated in Fig. 2.3. Generally, soils contain a dark organic-rich layer, designed as the O horizon, then a lighter colored layer, designated as the A horizon, where some humified organic matter accumulates. The layer that underlies the A horizon is called the E horizon because it is characterized by *eluviation*, which is the process of removal or transport of nutrients and inorganics out of the A horizon. Beneath the E horizon is the B horizon, which is characterized by *illuviation*. Illuviation is the deposition of the substances from the E horizon into the B horizon. Beneath the B horizon is the C horizon, which contains the parent material from which the soil was derived. The C horizon is generally unweathered or minimally weathered parent material. Although certain diagnostic horizons are common to most soils, not all soils contain each of these horizons.

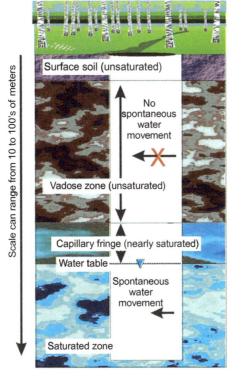

FIG. 2.1   Cross section of the subsurface showing surface soil, vadose zone, and saturated zone. *(Adapted from Environmental Microbiology, © 2000, Academic Press, San Diego, CA.)*

as the five *soil-forming factors* (Information Box 2.1). The biotic component consists of both microorganisms and plants. The *vadose zone* is the water-unsaturated and generally unweathered material between groundwater and the land surface. The major difference between a surface soil and a vadose zone is the fact that the vadose zone parent material has generally not been recently modified by climate. A model of a cross section of a typical subsurface environment is shown in Fig. 2.1.

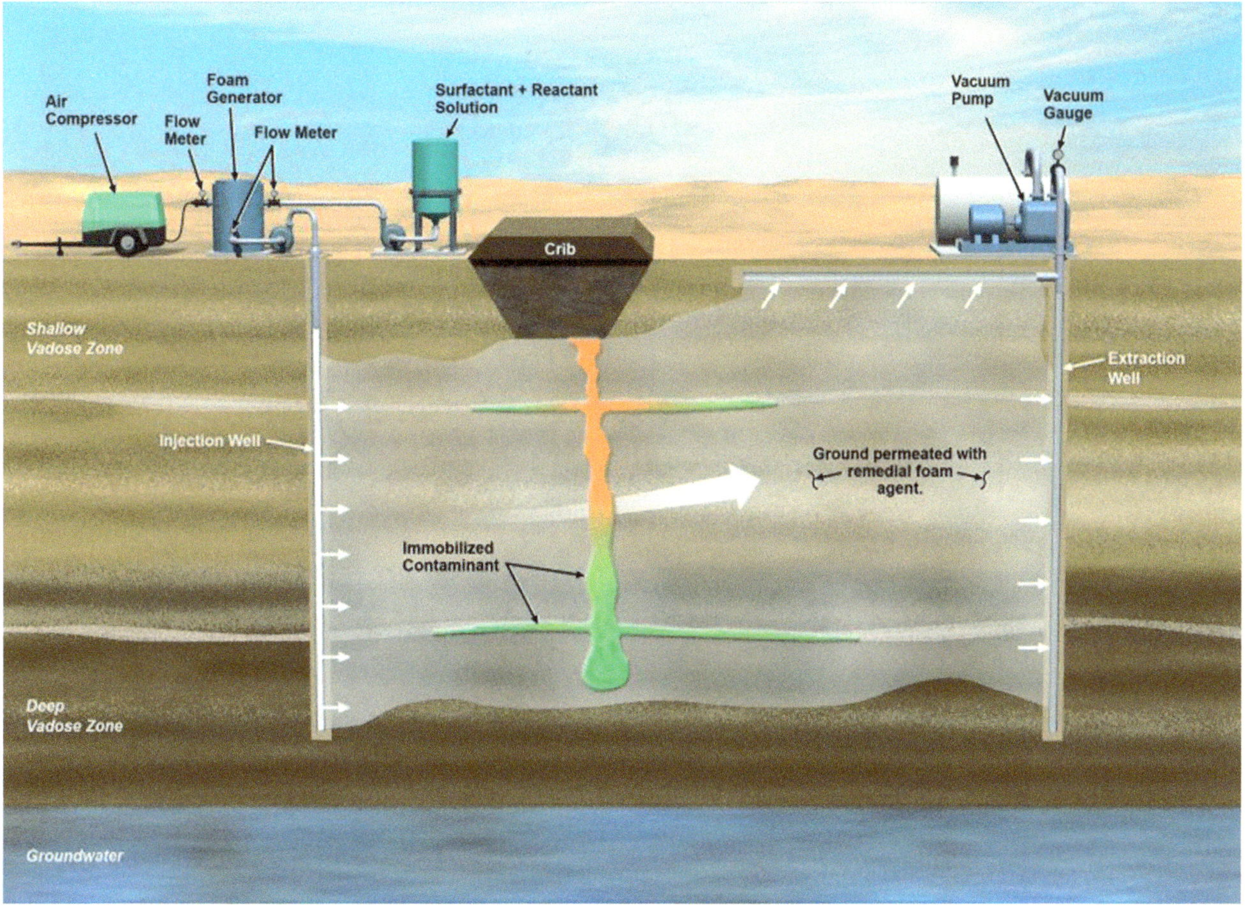

FIG. 2.2   Delivery of remedial solutions through a heterogeneous deep vadose zone to remove contaminants. *(Source: Pacific Northwest National Laboratory. From Environmental Microbiology,* © *2015, Academic Press, San Diego, CA.)*

2.2.2   Primary Particles and Soil Texture

Typically, a soil contains 45%–50% solids on a volume basis (Fig. 2.4). Of this solid fraction, 95% to >99.9% is the mineral fraction. Silicon (47%) and oxygen (27%) are the two most abundant elements found within the mineral fraction of the earth's crust. These two elements, along with lesser amounts of other elements, combine in a number of ways to form a large variety of minerals. For example, quartz is SiO_2 and mica is $K_2Al_2O_5[Si_2O_5]_3Al_4(OH)_4$. These are primary minerals that are derived from the weathering of parent rock. Weathering results in mineral particles that are classified on the basis of three different sizes: sand, silt, and clay (Information Box 2.2). The distribution (on a percent by weight basis) of sand, silt, and clay within a porous medium defines its *texture*. Soils predominated by sand are considered coarse textured while those with higher proportions of silt and clay are known as fine textured.

The differences in the size of the particles are due to the weathering of the parent rock. Table 2.1 presents the size fractionation of soil constituents including mineral,

organic, and biological constituents. This figure also illustrates the effect of size on specific surface area.

Soil texture affects many of the physical and chemical properties of the soil. Various mixtures of the three primary components result in different textural classes (Fig. 2.5). Of the three primary particles, clay is by far the dominant factor in determining a soil's properties. This is because there are more particles of clay per unit weight, than sand or silt, due to the smaller size of the clay particles. In addition, the clay particles are the primary soil particles that have an associated electric charge (see Chapter 8). The predominance of clay particles explains why any soil with >35% clay has the term *clay* in its textural class. In addition, because increases in soil clay concentrations result in increased surface area, this also increases the chemical reactivity of the soil (see Chapter 8).

2.2.3   Soil Structure

The three primary particles do not normally remain as individual entities. Rather, they aggregate to form secondary

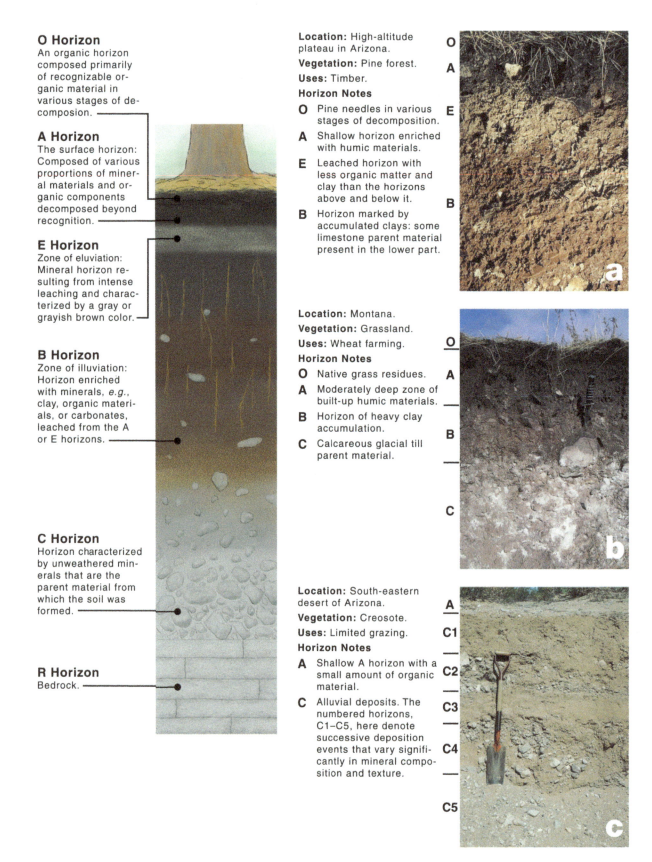

O Horizon
An organic horizon composed primarily of recognizable organic material in various stages of decomposion.

A Horizon
The surface horizon: Composed of various proportions of mineral materials and organic components decomposed beyond recognition.

E Horizon
Zone of eluviation: Mineral horizon resulting from intense leaching and characterized by a gray or grayish brown color.

B Horizon
Zone of illuviation: Horizon enriched with minerals, *e.g.*, clay, organic materials, or carbonates, leached from the A or E horizons.

C Horizon
Horizon characterized by unweathered minerals that are the parent material from which the soil was formed.

R Horizon
Bedrock.

Location: High-altitude plateau in Arizona.
Vegetation: Pine forest.
Uses: Timber.
Horizon Notes
O Pine needles in various stages of decomposition.
A Shallow horizon enriched with humic materials.
E Leached horizon with less organic matter and clay than the horizons above and below it.
B Horizon marked by accumulated clays: some limestone parent material present in the lower part.

Location: Montana.
Vegetation: Grassland.
Uses: Wheat farming.
Horizon Notes
O Native grass residues.
A Moderately deep zone of built-up humic materials.
B Horizon of heavy clay accumulation.
C Calcareous glacial till parent material.

Location: South-eastern desert of Arizona.
Vegetation: Creosote.
Uses: Limited grazing.
Horizon Notes
A Shallow A horizon with a small amount of organic material.
C Alluvial deposits. The numbered horizons, C1–C5, here denote successive deposition events that vary significantly in mineral composition and texture.

FIG. 2.3 Typical soil profiles illustrating different soil horizons. These horizons develop under the influence of the five soil-forming factors and result in unique soils. *(Source: Pepper et al. (2006). From* Environmental Microbiology, © 2015, Academic Press, San Diego, CA.)

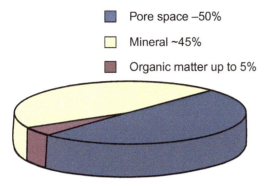

■ Pore space –50%

□ Mineral ~45%

■ Organic matter up to 5%

FIG. 2.4 Three basic components of a porous medium, such as a typical surface soil, on a volume basis. *(From* Environmental Microbiology, © *2015, Academic Press, San Diego, CA.)*

INFORMATION BOX 2.2 Primary mineral size classifications

Sand	0.05–2 mm
Silt	0.002–0.05 mm
Clay	<0.002 mm (2 μm)

structures, which occur because microbial gums, polysaccharides, and other microbial metabolites bind the primary particles together. This results in soil architecture (Fig. 2.6). In addition, particles can be held together physically by fungal hyphae and plant roots, or chemically by precipitates (e.g., cementation). These secondary aggregates, which are known as *peds* (Fig. 2.7), can be of different sizes and shapes, depending on the particular soil. Soils with even modest amounts of clay usually have well-defined peds, and hence a well-defined *soil structure*. These aggregates of primary particles usually remain intact as long as the soil is not disturbed, for example, by plowing. In contrast, sandy soils with low amounts of clay generally have less well-defined soil structure.

The phenomenon of soil structure has a profound influence on the physical properties of the soil. Because its particles are arranged in secondary aggregates, a certain volume of the soil includes voids that are filled with either air (the soil atmosphere) or soil water. Soils in which the structure has many voids within and between the peds offer favorable environments for soil organisms and plant roots, both of which require oxygen and water. Soils with no structure, that is, those consisting of individual primary particles are characterized as *massive*. Massive soils have very small void spaces and therefore little room for air or water.

TABLE 2.1 Size Fractionation of Soil Constituents

		Soil	
Specific Surface area Using a Cubic Model	Mineral Constituents	Size	Organic and Biologic Constituents
0.0003 m²/g	Sand Primary minerals: quartz, silicates, carbonates	2 mm	Organic debris
0.12 m²/g	Silt Primary minerals: quartz, silicates, carbonates	50 μm	Organic debris, large microorganisms Fungi Actinomycetes Bacterial colonies
3 m²/g	Granulometric clay Microcrystals of primary minerals Phyllosilicates Inherited; illite, mica Transformed: vermiculite, high-charge smectite Neoformed; kaolinite, smectite Oxides and hydroxides	2 μm	Amorphous organic matter Humic substances Biopolymers
30 m²/g	Fine clay Swelling clay minerals Interstratified clay minerals Low range order crystalline compounds	0.2 μm	Small viruses

From *Environmental Microbiology*, © 2000, Academic Press, San Diego, CA.

FIG. 2.5 A soil textural triangle showing different textural classes in the USDA system. These textural classes characterize soil with respect to many of their physical properties. *(From* Pollution Science, *© 1996, Academic Press, San Diego, CA.)*

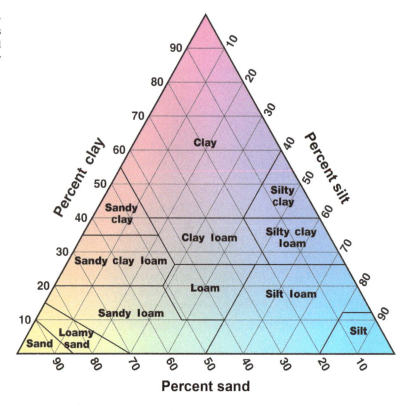

FIG. 2.6 Soil architecture resulting from secondary aggregate formation with intraaggregate and interaggregate pore space. *(Source: Pepper (2014). From* Environmental Microbiology *© 2015, Academic Press, San Diego, CA.)*

FIG. 2.7 Soil structure results from secondary aggregates known as peds. *(From* Environmental Microbiology *© 2015, Academic Press, San Diego, CA.)*

Void spaces are known as *pore space* which is made up of individual pores. These pores allow movement of air, water, and microorganisms through the soil. Pores that exist between aggregates are called *interaggregate pores*, whereas those within the aggregates are termed *intraaggregate pores* (Fig. 2.8). Although the average pore size is smaller in a clay soil, there are many more pores than in a sandy soil, and as a result, the total amount of pore space is larger in a fine-textured (clay) soil than in a coarse-textured (sandy) soil (Fig. 2.9). However, because small pores do not transmit water as fast as larger pores, a fine-textured soil will slow the movement of any material moving through it, including air, water, and microorganisms (see Chapter 3). Sometimes fine-textured layers of clay known as *clay lenses* can be found within volumes of coarser materials resulting in heterogeneous environments. In this case, water will move through the coarser material and flow around the clay lens. This has implications of remediative strategies such as pump and treat,

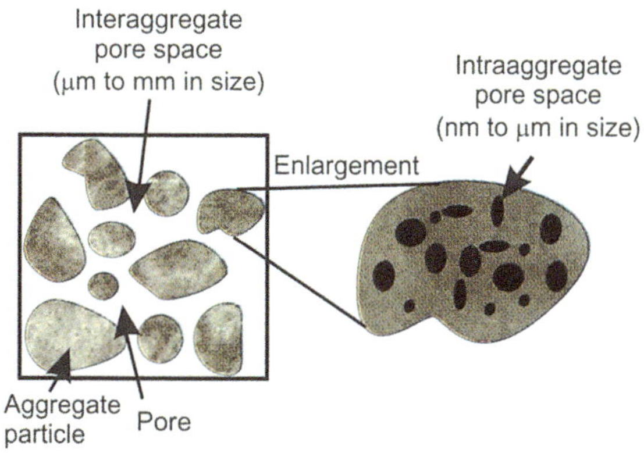

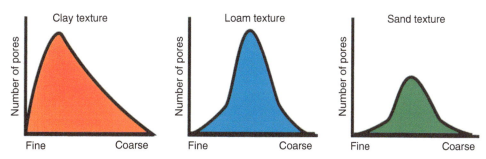

FIG. 2.8 Pore space. In surface soils, mineral particles are tightly packed together and even cemented in some cases with microbial polymers forming soil aggregates. The pore spaces between individual aggregates are called interaggregate pores and vary in size from micrometers to millimeters. Aggregates also contain pores that are smaller in size, ranging from nanometers to micrometers. These are called intraaggregate pores. *(From* Environmental Microbiology, *© 2015, Academic Press, San Diego, CA.)*

oxides. Substitution of a divalent magnesium cation (Mg^{2+}) for a trivalent aluminum cation (Al^{3+}) can result in the loss of one positive charge, which is equivalent to a gain of one negative charge. Other substitutions can also lead to increases in negative charge.

(II) *Ionization*: Hydroxyl groups (OH) at the edge of the lattice can ionize, resulting in the formation of negative charge:

$$Al + OH = Al + O^- + H^+$$

These are also known as *broken-edge bonds*. Ionizations such as these usually increase as the pH increases and are therefore known as *pH-dependent charge*. The functional groups of organic matter, such as carboxyl moieties, are also subject to ionization and can contribute to the total pH-dependent charge. The total amount of negative charge is usually measured in terms of equivalents of negative

FIG. 2.9 Typical pore size distribution for clay-, loam-, and sand-textured horizons. Note that the clay-textured material has the smallest average pore size, but the greatest total volume of pore space. *(From* Environmental Microbiology, *© 2015, Academic Press, San Diego, CA.)*

which is used to remove contaminants from the saturated zone. Pore space may also be increased by plant roots, worms, and small mammals, whose root channels, worm holes, and burrows create macro openings. These larger openings can result in significant aeration of surface and subsurface soils and sediments, as well as *preferential flow* of water through the soil.

Together texture and structure are important factors that control the movement of water, contaminants, nutrients, and microbes through soils, and hence affect contaminant transport and fate.

2.2.4 Cation-Exchange Capacity

The parameter known as cation-exchange capacity (CEC) arises because of the charge associated with clay particles and other soil components. Normally, this is a negative charge that occurs for one of two reasons:

(I) *Isomorphic substitution*: Clay particles exist as inorganic lattices composed of silicon and aluminum

charge per 100 g of soil and is a measure of the potential CEC of the soil. A milliequivalent (meq) is one-thousandth of an equivalent weight. Equivalents of chemicals are related to hydrogen, which has a defined equivalent weight of 1. The equivalent weight of a chemical is the atomic weight divided by its valence. For example, the equivalent weight of calcium is 40/2 = 20 g. A CEC of 15–20 meq per 100 g of soil is considered to be average, whereas a CEC > 30 is considered high. Note that it is the clays and organic particles that are negatively charged. Due to their small particle size, they are collectively called the *soil colloids*. The existence of CEC allows the phenomenon of cation exchange to occur (see also Chapter 8).

How does the process of cation exchange work? Common soil cations such as Ca^{2+}, Mg^{2+}, K^+, Na^+, and H^+ that exist in the soil solution are in equilibrium with cations on exchange sites. If the concentration of a cation in the soil solution is changed, for example, increased, then that cation is likely to occupy more exchange sites, replacing existing cations within the site (Fig. 2.10). Thus

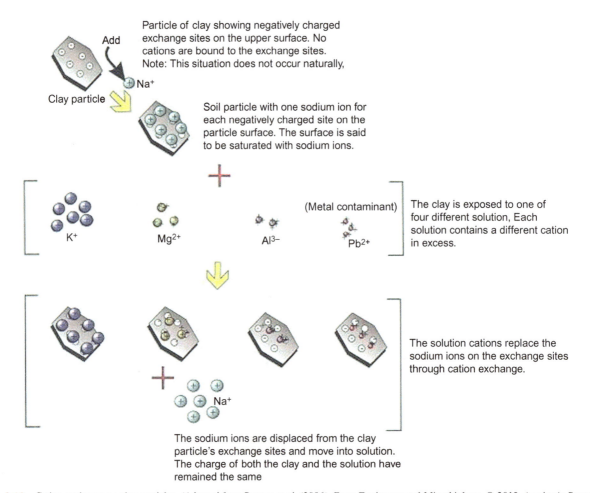

FIG. 2.10   Cation exchange on clay particles. *(Adapted from Pepper et al. (2006). From* Environmental Microbiology, © *2015, Academic Press, San Diego, CA.)*

a monovalent cation such as K^+ can replace another monovalent cation such as Na^+, or two K^+ can replace one Mg^{2+}. Note however, that when working with charge equivalents, one milliequivalent of K^+ replaces one milliequivalent of Mg^{2+}. Cation exchange ultimately depends on the concentration of the cation in soil solution and the *adsorption affinity* of the cation for the exchange site. The adsorption affinity of a cation is a function of its charge density, which in turn depends on its total charge and the size of the hydrated cation. The adsorption affinities of several common cations are given in the following series in decreasing order:

$$Al^{3+} > Ca^{2+} = Mg^{2+} > K^+ = NH_4^+ > Na^+$$

Highly charged small cations such as Al^{3+} have high adsorption affinities. In contrast, monovalent ions have lower affinities, particularly if they are highly hydrated such as Na^+, which increases the effective size of the cation. The extensive surface area and charge of soil colloids (clays +organic material) are critical to microbial activity since they affect both binding or sorption of solutes and microbial attachment to the colloids.

Sorption is a major process influencing the movement and bioavailability of essential compounds and pollutants in soil. The broadest definition of *sorption* is the association of organic or inorganic molecules with the solid phase of the soil. For inorganic charged molecules, cation exchange is one of the primary mechanisms of sorption (Fig. 2.10). Generally, positively charged ions, for example, calcium (Ca^{2+}) or lead (Pb^{2+}), participate in cation exchange. Since sorbed forms of these metals are in equilibrium with the soil solution, they can serve as a long-term source of essential nutrients (Ca^{2+}) or pollutants (Pb^{2+}) that are slowly released back into the soil solution as the soil solution concentration of the cation decreases with time.

Attachment of microorganisms can also be mediated by cation bridging (Fig. 2.11) or by the numerous functional groups on clays (Fig. 2.11). Although the clay surface and microbial cell surface both have net negative charges, clay surfaces are neutralized by the accumulation of

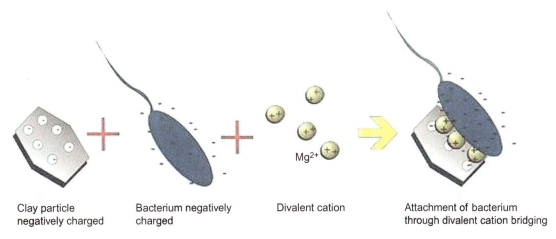

FIG. 2.11 Attachment of a bacterial cell to a clay particle via cation bridging. *(From* Environmental Microbiology, *© 2015, Academic Press, San Diego, CA.)*

positively charged counterions such as K^+, Na^+, Ca^{2+}, Mg^{2+}, Fe^{3+}, and Al^{3+}. Together, these negative and positive surface charges form what is called the *electrical double layer*. Similarly, microbes have an electrical double layer. The thickness of the clay double layer depends on the valence and concentration of the cations in solution. Higher valence and increased cation concentrations will shrink the electrical double layer. Because the double layers of the clay particles and microbial cells repel each other, the thinner these layers are, the less the repulsion between the clay and cell surfaces. As these repulsive forces are minimized, attractive forces such as electrostatic and van der Waals forces allow the attachment of microbial cells to the surface (Gammack et al., 1992). As a result, most microbes in terrestrial environments exist attached to soil colloids, rather than existing freely in the soil solution.

2.2.5 Soil pH

In areas with high rainfall, basic cations tend to leach out of the soil profile; moreover, soils developed in these areas have higher concentrations of organic matter, which contain acidic components and residues. Thus such soils tend to have decreased pH values and are acidic in nature. Soils in arid areas do not undergo such leaching, and the concentrations of organic matter are lower. In addition, water tends to evaporate in such areas, allowing salts to accumulate. These soils are therefore alkaline, with higher pH values.

Soil pH affects the solubility of chemicals in soils by influencing the degree of ionization of compounds and their subsequent overall charge. The extent of ionization is a function of the pH of the environment and the dissociation constant (pK) of the compound. Thus soil pH may be critical in affecting transport of potential pollutants through the soil and vadose zone.

We define pH as the negative logarithm of the hydrogen ion concentration:

$$pH = -\log[H^+]$$

Usually, water ionizes to H^+ and OH^-:

$$HOH = H^+ + OH^-$$

The *dissociation constant* (K_{eq}) is defined as

$$K_{eq} = \frac{[H^+][OH]}{[HOH]} = 10^{-14} \text{ mol L}^{-1}$$

Since the concentration of HOH is large relative to that of H^+ or OH^-, it is normally given the value of 1; therefore

$$[H][OH^-] = 10^{-14} \text{ mol L}^{-1}$$

For a neutral solution

$$[H^+] = [OH^-] = 1 \times 10^{-7}$$

and

$$pH = -\log[H^+] = -(-7) = 7$$

A pH value of <7 indicates acidity whereas a pH value >7 indicates alkalinity (or basicity) (Table 2.2).

2.2.6 Organic Matter

Organic compounds are incorporated into soil at the surface via plant residues, such as leaves or grassy material, and microorganisms. These organic residues are degraded microbially by soil microorganisms, which utilize the organics as food or microbial substrate. The main plant constituents, presented in Table 2.3, vary in degree of complexity and ease of breakdown by microbes. In general, soluble constituents are easily metabolized and break down rapidly, whereas lignin, for example, is very resistant to

TABLE 2.2 Soil pH Regimes

Soil	pH Regime
Acidic	>5.5
Neutral	6–8
Alkaline	>8.5

From *Pollution Science*, 8, 1996, Academic Press, San Diego, CA.

TABLE 2.3 Major Constituents of Plant Residues

Constituent	% Dry Weight
Cellulose	15–60
Hemicellulose	10–30
Lignin	5–30
Protein and nucleic acids	2–15
Soluble substances, for example, sugars	10

From *Pollution Science*, 8 1996, Academic Press, San Diego, CA.

microbial decomposition. The net result of microbial decomposition is the release of nutrients for microbial or plant metabolism, as well as the particle breakdown of complex plant residues. These microbial modified complex residues are ultimately incorporated into large macromolecules that form the stable basis of soil organic matter. This stable organic matrix is slowly metabolized by indigenous soil organisms, a process that results in about 2% breakdown of the complex materials annually. Owing to the slow but constant decomposition of the organic matrix and annual fresh additions of plant residues, an equilibrium is achieved in which the overall amount of soil organic matter remains constant. In humid areas with high rainfall, soil organic matter contents can be as high as 5% on a dry-weight basis. In arid areas with high rates of decomposition and low inputs of plant residues, values are usually <1%. The formation of soil organic matter is illustrated in Fig. 2.12, and terms used to define soil organic matter are given in Table 2.4.

The release of nutrients that occurs as plant residues degrade has several effects on soil. The enhanced microbial activity causes an increase in soil structure, which affects most of the physical properties of soil, such as aeration and infiltration. The stable humic substances contain many moieties that contribute to the pH-dependent CEC of the soil. In addition, many of the humic and nonhumic substances can complex or chelate heavy metals, and sorb organic contaminants. This retention affects their availability to plants and soil microbes as well as their potential for transport into the subsurface (see Chapter 6).

2.2.7 Vadose Zone Solid Phase

The *vadose zone* is defined as the unsaturated environment that lies between the surface soil and the saturated zone (groundwater). Physically, the vadose zone parent material may be very similar to that of the surface soil above it, except that it is less weathered and has very low organic matter content. In terms of texture, vadose zones normally contain larger rocks and cobbles than surface soils, but still have high amounts of sand, silt, and clay. The low organic content is due to the fact that organic material added to surface soils as vegetative leaf litter is usually degraded within the surface soil. Therefore the organic carbon content of vadose zones is usually very low. This leads to *oligotrophic* (low nutrient) conditions. Hence, microbial activity in vadose zones is normally lower than in surface soils (see also Chapter 5). This may affect the fate of subsurface organic contaminants since there may be decreased rates of biodegradation (see also Chapter 8).

2.3 GASEOUS PHASE

2.3.1 Constituents of Soil Atmosphere

Soil and the atmosphere are in direct contact; therefore most of the gases found in the atmosphere are also found in the air

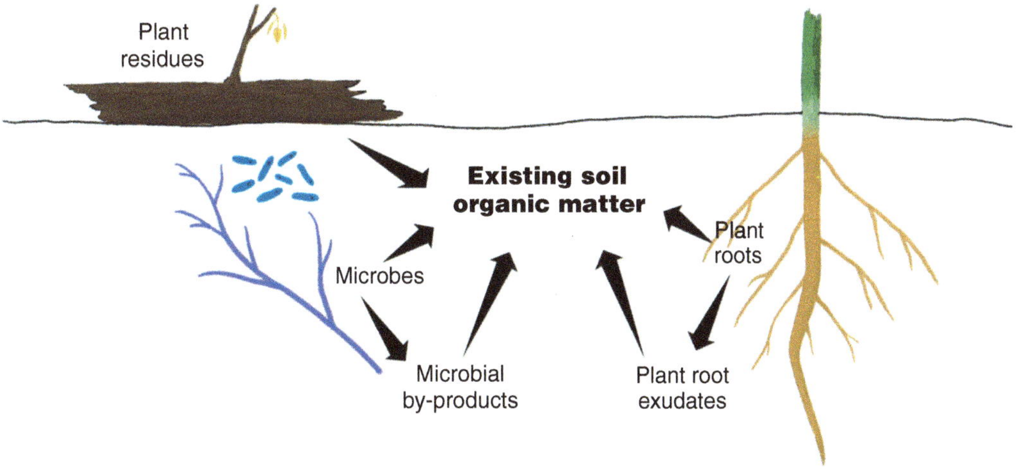

FIG. 2.12   Schematic representation of the formation of soil organic matter. *(From* Pollution Science, © *1996, Academic Press, San Diego, CA.)*

TABLE 2.4 Terms Used to Define Soil Organic Matter

Term	Definition
Organic residues	Undecayed plant and microbial biomass and their partial decomposition products
Soil biomass	Live microbial biomass
Soil organic matter or humus	All soil organic matter, except organic residues and soil biomass
Humic substances	High-molecular-weight complex stable macromolecules with no distinct physical or chemical properties. These substances are never exactly the same in any two soils because of variable inputs and environments. This is the stable backbone of soil organic matter and is degraded only slowly (2% per year)
Nonhumic substances	Known chemical materials such as amino acids, organic acids, carbohydrates, or fats. They include all known biochemical compounds and have distinct physical and chemical properties. They are normally easily degraded by microbes

From *Pollution Science*, 8, 1996, Academic Press, San Diego, CA.

phase within the soil (called the soil atmosphere), but at different concentrations. The main gaseous constituents are oxygen, carbon dioxide, nitrogen, and other volatile compounds such as hydrogen sulfide or ethylene. The concentrations of oxygen and carbon dioxide in the soil atmosphere are normally different than in the atmosphere (Table 2.5). This variable reflects the use of oxygen by aerobic soil organisms and subsequent release of carbon dioxide. In addition, the gaseous concentrations in soil are normally regulated by diffusion of oxygen into soil and of carbon dioxide from soil.

2.3.2   Availability of Oxygen and Soil Respiration

The oxygen content of soil is vital for *aerobic microorganisms*, which use oxygen as a terminal electron acceptor

during degradation of organic compounds (see Chapter 9). *Facultative anaerobes* can utilize oxygen or combined forms of oxygen (such as nitrate) as a terminal electron acceptor. *Anaerobes* cannot utilize oxygen as an acceptor. Strict anaerobes are lethally affected by oxygen because they do not contain enzymes that can degrade toxic peroxide radicals. Since microbial degradation of many organic compounds in soil, including xenobiotics, is carried out by aerobic organisms, the presence of oxygen in soil is necessary for such decomposition. Oxygen is found either dissolved in the soil solution or in the soil atmosphere, but soil oxygen concentrations in solution are much lower than in the soil atmosphere.

The total amount of pore space depends on soil texture and soil structure. Soils high in clays have more total pore space, but smaller pore sizes. In contrast, sandy soils have larger pore sizes, allowing more rapid water and air

TABLE 2.5 Characteristics of the Soil Atmosphere

Location	Composition (% volume basis)		
	Nitrogen (N_2)	Oxygen (O_2)	Carbon dioxide (CO_2)
Atmosphere	78.1	20.9	0.03
Well-aerated soil surface	78.1	18–20.5	0.3–3
Fine clay or saturated soil	>79	~0–10	Up to 10

From *Environmental Microbiology*, 8, 2000, Academic Press, San Diego, CA.

movement. In any soil, as the amount of soil structure increases, the total pore space of the soil increases. Aerobic soil microbes require both water and oxygen, which are both found within the pore space. Therefore the soil moisture content controls the amount of available oxygen in a soil. In soils saturated with water, all pores are full of water and the oxygen content is very low. In dry soils, all pores are essentially full of air, so the soil moisture content is very low. In soils at *field capacity*, that is, soils having moderate soil moisture, both air (oxygen) and moisture are readily available to soil microbes. In such situations, soil respiration via aerobic microbial metabolism is normally at a maximum. It is important to note, however, that low-oxygen concentrations may exist in certain isolated pore regions, allowing anaerobic microsites to exist even in aerobic soils, thereby supporting transformation processes carried out by facultative anaerobes and strict anaerobes. This is an excellent example of how soil can function as a discontinuous environment of great diversity.

2.3.3   Gaseous Phase Within the Vadose Zone

Vadose zones generally are primarily aerobic regions. However, due to the heterogeneous nature of the subsurface, anaerobic zones can occur particularly in clay lenses. Thus both aerobic and anaerobic microbial processes may occur.

At contaminated sites, volatile organic compounds can be found in the gaseous phase of the vadose zone. For example, chlorinated solvents, which are ubiquitous organic contaminants (see Chapters 12 and 15), are volatile and are often found in the vadose zone gaseous phase below hazardous waste sites. In such cases, soil venting is often used to remove the contamination (see Chapter 19). The porosity, structure, and water content of the vadose zone is critical to effective application of soil venting.

2.4   LIQUID PHASE

Water is, of course, essential for all biological forms of life, in part because of the unique nature of its structure. The fact that the oxygen moiety of the molecule is slightly more electronegative than the hydrogen counterparts results in a polar molecule. This polarity, in turn, allows water to hydrogen bond both to other water molecules and to other polar molecules. This capacity to bond with almost anything has a profound influence on biological systems, and it explains why water is a near-universal solvent (see Chapter 3). It also explains the hydration of cations and the adsorption of water to soil colloids (see Chapter 8).

By definition, the vadose zone is unsaturated and contains low moisture content. However, whenever rainfall or irrigation events occur at the soil surface, some moisture leaches into the vadose zone. Other avenues by which moisture can reach the subsurface is through burrowing animal holes or worm holes which results in preferential flow. Even so, significant moisture in the vadose zone is the exception rather than the rule. Basic properties of water in both surface and subsurface environments are discussed in Chapter 3.

2.5   BASIC SOIL PHYSICAL PROPERTIES

2.5.1   Bulk Density

Soil bulk density is defined as the ratio of dry mass of solids to bulk volume of the soil sample:

$$\rho_b = \frac{M_s}{V_T} = \frac{M_s}{V_s + V_w + V_a}$$

where:

ρ_b = Soil bulk density [M L^{-3}]
M_s = Dry mass of solid [M]
V_s = Volume of solids [L^3]
V_w = Volume of water [L]
V_a = Volume of air [L^3]
V_T = Bulk volume of soil [L^3]

The bulk volume of soil represents the combined volume of solids and pore space. In SI units, bulk density is usually expressed in $g\,cm^{-3}$ or $kg\,m^{-3}$. Bulk density is used as a measure of soil structure. It varies with a change in soil structure, particularly due to differences in packing. In addition, in swelling soils, bulk density varies with the water content. Therefore it is not a fixed quantity for such soils.

2.5.2 Porosity

Porosity (n) is defined as the ratio of void volume (pore space) to bulk volume of a soil sample:

$$n = \frac{V_v}{V_T} = \frac{V_v}{V_w + V_s + V_a}$$

where:

n is the total porosity [n];
V_v is the volume of voids [L^3];
V_T is the bulk volume of sample [L^3].

It is dimensionless and described either in percentages with values ranging from 0 to 100%, or as a fraction where values range from 0 to 1. The general range of porosity that can be expected for some typical materials is listed in Table 2.6.

Porosity of a soil sample is determined largely by the packing arrangement of grains and the grain-size distribution. Cubic arrangements of uniform spherical grains provide the ideal porosity with a value of 47.65%. Rhombohedral packing of similar grains presents the least porosity with a value of 25.95%. Because both packings

have uniformly sized grains, porosity is independent of grain size. If grain size varies, porosity is dependent on grain size as well as distribution. Total porosity can be separated into two types, primary and secondary, as discussed in Section 2.2.3. The porosity of a soil sample or unconsolidated sediment is determined as follows. First, the bulk volume of the soil sample is calculated from the size of the sample container. Next, the soil sample is placed into a beaker containing a known volume of water. After becoming saturated, the volume of water displaced by the soil sample is equal to the volume of solids in the soil sample. The volume of voids is calculated by subtracting the volume of water displaced from the bulk volume of the bulk soil sample.

In a saturated soil, porosity is equal to water content since all pore spaces are filled with water. In such cases, total porosity can also be calculated by weighing the saturated sample, drying it, and then weighing it again. The difference in mass is equal to the mass of water, which, using a water density of $1\,g\,cm^{-3}$, can be used to calculate the volume of void spaces. Porosity is then calculated as the ratio of void volume and total sample volume.

Porosity can also be estimated using the following equation:

$$n = 1 - \frac{\rho_b}{\rho_d}$$

where ρ_b is the bulk density of soil [ML^{-3}], and ρ_d is the particle density of soil [ML^{-3}].

A value of $2.65\,g\,cm^{-3}$ is often used for the latter, based on silica sand as a primary soil component. Void ratio (e),

TABLE 2.6 Porosity Values of Selected Porous Media	
Type of material	n (%)
Unconsolidated media	
Gravel	20–40
Sand	20–40
Silt	25–50
Clay	30–60
Rocks	
Karst Limestone	5–30
Sandstone	5–30
Shale	0–10
Fractured crystalline rock	0–20
Dense crystalline rock	0–10

From *Environmental Monitoring*, 8, 2004, Academic Press, San Diego, CA.

which is used in engineering, is the ratio of volume of voids to volume of solids:

$$e = \frac{V_V}{V_s}$$

The relationship between porosity and void is ratio is described as:

$$e = \frac{n}{1-n}$$

It is dimensionless. Values of void ratios are in the range of 0–3.

2.5.3 Soil Water Content

Soil water content can be expressed in terms of mass (θ_g) or volume (θ_v). Gravimetric (mass) water content is the ratio of water mass to soil mass, usually expressed as a percentage. Typically, the mass of dry soil material is considered as the reference state, thus:

$$\theta_g\% = [(\text{mass wet soil} - \text{mass dry soil})/\text{mass dry soil}] \times 100.$$

Volumetric water content expresses the volume (or mass, assuming a water density, ρ_w, of $1\,g\,cm^3$) of water per volume of soil, where the soil volume is comprised of the solid grains and the pore spaces between the grains. When the soil is completely saturated with water, θ_v should generally equal the porosity. The relationship between gravimetric and volumetric water contents is given by:

$$\theta_v = \theta_g \left[\frac{\rho_b}{\rho_w}\right]$$

A related term that is often used to quantify the amount of water associated with a sample of soil is "saturation", S_w, which describes the fraction of the pore volume (void space) filled with water:

$$S_w = \frac{\theta_v}{n}$$

2.5.4 Soil Temperature

Soil temperature is often a significant factor especially in agriculture and land treatment of organic wastes, since growth of biological systems is influenced by soil temperature. In addition, soil temperature influences the physical,

chemical, and microbiological processes that take place in soil. These processes may control the fate and transport of contaminants in the subsurface environment. The temperature of the soil zone fluctuates throughout the year in accordance with the above-ground temperature. Conversely, the temperature below the upper few meters of the subsurface remains relatively constant throughout the year.

QUESTIONS AND PROBLEMS

1. The hydrogen ion concentration of the soil solution from a particular soil is $3 \times 10^{-6}\,mol\,L^{-1}$. What is the pH of the soil solution?
2. What is the soil textural class of a soil with 20% sand, 60% silt, and 20% clay?
3. A 100-g sample of a moist soil initially has a moisture content of 15% on a dry weight basis. What is the new moisture content if 10 g of water is uniformly mixed into the soil?
4. Which factors within this chapter affect the cation-exchange capacity (CEC) of a soil? Explain why.
5. Which factors can potentially affect the transport of contaminants through soil and vadose zone? Explain why.
6. How does soil moisture content affect the activity of aerobic and anaerobic soil microorganisms?
7. Compare and contrast surface soils with the vadose zone.

REFERENCES

Gammack, S.M., Paterson, E., Kemp, J.S., Cresser, M.S., Killham, K., 1992. Factors affecting the movement of microorganisms in soils. In: Stotzky, G., Bollag, J.-M. (Eds.), Soil Biochemistry. 7, Marcel Dekker, New York, pp. 263–305.
Pepper, I.L., Gerba, C.P., Gentry, T.J., 2015. Environmental Microbiology, third ed. Academic Press, San Diego, CA.
Pepper, I.L., 2013. The soil health-human health nexus. Crit. Rev. Environ. Sci. Technol. 43, 2617–2652.
Roesch, L.F.W., Fulthorpe, R.R., Riva, A., Casella, G., Hadwin, A.K.M., Kent, A.D., Daroub, S.H., Camargo, F.A.O., Farmerie, W.G., Triplett, E.W., 2007. Pyrosequencing enumerates and contrasts soil microbial diversity. ISME J. 1, 283–290.
Young, I.M., Crawford, J.W., 2004. Interactions and self-organization in the soil-microbe complex. Science 304, 1634–1637.

Chapter 3

Physical-Chemical Characteristics of Water

M.L. Brusseau, D.B. Walker and K. Fitzsimmons

Apache Reservoir, Arizona. *Photo courtesy D. Walker.*

3.1 THE WATERY PLANET

3.1.1 Distribution

Ninety seven percent of water on the Earth is marine (salt-water), while only 3% is freshwater (Fig. 3.1). With regard to the freshwater, 79% is stored in polar ice caps and mountain glaciers, 20% is stored in aquifers or soil moisture, and 1% is surface water (primarily lakes and rivers). An estimated $110,000 \, \text{km}^3$ of rain, snow, and ice falls annually on land surfaces, and this is what replenishes fresh water resources. Possible effects of climate change, combined with continued increases in human population and economic development are resulting in critical concern for the future sustainability of freshwater resources.

The limited supplies of surface waters and groundwater receive significant amounts of the pollutants generated by humans. Lakes across the planet have an average retention time of 100 years, meaning it takes 100 years to replace that volume of water. Rivers, on the other hand, have a much shorter retention time. The relatively long retention time in lakes exacerbates the impacts of introducing pollutants that will be present for a long time (i.e., they are

"environmentally persistent"). The short retention time in rivers means that pollutants are transferred rapidly to other water bodies such as groundwater, lakes, or oceans. The retention time of groundwater is measured in hundreds if not thousands of years. In the groundwater environment, persistent pollutants may persist for extremely long periods because of constraints to transformation. The characteristics of oceans, surface water bodies, and groundwater are described in upcoming sections. Pollution of groundwater and surface water is discussed in Chapters 15 and 16, respectively.

3.1.2 The Hydrologic Cycle

Water covers much more of Earth's surface than does land. The continual movement of water across the Earth due to evaporation, condensation, or precipitation is called the hydrologic cycle (Fig. 3.2). The consistency of this cycle has taken millennia to establish, but can be greatly altered by human activities including climate change, desertification, or excessive groundwater pumping. Water, in its constantly changing and various forms, has been and continues to be an essential factor for all living things.

Environmental and Pollution Science. https://doi.org/10.1016/B978-0-12-814719-1.00003-3

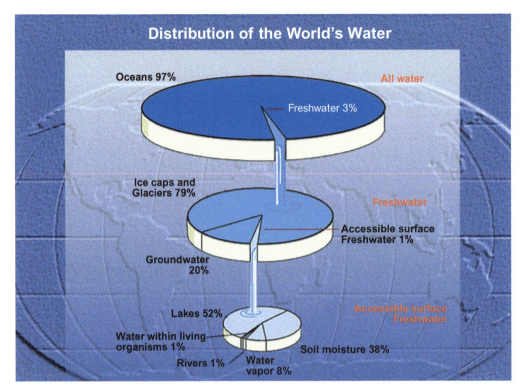

FIG. 3.1   Distribution of the world's water. *(Courtesy of Patricia Reiff and MTPE team, Rice University.)*

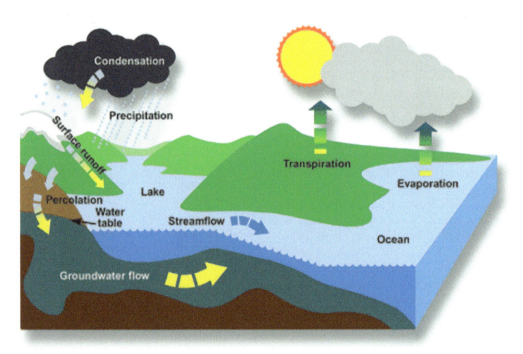

FIG. 3.2   The hydrologic cycle. *(Source: Environment Canada's Freshwater Website http://www.ec.gc.ca/water, 2004. Reproduced with the permission of the Minister of Public Works and Government Services, 2006.)*

Evaporating water moderates temperature; clouds and water vapor protect us from various forms of radiation; and precipitation spreads water to all regions of the globe, allowing life to flourish from the highest peaks to the deepest caves. Solar energy drives evaporation from open water surfaces as well as soil and plants. Air currents distribute this vaporized water around the globe. Cloud formation, condensation, and precipitation are functions of cooling. When vaporized, water cools to a certain temperature, condensation occurs, and often results in precipitation to the Earth's surface. Once back on the surface of the Earth, whether on land or water, solar energy then continues the cycle. The latent heat of water (the energy that is required or released as water changes states) serves to moderate global temperatures, maintaining them in a range suitable for humans and other living organisms.

Some processes involved with the hydrologic cycle aid in purifying water of the various contaminants accumulated during its cycling. For instance, precipitation reaching the soil will react with various minerals and neutralize the acidity that was generated from atmospheric processes. Suspended sediments entrained through erosion and runoff will settle out as the water loses velocity in ponds or lakes. Other solids will be filtered out as water percolates through soil and vadose zones and ultimately to an aquifer. Many organic compounds will be degraded by bacteria in soil or sediments. Salts and other dissolved solids will be left behind as water evaporates and returns to a gaseous phase or freezes into a solid phase (ice). These processes maintained water quality of varying degrees before human impacts on the environment; however, the current scale of these impacts often tends to overwhelm the ability of natural systems to cleanse water through the hydrologic cycle. Further, we have introduced many compounds that are resistant to normal removal or degradation processes (Chapters 14–17).

3.2 UNIQUE PROPERTIES OF WATER

3.2.1 Structure and Polarity

Water is an unusual molecule in that the structure of two hydrogen atoms and one oxygen atom provides several characteristics that make it a universal solvent. First is the fact that the two hydrogen atoms, situated on one side of the oxygen atom, carry positive charges, while the oxygen atom retains a negative charge (Fig. 3.3).

This induced polarity allows water molecules to attract both positive and negative ions to the respective poles of the molecule. It also causes water molecules to attract one another. This is termed hydrogen bonding. This contributes to the viscosity of water and to the dissolution capacity of water for different materials. This unique nature of water

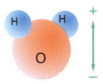

FIG. 3.3 Structure and charge distribution of water, http://faculty.uca.edu/~benw/biol1400/notes32.htm.

makes it an easy conduit for the dissolution and transport for any number of pollutants. Because so many materials dissolve so completely in water, their removal from water is often difficult.

3.2.2 Thermal Properties

Water has unique thermal properties that enable it to exist in three different states: vapor, solid, and liquid under environmentally relevant conditions. Changes in each phase have certain terminology, depending upon state changes, as described below:

Condensation: vapor → liquid
Evaporation: liquid → vapor
Freezing: liquid → solid
Melting: solid → liquid
Sublimation: solid → vapor
Frost Formation: vapor → solid

The fact that water becomes less dense in its solid state, compared to its liquid state, is yet another unusual characteristic. Most liquids contract with decreasing temperature. This contraction also makes these liquids denser (i.e., "heavier") as temperature decreases. Water is unique because its density increases only down to approximately 4°C, at which point it starts to become less dense (Fig. 3.4). Because of this, ice floats and insulates deeper water. This is important because without this unique property, icebergs and other solid forms of water would sink to the bottom of the ocean, displacing liquid water as they did so. Also, lakes and ponds would freeze from the bottom up with the same effect. This is critical to maintaining deep bodies of liquid waters on Earth rather than a thin layer of water on top of an increasingly deep bed of solid ice.

The *specific heat* of water is the amount of energy required to raise one gram of water, one degree C, and is usually expressed as *joules per gram-degree Celsius* ($J\,g^{-1}\,°C^{-1}$). Specific heat values for the different phases of water are given as follows.

Phase	$J\,g^{-1}\,°C^{-1}$
Vapor	2.02
Liquid	4.18
Solid	2.06

The *latent heat of fusion* is the amount of energy required to change 1 gram of ice, at its melting point temperature, to liquid. It is considered "latent" because there is no temperature change associated with this energy transfer, only a change in phase. The heat of fusion for water is $-333\,J\,g^{-1}\,°C^{-1}$.

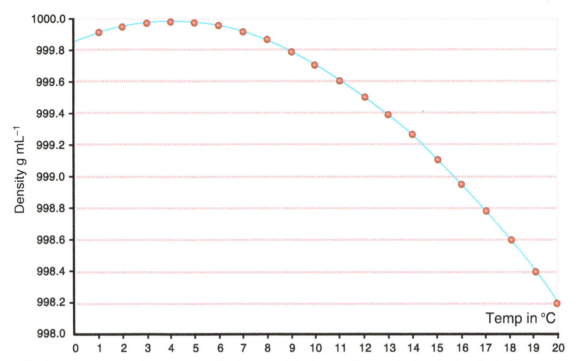

FIG. 3.4 The density of water at varying temperatures, http://www.cyberlaboratory.com/library/basicsofdensity/whatisdensity.htm.

The energy required for the phase changes of water is given in Table 3.1.

Earth is unique because it contains the necessary temperatures and pressures for all three states of water to exist. Water, under the correct combination of temperature and pressure, is capable of existing in all three states (solid, liquid, and vapor) simultaneously and in equilibrium. This is referred to as the *triple point,* where infinitesimally small increases or decreases in either pressure or temperature will cause water to be either a liquid, solid, or gas. Specifically, the triple point of water exists at a temperature and pressure of 273.16K (0.0098°C) and 611.73 pascals (0.00603 atm), respectively. Fig. 3.5 shows that decreasing temperature and increasing pressure causes water to pass directly from a gas to a solid. At pressures higher than the triple point, increasing temperature causes solid water (ice) to transform

into liquid and eventually gas (vapor). Liquid water cannot exist in pressures lower than the triple point and ice instantaneously becomes steam with increasing temperature. This process is known as sublimation.

3.3 MECHANICAL PROPERTIES

3.3.1 Interception, Evaporation, Infiltration, Runoff

Precipitation in a nonpolluted environment provides a fairly pure form of water. However, today precipitation may absorb pollutants in the environment to form acid rain (see also Chapter 17). Precipitation can also entrain fine particulates that were suspended in the air. As precipitation reaches the surface, raindrops are likely to fall upon and be

TABLE 3.1 Phase Changes of Water

Process	From	To	Energy Gained or Lost ($Jg^{-1} °C^{-1}$)
Condensation	Vapor	Liquid	2500
Deposition	Vapor	Ice	2833
Evaporation	Liquid	Vapor	−2500
Freezing	Liquid	Ice	333
Melting	Ice	Liquid	−333
Sublimation	Ice	Vapor	2833

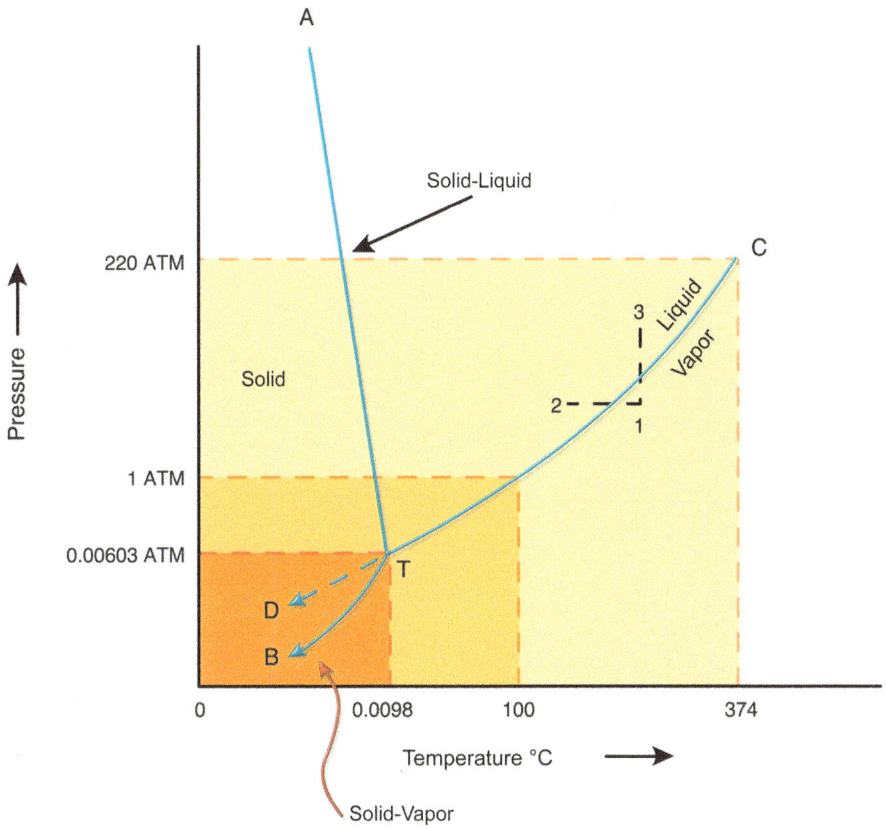

FIG. 3.5 Phase diagram of the triple point of water, http://www.sv.vt.edu/classes/MSE2094_NoteBook/96ClassProj/pics/trip_pt1.

intercepted by various types of vegetation. In many regions, much of the precipitation may fall on trees, shrubs, or grasses and never actually reach the ground. In others, the plants may slow the rate of fall of raindrops, break them into smaller drops, or channel them more gently to the surface. Interception can have multiple influences on the impact of rainfall on the land surface. First, the water may evaporate directly from the plant, never reaching the soil. Second, it may entrain materials settled on the plant surfaces. Third, by slowing the momentum and reducing the energy of falling rain, physical impacts on the soil and resulting erosion may be reduced.

Certain anthropogenic land use practices or natural events can lead to decreases in interception and subsequent increases in sediment suspended within water. Often, sediment may have other pollutants attached to it thereby polluting the water as well. Certain mining features, if not revegetated, can result in increased erosion and subsequent contamination of streams. Natural events, such as wildfires, can also result in substantial erosion and contamination of downstream areas (see also Chapter 14).

Evaporation of water is another crucial part of the hydrologic cycle. The rate of evaporation from a body of water, or mass of soil, is a function of the relative humidity, temperature, and wind speed. An important subcomponent

of evaporation is transpiration, the active transport and evaporation of water from plants. Plants transport nutrients in an aqueous solution and then dispose of the water through their leaves by evaporation. As water evaporates, it leaves a concentrated amount of compounds that were formerly dissolved in that water. This applies to nutrients left in plants, as well as to pollutants that were introduced with the water.

Water that is not contained in oceans is often referred to as "freshwater," implying that it is not saline. This is not always the case, and some inland waters can be much more saline than the world's oceans. This is especially true in arid environments or enclosed basins that have limited or no drainage. Often, salinity in inland waters reaches such high levels that it supports little, if any, life. Salinity in inland waters, and in the world's oceans, is largely a result of evaporation. As water is vaporized and once again enters the hydrologic cycle, salts accumulate on the Earth's surface, and in lieu of adequate dilution and flushing, can often make water increasingly saline (Fig. 3.6).

Precipitation that reaches the soil surface either infiltrates the ground or runs off the surface. Human uses of water also deliver enormous amounts of water onto soils or human-made structures that can either infiltrate or contribute to runoff. This is a major source of pollutants introduced into the environment. The infiltration rate of water

FIG. 3.6   This satellite image is of the Great Salt Lake in Utah. This is the largest lake in the United States west of the Mississippi River covering some 1700 square miles. It is also 3–5 times more saline than the world's oceans. It is a fishless lake with only the most saline-tolerant ("halophytic") organisms capable of surviving. The largest organisms inhabiting its waters are species of brine shrimp and brine flies, http://ut.water.usgs/gov/greatsaltlake/.

FIG. 3.7   The Grand Canyon of the Colorado River was formed by the dissolution and erosion of material over eons. Historically, most of this material was deposited in the Gulf of California. With the construction of large dams along the course of the Colorado River, most of this material is now deposited in storage reservoirs, http://www.kaibab.org/tr961/lg961110.jpg.

into the ground is an important measure used to determine how foundations and sewer systems are designed, how irrigation water should be applied, and how pollutants may migrate to a water supply.

How water runs off of surfaces is also a matter of interest to hydrology, fisheries, aquatic biology, and pollution science. Not only are pollutants entrained in flowing water, but erosion and flooding can also occur. Studies of runoff and surface flow focus upon the amounts of soil and pollutants that are transported and their eventual fate as they arrive into lakes or streams.

3.4   THE UNIVERSAL SOLVENT

One of the most unique properties of water is its ability to dissolve other substances. It is this ability that can lead to large-scale landscape transformations (Fig. 3.7) and the ability to carry contaminants relatively long distances. If it were not for the various substances dissolved in water, an organism's cells would quickly be deprived of essential nutrients, salts, and gasses, leading to eventual death. The dissolution of materials in water has shaped the nature of all living creatures on the planet.

3.4.1   Concentration Terminology

It is important to quantify the amount of material dissolved in water. Quantification requires a range of values so that we can determine high versus low concentrations for a given constituent. The values are always expressed as a ratio of quantity of solute to quantity of water (Table 3.2). The importance of very small concentrations should never be underestimated (see Information Box 3.1). This is especially true in toxicological studies where very small concentrations can lead to toxic impacts on organisms (Chapter 28).

There are two major expressions in concentration terminology.

- Mass/mass. An example would be milligrams/kilogram ($mg\,kg^{-1}$), which equals milligrams of dissolved matter per kg of solution.
- Mass/volume. An example would be milligrams/liter ($mg\,L^{-1}$), which equals milligrams of dissolved matter per liter of solution.

TABLE 3.2 Examples of Typical Concentrations of Solutes in Water

Percent		parts per hundred	10^2
Gram	$g L^{-1}$	parts per thousand	10^3
Milligram	$mg L^{-1}$	parts per million	10^6
Microgram	$\mu g L^{-1}$	parts per billion	10^9
Nanogram	$ng L^{-1}$	parts per trillion	10^{12}

INFORMATION BOX 3.1
Examples of Why Small Numbers are Important

- The solubility of trichloroethene in water is ~1000 mg/L or 1 g/L.
- The regulatory limit for its presence in drinking water is 5 μg/L.

  >>> the regulatory limit is 200,000 times lower than the solubility.

Another common concentration unit used is "parts per million" (ppm), representing the quantity of matter present out of a million. For dilute aqueous solutions, $mg L^{-1}$ and ppm will be the same number because the density of the solution will be close to 1. The following relationship can be used to convert units for cases where densities of the liquid or solution are significantly different than 1: specific gravity of solution × ppm = $mg L^{-1}$. Specific gravity is the ratio of the density of a substance to the density of a reference substance (typically water). Note that this same relationship holds true when using other concentrations such as parts per billion (ppb) and $\mu g L^{-1}$ or parts per trillion and $ng L^{-1}$.

Example Calculation 3.1

Using Concentration

Knowing the concentrations of constituents in water has many utilitarian uses. For example, an environmental scientist may need to determine the total mass of a contaminant present in a surface water reservoir to conduct a risk assessment and to develop a remediation strategy.

- Suppose the water contained 35 ppm of sulfate.
- The reservoir has a total storage volume of 1 million m^3.
- The reservoir is half full.
- How many kilograms of sulfate are contained in the reservoir?

  To calculate this we determine the following:

- Volume of water = 1,000,000 m^3 × 0.5 × 1000 L/m^3 = 500,000,000 L.
- Sulfate concentration = 35 ppm = 35 mg/L.
- Mass of sulfate in the water = 35 mg/L × 500,000,000 L × 1 g/1000 mg × 1 kg/1000 g = 17,500 kg.

3.4.2 Oxygen and Other Gases in Water

Just like terrestrial counterparts, aquatic organisms (other than anaerobic microbes) need dissolved oxygen and other gases in order to survive. Additionally, the world's oceans "absorb" an estimated ¼ to ⅓ of carbon dioxide emitted by human activity. If it were not for the ocean's ability to absorb carbon dioxide, an important greenhouse gas, global warming would proceed at an unprecedented rate. The amount of gas that an aqueous solution can hold is dependent upon several variables, the most important of which is atmospheric pressure. Simply stated, increasing atmospheric pressure causes a greater amount of gas to go into solution at a given temperature (Fig. 3.8). Generally, increasing water temperature will result in an increased solubility of gas.

There is a direct, linear relationship between the partial pressure and the concentration of gas in solution. For example, if the partial pressure is increased by ¼, the concentration of gas in solution is increased by ¼ and so on. This is because the number of collisions of gas molecules with the solvent molecules (water in this case) is directly proportional to increases or decreases in partial pressure. Since the concentration:pressure ratio remains the same, we can predict the concentration of gas in water under differing partial pressures. This relationship can be written as:

$$\frac{Concentration 1}{Pressure 1} = \frac{Concentration 2}{Pressure 2}$$

For example, 1 L of water under 1 atmosphere of pressure will contain 0.0404 grams of oxygen. What will the concentration of oxygen be if the partial pressure is increased to 15 atmospheres?

$C1 = 0.0404$ g O_2/L liter solution
$P1 = 1$ atm, $P2 = 15$ atm
$C2 = ?$

$$\frac{0.0404 g O_2}{1 atm} = \frac{C_2}{15 atm}$$

$C2 = (15 atm)(0.0404 g O_2$ per 1 L/1 atm)
$C2 = 0.606$ g O_2

In any body of water, there are sources and sinks of dissolved oxygen. Sources include atmospheric reaeration through turbulence, ripples and waves, and dams and waterfalls. Another potential source of dissolved oxygen is

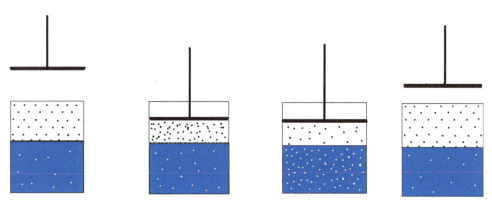

FIG. 3.8 The solubility of oxygen in water under different atmospheric pressures.

photosynthesis primarily by algae or submersed aquatic vegetation. During photosynthesis, plants convert CO_2 into oxygen in the process outlined as follows.

$$6CO_2 + 12H_2O + Light\ energy \rightarrow C_6H_{12}O_6 + 6O_2 + 6H_2O \tag{3.1}$$

All natural waters also have sinks of dissolved oxygen, which include:

Sediment Oxygen Demand (SOD): Due to decomposition of organic material deposited on bottom sediments.
Biological Oxygen Demand (BOD): The oxygen required for aerobic biodegradation (respiration) of contaminants by microorganisms.
Chemical Oxygen Demand (COD): The oxygen required for degradation or oxidation of *all* labile contaminants. Note that BOD is a subset of COD.
Respiration is the metabolic process by which organic carbon is oxidized to carbon dioxide and water with a net release of energy (see also Chapter 5). Aerobic respiration requires, and therefore consumes, oxygen.

$$C_6H_{12}O_6 + 6O_2 \rightarrow 6CO_2 + 6H_2O + energy \tag{3.2}$$

This is, essentially, the opposite of photosynthesis. In the absence of light, the CO_2 collected by plants via photosynthesis during the day is released back into the water at night, resulting in a net loss of dissolved oxygen. Depending upon the amount of nutrients, algae, and available light, this often results in large daily fluctuations in dissolved oxygen levels known as Diel patterns (Fig. 3.9).

The implications of dissolved oxygen sinks and sources on aquatic organisms and overall water quality are crucial in determining whether or not a river, lake, or stream is impaired and to what degree. If dissolved oxygen sinks are greater than sources for extended periods of time, it is safe to assume some degree of contamination has occurred.

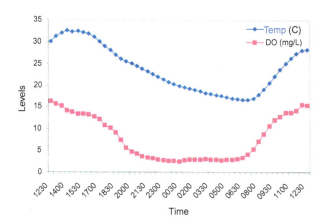

FIG. 3.9 Diel pattern of temperature and dissolved oxygen in Rio de Flag, an effluent-dominated stream in Flagstaff, Arizona. Data was collected every 30 minutes over a 24-hour period on December 08, 2003.

Examples of anthropogenic wastes that can cause dissolved oxygen impairment of receiving waters are sewage (raw and treated, human and nonhuman), agricultural runoff, slaughterhouses, and pulp mills.

3.4.3 Carbon Dioxide in Water

Carbon dioxide only accounts for approximately 0.033% of the gases in Earth's atmosphere, yet is abundant in surface water. The biggest reason for the abundance of carbon dioxide in water is due to its relatively high solubility—about 30 times that of oxygen. At room temperature, carbon dioxide has a solubility in water of 1.45 g/L. In the atmosphere, carbon dioxide is released when fossil fuels are burned for human uses, and as a result of large worldwide increases in the use of fossil fuels during the last century or so, the amount of carbon dioxide in the atmosphere has steadily increased.

Carbon dioxide dissociates and exists in several forms in water. First, carbon dioxide can simply dissolve into water going from a gas to an aqueous form. A very small portion of carbon dioxide (less than 1%) dissolved in water is hydrated to form carbonic acid (H_2CO_3). Equilibrium is then established between the dissolved carbon dioxide and carbonic acid.

$$CO_2 + H_2O \leftrightarrow H_2CO_3 \qquad (3.3)$$

Carbonic acid, a very weak acid, is then dissociated in two steps.

$$H_2CO_3 \leftrightarrow H^+ + HCO_3^- \qquad (3.4)$$

$$HCO_3^- \leftrightarrow H^+ + CO_3^{2-} \qquad (3.5)$$

As carbon dioxide is dissolved in water, equilibrium is eventually established with the bicarbonate (HCO_3^-) and carbonate (CO_3^{2-}) ions. Carbonate, being a largely insoluble anion, then reacts with cations in the water, causing these cations to precipitate out of solution. As a result, Ca^{2+} and Mg^{2+} often precipitate as carbonates. Calcium carbonate ($CaCO_3$), otherwise known as limestone, has been generated in large deposits as a result of this process. As limestone is once again dissolved, carbon dioxide is released back into the atmosphere. In addition, several aquatic organisms, such as corals and shelled creatures such as clams, oysters, and scallops, are capable of converting the carbon dioxide in water into calcium carbonate.

Diel fluctuations in oxygen and pH levels can occur during the day in waters where photosynthesis is taking place (see Fig. 3.10). Algae and plants convert carbon dioxide into carbohydrates to be used in metabolic processes. In very productive waters, this process can leave bicarbonate or carbonate in excess, leading to increased pH levels. In the absence of adequate light for photosynthesis, respiration predominates, resulting in carbon dioxide once again being restored to the water resulting in decreased pH levels.

Calcium carbonate, while insoluble at neutral to basic pH levels, readily dissolves in acidic conditions. In the initial step, carbonate acts as a base resulting in calcium ions and carbonic acid. In the next step, carbonic acid is dissociated releasing carbon dioxide as a gas.

$$CaCO_3 + 2H^+ \rightarrow Ca^{2+} + H_2CO_3 \qquad (3.6)$$

$$H_2CO_3 \leftrightarrow H_2O + CO_2 \qquad (3.7)$$

Rain is often slightly acidic due to the presence of atmospheric carbon dioxide. Recently, due to the burning of fossil fuels, other gases can also be dissolved in rain resulting in "acid rain." Atmospheric pollutants responsible for acid rain include sulfur dioxide (SO^2) and nitrous oxides (NO_x). More than 2/3 of these pollutants come from burning fossil fuels for electrical power generation, and prevailing winds can result in acid rain being deposited far from its original source. Acid rain has far-reaching environmental consequences including acidification of lakes and streams, making them uninhabitable by aquatic life; extensive damage to forests, plants, and soil; damage to building materials and automotive finishes; and human health concerns.

Natural conditions often exist that can result in the dissolution of limestone:

$$CO_2 + H_2O + CaCO_3 \leftrightarrow Ca^{2+} + 2HCO_3^- \qquad (3.8)$$

The remaining reaction is a three-step process:

$$CaCO_3 \leftrightarrow Ca^{2+} + CO_3^{2-} \qquad (3.9)$$

$$CO_2 + H_2O \leftrightarrow H_2CO_3 \qquad (3.10)$$

$$H_2CO_3 + CO_3^{2-} \leftrightarrow 2HCO_3^- \qquad (3.11)$$

This reaction can result in the formation of caves when naturally acidic rainwater reacts with a subterranean layer of limestone, dissolving the calcium carbonate and forming

FIG. 3.10 Profuse growth of attached algae ("periphyton") growing in the Rio de Flag stream at the time the data reported in Fig. 3.9 were recorded. Photosynthesis and respiration by these algae likely contributed to the large swings in dissolved oxygen levels within the water over the 24-hour period. *(Photo courtesy D. Walker.)*

FIG. 3.11 Cave formation in the Big Room, Carlsbad Caverns National Park, New Mexico, http://www.nps.gov/cave/home.htm.

openings. As slightly acidic water reaches the cave ceiling, the water evaporates and carbon dioxide escapes. It is this reaction that is responsible for the many elaborate formations in cave ecosystems (Fig. 3.11).

Total alkalinity is the total concentration of bases, usually carbonate and bicarbonate, in water and is expressed as mg/L of calcium carbonate. Analytically, total alkalinity is expressed as the amount of sulfuric acid needed to bring a solution to a pH of 4.2. At this pH, the alkalinity in the solution is "used up," and any further addition of an acid results in drastic decreases in pH levels. Total alkalinity, by definition, is the ability of a water body to neutralize acids. In other words, it is the "buffering capacity" of a water body, and it is influenced by the minerals in local soils. In areas of the northeastern United States, where parent material contributes little to the total alkalinity in the water, the cumulative effects of acid rain have been most devastating and have reduced levels of aquatic life in many streams. Mining activity and pulp mills can also add to reductions in total alkalinity and subsequent decreases in pH. Stopgap measures in watersheds, lakes, or streams where alkalinity has been depleted include such drastic actions as dropping lime from helicopters to increase the buffering capacity for aquatic life.

3.5 LIGHT IN AQUATIC ENVIRONMENTS

Just as in terrestrial systems, light at the water's surface marks the beginning of photosynthesis or "primary production." Light is the driving force behind almost all metabolic processes in aquatic ecosystems. Light carries heat energy to be used in many chemical and biological processes, and can simultaneously regulate and/or damage aquatic biota. Pollutants, especially suspended and dissolved substances, can have profound effects on both the amount of light available for photosynthesis and the heat energy needed for these processes.

3.5.1 Light at and Below the Water Surface

Several processes can affect both the intensity and quality of light reaching the Earth's surface. One such process is simple scattering of light by particles in the atmosphere, including water vapor. Refraction of light occurs when the speed of light changes going from one medium, such as air, into another, such as water. Light can also be reflected off of a water surface due to several factors such as the incident angle of light, wave height and frequency, or the presence of ice. Another factor affecting light intensity is absorption due to the decrease in light energy by its transformation into heat. Both atmospheric gases and water can cause absorption.

Absorption (i.e., "quenching") of light entering a body of water can be quantified using a *vertical absorption coefficient* expressed as:

$$k = \frac{\ln I_0 - \ln lz}{z}$$

where:

I_o is the natural log of the initial amount of light entering the water
I_z is the natural log of light remaining at any given depth
z is the thickness of the water in meters

The vertical absorption coefficient is somewhat analogous to the coefficient of extinction, except that the latter uses the base 10 logarithm. The vertical coefficient of absorption is therefore 2.3 times the coefficient of extinction. Another important variable when considering the fate of light in water is *the total coefficient of absorption*, which is the sum of all factors leading to the intensity (or "extinction") of light at any given depth. The total coefficient of absorption can be expressed as:

$$I_z = I_{o-}{}^{-kw} + I_{o-}{}^{-kp} + I_{o-}{}^{-kc}$$

where:

kw = the coefficient of absorption in pure water
kp = suspended particulate matter
kc = dissolved substances

Note that kp or kc can only be determined after filtration or centrifugation. The total coefficient of absorption is different for each body of water and is dependent upon the amount of dissolved or suspended material in the water. Dissolved substances are normally humic or fulvic acids, tannins, lignins, or anything that constitutes colored, dissolved, organic matter absorbing light strongly at relatively short wavelengths (e.g., blues and ultraviolet radiation < 500 nm). Suspended material includes fine clays, and phytoplankton, which absorb light evenly over the entire spectrum.

In standing water, vertical light penetration can be roughly estimated using a *secchi disk* which is standardized, 20-cm diameter, black and white, weighted disk lowered into the water using a calibrated line. The depth at which the disk almost, but not quite, disappears is recorded. This is known as the *secchi disk transparency* expressed as Z_{sd}. To be comparable, this has to be done between 10 am and 2 pm on any given day. Secchi disk depth is often mistakenly used as a proxy for primary production (the amount of standing algal biomass) in a water body. In reality there are many other mitigating factors besides algae that can cause either increases or decreases in transparency. The *photic zone* is the volume of water from the surface to where 99% of the light needed for photosynthesis has been extinguished. A very rough estimate of photic zone depth is anywhere from 2.7 to 3 times Z_{sd}.

3.6 OCEANS

As noted before, contaminants can enter oceans through transport in air, streams, and groundwater, as well as directly through disposal of wastes in the ocean. Contaminants in the ocean may be introduced into the food chain by filter-feeding organisms or possibly may be sequestered in cold, deep basins where they are resistant to degradation by natural processes. Much of the world's human population inhabits coastal areas, making oceans especially vulnerable to pollutants introduced directly or from surface water and groundwater drainage. In turn, ocean contamination can have both direct and indirect impacts on humans.

3.6.1 Salts

The oceans are saline due to the constant input of dissolved salts leached from rocks and soils on land surfaces. These dissolved solids consist of many salts including sodium, calcium, and magnesium salts. There are several common methods of determining total dissolved solids and/or salinity including electrical conductivity, density, light refraction, silver titration, and simple evaporation of a known volume. Each of these methods will provide a result that can be converted to a percentage, or more commonly parts per thousand, of salinity. Open ocean seawater will vary from 33 to 35 parts per thousand (ppt), while coastal waters may have less than 1 ppt.

3.6.2 Transport and Accumulation of Pollutants

The oceans tend to become the repository of many pollutants. Historically, one of the most common mind-sets was that the "solution to pollution is dilution." Over time

we have discovered that even the oceans are not vast enough to handle the volume of pollutants that are discharged by human activities. In many cases, the effects of pollutants are so toxic that even vast dilution is not effective. In other instances we have found that filter-feeding organisms bioaccumulate toxic compounds that ascend food chains and can affect grazers and top carnivores, as well as humans consuming various seafoods. Mercury in swordfish and certain sharks is such an example.

3.6.3 Wave Morphology and Currents

Waves and currents move pollutants within surface water. This is especially critical in marine systems as we attempt to control and/or track pollution movement. Waves are typically the result of winds blowing across the surface of a body of water. As the air friction pushes against the water, small ripples form. Continued breezes push against the sides of each ripple providing additional energy. The tops of the ripples may blow off, forming whitecaps. This releases some of the energy, but more will continue with the bulk of the water below. Continued wind energy transferred to the waves can store enormous amounts of energy. The size of a wave is a function of the average velocity of the wind, the period of time it blows, and the distance of open water across which it blows. The distance over which blowing winds create waves is called the *fetch*. Waves are usually described by the period (the time between two crests passing the same point), the wavelength (distance between two crests), and the wave height (vertical distance between a trough and the next crest).

Waves are important factors in the dispersion of pollutants, especially oil spills. Calm waters facilitate the recovery of oil and other floating pollutants. However, in cases where it cannot be collected, wave action can break up the thick mats of oil and spread the material so that bacterial degradation can break down the organic molecules. It will also allow the lighter fractions to volatilize.

Water motion in the oceans is a function of waves, which are wind driven, and currents, which are driven by a number of factors. The most important is the Coriolis effect caused by the spinning of the Earth (see Chapter 4). However, wind, runoff from rivers, density differences from temperature or salinity extremes, and tidal fluctuations can all drive currents. Currents are often compared to the circulatory system of a living organism. Trade winds will power currents that transport water and its constituents across vast distances along the surface. Cooling of surface waters in the high latitudes causes cold, dense water to sink into the depths, where it flows along the bottom until it upwells in lower latitudes to replace warmer surface waters that are blown away from coastlines. Tidal action will also drive currents in local situations.

Currents and waves effectively mix surface waters on a short time scale. Deep-water currents mix water on a much longer time scale. Together they effectively spread pollutants to every corner of the ocean. Water motion can be effectively measured in two ways. If the motion is fairly consistent in one direction, a current meter can be used to determine velocity. A more common situation is when water motion is not consistent, but varies in direction and speed in three dimensions over short periods of time. In this case, a better measure is to use a clod card. Clod cards consist of a block of calcium sulfate that slowly dissolves in water. The block is mounted on a card stock for easy attachment and handling. Its rate of dissolution can be measured by recording initial and final dry weights. These are normally used to compare two or more environments. The clod cards can be calibrated, if necessary, by placing controls in waters or tanks with known velocities of water motion.

3.7   LAKES AND RESERVOIRS—THE LENTIC SYSTEM

Lentic systems are closed ecosystems such as lakes. However, while lakes and reservoirs are relatively more "closed" than rivers and streams, they are far from isolated. Although some lakes and reservoirs have groundwater inputs, the majority of water entering them is a result of overland flow; therefore lakes and reservoirs are reflections of all processes that have occurred in the watershed up to that point. Both natural and anthropogenic watershed influences can have profound effects on both water quality for human use and aquatic communities living within lentic systems.

3.7.1   Lentic Typology

There are several ways to classify lakes and reservoirs: by origin, ecoregion, shape and size, regimen of mixing, and stratification. Detail about every different type of lake or reservoir is beyond the scope of this chapter, and the reader is instead referred to any of several available limnological texts. Rather, this section will discuss the differences

between two main types of lentic systems: lakes and reservoirs.

The main difference between lakes and reservoirs is that the former have natural origins, while the latter are manufactured by humans for anthropogenic needs. Both are lentic systems and therefore share some common attributes. Lakes are dominant where glaciers have scoured the landscape, as, for example, around the Great Lakes. In other cases, tectonic activity has formed rifts, allowing for the African Rift Lakes, or some other depression has been made through natural causes and there is adequate ground or surface water inputs to fill these depressions. Reservoirs, on the other hand, are constructed where lakes are not in abundance and water is needed for human use. Reservoirs are often found in greater abundance in arid and semiarid regions such as in the western United States, where large natural lakes are not abundant. The large number of reservoirs built in arid regions often means sacrificing lotic habitats through either direct impoundment (e.g., dams) or some change in water chemistry caused by impoundment. Endemic aquatic organisms living in streams and rivers of the western United States are among the most endangered species on the planet due to impoundment of habitat and changes in environmental conditions below large dams.

Some of the major differences between lakes and reservoirs are given in Table 3.3. The arrows are either increasing or decreasing as they relate to either lakes or reservoirs. For example, lakes generally have a much smaller watershed area than reservoirs.

3.7.2   Trophic State

Materials within lentic systems are generally classified as either *autochthonous* or *allochthonous* in origin. Allochthonous (from the Greek, meaning "other than from the earth or land itself") material is everything that has been imported to the lentic system from somewhere else in the watershed. This material can be thought of in terms of *loading*. Autochthonous (from the Greek, meaning "of or from the earth or land itself") material is that which is recycled within the lake or reservoir. Both sources play a role in a lentic system's *trophic state*.

TABLE 3.3 Comparisons of Natural Lakes and Reservoirs

Variable	Natural Lakes	Reservoirs
Watershed area	↓	↑
Maximum depth	↓	↑
Mean depth	↑	↓
Residence time	↑	↓

The trophic status of a lake or reservoir is largely a means to communicate the ecological condition of a water body. Trophic state is based upon the total weight of living biological material or biomass in a water body at a specific location and time. Time and location-specific measurements can be aggregated to produce waterbody-level estimations of trophic state. Trophic status is not equivalent to *primary production*, which is the rate of carbon fixed (usually expressed as g of C fixed $day^{-1} m^{-3}$). Trophic state, being a multidimensional phenomenon, has no single trophic indicator that adequately measures its underlying concept. Combining the major physical, chemical, and biological expressions of trophic state into a single index reduces the variability associated with individual indicators and provides a reasonable composite measure of trophic conditions in a water body.

Several trophic state indices have been devised. The selection of which one to use depends upon several different chemical and physical parameters. One of the most-used trophic state indices for lakes is the *Carlson's TSI* (Carlson, 1977). Especially important are the nutrients phosphorous and nitrogen, both essential macronutrients for algal growth and primary production. Other variables used include measures of chlorophyll *a* and secchi disk depth. Chlorophyll *a* is a pigment common to all algae and gives a measure of standing biomass. Phosphorous is often the nutrient that is most "limiting" in natural waters (although nitrogen limitation also does occur). Carlson's TSI relies upon three variables—chlorophyll a, secchi disk depth, and total phosphorous—to determine trophic status of any water body that is phosphorous limited. In the broadest sense, trophic status of a lake or reservoir is often divided into the categories presented in Table 3.4.

The advantage of a simple, fixed classification system is its easy application. However, trophic terminology has a history of being misused. For example, deeming a lake as eutrophic does not automatically mean it has poor water quality. Although the concepts of water quality and trophic state are related, they should not be used interchangeably. Trophic state is an absolute scale that describes the biological condition of a body of water. The trophic scale is a division of variables used in the definition of trophic state and is not subject to change because of the attitude or biases of the observer. An oligotrophic or a eutrophic lake has attributes of production that remain constant no matter what the use of the water or where the lake is located. For the trophic state terms to have meaning, they must be applicable in any situation and location, while keeping in mind that trophic status is just one of several aspects of the biology of the water body in question. Water quality, on the other hand, is a term used to describe the condition of a water body in relation to human needs or values. Quality is not an absolute; the terms "good" or "poor" water quality only have meaning relative to the user. An oligotrophic lake might have "good" water quality for swimming but have "poor" water quality for fishing. Confusion can ensue when trophic state is used to infer water quality.

3.7.3 Density and Layering

As light enters the water, different wavelengths are quenched exponentially (see Section 3.6.2). Wavelengths in the 620 to 740 nm range are absorbed first. This range also contains

TABLE 3.4 Trophic Categories for Lakes and Reservoirs

Trophic State	Chlorophyll A (μg/L)	Secchi Disk Depth (m)	Total *P* (μg/L)	Attributes
<30	<0.95	>8	<6	*Oligotrophy*: Clear water, dissolved oxygen throughout the year in the hypolimnion
30–40	0.95–2.6	8–4	6–12	Hypolimnia of shallower lakes may become anoxic
40–50	2.9–7.3	4–2	12–24	*Mesotrophy*: Water moderately clear; increasing probability of hypolimnetic anoxia during the summer
60–60	7.3–20	2–1	24–48	*Eutrophy*: Problems with excessive primary production begin. Anoxic hypolimnia in stratified lakes/reservoirs
60–70	20–56	0.5–1	48–96	Cyanobacteria dominate the phytoplankton. Increasing problems with anoxia
>70	56–155	0.25–0.5	96–192	*Hyper-eutrophy*: Primary production limited only by light. Dense growths of algae and/or aquatic plants. Increasing prevalence of anoxia throughout the water column. Fish kills possible
>80	>155	<0.25	192–384	Few aquatic plants or other forms of life. Sustained periods of anoxia

INFORMATION BOX 3.2
Stratification of Lakes and Reservoirs

The uppermost layer is called the *epilimnion* and is characterized by relatively warm water where most photosynthesis occurs. Depending upon environmental conditions, it is more oxygenated than layers below it. The middle layer is called the *metalimnion* and contains an area known as the *thermocline.* The thermocline is that area within the water column where the temperature gradient is the steepest. The metalimnion is that region surrounding the thermocline where the temperature gradient is steep compared to the upper and bottom layers. Due to the temperature gradient becoming increasingly steep within a correspondingly smaller volume of water, the thermocline becomes an infinitesimally small plane, whereas the metalimnion is a larger region encompassing the mean of the greatest rate of change. The *hypolimnion* is the bottom layer and is colder and denser than either the epilimnion or metalimnion. When a lake or reservoir is thermally stratified, the hypolimnion becomes largely isolated from atmospheric conditions and is often referred to as being stagnant. Additionally, the hypolimnion receives organic debris from the epilimnion, and as respiring bacteria begin the process of decomposition of this received material, consumption of dissolved oxygen (e.g., respiration) usually exceeds either production of oxygen from photosynthesis or atmospheric reaeration. The epilimnion is often referred to as the *trophogenic* area of lentic systems, where mixing through wind and wave action as well as photosynthesis exceeds respiration, whereas the hypolimnion is referred to as the tropholytic region, where organic material is synthesized and mineralization by bacteria occurs.

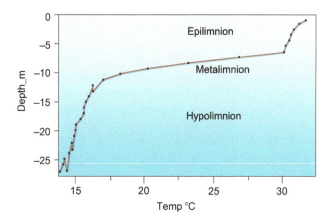

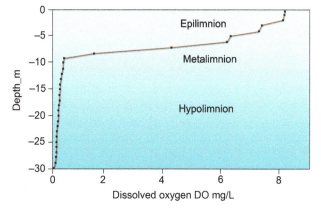

FIG. 3.12 Thermal stratification in Lake Pleasant, Arizona. Data collected on May 08, 2003. The upper graph plots depth and temperature and the bottom graph depth and dissolved oxygen.

those wavelengths of light that carry the most amount of heat energy. The relative density of water is temperature dependent. Water becomes increasingly dense down to about 4°C, at which point it becomes less dense (see Section 3.2.2). Thus differential heating leads to differential vertical layering of the water column usually beginning in spring and early summer. The definitions of the various layers are given in Information Box 3.2. An example of thermal stratification in a lake is presented in Fig. 3.12.

Interaction between the epilimnion and hypolimnion often results in the formation of autochthonous feedback mechanisms. In almost every case, anoxia within the hypolimnion mirrors epilimnetic production so that increases in trophic state result in increased hypolimnetic anoxia. Prolonged anoxia within the hypolimnion often results in the prevalence of reducing conditions (see Section 3.5). Under these reducing conditions, nutrients that would otherwise be bound to material within sediments become unbound and once again available for biological uptake. The density differences between the epilimnion and hypolimnion mean that relatively few of these nutrients are available for uptake

by phytoplankton until the lake or reservoir destratifies or "turns over," which occurred when the epilimnion cools to a temperature similar to the hypolimnion (usually during the fall). Algal "blooms" or sudden increases in biomass of algae are common during this time.

Recycling of nutrients through the thermocline from the hypolimnion to the trophogenic epilimnion does occur, although this is relatively small compared to overall nutrient levels within the hypolimnion. Increases in algal biomass mean that more organic material is available for transport back through the thermocline into the hypolimnion, adding to anoxia and the potential release of more nutrients from sediments, so that a positive feedback loop is established. This autochthonous cycling of nutrients can keep a lake or reservoir locked into a eutrophic state even after other sources of pollutants from the watershed have been reduced.

3.8 STREAMS AND RIVERS—THE LOTIC SYSTEM

Lotic systems consist of running water such as rivers or streams and are the great transporter of material to oceans, lakes, and reservoirs. Rivers and streams are also vulnerable to both natural and anthropogenic sources of pollution and have a history of being used and misused.

3.8.1 Stream Morphometry

Stream morphometry was initiated by R.E. Horton and A.E. Strahler in the 1940s and 50s to find suites of holistic stream properties from the measurement of various attributes. This was designed to allow some type of classification system that could be used as a communicative tool for hydrologists (Fig. 3.13).

The original idea was to develop a hierarchical classification system of stream segments. These segments were ordered numerically from headwaters so that individual tributaries at the headwaters were given the order of "1." The joining of two 1st order streams were given the order of "2" and the joining of two 2nd order streams, the order of "3," and so on.

Horton found that the ratio between number of stream segments in one order and the next was consistently around three. This is called the *bifurcation ratio*. This ratio has also been discovered in the rooting system of plants, the branching structure of woody plants, leaf venation, and the human circulatory system.

Order	# of Segments	Bifurcation Ratio
1	30	3.0
2	10	3.3
3	3	3.0

Horton called this association the *Law of Stream Numbers*, which is defined as the "morphometric relationship observed in the number of stream segments of a particular classification order in stream order branching." Horton combined the information that he obtained to define the *Laws of Stream Lengths* and *Basin Areas*. The Law of Stream Lengths states that a geometric relationship exists between the numbers of stream segments in successive stream orders, whereas the Law of Basin Areas indicates that the mean basin area of successively ordered streams form a linear relationship.

A quantifiable measure of the morphometry of drainage networks is *drainage density*, which is the length of stream channel per unit area of drainage basin expressed as:

$$\text{Drainage density (Dd)} = \frac{\text{stream length}}{\text{basin area}}$$

Drainage density is a useful numerical measure of landscape dissection and runoff potential.

Rivers are often divided into relatively homogenous units or *reaches*. A river has distinct chemical, physical, and biological attributes, depending upon stream order and overall size of the channel. General physical characteristics based upon channel size are given in Table 3.5.

3.8.2 Stream Hydraulics

The flow of fluids is generally classified into two types: *laminar* and *turbulent*. For laminar flow, which occurs at lower velocities, the individual fluid (e.g., water) molecules move uniformly in the direction of the mean gradient, with minimal mixing. Conversely, at higher velocities, the fluid molecules do not always move uniformly; instead, they may also cross the paths of other molecules, mixing and forming eddies. This is referred to as turbulent flow. The Reynolds number, a dimensionless parameter, is used to characterize conditions for which flow will be laminar or turbulent.

The volume of water flowing in a stream at a given time is referred to as the stream discharge. Discharge has units of volume per time (e.g., cubic meters per second). It is calculated as $Q = A * v$, where Q is discharge, A is cross-sectional area, and v is velocity. The cross-sectional area of the stream, which has units of length squared, is controlled by the shape and size of the channel and the height of water in the channel. This latter term is often referred to as the stream stage. The velocity (units of length per time) of water in a stream is controlled by the gradient (slope) of the stream and roughness of the stream channel surfaces. Water velocities are generally not uniform across the stream cross-section. Rather, the highest velocities typically occur in the center of the channel just below the surface, and the lowest velocities occur along the channel surfaces (where friction is greatest).

An empirical equation, called the *Gauckler-Manning formula*, is often used to determine the mean velocity of water flowing in an open channel. This is given as:

$$v = \frac{1}{n} R^{2/3} S^{1/2}$$

where:

v: mean stream velocity [m/s or ft/s]
R: hydraulic radius [m or ft]
S: slope of water surface
n: Manning roughness coefficient

FIG. 3.13 Example of stream ordering. Notice that it takes 2 stream orders of the same magnitude to increase downstream ranking. For example, the confluence of a 2nd order and a 1st order stream does not equal a third order, but the joining of two 2nd order streams does, http://www.cotf.edu/ete/modules/waterq/streamorder.html.

TABLE 3.5 Properties of Stream Channels

Channel Size	Physical Characteristics	Order
Small	Cobbles (>64 mm dia.) And boulders (>256 mm dia.) Dominate the substrate. Pools form behind rocks or logs (step-pool formations) Relatively steep gradients (2° to 20°) Banks composed of bedrock, boulders, and roots. Highly erosional areas.	Low
Intermediate	20–30 m max. width Dominated by pool-riffle-bar units Riffles are zones of relatively shallow, rapid flow Rapids, cascades, and glides (extended riffles) may be prominent Major pool types include *backwater* (formed from either obstructions in the main channel or from periodic flooding of banks), *dammed* (found upstream of boulder lies and gravel bars), and *scour* (where flow converges past an obstruction)	Intermediate
Large	Dominated by pool-riffle sequences, bar formations, and meanders Reach gradient largely determined by valley gradient Transport of sediment may increase *sinuosity* Sinuosity is defined as river length/valley length Braided channels may form when the river can no longer carry its sediment load Increased deposition of sediment in large channels	High

Data from: Church, M., 1992. Channel morphology and typology. In: Callow, P., Petts, G.E. (Eds.), The Rivers Handbook: Hydrological and Ecological Principles, vol. 1. Blackwell Scientific Publications, Oxford, 126–143.

A depiction of an open channel with the relevant variables is presented in Fig. 3.14. The hydraulic radius, R, is the ratio of the cross-sectional area of flow (A) to the wetted perimeter (P). The water surface slope, S, is the ratio of the water level change (h_L) to the length of the channel segment (L). The roughness coefficient represents the properties of the channel surfaces and accounts for the degree of friction associated with the different types of surfaces. The values range from a low of 0.012 for smooth concrete, used for example in aqueducts, to 0.04–0.05 for rocky mountain streams. Inspection of the previous equation shows that water velocity will be lower for a channel with a rocky surface compared to one with smooth concrete, with all other factors being equal. Once you have calculated the mean velocity of water flow, you can calculate the residence time of water in the system as L/v.

3.9 GROUNDWATER—WATER IN THE SUBSURFACE

Water in the subsurface serves as a critical resource for human consumption, both directly and indirectly (see Chapter 15). Groundwater resources serve as one of the two primary sources of potable water supply in the world (the other being surface water). In addition, water in the soil profile supports plant life, upon which humans are dependent in several ways. Water is also central to the transport and fate of contaminants in the subsurface. We will examine the distribution and movement of groundwater in this section. The impact of water flow on transport of contaminants in the subsurface is discussed in Chapters 7 and 15.

3.9.1 Water in the Subsurface

We can observe a cross-section of water distribution in the subsurface by drilling a borehole or excavating a pit. A schematic of a typical subsurface profile was presented in Fig. 2.1. The vadose zone, also known as the zone of aeration or unsaturated zone, represents a region extending from near the ground surface to a water table. The water table is defined as a water surface that is at atmospheric pressure. In the soil and vadose zones, all pores are usually not filled with water; many pores will also contain air. In such cases, the porous medium is considered to be unsaturated. Water pressure in the soil and vadose zones is less than atmospheric pressure ($P < 0$). The thickness of the vadose zone varies from a meter or less in tropical regions to a few hundred meters in arid regions, depending upon the climate (e.g., precipitation), soil texture, and vegetation (see also Chapter 2).

Water stored in the soil and vadose zones is retained by surface and capillary forces acting against gravitational forces. Molecular forces hold water in a thin film around soil grains. Capillary forces hold water in the small pores between soil grains. Gravity forces are not sufficient to

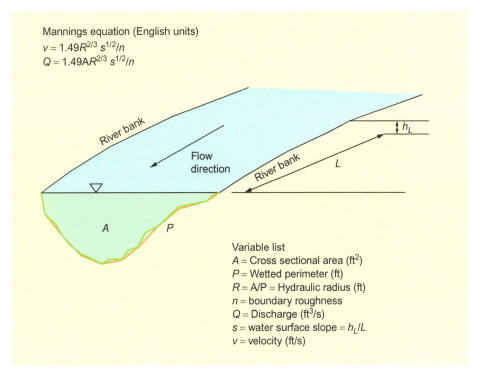

Mannings equation (English units)
$$v = 1.49R^{2/3} s^{1/2}/n$$
$$Q = 1.49AR^{2/3} s^{1/2}/n$$

River bank

Flow direction

River bank

h_L

L

A

P

Variable list
A = Cross sectional area (ft^2)
P = Wetted perimeter (ft)
R = A/P = Hydraulic radius (ft)
n = boundary roughness
Q = Discharge (ft^3/s)
s = water surface slope = h_L/L
v = velocity (ft/s)

FIG. 3.14 Diagram of components of an open channel.

force this water to percolate downward. Thus there is very little movement of water in the vadose zone when water contents are relatively low. For higher water-content conditions, some of the water is free to move under the influence of gravity. When this occurs, water movement would generally be vertically downward. If this water moves all the way to the water table, it serves to replenish (or recharge) ground water. The *capillary fringe* is the region above the water table where water is pulled from the water table by capillary forces. This zone is also called the *tension-saturated zone*. The thickness of this zone is a function of grain-size distribution and varies from a few centimeters in coarse-grained soils to a few meters in fine-grained soils. The water content in this zone ranges from saturated to partially saturated, but fluid pressure acting on the water is less than atmospheric pressure ($P < 0$).

The region beneath the water table is called the saturated or phreatic zone. In this zone, all pores are saturated with water and the water is held under positive pressure. Because all pores are filled with water, soil-water content is equal to porosity, except when liquid organic contaminants are also present in the pore spaces (see Chapter 15). The water in the saturated zone is usually referred to as groundwater. Water movement in the saturated zone is generally horizontal. Specific sections of the saturated zone, particularly those comprised of sands and gravels, are called aquifers, which are geologic units that store and transmit significant quantities of groundwater.

3.9.2 Principles of Subsurface Water Flow

Water at any point in the subsurface possesses energy in mechanical, thermal, and chemical forms. The energy status of water, for example, the effort required to move water from one point to another, is a critical aspect of quantifying water flow. For groundwater flow, the contributions of chemical and thermal energies to the total energy of water are generally relatively minor and thus are usually ignored. Therefore we consider water flow through porous media to be primarily a mechanical process. Fluid flow through porous media always occurs from regions where energy per unit mass of fluid (fluid potential) is higher, to regions where it is lower.

From fluid mechanics, the mechanical energy of water at any point is composed of the kinetic energy of the fluid, the potential (or elevation) energy, and the energy of fluid pressure. For water flow in porous media, kinetic energy is generally negligible because pore-water velocities are usually small. Thus total energy of water is considered to consist of potential and pressure energies.

The potential or elevation energy results from the force of gravity acting on the water. In the absence of pressure energy considerations, water always flows from regions of higher elevation potential to lower. This is why surface water flows "downhill." Similarly, groundwater usually flows downward from higher elevations (underneath mountain peaks) to lower elevations (underneath the valley floor).

In the vadose zone, the pressure potential is negative, indicating that energy is required to "pull" water away from the soil surfaces and small pores. In the saturated zone, the pressure potential is positive due to the pressure exerted by overlying water. The water pressure increases with depth in the saturated zone. This condition occurs in all water bodies—including swimming pools—which is manifested by the increasing pressure one feels on their sinuses and eardrums, as one swims deeper below the surface.

The energy potentials are commonly expressed in terms of length to simplify their use. In length terms, the energy potentials are referred to as "heads." The total energy potential head for water is called the *hydraulic head*; sometimes, particularly for vadose-zone applications, it is called the *soil-water potential head*. The equation we use to relate the hydraulic head (h) to its two parts, elevation head (z) and pressure head (Ψ), is $h = \Psi + z$. Each head has a dimension of length (L) and is generally expressed in meter or feet.

Hydraulic head measurements are essential pieces of information that are required for characterizing groundwater flow systems (i.e., direction and magnitude of flow), determining hydraulic properties of aquifers, and evaluating the influence of pumping on water levels in a region. Piezometers are used to measure the hydraulic head at distinct points in saturated regions of the subsurface. A piezometer is a hollow tube or pipe drilled or forced into a profile to a specific depth. Water rises inside the tube to a level corresponding to the pressure head at the terminus. The level to which water rises in the piezometer with reference to a datum such as sea level is the hydraulic head (Fig. 3.15).

The relationship between the three head components is illustrated in a piezometer in Fig. 3.15. The value of z represents the distance between the measurement point in the profile and a reference datum. Sea level is often taken as the reference point where $z = 0$, although some people use the elevation of land surface as the reference datum. The value of Ψ represents the distance between the measurement inlet point and the water level in the well. The value of h represents the elevation of water from the reference datum. This basic hydraulic head relationship is essential to an understanding of groundwater flow. Water flow in porous media always occurs from regions in which hydraulic head is higher to regions in which it is lower.

Groundwater level measurements in a network of production or monitor wells are used to define the potentiometric (or water table) surface of regional groundwater. Knowledge of this surface is required to define hydraulic gradients and flow directions. An example of a groundwater level contour map is presented in Fig. 3.16. In an unconfined aquifer, the water table represents the potentiometric surface of the aquifer. However, in a confined aquifer (one where a low-permeability unit resides on top of the aquifer), the potentiometric surface is an imaginary surface connecting the water levels in wells.

Data from piezometers terminating in depth-wise increments provide information about the vertical flow direction of water in saturated regions (Fig. 3.17). In the first (left side) diagram, the water level is uniform, indicating no vertical flow. In the second (middle) diagram, the water level is highest in the shallowest piezometer and lowest in the deepest one, indicating downward flow. The reverse is true for the third (right side) diagram.

3.9.3 Darcy's Law

In 1856, a French hydraulic engineer, Henry Darcy, established a relationship that bears his name to this day. The relationship is based on studies of water flow through columns of sand, similar to the schematic shown in Fig. 3.18. In Darcy's experiment, the column is packed with sand and plugged on both ends with stoppers. Water is introduced into the column under pressure through an inlet in the stopper and allowed to flow through it until all the pores are fully saturated with water and inflow and outflow rates are equal. Water pressures along the flow path are measured by the manometers installed at the ends of the column.

In his series of experiments, Darcy studied the relationship between flow rate and the head loss between the inlet and outlet of the column. He found that:

1. The flow rate is proportional to the head loss between the inlet and outlet of the column:

$$Q \propto (H_a - h_b)$$

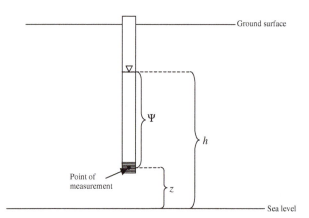

FIG. 3.15 Concept of hydraulic head (h), elevation head (z), and pressure head (Ψ) in a piezometer. The cross-hatched section at the bottoms of the tube (terminus) represents the screened interval, which allows water to flow into the piezometer. Water level is denoted by the "∇" symbol. *(From Yolcubal, I., Brusseau, M.L., Artiola, J.F., Wierenga, P., Wilson, L.G., 2004. Environmental physical properties and processes. In: Artiola, J.F., Pepper, I.L., Brusseau, M. (Eds.), Environmental Monitoring and Characterization. Elsevier Academic Press, San Diego, California, pp. 207–237.)*

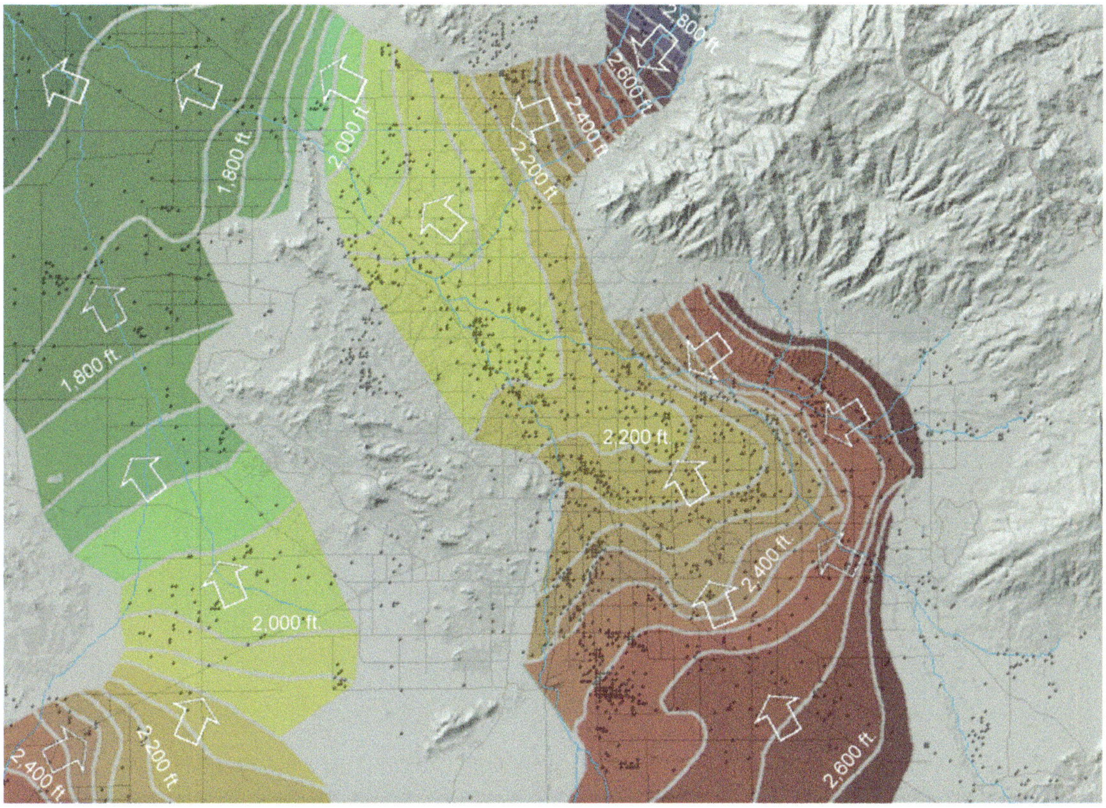

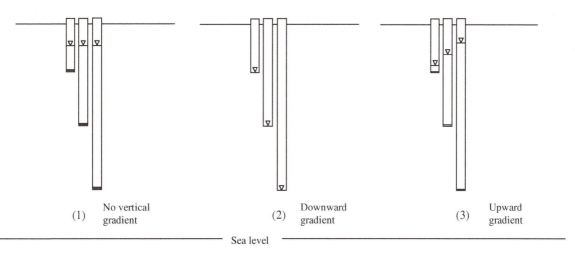

FIG. 3.16 Example of a groundwater contour map. *(Image courtesy of the Arizona Department of Water Resources and the University of Arizona Water Resources Research Center.)*

(1) No vertical gradient	(2) Downward gradient	(3) Upward gradient

Sea level

FIG. 3.17 Three hypothetical situations for vertical hydraulic gradient water level are denoted by the "∇" symbol. *(From Yolcubal, I., Brusseau, M.L., Artiola, J.F., Wierenga, P., Wilson, L.G., 2004. Environmental physical properties and processes. In: Artiola, J.F., Pepper, I.L., Brusseau, M. (Eds.), Environmental Monitoring and Characterization. Elsevier Academic Press, San Diego, California, pp. 207–237.)*

The flow rate is inversely proportional to the length of flow path:

$$Q \propto \frac{1}{dl}$$

2. The flow rate is proportional to the cross-sectional area of the column:

$$Q \propto A$$

Mathematically, these experimental results can be written as:

$$Q = KA\left[\frac{h_a - h_b}{\Delta l}\right] = KA\left[\frac{\Delta h}{\Delta l}\right]$$

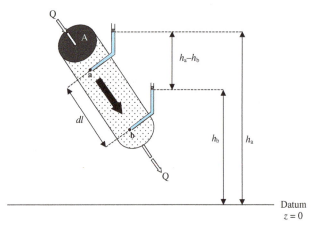

FIG. 3.18 Schematic of Darcy's experimental apparatus (original apparatus was vertically oriented). *(From Yolcubal, I., Brusseau, M.L., Artiola, J.F., Wierenga, P., Wilson, L.G., 2004. Environmental physical properties and processes. In: Artiola, J.F., Pepper, I.L., Brusseau, M. (Eds.), Environmental Monitoring and Characterization. Elsevier Academic Press, San Diego, California, pp. 207–237.)*

where:

Q = flow rate or discharge [$L^3 T^{-1}$]
A = cross-sectional area of the column [L^2]
$h_{a,b}$ = hydraulic head [L]
dh = head loss between two measurement points [L]
dl = the distance between the measurement locations [L]
dh/dl = hydraulic gradient
K = proportionality constant or hydraulic conductivity [LT^{-1}]

We can rewrite Darcy's Law as:

$$\frac{Q}{A} = q = \frac{K\,dh}{dl}$$

where q is called specific discharge or Darcy velocity with units designated as [LT^{-1}]. This is an apparent velocity because Darcy velocity represents the total discharge over a cross-sectional area of the porous medium.

The cross-sectional area of a porous medium includes both void space and solids. However, water flow occurs only in the connected pore spaces of the cross-sectional area (see Fig. 3.19). Therefore, to determine the actual mean pore-water velocity, specific discharge is divided by the porosity of the porous medium: $v = q/n$, where v is the pore-water velocity or average linear velocity. Pore-water velocity is always greater than Darcy velocity. With this v, we can calculate the mean residence time for groundwater flow in a specified zone as L/v, where L is the length of the subsurface zone of interest.

Measurements of groundwater velocity are important for hydrogeological studies and environmental monitoring applications including predicting the rate of contaminant

movement. There are several techniques available for determining groundwater flow velocities in the field. These include using Darcy's Law, borehole flowmeters, and tracer tests. Using Darcy's Law requires knowledge of hydraulic conductivity and the hydraulic gradient, which are obtained as described before. Borehole flowmeters are used for many applications including well-screen positioning, recharge zone determination, and estimation of hydraulic conductivity distribution. In addition, they can be used for measurements of horizontal and vertical flow characteristics in a cased well or borehole. Tracers may also be used in field studies to determine velocities of groundwater. A tracer test is conducted by injecting a tracer solution into the aquifer and monitoring tracer concentrations at downgradient locations. The time required for the tracer to travel from the injection point to the monitoring point can be used to calculate groundwater velocity. A suitable tracer should not react physically or chemically with groundwater or aquifer material and must not undergo transformation reactions. These types of substances are generally called conservative tracers.

3.9.4 Hydraulic Conductivity

The proportionality constant in Darcy's law, which is called *hydraulic conductivity (K)* or *coefficient of permeability*, is a measure of the fluid transmitting capacity of a porous medium and is expressed as:

$$K = \frac{k\rho g}{\mu}$$

where:

k is the intrinsic or specific permeability [L^2]
ρ is the fluid density [ML^{-3}]
g is the acceleration due to gravity [LT^{-2}]
μ is the dynamic viscosity of the fluid [$MT^{-1}L^{-1}$]

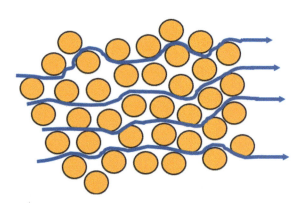

FIG. 3.19 Groundwater flow through the connected pores of a porous medium.

Hydraulic conductivity has a dimension of velocity $[L\,T^{-1}]$ and is usually expressed in $m\,s^{-1}$, $cm\,s^{-1}$, or $m\,day^{-1}$ in SI units, or $ft\,s^{-1}$, $ft\,day^{-1}$, or $gal\,day^{-1}\,ft^{-2}$.

As indicated by the previous equation, hydraulic conductivity depends on properties of both the fluid and porous medium. The two fluid properties are density and dynamic viscosity. Intrinsic permeability is a property that in most cases depends solely on the physical properties of the porous medium. This relationship can be illustrated using the expression called the *Hazen approximation*:

$$k = C(d_{10})^2$$

where:

C is a shape factor [dimensionless]
d_{10} is the effective grain diameter [L]

C is a constant that represents the packing geometry, grain morphology (size and shape), and grain-size distribution of the porous medium. The value of C ranges between 45 for clays and 140 for sand. A value of $C = 100$ is often used as an average. d_{10} is the diameter for which 10% (by weight) of the sample has grain diameters smaller than that diameter, as determined by sieve analysis. The Hazen approximation is applicable to sand with an effective mean diameter between 0.1 and 3.0 mm. Intrinsic permeability (k) has dimensions of square feet (ft^2), square meter (m^2), or square centimeter (cm^2).

As noted before, porous media properties that control K include pore size, grain-size distribution, grain geometry, and packing of grains. Among those properties, the influence of grain size on K is dramatic, since K is linearly proportional to the square of grain diameter. The larger the grain diameter, the larger is the hydraulic conductivity. For example, hydraulic conductivity of sands ranges from 10^{-4} to $10^{-1}\,cm\,s^{-1}$, whereas the hydraulic conductivity of clays ranges from 10^{-9} to $10^{-7}\,cm\,s^{-1}$. The values of saturated hydraulic conductivity vary by several orders of magnitude, depending on the material. The range of values of hydraulic conductivity and intrinsic permeability for different media is illustrated in Fig. 3.20.

In the vadose zone, hydraulic conductivity is not only a function of fluid and media properties, but also the soil-water content (θ), and is described by the following equation:

$$K(\theta) = Kk_r(\theta)$$

where:

$K(\theta)$ is the unsaturated hydraulic conductivity
K is the saturated hydraulic conductivity
$k_r(\theta)$ is the relative permeability or relative hydraulic conductivity

Relative permeability is a dimensionless number that ranges between 0 and 1. The $k_r(\theta)$ term equals 1 when all the pores are fully saturated with water, and equals 0 when the porous medium is dry. Unsaturated hydraulic conductivity is always lower than saturated hydraulic conductivity.

Unsaturated hydraulic conductivity is a function of soil-water content. As the soil-water content decreases, so does $K(\theta)$. In fact, a small drop in the soil-water content of a porous medium, depending upon its texture, may result in a dramatic decrease (e.g., 10^3, 10^6) in the unsaturated hydraulic conductivity. As we discussed earlier, the hydraulic conductivity of sands is always greater than that of clays for saturated porous media. However, in the vadose zone, this relationship may not always hold true. For example, during drainage of a soil, larger pores drain first and the residual water remains in the smaller pores. Since sand has larger pores than clay, it will lose a greater proportion of water for a given suction. Consequently, at relatively low soil-water contents (high suctions), most of the pores of a sand will be drained, while many for a clay will remain saturated. Therefore the unsaturated hydraulic conductivity of a clay unit may become greater than that of a sand unit at lower soil-water contents.

Hydraulic conductivity is a critical piece of information required for evaluating aquifer performance, characterizing contaminated sites for remediation, and determining the fate and transport of contaminant plumes in subsurface environments. For example, for water management issues, one needs to know the hydraulic conductivity to calculate the water-transmitting and storage capacities of the aquifers. For remediation applications, knowledge of K distribution of contaminated soils is necessary for calculating plume velocity and travel time, to determine if the plume may reach a downgradient location of concern. Hydraulic conductivity can be measured in the laboratory as well as in the field. Laboratory measurements are performed on either disturbed or undisturbed samples that are collected in the field. Laboratory measurements are relatively inexpensive, quick, and easy to make compared to field measurements. They are often used to obtain an initial characterization of a site before on-site characterization is initiated. However, measurements made for a sample represent that specific volume of media. A single sample will rarely provide an accurate representation of the field because of the heterogeneity inherent to the subsurface. Thus a large number of samples would usually be required to characterize the hydraulic conductivity distribution present at the site. Thus field tests, while more expensive, are generally preferred to characterize large sites.

3.10 A WATERSHED APPROACH

Strictly defined, a watershed is a bound hydrologic unit where all drainages flow to a common water source.

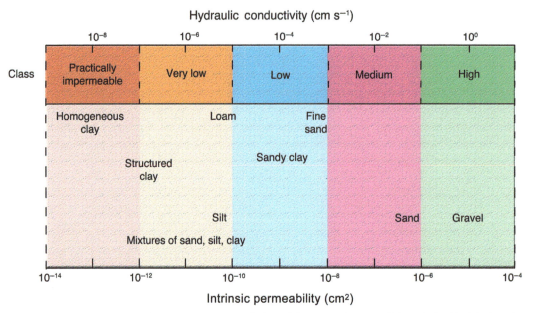

FIG. 3.20 Hydraulic conductivity and permeability at saturation. *(Modified from Klute, A., Dirksen, C., 1982. Methods of Soil Analysis. Part I—Physical and mineralogical methods. Soil Science Society of America, Madison, WI.)*

Referring back to the section on stream order, we can see that a watershed encompasses all stream orders from beginning to end. No matter where one goes on the Earth (excluding the oceans), that person would be in a distinct watershed. Watersheds are usually delineated by sharp gradients and precipitation or snowmelt flows along one side of the gradient or the other, depending upon local topography and natural hydrology. It is also important to recognize that groundwater is also part of the watershed system.

Increasingly, resource managers are using watersheds as the unit of measure for management and delineation purposes on a landscape scale. In hydrologic terms, it makes perfect sense to focus management efforts in these terms, because a land use activity in one part of a watershed often has an effect on downstream areas. The more traditional approach of addressing specific sources of pollution in a specific area has been successful in addressing problems that are readily noticeable; however, this approach has met with limited success in addressing the chronic and subtle stressors that often contribute to impairment within a watershed. A watershed framework is better able to capture these subtle changes over time.

It is important to understand that watersheds, and the organisms they contain, pay no attention to human-imposed political boundaries. Natural physical, chemical, and biological processes within watersheds provide civilization-sustaining benefits when functioning properly, which are referred to as ecosystem services as discussed in Chapter 6. Unfortunately, the same community to which watersheds provide these benefits is often the source of impairment and disruption of natural processes within them. Acknowledging this fact while understanding that political boundaries do exist is a necessary prerequisite toward cooperative management of the world's most precious natural resource.

While water within watersheds follows the physico-chemical "laws" previously described in this chapter, watershed management is equal parts hydrology and social science. Watershed management acknowledges that humans are a part of, not separated from, watershed processes and functioning. Watershed management, as a component of ecosystem management, tries to unify communities, managers, water quality experts, and as many stakeholders as possible, with the overall goal of increasing or sustaining water quality over large geographic areas for the common good of not only affected communities, but also of the watershed itself.

This ideology is far from revolutionary, and perhaps the best definition of a watershed was by the geologist, explorer, and teacher John Wesley Powell, who stated over 100 years ago that:

"A watershed is that area of land, a bounded hydrologic system within which all living things are inextricably linked by their common water course and where, as humans settled, simple logic demanded that they become part of a community."

QUESTIONS AND PROBLEMS

1. **A.** Determine pressure head (Ψ) elevation head (z), and hydraulic head (h) for both wells.
   B. Calculate the hydraulic gradient.
   C. Does groundwater flow from well #1 to #2, or from #2 to #1?

The horizontal distance between the wells is 250 m Well 1: The elevation at the ground surface is 100 m; the measured water level in the well is 40 m below the ground surface. The well casing length is 60 m.

Well 2. The elevation at the ground surface is 115 m; the measured water level in the well is 50 m below the ground surface. The well casing length is 80 m.

2. Why is hydraulic conductivity a function of the fluid as well as the porous medium? Is the conductivity for water greater than or less than that for air (for a given porous medium)?

3. Water has been described as a universal solvent. Describe three properties of water that account for its ability to dissolve so many types of compounds.

4. Describe some physical, chemical, and biological characteristics within the epilimnion and hypolimnion of a thermally stratified lake.

5. How do you calculate residence times for streams and for groundwater? Are typical residence times for the two similar or different?

REFERENCE

Carlson, R.E., 1977. A trophic state index for lakes. J. Limnol Ocean. 22, 361–369.

FURTHER READING

Birge, E.A., 1898. Plankton studies on Lake Mendota II. The crustacea of the plankton from July, 1894, to December, 1896. Trans. Wis. Acad. Sci. Arts Lett. 11, 274–448.

Brezonik, P.L., 1994. Chemical Kinetics and Process Dynamics in Aquatic Systems. Lewis-CRC Press, Boca Raton, FL, p. 754.

Callow, P., Petts, G.E., 1992. The Rivers Handbook. 1,Blackwell Sci. Publ., Cambridge 526. p.

Clesceri, L.S., Greenberg, A.E., Eaton, A.D.e., 1999. Standard Methods for the Examination of Water and Wastewater, 20th ed. American Public Health Association, Washington, DC.

Cole, G.A., 1994. Textbook of limnology. In: Prospect Heights. fourth ed. Waveland Press Inc., Illinois

Edmondson, W.T., 1994. What is limnology? In: Margalef, R. (Ed.), Limnology Now: A Paradigm of Planetary Problems. Elsevier, New York, NY.

Frey, D.G., 1963. Wisconsin: The Birge and Juday years. (Chapter 1) In: Frey, D.G. (Ed.), Limnology in North America. University of Wisconsin Press, Madison, Wisconsin 734. pp.

Horton, R.E., 1931. The field, scope, and status of the science of hydrology. In: Reports and Papers, Hydrology, Trans. AGU. National Research Council, Washington, DC, pp. 189–202.

Hutchinson, G.E., 1957. A Treatise on Limnology. Vol. I. John Wiley & Sons, New York.

Hynes, H.B.N., 1970. The Ecology of Running Waters. University of Toronto Press, Toronto.

Kelly, C.A., Rudd, J.W.M., Hesslein, R.H., Schindler, D.W., Dillon, P.J., Driscoll, C.T., Gherini, S.A., Hecky, R.E., 1988. Prediction of biological acid neutralization in acid sensitive lakes. Biogeochemistry 3, 129–140.

Klute, A., Dirkson, C., 1982. Methods of Soil Analysis. Part I—Physical and mineralogical methods. Soil Science Society of America, Madison, WI.

Mitsch, W., Gosselink, J.G., 1993. Wetlands, second ed. Van Nostrand Reinhold, New York, NY, p. 722.

Reynolds, O., 1883. An experimental investigation of the circumstances which determine whether the motion of water shall be direct or sinuous, and of the law of resistance in parallel channels. Phil. Trans. R. Soc. London, 84–99.

Schlesinger, W.H., 1991. Biogeochemistry: An Analysis of Global Change. Academic Press, Orlando, FL.

Streeter, H.W., Phelps, E.B., 1925. A study of the pollution and natural purification of the Ohio River. III. Factors concerned in the phenomena of oxidation and reaeration. U.S. Public Health Serv. Public Health Bull. 146.

Swain, E.B., Engstrom, D.R., Brigham, M.E., Henning, T.A., Brezonik, P.L., 1992. Increasing rates of atmospheric mercury deposition in midcontinental North America. Science 257, 784–787.

van der Leeden, F., Troise, F.L., Todd, D.K., 1990. The Water Encyclopedia. Lewis Pubs, Chelsea, MI.

Vannote, R.L., Minshall, G.W., Cummins, K.W., Sedell, J.R., Cushing, C.E., 1980. The river continuum concept. Can. J. Fish. Aquat. Sci. 37, 130–137.

Vollenweider, R.A., 1975. Input-output models with special reference to the phosphorous loading concept in limnology. Schweiz. Z. Hydrol. 37, 53–84.

Wetzel, R.G., Likens, G.E., 1991. Limnological Analyses. Springer-Verlag, New York, NY.

Williams, J.E., Wood, C.A., Dombeck, M.P. (Eds.), 1997. Watershed Restoration: Principles and Practices. American Fisheries Society, Bethesda, MD.

Yolcubal, I., Brusseau, M.L., Artiola, J.F., Wierenga, P., Wilson, L.G., 2004. Environmental physical properties and processes. In: Artiola, J.F., Pepper, I.L., Brusseau, M. (Eds.), Environmental Monitoring and Characterization. Elsevier Academic Press, San Diego, California, pp. 207–237.

Chapter 4

Physical-Chemical Characteristics of the Atmosphere

M.L. Brusseau, A.D. Matthias, S.A. Musil and H.L. Bohn

Wildland fires, such as this one outside of Tucson, Arizona, can contribute particulate pollutants to the atmosphere. *Photo courtesy: Janick F. Artiola.*

4.1 INTRODUCTION

The hydrologic cycle, climate, weather patterns, and other large-scale functions are influenced by atmospheric processes. In addition, the transport and fate of contaminants in the atmosphere is strongly related to its various physical and dynamic properties. Atmospheric winds, for example, determine the pathways and speeds at which pollutants are transported away from sources such as cars and smoke stacks. Another physical process, the condensation of water vapor into rain and fog droplets, scavenges water-soluble pollutants from the atmosphere, ultimately determining the rate of their removal. In addition, the vertical variation of temperature greatly influences atmospheric stability and hence the turbulent mixing of polluted air with clean air. Temperature also affects reaction rates between chemical species, such as those involved in ozone formation in polluted urban environments. The chemical composition of the atmosphere is also critically important for both small- and global-scale pollution issues. The following sections provide a brief introduction to the physical and chemical

properties of the atmosphere that are most relevant to our understanding of air pollution processes. The purpose here is to gain an overall understanding of air density, pressure, wind, water vapor, precipitation, radiation transfer, and temperature. We will not cover several important topics concerning the atmosphere, such as large-scale weather disturbances (e.g., hurricanes) or forecasting weather conditions for air pollution advisories. Interested readers should consult more comprehensive textbooks in atmospheric science for detailed information. Suggested references are listed at the end of this chapter.

4.2 PHYSICAL PROPERTIES

4.2.1 Density, Pressure, and Wind

Air is a multicomponent mixture of gaseous molecules and atoms, which are constantly moving about and undergoing frequent collisions. The mass and kinetic energy of each of these moving molecules imparts a force upon collision, which gives rise to *atmospheric pressure*. The horizontal variations of pressure, which result in air flow (winds) across the earth's surface, are an important factor in air pollution dispersal.

Pressure is the force per unit area exerted by air molecules. Usually expressed in units of Newtons per square meter ($N\,m^{-2}$) or Pascals (Pa), pressure is exerted equally in all directions because molecular scale motion is uniformly distributed in all directions. Thus, at any height in the atmosphere, pressure is the cumulative force (weight) per unit area exerted by all molecules above that height. Under static equilibrium conditions, the weight of the atmosphere pushing down on an air parcel at any height is exactly balanced by a pressure gradient force pushing upward. The weight of the atmosphere compresses air molecules near the earth's surface. In fact, nearly two-thirds of all atmospheric molecules are contained within a one scale-height distance of about 8.4 km above the surface. Air density (measured in $kg\,m^{-3}$) and pressure are both highest at sea level, with values of about $1.2\,kg\,m^{-3}$ and $101.3\,kPa$

Environmental and Pollution Science. https://doi.org/10.1016/B978-0-12-814719-1.00004-5

(1013 mbar) on average, respectively. From sea level upward, both decrease exponentially with height, as illustrated in Fig. 4.1. At heights greater than about 60 km, so few molecules (and atoms and ions) are present that both density and pressure become almost negligible.

Sea-level pressure varies both temporally and spatially across the earth's surface. For example, at any time of day, the pressure at Seattle, Washington may be several millibars higher or lower than the pressure at Miami, Florida, even though both cities are at sea level.

Surface pressure differences (gradients) can result from several factors, such as variations in the heating of air molecules by the sun, fluctuations in atmospheric water vapor and cloud cover, and rotation of the earth. Belts of semipermanent high (H) and low (L) surface pressure circle the earth at various latitudes. These belts are a result of the general circulation of the atmosphere, as illustrated simplistically in Fig. 4.2A. Large-scale circulation is composed, on average, of six main convective cells (three each in the northern and the southern hemispheres). The two main large equatorial cells shown in Fig. 4.2A are known as *Hadley cells*. The six convective cells result primarily from the heating of the earth's surface by sunlight. More sunlight is absorbed per unit area over the equator than at higher latitudes, thus heated air ascends over the equator and cooler air descends at higher latitudes. If the earth did not spin on its axis of rotation, there would likely be only one large convective cell in the northern hemisphere and one large cell in the southern hemisphere—extending between the equator

and the poles. There are six cells instead of two cells, however, because of the earth's spin on its axis, which deflects the winds. The six convective cells shown in Fig. 4.2A are important because they carry heat away from the warm equatorial region toward the poles and they transport air pollutants long distances over the earth's surface.

Low surface pressure results when warm, moist, buoyant air, such as that over the equator, ascends from the earth's surface. Moist air rises above the equator because the temperature is relatively high and because moist air, containing perhaps 2%–4% water vapor, is less dense (and lighter) than dry air. The relatively low molecular weight of water vapor (18) relative to dry air (29) lowers the average molecular weight of moist air. The lighter air moves upward and exerts relatively less pressure at the surface.

Surface flows of moist air associated with the trade winds converge at the *intertropical convergence zone* (ITCZ) near the equator (see Fig. 4.2A). As the flows converge from the northern and southern hemispheres, the air is heated by the equatorial sun and thus rises. As the air rises, it cools and loses its moisture by condensation and precipitation. At high altitudes the rising air current diverges northward and southward. At subtropical latitudes (about 30° north and south), the dry upper-level air flow subsides (sinks) toward the (mostly) ocean surfaces where it again becomes moist and flows back to the ITCZ. The subsiding air compresses (and thus warms) the atmosphere and increases pressure. Subsidence may occur at a rate of about 1 km per day at subtropical latitudes. High pressure is therefore associated with relatively warm, dry, subsiding air. This process has important implications for dispersal of air pollutants.

Surface wind patterns associated with atmospheric circulation are more complex than the simple idealized flow patterns shown in Fig. 4.2. In general, surface winds are influenced by several forces acting on air masses, including pressure gradient (flow from high to low pressure), Coriolis (deflection of air flow to the right in the northern hemisphere due to the rotation of the earth), frictional, and centrifugal forces. In the northern hemisphere, these forces combine to cause air to flow counterclockwise around low pressure and clockwise around high pressure.

Flow around low pressure is called *cyclonic flow*. Flow around high pressure is termed *anticyclonic flow*. At low latitudes in the northern hemisphere, prevailing surface *trade winds* are generally from northeast to southwest. At midlatitudes in the northern hemisphere, prevailing surface winds are generally from southwest to northeast—the *westerlies* shown in Fig. 4.2A. At high latitudes over the Arctic, air flow is generally northeast to southwest. Because of convergence at the ITCZ, winds tend to be light over the equator.

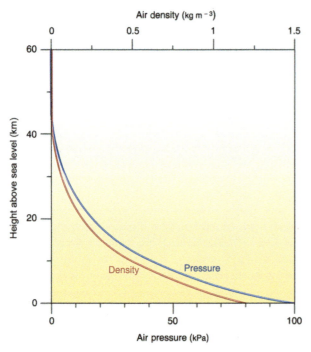

FIG. 4.1 Approximate density and pressure variations with altitude in the earth's atmosphere.

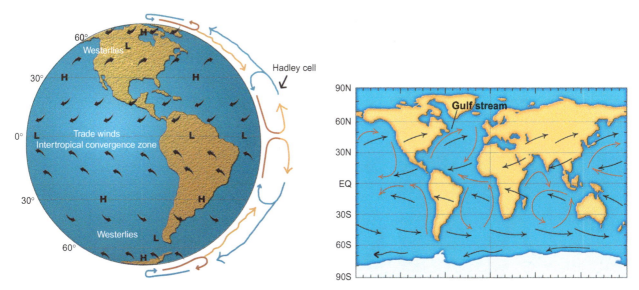

FIG. 4.2   (A) General circulation of the earth's atmosphere showing large-scale convective cells, predominant surface wind directions, and long-term surface pressure features. H = high pressure, L = low pressure. Black arrows indicate the direction of surface winds. (B) General surface wind patterns (black arrows) and ocean currents (red arrows). *(From* Pollution Science, *© 1996, Academic Press, San Diego, CA.)*

The surface wind patterns shown in Fig. 4.2A are also illustrated by the black arrows on the world map in Fig. 4.2B. Frictional drag between the winds and the ocean surface tends to cause the ocean currents (red arrows in Fig. 4.2B) to generally follow the wind patterns. Note, for example, in Fig. 4.2B how the westerly winds off the east coast of North America coincide with the direction of the Gulf Stream, which brings warm tropical ocean water to northern Europe. Although the British Islands, for example, are at a relatively high latitude range (~50–60° north), the Gulf Stream helps moderate the climate there.

The general circulation and resultant spatial pattern of high and low pressure influence long-range pollutant transport such as dust (see Fig. 4.3). In regions with semipermanent high pressure features (such as those within the subtropical high-pressure belt at about 30° latitude), calm, stagnant conditions often persist for long periods, thereby amplifying air pollutant concentrations near the surface. For example, air subsidence associated with high pressure over the eastern Pacific Ocean (see Fig. 4.2A) markedly influences air quality in the coastal cities of California. Similarly, high pressure over the southwestern United States also adversely affects air quality in the region, particularly over large urban areas such as Phoenix, Arizona. Similar high pressures can cause air quality problems over Cairo, Egypt.

4.2.2   Temperature

Temperature is a measure of the kinetic energy (heat content) of molecules and atoms. Air temperature affects nearly all physical, chemical, and biological processes within the earth–atmosphere system. A good example is the influence it has on the atmosphere of polluted urban environments, where high temperature greatly increases the rate of photochemical smog formation. Furthermore, once smog is formed, it may be dispersed upward and downward by buoyancy-generated atmospheric turbulence resulting from temperature (and hence density) differences between individual air parcels and their surrounding environment. We can see the effects of buoyancy on air motion by watching the erratic motion of a helium-filled balloon once it is released into the atmosphere.

Air temperature near the earth's surface varies markedly over different time scales, ranging from seconds to years. By midsummer, for example, air temperature at 2 m above the Sonoran Desert floor in Arizona may vary diurnally from about 45°C maximum (at midafternoon) to 20°C minimum (at dawn). In midwinter the daily variation in the desert may range about 10–30°C. Temporal variations of air temperature are caused mainly by varying solar energy input to the surface.

Air temperature at a given height, say, 2 m, also varies markedly across the earth's surface owing to spatial variations in energy input. Obviously, the lowest temperatures occur in the polar regions where solar energy input per unit surface area is small. The highest temperatures occur in low-latitude deserts, such as the Sahara in Africa, where solar energy input per unit surface area is very large and little water is available for evaporative cooling of the ground.

The question of how air temperature changes with increasing height above the ground is important when considering how air pollution is dispersed near the ground. To answer this question, we must recognize that energy exchange takes place almost continuously between the surface and the atmosphere. Some heat exchange occurs by *conduction* through a very thin layer of air over the surface; however, most heat

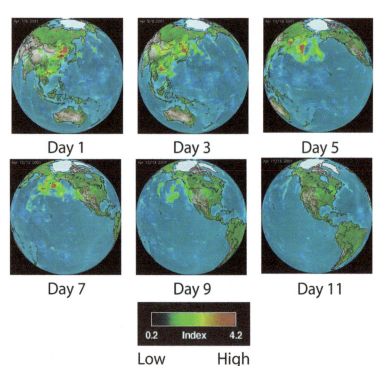

Day 1   Day 3   Day 5

Day 7   Day 9   Day 11

0.2   Index   4.2

Low   High

FIG. 4.3 A large dust storm moving from east Asia across North America in April 2001. Note the storm is intense enough to carry aerosol particles from China to northeastern North America in 11 days. Atmospheric circulation can move air pollutants on a global scale. *(Data from the Earth Probe TOMS (Total Ozone Mapping Spectrometer); Images: U.S. National Aeronautics and Space Administration.)*

exchange occurs by means of *convection*, which is the turbulent exchange caused by buoyancy and shear stress. Convection becomes increasingly more efficient with increasing height. This increase is due to the decreased effect of surface frictional drag at greater heights. At midday the change in temperature with height above the surface is often very large. Temperature gradients within the first few millimeters above a hot desert soil surface may be as high as $-1°C$ per millimeter. Note that a negative temperature gradient means that air temperature decreases with increasing height from the ground surface. Because convection quickly becomes very efficient in mixing air with increased height, the temperature gradients within turbulent air rapidly decrease in magnitude with increasing height. It is important to remember that air in contact with the earth's surface during daylight hours is generally warmer than air aloft because of strong surface heating. Thus air temperature generally decreases with increased height within the lower part of the atmosphere during the daytime.

Atmospheric scientists use the term *lapse rate* to describe the observed change (generally a decrease) of air temperature with height ($\Delta T/\Delta z$). The lapse rate at a given height (z, in meters) and location may vary greatly throughout the day in response to changes in heat flow between the surface and the atmosphere. At a height of 2 m at midday above a hot desert soil surface, for example, the lapse rate may range from about $-0.01°C$ to $-0.2°C$ per meter. Within the first few kilometers of the lower atmosphere, however, the lapse rate is, on average, about $-0.0065°C$ per meter.

During the night, however, the situation is generally the reverse of daytime conditions. The ground surface may quickly lose energy to space by infrared radiation emission (see Section 4.4) and become relatively cool. This cooling process also cools the air in direct contact with the surface. Thus at night, air temperature often increases with increasing height, typically on the order of $0.1–1°C$ per meter. An *air temperature inversion* occurs when temperature increases with height up to a level (called the *inversion height*) of maximum air temperature. Above the inversion height, the temperature decreases with height. Radiation inversions are particularly common in the dry desert environment of the southwestern United States and northeastern Africa, where nocturnal loss of radiant energy from the ground to space causes cooling of the air in contact with the ground. Inversions can also occur as a result of subsidence associated with anticyclonic flow, which is also common over the southwestern United States. (These and other causes of inversions are discussed further in Chapter 17.4.1) As discussed in the following paragraphs, the stable atmospheric conditions associated with inversions tend to trap pollutants near their source. The stability of air defines its ability to mix and disperse pollutants. Air can be unstable, stable, or neutral.

Unstable air results in turbulent motion associated with free convection due to buoyancy within the atmosphere. Buoyant motion enhances upward penetration of air parcels into the atmosphere, thus helping to disperse pollutants. Under unstable conditions, an air parcel that is displaced adiabatically (without heat exchange with its surroundings) upward or downward a short distance is accelerated away from its initial position by buoyancy. Air is unstable because the net buoyancy force acting on the parcel accelerates it either upward or downward, depending upon the temperature (density) difference between the parcel and its surrounding environment. Unstable conditions are prevalent during daytime when convection carries heat upward from the soil surface.

In *stable air*, turbulence is suppressed or even absent. In stable conditions, buoyancy tends to restore an adiabatically displaced parcel to its original height. In other words, the buoyancy force acts in the direction opposite to the motion of the displaced parcel. Stable conditions occur most often at night, when convective heat flow is downward from the atmosphere to the soil surface.

Neutral stability means that the buoyancy force is zero and that a balance exists between gravity (acting downward) and the pressure gradient force (acting upward) on the parcel. The pressure gradient force is the difference between the pressures at the top and bottom of the parcel divided by the distance between top and bottom. Thus under neutral conditions an air parcel displaced upward or downward from its initial height remains at its new height unless acted upon by an external force. Neutral conditions often occur briefly after sunrise, and before sunset, when convective heat flow is zero. Cloudy, windy days are also favorable for neutral stability.

We base the assessment of the pollutant-dispersal ability of the atmosphere on quantification of the stability of the atmosphere. Stability is largely determined by the value of the measured lapse rate $\Delta T/\Delta z$ relative to the adiabatic lapse rate (Γ). The constant Γ is defined as the change in the temperature of the air parcel when the parcel is displaced upward or downward adiabatically from a base height z_b (see Fig. 4.4). This change in temperature results from a change in pressure, as described by the ideal gas law. When the atmosphere is relatively dry, Γ is equal to $-g/c_p$ where g is the acceleration due to gravity, c_p is the specific heat of air at constant pressure.

The Γ thus has a value of about $-0.01°C$ per meter. This means that the temperature (T_b) of an air parcel adiabatically lifted from height z_b will decrease by $0.01°C$ per meter of displacement. Likewise, if the parcel is lowered from z_b its temperature will increase by $0.01°C$ per meter of displacement.

- When $\Delta T/\Delta z < \Gamma$ (e.g., $-0.05°C$ per meter is less than $-0.01°C$ per meter), the atmosphere is unstable, as shown in Fig. 4.4 *(center)*. It is unstable because a parcel adiabatically displaced upward from its initial position at z_b is always warmer than its surroundings. When it is adiabatically displaced downward, it is cooler than its surroundings. Thus it may be accelerated up or down from z_b by the buoyancy force, causing turbulence.

- When $\Delta T/\Delta z < \Gamma$, conditions are said to be stable (Fig. 4.7, *bottom*). During stable conditions, a parcel adiabatically displaced upward from z_b becomes cooler than its surroundings. When it is displaced downward, it becomes warmer than its surroundings. Thus buoyancy restores the parcel to z_b, suppressing turbulent motion.

- When $\Delta T/\Delta z < \Gamma$, neutral conditions are present (Fig. 4.7, *top*), and the buoyancy force acting on the parcel is zero at any height.

The criteria for characterizing stability are summarized as follows:

Unstable conditions: $\Delta T/\Delta z < \Gamma$.
Stable conditions: $\Delta T/\Delta z < \Gamma$.
Neutral conditions: $\Delta T/\Delta z = \Gamma$.

4.3 ATMOSPHERE STRUCTURE

4.3.1 Lower Atmosphere

Temperature variation with height defines the various layers of the atmosphere. The major atmospheric layers are shown in Fig. 4.5. The *troposphere* is the lowest major layer. Within the troposphere, vertical variation of temperature is characterized by lapse-rate conditions; thus the troposphere is generally unstable and well mixed. The troposphere extends upward from the surface to a height of about 10–15 km, depending upon latitude and season of the year. The troposphere is certainly familiar to us, since it is tropospheric air that we breathe. Also, most of our weather occurs in the troposphere, including cloud formation, rain, winds, and other meteorological processes. The *tropopause* is the upper limit of the troposphere, which separates the troposphere from the stratosphere above.

The *atmospheric boundary layer*, which is an important sublayer at the bottom of the troposphere, forms the atmospheric interface between the troposphere and the ground surface. In this region of the atmosphere, airflow patterns are strongly affected by buoyancy (free convection) and surface shear forces (forced convection). Within the first few meters above the ground surface, vertical gradients of air temperature, wind speed, humidity, and other scalar quantities are often large and variable with time. These gradients are due mainly to the temporal variability of energy and mass exchanges (e.g., evaporation of water) between the surface and the atmosphere. The depth of the boundary layer varies over the course of the day. By midafternoon, rising air from the heated ground may extend the boundary layer up to the 1 km height. This height is often referred to as the *mixing depth or mixed layer*. The turbulent air parcels, often referred to as *eddies*, undergo eddying motion. By night however, the atmosphere cools, and the boundary may shrink to a thickness of only about 0.1 km. Most of the important atmospheric pollutant transport and

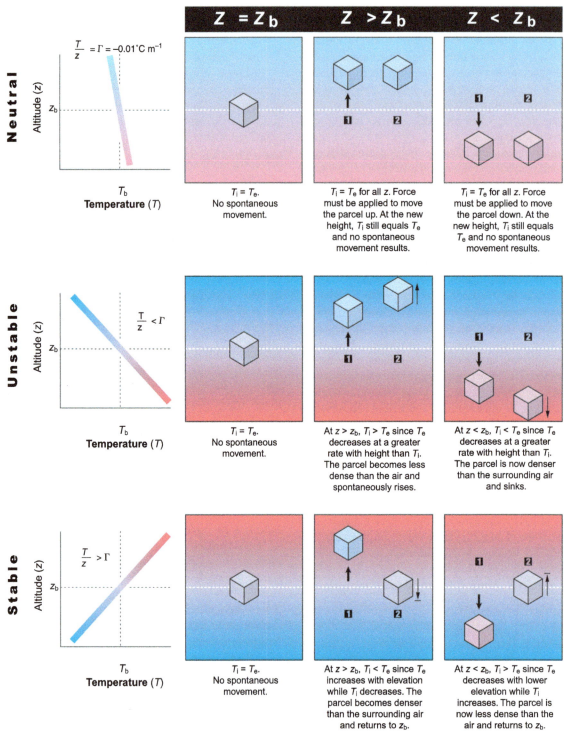

FIG. 4.4 Air-temperature variations with height during neutral, unstable, and stable conditions. *(From Pollution Science, © 1996, Academic Press, San Diego, CA.)*

transformations occur within the boundary layer. However, some chemically stable gases are dispersed upward throughout much of the troposphere. Some, such as N_2O and CFCs, eventually diffuse upward into the stratosphere.

At the very bottom of the boundary layer, directly above the earth's surface, is a sublayer known as the *surface layer*. This sublayer generally extends upward to about one-tenth of the boundary-layer depth. The properties of the surface layer are most directly affected by surface roughness and surface heat exchange. Energy and mass fluxes are nearly constant with height in the surface layer; thus this sublayer is sometimes called the *constant flux layer*.

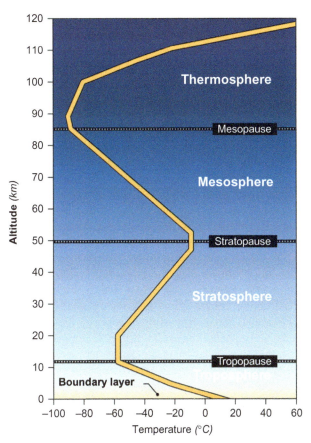

FIG. 4.5 The structure of the atmosphere as defined by average variation of temperature with altitude. *(From* Pollution Science, © *1996, Academic Press, San Diego, CA.)*

4.3.2 Upper Atmosphere

The *stratosphere* is the stable (stratified) layer of atmosphere extending from the tropopause upward to a height of about 50 km (Fig. 4.5). The stratosphere is highly stable because the air temperature increases with height up to the stratopause, which is the height of the temperature inversion. The increased temperature in this layer is due mainly to UV absorption by various chemical species, including ozone and molecular oxygen present in the stratosphere. Maximum heating takes place in the upper part of the stratosphere. Because of the stable air, pollutant mixing is suppressed within this layer. Thus natural (e.g., N_2O) and synthetic (e.g., CFC) chemicals that reach the stratosphere from the troposphere tend to diffuse upward very slowly within the stratosphere.

Ozone is formed naturally and photochemically within the stratosphere. Ozone is considered a pollutant in the troposphere, but in the stratosphere it is essential to life on earth because it absorbs biologically harmful UV radiation.

The *mesosphere* and the *thermosphere* are two additional atmospheric layers above the stratosphere. These layers are largely decoupled from the stratosphere and troposphere below; therefore they exert little influence on our weather and on pollutant transport processes. Likewise, pollution has little or no effect upon these two upper layers.

4.4 ELECTROMAGNETIC RADIATION

The energy associated with electromagnetic radiation is proportional to the wavelength of the light waves. Shorter wavelength light, such as the visible light produced by the sun, has higher energy per unit wavelength than longer wavelength light, such as infrared radiation emitted by the ground and the earth's atmosphere. Quantum theory states that electromagnetic energy, such as light, is transmitted in discrete amounts or "packets" called *quanta*. A single quanta (i.e., "quantum") of electromagnetic energy is also referred to as a *photon*, and the energy carried by each photon is proportional to its frequency. The quantity of electromagnetic energy flow over time is measured as a rate, i.e., $quanta\,s^{-1}$ known as the *radiant flux of light*.

The arrangement of light based upon differing wavelengths, frequencies, and energies is described by *spectra*. For example, the spectrum formed by white light contains all colors and is therefore said to be continuous (Fig. 4.6). Certain biological, chemical, and physical processes occur only at specific frequencies of spectra, some of which can be seen with the human eye and several of which cannot. Light that is divided over a certain range of spectra is divided by color and measured by its frequency in nanometers.

Light travels at $299,792\,km\,s^{-1}$. It takes approximately 400 trillion waves of red light at 750 nm to span the distance light travels every second. It takes almost twice as many violet waves, at 380 nm, to fill the same volume of a light second. The amount of radiation emitted by the sun that reaches the earth's outer atmosphere is $1.94\,cal\,cm^{-2}\,min^{-1}$ and is known as the *solar constant*. The most common wavelength that makes it to our outer atmosphere is ~480 nm.

Light wave frequency and energy are interrelated as explained by *Planck's equation*, which is expressed as:

$$E = h\nu$$

where

E is the energy in a photon of light.

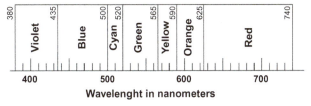

FIG. 4.6 Wavelengths of light and associated spectra (http://hyperphysics.phy-astr.gsu.edu/hbase/vision/specol.html).

h is Planck's constant of 6.6255×10^{-34} J s^{-1}.
v is the wave frequency.

Frequency of a light wave is given by

$$v = c/\lambda$$

where

c is the speed of light (3×10^8 m/s), and.
λ is the wavelength.

For example, the frequency of red light (750 nm) is calculated from.

$$v = \frac{3 \times 10^8}{7.5 \times 10^{-7}} = \frac{4.00 \times 10^{14}}{\text{s}^{-1} \text{ wavelengths light}}$$

Substituting this number into Planck's equation ($E = hv$), 26.5×10^{-20} joules are contained in a single photon of light. Compare this with the fact that it takes 6.024×10^{23} photons just to initiate a photosynthetic reaction (otherwise known as *Avogadro's number*). Thus it can be seen that it takes a large amount of light energy to perform what we would consider a simple biological process. Photosynthetically active radiation, the spectrum of light needed for photosynthesis by most plants, is approximately 400–700 nm. Specific types of chlorophylls and accessory pigments in plants have narrowed the requirements of wavelength ranges. For example, chlorophyll *a*, a photosynthesizing pigment common to all algae, absorbs light in two peaks, 670–680 nm and again at 435 nm.

The transfer of energy within the earth–atmosphere system by electromagnetic radiation (light) is very important for maintaining the climate of the earth. The radiation environment of the earth is largely a function of three main radiative transfer processes:

1. The flux of radiant energy reaching the earth's surface from the sun,
2. The redistribution of the radiant energy between the ground and the atmosphere,
3. The loss of radiant energy to space.

The overall radiation balance of the earth–atmosphere system involves absorption, scattering, transmission, and emission processes, which are described briefly in this section. Two wavelength intervals of the electromagnetic spectrum are of primary importance to the overall radiation balance: shortwave (solar) and longwave radiation. Shortwave consists of the wavelength interval from about 0.15 to 3.0 μm. Within this portion are the components of *ultraviolet* (0.15–0.36 μm), *visible* (0.36–0.75 μm), and *near infrared* (0.75–3 μm) radiation. Shortwave energy is emitted by the sun, which is an almost perfect blackbody radiator with a temperature of about 6000 K. A perfect blackbody radiator emits the maximum possible radiant energy per unit wavelength at a given temperature.

Ultraviolet and visible light are the shortwave components most significant to the global environment.

On a per-unit-wavelength basis, ultraviolet (UV) is very high-energy radiation. Fortunately, most of the high-energy UV wavelengths are selectively absorbed by ozone, oxygen, and other constituents in the earth's upper atmosphere. Some UV, however, reaches the earth's surface, where it can be harmful to life and contributes to the production of photochemical smog in the lower atmosphere.

Most shortwave radiation is within the visible portion of the spectrum. As noted above, the wavelength of maximum energy flux from the sun is at 0.48 μm, which is visible to the human eye as green light. Most visible light from the sun passes through clear air without significant loss. Scattering of visible light by atmospheric molecules, clouds, and aerosols does occur, however. Visible light reaching the land surface is either absorbed (about 75%) or reflected (about 25%) by surface matter (e.g., plants, water, and soil). The absorbed radiant energy heats the soil and air and evaporates water. Some of the absorbed energy is reemitted back to the atmosphere in the form of longwave radiation.

Longwave radiation encompasses the spectrum from about 3.0 to 100 μm and is emitted by matter within the earth–atmosphere system. Since terrestrial absolute temperatures are about 288 K (kelvin) (~15°C), the wavelength of maximum longwave emission is about 10 μm. Longwave radiation is commonly referred to as terrestrial or *infrared* radiation.

The earth's atmosphere is largely opaque to most of the longwave spectrum. A window exists, however, between about 8 and 11 μm that permits escape of a portion of longwave energy to space. Absorption and reemission of longwave energy within the atmosphere occur within the vibrational energy mode of various molecular species, including water vapor, carbon dioxide, nitrous oxide, and methane. These and a few other species are the well-known "greenhouse" gases and are responsible for the earth's *greenhouse effect*.

Without the warming by the natural greenhouse effect, the earth surface would be about 33°C colder than its present mean temperature of about 15°C. There is, however, much public and scientific concern that an additional 1.4–5.8°C global warming will occur by the end of this century owing to gradual accumulation of atmospheric greenhouse gases from anthropogenic sources.

The major greenhouse gas is water vapor because of its high concentration relative to the other greenhouse gases. Dust and aerosol clouds also affect radiative transfer. The enormous Tambora volcanic explosion April 10–11, 1815, on the island of Sumbawa in Indonesia caused the "year without a summer" in 1816 and crop failures in the northern hemisphere due to cooling by dust in the atmosphere (see Information Box 4.1). However, some volcanos can cause cooling by injecting large amounts of SO_2, in

INFORMATION BOX 4.1 Effects of a Large Volcanic Explosion

The Tambora explosion of 1815 is an example of how dust and aerosols from a volcano can affect the earth's climate. The eruption of the Tambora volcano on the island of Sumbawa in Indonesia during April 10–11, 1815, took the lives of about 92,000 people. It was a "supercolossal" eruption that rated a 7 out of 8 (8 being the largest) on the Volcanic Explosive Index scale. It was the largest volcanic eruption in the past 500 years. The energy of the eruption (equivalent to about 20,000 MTons of TNT, which is far larger than the largest nuclear bomb ever detonated (58 MTons TNT)), injected about $150 \, km^3$ ($\sim 36 \, miles^3$) of ash and dust presumably >25 km into the stratosphere that darkened the earth for many days. Dust and aerosols from the volcano remained suspended in the atmosphere for several months and ultimately significantly cooled the earth on average by $-0.7°C$ by reflecting sunlight back to space. The cooling led to "the year without a summer," causing crop failure throughout North America and Northern Europe in 1816. The anomalous cold weather brought snow every month of the year to New England. The crop failures there may also have caused an increased migration of people from the New England states to the new territories in the Midwest.

FIG. 4.7 A major eruption of Mount Pinatubo, Philippines, in June 1991. The eruption sent a cloud of ash and gases into the stratosphere that circled the world multiple times. It is estimated that the cloud cooled annual temperatures in some regions by as much as 0.5°C. *(Photograph: Karin Jackson, courtesy of the U.S. Geological Survey.)*

addition to or in place of dust, into the upper atmosphere. The 1991 eruption of Mt. Pinatubo is a good example (see Fig. 4.7).

4.5 WATER VAPOR AND PRECIPITATION

Water vapor is a highly variable part of the atmosphere and is a major component of the hydrologic cycle. In warm, humid, tropical rain forests, high rates of evaporation of water from the earth's surface keep the lower atmosphere almost continuously saturated. On the other hand, in dry, hot deserts, there is usually little water to evaporate, and the amount of water vapor in the atmosphere is almost negligible.

Atmospheric water vapor is characterized by various parameters, including vapor pressure, relative humidity, dew point temperature, water vapor density, and specific humidity. *Relative humidity* is probably the most familiar. It is defined as the ratio of the actual vapor pressure to the saturation vapor pressure of the air, which is solely a function of air temperature.

Condensation of water vapor into cloud and fog droplets occurs when air is cooled to saturation at the *dew point temperature*. Cooling occurs by various processes, such as the radiational cooling of the surface at night, the upward convective movement, advective motion of the atmosphere (in which a cold air front displaces warm moist air upward),

and orographic lifting (in which air rises over mountain ranges).

Precipitation in the form of rain or snow rids the air of many types of particulate matter and gaseous pollutants. This removal by scavenging is known as *wet deposition*. Pollutants may dissolve directly in the water droplets or they may be adsorbed on the droplets. Soluble pollutants include, among other compounds, dioxides of sulfur and nitrogen. Removal of these pollutants increases the acidity of precipitation, producing acid rain.

4.6 CHEMICAL COMPOSITION

By mass and by volume, >99% of the atmosphere consists of nitrogen (N_2), oxygen (O_2), and argon (Ar) gases (Table 4.1). The concentrations of these atmospheric gases, together with neon (Ne), helium (He), and krypton (Kr), have probably been constant for many millions of years and are unlikely to change markedly, either by natural or anthropogenic means.

The trace gas concentrations, conversely, are variable. These gases, which are the subject of considerable concern, are listed in Table 4.2. [*Note:* Although water vapor is listed

TABLE 4.1 Constant Atmospheric Components

Gas	Percent by Volume of Dry Air	Concentration ($\mu L L^{-1}$)
Nitrogen (N_2)	78.1	780,840
Oxygen (O_2)	20.9	209,460
Argon (Ar)	0.9	9340
Neon+helium+krypton (Ne+He+Kr)	0.002	24

From *Pollution Science*, © 1996, Academic Press, San Diego, CA.

TABLE 4.2 Variable Gas Concentrations in the Atmosphere

Gas	Concentration ($\mu L L^{-1}$)
Water vapor (H_2O)	<10,000
Carbon dioxide (CO_2)	380
Methane (CH_4)	1.5
Hydrogen (H_2)	0.50
Nitrous oxide (N_2O)	0.31
Ozone (O_3)	0.02
Carbon monoxide (CO)	<0.05
Ammonia (NH_3)	0.004
Nitrogen dioxide (NO_2)	0.001
Sulfur dioxide (SO_2)	0.001
Nitric oxide (NO)	0.0005
Hydrogen sulfide (H_2S)	0.00005

Adapted from *Pollution Science*, © 1996, Academic Press, San Diego, CA.

among the variable components, it will not be discussed here.] We know that all of these gases are affected by human activities as well as by reactions with the soil, biosphere, and oceans. But how much change in trace gas concentrations is due to human activity and how much to lesser known natural causes is still unclear.

Soils serve as both source and sink for virtually all of the gases in Table 4.2. Whether soils function as a source or sink can vary between day and night, with the season, with water content, with cultivation, with fertilization, and with the gas being considered, as can the strength of that function. The oceans and biosphere also fluctuate in their source/sink behavior. This chapter emphasizes the role of soils in controlling trace gas concentrations within the atmosphere.

The global average concentration[1] of carbon dioxide (CO_2) in the atmosphere was ~403 $\mu L L^{-1}$ as of 2016, which is 100 $\mu L L^{-1}$ larger than at any time in the past 800,000 years. The annual rate of increase in atmospheric carbon dioxide over the past 60 years is about 100 times greater than any time prior. The significant increase observed over the past century is primarily due to the combustion of fossil fuels. Superimposed on the long-term rising trend are smaller scale changes that occur over different time periods. For example, the CO_2 concentration decreases by a few parts per million each summer because of increased photosynthesis by terrestrial plants. From fall through spring, CO_2 rises because of microbial decomposition of organic matter and plant respiration. The heights

---

1. Various units are used to express gas concentrations in the atmosphere. Here, $\mu L L^{-1}$ and percentages by mass and volume will be used. The unit $\mu L L^{-1}$ corresponds to the commonly used unit of ppmv (parts per million by volume). We can also interpret both $\mu L L^{-1}$ and ppmv as a mol fraction $\times 10^6$, or the number of gas molecules (or atoms) of a component gas in one million air molecules (or atoms) (as we can show through derivation). For air pollutants and regulatory purposes, a mass per volume unit, $mg\,m^{-3}$, is also used.

of the annual peaks and valleys are buffered by the less seasonal photosynthesis-degradation cycle in the oceans. In the southern hemisphere, the CO_2 peaks and valleys are six months out of phase with those of the northern hemisphere. The amplitude of the annual CO_2 variation in the southern hemisphere is also much smaller because the southern land areas are much smaller and a larger percentage is arid.

Another carbon-based gas is methane (CH_4), whose major sources are swamps, natural gas seepage, and termite activity. Methane concentrations in the atmosphere have been increasing over the past several decades. Hydrogen is also liberated in small amounts from wetlands.

Nitrous oxide (N_2O) and ammonia (NH_3) are released from and absorbed by soils naturally, and releases of these gases are higher after fertilization. Nitrous oxide, whose concentration has also been increasing slowly, is a rather unreactive gas that has a long residence time ($\sim$150 years) in the atmosphere. Atmospheric ammonia has also been observed in higher concentrations, especially in industrial regions, where it often takes the form of ammonium sulfate. Ammonia reacts rapidly with soils, plants, and the ocean.

Nitrogen dioxide (NO_2), its dimer N_2O_4, and nitric oxide (NO) (which are often combined and written as NO_x) are produced by combustion of coal and petroleum fuels, as is sulfur dioxide (SO_2). While sulfur is a constituent of coal and oil, the nitrogen oxides are by-products of high-temperature furnaces and internal combustion engines. In addition, lightning produces NO_x. These gases are highly reactive in air: they rapidly oxidize to nitric and sulfuric acid, which quickly dissolve in water and wash out as *acid rain*. These gases are also absorbed directly from the air by plants and calcareous soils.

Rain is naturally acidic (pH 5–6) because it absorbs atmospheric carbon dioxide to form carbonic acid. Rain becomes even more acidic as it absorbs SO_2 and NO_x and forms nitric and sulfuric acids. The effect of this pollutant absorption is particularly evident in the low pH of rain downwind of the major industrial centers of North America and Europe (see Information Box 4.2). The low pH of rain can have a number of deleterious effects on living organisms in freshwater and terrestrial ecosystems, as most organisms do best in a rather narrow range of pH levels

INFORMATION BOX 4.2 Acid Rain Effects on Built Environments

Acid rain can cause significant surface damage to man-made items such as buildings, metals, and glass (see Fig. 4.8). The Acropolis in Athens, Greece, is a good example of an ancient structure that has survived earthquakes, wars, and time, only to be severely damaged by acid deposition. SO_2 and sulfuric acid also attack medieval stained glass, which is causing damage at the Canterbury Cathedral in England and the Chartres Cathedral in France.

Continued

INFORMATION BOX 4.2 Acid Rain Effects on Built Environments—cont'd

FIG. 4.8 Limestone pillar from an old (circa 1200 CE) church in central Paris. Acid rain accelerates weathering of building materials such as marble and limestone. *(Photo courtesy: M.A. Crimmins.)*

(near neutral, pH 5–9). At its worst, acid rain can gradually decrease the pH of the local water and soil to the point of indirectly killing many organisms. The strength of the effect of acid rain is dependent on the overall amount of acid deposited and the buffering capacity (acid-neutralizing capacity) of the soil and the underlying bedrock. For instance, areas with limestone tend to be less sensitive to acid rain because the limestone reacts with the acids to keep pH levels more neutral. Large regions of eastern Canada are strongly affected by acid rain, in part because the parent soil material is granite, which has very little buffering capacity.

Air pollutants, including SO_2 and NO_x, can be transported long distances from their sources by wind. Reducing acid rain requires a regional approach, often including multiple countries. The United States and Canada have joint agreements on pollutant controls and emissions trading programs. The United Nations also has developed protocols that are followed by many European countries. As a result, recent studies show that acid deposition has decreased in many parts of North America and Europe due in part to new emissions regulations over the last 20 years. With this decrease, some regions are showing signs of relatively rapid recovery. Other areas continue to acidify, perhaps in part because the local buffering capacity is poor. However, most of the reduced emissions are in sulfur compounds, with NO_x continuing to contribute to acid deposition. Over 85% of SO_2 production in the United States in 2002 (see Fig. 4.9) was from stationary fuel combustion (power and industrial plants). About 54% of NO_x production in the United States was from transportation, which is more difficult to control due to the large number of sources. As newer vehicles with improved catalytic converters become

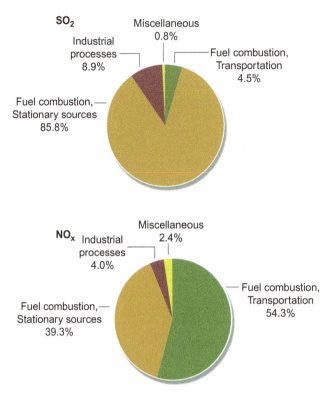

FIG. 4.9   Sources of SO_2 and NO_x in the United States in 2002. *(Data: U.S. Environmental Protection Agency.)*

common, there should be some reduction of NO_x emissions. Finally, some areas of the world, such as China, are currently undergoing rapid industrialization and are experiencing an increase in acid rain.

On the global scale, carbon monoxide (CO) is not considered an air pollutant because soil microorganisms adsorb it relatively rapidly and oxidize it to CO_2. In urban areas, however, carbon monoxide can accumulate during rush hour traffic. Long-term exposure to low levels of carbon monoxide can affect cardiovascular health. Exposure to high levels of CO is toxic.

To decrease air pollution, regulatory agencies worldwide have put increasing restrictions on SO_2, NO_x, and organic chemical emissions. London, Pittsburgh, Salt Lake City, Los Angeles, and many other North American and European cities have already shown obvious improvements. Fig. 4.10 illustrates smog in Salt Lake City, Utah, during the early 1970s, before strict emissions standards were promulgated. Visibility and air quality in Salt Lake City have improved because of regulatory restrictions on emissions. Unfortunately, in other cities, such as Mexico City, the situation will probably worsen before it improves.

Ozone (O_3) is considered an air pollutant in the lower atmosphere because it is harmful to plants and humans. Ozone is produced by the action of ultraviolet (UV) sunlight on polluted air that contains nitrogen oxides (NO and NO_2) and organic gases such as industrial solvents, fuels, and partially oxidized hydrocarbons. The ozone concentration has

therefore been adopted by the U.S. Environmental Protection Agency (EPA) as an index of air pollution.

In the upper atmosphere, ozone is beneficial to life on earth in that it absorbs much of the UV fraction of sunlight. The UV fraction of sunlight can cause skin cancer in humans and animals and stunt plant growth. A distinctive pollution problem that increases harmful UV light at the earth's surface is the *ozone hole*, a seasonal (springtime) decrease in stratospheric ozone measured over the South

FIG. 4.10   Smog covers Salt Lake City in July 1972 prior to increased emissions restrictions. Visibility and air quality have improved because of regulatory restrictions on emissions. *(Photo courtesy: U.S. Environmental Protection Agency.)*

Pole. The hole was discovered by British scientists in the 1980s. During the 1980s and 1990s the size of the hole generally increased, but recent observations indicates that the size is decreasing (see Fig. 20.10).

At higher altitudes, ozone is attacked by chlorine (Cl) and to a lesser extent by NO_x. In the more intense UV of that region, chlorofluorocarbons (CFCs, such as CCl_2F_2) decompose to Cl_2, which degrades O_3 to oxygen. NO_x degrades O_3 as well, but to a lesser extent; N_2O at higher altitudes is converted by UV light to NO_x (see Chapter 17.2.4.2).

Hydrogen sulfide is a colorless gas that smells like rotten eggs. There are numerous natural sources of hydrogen sulfide, including thermal springs and swamps. Anthropogenic sources include oil production, pulp and paper mills, municipal sewer plants, and large livestock operations. Since the odor can be perceived at levels as low as 10 ppb, in low quantities it can be a nuisance gas. High exposures (>300 ppm) can cause severe respiratory distress, with exposures over 700–800 ppm usually resulting in death. Hydrogen sulfide gas reacts with water vapor to form sulfuric acid, which can contribute to acid rain.

4.7 QUESTIONS AND PROBLEMS

1. What causes wind? How does wind affect the movement of heat, water vapor, and pollution in the atmosphere?
2. How can surface pressures differ between Venice and New York even though both cities are at sea level?
3. Describe how and why air temperature varies with increasing height in the troposphere and stratosphere.
4. Why is air generally well mixed in the troposphere but not in the stratosphere?
5. The dry adiabatic lapse rate of the atmosphere is given by $\Gamma = g/c_p$ where g is acceleration due to gravity $(9.8\,m\,s^{-2})$ and c_p is the specific heat of air $(1010\,J\,kg^{-1}\,K^{-1})$. Calculate the actual numerical value of Γ and show that its units are $K\,m^{-1}$. [Hint: Recall that Joules (J) of energy can be represented in terms of kinetic energy $(kg\,m^2\,s^{-2})$.]
6. Suppose a small parcel of air is lifted adiabatically (i.e., it is insulated from its surroundings) from the ground upward to a height of 100 m. How much cooler or warmer will the parcel be at 100 m than at the ground surface?
7. Suppose the small parcel of air in question #6 is 1° Celsius warmer than the air surrounding it at 100 m height. Will the parcel move upward, downward, or remain stationary at 100 m? Is the atmosphere unstable or stable at 100 m height? Explain.
8. The concentration of atmospheric carbon dioxide is increasing about $1.5\,\mu L\,L^{-1}$ per year. If the current rate of increase continues, when (at about what year) will its concentration be double the pre-Industrial Revolution concentration of $280\,\mu L\,L^{-1}$?
9. Why is there environmental concern about the increasing concentration of carbon dioxide in the atmosphere? What is causing the increase?

FURTHER READING

Ahrens, C.D., 2013. Meteorology Today An Introduction to Weather. In: Climate and the Environment. 10th ed. Brooks/Cole-Thomson Learning Inc., Pacific Grove, CA

Global Climate Change, NASA, https://climate.nasa.gov/causes/.

Chapter 5

Biotic Characteristics of the Environment

I.L. Pepper

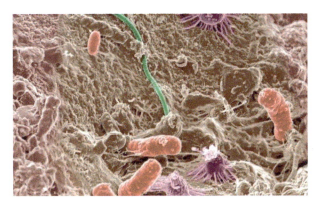

Microbes in the soil are central players converting carbon into greenhouse gases. *Credit: Alice Dohnalkova/PNNL appearing in PLoS ONE. https://phys.org/news/2016-03-microbes-climate.html*

5.1 MAJOR GROUPS OF ORGANISMS

Microorganisms in the environment are diverse in origin and ubiquitous. Environmental microorganisms are also fundamentally different from laboratory-maintained or clinical isolates of microbes, because they are adapted to harsh and often widely fluctuating environments. The smallest organisms are the viruses, which do not carry out metabolic reactions and thus require a host for self-replication. Viruses are unique in that they consist solely of nucleic acids and proteins, and are not technically viable, living organisms. To date, all biological entities contain nucleic acids as DNA or RNA. In environmental microbiology, we can also categorize microbes as prokaryotes or eukaryotes, both of which clearly affect human health and welfare, and are essential for maintaining life, as we know it (Table 5.1). Bacteria and actinomycetes are prokaryotic. Larger and more complex organisms include the eukaryotic fungi, algae, and protozoa (Fig. 5.1).

Viruses are significant because of their ability to infect other living organisms and cause disease. Bacteria can also cause infections, but in addition are important because of their ability to transform organic and inorganic compounds. Fungi are also involved in biochemical transformations. Algae affect surface water pollution and can also produce microbial toxins, but their overall impact on pollution is generally not as significant as that of the bacteria or fungi. Finally, note that the protozoa are also significant sources of pathogens that directly affect human health. A classic example of this would be the outbreak of *Cryptosporidium* contamination in potable water supplies that resulted in over 100 deaths in 1993 in Milwaukee, Wisconsin (Gerba, 2015). Protozoa are also important entities in surface environments as grazers of bacteria, helping to control bacterial populations.

In this chapter, the basic structure and function of key environmental microbes will be described. We will focus primarily on viruses, bacteria, and fungi because of their importance and relevance to environmental issues. The objective of this chapter is to outline the key characteristics of these microbes, their roles in the environment, and their requirements to function successfully in the environment. Special emphasis is placed on the bacteria because of their multiple roles in environmental microbiology.

Microorganisms in the environment are ubiquitous and are represented by diverse populations and communities. Microscopic organisms, including viruses, bacteria, fungi, algae, and protozoa, are too small to be seen with the naked eye but large enough to be studied under a microscope. However, there are certain genera of fungi, algae, and protozoa that are macroscopic in nature, and therefore only viruses and bacteria are totally microscopic (Fig. 5.2).

Viruses are the smallest microorganisms and are considered to be ultramicroscopic. Viruses consist primarily of nucleic acid material surrounded by a protein coat. Electron microscopes are required to visualize viruses because their size is below the resolution capacity of light microscopes. Ranging in size from 20 to several hundred nanometers, virus particles are very simple in organization, and normally consist of only an inner nucleic acid genome, an outer protein capsid, and sometimes an additional membrane envelope (Fig. 5.3). Unlike other microorganisms, viruses do not necessarily fulfill the requirements for being alive as they have no ribosomes and do not metabolize. Viruses must infect and replicate inside other living cells or hosts. For example, noroviruses can be transmitted via ingestion of contaminated water. When a human ingests contaminated water, norovirus enter the intestines, attach,

Environmental and Pollution Science. https://doi.org/10.1016/B978-0-12-814719-1.00005-7

TABLE 5.1 Microbial Influences on Our Daily Lives

Activity	Environmental Matrix	Impact	Microorganisms
Municipal wastewater treatment	Wastewater	Waterborne disease reduction	*E. coli* *Salmonella*
Water treatment	Water	Waterborne disease reduction	Norovirus *Legionella*
Food consumption	Food	Foodborne disease	*Clostridium botulinum* *E. coli* O157:H7
Indoor activities	Fomites	Respiratory disease	Rhinovirus
Breathing	Air	Legionellosis	*Legionella pneumophila*
Enhanced microbial antibiotic resistance	Hospitals	Antibiotic-resistant microbial infections	Methicillin Resistant *Staphylococcus aureus*
Nutrient cycling	Soil	Maintenance of biogeochemical cycling	Soil heterotrophic bacteria
Rhizosphere/Plant interactions	Soil	Enhanced plant growth	Rhizobia Mycorrhizal fungi
Bioremediation	Soil	Degradation of toxic organics	*Pseudomonas* spp.

From Environmental Microbiology, third edition, © 2015 Elsevier, San Diego, CA. (Table 1.1, pg. 5)

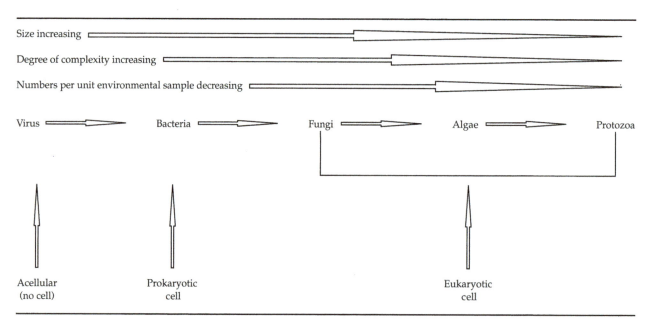

FIG. 5.1   Scope and diversity of microbes found in the environment. *(From Environmental and Pollution Science, Second Edition © 2006 Elsevier, San Diego, CA. (Fig. 5.1, pg. 59).)*

infect, and replicate in intestinal cells, ultimately causing the symptoms of disease such as gastroenteritis.

Phage are viruses that infect bacteria and are thought to be the most abundant biological entity on the planet. Global estimates of the virosphere are 10^{31} (Suttle, 2005). Phage are also very prevalent in soils with estimates being $10^8–10^9$/g

(Williamson et al., 2007). Mean *virus:bacteria (prokaryote) ratios (VPR)* for a variety of ecosystems vary from 5.6 to 28.5. However, for soil ecosystems the mean VPR value has been reported to be 704 (Parikka et al., 2017) Fig. 5.4). The significance of phage worldwide is illustrated in Information Box 5.1. With respect to environmental and

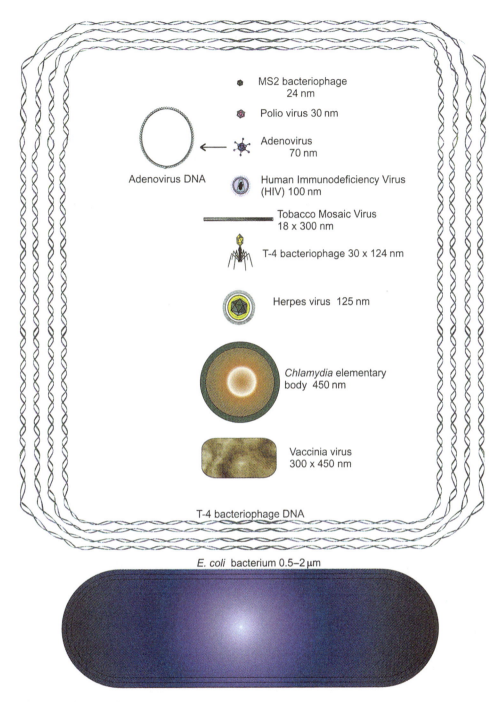

FIG. 5.2 Comparative sizes of selected bacteria, viruses, and nucleic acids. *(From Environmental and Pollution Science, Second Edition © 2006 Elsevier, San Diego, CA. (Fig. 5.2, pg. 60).)*

pollution science, phage are important as surrogates for human pathogenic viruses and as such are used in fate and transport studies in both soil and water (see also Chapter 13).

Bacteria are typically 1–2 μm, relatively simple, single celled organisms whose genetic material is not enclosed in a nuclear membrane (Fig. 5.5A and B). Based upon this cellular organization bacteria are classified as prokaryotes and include the eubacteria and the archaea. Although they are

classified as bacteria, special mention should be made of the *actinomycetes*.

These prokaryotic microbes consist of long chains of single cells, which allow them to exist in a filamentous form. Structurally, from the exterior, actinomycetes resemble miniature fungi, and this is the reason why they are often reported as a specialized subgroup of bacteria. Actinomycetes are important producers of antibiotics such as streptomycin,

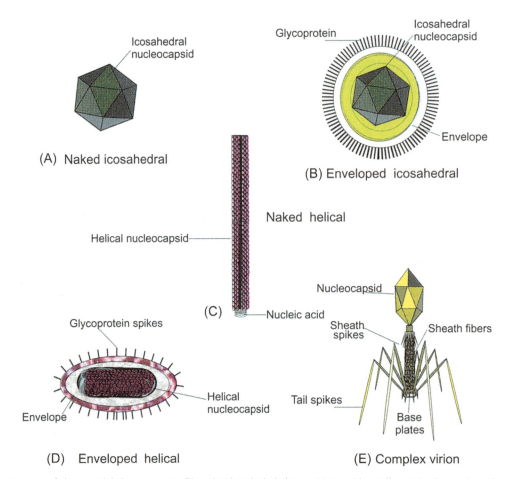

FIG. 5.3 Simple forms of viruses and their components. The naked icosahedral viruses (A) resemble small crystals; the enveloped icosahedral viruses (B) are made up of icosahedral nucleocapsids surrounded by the envelope; naked helical viruses (C) resemble rods with a fine, regular helical pattern in their surface; enveloped helical viruses (D) are helical nucleocapsids surrounded by the envelope; and complex viruses (E) are mixtures of helical and icosahedral and other structural shapes. *(From Environmental and Pollution Science, Second Edition © 2006 Elsevier, San Diego, CA. (Fig. 5.3, pg. 61).)*

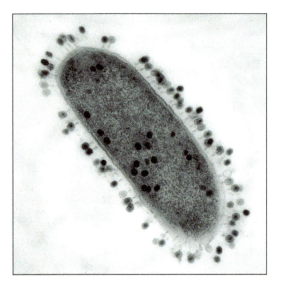

FIG. 5.4 Transmission electron micrograph of an *E. coli* cell being infected by several T4 phages. *(Courtesy of J. Wertz. From Environmental Microbiology, Third Edition © 2015 Elsevier, San Diego, CA. (Fig. 2.18, pg. 30).)*

are important in the biodegradation of complex organics, and also produce geosmin, which is the compound that gives soil its characteristic odor. Geosmin can also result in taste and odor problems in potable waters (see also Chapter 24).

Prokaryotes lack developed internal structures, including a membrane-enclosed nucleus, which distinguishes them from the more complex microorganisms such as fungi.

INFORMATION BOX 5.1 Role of Phage in Environmental Microbiology

- Control of bacterial populations
- Control of specific bacterial pathogens
- Control of marine cyanobacteria
- Interactions with food web processes
- Interactions with biogeochemical cycles
- Enhanced prokaryotic diversity via horizontal gene transfer

(From Environmental Microbiology, Third Edition, © 2015 Elsevier, San Diego, CA. (Information Box 2.9, pg. 31))

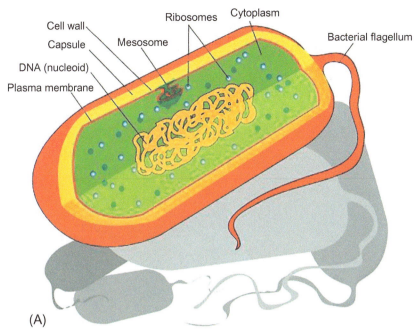

Cell wall
Capsule
DNA (nucleoid)
Plasma membrane
Mesosome
Ribosomes
Cytoplasm
Bacterial flagellum

(A)

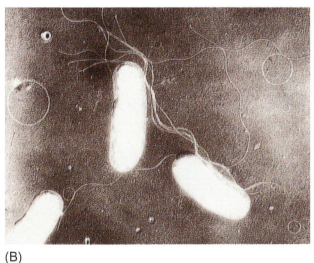

(B)

FIG. 5.5 (A) The components of a bacterial cell are illustrated. All bacteria have the cellular DNA dispersed in the cytoplasm, ribosome, cell membranes, a cell wall and a capsule. Some bacteria do not have flagella, some have a single flagellum and some have multiple flagella. Not all bacteria have pili, but bacterial pathogens do, and the pili allow them to attach to host cells. (B) Scanning electron micrographs of a soil bacterium with multiple flagella. The circles are detached flagella that have spontaneously assumed the shape of a circle. *A. (From Environmental Microbiology, Third Edition © 2015 Elsevier, San Diego, CA. (Fig. 2.2, pg. 13)). B. (From Environmental Microbiology, Third Edition © 2015 Elsevier, San Diego, CA. (Fig. 2.6, pg. 16).)*

Also lacking are internal cell membranes and complex internal cell organelles involved in growth, nutrition, or metabolism. Bacteria are ubiquitous organisms and can be found in even the most extreme environments as they have evolved a capacity for rapid growth, metabolism, and reproduction, as well as the ability to use a diverse range of organic and inorganic substances as carbon and energy sources. Bacteria are considered to be the simplest form of life, as viruses are not considered to be alive. A single gram of soil can contain from 10^5 to 10^{10} bacteria depending on the particular soil. Bacteria are especially vital to life on earth because of their functioning in all major environments. They play an important role in biogeochemical processes, nutrient cycling in soils, bioremediation, human and plant diseases, plant-microbe interactions,

municipal waste treatment, and the production of important drug agents including antibiotics (see Chapter 27).

Fungi, like protozoa, are also eukaryotic organisms (Fig. 5.6). They are very ubiquitous in the environment and also critically affect human health and welfare since they can be both beneficial and harmful to plants, animals, and humans. Certain species of fungi promote the health of many plants through mycorrhizal associations, while other species are phytopathogenic and capable of destroying plant tissue and even whole crops. Fungi are also important in the cycling of organics and in bioremediation. One of the most important fungi is the yeasts, which are utilized in the fermentation of sugars to alcohol in the brewing and wine industries. Fungi range from microscopic, with a single cell,

FIG. 5.6 Various types of fungi. (A)–(D) show the fruiting bodies of various molds. In (E) is illustrated the Maiden Veil fungus or *Dictyophora indusiata* within the *Basidiomycota* (mushroom). In (F) shown the building yeast *Saccharomyces cerevisiae*. *(From Environmental Microbiology, Third Edition © 2015 Elsevier, San Diego, CA. (Fig. 2.11, pg. 23).)*

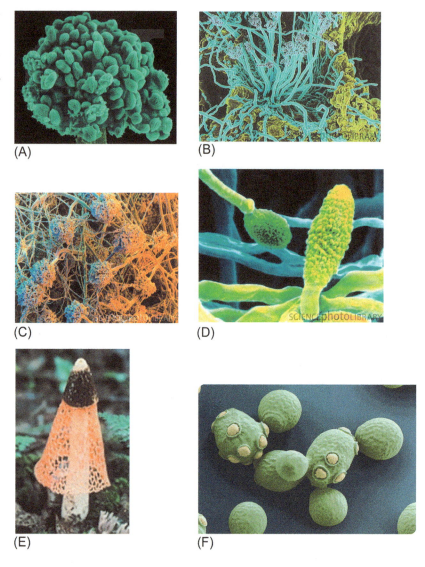

(A)

(B)

(C)

(D)

(E)

(F)

to macroscopic, with filaments of a single fungal cell being several cm in length. Overall, fungi are heterotrophic in nature, with different genera metabolizing everything from simple sugars to complex aromatic hydrocarbons. Fungi are also important in the degradation of the plant polymers cellulose and lignin. For the most part, fungi are aerobic, though some such as the yeasts are capable of fermentation. Fungi are distinguished from algae by their lack of photosynthetic ability.

Algae are a group of photosynthetic organisms that can be macroscopic, as in the case of seaweeds and kelps, or microscopic in size. Algae are aerobic eukaryotic organisms and as such exhibit structural similarities to fungi with the exception of the algal chloroplast. Chloroplasts are photosynthetic cell organelles found in algae that are capable of converting the energy of sunlight into chemical energy through photosynthesis. Algae are abundant in fresh and salt water, in soil, and in association with plants. When excessive concentrations of N and P occur in aquatic environments, they can cause eutrophication. The capacity of algae to photosynthesize is critical because these primary producers provide the basis of aquatic food chains and can also be important in the initial colonization of disturbed terrestrial environments. Overall, the algae are sometimes known by common names such as green algae, brown algae, or red algae based on their predominant color.

Blue-green algae are actually classified as bacteria known as *Cyanobacteria*. The *Cyanobacteria* are photosynthetic and some can also fix atmospheric nitrogen. Large blooms of freshwater *Cyanobacteria* may produce toxins, which are harmful to animals and humans. Conversely, *Spirulina* is a blue-green algae grown commercially and sold as a natural dietary supplement.

Protozoa are unicellular eukaryotes with characteristic organelles (mitochondria, plasma membrane, nuclear envelope, eukaryotic ribosomal RNA, endoplasmic membranes, chloroplast, and flagella). Protozoa are relatively large and can be visible to the naked eye. Their sizes can range from 2 μm to several cm. Although they are single-celled organisms, they are by no means simple in structure, and many diverse forms can be observed. Morphological variability, evolved over hundreds of millions of years, has enabled protozoan adaptation to a wide variety of environments. Protozoa can be found in nearly all terrestrial and aquatic environments and are thought to play a valuable role in ecological cycles by in part controlling bacterial populations. Many species are able to exist in extreme environments from polar regions to hot springs and desert soils. In recent years, protozoan pathogens such as *Giardia*, *Cyclospora*, *Cryptosporidium*, and the *Microsporidia*, have emerged as major issues with respect to safe potable water (see Chapters 13 and 24). Protozoa may be free living, capable of growth and reproduction outside any host, or parasitic, meaning that they colonize host cell tissues. Some are opportunists, adapting either a free-living or parasitic existence as their environment dictates.

5.2 MICROORGANISMS IN SURFACE SOILS

Surface soils are predominantly occupied by indigenous populations of phage, bacteria, actinomycetes, fungi, algae, and protozoa. In general, as the size of these organisms increases from bacteria to protozoa, the number present decreases (Fig. 5.1). Most environmental monitoring of microbes focuses on bacteria, actinomycetes, and fungi, particularly in soil and water environments. Estimates of soil microbial biomass are shown in Information Box 5.2. A comparison of soil microbial properties is presented in Table 5.2. In addition to these indigenous populations, specific microbes can be introduced into soil by human or animal activity. Human activities include the deliberate introduction of bacteria as biological control agents or as biodegradative agents. Microbes are also introduced indirectly as a result of application of biosolids to agricultural fields (see Chapter 23). Animals introduce microbes through bird droppings and animal excrement. Regardless of the source, introduced organisms usually die within weeks or months, and rarely significantly affect the abundance and distribution of indigenous populations, which usually outcompete introduced organisms.

The following discussion is an overview of the dominant types of microbes found in surface soils, including their occurrence, distribution, and function.

5.2.1 Bacteria

Bacteria are prokaryotic organisms lacking a nuclear membrane. Only phage are more prevalent on Earth than bacteria. Surface soils routinely contain 10^8 bacteria per gram of soil, whereas total counts including those that cannot be cultured on media are frequently 10^{10} per gram of soil. Logically it can be deduced that perhaps 99% of all soil bacteria cannot be cultured. Shade et al. (2012) used conventional culturing methods to enumerate soil bacteria as well as high throughput sequencing on the same soil samples (culture independent). Analyses showed that soil bacteria captured by culturing were in very low abundance or absent from the culture-independent community (Fig. 5.7). Thus culturable soil bacteria may actually be atypical members of the community.

Bacteria are characterized by a complex cell envelope, which contains cytoplasm but no cell organelles. Bacteria are capable of rapid growth and reproduction, both of which occur by binary fission. Genetic exchange occurs predominantly by *conjugation* (cell-to-cell contact) or *transduction* (exchange via viruses), although *transformation* (transfer of naked DNA) also occurs (Keen et al., 2017). The size of bacteria generally ranges from 0.1 to 2 μm. Soil bacteria can be rod-shaped, coccoidal, helical, or pleomorphic (Fig. 5.8). Soil bacteria also exhibit great diversity with respect to colony morphology (Fig. 5.9).

5.2.1.1 Mode of Nutrition

Bacteria can be classified according to their mode of nutrition:

1. *Autotrophic mode*: Strict soil autotrophs obtain energy from inorganic sources and carbon from carbon dioxide. These kinds of organisms generally have few growth factor requirements. Chemoautotrophs obtain energy from the oxidation of inorganic substances, whereas photoautotrophs obtain energy from photosynthesis.
2. *Heterotrophic mode*: Heterotrophs obtain energy and carbon from organic substances. Chemoheterotrophs obtain energy from the oxidation of organic compounds; however, photoheterotrophs obtain energy from photosynthesis, which requires organic electron donors.

INFORMATION BOX 5.2 Estimates of Soil Microbial Biomass

Microbe	Soil Biomass (μg/g)
Phage	10*
Bacteria	1000**
Fungi	5000***

*Based on estimates of 10^8 phage/g soil (Williamson et al., 2007) and weight of 1 virus ≃ 1 fg (Gupta et al., 2004).

**Based on estimates of 10^9 bacteria/g soil and weight of a bacterial cell = 1 pg (Ingraham et al., 1983).

***Based on estimates from soil ergosterol analyses (Montgomery et al., 2000).

TABLE 5.2 Characteristics of Bacteria, Actinomycetes, and Fungi

Characteristic	Bacteria	Actinomycetes	Fungi
Population	Most numerous	Intermediate	Least numerous
Biomass	Bacteria and actinomycetes have similar biomass		Largest biomass
Degree of branching	Slight	Filamentous, but some fragment to individual cell	Extensive filamentous forms
Aerial mycelium	Absent	Present	Present
Growth in liquid culture	Yes—turbidity	Yes—pellets	Yes—pellets
Growth rate	Exponential	Cubic	Cubic
Cell wall	Murein, teichoic acid, and lipopolysaccharide	Murein, teichoic acid, and lipopolysaccharide	Chitin or cellulose
Complex fruiting bodies	Absent	Simple	Complex
Competitiveness for simple organics	Most competitive	Least competitive	Intermediate
Fix N	Yes	Yes	No
Aerobic	Aerobic, anaerobic	Mostly aerobic	Aerobic except yeast
Moisture stress	Least tolerant	Intermediate	Most tolerant
Optimum pH	6–8	6–8	6–8
Competitive pH	6–8	>8	<5
Competitiveness in soil	All soils	Dominate dry, high-pH soils	Dominate low-pH soils

From *Environmental and Pollution Science, Second Edition* © 2006, Academic Press, San Diego, CA (Table 5.2, pg. 64)

In soil, chemoheterotrophs and chemoautotrophs predominate; phototrophs of either variety are not as numerous because soil is not permeable to sunlight.

5.2.1.2 Type of Electron Acceptor

Aerobic bacteria use oxygen as a terminal electron acceptor and possess superoxide dismutase or catalase enzymes that are capable of degrading peroxide radicals. *Anaerobic bacteria* do not utilize oxygen as a terminal electron acceptor. Strict anaerobes do not possess superoxide dismutase or catalase enzymes, and are thus poisoned by the presence of oxygen. Although other kinds of anaerobes do possess these enzymes, they utilize terminal electron acceptors other than oxygen, such as nitrate or sulfate. *Facultative anaerobes* can use oxygen or combined forms of oxygen as terminal electron acceptors. In unsaturated soils aerobic bacteria normally outnumber anaerobic bacteria. Anaerobic populations increase with increasing soil, depth, but rarely predominate unless soils are saturated (Pepper et al., 2015).

5.2.1.3 Soil Bacterial Classification

Bacteria can be classified based on the concept of growth characteristics resulting in *r and K selection*. Organisms adapted to living under conditions in which substrate is plentiful are designated as K-selected. *Rhizosphere* organisms living off root exudates are examples of K-selected organisms. Organisms that are r-selected live in environments in which substrate is the limiting factor except for occasional flushes of substrate. r-Selected organisms rely on rapid growth rates when substrate is available, and generally occur in uncrowded environments. In contrast, K-selected organisms exist in crowded environments and are highly competitive.

Soil bacteria can also be classified according to diversity using next-generation sequencing in combination with statistical approaches. From a soil community perspective, microbial diversity can be defined as the amount of genetic, morphological, and functional differences in the microbial populations occupying a given environment. For soil bacteria, this typically involves the enumeration of the number of bacteria species per unit weight of soil. The criteria for what constitutes a bacterial species and how to distinguish between two different species are now based on 16S rRNA gene sequence identity. Bacteria with at least 97% sequence similarity are considered to be within the same species and are classified as an *operational taxonomic unit (OTU)* (Gentry et al., 2015).

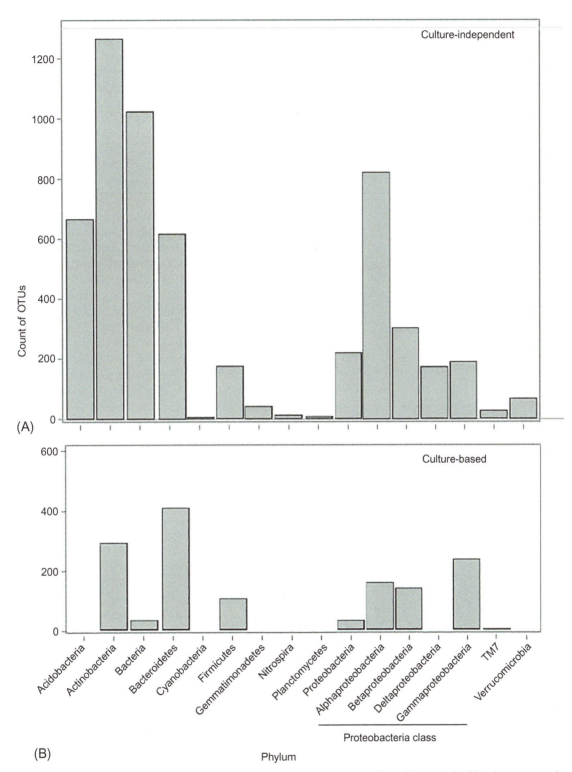

FIG. 5.7 Impact of culture-based and culture-independent methods on characterization of a soil bacterial community. Note the greater number of phyla and operational taxonomic units (OTUs) detected using the culture-independent approach. *(From Shade, A., Hogan, C.S., Klimowiez, A.K., Linske, M., McManus, P.S., Hendelsman, J., 2012. Culturing captures members of the soil rare biosphere. Environ. Microbiol. 14:2247-2252. From Environmental Microbiology, Third Edition © 2015, Academic Press, San Diego, CA. (Fig. 4.21, pg. 79).)*

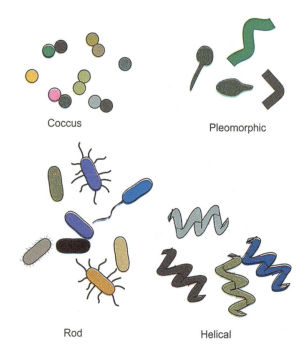

Coccus

Pleomorphic

Rod

Helical

FIG. 5.8 Typical shapes of representative bacteria. *(From Environmental Microbiology © 2000, Academic Press, San Diego, CA. (Fig. 5.8, pg. 64).)*

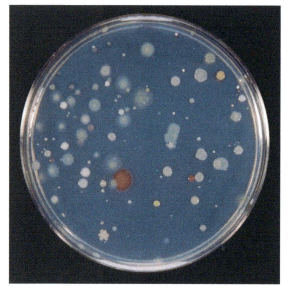

FIG. 5.9 Here, a variety of bacteria with some actinomycetes are isolated from a field soil on a petri dish on a general heterotrophic medium. *(From Pollution Science, First Edition © 1996, Academic Press, San Diego, CA. (Fig. 5.9, pg. 64).)*

Estimates of soil bacterial diversity have steadily increased over the past 20 years, with the advent of high throughput sequence analysis. It is now well recognized that many soil samples must be sequenced to account for environmental variation at a particular location to obtain the total species richness. A 2017 study concluded that there were 42,866 bacterial species within a 50-ha forest soil

(Chen et al., 2017). This value was derived from a massive sequencing data set obtained from 550 soil samples. Interestingly, a total of over 8 million sequences were generated from all 550 samples with a range of $\simeq 8000$ to 21,500 sequences per sample. Soil fungal diversity appears to be less than that of soil bacteria with $\simeq 2000$ fungal OTUs reported from within 66 soil samples (Song et al., 2015).

5.2.1.4 Dominant Culturable Soil Bacteria

Examples of dominant culturable heterotrophic bacteria include *Arthrobacter*, *Bacillus*, and *Pseudomonas* species, which are important with respect to degradation of organic compounds including toxic anthropogenic compounds. Examples of other important heterotrophic bacteria are presented in Table 5.3. Examples of important autotrophic soil bacteria are presented in Table 5.4.

Bacteria are critically involved in almost all soil biochemical transformations, including both organic and inorganic compounds. The importance of soil bacteria in the fate and mitigation of pollutants cannot be overestimated. Because of their prevalence and diversity, as well as fast growth rates and adaptability, they have an almost unlimited ability to degrade most natural products and many xenobiotics. We will examine, in detail, the influence of bacteria on waste disposal and pollution mitigation in the succeeding chapters of this text (Chapters 9, 19, and 22).

Actinomycetes are organisms that technically are classified as bacteria, but are unique enough to be discussed as an individual group. They have some characteristics in common with bacteria but are also similar in some respects to fungi. For the most part, they are aerobic chemoheterotrophic organisms consisting of elongated single cells. They display a tendency to branch into filaments, or hyphae, that resemble fungal mycelia; these hyphae are morphologically similar to those of fungi, but are smaller in diameter (about 0.5–2 μm). The diversity of soil bacteria including actinomycetes is shown in Fig. 5.9, where the actinomycetes have a dull chalky appearance. Finally, the spatial relationship of soil bacteria, actinomycetes, and fungi is illustrated in Fig. 5.10. Generally, the population of actinomycetes is 1–2 orders of magnitude less than that of other bacteria in soil around 10^7 per gram of soil (Vieira and Hahas, 2005). They are not known to reproduce sexually, but all produce asexual spores called *conidia*. The genus *Streptomyces* is prominent within the actinomycete population, and these Gram-positive organisms produce Streptomycin.

Actinomycetes are an important component of bacterial populations, especially under conditions of high pH, high temperature, or water stress (Information Box 5.3). One distinguishing feature of this group of bacteria is that they are able to utilize a great var iety of substrates found in soil, especially some of the less degradable insect and plant polymers such as chitin, cellulose, and hemicellulose.

TABLE 5.3 Examples of Important Heterotrophic Soil Bacteria

Organism	Characteristics	Function
Actinomycetes, e.g., *Streptomyces*	Gram positive, aerobic, filamentous	Produce geosmins, "earthy odor," and antibiotics
Bacillus	Gram positive, aerobic, spore former	Carbon cycling, production of insecticides and antibiotics
Clostridium	Gram positive, anaerobic, spore former	Carbon cycling (fermentation), toxin production
Methanotrophs, e.g., *Methylosinus*	Aerobic	Methane oxidizers that can cometabolize trichloroethene (TCE) using methane monooxygenase
Cupriavidus pinatubonesis	Gram negative, aerobic	2,4-D degradation via plasmid pJP4
Rhizobium	Gram negative, aerobic	Fixes nitrogen symbiotically with legumes
Frankia	Gram positive, aerobic	Fixes nitrogen symbiotically with nonlegumes
Agrobacterium	Gram negative, aerobic	Important plant pathogen, causes crown gall disease

From *Environmental & Pollution Science, Second Edition* © 2006, Academic Press, San Diego, CA. (Table 5.4, pg. 66)

TABLE 5.4 Examples of Important Autotrophic Soil Bacteria

Organism	Characteristics	Function
Nitrosomonas	Gram negative, aerobe	Converts $NH_4^+ \rightarrow NO_2^-$ (first step of nitrification)
Nitrobacter	Gram negative, aerobe	Converts $NO_2^- \rightarrow NO_3^-$ (second step of nitrification)
Thiobacillus	Gram negative, aerobe	Oxidizes $S \rightarrow SO_4^{2-}$ (sulfur oxidation)
Thiobacillus denitrificans	Gram negative, facultative anaerobe	Oxidizes $S \rightarrow SO_4^{2-}$; functions as a denitrifier
Thiobacillus ferrooxidans	Gram negative, aerobe	Oxidizes $Fe^{2+} \rightarrow Fe^{3+}$

From *Environmental & Pollution Science, Second Edition* © 2006, Academic Press, San Diego, CA (Table 5.3, pg. 65).

Like all bacteria, actinomycetes are prokaryotic organisms. In addition, the adenine–thymine and guanine–cytosine contents of bacteria and actinomycetes are similar, as are the cell wall constituents of both types of organisms. Actinomycete filaments are also about the same size as those of bacteria.

Like fungi, however, actinomycetes display extensive mycelial branching, and both types of organisms form aerial mycelia and conidia. Moreover, growth of actinomycetes in liquid culture tends to produce fungus-like clumps or pellets rather than the uniform turbidity produced by bacteria. Finally, growth rates in fungi and actinomycetes are not exponential as they are in bacteria; rather, they are cubic (see also Table 5.2).

Actinomycetes can metabolize a wide variety of organic substrates including organic compounds that are not easily metabolized such as phenols and steroids. They are also important in the metabolism of heterocyclic compounds such as complex nitrogen compounds and pyrimidines. The breakdown products of their metabolites are frequently aromatic, and these metabolites are important in the formation of humic substances and soil humus. The earthy odor associated with most soils is due to *geosmin*, a compound produced by actinomycetes.

Actinomycetes often comprise about 10% of the total bacterial population. Actinomycetes are tolerant of alkaline soils (pH > 7.5) than other bacteria but less tolerant of acidic soils (pH < 5.5); actinomycetes are also more tolerant of low soil moisture contents than other bacteria. Because of this and their tolerance of alkaline soils, actinomycete populations tend to be higher in desert soils, such as the southwestern U.S.A.

Many actinomycetes interact with plants in symbiotic and pathogenic associations. For example, the genus *Frankia* initiates root nodules with nonleguminous, nitrogen-fixing plants, whereas the species *Streptomyces scabies* is the causative agent of potato scab. On the other hand, many streptomycetes produce antibiotics, including of course streptomycin (deLima Procopio et al., 2012).

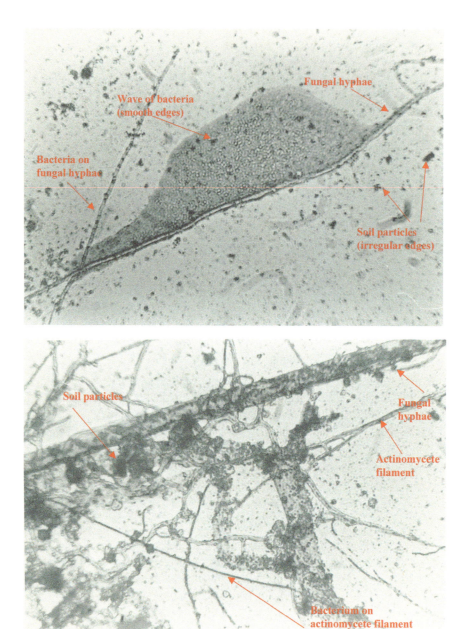

FIG. 5.10 Comparison of soil bacteria, actinomycetes, and fungi viewed under a light. *(Photos courtesy W.H. Fuller. From Environmental and Pollution Science, Second Edition © 2006, Academic Press, San Diego, CA. (Fig. 5.10, pg. 67).)*

5.2.2 Archaea

The archaeans are microbes that look similar to bacteria under the microscope, but they are actually quite different genetically and biochemically. They may be the oldest form of life on Earth and were originally thought to only inhabit extreme environments (Pepper and Gentry, 2015). Now it is recognized that archaea are actually widespread in nature. Some archaea, the Planctomycetes are capable of intracellular compartmentalization, utilizing internal membranes and even membrane-bound nucleotides (Fuerst and Sagulenkol, 2011). Thus Planctomycetes could be an evolutionary intermediate between prokaryotes and eukaryotes.

Archaeal populations can be as large as 10^8 per gram of soil, but they are typically at least two orders of magnitude less numerous than bacteria around 10^5–10^6 per gram of soil (Zhang et al., 2017). Archaea contribute to biogeochemical cycling of soil macronutrients including nitrification. Ammonia-oxidizing archaea (AOE) can be important in environments that have lower levels of ammonia and also in subsurface soils (Verhamme et al., 2011).

INFORMATION BOX 5.3 Characteristics and Functions of Actinomycetes

Characteristics

Structure	Prokaryotic
Size	1–2 µm diameter
Morphology	Filamentous lengths of cocci
Gram stain	Gram positive
Respiration	Mostly aerobic, can be
Habitat	anaerobic
Abundance, marine	Soil or marine
isolates	5–40 CFU/ml
Abundance, soils	$10^6–10^8$/g

Functions

- Source of natural products and antibiotics, e.g., streptomycin
- Produce geosmin the compound which gives soil and water a characteristic earthy odor
- Capable of degradation of complex organics
- Capable of biological nitrogen fixation with the non-legume-associated *Frankia* spp.

From Environmental Microbiology, Third Edition, *Elsevier, San Diego, CA. (Information Box 4.8, pg. 81)*

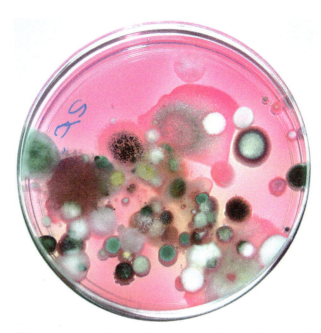

FIG. 5.11 Soil fungi isolated from a surface soil grown in a petri dish containing Rose Bengal Agar. *(Photo courtesy K.L. Josephson. From Environmental & Pollution Science, Second Edition © 2006, Academic Press, San Diego, CA. (Fig. 5.11, pg. 68).)*

5.2.3 Fungi

Fungi other than yeasts are aerobic and are abundant in most surface soils. Numbers of fungi usually range from 10^5 to 10^6 per gram of soil. Despite their lower numbers compared with bacteria, fungi usually contribute a higher proportion of the total soil microbial biomass. This is due to their comparatively large size; a fungal hypha can range from 2 to 10 µm in diameter. Fig. 5.11 shows an example of the diverse fungal population that can be isolated and cultured from surface soil. Because of their large size, fungi are more or less restricted to the interaggregate regions of the soil matrix. Yeasts can metabolize anaerobically (fermentation) and are less numerous than aerobic mycelium-forming fungi. Generally, yeasts can be found at populations of up to $10^3–10^5$ per gram of soil (Maksimova et al., 2016). Because of their reliance on organic sources for substrates, fungal populations are greatest in the surface O and A horizons, and numbers decrease rapidly with increasing soil depth. As with bacteria, soil fungi are normally found associated with soil particles or within plant rhizospheres.

Fungi are important components of the soil with respect to nutrient cycling, especially organic matter decomposition. They decompose both simple sugars and complex polymers such as cellulose and lignin. The role of fungi in decomposition is increasingly important when the soil pH declines, because fungi tend to be more tolerant of acidic conditions than bacteria. Some of the common genera of soil fungi involved in nutrient cycling are *Penicillium* and *Aspergillus*. These organisms are also important in the development of soil structure because they physically entrap soil particles with fungal hyphae. Fungi are critical in the degradation of complex plant polymers such as cellulose and lignin, and some fungi can also degrade a variety of pollutant molecules. The best-known example of such a fungus is the white rot fungus *Phanerochaete chrysosporium*. Other fungi such as *Fusarium* spp., *Pythium* spp., and *Rhizoctonia* spp. are important plant pathogens. Still, others cause disease; for example, *Coccidioides immitis* causes a chronic human pulmonary disease known as valley fever in the southwestern deserts of the United States. Finally, note that mycorrhizal fungi are critical for establishing plant–fungal interactions that act as an extension of the root system of almost all higher plants. Without these mycorrhizal associations, plant growth as we know it would be impossible.

5.2.4 Algae

Algae are typically phototrophic, and thus would be expected to survive and metabolize in the presence of a light energy source and CO_2 carbon source. Therefore one would expect to find algal cells predominantly in areas where sunlight can penetrate the surface of the soil. One can actually find algae to a depth of 1 m because some algae, including the green algae and diatoms, can grow heterotrophically as well as photoautotrophically. In general, though, algal populations are highest in the surface 10 cm of soil. Typical algal populations close to the soil surface can range from 10^3 to 10^5 per gram of soil (Zancan et al., 2006). Note that a surface soil where a visible algal bloom has developed can contain millions of algal cells per gram of soil.

Algae are often the first to colonize surfaces in a soil that are devoid of preformed organic matter. Colonization by this group of microbes is important in establishing soil formation processes, especially in barren volcanic areas, desert soils, and rock faces. Algal metabolism is critical to soil formation in two ways: algae provide a carbon input through photosynthesis, and as they metabolize they produce and release carbonic acid, which aids in weathering the surrounding mineral particles. Further, algae produce large amounts of extracellular polysaccharides, which also aid in soil formation by causing aggregation of soil particles (Killham, 1994).

Populations of soil algae generally exhibit seasonal variations with numbers being highest in the spring and fall. This is because desiccation caused by water stress tends to suppress growth in the summer and cold stress affects growth in the winter. Some of the major groups of algae found in soil include: the Cyanophyceae (cyanobacteria) formerly called blue-green algae; the Clamydophyceae including *Chlamydomonas* and the Bacillariophyceae including diatoms (Zancan et al., 2006). Diatoms are found primarily in neutral and alkaline soils. Diatoms are characterized by the presence of a silicon dioxide cell wall. The cyanobacteria, often called blue-green algae, (e.g., *Nostoc* and *Anabaena*), are actually classified as bacteria but have many characteristics in common with algae. The cyanobacteria participate in the soil-forming process discussed in the previous paragraph, and some cyanobacteria also have the capacity to fix nitrogen, a nutrient that is usually limiting in a barren environment.

5.2.5 Protozoa

Protozoa are unicellular, eukaryotic organisms that can be several mm in length, although most are much smaller. Most protozoa are heterotrophic and survive by consuming bacteria, yeast, fungi, and algae. There is evidence that they may also be involved, to some extent, in the decomposition of soil organic matter. Because of their large size and requirement for large numbers of smaller microbes as a food source, protozoa are found mainly in the top 15–20 cm of the soil. Protozoa are usually concentrated near root surfaces that have high densities of bacteria or other prey.

There are three major categories of protozoa: the flagellates, the amoebae, and the ciliates. The flagellates are the smallest of the protozoa and move by means of one to several flagella. Some flagellates (e.g., Euglena) contain chlorophyll, although most do not. The amoebae, also called rhizopods, move by protoplasmic flow, either with extensions called pseudopodia or by whole-body flow. Amoebae are usually the most numerous types of protozoan found in a given soil environment. Ciliates are protozoa that move by beating short cilia that cover the surface of the cell. The

protozoan population of a soil is often correlated with the bacterial population, which is the protozoan's major food source. Numbers of protozoa can vary around 10^4 per gram of soil (Takenouchi et al., 2016).

5.3 MICROORGANISMS IN THE SUBSURFACE

Although the microorganisms of surface soils have been studied extensively, the study of subsurface microorganisms is relatively new, beginning in earnest in the 1980s. Complicating the study of subsurface life are the facts that sterile sampling is problematic, which can result in contamination of samples with surface soil microbes. In addition, subsurface microorganisms are difficult to culture.

Because subsurface microbiology is still a developing field, information is limited in comparison with that for surface microorganisms. Yet there is enough information available to document that subsurface environments, once thought to contain very few if any microorganisms, actually have a significant and diverse population of microorganisms. In particular, shallow subsurface zones, specifically those with a relatively rapid rate of water recharge, have high numbers of microorganisms. The majority of these organisms are bacteria, but protozoa and fungi are also present. As a general rule, total numbers of bacteria, as measured by direct counts, remain fairly constant, ranging between 10^5 and 10^7 cells per gram throughout the profile of a shallow subsurface system. For comparison, numbers in surface soils range from 10^9 to 10^{10} cells per gram (Pepper and Gentry, 2015). This decrease in numbers is directly correlated with the low amounts of inorganic nutrients and organic matter in subsurface materials. Subsurface eukaryotic counts are also lower than surface counts by several orders of magnitude. Low eukaryotic counts can be due to low organic matter content or due to removal by physical straining by small soil pores. Both prokaryotic and eukaryotic counts are highest in portions of the subsurface that contain sandy sediments. This does not mean that clayey regions are not populated, but the numbers tend to be lower due to exclusion and physical straining of microorganisms by small pores in clay-rich media.

Numbers of culturable bacteria in subsurface environments show more variability than direct counts, ranging from zero to almost equal numbers of direct counts. Thus, in general, the difference between direct and viable counts in the subsurface is greater than the difference in surface soils which is normally one to two orders of magnitude. The larger difference between direct and viable counts in the subsurface may be because nutrients are much more limiting in the subsurface, and therefore a greater proportion of the population may be in a nonculturable state.

The dominant populations are aerobic, heterotrophic bacteria, although there are small populations of eukaryotic organisms as well as anaerobic and autotrophic organisms. Subsurface microbes are diverse in type, although not as diverse as surface organisms. Great diversity in heterotrophic activity has been found; subsurface microbes have shown the capacity to degrade simple substrates such as glucose as well as more complex substrates such as aromatic compounds, surfactants, and pesticides. It has also been shown that subsurface bacteria are capable of denitrification activity (Artiola and Pepper, 1992).

5.4 BIOLOGICAL GENERATION OF ENERGY

Biological activity requires energy, and all microorganisms generate energy. This energy is subsequently stored as *adenosine triphosphate* (ATP), which can then be used for growth and metabolism as needed, subject to the second law of thermodynamics.

The Second Law of Thermodynamics:

Can be paraphrased as- *In a chemical reaction, only part of the energy is used to do work. The rest of the energy is lost as entropy.*

Gibbs free energy ΔG is the amount of energy available for work for any chemical reaction. For the reaction

$$A + B \rightleftarrows C + D$$

the thermodynamic equilibrium constant is defined as

$$K_{eq} = \frac{[C][D]}{[A][B]}$$

Case 1. If product formation is favored:

That is, if

$$[C][D] > [A][B]$$

then $K_{eq} > 1$ and $\ln K_{eq}$ is positive.

For example, if $K_{eq} = 2.0$, then

$$\ln K_{eq} = 0.69$$

Case 2. If product formation is *not* favored:

$$[C][D] < [A][B]$$

then $K_{eq} < 1$ and $\ln K_{eq}$ is negative.

For example, if $K_{eq} = 0.20$, then

$$\ln K_{eq} = -1.61$$

The relationship between the equilibrium coefficient and the free energy ΔG is given by

$$\Delta G = -RT \log K_{eq}$$

where R is the universal gas constant and T = absolute temperature (K).

If ΔG *is negative*, energy is released from the reaction due to a spontaneous reaction. This is because if ΔG is negative, $\log K_{eq}$ must be positive. Therefore $K_{eq} > 1$, which means energy must be released.

If ΔG *is positive*, energy is needed to make the reaction proceed. This is because if ΔG is positive, $\log K_{eq}$ must be negative. Therefore $K_{eq} < 1$, which means energy must be added to promote the reaction.

Thus we can use ΔG values for any biochemical reaction mediated by microbes to determine whether energy is liberated for work, and how much energy is liberated.

Soil organisms can generate energy via several mechanisms, which can be divided into two main categories.

1. Photosynthesis:

$$2\,H_2O + CO_2 \xrightarrow{\text{light}} \underset{\text{biomass}}{CH_2O} + O_2 + H_2O$$

$$\Delta G \simeq +115\,\text{Kcal}\,\text{mol}^{-1}$$

For this reaction to proceed, energy supplied by sunlight is necessary. The fixed organic carbon is then used to generate energy via respiration. Examples of soil organisms that undergo photosynthesis are *Rhodospirillum*, *Chromatium*, and *Chlorobium*.

2. Respiration
   (a) *Aerobic heterotrophic respiration*: Many soil organisms undergo aerobic, heterotrophic respiration, for example, *Pseudomonas* and *Bacillus*.

$$C_6H_{12}O_6 + 6O_2 \rightarrow 6CO_2 + 6H_2O$$

$$\Delta G = -686\,\text{Kcal}\,\text{mol}^{-1}$$

   (b) *Aerobic autotrophic respiration*: The reactions carried out by *Nitrosomonas* and *Nitrobacter* are known as *nitrification*:

$$NH_3 + 1\tfrac{1}{2}O_2 \rightarrow HNO_2 + H_2O \ (\textit{Nitrosomonas})$$

$$\Delta G = -66\,\text{Kcal}\,\text{mol}^{-1}$$

$$KNO_2 + \tfrac{1}{2}O_2 \rightarrow KNO_3 \ (\textit{Nitrobacter})$$

$$\Delta G = -17.5\,\text{Kcal}\,\text{mol}^{-1}$$

The following two reactions are examples of *sulfur oxidation*:

$$2H_2S + O_2 \rightarrow 2H_2O + 2S \ (\textit{Beggiatoa})$$

$$\Delta G = -83\,\text{Kcal}\,\text{mol}^{-1}$$

$$2S + 3O_2 \rightarrow 2H_2SO_4 \ (\textit{Thiobacillus thiooxidans})$$

$$\Delta G = -237\,\text{Kcal}\,\text{mol}^{-1}$$

The next reaction involves the degradation of cyanide:

$$2KCN + 4H_2O + O_2 \rightarrow 2KOH$$

$$+ 2NH_3 + 2CO_2 \; (Streptomyces)$$

$$\Delta G = -56 \, \mathrm{Kcal \, mol^{-1}}$$

All of the prior reactions illustrate how soil organisms mediate reactions that can cause or negate pollution. For example, nitrification and sulfur oxidation can result in the production of specific pollutants, that is, nitrate and sulfuric acid, whereas the destruction of cyanide is obviously beneficial with respect to the mitigation of pollution.

(c) *Facultative anaerobic, heterotrophic respiration*: *Pseudomonas denitrificans* undertakes this kind of metabolism by using nitrate as a terminal electron acceptor rather than oxygen. Note that these organisms can use oxygen as a terminal electron acceptor if it is available, and that aerobic respiration is more efficient than anaerobic respiration.

$$5C_6H_{12}O_6 + 24KNO_3 \rightarrow 30CO_2 + 18H_2O + 24KOH + 12N_2$$

(*P. denitrificans*)

$$\Delta G = -36 \, \mathrm{Kcal \, mol^{-1}}$$

(d) *Facultative anaerobic autotrophic respiration*:

$$S + 2KNO_3 \rightarrow K_2SO_4 + N_2 + O_2 \; (\; Thiobacillus \, denitrificans)$$

$$\Delta G = -66 \, \mathrm{Kcal \, mol^{-1}}$$

(e) *Anaerobic heterotrophic respiration*: *Desulfovibrio* is an example of an organism that carries out this type of metabolism.

Lactic acid $+ H_2SO_4 \rightarrow$ Acidic acid $+ CO_2 + 2H_2S + 2H_2$

(*Desulfovibrio* spp.)

$$\Delta G = -40 \, \mathrm{Kcal \, mol^{-1}}$$

Soil microorganisms can also conduct fermentation, which is also an anaerobic process. But fermentation is not widespread in soil. Overall, there are many ways in which soil organisms can and do generate energy. The above-listed mechanisms illustrate the diversity of soil organisms and explain the ability of the soil community to break down or transform almost any natural substance. In addition, enzyme systems have evolved to metabolize complex molecules, be they organic or inorganic. These enzymes can also be used to degrade xenobiotics with similar chemical structures. Xenobiotics, which do not degrade easily in soil, are normally chemically different from any known natural substance; hence soil organisms have not evolved enzyme systems capable of metabolizing such compounds. There are many other factors that influence the breakdown or degradation of chemical compounds. These will be discussed in Chapter 9.

5.5 SOIL AS AN ENVIRONMENT FOR MICROBES

5.5.1 Biotic Stress

Since indigenous soil microbes are in competition with one another, the presence of large numbers of organisms results in biotic stress factors. Competition can be for substrate, water, or growth factors. In addition, microbes can secrete inhibitory or toxic substances, including antibiotics, that harm neighboring organisms. Finally, many organisms are predatory or parasitic on neighboring microbes. For example, phages infect both bacteria and fungi. Because of biotic stress, nonindigenous organisms that are introduced into a soil environment often survive for only fairly short periods of time (days to several weeks). This effect has important consequences for pathogens (see Chapter 13) and for other organisms introduced to aid biodegradation (see Chapter 9). This process of adding organisms to assist biodegradation is called *bioaugmentation*.

5.5.2 Abiotic Stress

Light. Soil is impermeable to light, that is, no sunlight penetrates beyond the top few centimeters of the soil surface. Phototrophic organisms are therefore limited to the top few centimeters of soil. At the surface of the soil, however, such physical parameters as temperature and moisture fluctuate significantly throughout the day and also seasonally. Hence most soils tend to provide a harsh environment for photosynthesizing organisms. A few phototrophic organisms, including algae, have the ability to switch to a heterotrophic respiratory mode of nutrition in the absence of light. Such "switch-hitters" can be found at significant depths within soils. Normally, these organisms are not competitive with other indigenous heterotrophic organisms for organic substrates.

Soil Moisture. Typically, soil moisture content varies considerably in any soil, and soil organisms must adapt to a wide range of soil moisture contents. Soil aeration is dependent on soil moisture: saturated soils tend to be anaerobic, whereas dry soils are usually aerobic. But soil is a heterogeneous environment; even saturated soils contain pockets of aerobic regimes, and dry soils harbor anaerobic microsites that exist within the centers of secondary aggregates. Although the bacteria are the least tolerant of low soil moisture, as a group they are the most flexible with respect to soil aeration. They include aerobes, anaerobes, and facultative anaerobes, whereas the actinomycetes and fungi are predominantly aerobic.

Soil Temperature. Soil temperatures vary widely, particularly near the soil surface. Most soil populations are resistant to wide fluctuations in soil temperature although soil populations can be psychrophilic (0–10°C), mesophilic (20–40°C), or thermophilic (>40°C), depending on the geographic location of the soil. Most soil organisms are mesophilic because of the buffering effect of soil on soil temperature, particularly at depths beneath the soil surface.

Soil pH. Undisturbed soils usually have fairly stable soil pH values within the range of 6–8, and most soil organisms have pH optima within this range. There are, of course, exceptions to this rule, as exemplified by *Thiobacillus thiooxidans*, an organism that oxidizes sulfur to sulfuric acid and has a pH optimum of 2–3. Microsite variations of soil pH can also occur owing, perhaps, to local decomposition of an organic residue to organic acids. Here again, we see that soil behaves as a heterogeneous or discontinuous environment, allowing organisms with differing pH optima to coexist in close proximity. Normally, soil organisms are not adversely affected by soil pH unless drastic changes occur. Drastic change can happen, for example, when lime is added to soil to increase the pH, or when sulfur is added to decrease the pH.

Soil Texture. Almost all soils contain populations of soil organisms regardless of the soil texture. Even soils whose textures are extreme, such as pure sands or clays, usually contain populations of microbes, albeit in lower numbers than in soils with less extreme textures. Most nutrients are associated with clay or silt particles, which also retain soil moisture efficiently. Thus soils with at least some silt or clay particles offer a more favorable habitat for organisms than do soils without these materials.

Soil Carbon and Nitrogen. Carbon and nitrogen are both nutrients that are found in soils. However, since both of these nutrients are typically present in low concentrations, growth and activity of soil organisms are limited. In fact, many organisms exist in soil under limited starvation conditions and hence are dormant. Without added substrate or amendment, soil organisms generally metabolize at low rates. The major exception to this is the rhizosphere, whose root exudates maintain a high population level. In most cases, all available soil nutrients are immediately utilized. Soil humus represents a source of organic nutrients that is mineralized slowly by autochthonous organisms. Similarly, specific microbial populations can utilize xenobiotics as a substrate, even though the rate of degradation is sometimes quite low.

Soil Redox Potential. Redox potential (E_h) is the measurement of the tendency of an environment to oxidize or reduce substrates. In a sense, we can think of it as the availability of different terminal electron acceptors that are necessary for specific organisms. Such electron acceptors exist only at specific redox potentials, which are measured in millivolts (mV). An aerobic soil, which is an oxidizing environment, has a redox potential or E_h of +800 mV; an

TABLE 5.5 Redox Potential at Which Soil Substrates Are Reduced

Redox Potential (mV)	Reaction	Type of Organism and Metabolism
+800	$O_2 \rightarrow H_2O$	Aerobes, aerobic respiration
+740	$NO_3^- \rightarrow N_2$, N_2O	Facultative anaerobes, nitrate reduction
−220	$SO_4^{2-} \rightarrow S^{2-}$	Anaerobes, sulfate reduction
−300	$CO_2 \rightarrow CH_4$	Anaerobes, methanogenesis

From *Environmental Microbiology, Third Edition*, © 2015, Elsevier, San Diego, CA (Table 4.4, pg. 76).

anaerobic soil, which is a reducing environment, has an E_h of about to −300 mV. Oxygen is found in soils at a redox potential of about +800 mV. When soil is placed in a closed container, oxygen is used by aerobic organisms as a terminal electron acceptor until all of it is depleted. As this process occurs, the redox potential of the soil decreases, and other compounds can be used as terminal electron acceptors. Table 5.5 presents the redox potential at which various substrates are reduced and the activity of different types of organisms in a soil.

The fact that different terminal electron acceptors are available for various organisms having diverse pH requirements means that some soil environments are more suitable

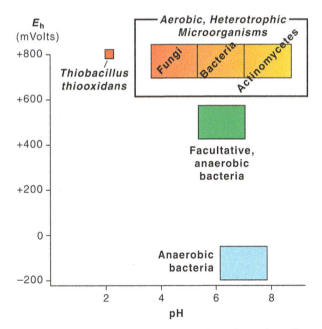

FIG. 5.12 Optimal E_h/pH relationships for various soil organisms. *(From Environmental and Pollution Science, Second Edition, © 2006, Academic Press, San Diego, CA. (Fig. 5.12, pg. 72).)*

than others for various groups of organisms. Fig. 5.12, which illustrates optimal E_h/pH relationships for various groups of organisms, shows that redox potential affects the activity of all organisms.

5.6   ACTIVITY AND PHYSIOLOGICAL STATE OF MICROBES IN SOIL

Is the soil environment a favorable one for soil microbes? There are actually two possible answers to this question. On the one hand, soil is a very harsh environment; on the other hand, soil contains very large populations of microbes. The reason for this is that some soil organisms are spore formers such as *Bacillus* or *Clostridium*. Other bacteria reduce in size to provide a more favorable volume:surface area ratio, a process known as rounding. However, many organisms apparently just go dormant and survive via a variety of mechanisms including reduced metabolic activity, adjustments of the cell wall, and production of extracellular slime (Alvarez et al., 2004). These mechanisms also allow organisms to survive long periods of time when resources are not available. Viewing soil as a community having large numbers of organisms and great microbial diversity, we can infer that each species has a habitat and a niche, rather like a home and a job. Because soil is a heterogeneous environment, many different kinds of organisms can coexist. Diversity also ensures that all available nutrients are utilized in soil.

Owing to the harsh physical soil environment and the fact that nutrients are usually limiting, soil organisms do not actively metabolize most of the time. In fact, they exist under stress, so they may exist injured or even be killed. Fig. 5.13 shows the various potential states of soil organisms, the two extremes being live microbes and dead ones. Between these two extremes, other physiological states are possible, including metabolically active and dormant states. Because many organisms are in fact injured in soil, and because they have diverse specific nutritional needs, many soil organisms cannot be cultured by conventional methods. These are the so-called *viable but nonculturable organisms*. In practice, perhaps 99% of all soil organisms may be nonculturable (Roszak and Colwell, 1987). Thus any methodology that relies on obtaining soil organisms via a culturable procedure may in fact be sampling a very small subsection of the soil population (see also Section 5.2.1).

Based on our discussion of the soil as an environment for microbes, we may reasonably infer indigenous organisms within a particular soil are selected by the specific environment in that soil. Indigenous organisms are often capable of surviving wide variations in particular environmental parameters. Introduced organisms, however, are unlikely to be as well adapted and cannot be expected to compete with indigenous organisms unless a specific niche is available.

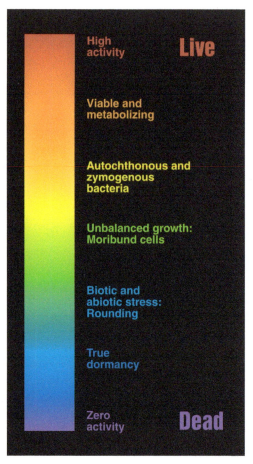

FIG. 5.13   Physiological states of soil organisms. *(From Environmental and Pollution Science, Second Edition © 2006, Academic Press, San Diego, CA. (Fig. 5.13, pg. 72).)*

5.7   ENUMERATION OF SOIL BACTERIA VIA DILUTION AND PLATING

Because of the importance of soil bacteria, there is extensive interest in estimating soil bacterial numbers. Perhaps the most common method of bacterial enumeration is the culturable technique known as "dilution and plating." The following section provides an overview of this technique and an example calculation (Example Calculation 5.1).

Since soils generally contain millions of bacteria per gram, normally a dilution series of the soil is made by suspending a given amount of soil in a dispersing solution (often a dilute buffered peptone or saline solution) and transferring aliquots of the suspensions to fresh solution until the suspension is diluted sufficiently to allow individual discrete bacterial colonies to grow on agar plates.

After inoculation on several replicate agar plates, the plates are incubated at an appropriate temperature and counted after they have formed macroscopic fungal colonies. Because the assumption is that one colony is derived from one organism, the term *Colony Forming Units (CFUs)*

EXAMPLE CALCULATION 5.1 Dilution and Plating Calculations

A 10-gram sample of soil with a moisture content of 20% on a dry weight basis is analyzed for viable culturable bacteria via dilution and plating techniques. The dilutions were made as follows:

Process	Dilution
10 g soil→95 mL saline (solution A)	10^{-1} (weight/volume)
1 mL solution A→9 mL saline (solution B)	10^{-2} (volume/volume)
1 mL solution B→9 mL saline (solution C)	10^{-3} (volume/volume)
1 mL solution C→9 mL saline (solution D)	10^{-4} (volume/volume)
1 mL solution D→9 mL saline (solution E)	10^{-5} (volume/volume)

1 mL of solution E is pour plated onto an appropriate medium and results in 200 bacterial colonies.

$$= \frac{1}{\text{dilution factor}} \times \text{number of colonies}$$

$$\text{Number of } CFU = \frac{1}{10^{-5}} \times 200 \, CFU/g \text{ moist soil}$$

$$= 2.00 \times 10^7 \, CFU/g \text{ moist soil}$$

But, for 1 g of moist soil,

$$\text{Moisture content} = \frac{\text{Moist weight} - \text{Dry weight (D)}}{\text{Dry weight (D)}}$$

Therefore

$$0.20 = \frac{1-D}{D} \text{ and}$$

$$D = 0.833 \, g$$

$$\text{Number of CFU per g dry soil} = 2.00 \times 10^7 \times \frac{1}{0.833} = 2.4 \times 10^7$$

From *Environmental and Pollution Science, Second Edition*, © 2006, Academic Press, San Diego, CA. (Example Calculation Box 5.1, pg. 75).

is used in the final analysis, with the results expressed in terms of CFUs per gram of dry soil.

Fig. 5.14 describes a dilution and plating protocol procedure. Beginning at step 1, a 10-fold dilution series is performed. A 10-fold series is very common as the calculations for the determinations of the organism count are very simple. Here, 10 g of moist soil is added to 95 mL (solution A) of deionized water and shaken well to disperse the organisms. The reason that 10 g of soil are used is that 10 g of soil occupies approximately 5 mL. Thus we have 10 g of soil in 100 mL total volume, thereby forming a 1:10 w/v dilution.

Next, 1.0 mL of suspension is removed from the bottle and added to a tube (b) containing 9.0 mL of the same dispersion solution as in A. The tube is capped and vortexed. Working diligently, the dilution series is continued to the highest desired dilution (tubes C, D, and E). Three different dilutions (tubes C, D, and E) are plated so as to increase the chance of obtaining a dilution that will result in a countable number of organisms.

Here, pour plates are utilized for the plating procedure. The dilution of interest is vortexed and 1.0 mL of suspension is removed from the tube and added to each of two sterile petri dishes. Before the soil particles in the inoculum can settle, pour plates are made (step 3a). Here, a suitable agar such as R_2A medium is poured into the plate with one mL of inoculum. The agar is at a temperature warm enough to keep the agar fluid, but cool enough not to kill the organisms.

Then the plate is *gently* swirled (step 3b) to distribute the agar and inoculum across the bottom of the plate (*without splashing agar on the sides or lid of the dish*). Finally, the agar is allowed to solidify, and the plates are incubated upside down to prevent condensation from falling on the growing surface of the agar (step 4). Counting takes place after an incubation period suitable for the organism(s) of interest (often 5–7 days). It is also useful to recount after 14 days to allow for slow-growing bacteria to form additional colonies.

5.8 MICROORGANISMS IN AIR

5.8.1 The Atmosphere as a Microbial Environment

The atmosphere is a harsh environment for microorganisms to survive due to abiotic stress including UV radiation, temperature, desiccation, and lack of nutrients (Fig. 5.15). This is particularly true for vegetative cells such as bacteria. Thus many of the bacteria capable of surviving harsh atmospheric conditions are capable of forming endospores such as *Bacillus* spp. (Wood et al., 2010). Some fungi are also spore formers and capable of resisting environmental stress. *Aspergillus fumigatus* is a major airborne fungal pathogen capable of causing human disease when the conidia are inhaled into the lungs (McCormick et al., 2010). In general, microbes in the atmosphere survive better when they are surrounded by water droplets or colloidal particles that act as rafts. For example, atmospheric movement of microorganisms can occur via clouds of desert dust (Griffin, 2007). Microbes can be found in different areas of the atmosphere including the atmospheric boundary layer and the troposphere itself (see also Chapter 4).

Microbes that are found in air are generally known as *bioaerosols* and the process that moves microbes into air is known as *aerosolization*. Bioaerosols can be whole microorganisms including viruses, bacteria, fungi, or spores, or they can be biological remnants such as endotoxin and cell wall constituents.

Step 1. Make a 10-fold dilution series.

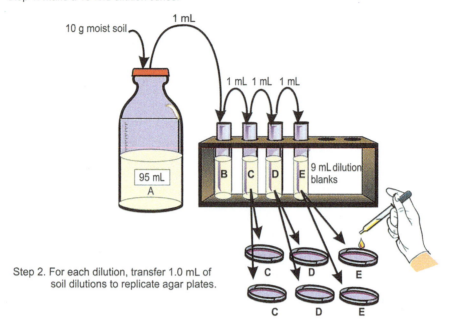

Step 2. For each dilution, transfer 1.0 mL of
soil dilutions to replicate agar plates.

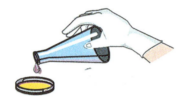

Step 3a. Add molten agar cooled to 45°C
to the dish containing the soil
suspension.

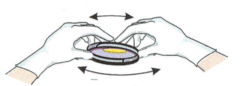

Step 3b. After pouring each plate, replace
the lid on the dish and gently swirl
the agar to mix in the inoculum and
completely cover the bottom of the
plate.

Step 4. Incubate plates under specified conditions.

Step 5. Count dilutions yielding 30–300 colonies
per plate. Express counts as CFUs per
g dry soil.

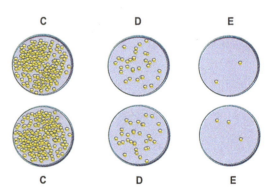

FIG. 5.14   Dilution and pour plating technique. Here, the diluted soil suspension is incorporated directly in the agar medium rather than being surface applied as in the case of spread plating. *(From Environmental and Pollution Science, Second Edition, © 2006, Academic Press, San Diego, CA. (Fig. 5.14, pg. 74).)*

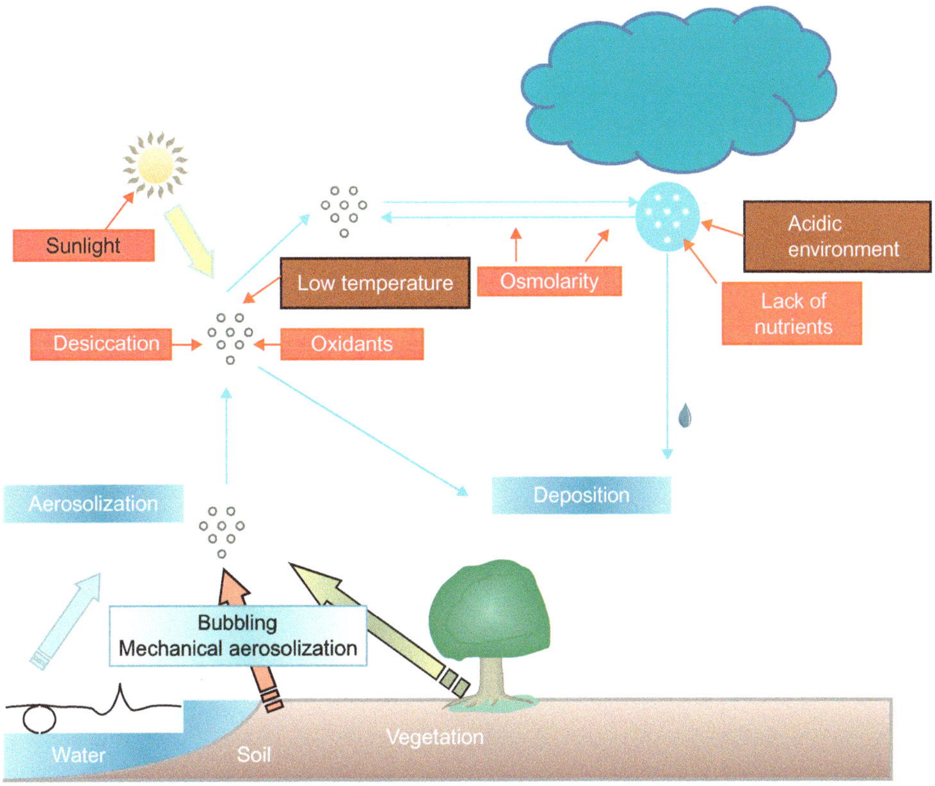

FIG. 5.15 Atmospheric environmental factors that can adversely influence microbes. *(From Environmental Microbiology, Third Edition, © 2015 Elsevier, San Diego, CA.)*

5.8.2 Bioaerosols Within the Atmospheric Boundary Layer

Here, microbes that are aerosolized can be pathogenic or nonpathogenic depending on the source of microorganisms. Soil itself can be a large source of nonpathogenic indigenous soil microbial inhabitants that do not normally pose adverse human health threats. Numbers of aerosolized microbes are frequently a function of wind duration and speed, as well as processes involving agriculture, mining, or road construction. They can range from essentially zero to 10^4–10^6 per m^3 of air. Microbial numbers in near-surface air samples have been reported as 10^4 per m^3 of air for bacteria, and 10^3–10^4 per m^3 for fungal spores (Frölich-Nowoisky et al., 2016). Numbers of microbes in air can be estimated by use of specialized air samplers (Fig. 5.16). In this technique, vacuum pumps are used to deliver known volumes of air to a sampler containing a microbial trapping solution. After air collection, microbes trapped in the liquid are identified and enumerated via cultural or molecular assays.

There are also numerous airborne pathogenic microorganisms that can infect plants, animals, or humans (Information Box 5.4). Clearly disease caused by airborne microorganisms can be devastating, as in the case of foot and mouth disease of animals, or common cold infections in humans (Information Box 5.5). Generally these

pathogens are released from infected animals or humans, or they can be of soil borne origin as in the case of valley fever, which is particularly prevalent in the southwest USA. Plant pathogens can also adversely affect humans as illustrated by the Irish potato famine (Information Box 5.6). Bioaerosols have also been of concern to residents close to biosolid land application sites. A discussion of this issue is presented in Chapter 23. Indoor air quality within homes and buildings is also an issue and discussed in Chapter 18.

5.8.3 Bioaerosols in the Clouds in the Troposphere

Recent studies have suggested that airborne microbes can affect meteorological processes. In particular, some microbes known as ice nucleators efficiently catalyze ice formation and also play a role in the formation of precipitation from clouds (Christner, 2012). This is thought to occur because bacteria can act as cloud concentration nuclei (CCN), which can combine to form "giant CCN" (Amato, 2012). Giant CCN in turn can promote the formation of precipitation by acting as collector drops that form drizzle.

DeLeon-Rodriguez et al. (2013) used culture independent analyses to evaluate the microbiome of the middle to upper troposphere (8–15 km altitude above oceans). Use

FIG. 5.16 Air samplers used to estimate bioaerosol concentrations. *(Photo courtesy J. Brooks. From Environmental and Pollution Science, Second Edition © 2006, Elsevier, San Diego, CA. (Fig. 5.15, pg. 75).)*

INFORMATION BOX 5.4 Examples of Important Airborne Pathogens

Plant Disease	Fungal Pathogen
Dutch Elm Disease	*Ceratocystis ulmi*
Stem rust of wheat	*Puccinia graminis*
Potato blight	*Phytophthora infestans*

Animal Disease	Pathogen
Brucellosis	*Brucella* spp. (bacteria)
Aspergillosis	*Aspergillus* spp. (fungi)
Rabies	*Rhabdoviridae* (virus)
Foot and mouth disease	Aphthovirus (virus)

Human Disease	Pathogen
Pulmonary anthrax	*Bacillus anthracis* (bacteria)
Typhoid fever	*Salmonella typhi* (bacteria)
Valley fever	*Coccidioides immitis* (fungus)
Common cold	Picornavirus (virus)
Hepatitis	Hepatitis (virus)

(From Environmental and Pollution Science, Second Edition, © 2006, Academic Press, San Diego, CA. (Information Box 5.1, pg. 75).)

INFORMATION BOX 5.5 The United Kingdom Foot and Mouth Crisis 2001

Outbreaks of foot and mouth disease have occurred worldwide including multiple occurrences within the United States. However, one of the more devastating outbreaks occurred in the U.K. in the spring and summer of 2001. Foot and mouth viruses are from the *aphthovirus* genus of the family Picornaviridae and are single-stranded RNA viruses. The 2001 U.K. outbreak was due to a type O pan-Asia strain that was thought to have arisen from infected meat illegally importuned into the U.K. Transmission of the virus can occur via direct contact or via bioaerosols (Grubman and Baxt, 2004). Overall, 2000 cases were reported throughout Britain, resulting in the culling of 7 million sheep and cattle, costing the U.K. $\cong$ $16 million.

(From Environmental Microbiology, Third Edition, © 2015, Elsevier, San Diego, CA. (Case Study 5.3, pg. 100).)

of culture-independent techniques was important since only 1% of bacteria and 50% of fungi found in the atmosphere have been reported as culturable (Delort et al., 2010). Multiple air samples were taken under a variety of conditions including cloudy and cloud free air masses, and during and after two major hurricanes. Quantitative PCR (qPCR) of the gene SSU rRNA and the assumption of four rRNA copies per bacterial genome resulted in an average bacterial concentration of 5.1×10^3 cells per m^3 of air. Using the same approach fungal cells were also present with an average concentration two orders of magnitude less than the bacterial concentrations. Overall, 17 bacterial taxa were found in all samples, including taxa known to utilize C1–C4

INFORMATION BOX 5.6 Examples of Airborne Plant Pathogens

This figure shows the airborne spread of late blight of potato that caused the 1845 epidemic known as the Irish potato famine. *Phytophthora infestans* spread from Belgium (mid-June) throughout Europe by mid-October. Famine-related deaths are estimated from 750,000 to 1,000,000. Economic devastation from this famine caused the population of Ireland to decrease from approximately 8 to 4 million from 1840 to 1911.

(From Environmental Microbiology, Third Edition, © 2015 Elsevier, San Diego, CA. (Information Box. 5.3, pg. 99).)

carbon compounds present in the atmosphere. These data suggest that these organisms possessed traits that allowed survival in the atmosphere. These findings indicate that airborne microbial cells may quantitatively be more important for cloud formation and precipitation than previously thought, and play an important role in the biochemistry of the atmosphere, and the hydrological cycle.

5.9   MICROORGANISMS IN SURFACE WATERS

The study of microorganisms in water is known as aquatic microbiology. Seventy percent of the earth's surface is covered with water in the form of oceans, estuaries, lakes, wetlands, and streams. Clearly the diversity of microbial communities within these environments is enormous, and an in-depth review is beyond the scope of this book. However, microbes that are known to inhabit surface waters include phage, bacteria, fungae, and protozoa. With respect to pollution, we are often concerned with either microbes that are pathogenic or produce toxins that adversely affect human health and welfare (see Chapter 13). In other situations, we are concerned with the transport and fate of pathogens introduced into surface waters via, for example, waste disposal or contamination of surface waters with animal wastes. Contamination of surface waters with animal wastes resulted in a large outbreak of waterborne disease in Milwaukee in 1993. In this outbreak, more than 100 individuals died of infection with *Cryptosporidium*—a protozoan parasite (see Case Study, Chapter 13).

5.9.1   Microbial Lifestyles in Aquatic Environments

Primary production via photosynthesizing microbes has been estimated at 50–60 petagrams ($Pg = 10^{15}$ g) of carbon per year (DeLaRocha, 2006). When primary production from freshwater is also taken into account, this represents $\simeq 50\%$ of the total primary production globally (Rich and Maier, 2015). The amount of primary production within any aquatic environment can vary considerably depending on a variety of environmental factors. Of these high turbidity and lack of essential inorganic nutrients, particularly nitrogen and phosphorus, limit primary production. Turbidity and increasing depth of water both limit photosynthesis by decreasing light transmission through the water. Phototrophic organisms include prokaryotic cyanobacteria and eukaryotic algae. Secondary production relies on heterotrophic microbes in water including bacteria, fungi, and protozoa. Bacteria and fungi rely on the absorption of dissolved organic compounds as a mode of nutrition. In contrast, protozoa consume other organisms including algae and bacteria.

Primary and secondary production within aquatic environments can vary tremendously depending on the particular water type and environmental conditions. Freshwaters are those aquatic environments not directly influenced by marine waters. True marine waters are characterized by high salinity with oceans consisting of approximately 3% salt. Estuaries are transitional between fresh and marine waters and are less saline than marine waters. For any given aquatic environment, microbial numbers will likewise vary as a function of light penetration of the water and the amount of dissolved organic matter within the water (Information Box 5.7).

Overall aquatic environments exist as complex ecosystems containing phototrophic and heterotrophic microbes that control the oxygen content of surface waters and the degradation of organic compounds (see Chapter 16). Aquatic microorganisms are also important in urban settings through the formation of biofilms in potable water distribution systems (see Chapter 24).

INFORMATION BOX 5.7 Environmental Influences on Aquatic Microbial Communities

Parameter	Microbial Community Influence	Effect
Depth of water	Primary producers	As depth increases photosynthesis and primary producer population decreases
Proximity to habitat	Heterotrophic populations	The greater the proximity to terrestrial habitat, the more dissolved organic matter and heterotrophic activity
Size of stream	Primary producers and heterotrophic populations	As the size increases, more dissolved organic matter is acquired. Heterotrophic activity increases, while light penetration and phototrophic activity decreases.

(From Environmental and Pollution Science, Second Edition, © 2006, Academic Press, San Diego, CA. (Information Box 5.2, pg. 76).)

5.10 BIOINFORMATION AND 'OMIC-BASED APPROACHES TO CHARACTERIZE MICROBIAL COMMUNITIES

The term "ome" refers to the totality of a collection of specific things, and hence a "biome" is a collection of living organisms. In contrast, 'omics involve the characterization of multiple biological molecules simultaneously. An overview of 'omics-based approaches for characterizing environmental microorganisms is shown in Fig. 5.17. With the advent of next generation or high throughput sequencing, the field of *bioinformatics* has developed to provide the statistical and computational approaches needed to evaluate very large complex data sets.

5.10.1 Genomics and Comparative Genomics

Genomes encompass all of an organisms' hereditary information as DNA for bacteria. For many viruses such as enteroviruses the information is stored as RNA, whereas for other viruses such as adenovirus it is DNA. As of 2016 close to 30,000 genomes had been sequenced (in complete or draft stage) according to the Genomes Online Database. This sequence information has allowed a new field to develop known as *comparative genomics*. This new field involves examining the similarities and differences among genomes to draw inferences about the functions of particular genes and evolutionary relationships between organisms. Thus organisms that share the same habitats may be related. As an example bacterial and archaeal thermophiles often occur in the same habitat and undergo lateral gene transfer that has resulted in 20% gene homology for the bacterial genome of *Thermotoga maritima* with the archaeal *Pyrococcus* spp. despite the fact that they are in different domains (Hollister et al., 2015).

5.10.2 Metagenomics

This technique relies on the analysis of DNA extracted from environmental samples to allow for subsequent analysis of the gene content of microbial communities. This technique has become the gold standard for microbial community analysis, with 16S rRNA sequencing being used to estimate bacterial diversity in a variety of environments (Table 5.6).

FIG. 5.17 Overview of 'omics-based approaches for characterizing environmental microorganisms. *(Adapted from Zhang, W., Li, F., Nie, L., 2010. Integrating multiple "omics" analysis for microbial biology: applications and methodologies, Microbiology 156, 287–301. From Environmental Microbiology, Third Edition, © 2015, Elsevier, San Diego, CA. (Fig. 21.1, pg. 484).)*

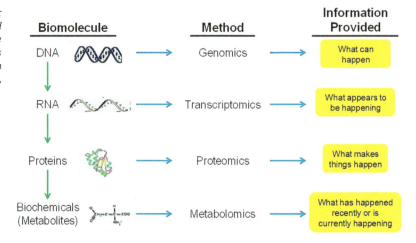

TABLE 5.6 Estimates of Bacterial Diversity (Based Upon 16S rRNA Sequences) in Different Environments

Environment	Diversity Estimate (Species Richness)	Source
Polar desert soil	2935	Fierer et al. (2012)
Agricultural soil	3409	Hollister et al. (2013)
Diesel-contaminated soil	3259	Sutton et al. (2013)
Hypersaline soil	5285	Hollister et al. (2010)
Arctic tundra soil	6965	Fierer et al. (2012)
North Atlantic ocean	6997	Sogin et al. (2006)
Tropical forest soil	8772	Fierer et al. (2012)
Temperate grassland soil	10,253	Fierer et al. (2012)
Temperate forest soil	12,150	Fierer et al. (2012)

From *Environmental Microbiology, Third Edition*, © 2015, Elsevier, San Diego, CA (Fig. 19.1, pg. 444).

This in turn allows for the evaluation of environmental factors such as the addition of organic or metal pollutants on bacterial diversity and an estimate of the toxicity of the contaminant. This is particularly useful since these assays are culture independent and avoid the problems associated with viable but nonculturable bacteria, which dominate all soils. Following sequencing and processing sequences for quality control, sequences or "reads" are frequently compared to existing databases such as the National Center for Biotechnological Information (NCBI) or the Metagenomics Analysis Server (MG-RASI). Additionally,

INFORMATION BOX 5.8 Applications of Metagenomics Relevant to Pollution Science

- Microbial characterization of polluted soil environments
- Characterization of microbial megacommunities found within biofilms
- "Sewage mining": characterization of communities in sewage to identify incidence of microbial pathogens prior to outbreaks
- Infectious disease diagnosis
- Microbial characterization of extreme environments
- Environmental remediation and the need for bioaugmentation

"reads" can be assembled into longer stretches of DNA to give better information since they represent larger portions of the genome.

Currently, diversity estimates of communities via "biome" analysis are extremely popular and useful with respect to Pollution Science and various applications of metagenomics are illustrated in Information Box 5.8.

5.11 TRANSCRIPTONOMICS AND PROTEOMICS

Whereas metagenomics provides data on the DNA potential of a community, other techniques are necessary to identify which genes are actually being expressed. This includes *transcriptonomics* which identifies mRNA, the product of DNA transcription, and *proteomics* which identifies protein expression. For transcriptonomics, mRNA must be collected and preserved immediately since mRNA is extremely labile and is only stable for a few minutes. To do this, commercial kits are available and mRNA is stabilized either by immediate freezing in liquid nitrogen or the addition of RNA stabilization buffers such as the MO BIO Lifeguard Soil Preservation Solutions (MO BIO Laboratories, Inc., Carlsbad, CA) (Hollister et al., 2015).

These buffers preserve the total RNA in the biological system, but the RNA still has to be extracted from intact cells using commercial kits and bead-beating technology with a stable buffer. Finally, the mRNA must be converted to cDNA prior to sequence analysis. For sequencing, Illumina systems are frequently used, followed by quality control procedures to remove short read sequences. Sequences are then compared to available databases which assign gene function and identification.

Although metagenomics and transcriptonomics provide great insight into the environmental roles of microorganisms, it is proteins not genes that conduct microbial processes, and thus proteomic analyses are vital. Measurement of proteins produced by a given community gives a more direct measurement of microbial activity, and the proteins produced by a given microbe within a certain environment are known as the *proteome*. Typically proteomic-based studies involve the exposure of the microbes to specific environmental conditions, followed by the isolation, separation, and identification of the proteins. Separation and identification of proteins is not trivial and can involve liquid chromatography mass spectrometry (LC-MS).

Overall, transcriptonomics and proteomics can be utilized in the same applications as many of the metagenomics studies and are utilized to link microbial species to function. Combined together, these "omic approaches to characterizing environmental communities provide a whole new "toolbox" for environmental scientists.

QUESTIONS AND PROBLEMS

1. If you could be a soil microorganism, what would you be? Why?
2. The use of metagenomics overcomes many of the problems associated with cultural classification of soil microorganisms. Explain what these advantages are. What are the major limitations of metagenomics analysis?
3. A soil is saturated with water and subsequently allowed to dry out. Discuss the changes in microbial populations that occur during the process.
4. Briefly discuss the statement "Soil is a favorable environment as a habitat for microorganisms."
5. For the chemical reaction

$$A + B \rightleftarrows C + D$$

the thermodynamic equilibrium constant $K_{eq} = 0.38$.

Deduce whether $\triangle G$ for the reaction is negative or positive. Is energy liberated from the reaction or must energy be added to promote the reaction?

6. A 10-g sample of soil with a moisture content of 25% on a dry weight basis is analyzed for viable culturable bacteria via dilution and plating techniques. A 10^{-5} dilution of the soil suspension resulted in 421 bacterial colonies. How many culturable bacteria are there per gram of dry soil?
7. For the following organisms identify whether the microbe is heterotrophic or autotrophic. Also identify the substrate that can be oxidized as well as the terminal electron acceptor utilized by the organism.
   a. *Nitrosomonas* spp.
   b. *Thiobacillus thiooxidans*
   c. *Thiobacillus denitrificans*
   d. *Desulfovibrio* spp.
8. Discuss the likely microbial populations for the following scenarios. Identify modes of nutrition and whether or not populations are likely to be aerobic or anaerobic.
   a. peat bog
   b. pristine alpine lake
   c. clouds
   d. the air in a city on a rainy day
   e. the air surrounding a farm during a windy day when a dry bare field is being plowed

REFERENCES

Alvarez, H.M., Silva, R.A., Cesari, A.C., Zamit, A.L., Peressutti, S.R., Reichelt, R., Keller, U., Mulkus, U., Rasch, C., Maskow, T., Mayer, F., Steinbchel, A., 2004. Physiological and morphological responses of the soil bacterium *Rhodococcus opacus* strain PD630 to water stress. FEMS Microbiol. Ecol. 50, 75–86.

Amato, P., 2012. Clouds provide atmospheric oases for microbes. Microbe 7, 119–123.

Artiola, J.F., Pepper, I.L., 1992. Denitrification activity in the root zone of a sludge amended desert soil. Biol. Fertil. Soils 13, 200–205.

Chen, Y., Kuang, J., Jia, P., Cadotle, M.W., Huang, L., Li, J., Liao, B., Wang, P., Shu, W., 2017. Effect of environmental variation on estimating the bacterial species richness. Front. Microbiol. 8, https://doi.org/10.3389/fmicb.2017.00690. Article Number 690.

Christner, B.C., 2012. Cloudy with a chance of microbes. Microbe 7, 70–75.

DeLaRocha, 2006. The biological pump. In: Elderfield, H. (Ed.), The Oceans and Maine Geochemistry. Elsevier, New York, pp. 83–111.

DeLeon-Rodriguez, N., Lathem, T.L., Rodriguez-R, L.M., Barazesh, J.M., Anderson, B.E., Beyersdorf, A.J., Ziemba, L.D., Bergin, M., Nenes, A., Konstantinidis, K.T., 2013. Microbiome of the upper troposphere: species composition and prevalence, effects of tropical storms; and atmospheric implications. PNAS 110, 2575–2580.

deLima Procopio, R.E., DaSilva, I.R., Martins, M.K., deAzevedo, J.L., deAraujo, 2012. Antibiotics produced by Streptomyces. Braz. J. Infect. Dis. 16, 466–471.

Delort, A.M., Vatilingom, M., Amato, P., Sancelme, M., Prazols, M., Mailot, G., et al., 2010. A short overview of the microbial population in clouds: potential in atmospheric chemistry and nucleation process. Atmos. Res. 98, 249–260.

Fierer, N., Leff, J.W., Adams, B.J., Nielsen, U.N., Bates, S.T., Lauber, C.L., Owens, S., Gilbert, J.A., Wall, D.H., Caporaso, J.G., 2012. Cross-biome metagenomic analyses of soil microbial communities and their functional attributes. Proc. Natl. Acad. Sci. U.S.A. 109, 21390–21395.

Frölich-Nowoisky, J., Kampf, C.J., Weber, B., Hutfman, J.A., Pőhlker, C., Andreae, M.D., Lang-Yona, N., Burrows, S.M., Gunthe, S.S., Ebert, W., Su, H., Hoor, P., Thines, E., Hoffman, T., Deprés, U.R., Pőschl, U., 2016. Bioaerosols in the Earth system: climate health and ecosystem interactions. Atmos. Res. 182, 346–376.

Fuerst, J.A., Sagulenkol, E., 2011. Beyond the bacterium: planctomycetes challenge our concepts of microbial structure and function. Nat. Rev. Microbiol. 9, 403–413.

Gentry, T.J., Pepper, I.L., Pierson III, L.S., 2015. Microbial diversity and interactions in natural ecosystems. In: Pepper, I.L., Gerba, C.P., Gentry, T.J. (Eds.), Environmental Microbiology. third ed. Elsevier, San Diego, CA.

Gerba, C.P., 2015. Environmentally transmitted pathogens. In: Pepper, I.L., Gerba, C.P., Gentry, T.J. (Eds.), Environmental Microbiology. third ed. Elsevier, San Diego, CA.

Griffin, D.W., 2007. Atmospheric movement of microorganisms in clouds of desert dust and implications for human health. Clin. Microbiol. Rev. 20, 459–477.

Grubman, M.J., Baxt, B., 2004. Foot and mouth disease. Clin. Microbiol. Rev. 17, 465–493.

Gupta, A., Akin, D., Bashir, R., 2004. Single virus particle mass detection using microresonators with nanoscale thickness. Appl. Phys. Lett. 84, 1976–1978.

Hollister, E.B., Brooks, J.P., Gentry, T.J., 2015. Bioinformation and "omic" approaches for characterization of environmental microorganisms. In: Pepper, I.L., Gerba, C.P., Gentry, T.J. (Eds.), Environmental Microbiology. Elsevier, San Diego, CA.

Hollister, E.B., Engledow, A.S., Hammett, A.J., Provin, T.L., Wilkinson, H.H., Gentry, T.J., 2010. Shifts in microbial community structure along an ecological gradient of hypersaline soils and sediments. ISME J. 4, 829–838.

Hollister, E.B., Hu, P., Wang, A.S., Hons, F.M., Gentry, T.J., 2013. Differential impacts of brassicaceous and non-brassicaceous oilseed meals on soil bacterial and fungal communities. FEMS Microbiol. Ecol. 83, 632–641.

Ingraham, J.L., Maalte, Neidhardt, F.C., 1983. Growth of the Bacterial Cell. Sinawer Associates, Inc., Sunderland, MA

Keen, E.C., Bliskovsky, V.V., Malagon, F., Baker, J.D., Prince, J.S., Klaus, J.S., Adnya, S.L., 2017. Novel "superspreader" bacteriophages promote horizontal gene transfer by transformation. MBio 8, e02115–e02116. https://doi.org/10.1128/mbio.02115-16.

Killham, K., 1994. Soil Ecology. Cambridge University Press, Cambridge.

Maksimova, I.A., Glushakova, A.M., Kachalkin, A.V., Chernov, I.Y., Panteleeva, S.N., Reznikova, Z.I., 2016. Yeast communities of Formica aquilonia colonies. Microbiology 85, 124–129.

McCormick, A., Loeffler, J., Ebel, F., 2010. *Aspergillus fumigatus*: contours of an opportunistic human pathogen. Cell. Microbiol. 12, 1535–1543.

Montgomery, H.J., Monreal, C.M., Young, J.C., Seifert, K.A., 2000. Determination of soil fungal biomass from soil ergosterol analyses. Soil Biol. Biochem. 32, 1207–1217.

Parikka, K.J., LeRomancer, M., Wautero, N., Jacquet, S., 2017. Deciphering the virus-to-prokaryote ratio (VPR): insights into virus-host relationships in a variety of ecosystems. Biol. Rev. 92, 1081–1100.

Pepper, I.L., Gentry, T.J., 2015. Earth environments. In: Pepper, I.L., Gerba, C.P., Gentry, T.J. (Eds.), Environmental Microbiology. third ed. Elsevier Inc., San Diego, CA

Rich, V.I., Maier, R.M., 2015. Aquatic environments. In: Pepper, I.L., Gerba, C.P., Gentry, T.J. (Eds.), Environmental Microbiology. third ed. Elsevier, San Diego, CA.

Roszak, D.B., Colwell, R.R., 1987. Survival strategies of bacteria in the natural environment. Microbiol. Rev. 51, 365–379.

Shade, A., Hogan, C.S., Klimowiez, A.K., Linske, M., McManus, P.S., Hendelsman, J., 2012. Culturing captures members of the soil rare biosphere. Environ. Microbiol. 14, 2247–2252.

Sogin, M.L., Morrison, H.G., Huber, J.A., Welch, D.M., Huse, S.M., Neal, P.R., Arrieta, J.M., Herndl, G.J., 2006. Microbial diversity in the deep sea and the underexplored "rare biosphere.". Proc. Natl. Acad. Sci. U.S.A. 103, 12115–12120.

Song, Z., Schlatter, D., Kennedy, P., Kinkel, L.L., Kistler, H.C., Nguyen, H., Bates, S.T., 2015. Effort versus reward: preparing samples for fungal community characterization in high throughput sequencing surveys of soils. PLoS ONE. 10 (5). https://doi.org/10.1371/journal.pone.0127234. Article Number: UNSPe0127234.

Suttle, C.A., 2005. Viruses in the sea. Nature 437, 356–361.

Sutton, N.B., Maphosa, F., Morillo, J.A., Al-Soud, W.A., Langenhoff, A.A.M., Grotenhuis, T., Rijnaarts, H.H.M., Smidt, H., 2013. Impact of long-term diesel contamination on soil microbial community structure. Appl. Environ. Microbiol. 79, 619–630.

Takenouchi, Y., Iwasaki, K., Murase, J., 2016. Response of the protistan community of rice field soil to different oxygen tensions. FEMS Microbiol. Ecol. https://doi.org/10.1093/femsec/fiw104.

Verhamme, D.T., Prosser, J.I., Nicol, G.W., 2011. Ammonia concentration determines differential growth of ammonia-oxidizing archaea and bacteria in soil microcosms. ISME J. 5, 1067–1071.

Vieira, F.C.S., Hahas, E., 2005. Comparison of microbial numbers in soils by using various culture media and temperatures. Microbiol. Res. 160, 197–202.

Williamson, K.E., Radosevich, M., Smith, D.W., Wommack, K.E., 2007. Incidence of lysogeny within temperate and extreme soil environments. Environ. Microbiol. 9, 2563–2574.

Wood, J.P., Lemieux, P., Betancourt, D., Kariher, P., Gatchalian, N.G., 2010. Dry thermal resistance of *Bacillus anthracis* (Sterne) spores and spores of other *Bacillus* species: implications for biological agent destruction via waste incineration. J. Appl. Microbiol. 109, 99–106.

Zancan, S., Trevisan, R., Paoletti, M.G., 2006. Soil algae composition under different agro-ecosystems in North-Eastern Italy. Agric. Ecosyst. Environ. 112, 1–12.

Zhang, G.-Y., Zhang, L.-M., He, J.-Z., Lui, F., 2017. Comparison of Archael populations in soil and their encapsulated iron-manganese nodules in four locations spanning from North to South China. Geomicrobiol. J. https://doi.org/10.1080/01490451.2016.1278056.

FURTHER READING

Deshmukh, S.K., Sing, A.K., Datta, S.P., Annapurna, K., 2011. Impact of long-term wastewater application on microbiological properties of vadose zone. Environ. Monit. Assess. 175, 601–612.

Fuhrman, J., 1992. Bacterioplankton roles in cycling of organic matter. The microbial food web. In: Falkowski, P.G., Woodhead, A.D. (Eds.), Primary Productivity and Biogeochemical Cycles in the Sea. Plenum, New York, pp. 361–383.

Maier, R.M., Pepper, I.L., Gerba, C.P., 2000. A Textbook of Environmental Microbiology. Academic Press, San Diego, CA.

Turco, R.F., Sadowsky, M., 1995. The microflora of bioremediation. In: Skipper, H.D., Turco, R.F. (Eds.), Bioremediation: Science and Applications. In: Special Publ. No. 43Soil Science Society of America, Madison, WI, pp. 87–102.

Zhang, W., Li, F., Nie, L., 2010. Integrating multiple "omics" analysis for microbial biology: applications and methodologies. Microbiology 156, 287–301.

Chapter 6

Ecosystems and Ecosystem Services

M.L. Brusseau

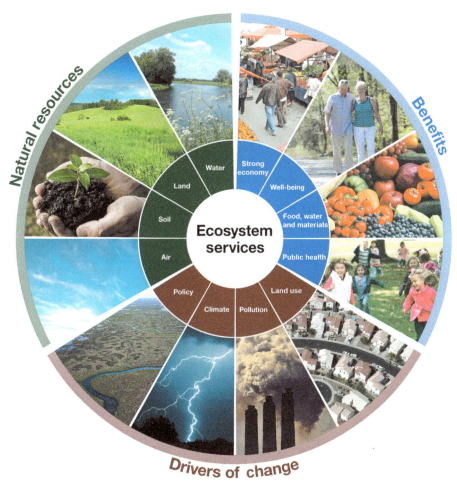

Chapter cover art: Ecosystem services—sources, benefits, and drivers of change *From: https://www.epa.gov/enviroatlas/ecosystem-services-enviroatlas*

The environment is much more than its physical components such as soil, rocks, water, and air. Lifeforms are present in the environment, and these lifeforms are supported by the environment and have myriad impacts on the environment. These impacts can in turn affect the ability of the environment to support life. This integrated system of organisms and the physical components of the environment is called an ecosystem. Organisms such as humans receive many benefits from ecosystems and can also be negatively impacted by environmental components or processes. The

region at the Earth's surface (called the critical zone), and particularly soil, is a key component of ecosystems that support all terrestrial life. These concepts are presented in this chapter.

6.1 ECOSYSTEMS

Chapters 2–4 covered the physical and chemical properties of the environment. The biological component of the

Environmental and Pollution Science. https://doi.org/10.1016/B978-0-12-814719-1.00006-9

environment—life, is covered in Chapter 5 specifically for microorganisms and more generally in this chapter.

The fundamental operational unit for characterizing the presence, activities, and impacts of lifeforms in the environment is an *ecosystem*. An ecosystem consists of a community of organisms living in a discrete designated area, together with the nonliving components of that area (the local physical environment). It is important to understand the difference between "community" and "ecosystem." A *community* is the aggregate of all of the populations of the different species that live together in the designated area. In other words, a community is the biotic, or living, component of an ecosystem. Conversely, an ecosystem includes the physical environment in addition to the biotic component.

The sizes of ecosystems vary across a wide range. They can be small, such as a tidal pool, a decaying tree trunk, or the earth beneath a boulder. They can also be very large, such as the Sahara desert or the Amazon rainforest. Very large ecosystems are sometimes referred to as "Ecoregions." In addition, ecosystems can often be subdivided into smaller individual subecosystems. For example, the Amazon rainforest basin has several subecosystems, at different scales. The largest set of these subecosystems include (i) floodplain forests or varzea (the areas close to rivers that are flooded during the rainy season); (ii) terra firma rainforest (forest that is off the floodplain); (iii) igapo (forests around freshwater lakes and lower reaches of rivers); (iv) the Cerrado savanna, significantly drier than the interior Amazon forest; and (v) the Amazon River aquatic ecosystem. Each one of these major subecosystems can be divided into smaller individual ecosystems, such as the tree canopy, the forest land surface, and streams and ponds within a single forest area. Thus we recognize that ecosystems can be defined at different telescoping scales. Artificial boundaries are imposed to define a specific ecosystem when scientists wish to examine that particular ecosystem.

Ecosystems also vary greatly in their properties, both in the physical nature of the environment and in the types of lifeforms present. The major types of large-scale ecosystems are categorized into *biomes*. These are very large regions of the Earth's surface that have unique fauna and flora that have developed because of the specific geologic factors present, such as climate (temperature and precipitation). The major biomes are typically designated as Desert, Aquatic, Grasslands, Forests, and Tundra. These are briefly described in Information Box 6.1. A map showing the distribution of major biomes on Earth is presented in Fig. 6.1.

Ecology is the science discipline that studies how organisms interact with each other and with their physical environment. Ecological studies can be conducted at six general levels: organism, population, community, ecosystem, biome, and biosphere (Fig. 6.2). These levels are defined as follows:

- *Organism*: An individual organism (single cell or animal). Organismal ecologists study *adaptations*, beneficial features arising by natural selection, that allow organisms to live in specific habitats. These adaptations can be morphological, physiological, or behavioral.
- *Population*: A *population* is a group of organisms of the same species that live in the same area at the same time. Population ecologists study the size, density, and structure of populations and how they change over time.
- *Community*: A biological *community* consists of all the populations of different species that live in a given area. Community ecologists focus on interactions between populations and how these interactions shape the community.
- *Ecosystem*: An *ecosystem* consists of all the organisms in an area, the community, and the abiotic factors that influence that community. Ecosystem ecologists often focus on flow of energy and recycling of nutrients.
- *Biome*: A *biome* is a set of ecosystems sharing similar characteristics (Information Box 6.1).
- *Biosphere*: The *biosphere* is planet Earth, viewed as the aggregation of all of the ecosystems and biomes into one integrated and interdependent system. Ecologists working at the biosphere level may study global patterns—for example, climate or species distribution—interactions among ecosystems, and phenomena that affect the entire globe, such as climate change.

The different levels of ecological investigation presented earlier are listed from small to large. They build progressively—populations are comprised of individual organisms, while communities are comprised of populations, and similarly a biome consists of a set of ecosystems. Each level has emergent properties, meaning new properties that are not a part of any of the individual components comprising that level, but instead emerge from the various synergistic and antagonistic interactions and relationships among the components. Ecological studies will often incorporate more than one level to examine interactions among the levels.

There are several different types of interactions among individuals and populations within a community. *Competition* occurs when more than one organism or population uses a resource at the same time. *Predation* is one organism or population feeding on another. Many species obtain their food by eating other organisms. *Symbiotic Relationships* are the close relationships that exist when two or more species live together and each provides benefits to the other. A *habitat* is an area where an organism lives. A *niche* is the role or position that an organism has in its ecosystem.

The feeding relationship in an ecosystem is called a *food chain*. Food chains are usually in a sequence, with an arrow used to show the flow of food. A food web is a network of many food chains and is more complex, detailing the feeding interactions within the whole ecosystem

INFORMATION BOX 6.1 Major Biomes

Aquatic:

Aquatic biomes contain numerous species of plants and animals, both large and small. Although water temperatures can vary widely, aquatic areas tend to be more humid and the air temperature on the cooler side. The aquatic biome can be divided into two basic types, freshwater (i.e., lakes, ponds, streams, and wetlands) and marine (i.e., oceans and estuaries). The major property differentiating these two types of systems is the salinity of the water. See Chapter 3 for discussion of these two systems.

The plants and animals that live in each type of system have evolved the ability to thrive under the respective conditions. The physical conditions of aquatic systems have led to great differences in the morphology of lifeforms living within these systems versus those of terrestrial lifeforms. Examples include massively large whales and the general streamlined shapes of fish.

Deserts:

Deserts cover about one-fifth of the Earth's surface and occur where precipitation is less than 50 cm/year. Low magnitude of rainfall is the defining characteristic of deserts, and it has a significant impact on the vegetation and animals present. Vegetation is usually sparse, with a preponderance of smaller, hardy plants. There are relatively few large animals in deserts because most are not capable of storing sufficient water and withstanding the heat (or cold). The dominant animals of warm deserts are nonmammalian vertebrates, such as reptiles. Mammals are usually small, like the kangaroo mice of North American deserts.

Although most deserts, such as the Sahara of North Africa and the deserts of the southwestern United States, Mexico, and Australia, occur at low latitudes, another kind of desert, cold deserts, occur in the basin and range area of Utah and Nevada and in parts of western Asia. More generally, desert biomes can be classified according to several characteristics. There are four major types of deserts: Hot and Dry (e.g., Chihuahuan, Sonoran, and Mojave in North America), Semiarid (e.g., the Great Basin in North America), Coastal (e.g., the Atacama of Chile), and Cold (e.g., regions in Antarctic and Greenland).

Forests:

Today, forests occupy approximately one-third of the Earth's land area, account for over two-thirds of the leaf area of land plants, and contain about 70% of carbon present in living things. Forest biomes have relatively high rates of precipitation and are dominated by trees and other woody vegetation. Soils are generally rich, with high acidity and decaying organic matter. Forests typically have very large and diverse communities (large biodiversity).

Forest biomes are classified according to numerous characteristics, with seasonality being the most widely used. Distinct forest types also occur within each of these broad groups. There are three major types of forests, classified according to latitude: tropical (e.g., the Amazon rain forest in Brazil), temperate (e.g., the forests of eastern United States), and boreal or taiga (e.g., the Scandinavian forest). The types of trees and other vegetation present differ greatly among these three forests.

Grasslands:

Grasslands are characterized as lands dominated by grasses rather than large shrubs or trees. Grasslands occur between the desert and mountain biomes, and are considered a transitional biome. They typically receive moderate amounts of precipitation.

There are two main types of grasslands, tropical grasslands (called savannas) and temperate grasslands. Savannas cover almost half of the African continent (about 5 million square miles, generally in central Africa) and large areas of Australia, South America, and India. The major examples of temperate grasslands are the veldts of South Africa, the puszta of Hungary, the pampas of Argentina and Uruguay, the Eurasian steppes, and the plains and prairies of central North America.

Tundra:

Tundra is the coldest of all of the terrestrial biomes. Tundra originates from the Finnish word *tunturia*, meaning treeless plain. This biome has very little precipitation, freezing temperatures, and covers about a fifth of the Earth's land surface. It has the lowest biodiversity of all of the biomes. A key property of this biome is *permafrost*, which is the frozen ground of the tundra. Tundra is separated into two types, arctic tundra (the Arctic region of the northern hemisphere) and alpine tundra (the tops of mountainous areas).

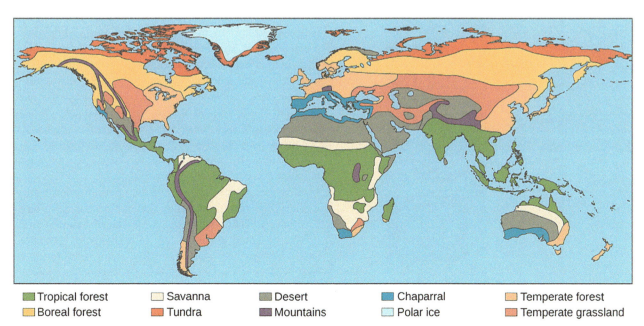

| ■ Tropical forest | □ Savanna | ■ Desert | ■ Chaparral | ■ Temperate forest |
| ■ Boreal forest | ■ Tundra | ■ Mountains | □ Polar ice | ■ Temperate grassland |

FIG. 6.1 Major biomes on Earth. *Reproduced from OpenStax College, Terrestrial Biomes. OpenStax CNX. 21 Jun 2013. http://cnx.org/contents/ cf87c2b7-d8ec-4cba-adba-6c30f5736f63@4.*

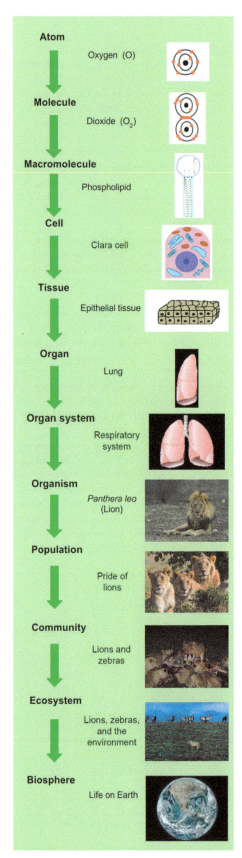

FIG. 6.2 Levels of ecology. *From https://en.wikipedia.org/wiki/Bio logical_organisation*

(Fig. 6.3). The levels of a food chain are called *Trophic levels*. The trophic level of an organism is the level it holds in the overall food pyramid (Fig. 6.4). The sun is the source of energy in the vast majority of food chains. Green plants, usually the first level of any food chain, absorb some of the Sun's light energy to make their own food via photosynthesis. Green plants are therefore known as "Primary Producers" in a food chain. The second level of the food chain is called the Primary Consumer. These organisms consume the green plants. Examples include insects, sheep, caterpillars, and cows. The third level in the chain is Secondary Consumers, which eat the primary consumers and other animal matter. They are commonly called carnivores and examples include lions, snakes, and cats. The fourth level is called Tertiary Consumers. These are animals that eat secondary consumers. Quaternary Consumers eat tertiary consumers. At the top of the levels are the apex Predators.

Food chains and webs are an example of the flow of matter and energy within an ecosystem. Other examples of matter and energy flow are the various biogeochemical cycles that occur within ecosystems. Important examples are the carbon and nitrogen cycles, covered in Section 6.4. Finally, physical cycles of geological and hydrological processes are also important. These physical and biogeochemical cycles are critical to the existence of life, transforming energy and matter into usable forms to support the functioning of ecosystems. A few general principles govern energy and matter flow in ecosystems. Energy flows through the ecosystem, usually entering as sunlight and exiting as heat, and is not recycled. Conversely, matter is constantly recycled within an ecosystem.

6.2 ECOLOGICAL PRINCIPLES

Ecosystems are dynamic systems—energy is constantly flowing through an ecosystem, and materials are continually being recycled; individual organisms die while others are being born; populations of different species are fluctuating in their numbers; and climate is varying seasonally. In essence, a static ecosystem is normally a dead ecosystem—just as a static cell would normally be a dead cell or at least dormant. *An ecosystem* whose composition and identity remain generally constant despite fluctuations in physical conditions and community composition is at steady state and is said to be in equilibrium. A system in equilibrium is still a dynamic system; it is not static.

Ecosystems may be knocked out of equilibrium by disturbances, disruptive events that affect their composition or functioning. These disturbances may be of natural (earthquake, volcanic eruption) or anthropogenic (pollution, excessive resource use) origin. If a disturbance is severe enough, it may change an ecosystem beyond the point of recovery, to where it is no longer resilient. This could lead to permanent impairment or loss of the ecosystem and the

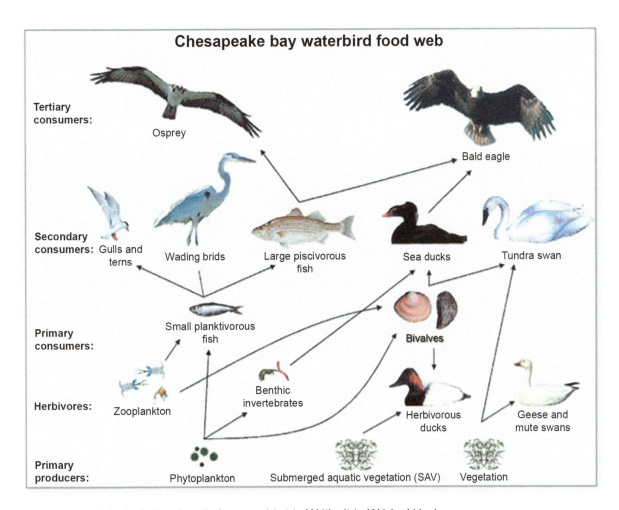

FIG. 6.3 Example of a food web. *From https://pubs.usgs.gov/circ/circ1316/html/circ1316chap14.html*

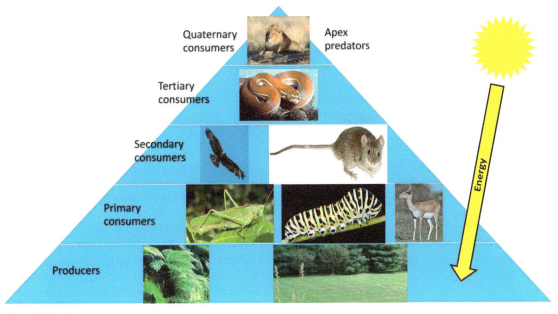

FIG. 6.4 Trophic levels.

services they provide. Ecosystems may have more than one state of equilibrium. Hence a perturbed ecosystem may be able to transition to another state of equilibrium that is optimal for the current, postdisturbance, conditions.

The state of an ecosystem with respect to its stability or instability is measured by two parameters—latitude and precariousness. *Latitude* is the maximum extent to which a system can be altered before losing its ability to recover. This point is designated as the critical threshold, or tipping point, beyond which recovery is difficult or impossible. *Precariousness* describes how close the ecosystem currently is to the critical threshold.

The response of ecosystems to disturbances, and their ability to withstand or adapt to said disturbances, is determined by their resilience. *Ecological resilience*, then, represents how readily and quickly an ecosystem returns to equilibrium after being disturbed. This is a critical concept in ecology for understanding the impacts of human activities on ecosystems. In addition, ecological resilience has become a central concept for conservation practices and ecosystem restoration and management efforts.

A resilient ecosystem is one that has the capability to maintain key functions and processes in the face of stresses, either by resisting or adapting to change. Possessing a high degree of resilience is a positive attribute in the vast majority of cases. However, resilience is not always a positive feature of an ecosystem. An ecosystem may be locked into an undesirable state, thus operating at suboptimal levels. An example of such a condition is a lake under permanent eutrophic conditions, where an overabundance of nutrients results in depleted oxygen levels that in turn lead to loss of desirable aquatic species and the proliferation of undesirable ones.

Two critical characteristics determine the resilience of a particular ecosystem, resistance and adaptive capacity. Resistance is the ability of an ecosystem to remain at equilibrium in spite of a disturbance. It represents the ease or difficulty of changing the system. An ecosystem that is more resistant will likely exhibit fewer deleterious effects of a perturbation. Adaptive capacity is the ability of an ecosystem to adapt to the change in conditions brought about by the disturbance.

Many factors influence the resistance and adaptive capacity of an ecosystem and thus its resilience. One major factor that is generally considered essential for resilience is *biodiversity*. There are several types or levels of biodiversity. Genetic diversity represents biological diversity within a single population of a species. Community diversity refers to the numbers and types of populations or species present in the ecosystem. Functional diversity refers to the differentiation of populations based on their function or role in the ecosystem. *Redundancy* refers to overlap in populations for a particular role or niche. Possessing greater levels of biodiversity provides the ecosystem with more opportunities and abilities to resist and adapt to changes.

The degree to which individual components and processes are connected and interact among the different scales and levels of an ecosystem is another major factor influencing resilience. This has been referred to as *panarchy*. Systems with higher levels of interconnectedness are generally considered to be more resilient.

Abiotic factors may also influence the resilience of an ecosystem. Climate is one primary abiotic factor, acting through, for example, its control of water availability. The properties and conditions of soil are another example. Third, an ecosystem that comprises a higher diversity of habitats will generally be more resilient. Finally, the type of disturbance itself can influence resilience, as a particular ecosystem may respond differently to different types of disturbances.

If an ecosystem's resilience is overcome, it will then undergo a longer term series of changes through the process of *ecological succession*. Think of a lava flow, such as those that still occur on the island of Hawaii on the slopes of Mauna Loa and Kilauea, burning through portions of the native rain forest. The lava cools quickly, but lays barren for many years, too hostile to support life. Eventually rain and wind erosion create tiny fissures in the lava, where life can establish a foothold. Microorganisms, usually bacteria, are often the first forms of life to become established. In fact, very few disturbed sites are microbiologically sterile. Microorganisms are a prerequisite for plant growth. Lichens and cyanobacteria are the next colonists to arrive on the flow. Continual breakdown of the parent lava by acids secreted by the lichens produces a thin layer of soil in which the first higher order plants can root. Generally, small ferns appear first, followed by grasses and shrubs as the fissures widen due to the action of the plant roots. Cyanobacteria fix nitrogen, which also supports the plant life. Each stage of succession conditions the lava substrate to favor the next stage, and finally the rain forest is restored. This process may take many tens to hundreds or even thousands of years. These long time scales for recovery make it imperative that anthropogenic disturbances are minimized and that ecosystems are protected and managed to maintain their resilience.

6.3 THE CRITICAL ZONE

The Critical Zone is the upper surface layer of Earth—from the tops of the trees to the bottom of the groundwater (Fig. 6.5). The term was coined for this purpose in 1998 by Gail Ashley and represents the concept of ecosystems applied to that region of Earth that supports all terrestrial life, including humans. It is a living, breathing, constantly evolving boundary layer where rock, soil, water, air, and living organisms interact. It also borders the most critical zone of the atmosphere that affects human life—the atmospheric boundary layer (see Chapter 4). These complex interactions regulate the natural habitat and determine the

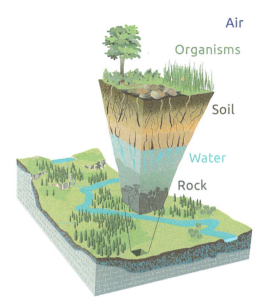

Air

Organisms

Soil

Water

Rock

FIG. 6.5 The critical zone. *Reproduced from http://criticalzone.org/ national/research/the-critical-zone-1national/ based on a figure from Chorover, J., Kretzschmar, R., Garcia-Pichel, F., Sparks, D.L., 2007. Soil biogeochemical processes in the critical zone. Elements 3, 321–326.*

availability of life-sustaining resources, including food production and water quality.

The critical zone is imprinted by different types of events over an extremely wide range of time scales, from seconds, hours, years, millennia, and geologic time. The present structure and functioning of the critical zone reflects the aggregate impacts of these various events. The critical zone and human society are closely intertwined, impacting each other in many ways. Human activities can have a major impact on the conditions and functioning of the critical zone. Various examples of these impacts are covered in many later chapters of this book. Understanding the complex web of physical, chemical, and biological processes of the critical zone requires a **systems approach** across a broad array of sciences: hydrology, geology, soil science, biology, ecology, geochemistry, geomorphology, and more.

The critical zone provides many benefits to humans. These have been termed "ecosystem services" and will be covered in a following section. Certain interactions between humans and the critical zone can lead to negative health outcomes. This is covered in Chapter 27. An essential component of the critical zone is soil. Soil health and the interaction between soil and humans are also covered in Chapter 27.

6.4 BIOGEOCHEMICAL CYCLES

Biogeochemical cycles are critical to the existence of life, transforming energy and matter into usable forms to support the functioning of ecosystems, as noted previously. These cycles describe the movement of matter between the major

reservoirs of Earth—the atmosphere, the terrestrial biosphere, the oceans, and the geosphere (soil, sediments, and rocks). The six most common elements comprising organic molecules—the backbone of lifeforms—are carbon, nitrogen, hydrogen, oxygen, phosphorus, and sulfur. These elements reside in the different reservoirs to different extents and consist of a variety of chemical forms both organic and inorganic.

Carbon is an essential element in the bodies of all living organisms and an essential source of energy for many organisms. As discussed in Chapter 4, carbon in the form of CO_2 in the atmosphere is a greenhouse gas that helps regulate Earth's temperature to make it hospitable for life. Carbon, in the form of oil, natural gas, and coal, is a critical source of energy for human activities. Carbon is also a key element for the manufacture of everyday products such as plastics. Carbon is central to current issues of global climate change (Chapter 25). The *carbon cycle* describes the movement of carbon among the various reservoirs (Fig. 6.6).

Carbon cycles through the environment at different spatial and temporal scales. The carbon cycle can be characterized as two interconnected subcycles. One subcycle consists of carbon exchange among living organisms, which occurs over short time scales of days–weeks–months. The second subcycle comprises long-term cycling of carbon through geologic processes. It is important to note that these two subcycles are linked.

The movement of carbon through the food chain represents the rapid carbon subcycle. Plants move carbon from the atmosphere into the terrestrial biosphere through photosynthesis. These so-called autotrophs use energy from the sun to convert CO_2 into complex organic molecules such as glucose (sugar). Animals that eat these plants digest the sugar molecules to obtain energy for their bodies in the process termed respiration. These organisms are called heterotrophs. Respiration, excretion, and decomposition release the carbon back into the atmosphere or soil, continuing the carbon cycle. A similar cycle occurs in the ocean, except that the marine autotrophs obtain their carbon in the dissolved form (HCO_3^-).

The movement of carbon between the primary reservoirs of Earth, the atmosphere, ocean, terrestrial biosphere (land), and the geosphere is generally very slow, occurring over scales of hundreds to thousands of years or more. Carbon dioxide is the major form of carbon in the atmosphere. The level of carbon dioxide in the atmosphere is greatly influenced by exchange between it and the ocean and land reservoirs.

Carbon dioxide in the atmosphere dissolves in the ocean, after which it is transformed into different forms (Fig. 6.7). Phytoplankton convert the carbon through photosynthesis and become food sources for higher level organisms. Some organisms combine carbon with calcium ions in the seawater to form calcium carbonate ($CaCO_3$), a

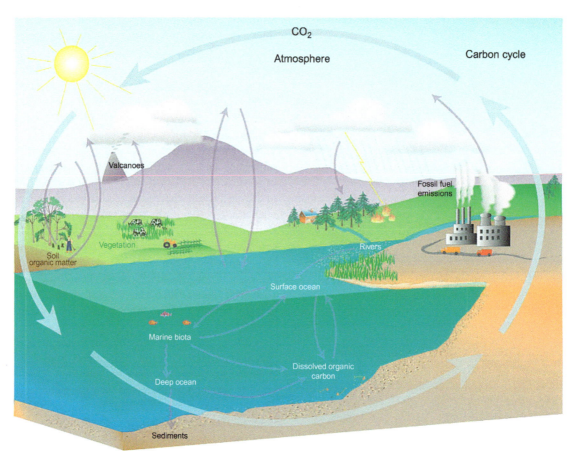

FIG. 6.6   The carbon cycle. *From http://www.noaa.gov/resource-collections/carbon-cycle*

major component of the shells of marine organisms. Marine organisms die and eventually form sediments on the ocean floor. Over many hundreds of thousands of years, the sediment is subjected to geological processes to form limestone and other marine sedimentary rock. These solid forms of carbon at the ocean bottom represent the largest carbon reservoir on Earth.

On land, carbon exists in living plants and animals generated through the food web as discussed previously. Transfer of carbon between land and the atmosphere occurs through photosynthesis–respiration processes associated with these living organisms. In addition, human activities such as fossil-fuel consumption are another source of carbon to the atmosphere. Both of these sources are depicted in Fig. 6.8. These transfer processes occur over relatively short time scales. Another way for carbon to transfer from land to the atmosphere is through volcanoes and other geothermal systems. Carbon-containing sediments and rocks on the ocean floor move deep within the Earth by the process of subduction, which is the movement of one tectonic plate beneath another. These materials can be transformed into molten rock (magma) under the high temperatures and pressures present at great depth. This process can lead to the formation of volcanoes, which releases

carbon in the form of CO_2 upon eruption. Clearly, this process occurs over many hundreds of thousands or millions of years.

Carbon is stored in soil as a result of the decomposition of living organisms or from weathering of terrestrial rock and minerals. Carbon cycling can be quite complex in the shallow subsurface as shown in Fig. 6.9. Carbon is stored deeper underground in the form of decomposed remains of plants and animals that have been converted into petroleum, natural gas, coal, and oil shale.

The *nitrogen cycle* is another biogeochemical cycle critical to life (Fig. 6.10). Nitrogen is especially important to ecosystem dynamics because many ecosystem processes, such as primary production and decomposition, are limited by the available supply of nitrogen. While it is an essential element, the vast majority of organisms cannot make direct use of the primary source of nitrogen, nitrogen gas (N_2) in the atmosphere. The conversion of N_2 into a more useable form of nitrogen, ammonia, is carried out by nitrogen-fixing bacteria. Once this occurs, various processes transfer the nitrogen within the ecosystem, and ultimately back to the atmosphere, as depicted in Fig. 6.10.

Human activity can release nitrogen into the environment by two primary means. One is through the

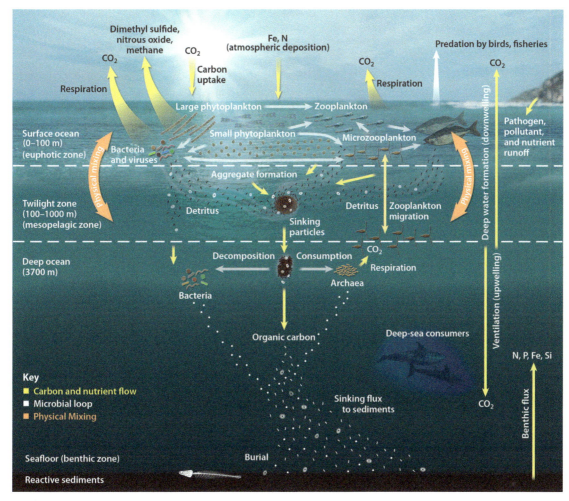

FIG. 6.7 The oceanic carbon cycle. *From Office of Biological and Environmental Research of the U.S. Department of Energy Office of Science. https:// science.energy.gov/ber/*

combustion of fossil fuels, which releases nitrogen oxides into the atmosphere (Chapter 17). The second is through the use of fertilizers that contain nitrogen and phosphorus compounds in agriculture. Nitrogen can be taken up by surface runoff during precipitation events and eventually deposit into lakes, streams, and rivers. A major effect from fertilizer runoff is eutrophication, a process whereby nutrients from the runoff cause the overgrowth of algae and a number of resulting problems (Chapter 16). The phosphorus and sulfur cycles are other critical biogeochemical cycles. More discussion of the nitrogen and phosphorous cycles is presented in Chapter 16.

6.5   ECOSYSTEM SERVICES

It has long been understood that human life and well-being depend upon resources and other benefits provided by the environment. However, to this day, the value of these benefits, as well as the reduction or loss of benefits due to human impacts on the environment, is not fully recognized

by most people. This situation has resulted in the fact that the full value and importance of the benefits are not considered in decision-making for most human activities. Concomitantly, the full costs of negative impacts to the environment, and the subsequent loss of benefits, are not fully considered. The concept of *Ecosystem Services* was developed to improve our ability to recognize and assign value to the benefits provided by the environment and to better assess the impacts of human activities on the systems that provide these benefits.

A landmark event in the development of this concept was the Millennium Ecosystem Assessment (MEA), conducted from 2001 to 2005 under the auspices of the United Nations (WRI, 2005). It was governed by a multistakeholder board that included representatives of international institutions, governments, business, NGOs, and indigenous peoples. Its purpose was to assess the consequences of ecosystem change for human well-being and provided an appraisal of the condition and trends in the world's ecosystems and the services they provide, as well as the

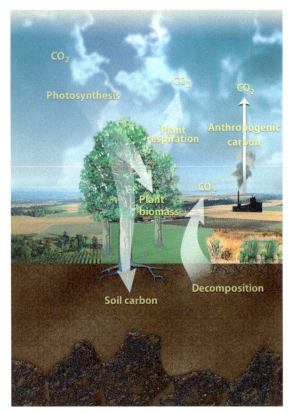

FIG. 6.8 The terrestrial carbon cycle. *From Office of Biological and Environmental Research of the U.S. Department of Energy Office of Science. https://science.energy.gov/ber/*

scientific basis for action to conserve and use them sustainably. The main findings of the assessment are:

1. Over the past 50 years, humans have changed ecosystems more rapidly and extensively than in any comparable period of time in human history, largely to meet rapidly growing demands for food, fresh water, timber, fiber, and fuel due to population growth and economic development.

2. The changes that have been made to ecosystems have contributed to substantial net gains in human well-being and economic development. However, these gains have been achieved at growing costs in the form of the degradation of many ecosystem services, increased risks of nonlinear changes, and the exacerbation of poverty for some groups of people. These problems, unless addressed, will substantially diminish the benefits that future generations obtain from ecosystems.

3. The degradation of ecosystem services could grow significantly worse during the first half of this century and is a barrier to achieving the Millennium Development Goals (see Chapter 28).

4. The challenge of reversing ecosystem degradation while meeting increasing demands for services can be partially met under some scenarios considered by the MEA but will involve significant changes in policies, institutions, and practices that are not currently under way.

Ecosystem services are classified into four groups—Provisioning, Regulating, Cultural, and Supporting (Fig. 6.11). Humans receive both direct and indirect benefits from these services. *Provisioning services* are material and energy resources obtained from ecosystems: food (such as crops, fruit, fish), fuel, building materials, fiber (such as timber, cotton, and wool), biochemicals (such as natural medicines, pharmaceuticals), and ornamental resources (such as flowers, shells, stone). *Regulating services* are benefits obtained from processes carried out by ecosystems: maintenance of air quality, climate and water regulation,

FIG. 6.9 Carbon cycling in the shallow subsurface environment. *From Office of Biological and Environmental Research of the U.S. Department of Energy Office of Science. https://science.energy.gov/ber/*

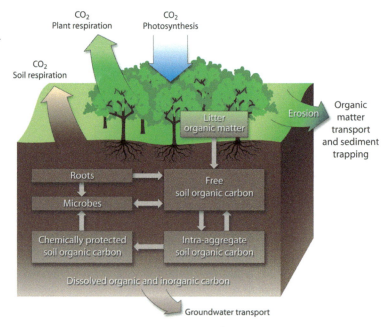

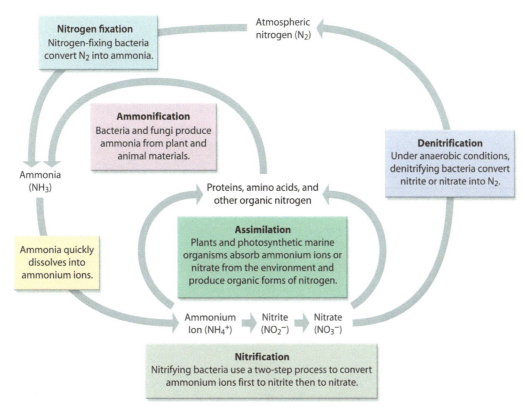

FIG. 6.10 The nitrogen cycle. *From Office of Biological and Environmental Research of the U.S. Department of Energy Office of Science. https:// science.energy.gov/ber/*

erosion control, water purification and detoxification, natural hazard protection, and bioremediation of waste. *Cultural services* are the nonmaterial benefits that people obtain from the environment that support physical and mental well-being. These include spiritual enrichment, cognitive development, recreation, art, social relations, esthetic values, cultural heritage values, and ecotourism. *Supporting services* are the services that are necessary for the production of all other ecosystem services, in other words, the services that maintain ecosystem health and functioning (soil formation and retention, water cycling, nutrient

cycling, primary production, production of atmospheric oxygen, and provision of habitat).

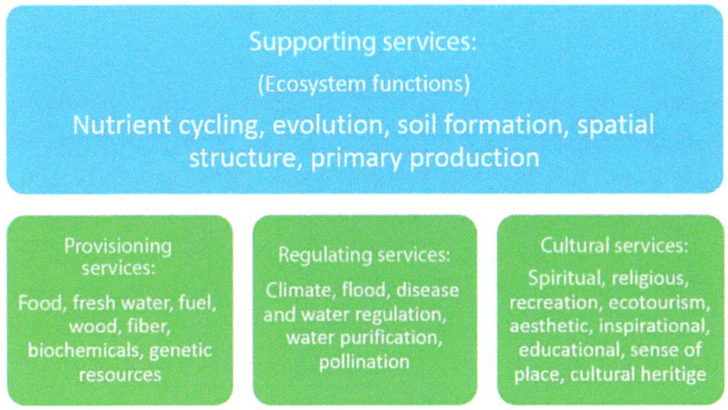

The current status of ecosystems and the services they provide based on the MEA is presented in Information Box 6.2. Inspection shows that many services are in decline, due mostly to human impacts. A primary concept of ecosystem services is that human impacts on ecosystems can impair the condition and functioning of ecosystems, and that this impairment can lead to reduced or lost services, which in turn can negatively impact human well-being. In other words, it recognizes that humans are an integral part of ecosystems and that a dynamic interaction exists between them and other parts of ecosystems, with human activities directly and indirectly driving changes in ecosystems that in turn cause changes in human well-being. The connection between human well-being and ecosystem services is illustrated in Fig. 6.12. Illustrations of how human activities create "drivers for change" of ecosystems, and how these changes impact services and human well-being are presented in Fig. 6.13.

Metro nature is a relatively new term that can be defined as the integration of nature into urban areas or cities (Wolf and Robbins, 2015). An example of metro nature would be the iconic Central Park in New York, which is critically important to the community for recreational and other

Supporting services:

(Ecosystem functions)

Nutrient cycling, evolution, soil formation, spatial structure, primary production

Provisioning services:	Regulating services:	Cultural services:
Food, fresh water, fuel, wood, fiber, biochemicals, genetic resources	Climate, flood, disease and water regulation, water purification, pollination	Spiritual, religious, recreation, ecotourism, aesthetic, inspirational, educational, sense of place, cultural heritige

FIG. 6.11 Ecosystem services. *Reproduced from http://www.ceeweb.org/ work-areas/priority-areas/ecosystem-services/what-are-ecosystem-services/.*

INFORMATION BOX 6.2 Status of Ecosystems and Ecosystem Services

From the Millennium Ecosystem Assessment (World Resources Institute (WRI), 2005. Millennium ecosystem assessment, ecosystems and human well-being: synthesis. Island Press, Washington, DC.

Legend: ▲ Enhanced, ▼ Degraded, +/− Both.

Status indicates whether the condition of the service globally has been enhanced (if the productive capacity of the service has been increased, for example) or degraded in the recent past. Definitions of "enhanced" and "degraded" are provided in the following note. A fourth category, supporting services, is not included here as they are not used directly by people.

Service	Subcategory	Status	Notes
Provisioning Services			
Food	Crops	▲	Substantial production increase
	Livestock	▲	Substantial production increase
	Capture Fisheries	▼	Declining production due to overharvest
	Aquaculture	▲	Substantial production increase
	Wild foods	▼	Declining production
Fiber	Timber	+/−	Forest loss in some regions, growth in others
	Cotton, hemp, silk	+/−	Declining production of some fibers, growth in others
	Wood fuel	▼	Declining production
Genetic resources		▼	Lost through extinction and crop genetic resource loss
Biochemicals, natural medicines, pharmaceuticals		▼	Lost through extinction, overharvest
Fresh water		▼	Unsustainable use for drinking, industry, and irrigation; amount of hydro energy unchanged, but dams increase the ability to use that energy
Regulating Services			
Air quality regulation		▼	Decline in ability of atmosphere to cleanse itself
Climate regulation	Global	▲	Net source of carbon sequestration since mid-century
	Regional and local	▼	Preponderance of negative impacts
Water regulation		+/−	Varies depending on ecosystem change and location
Erosion regulation		▼	Increased soil degradation
Water purification and waste treatment		▼	Declining water quality
Disease regulation		+/−	Varies depending on ecosystem change
Pest regulation		▼	Natural control degraded through pesticide use
Pollination		▼[a]	Apparent global decline in abundance of pollinators
Natural hazard regulation		▼	Loss of natural buffers (wetlands, mangroves)
Cultural Services			
Spiritual and religious values		▼	Rapid decline in sacred groves and species
Esthetic values		▼	Decline in quantity and quality of natural lands
Recreation and ecotourism		+/−	More areas accessible but many degraded

[a]*Indicates* low to medium certainty. *All other trends are* medium to high certainty.

Note: For provisioning services, we define enhancement to mean increased production of the service through changes in area over which the service is provided (e.g., spread of agriculture) or increased production per unit area. We judge the production to be degraded if the current use exceeds sustainable levels. For regulating and supporting services, enhancement refers to a change in the service that leads to greater benefits for people (e.g., the service of disease regulation could be improved by eradication of a vector known to transmit a disease to people). Degradation of regulating and supporting services means a reduction in the benefits obtained from the service, either through a change in the service (e.g., mangrove loss reducing the storm protection benefits of an ecosystem) or through human pressures on the service exceeding its limits (e.g., excessive pollution exceeding the capability of ecosystems to maintain water quality). For cultural services, enhancement refers to a change in the ecosystem features that increase the cultural (recreational, esthetic, spiritual, etc.) benefits provided by the ecosystem.

activities (Fig. 6.14). Potential benefits of exposure to nature are presented in Table 6.1. The concept of metro nature is an example of applying ecosystem services concepts to a specific case.

Soil is a central component of the environment. It provides many critical services to humans. However, it can also have deleterious impacts on humans, affecting their health and well-being. The complex interactions between soil health and human health are discussed in Chapter 27.

As will be discussed in several upcoming chapters in this text, many human activities can disturb ecosystems, disrupting their functioning and thereby reducing the services they provide. This can happen via pollution events, impairment or destruction of habitat, or by overuse of resources. Methods to restore ecosystems and thus ecosystem services are discussed in Chapter 20.

The *Tragedy of the Commons* is a concept related to overuse of resources provided by the environment, that is, an ecosystem service. Central to the concept is the idea that individuals will put their own personal interests above those of the common good (the general public). Thus when multiple individuals share a common resource, each will attempt to maximize their use of the resource and the benefits they receive. However, the cumulative impact of these

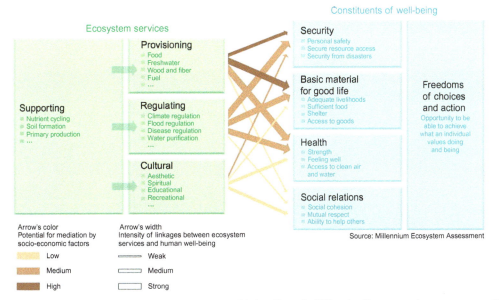

FIG. 6.12 The connections between ecosystem services and human well-being. *From the Millennium Ecosystems Assessment report, World Resources Institute (WRI), 2005. Millennium Ecosystem Assessment, Ecosystems and Human Well-Being: Synthesis. Island Press, Washington, DC.*

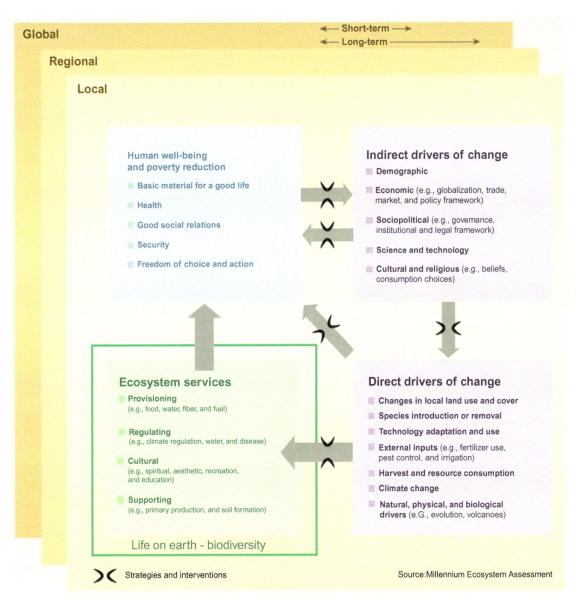

FIG. 6.13 The connections between drivers for change in ecosystems and the impacts on services and human well-being. *From the Millennium Ecosystems Assessment report, World Resources Institute (WRI), 2005. Millennium Ecosystem Assessment, Ecosystems and Human Well-Being: Synthesis. Island Press, Washington, DC.*

FIG. 6.14 An example of "metro nature," the iconic Central Park in New York. *Reproduced from Jean-Christophe BENOIST [CC BY 3.0 (https:// creativecommons.org/licenses/by/3.0)], from Wikimedia Commons.*

TABLE 6.1 Health Benefits From Contact With Nature

Health/Well-Being Benefits
Reduced stress
Better sleep

Improved mental health
   Reduced depression
   Reduced anxiety

Greater happiness, well-being
Reduced aggression
Lower blood pressure
Improved congestive heart failure
Reduced obesity
Reduced diabetes
Better eyesight
Reduced mortality

Modified from Table 1, Frumkin, H., Bratman, G.N., Breslow, S.J., et al., 2017. Nature contact and human health: a research agenda. Environ. Health Perspect. 125 (7), 075001. doi:10.1289/EHP1663.

individual decisions results in overuse of the resource. This in turn ultimately results in a reduction or loss of benefits for all of the users. This concept illustrates the need for sustainable management practices to regulate resource use to preserve the long-term benefits accruing from ecosystem services. Sustainable management and development concepts are discussed in Chapter 32.

QUESTIONS AND PROBLEMS

1. Define ecosystems and biomes and discuss the differences
2. What biome do you live in?
3. Identify some examples of ecosystem services important to you.
4. What is the critical zone?
5. Discuss the different types of biodiversity, using a specific example ecosystem.
6. Discuss a specific example of the tragedy of the commons.
7. Describe the critical components and transfer processes for the sulfur or phosphorous cycle.

REFERENCES

Wolf, K.L., Robbins, A.S.T., 2015. Metro nature, environmental health and economic value. Environ. Health Perspect. 123 (5), 390–398.
World Resources Institute (WRI), 2005. Millennium Ecosystem Assessment, Ecosystems and Human Well-Being: Synthesis. Island Press, Washington, DC.

Chapter 7

Physical Processes Affecting Contaminant Transport and Fate

M.L. Brusseau

Plate 1: Transport of dyed water in a flow cell containing porous media.

7.1 CONTAMINANT TRANSPORT AND FATE IN THE ENVIRONMENT

An understanding of the transport and fate of contaminants in the environment is required to evaluate the potential impact of contaminants on human health and the environment. For example, such knowledge is needed to conduct risk assessments, such as evaluating the probability that a contaminant spill would result in groundwater pollution, or if an existing groundwater contaminant plume poses a hazard to humans living nearby. Such knowledge is also required to develop and evaluate methods for remediating environmental contamination. Just as important, knowledge of contaminant transport and fate is necessary to help design chemicals and processes that minimize adverse impacts on human health and the environment to enhance pollution prevention.

The four general processes that control the transport and fate of contaminants in the environment are advection, dispersion, interphase mass transfer, and transformation reactions. These are defined in Information Box 7.1.

Many of the processes influencing contaminant transport and fate in the subsurface are illustrated in Fig. 7.1, which

shows the disposition of an organic liquid contaminant spilled into the ground (DNAPL is a "denser than water" nonaqueous phase liquid).

The fate of a specific contaminant in the environment is a function of the combined influences of these four general processes. The combined impact of the four processes determines the "migration potential" and "persistence" of a contaminant in the environment. These two properties control what is, in essence, the "ability" of the chemical to contaminate the medium of interest (soil, water, air), or in other words the "*pollution potential*." Constituents that have higher aqueous solubilities and low sorption are generally transported more readily, which means that the contaminant can readily spread from the site where it first entered the environment. Thus it has a larger *migration potential*. Constituents that are resistant to transformation or degradation (i.e., are *persistent*) can remain in the environment, and thus remain hazardous, for longer times. A constituent that has a larger migration potential and that is persistent will generally pose a much greater risk for contamination compared to those that have lower migration potentials and/or that are not persistent.

The health risk posed by a specific contaminant to humans or other organisms is, of course, a function of its toxicological characteristics (as discussed in Chapter 28), as well as its contamination potential. Thus it is important to understand both types of properties. For example, the greatest potential health risk will generally be associated with contaminants that are persistent and highly toxic. However, actual harmful effects will occur only if the organism is exposed to the contaminant. Thus the pathways of exposure are critical to assessing risks posed by contaminants. The processes influencing the transport and fate of contaminants in the environment have a significant impact on pathways and levels of exposure.

Once a chemical is applied to (or spilled onto) the land surface, it may remain in place or it may transfer to the air, surface runoff, or the subsurface. Chemicals with moderate to large vapor pressures may evaporate or volatilize into the

Environmental and Pollution Science. https://doi.org/10.1016/B978-0-12-814719-1.00007-0

INFORMATION BOX 7.1 Processes Influencing Contaminant Transport and Fate

Advection is the transport of matter via the movement of a fluid; as discussed in Chapter 3, a fluid such as groundwater moves in response to a gradient of fluid potential. For example, contaminant molecules dissolved in water will be carried along by the water as it flows through (infiltration) or above (runoff) the soil. Similarly, contaminant molecules residing in air will be carried along as the air flows. *Dispersion* represents spreading of matter about the center of the contaminant mass. Spreading is caused by molecular diffusion and nonuniform flow fields. Contaminant molecules can reside in several phases in the environment, such as in air (atmosphere and soil gas phase), in water, and associated with soil particles. So-called *mass transfer* processes, such as sorption, evaporation, and volatilization, involve the transfer of matter between phases in response to gradients of chemical potential or, more simply, concentration gradients. *Transformation reactions* include any process by which the physicochemical nature of a chemical is altered. Examples include biotransformation (metabolism by organisms), hydrolysis (interaction with water molecules), and radioactive decay.

gas phase, thus becoming subject to atmospheric transport and fate processes including advection (carried along by wind), dispersion, and transformation reactions (Chapter 17). Breathing contaminated air is one potential source of exposure to hazardous contaminants. Transfer of contaminants to surface runoff during precipitation or irrigation events is a major concern associated with the nonpoint source pollution issue. Once entrained into surface runoff, the contaminant may then be transported to surface water bodies. Consumption of contaminated surface water is another potential route of human or animal exposure to toxic chemicals. Another route of toxic chemical exposure for humans is consumption of contaminated groundwater. Once applied to a land surface, a contaminant can partition to the soil pore water. The contaminant then has the potential to move downward to a saturated zone (aquifer), thereby contaminating groundwater. Whether or not this will occur, as well as the time it will take, and the resulting magnitude of contamination, depends on numerous factors. These include the magnitude and rate of infiltration or recharge, the soil type, the depth to the aquifer, and the quantity of contaminant and its physicochemical properties (e.g., solubility, degree of sorption, and transformation

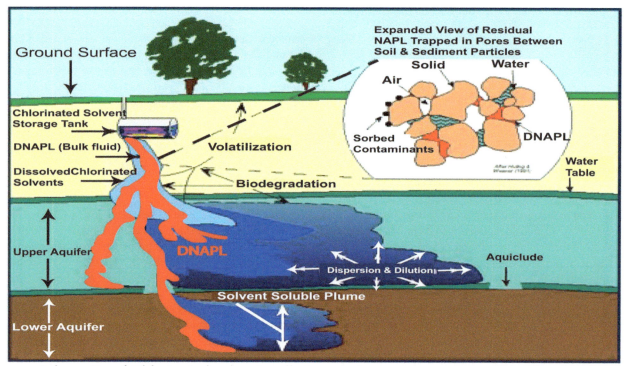

Schematic of Chlorinated solvent pollution as dense non-aqueous phase liquids migrating downward in an aquifer and serving as a source for a solvent soluble plume. Also shown are natural attenuation processes (Modified USEPA 1999).

FIG. 7.1   Disposition of organic liquid spilled into the subsurface. Schematic of chlorinated solvent pollution: dense nonaqueous phase liquids migrate downward in the subsurface, and serve as a source of contamination for groundwater. Also shown are natural attenuation processes. *DNAPL*, Dense nonaqueous phase liquid; *NAPL*, nonaqueous phase liquid. *(Modified from U.S. EPA, 1999)*.

potential). Volatile contaminants can move by advection and diffusion in the soil-gas phase in addition to the pore-water phase, which provides an additional means to travel to an aquifer. The physicochemical properties of the contaminant control its transport and fate behavior. This is discussed in Chapters 12, 14–17.

7.2 ADVECTION

As introduced before, advection is the transport of matter by the movement of a fluid. Any contaminant dissolved in water (which we call a "solute") will be carried along by the water as it flows. Similarly, contaminant molecules residing in air (vapor phase) will be carried along as the air flows. Matter suspended in water or air, such as suspended solids, will also move with fluid flow. An example of a plume emanating from a smokestack is shown in Fig. 7.2. The plume is moving in the direction of mean air flow (the direction of the prevailing winds). Similarly, a plume of groundwater contamination moves in the mean direction of groundwater flow. For example, the mean direction of groundwater flow is NNE for Fig. 7.4. Hence, the groundwater contaminant plume is migrating in that direction. Advection is generally the single most important mode of transport of contaminants in the environment. Advective transport by water or air is the primary reason for large-scale movement of contaminants in the environment.

Transport of contaminant mass by advection is proportional to the rate of fluid movement. Thus characterizing the advective transport potential of a contaminant at a specific location requires one to determine the direction and rate of fluid flow at that location. Monitoring air flow in the atmosphere was discussed in Chapter 4. Determining the direction and rate of surface water flow is relatively straightforward, e.g., for streams, the direction is defined by the stream channel and the rate is related to the water level (see Chapter 3). Conversely, characterizing the direction and rate of fluid flow in the subsurface is much more complex, as discussed in Chapter 3.

7.3 DISPERSION

Dispersion represents spreading of matter about the center of the contaminant mass. In essence, as a mass of contamination (called a "pulse" or "plume") moves by advection, the size of the plume increases due to dispersion. In other words, the plume "grows" as it moves. This spreading is caused by molecular diffusion and nonuniform flow fields. It is important to note that spreading occurs in both water flow and air flow systems.

Molecular diffusion is the result of random motion of individual molecules. Every molecule vibrates and moves due to its individual kinetic energy. At any given time, each molecule may move in any given direction; thus we call this random motion. The net result of this series of individual movements is that molecules will spread from regions containing greater numbers of molecules (i.e., higher concentrations) to regions with fewer numbers (lower concentrations). The effect of diffusion is readily seen by adding a drop of food coloring dye to a beaker of water—the dye will spread and eventually color the entire volume of water. The contribution of molecular diffusion to overall transport and spreading of a contaminant is generally small in the subsurface. It becomes significant primarily only in systems where advection is minimal, such as clay units in the saturated zone of the subsurface.

The major cause of dispersion is nonuniform flow fields. When a fluid flows through the environment, it does not move as one uniform body. Rather, sections of the fluid move at different rates or velocities. For example, in a river, water flowing along the channel walls moves slower due to friction than water located in the center of the channel. For the atmosphere, local variations in flow directions and rates compared to the mean flow can cause dispersion. In the subsurface, the dispersion effect occurs at a range of spatial scales. For smaller scales (<1 m), dispersion is caused by nonuniform flow in the porous-medium pores and occurs in three major ways (Fig. 7.3A). First, fluid flow in a single pore is faster in the center of the pore because friction slows

FIG. 7.2 Example of a plume emanating from a smokestack. *(Image courtesy of NASA).*

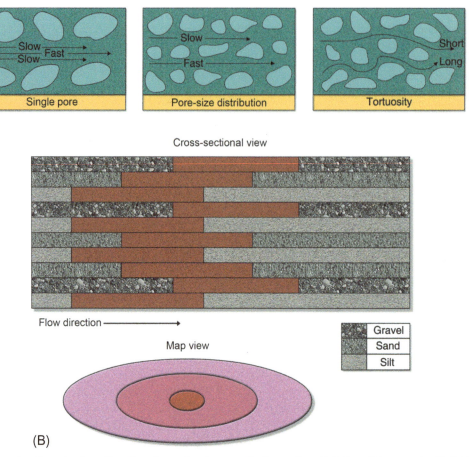

FIG. 7.3 Processes causing dispersion (spreading of a pulse or plume) as a result of nonuniform fluid flow: (A) pore scale; (B) field scale.

the fluid near the pore walls. Second, fluid flow is faster in larger diameter pores than it is in smaller diameter pores (a smaller proportion of fluid is influenced by friction for larger pores). Third, the time it takes a fluid or contaminant molecule to travel from one location to another is less for pore sequences that have fewer twists and turns (less tortuous flow path). For larger scales (field scale), fluid moves faster in larger permeability units such as sand and slower in lower permeability units such as clay (see Chapter 3). As a result, the rates of advection differ for different portions of the fluid. Thus, as a plume of dissolved contaminant moves, different sections of the plume will be moving at different velocities. This results in spreading or "growth" of the plume in the direction of travel. This dispersion effect is illustrated in Fig. 7.3B. An analogy to help picture the dispersion process is to think of runners at a marathon. At the start of the event, all of the participants will be tightly grouped in a small area at the starting point. After the event starts, the runners will become more spread out over the course because some runners are faster than other runners. By the time the first runner finishes, there may be a mile or more separation between the lead runner and the slowest runners.

7.4   MASS TRANSFER

As noted in a previous section, a contaminant may reside in the air, in water, or associated with a solid phase. The transfer of contaminants from their original phase to other phases is an important aspect of contaminant behavior. The primary mass transfer processes of interest for contaminant transport in the environment are dissolution, evaporation, volatilization, and sorption. These four processes will be discussed briefly below; a more detailed discussion is presented in Chapter 8.

The transfer of molecules from their pure state to water is called *dissolution*. For example, placing salt crystals in water will result in dissolution—the salt will dissolve in the water. Similarly, placing an immiscible organic liquid (such as a cleaning solvent) in contact with water will result in the transfer of some of the organic liquid molecules to the water. This is how water becomes polluted by a contaminant. The extent to which contaminant molecules or ions will dissolve in water is governed by their aqueous solubility (see Chapter 8). The solubilities of different contaminants vary by orders of magnitude. A critical aspect of solubility is how it compares to pollution action levels for the contaminant. For example, the aqueous solubility of trichloroethene (a chlorinated cleaning solvent) is approximately 1 g/L. This is very low compared to many other compounds. However, the maximum contaminant level for trichloroethene is 5 µg/L, 200,000 times smaller. Thus it does not take very much trichloroethene to pollute water.

Evaporation of a compound involves transfer from the pure liquid or pure solid phase to the gas phase. The vapor pressure of a contaminant, then, is the pressure of its gas phase in equilibrium with the solid or liquid phase, and is an index of the degree to which the compound will evaporate. In other words, we can think of the vapor pressure of a compound as its "solubility" in air. Evaporation can be an important transfer process when pure-phase contaminant is present in the vadose zone, such that contaminant molecules evaporate into the soil gas.

Volatilization is the transfer of contaminant between water and gas phases. Volatilization is different from evaporation, which specifies a transfer of molecules from their pure phase to the gas phase. For example, the transfer of benzene molecules from a pool of gasoline to the atmosphere is evaporation, whereas the transfer of benzene molecules from water (where they are dissolved) to the atmosphere is volatilization. The vapor pressure of a compound gives us a rough idea of the extent to which a compound will volatilize, but volatilization also depends on the solubility of the compound as well as environmental factors. Volatilization is an important component of the transport and remediation of volatile organic contaminants in the vadose zone (see Chapters 15 and 19).

Sorption also influences the transport of many contaminants in the environment. The broadest definition of sorption (or retention) is the association of contaminant molecules with the solid phase of the porous medium (e.g., soil particles). Sorption can occur by numerous mechanisms, depending on the properties of the contaminant and of the sorbent (see Chapter 8). A critical impact of sorption is that it slows or retards the rate of movement of contaminants (referred to as "retardation"). As long as the soil grains are immobile, contaminant molecules will also be immobile when they are sorbed to the grains. Thus contaminants that sorb will move more slowly than the mean water velocity. Conversely, contaminants that do not sorb will generally move at the same velocity as water.

7.5 TRANSFORMATION REACTIONS

The transport and fate of many contaminants in the environment is influenced by transformation reactions, such as biodegradation, hydrolysis, and radioactive decay. The susceptibility of a contaminant to such transformation reactions is very dependent on their molecular structure (organics) or speciation (inorganics). For example, chlorinated hydrocarbons are generally more resistant to biodegradation than are nonchlorinated hydrocarbons. Abiotic and

biotic transformation reactions are discussed in Chapters 8 and 9, respectively.

A critical impact of transformation reactions on transport and fate is that the mass of original contaminant is reduced. This is usually a positive result, as it leads to reducing the amount of potential pollution present in the environment. However, in some cases it may produce a negative result, such as when the transformation reaction produces a more hazardous compound.

7.6 CHARACTERIZING SPATIAL AND TEMPORAL DISTRIBUTIONS OF CONTAMINANTS

Characterizing the transport and fate behavior of contaminants in the environment is based on evaluating their distribution in the environment in terms of phase distribution (are they in groundwater, sorbed to soil, etc.), spatial distribution (where are they located at the site), and temporal distribution (i.e., how concentrations change with time). The information needed to assess these distributions is obtained from sampling and monitoring programs, as described in Chapter 10 (detailed information on environmental monitoring is available in Artiola et al. (2004).

Examining spatial distributions of contamination is based on developing contaminant contour maps. This is done by plotting contaminant concentrations, obtained from analysis of samples collected from discrete points in space, on a map of the site. The points of equal concentrations are connected by contour lines. This procedure produces a diagram of the contaminant plume, which provides a means to visualize the size and distribution of contamination at the site (see Fig. 7.4). These contaminant plume maps can be produced periodically (at different times), which allows examination of the transport behavior of the contamination—is the contaminant plume moving or staying in place?

Simple examples of spatial distributions of contaminant plumes as affected by the four transport and fate processes are presented in Fig. 7.5. The impact of fluid velocity on advective transport is illustrated by comparing the locations of plumes 1 and 2—plume 1 has traveled a greater distance downgradient from the point of origin because of the greater fluid velocity for that case. The impact of dispersion is shown for plume 3, which is longer than plume 2 (which has no dispersion). The impact of sorption and retardation on transport is observed for plume 4, which has not traveled as far as plume 3. The impact of a transformation reaction that causes loss of contaminant mass is illustrated for plume 5, which is smaller than plume 3.

The change in contaminant concentrations at a site over time is an important aspect of the risk posed by the contamination. The temporal variability of contaminant

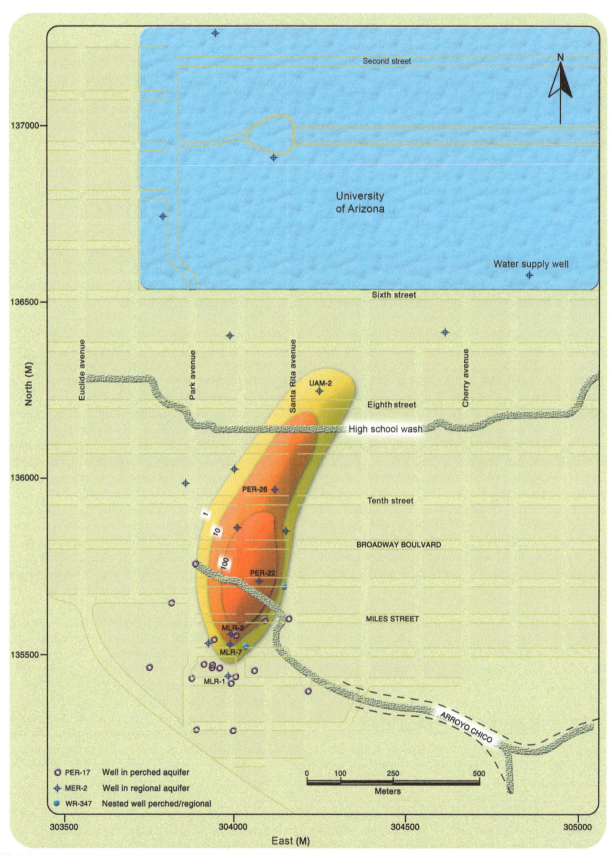

FIG. 7.4   Groundwater contaminant plume contour map for a site in Tucson, Arizona. The contours represent tetrachloroethene concentrations in µg/L. The contamination source is located near the south end of the plume. *(Drawn by Concepción Carreón Diazconti).*

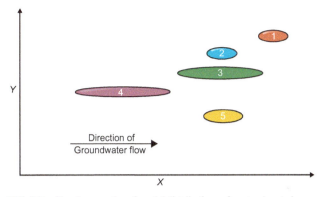

FIG. 7.5 Simple examples of spatial distributions of contaminant plumes as affected by the four transport and fate processes. Case 1: Advection (larger fluid velocity, *v*); Case 2: Advection (small *v*); Case 3: Advection (smaller *v*) and Dispersion; Case 4: Advection (smaller *v*), Dispersion, and Retardation; Case 5: Advection (smaller *v*), Dispersion, and Transformation. The plumes are drawn with a single contour representing the detectable concentration. The plumes all originated at the *x* = 0 plane.

concentrations is characterized by plotting concentration histories, in which concentrations of samples collected from one location at different times are plotted as a function of time. A special case of these concentration histories is obtained when we are monitoring a location at which contamination is just beginning to arrive. These plots are called breakthrough curves.

Simple examples of breakthrough curves as affected by the four transport and fate processes are presented in Fig. 7.6. The impact of fluid velocity on advective transport is illustrated by comparing the locations of breakthrough curves 1 and 2; breakthrough curve 1 appears earlier than breakthrough curve 2 because of the greater fluid velocity for case 1. The impact of dispersion is shown for breakthrough curve 3, which is more spread out (rotated clockwise) than breakthrough curve 2 (which has no dispersion). The impact of sorption and retardation on transport is observed for breakthrough curve 4, which appears later than breakthrough curve 3. The impact of a transformation reaction that causes loss of contaminant mass is illustrated for breakthrough curve 5, which peaks at a lower concentration than breakthrough curve 3.

7.7 ESTIMATING PHASE DISTRIBUTIONS OF CONTAMINANTS

As discussed in the beginning of this chapter, contaminants can reside in multiple phases of the environment—as vapor in air, dissolved in water, and sorbed to porous medium particles. It is often important to know how a contaminant will distribute among these phases. For example, conducting risk assessments requires an evaluation of potential routes of exposure. To do this, we would need to know if and how much contaminant is present in the various phases. In addition, we would like to know how much contaminant is present in the various phases when we design systems to clean up contaminated sites.

A simple way to estimate contaminant distributions is to use phase distribution coefficients. These coefficients provide information about the distribution of contaminant between two phases and are presented as ratios of concentrations in the two phases. Two key distribution coefficients are the sorption coefficient K_d ($K_d = S/C_w$), which describes the distribution of a contaminant between porous medium particles and water (i.e., sorption), and Henry's Constant, H ($H = C_g/C_w$), which describes the distribution between gas (air) and water phases (i.e., volatilization). In these equations,

- C_w is the concentration of contaminant in water
- C_g is the concentration of contaminant in air
- S is the concentration of contaminant in the sorbed phase.

The first piece of information of interest is the total amount of contaminant in the system. For example, the total mass of contaminant contained in a given volume of soil can be defined as:

$$M = V_w C_w + M_s S + V_g C_g \qquad (7.1)$$

where

M is total contaminant mass
V_w is the volume of the water phase
M_s is the mass of soil particles
V_g is the volume of soil gas.

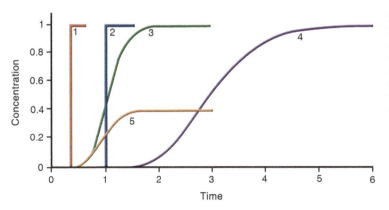

FIG. 7.6 Simple examples of breakthrough curves as affected by the four transport and fate processes. Case 1: Advection (larger fluid velocity, *v*); Case 2: Advection (smaller *v*); Case 3: Advection (smaller *v*) and Dispersion; Case 4: Advection (smaller *v*), Dispersion, and Retardation; Case 5: Advection (smaller *v*), Dispersion, and Transformation. The input concentration is equal to 1 concentration unit.

We can define a unit mass of contaminant (M^*) by dividing Eq. (7.1) by V_T, the total volume of the soil system. We can also define M^* in terms of one concentration by substituting the distribution equations [$S = K_d C_w$, $C_g = H C_w$] into Eq. (7.1). We use C_w as the key concentration since it is usually the concentration measured in subsurface monitoring programs. The modified equation for estimating contaminant mass is:

$$M^* = \left[\theta_w + \rho_b K_d + \theta_g H\right] C_w \tag{7.2}$$

where

θ_w is the soil-water content (volume of water per volume of soil)

θ is the soil-gas content (volume of soil gas per volume of soil)

ρ_b is soil bulk density (mass of soil per volume of soil).

Remember that M^* is mass of contaminant per soil volume element; thus total contaminant mass (M) is calculated by $M^* \times V_T$.

The fraction of contaminant residing in each phase can be calculated by the following equations:

$$\text{Fraction in water} = \frac{\theta_w C_w}{M^*} \tag{7.3}$$

$$\text{Fraction retained by soil particles} = \frac{\rho_b K_d C_w}{M^*} \tag{7.4}$$

$$\text{Fraction in soil gas} = \frac{\theta_g H C_w}{M^*} \tag{7.5}$$

It is important to remember that this approach is based on assuming that the distribution processes have reached equilibrium. If this is not true, the estimates obtained with Eqs. (7.1)–(7.5) can be erroneous.

Example Calculation 7.1

The calculation of contaminant phase distributions will be illustrated with the following example. We will assume the following values for the soil properties: $\theta_w = 0.25$, $\theta_g = 0.25$, and $\rho_b = 1.5\,g/cm^3$. To simplify the calculations, we will assume

$$K_d = 1\,mL/g$$
$$H = 1$$
$$C_w = 1\,mg/L$$

Using these values, the unit mass of contaminant is calculated to be:

$$M^* = \left[0.25 + 1.5\frac{g}{mL} \times 1\frac{mL}{g} + 0.25 \times 1\right] \times 1\frac{mg}{L} = 2\frac{mg}{L}$$

The fraction of pollutant residing in each phase is:

$$\text{Fraction in water} = \frac{0.25 \times 1\frac{mg}{L}}{2\frac{mg}{L}} = 0.125$$

$$\text{Fraction in soil gas} = \frac{0.25 \times 1 \times 1\frac{mg}{L}}{2\frac{mg}{L}} = 0.125$$

Fraction retained by soil particles

$$= \frac{1.5\frac{g}{mL} \times 1\frac{mL}{g} \times 1\frac{mg}{L}}{2\frac{mg}{L}} = 0.75$$

Thus 75% of the contaminant mass is associated with the soil particles, while 12.5% is associated each with water and with air.

7.8 QUANTIFYING CONTAMINANT TRANSPORT AND FATE

The transport and fate of contaminants in the environment is quantified by developing a set of governing equations. These equations are used to represent the various processes that influence transport and fate, as discussed before. The set of equations are developed based upon a conceptual model of the transport system (i.e., an idea of what is happening in the system). This set of equations constitutes a mathematical model of the system. It is important to recognize that a mathematical model is an inexact representation of reality, built upon a suite of assumptions and simplifications. The accuracy of a mathematical model will depend on the validity of the assumptions, representativeness of the simplifications, and the quality of the input data needed to implement the model.

The most widely used model to represent the transport and fate of contaminants in the environment is the advection-dispersion equation. A simplified version of this equation for solute transport in porous media is given by:

$$R\frac{\partial C}{\partial t} = -v\frac{\partial C}{\partial x} + D\frac{\partial^2 C}{\partial x^2} - T \tag{7.6}$$

where

C is contaminant concentration
D is the dispersion coefficient
R is the retardation factor
T represents possible transformation reactions
v is mean fluid velocity
t is time, and x is distance.

This equation is developed with numerous assumptions, including uniform, one-dimensional fluid flow, homogeneous conditions (e.g., a homogeneous porous medium), and that sorption is the only mass transfer process.

The term on the left-hand side of the equation represents the change in the amount of contaminant present at a given location. Changes in the amount of contaminant present at a given location are caused by transport of contaminant to and from that location, which occurs by advection and dispersion, and by mass transfer and transformation processes. The first term on the right-hand side of the equation represents advection, while the second term represents dispersion. The term T represents potential transformation reactions. A specific equation would be used for each given transformation process. One widely used equation is the "first-order" reaction equation, which is discussed in Chapter 8.

The effect of sorption on transport, namely, retardation, is represented by the retardation factor, R. The retardation factor is defined as:

$$R = 1 + \frac{\rho_b}{\theta} K_d = \frac{v_w}{v_p} = \frac{d_w}{d_p} \quad (7.7)$$

where

v_w is the velocity of the fluid
v_p is velocity of the contaminant
d_w is the distance traveled by the fluid
d_p is the distanced traveled by the contaminant.

Inspection of Eq. (7.7) shows that as sorption increases (i.e., K_d increases), the magnitude of the retardation factor increases. When the contaminant is not sorbed by the soil (i.e., $K_d = 0$), the retardation factor is equal to 1. This means that the contaminant will move at the same velocity as the mean velocity of the fluid $[v_p = v_w]$. When $R = 2$, the contaminant moves at an effective velocity that is half that of the fluid. In other words, the contaminant moves only half as fast as the fluid. When $R = 10$, the contaminant moves at one-tenth the velocity of water. Contaminants that have small retardation factors (<10) are considered to be relatively mobile. They can move rapidly from a spill site and thus quickly contaminate a large area. Conversely, contaminants that have very large retardation factors (>100) move very slowly with respect to fluid flow. Thus they probably will not create a contaminated zone as extensive as that created by mobile contaminants.

A special case exists for some systems wherein the retardation factor is less than 1, which would mean that the solute moves *faster* than the fluid by which it is being carried. A major example of this behavior is the transport of anionic substances such as Cl^- in the subsurface. As discussed in Chapter 2, the surfaces of many soils have a net negative charge. This leads to a repulsion of negatively charged solutes. For soil domains characterized by very small pores, this expulsion could prevent the solute from entering the water residing in the pores. Thus the solute would travel through only a portion of the soil, which results in transport that appears more rapid than the rate of water movement. This effect is termed anion exclusion. Of course, the individual solute molecules are not actually moving faster than the water molecules; they only appear to do so because they travel through a smaller portion of the porous medium than do the water molecules. Another

cause of $R < 1$ behavior is called size exclusion. In this case, extremely large molecules may be too large to pass through the smallest pores comprising a soil. Thus, just like the case before, the solute would travel through only a portion of the soil, resulting in transport that appears more rapid than the rate of water movement. Size exclusion has been observed in transport experiments for large colloids such as bacteria and protozoa.

Contaminant transport in the environment is usually much more complicated than what is represented by the simple advection-dispersion equation presented above. More complicated models have been developed to attempt to quantify transport and fate in real systems. A major factor limiting our ability to use these complex models is the difficulty in determining values for all of the unknown parameters in the model. This requires extensive characterization of the site, which is expensive and time consuming.

QUESTIONS AND PROBLEMS

1. A tanker truck containing a large volume of a liquid contaminant has overturned off the side of the freeway, and the contents of the tank have spilled onto the ground surface.
   (a) How will the compound move through the environment?
   (b) Identify and define the major processes that will control the transport and fate of the contaminant.
   (c) What contaminant properties will affect the transport of the compound in the environment?
2. (a) Calculate the total contaminant mass given the following data:

$$V_T = 1\,m^3 \; H = 0.5$$
$$\theta_w = 0.3 \; K_d = 2\,mL/g$$
$$\theta_g = 0.1 \; \rho_b = 1.5\,g/cm^3$$
$$C_w = 0.1\,mg/L$$

   (b) Calculate the fraction of contaminant present in water, in soil atmosphere, and sorbed to the soil.

3. (a) Calculate retardation factors for the following chemicals, given the following data:

Chemical	K_d (mL/g)	R
Benzene	0.1	
Trichloroethane	0.2	
Chlorobenzene	0.4	
Naphthalene	0.6	
PCB	10	

$[\rho_b = 1.5\,g/cm^3, \; \theta_w = 0.3]$

(b) Calculate the distances traveled by each chemical given the following information: the velocity of water moving through the soil is 1 cm/day; the elapsed time is 300 days.

REFERENCE

Artiola JF, Pepper IL, Brusseau ML: *Environmental Monitoring and Characterization,* San Diego, CA, 2004, Academic Press.

Chapter 8

Chemical Processes Affecting Contaminant Transport and Fate

M.L. Brusseau and J. Chorover

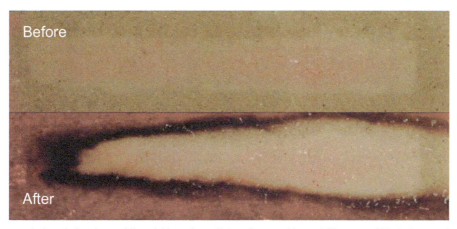

The use of permanganate solution (dark red) to oxidize trichloroethene (light red) trapped in sand. The zone of black discoloration in the "after" photo results from the formation of MnO_2 solids associated with the degradation of trichloroethene when it reacts with permanganate. *Photo courtesy Justin Marble.*

8.1 INTRODUCTION

From the perspective of chemistry, the environment is a heterogeneous system, meaning that it contains solid, liquid, and gaseous phases. As discussed in Chapter 7, the transport behavior of contaminants in the environment is influenced by their partitioning or transfer among these phases and also by transformation reactions. Major mass-transfer processes will be discussed in this chapter, including precipitation-dissolution and sorption-desorption. We will examine physical and chemical properties of contaminants that influence their speciation and solubility in water, and the magnitude of their sorption by porous media. The mass-transfer behavior of inorganic and organic contaminants can be quite different; therefore we will discuss them separately. In both cases, however, we rely on the principles of thermodynamic equilibrium to assist in the prediction of phase-transfer processes. This will be followed by a brief discussion of selected transformation reactions that alter the physical and chemical properties of contaminants.

8.2 BASIC PROPERTIES OF INORGANIC CONTAMINANTS

8.2.1 Speciation of Inorganic Pollutants

The precise chemical form of an element at a given point in time and space is defined as its speciation. Aqueous-phase speciation, which can be predicted on the basis of thermodynamics, is strongly affected by the environmental redox status and pH. Whereas redox status affects an element's most stable oxidation state, the solution-phase acidity (pH) affects its charge and degree of hydrolysis.

For example, the trace element arsenic (As) occurs in two different principal oxidation states [As(III) and As(V)], and each of these oxidation states is represented by various species with differing reactivities. The reduced form—As(III)—is favored under oxygen-depleted conditions, such as might be found in wetland sediments, and it occurs principally as the neutral species arsenite, $As(OH)_3^0$, in the pH range (2–9) of most natural waters. The oxidized form—As(V), known as arsenate—is favored

Environmental and Pollution Science. https://doi.org/10.1016/B978-0-12-814719-1.00008-2

under well-oxygenated conditions, and occurs dominantly as the monovalent anion $H_2AsO_4^-$ between pH 3 and 6, and as the bivalent anion $HAsO_4^{2-}$ at pH $>$ 7. Aqueous-phase speciation is also affected by the availability of complexing agents that can form *soluble complexes* with the ion or molecule of interest.

In addition to influencing the aqueous-phase speciation of solutes, pH and redox status also affect the solubility and charge of mineral solids that serve to sequester inorganic contaminants in soils. Solid-phase speciation of a contaminant can be diverse, as it is for aqueous systems. Contaminants become part of the solid phase via various mechanisms of *adsorption* to the surfaces of natural particles and/or *precipitation* within existing or newly formed solids. Thus solution-phase speciation controls the total amount of a contaminant element in equilibrium with these adsorbed or solid phases. The various interrelated chemical processes affecting the disposition of inorganic compounds in heterogeneous environmental systems are depicted in Fig. 8.1. They are discussed in more detail in the following sections.

8.2.2 Aqueous Phase Activities and Concentrations

Prediction of contaminant behavior requires working with balanced chemical reactions that represent changes in speciation both within a phase (e.g., within the aqueous solution) and between phases (e.g., between the solution and solid phases). A *balanced chemical reaction* is one that contains an equal number of moles of every element (*mass balance*) and an equal number of moles of charge (*charge balance*) on both sides of the equation. Since water is

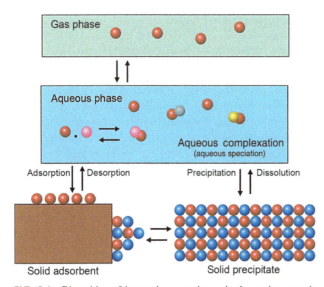

FIG. 8.1 Disposition of inorganic contaminants in the environment is controlled by aqueous-phase complexation reactions, precipitation-dissolution reactions, and sorption-desorption reactions.

ubiquitous in the environment, many chemical reactions of inorganic contaminants involve aqueous species as reactants or products.

When we use thermodynamic equilibrium constants to predict speciation, the reactions must be written in terms of *effective concentration* or *activities* of dissolved species, not their concentrations. The activity and concentration of a species, i, are related through the *activity coefficient* (γ_i):

$$(i) = \gamma_i[i] \tag{8.1}$$

where the term in parentheses on the left side represents activity and the term in square brackets on the right side denotes aqueous-phase concentration in either *molal* ($mol\,kg^{-1}$) or *molar* ($mol\,L^{-1}$) concentration units. *Molal* is used when a mass-based concentration is desired and *molar* is used when a volume-based concentration is needed. Since activity itself is unitless, effective units for γ_i are inverse to those of concentration.

In very dilute solutions similar to pure water, the activity coefficients of all dissolved species approach the value of 1.0 and therefore concentration is equal to activity. As the concentration of electrolytes increases, however, the behavior of individual ions is affected by the presence of others that are in close proximity. This results in a decrease in the value of γ_i with increasing electrolyte concentration over the range observed for fresh waters *and a corresponding decrease in activity relative to concentration* of ion i. In essence, an activity coefficient that deviates from 1.0 indicates that the impact of that particular species on the aqueous system (i.e., its activity) is not directly proportional to the amount of the species present (i.e., its concentration). An activity coefficient less than 1, as is observed for inorganic species, represents a case wherein the activity is reduced relative to the amount of the species present. Conversely, as will be discussed later, many organic compounds have activity coefficients greater than 1. Knowledge of activity coefficients is important for evaluating the behavior of contaminants in aqueous systems.

The reduced-activity effects observed for inorganics arise from mutual electrostatic interactions that are proportional to the charges of the ions involved. They are embodied in the definition of the *ionic strength* (I) of an aqueous solution, which is given by

$$I = \tfrac{1}{2}\Sigma_i[i]Z_i^2 \tag{8.2}$$

where Z_i is the valence of species i and the sum is over all charged species in solution. For example, the ionic strength of a 0.05-M NaCl solution is given by (note the valence of both Na and Cl is 1):

$$I(0.05\,M\,NaCl) = \tfrac{1}{2}[(0.05)1^2(\text{From Na})$$
$$+ (0.05)1^2]) (\text{From Cl}) = 0.05\,M \tag{8.3a}$$

whereas the ionic strength of a 0.05-M CaCl$_2$ solution is substantially higher (note the valence of Ca is 2):

$$I(0.05\,\text{M CaCl}_2) = \tfrac{1}{2}\left[(0.05)2^2\,(\text{From Ca})\right.$$
$$\left.+2(0.05)1^2\right]\,(\text{From Cl}_2) = 0.15\,\text{M} \quad (8.3b)$$

Activities of individual, charged aqueous species are then estimated from empirical relationships between γ_i and I as given by the *Davies equation*:

$$\log\gamma_i = -0.512\,Z_i^2\left[\frac{\sqrt{I}}{1+\sqrt{I}} - 0.3I\right] \quad (8.4)$$

This equation is valid for natural waters with ionic strengths approaching that of seawater (–0.7 M). The Davies equation shows that the influence of ionic strength on γ_i increases with charge (Z) of the dissolved species (Fig. 8.2). This reflects the fact that ions of higher charge interact more strongly with each other, as is expected for Coulombic attraction (or repulsion) between ions of opposite (or like) charge. It is important to understand the influence of ionic strength on speciation of inorganics in aqueous solution, because for example, the transport behavior of metals will be affected by ionic strength. Thus metal transport in sites with highly saline conditions may be different from that for sites with low-salinity (i.e., lower ionic strength) water.

8.2.3   Ion Hydration, Ion Hydrolysis, and Acid-Base Reactions

The activities of solutes impact the rate and extent of various phase-transfer processes and biotic uptake (by plants and microbes). Inorganic solutes are present as cations, anions, or neutral molecules. Major species present in natural waters affect the fate of contaminants via their mutual participation in chemical reactions. Major species include: (1) *nonhydrolyzing cations* (Na$^+$, Ca^{2+}, Mg^{2+}, K$^+$), (2) *hydrolyzing cations* (Al^{3+}, H$^+$), (3) *strong acid anions* (Cl$^-$, NO$_3^-$, SO$_4^{2-}$), and (4) *weak acid anions* (CO$_3^{2-}$, HCO$_3^-$, organic acids). Inorganic pollutants also fall into these four categories (see Table 8.1), but are often present at much lower concentrations than the major species.

The distinction between hydrolyzing and nonhydrolyzing cations relates to the strength of their bonding to water molecules. When ions are dissolved in solution, they become *hydrated* by forming *ion-dipole bonds* with the dipolar solvent water. Cations coordinate with the oxygen atoms in water, each of which has a partial negative charge, whereas anions coordinate with the hydrogen atoms (H), each of which has a partial positive charge (Fig. 8.3). The number of water molecules that surround a given ion depends on its ionic radius (r), with larger ions coordinating with a larger number of water molecules. A useful parameter for prediction of ion behavior in water is termed the *ionic potential:* the ratio of charge (Z) to radius (r) of a given ion. The force of attraction between an ion and its hydration waters increases as the charge of the ion becomes more concentrated or, in other words, as ionic potential (Z/r) increases. The chemical behavior of main-group cations in aqueous solution depends strongly on the ionic potential of the cation. Small values of ionic potential give rise to hydrated cations, whereas higher values give rise to *hydrolysis products* of these hydrated cations, which can result in the formation of both *insoluble oxides or hydroxides* and *soluble oxyanions*.

The force of ion-dipole attraction influences the *hydrolysis* of ions, which strongly impacts their transport and fate in the environment. Ion hydrolysis—defined as the breaking of O–H bonds in water molecules that are attached to the ions—occurs in response to changes in the pH of the

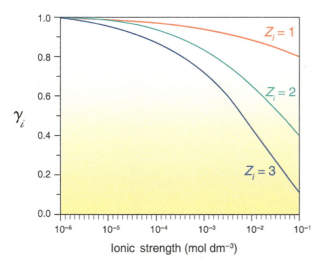

FIG. 8.2 The influence of ionic strength on activity coefficients of charged (Z_i = 1, 2, or 3) aqueous species, as calculated from the Davies Equation.

TABLE 8.1 Classification of Several Important Inorganic Contaminants on the Basis of Their Hydrolysis and Acid-base Behavior in Aqueous Solution[a]

Nonhydrolyzing Cations:	^{137}Cs$^+$, ^{90}Sr^{2+}
Hydrolyzing Cations:	Al^{3+}, Cr^{3+}, Co^{2+}, Ni^{2+}, Cu^{2+}, Zn^{2+}, Cd^{2+}, Hg^{2+}, Pb^{2+}, U^{VI}O$_2^{2+}$
Strong Acid Anions:	NO$_3^-$, Se^{VI}O$_4^{2-}$, ^{129}I$^-$
Weak Acid Anions:	PO$_4^{3-}$, Cr^{VI}O$_4^{2-}$, As^{V}O$_4^{3-}$, As^{III}(OH)$_4^-$, Se^{IV}O$_3^{2-}$, MoO$_4^{2-}$

[a]Radioactive contaminants are indicated by the atomic weight of the relevant isotope. The oxidation state of redox active elements is shown in Roman numerals in the superscript.

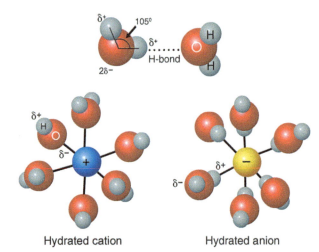

Hydrated cation   **Hydrated anion**

FIG. 8.3 The dipole structure of water, which arises from the H-O-H bonding geometry of the water molecule, influences H-bonding between water molecules in solution (top), and the coordination of water molecules to form the primary hydration sphere around cations and anions in solution (bottom). The bond to an ion from the negative (for cations) and positive (for anions) poles of the H_2O molecule is called an ion-dipole bond.

aqueous solution. These reactions are important for cations of high ionic potential (all the hydrolyzing cations listed in Table 8.1). The strong ion-dipole interaction (and also covalent bonding for transition metals) makes one or more of the protons (H^+ ions) in the coordinating water molecules susceptible to release to solution. This process is termed *proton dissociation*. Hydrolysis of the hydrated Pb^{2+} ion, for example, is given by:

$$Pb(H_2O)_6^{2+}(aq) \leftrightarrow PbOH(H_2O)_5^{+}(aq) + H^+(aq) \quad (8.5a)$$

where (aq) indicates the species is dissolved in aqueous solution. This reaction clearly shows that the H^+ released to solution comes from one of the water molecules attached to the Pb^{2+} ion. More typically, the reaction depicted in Eq. (8.5a) is written in a simpler notation, leaving out the hydration waters:

$$Pb^{2+}(aq) + H_2O(l) \leftrightarrow PbOH^+(aq) + H^+(aq) \quad (8.5b)$$

where (l) represents liquid water. Reactions 7.5 indicate that an increase in pH (i.e., *a decrease in H^+ activity*) will favor the hydrolysis reaction. [Recall: pH = $-\log(H^+)$]. An *equilibrium constant* for this reaction may be written in terms of activities of reactants and products, and its magnitude depends on the temperature (T) and pressure (P) of the system. At standard temperature (25°C) and pressure (0.1 MPa) (as will be the case throughout this chapter unless otherwise noted):

$$K_{h1} = \left(\frac{(PbOH^+)(H^+)}{(Pb^{2+})(H_2O)} \right) = 10^{-7.7} \quad (8.6)$$

where K_{h1} is the first hydrolysis constant for aqueous Pb^{2+}. The value of this constant indicates that (assuming the

activity of solvent $H_2O = 1$) the ratio of $(PbOH^+)/(Pb^{2+}) \sim 1$ at pH = 7.7, and it increases with pH. Further increases in pH likewise result in further hydrolysis:

$$PbOH^+(aq) + H_2O(l) \leftrightarrow Pb(OH)_2^{0}(aq) + H^+(aq)$$
$$K_{h2} + 10^{-9.4} \quad (8.7)$$

The uncharged $Pb(OH)_2^0$ aqueous species (favored at pH 9.4) is relatively insoluble, which results in the formation of an insoluble hydroxide [$Pb(OH)_2$] solid. Similar pH-dependent hydrolysis equilibria occur for all of the hydrolyzing cations in Table 8.1.

These effects of pH on the speciation of inorganics are critical to their transport and fate behavior in aqueous systems (surface water, soil, groundwater). For example, as shown in Eq. (8.7), an uncharged species of Pb is formed at higher pH values. This species has a low solubility in water, so it will tend to precipitate into a solid form. Lead in solid form is unlikely to be transported, which means a low probability of lead pollution for the surrounding area. In addition, because of their low solubilities at these higher pHs, lead and other metals are not as available to organisms (termed bioavailability). Thus they do not pose as great a risk as they would in more soluble forms that occur at lower pH values (see Chapter 28).

Contaminant elements that occur in high oxidation states tend to form stronger covalent bonds with oxygen, resulting in the formation of *oxyanions*—compounds composed of another element combined with oxygen (see Table 8.1). These oxyanions can be considered as hydrolysis products that have undergone significant H^+ dissociation. Since this dissociation is pH dependent, some of these species (the weak acid oxyanions) form complexes with aqueous phase H^+ ions in the pH range of natural waters, thereby altering their charge and behavior.

The tools of acid-base chemistry permit us to evaluate the extent of protonation or proton dissociation of any acid as a function of pH. For example, the acid-base chemistry of arsenate (which is very similar to that of phosphate) is given by:

$$H_3AsO_4^{0}(aq) \leftrightarrow H_2AsO_4^{-}(aq) + H^+(aq)$$
$$K_{a1} = 10^{-2} \quad (8.8a)$$

$$H_2AsO_4^{-}(aq) \leftrightarrow HAsO_4^{2-}(aq) + H^+(aq)$$
$$K_{a2} = 10^{-7} \quad (8.8b)$$

$$HAsO_4^{2-}(aq) \leftrightarrow AsO_4^{3-}(aq) + H^+(aq)$$
$$K_{a3} = 10^{-12} \quad (8.8c)$$

where K_{a1}, K_{a2}, and K_{a3} are the first, second, and third acid dissociation constants. The equilibrium constant for Eq. (8.8a is given by:

$$K_{a1} = \frac{(H_2AsO_4^{-})(H^+)}{(H_3AsO_4^{0})} \quad (8.9a)$$

Taking $-\log_{10}$ of both sides and rearranging gives:

$$-\log\left(H^{+}\right) = -\log K_{a1} - \log\left[\frac{\left(H_2AsO_4^-\right)}{\left(H_3AsO_4^0\right)}\right] \qquad (8.9b)$$

Eq. (8.9b) is termed the *Henderson-Hasselbalch Equation,* and it provides a useful index for the speciation of weak acids. When $(H_3AsO_4^0) = (H_2AsO_4^-)$ (i.e., activity of acid equals that of the conjugate base), the second term on the right side vanishes (log 1 = 0), and pH = pK_a. Thus, given a knowledge of solution pH and the pK_a values of oxyanions, we can determine the relative predominance of the various weak acid species. In this case, the predominant species are $H_3AsO_4^0$ below pH 2, $H_2AsO_4^-$ between pH 2 and 7, $HAsO_4^{2-}$ between pH 7 and 12, and AsO_4^{3-} at pH above 12. The transport behavior and bioavailability of As and other similar elements will be greatly affected by the effect of pH on speciation.

8.2.4   Aqueous-Phase Complexation Reactions

Dissolved ions can form stable bonds to other ions and molecules in solution giving rise to *aqueous-phase complexes.* Bonding between cations and anions may occur as *inner-sphere* or *outer-sphere complexes.* The former case involves direct ion-ion contact with no hydration waters interposed and may involve some degree of covalent bonding, whereas in the latter case, hydration waters are retained around each ion and the two are attracted electrostatically (Fig. 8.4). In either case, the stable unit is termed a *metal-ligand complex,* where the cation is the *central metal group* and the coordinating anion is termed a *ligand.* The formation of aqueous phase complexes influences the solubility, fate, and transport of the constituent molecules.

The tendency for metal-ligand complexes to form is characterized by a *thermodynamic stability constant* (K_{stab}). The K_{stab} is the equilibrium constant for a complex formation reaction written in terms of aqueous phase activities. For example, the mercuric (Hg^{2+}) cation forms a stable complex with the chloride (Cl^-) anion. The aqueous phase complexation of these two ions is given by:

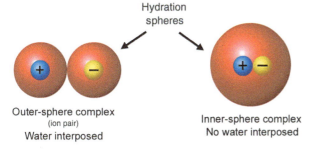

Outer-sphere complex
(ion pair)
Water interposed

Hydration spheres

Inner-sphere complex
No water interposed

FIG. 8.4   Aqueous phase complexation of inorganic ions can result in the formation of outer-sphere or inner-sphere complexes.

$$Hg^{2+}(aq) + Cl^-(aq) \leftrightarrow HgCl^+(aq)$$
$$K_{stab} = 10^{6.7} \qquad (8.10)$$

According to Eq. (8.10), if Cl^- (aq) is present in solution at $1\,mmol\,dm^{-3}$, then the activity of $HgCl^+$ (aq) will exceed that of the free Hg^{2+} (aq) cation by more than 5000 times (assuming activity coefficients = 1). Since $HgCl^+$ (aq) is a monovalent cationic species, it exhibits a lower affinity for negatively charged sites on soil particle surfaces (see Section 8.2.5) and, therefore a greater mobility in soils and sediments than the Hg^{2+} cation. Hence, aqueous-phase speciation is an important determinant of environmental fate. Aqueous-phase complexation reactions are quite rapid, occurring on time scales ranging from μs (10^{-6} s) to minutes, and therefore can be considered to achieve equilibrium during the time scales of water movement through porous media.

The formation of dissolved complexes increases the total aqueous-phase concentration of a given element in equilibrium with solid-phase precipitates (see Section 8.2.5). In addition to inorganic anions (e.g., Cl^- and SO_4^{2-}), organic molecules, including the low molecular weight organic acids (e.g., oxalate, citrate, and salicylate) and fulvic acids, are important ligands in natural waters. These organic anions are particularly important to forming metal-ligand complexes in zones of high biological activity such as surficial soils and wetland sediments. Prediction of transport and fate of inorganic contaminants must include consideration of their aqueous-phase speciation, including complexation with organic and inorganic molecules.

8.2.5   Precipitation-Dissolution Reactions

The solubility of minerals can regulate the partitioning of inorganic contaminants between the aqueous and solid phases (see Fig. 8.1). Soils and sediments contain a mixture of mineral solids, all of which may be subjected to dissolution or precipitation reactions, depending on aqueous-phase conditions at a given point in time. Although these processes typically proceed more slowly than adsorption-desorption reactions, they are very important over long time frames (weeks to decades). Precipitation can result in contaminant immobilization into sparingly soluble solids, whereas mineral dissolution can result in replenishment of the bioavailable pool of dissolved and exchangeable elements. Contaminants are incorporated into precipitated minerals both as *major elements* (>100 mg contaminant per kg of solid) and also as minor or *trace elements* ($<100\,mg\,kg^{-1}$).

The solubility of chromium (III) hydroxide, for example, is governed by the following reaction:

$$Cr(OH)_3(s) + 3H^+(aq) \leftrightarrow Cr^{3+}(aq) + 3H_2O(l) \qquad (8.11a)$$

where dissolution of the $Cr(OH)_3$ solid proceeds from left to right, releasing $Cr^{3+}(aq)$ into solution. Precipitation proceeds from right to left, as the solid phase removes $Cr^{3+}(aq)$ from solution. This reaction can be characterized by the thermodynamic *dissolution equilibrium constant*:

$$K_{dis} = \frac{(Cr^{3+})(H_2O)^3}{(Cr(OH)_3)(H^+)^3} = 10^{12} \qquad (8.11b)$$

In considering the dissolution-precipitation of pure mineral solids in water-saturated systems, the activities of solvent H_2O and mineral solid ($Cr(OH)_3$ in this case) are assumed to be unity, and the dissolution equilibrium constant is readily transformed into a *solubility product constant* (K_{so}), written in terms of activities of only aqueous phase species:

$$K_{so} = \frac{(Cr^{3+})}{(H^+)^3} = 10^{12} \qquad (8.11c)$$

The solubility product gives the activities of Cr^{3+} (aq) and H^+ (aq) in equilibrium with the solid phase $Cr(OH)_3$.

One can readily determine whether precipitation or dissolution will occur in a given aqueous environmental system by comparing the *actual values* of species activities in that system with those that occur at equilibrium. To accomplish this, actual values for a given natural system (which may not be at equilibrium) are incorporated into a term, comparable to that of Eq. (8.11c), called the *ion activity product (IAP)*:

$$IAP = \frac{(Cr^{3+})}{(H^+)^3} = \frac{\gamma_{Cr^{3+}}[Cr^{3+}]}{\gamma_{H+}^3[H^+]^3} \qquad (8.12)$$

where activities in this case are determined from *measured* species concentrations, and activity coefficient terms are calculated from a model such as the Davies equation (Eq. 8.4). The tendency for precipitation or dissolution to occur is then assessed on the basis of the *relative saturation* (Ω) of the system:

$$\Omega = \frac{IAP}{K_{so}} \qquad (8.13)$$

If $\Omega > 1$ ($IAP > K_{so}$), the solution is termed *supersaturated* with respect to the selected solid phase (e.g., Cr $(OH)_3$) and precipitation will occur. If $\Omega < 1$ ($IAP < K_{so}$), the solution is termed *undersaturated* with respect to the selected solid phase and (if the solid phase is present) dissolution will occur. If $\Omega = 1$ ($IAP = K_{so}$), the solution is termed *saturated* with respect to the selected solid phase; the solution and solid are at equilibrium and neither precipitation nor dissolution of the solid will occur.

The preceding example pertains to a single Cr(III) solid phase in which Cr^{3+} is the major cationic constituent. However, the natural environment contains numerous potential contaminant-bearing minerals, including those

in which the contaminant exists as a minor species. For example, Cr^{3+} can be incorporated into iron oxides, such as the mineral goethite (α-FeOOH), where it substitutes for a very small fraction of the Fe^{3+} cations:

$$Fe_{(1-x)}Cr_xOOH(s) + 3H^+(^{aq}) \leftrightarrow (1-x)Fe^{3+}(aq)$$
$$+ x\, Cr^{3+}(aq) + 2H_2O(l) \qquad (8.14)$$

Since goethite is very insoluble (i.e., low K_{so}), natural waters may become supersaturated with the Cr-containing goethite phase (Eq. (8.14) proceeds from right to left) at Cr concentrations much lower than those that would result in precipitation of $Cr(OH)_3$ (s). This type of reaction, called *coprecipitation*, is very important for controlling the fate of inorganic contaminant species in natural waters, since contaminant concentrations are often much lower those of major mineral-building elements such as Si, Al, and Fe. Prediction of geochemical fate of inorganic contaminants, therefore requires the assessment of Ω for each of the potential contaminant-bearing solids. As discussed previously, the transport potential and bioavailability of metals and other inorganics is much lower for solid species than for aqueous (dissolved) species.

It is important to note that K_{so}, *IAP*, and Ω values are *written in terms of activities of particular aqueous species, normally the free ion of interest* (e.g., Cr^{3+} (aq) in this case). Therefore speciation of solution for the mineral-building constituent is a prerequisite for calculating the relative saturation with respect to any mineral phases. In cases where numerous potential solid phases exist for a given inorganic contaminant, speciation of the solution phase and calculation of Ω values for each of the solids is normally accomplished by using a computer equipped with a geochemical speciation program.

8.3 BASIC PROPERTIES OF ORGANIC CONTAMINANTS

8.3.1 Phases—Solids, Liquids, Gases

As with inorganic contaminants, a critical property to consider when evaluating the transport and fate behavior of an organic contaminant is its phase state. Under "natural" conditions (temperature $T = 25°C$, pressure $P = 0.1\,MPa$), organic chemicals in their pure form exist as solids, liquids, or gases (see Table 8.2). The phase state of a contaminant has a significant impact on mass-transfer processes such as dissolution and evaporation (Fig. 8.5).

8.3.2 Dissolution and Aqueous Solubilities of Organic Contaminants

The transfer of molecules from their pure state to water is called dissolution. The extent to which molecules of a

TABLE 8.2 Aqueous Solubilities, Vapor Pressures, and Henry's Coefficients for Selected Organic Compounds

	Aqueous Solubility (mg/L)	Vapor Pressure (atm)	Henry's Constant	Phase State (Standard Conditions: $T = 20°C$, P = Atmospheric)
Benzene	1780	0.1	0.18	Nonpolar liquid
Toluene	515	0.03	0.23	Nonpolar liquid
Naphthalene	30	~0.00066	0.02	Nonpolar solid
Phenol	82,000	0.00026	0.00005	Polar liquid
Methane	24	275	27	Nonpolar gas

Data from Verschueren, K., 1983. Handbook of Environmental Data on Organic Chemicals, and Schwarzenbach, R.P., Gschwend, P.M., Imboden, D.M., 2003. Environmental Organic Chemistry.

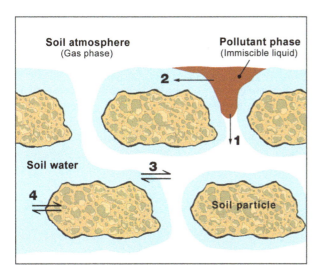

FIG. 8.5 Phase transfer of pure pollutant with water and air phases: (1) evaporation, (2) solubilization, (3) volatilization, (4) sorption. *(From Pollution Science © 1996, Academic Press, San Diego, CA.)*

compound will transfer from its pure phase into water is the aqueous solubility. The solubility of organic compounds depends strongly on the degree to which water and contaminant molecules interact. The well-known rule of thumb that "likes dissolve likes" generally holds for solubility. Water is a very polar solvent (see Fig. 8.3) and, therefore the solubility of organic compounds depends strongly on the degree of polarity of the molecules. Water can interact easily with other polar compounds. Therefore the aqueous solubilities of ionic or polar organic compounds are relatively large. Conversely, it requires much more energy for water to interact with or solvate nonpolar organic compounds. Therefore the solubilities of nonpolar compounds are generally much smaller than those of the polar and ionic compounds. This is presented in Table 8.2, where solubilities for several representative organic compounds are presented. Compare the solubilities of phenol, a polar liquid, with that of toluene, a nonpolar liquid. The solubility of phenol is 159 times greater than that of toluene.

Organic compounds in their pure form exist as either a solid (e.g., naphthalene), liquid (benzene, toluene, phenol), or gas (methane). Because dissolution (solubilization) requires breaking bonds between contaminant molecules, the solubility of organic compounds also depends on the form of the compound. For example, more energy is required to break bonds in solids than in liquids; the solubilization of solids can be thought of as a two-step process, where the solid must first "melt" (convert to a liquid), and then dissolve. Therefore the solubilities of solid organic compounds are usually smaller than those of liquid compounds (see Table 8.2, compare naphthalene to toluene or benzene).

The concept of activity coefficients was introduced in the prior section. For organic compounds, the aqueous solubility is inversely proportional to the activity coefficient. In other words, compounds with larger activity coefficients will have lower solubilities, and vice versa. In contrast to inorganics, for which activity coefficients are generally less than one, the activity coefficients of organic compounds are greater than one. This reflects the fact that it is more difficult (requires more energy) for water to interact with most organics than most inorganics.

An important property of liquid organic compounds is their miscibility, or lack thereof, with water. A miscible organic liquid is one that can be mixed with water such that a single liquid phase results. Alcohols such as methanol and ethanol are examples of miscible liquids. They can be considered as having an infinite solubility in water. Conversely, an immiscible liquid is one that cannot be mixed with water. Benzene, an aromatic hydrocarbon and a major component of gasoline, is an example of an immiscible liquid. If a volume of pure liquid benzene is mixed with water, the two liquids will separate quickly after cessation of mixing because they are immiscible. However, a relatively small fraction of the benzene molecules will transfer into the water phase, thus becoming dissolved. The maximum amount of benzene that can be dissolved in water is the aqueous solubility of benzene. Even though the amount of benzene that can dissolve in water is very small ($<2\,\mathrm{g\,L^{-1}}$), it can be very significant because the federal maximum contaminant level for benzene is $5\,\mathrm{\mu g\,L^{-1}}$.

The solubility of organic compounds in water is a function of water-contaminant and contaminant-contaminant molecular interactions, and depends primarily on chemical properties of the compounds. However, solubility is also affected by environmental factors such as temperature. Many organic compounds become more soluble as the temperature increases, but a few behave in the opposite way. Generally, solubility changes by less than a factor of two in the temperature range of most natural systems (e.g., 0–35°C). Salinity or ionic strength usually causes a small decrease in the solubility of nonpolar organic compounds ("salting-out effect"). This effect is also small for typical environmental conditions (less than a factor of two change in solubility).

The presence of other contaminants can influence the solubilities of organic contaminants. For example, the presence of alcohols or detergents (surfactants) can increase the amount of contaminant in solution. Alcohols such as ethanol are added to gasoline, often at quantities of 10% or more, to boost their oxygen content. If this gasoline mixture leaks from a storage tank into the subsurface, the presence of the alcohol can result in greater transport of the gasoline components. Indeed, the ability of alcohols and surfactants to increase the solubilities of contaminants is also the basis for the enhanced-solubilization flushing method for subsurface remediation (see Chapter 19).

8.3.3 Evaporation of Organic Contaminants

Evaporation of a compound involves transfer from the pure liquid or solid contaminant phase to the gas phase. The vapor pressure of a compound is the pressure of contaminant gas in equilibrium with the solid or liquid contaminant phase and is an index of the degree to which a compound will evaporate. The vapor pressure can be thought of as the "solubility" of the compound in air. Evaporation can be an important transfer process when pure-phase contaminant exists in the vadose zone, such that contaminant molecules can evaporate into the soil atmosphere.

In contrast to solubility, which is governed by contaminant-contaminant and water-contaminant molecular interactions, evaporation is controlled primarily by contaminant-contaminant interactions (i.e., the energy of bonding) in the solid or liquid phase. This is because intermolecule interactions are very minor for most gases, where the space between molecules is relatively large compared to that for liquids or solids. Simply put, the greater the bonding energy between contaminant molecules, the lower will be the vapor pressure. The vapor pressures of liquids are therefore typically larger than those of solids (see Table 8.2). The vapor pressures for benzene and toluene

are approximately 100 times larger than naphthalene's vapor pressure. However, the vapor pressure for phenol is smaller than naphthalene—this is because phenol is a polar compound. The vapor pressures for gases are very large (see methane in Table 8.2). The vapor pressure is a strong function of temperature because of the strong influence of temperature on gas-phase interactions.

8.3.4 Volatilization of Organic Contaminants

Transfer of contaminant molecules between water and gas phases is an important component of the transport of many organic compounds in the vadose zone and the atmosphere. Volatilization is different from evaporation; we use the latter term to specify a transfer of contaminant molecules from their pure phase to the gas phase. For example, the transfer of benzene molecules from a pool of gasoline to the atmosphere is evaporation, whereas the transfer of benzene molecules from water (where they are dissolved) to the atmosphere is volatilization. The vapor pressure provides a rough idea of the extent to which a compound will volatilize. However, volatilization depends also on the aqueous-phase solubility of the compound and on environmental factors.

At equilibrium, the distribution of a contaminant between gas and aqueous phases is described by *Henry's law*:

$$C_\mathrm{g} = HC_\mathrm{w} \qquad (8.15)$$

where:

C_g is concentration of pollutant in the gas phase ($\mathrm{ML^{-3}}$)
C_w is concentration of pollutant in the water phase ($\mathrm{ML^{-3}}$)
H is Henry's Constant (dimensionless). Henry's law can be used to evaluate the preference of a contaminant for aqueous and gas phases (Information Box 8.1)
M and $\mathrm{L^3}$ represent any consistent set of mass and volume units (or molar/molal units).

8.3.5 Multiple-Component Organic Phase

The preceding discussion dealt with the behavior of single organic contaminants. However, many important contaminants contain multiple components. The primary examples of this type of contamination are multicomponent immiscible liquids such as gasoline, diesel fuel, and coal tar. Knowledge of the partitioning behavior of multicomponent contaminants is essential to the prediction of their impact on environmental quality.

The transfer of individual components of a multiple-component contaminant into water is controlled by the aqueous solubility of the component and the composition of the liquid. A simple approach to estimating the solubility

Consider the preference of three contaminants in three separate closed containers in which reside equal volumes of water and air.

The first contaminant has a Henry's Constant of 1. A value of one means that the contaminant concentration in the air is equal to the concentration in the water. Thus this contaminant "likes" water and air equally.

The second contaminant has a Henry's Constant of 0.1. This means that the concentration of the contaminant in the air is ten times less than the concentration in the water; this contaminant prefers the water.

The third contaminant has a Henry's Constant of 10, which means that its concentration in air is ten times greater than it is in water.

Henry's law describes the distribution of contaminant mass at equilibrium. Instantaneous transfer and the achievement of equilibrium is not guaranteed in natural systems. In soils, however, the rate of contaminant transfer between water and gas phases is relatively rapid in comparison to other transport processes. Thus assuming instantaneous transfer is often not a major problem. The magnitude of the Henry's Constant is influenced by temperature, with larger values obtained at higher temperatures.

of multiple-component liquids involves an assumption of ideal behavior in both aqueous and organic phases and the application of *Raoult's Law*:

$$C_w^i = X_o^i \, S_w^i \qquad (8.16)$$

where:

C_w^i is aqueous concentration ($mol\,L^{-1}$) of component i
S_w^i is aqueous solubility ($mol\,L^{-1}$) of component i
X_o^i is mole fraction of component i in the organic liquid

The mole fraction represents the concentration of component i in the immiscible liquid. In essence, Raoult's Law states that the aqueous concentration obtained for any given component is proportional to the amount of that component in the immiscible liquid.

For example, assume we have a two-component immiscible liquid, with the mole fraction of each component equal to 0.5. This means that there is an equal amount of each component in the liquid contaminant. Let us further assume that the aqueous solubility of component A is $100\,mg\,L^{-1}$ and that of component B is $10\,mg\,L^{-1}$. We now wish to calculate the concentrations of A and B in a volume of water that is in contact with the immiscible liquid. Using Raoult's Law we find that the aqueous concentration for component A is $50\,mg\,L^{-1}$ and for component B is $5\,mg\,L^{-1}$. Thus the aqueous concentrations for the two components are half of their aqueous solubilities because they are not dissolving from their pure state, but rather from a mixture. One can

think of this as the two components "competing" with each other to dissolve into water. Inspection of Eq. (8.16) shows that when the mole fraction is equal to one (i.e., a single-component liquid), the aqueous concentration is equal to the aqueous solubility. Raoult's Law can also be used to determine the concentrations of components in air that are in equilibrium with a multiple-component contaminant liquid.

8.4 SORPTION PROCESSES

Sorption is a major process influencing the transport and fate of many contaminants in the environment. The broadest definition of sorption (or retention) is the association of contaminant molecules with the solid phase (soil or sediment particles). The solid phase to which the contaminants sorb is often referred to as the sorbent; we will use this term for the following discussion.

A critical impact of sorption is that it generally slows or retards the rate of movement of contaminants (see Chapter 7). In addition, sorption can influence the magnitude and rate of other processes. For example, sorption can influence biodegradation rates by affecting the bioavailability of contaminants to microorganisms. Sorption processes are also a basic component of traditional water treatment technologies, such as the use of granular activated carbon beds to remove organic contaminants from water or the use of ion-exchange beds to remove inorganics from water (see Chapter 22). Sorption can occur by numerous mechanisms, depending on the properties of the contaminant and of the sorbent; this will be discussed in the following sections.

8.4.1 Inorganics

As noted in Chapter 2, the surfaces of soil and sediment particles are important for the uptake and release of contaminants. Inorganic pollutants are attracted to particle surfaces largely because of their charge properties. Soil particles contain positively charged sites and negatively charged sites, which results in the uptake of anions and cations, respectively. *Isomorphic substitutions* in clay minerals give rise to *permanent charge*, which is dominantly negative. *pH-dependent ionization reactions* at surface hydroxyl groups of minerals and organic matter give rise to *variable charge*, which can be positive or negative, depending on acid-base reactions at surface sites. For example, many Fe and Al oxides are positively charged at pH < 8 and negatively charged at pH > 8. At any given point in time, a volume of soil contains some number of both positive-charged and negative-charged sites that contribute to the uptake of ionic contaminant species. As shown in Fig. 8.1, the accumulation of ions or molecules at these particle surfaces during their removal from aqueous solution is termed *adsorption*. The release of adsorbed molecules from surfaces to aqueous solution is termed desorption.

A *surface complex* is formed when an *adsorbate* (the ion or molecule adsorbed at a surface) forms a stable bond with an *adsorbent* (the solid phase whose surface provides the site for contaminant adsorption). Mechanisms of ion adsorption include the formation of inner-sphere and outer-sphere *surface complexes* that are analogous to those that form in aqueous solution. Outer-sphere surface complexes are dominantly the result of electrostatic interaction. Inner-sphere complex formation involves some degree of covalent bonding and, therefore chemical *specificity* between the adsorbate and adsorbent. For this reason, formation of outer-sphere complexes is often referred to as *physisorption*, whereas inner-sphere complex formation is termed *chemisorption*. Ions adsorbed in inner-sphere coordination are bound more strongly and are less likely to desorb from the surface because of displacement by other ions in solution.

Adsorption mechanisms are dependent on the adsorbate and adsorbent composition. For example, the weak acid species of arsenate $(H_2AsO_4^-, HAsO_4^{2-})$ and phosphate $(H_2PO_4^-, HPO_4^{2-})$ are known to form strong, inner-sphere complexes with surface sites on oxides and hydroxides, whereas the oxyanions selenate (SeO_4^{2-}), sulfate (SO_4^{2-}) and nitrate (NO_3^-) form weaker, outer-sphere complexes. Likewise, the transition metal cations Cu^{2+} and Pb^{2+} form strong inner-sphere complexes at variable-charge surface hydroxyl groups, whereas Ni^{2+} and Mn^{2+} tend to form outer-sphere complexes (Fig. 8.6).

Ions adsorbed as outer-sphere complexes are termed *exchangeable* because their electrostatic attraction to the surface is readily disrupted, leading to subsequent desorption, by introduction of an ion of similar or greater charge and surface affinity. Exchangeable ions are also considered to be "bioavailable," since they are a principal source for replenishment of ions that are removed from the solution phase by microbial or plant uptake. The number of moles of exchangeable cation charge that can be adsorbed per unit mass of soil or sediment is termed the *cation exchange capacity* (CEC). The corresponding value for anions is termed the *anion exchange capacity* (AEC).

In temperate zone soils dominated by permanent-charged silicate clays and organic matter, the CEC is typically much higher than the AEC, and adsorption of exchangeable cations exceeds that of exchangeable anions. The mobility of cations relative to anions is diminished as a result of adsorption processes. Cation exchange reactions are assumed to be reversible, and they conserve the number of moles of charge adsorbed to the negatively charged exchange sites, even when the exchanging cations have different valence.

For example, consider the exchange of adsorbed *bivalent* Cd^{2+} for adsorbed *monovalent* Na^+:

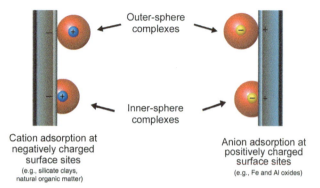

FIG. 8.6 Mechanisms of inorganic contaminant adsorption: Outer-sphere and inner-sphere surface complexation.

$$X_2Cd\,(s) + 2Na^+(aq) \leftrightarrow 2XNa(s) + Cd^{2+}(aq) \qquad (8.17)$$

where X represents one mole of negative charge on the exchanger and (s) refers to the solid phase. Thus one mole of Cd^{2+} occupies two moles of exchanger charge, and $Cd^{2+} \rightarrow Na^+$ exchange involves two moles of Na^+ to maintain charge balance. The reaction 8.17 can be considered from the perspective of *Le Chatelier's Principle of Equilibrium Chemistry*. This principal states that if a system is perturbed from its equilibrium state, chemical reactions will proceed to return the system to equilibrium. Thus an increase in aqueous-phase Na^+ concentration will result in the release of Cd^{2+} from exchange sites, enhancing their mobility in the dissolved state. The charge-based stoichiometry of cation exchange is further illustrated in Fig. 8.7.

Highly weathered, silicate-clay-depleted soils, such as Oxisols and Ultisols of the humid tropics, contain large amounts of variable-charge iron and aluminum oxides. These soils can have AEC > CEC, particularly under acidic conditions when pH is low and variable-charge sites become positive charged because of proton adsorption. In such systems, the mobility of anions is diminished relative to that of cations. Whereas ion exchange reactions involving outer-sphere complex formation can be quite rapid (time scales ranging from microseconds to minutes), longer term diffusion into small pores of adsorbent solids and the formation of inner-sphere complexes can extend equilibration time frames into weeks or even months.

8.4.2 Organics

The mechanisms by which many organic contaminants are sorbed or retained by porous media particles are usually quite different from those involved for inorganic contaminants. We can use the "likes dissolve likes" rule to help explain the sorption of nonpolar organic contaminants by sorbents. It is now generally accepted that for many natural sorbents (soil, sediment), organic contaminants interact primarily with organic material associated with the sorbent. This organic material is generally less polar than water

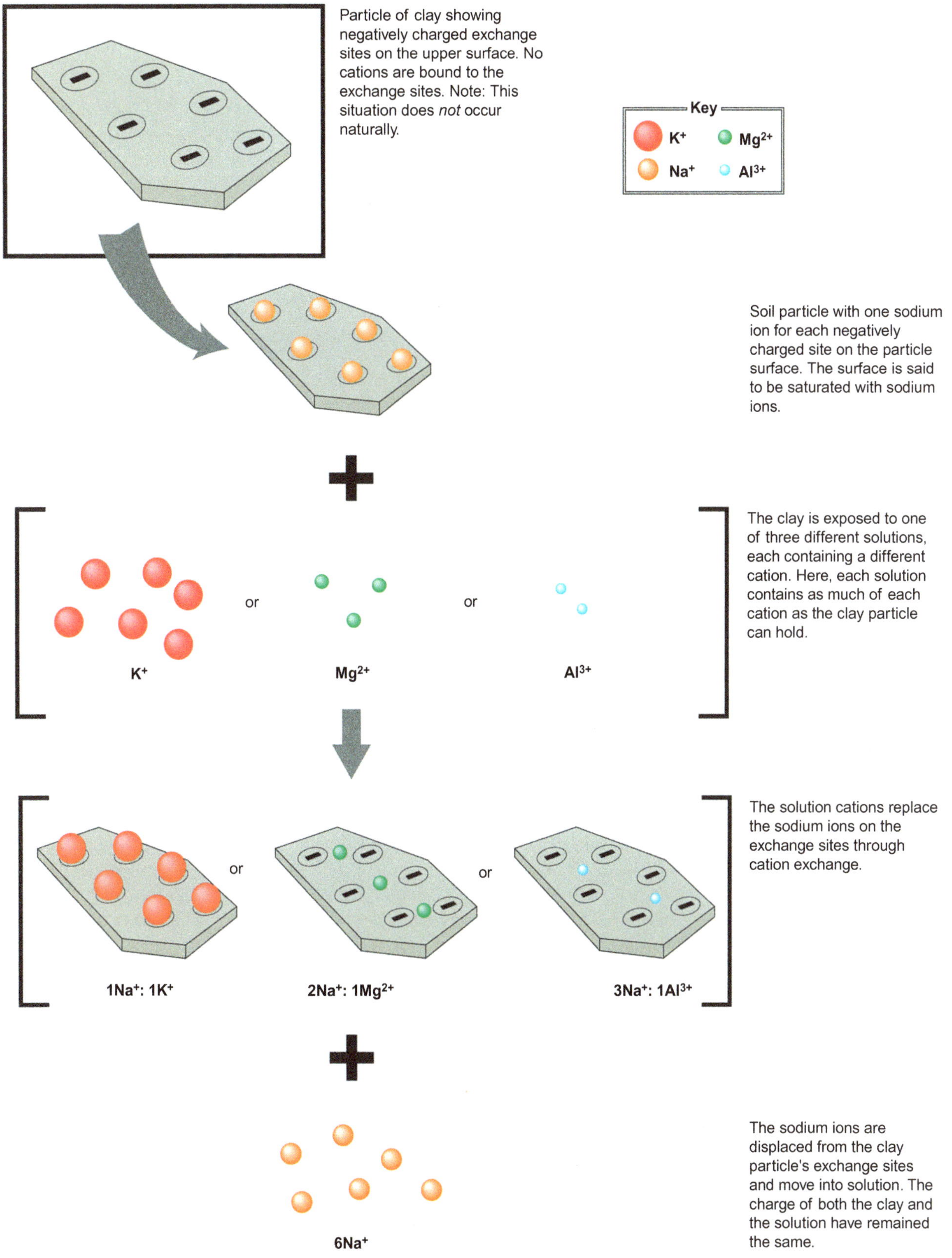

Particle of clay showing negatively charged exchange sites on the upper surface. No cations are bound to the exchange sites. Note: This situation does *not* occur naturally.

Key

K^+ Mg^{2+}
Na^+ Al^{3+}

Soil particle with one sodium ion for each negatively charged site on the particle surface. The surface is said to be saturated with sodium ions.

+

or   or

K^+ Mg^{2+} Al^{3+}

The clay is exposed to one of three different solutions, each containing a different cation. Here, each solution contains as much of each cation as the clay particle can hold.

or   or

$1Na^+: 1K^+$ $2Na^+: 1Mg^{2+}$ $3Na^+: 1Al^{3+}$

The solution cations replace the sodium ions on the exchange sites through cation exchange.

+

$6Na^+$

The sodium ions are displaced from the clay particle's exchange sites and move into solution. The charge of both the clay and the solution have remained the same.

FIG. 8.7 Mechanisms of inorganic contaminant adsorption: Outer-sphere and inner-sphere surface complexation. *(From Pollution Science © 1996, Academic Press, San Diego, CA.)*

and provides a more favorable environment for nonpolar organic contaminants. Thus the sorption of nonpolar organics is driven primarily by the incompatibility between water and the organic compound, a phenomenon known as the "hydrophobic effect." In essence, this effect involves the expulsion of a nonpolar organic compound from the aqueous phase. This happens because its presence there requires the breaking up of the hydrogen-bonded structure of liquid water (see Fig. 8.3) and there are no favorable bonds formed between water molecules and nonpolar organic molecules in return. Conversely, association with hydrophobic portions of organic matter does not require such disruption, and there are also weak bonding interactions (e.g., van der Waals associations) that can contribute to the favorable overall energetics of sorption.

The mechanisms governing the sorption of polar or ionic organics are similar to some of those governing retention of inorganics (e.g., electrostatic attraction, surface complex formation). Many of the important ionizable organic contaminants are negatively charged under environmental conditions (e.g., phenols, chlorophenols, carboxylic acids). Because most soil and sediment particles have a net negative charge, there is often repulsion between negatively charged organic contaminants and the sorbent, which results in little sorption to such particles. However, sorption to positive charged particles (e.g., oxides of Fe and Al) can be significant, so the amount of surface area composed of such particles can be an important determinant of contaminant sorption.

Sorption processes are usually considered in terms of aqueous systems, such as sorption by soil or aquifer material in subsurface systems, sorption by sediments in surface water bodies, and sorption by porous media in packed-bed wastewater treatment systems. In all these cases, the contaminant is dissolved in water that is flowing through or over the porous media. However, sorption processes may also occur in gas-phase systems, wherein volatile organic contaminants are associated with the atmosphere or soil atmosphere that is in contact with porous media. This process is often referred to as *vapor adsorption*. This issue is of particular interest for gas-phase contaminant transport in the vadose zone, and for potential transport of contaminants sorbed to particles suspended in air, which can be transported great distances by atmospheric processes. Vapor adsorption is strongly influenced by the amount of water present at the surfaces of the porous-medium grains. For example, vapor adsorption by oven-dry soil has been observed to be orders of magnitude larger than adsorption by water-saturated soil. However, once water starts coating the soil surfaces, the magnitude of observed adsorption decreases greatly. This effect is related to the ability of water to "out compete" organic contaminants for adsorption at the soil particle surfaces.

The properties of the contaminant (e.g., nonpolar or polar organic, inorganic charge sign and magnitude) are key determinants of the degree to which the contaminant will be sorbed or retained by a sorbent. However, the physical/chemical properties of the sorbent are also important. For example, sorption of nonpolar organic contaminants is often controlled by sorbent organic matter. Therefore the amount of organic matter associated with the sorbent is very important. The cation exchange capacity, clay content, and metal-oxide content are important properties for sorption of ionizable and ionic organic contaminants, as they are for inorganics.

8.4.3 Magnitude and Rate of Sorption

Sorption is quantified by measuring a sorption isotherm, which is simply a description of the relationship between the concentration of contaminant in the sorbed state and the concentration in the aqueous phase (or air for vapor-phase sorption). Many different forms of isotherms have been proposed and used to describe sorption. The simplest is the *Linear Isotherm* (the same idea as Henry's Law), which is given by:

$$S = K_d C_w \tag{8.18}$$

where:

S is the concentration of contaminant sorbed by the soil (ML^{-3})
K_d is the sorption coefficient $(L^3 M)$

The larger the value of K_d, the greater the degree to which a contaminant is sorbed by the sorbent.

A linear isotherm signifies that for all concentrations of contaminant in water, there will always be proportionally the same sorbed concentration. The sorption of many nonpolar organic contaminants is linear or close to linear. An example of a linear isotherm is shown in Fig. 8.8. Some organic and many inorganic contaminants exhibit nonlinear sorption. An example of a widely used nonlinear isotherm is the *Freundlich Isotherm* given by:

$$S = K_f C_w^n \tag{8.19}$$

where K_f is the Freundlich sorption coefficient and n is, in essence, a power function related to the sorption mechanism (s). Examples of nonlinear isotherms for the cases of $n > 1$ and $n < 1$ are presented in Fig. 8.8. For a nonlinear isotherm, the distribution of contaminant between the sorbed and aqueous phases is proportionally different at different concentrations. For example, at higher concentrations, the proportional distribution is less than it is at lower concentrations for the $n < 1$ isotherm. This is noted by observing the slope of the isotherm line; the magnitude of sorption for a given aqueous concentration correlates to the slope at that concentration.

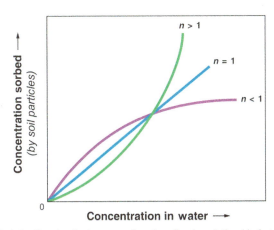

FIG. 8.8 Sorption isotherms used to describe the relationship between sorbed concentrations and aqueous concentrations of a contaminant; linear ($n=1$) and nonlinear isotherms are shown. *(From Pollution Science © 1996, Academic Press, San Diego, CA.)*

As the name implies, an isotherm is measured at one temperature. Temperature has a small but measurable effect on the sorption of contaminants. The effect of temperature will depend on the type of sorption mechanism involved. For low-polarity organic compounds, the sorption is governed by aqueous-solubility interactions as discussed before. So, if a change in temperature causes a change in solubility, it might be expected that a change in temperature may also cause a change in sorption. For relatively large organic compounds like anthracene, an increase in temperature causes an increase in solubility. Thus the sorption of anthracene may decrease with increasing temperature. The sorption mechanisms for inorganic contaminants involve strong contaminant-sorbent interactions. An increase in temperature could increase the energetics of this interaction and therefore produce an increase in sorption. The effect of salinity or ionic strength on sorption is also dependent on the type of sorption mechanism involved. For nonpolar organic contaminants, the effect is usually relatively small, whereas it can be significant for inorganic contaminants.

The use of sorption isotherms presumes the existence of equilibrium between the sorbent and aqueous or gas phases. Research has shown that the rate of sorption of many nonpolar organic contaminants is very slow, taking anywhere from several hours to several months to reach equilibrium. This slow rate of sorption is often due to the slow diffusion of contaminant molecules into components of the soil that have very small openings or pores. We know that the sorption of many organic contaminants is dominated by sorbent organic matter. This organic matter has a polymeric type of structure. It can take a long time for contaminant molecules to move from the surface to the inside of the organic material. In addition, some sorbents have solid particles that have very small pore spaces (microporosity). It can take considerable time for contaminant molecules to diffuse into these pores.

8.4.4 Liquid-Liquid Partitioning

We have previously covered important phase-transfer or partitioning processes, namely, evaporation, volatilization, dissolution, and sorption from solution or vapor to solid phases. Another important phase-transfer process is the partitioning of constituents between two immiscible liquids. For example, consider a site that has a leaking underground gasoline storage tank. A quantity of gasoline, which is immiscible with water, has leaked into the aquifer. Now we have a situation similar to that depicted in Fig. 8.5, but without the air phase present. The transfer of constituents of the gasoline, such as benzene, from the gasoline liquid to water is the process of dissolution discussed previously. Let us further assume that the site also contains other types of contaminants, such as chlorinated solvents (trichloroethene, TCE, for example). In such a case, the groundwater will likely contain a certain amount of dissolved TCE. As that contaminated groundwater passes by the gasoline trapped in the aquifer, some of the TCE molecules will transfer to the gasoline—because gasoline and TCE are low-polarity compounds compared to water, the TCE prefers to reside within the gasoline rather than water. This process is termed liquid-liquid partitioning.

This process can be important for transport of contaminants at sites that have multiple types of wastes. For example, employing a contaminant transport model that accounts explicitly for liquid-liquid partitioning and solid-phase sorption, it is demonstrated that the retardation factor for a TCE-like solute increased from 2 for the case of only solid-phase sorption to 17 for a case including partitioning to an immiscible liquid present at a volume saturation of 1% (i.e., the volume of organic liquid present fills only 1% of the pore space). This means that the TCE would migrate in groundwater 8.5 times slower because of partitioning to the organic liquid. The R increases to 152 when the volume saturation is increased to 10%.

Note that liquid-liquid partitioning is a reversible phase-transfer process, meaning that solutes that transfer into the immiscible liquid can transfer back into aqueous solution upon a change in the concentration gradient. The process is relatively rapid compared to advective-dispersive transport processes. The octanol-water partition coefficient is widely used to represent the propensity of an organic compound to partition from water to an immiscible organic liquid.

8.4.5 Adsorption to Fluid-Fluid Interfaces

In many environmental systems, two or more fluids will exist simultaneously. A fluid-fluid interface is formed when two fluids come into contact. This is observed in Fig. 8.5. The red body represents an immiscible organic liquid and the blue represents water. The surface wherein the two liquids contact each other is the oil-water interface. The surface wherein air and the organic liquid contact is the air-oil interface. Finally, the surface where the air and water contact is the air-water interface. The presence of these interfaces can be critical for transport and fate of contaminants.

There are three scenarios for which adsorption at fluid-fluid interfaces may be particularly relevant. It is well known that water droplets and hydrated aerosols in the atmosphere provide air-water interface that can serve as significant retention domains for contaminants. Extensive research has demonstrated that a wide range of organic compounds undergo adsorption at air-water interfaces. Thus it is likely that air-water interfacial adsorption can influence the atmospheric transport of contaminants.

A second scenario involves transport of contaminants in vadose-zone systems, in which both air and water are present in the soil pores. The air-water interface is a primary potential retention domain for constituents in the vadose zone. Transport studies have shown that air-water interfacial adsorption can significantly influence the transport of different types of contaminants.

The transport of contaminants in source zones containing immiscible organic liquids represents a third scenario. The previous section described the case where TCE dissolved in groundwater partitioned into an organic liquid trapped in the subsurface. In addition to transferring into the bulk organic liquid, the TCE can also adsorb at the interface between the organic liquid and water. Prior research has demonstrated that the transport of contaminants in organic-liquid-contaminated porous media can be influenced by adsorption at the oil-water interface.

It should be noted that fluid-fluid interfaces are relevant beyond their influence on retention of constituents. The interface between fluids represents the surface through which transfer of matter occurs. Therefore the amount of interfacial area present will control the magnitude and rate of mass transfer. Examples of processes for which the magnitude of interfacial area is critical include the dissolution of immiscible organic liquids, volatilization and evaporation, and the transfer of oxygen into water to support biodegradation (Chapter 9). The magnitude of interfacial area depends upon properties of the soil and the amount of the individual fluids present.

8.5 ABIOTIC TRANSFORMATION REACTIONS

The transformation of contaminants by microorganisms is discussed in Chapter 9. Some contaminants can also be transformed by abiotic (physical/chemical) processes. We will briefly discuss some major abiotic transformation processes in this section.

8.5.1 Hydrolysis

Water is a ubiquitous component of the environment. As a result, most contaminants will come into contact with water to some extent. It is important, therefore to understand if and when a contaminant will react with water when they are in contact. Reaction of a contaminant with water is termed hydrolysis. A generalized example of this reaction for organic compounds is given by:

$$R - X + H_2O \rightarrow R - OH + X^- + H^+ \qquad (8.20)$$

where R-X is an organic compound with X representing a functional group such as a halide (e.g., Cl). By reacting with water, the original compound (R-X) has been transformed to another compound (R-OH). The hydrolysis of inorganics was covered in Section 8.2.3.

The two key factors in hydrolysis reactions are the charge properties of the contaminant molecules and the pH. Hydrolysis is essentially an interaction between nucleophiles (substance with excess electrons, such as OH^-) and electrophiles (substance deficient in electrons, such as H^+). Thus the charge properties of the molecules will govern its reactivity with water. For many compounds, hydrolysis may be catalyzed or enhanced under acidic or basic conditions. This means that the occurrence and rate of hydrolysis is often pH dependent. For example, a hydrolysis reaction catalyzed by OH^- would occur more rapidly at higher pH values because of larger OH^- concentrations. Hydrolysis can also be influenced by sorption interactions. For example, the pH at the surface of many soils is lower than the pH of the water surrounding the soil particles. Thus an acid-catalyzed hydrolysis reaction could be enhanced when the contaminant is associated with the soil.

8.5.2 Oxidation-Reduction Reactions

Oxidation-reduction (*redox*) reactions are chemical reactions that involve the transfer of electrons between two molecular species. The two species involved can be organic or inorganic, and they may be present in any environmental phase (gas, liquid, or solid). In a full redox reaction, one species begins the reaction in its more reduced form and this species is *oxidized* (i.e., loses one or more electrons) during the reaction. Conversely, the other species enters the reaction in its more oxidized form and is *reduced* (accepts one or more electrons). Fig. 8.9 depicts this process schematically. Many of the environmentally important redox reactions are *catalyzed* (i.e., made to proceed faster) by microorganisms, but they only proceed when favorable thermodynamically.

Take the oxidation of zinc solid as an example: $Zn(s) + 2H^+(aq) \leftrightarrow Zn^{2+}(aq) + H_2(g)$. In this, the oxidation number of Zn has changed from 0 to +2, producing Zn^{2+}, and the oxidation number of H^+ has changed from +1 to 0,

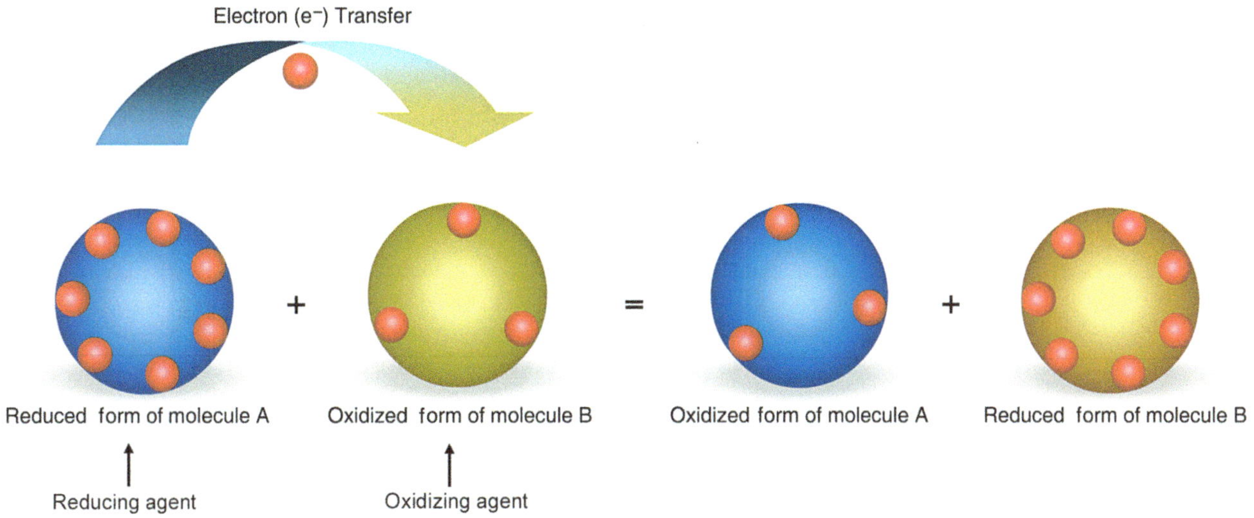

FIG. 8.9 A full oxidation-reduction reaction involves the transfer of electrons from one species (the *reducing agent*) to another (the *oxidizing agent*).

producing H_2 gas. In this reaction, Zn has been oxidized and H^+ has been reduced. Since Zn(s) was oxidized, it caused the reduction of H^+(aq) and is therefore the reducing agent. Likewise H^+(aq) caused the oxidation of Zn(s), making H^+ the oxidizing agent.

Loss of electrons from one substance must simultaneously be accompanied by the gain of electrons from another. Electrons are neither created nor destroyed in chemical reactions, and thus oxidation-reduction reactions occur in pairs. Just as the transfer of hydrogen ions determines the pH of a solution, the transfer of electrons between species determines the redox potential of an aqueous solution. Redox potential is also referred to as "ORP" for oxidation-reduction potential and is measured in volts or *Eh* ($1 V = 1 Eh$). ORP specifically measures the tendency for a solution to either gain or lose electrons when it is subject to change by the introduction of a new species. A solution with a higher ORP will have a tendency to gain electrons (i.e., oxidize them) and a solution with a lower ORP will have a tendency to lose electrons to new species (i.e., reduce them).

Perhaps the best known example of a redox reaction is aerobic, heterotrophic *respiration,* with molecular oxygen (O_2) acting as an electron acceptor during the oxidation of carbohydrate (see also Chapter 5):

$$CH_2O(aq) + O_2(g) \leftrightarrow CO_2(g) + H_2O(l) \qquad (8.21)$$

In this reaction, one mole of carbon (C) is reduced from the 0 oxidation state in CH_2O to the +4 oxidation state in CO_2 while two moles of oxygen (O) are reduced from the 0 oxidation state in O_2 to the -2 oxidation state (one mole ends up in H_2O and the other in CO_2; the third mole of O was already in the -2 oxidation state in CH_2O). Thus a total of four moles of electrons are transferred per mole of CH_2O oxidized. When microorganisms catalyze the respiration of

carbohydrates, they capture some of the energy released in the reaction. In a similar way, microbes can catalyze the oxidation of other organic compounds that contain reduced C, including many organic contaminants. Although many organic contaminants are oxidized much more slowly than "labile" forms of C, such as carbohydrates, they are eventually subjected to oxidation, and the process is most favorable energetically when oxygen is available to act as the electron acceptor (i.e., in *oxic* environments).

Many subsurface environmental systems, including biologically active soils or sediments, are depleted of gaseous or dissolved O_2. This occurs when respiration consumes O_2 faster than it can be replenished by diffusion from the atmosphere. In these *anoxic* systems, alternative oxidizing agents must be used as electron acceptors in respiration. The major *alternative electron acceptors* in aqueous environments include reducible solutes and mineral solids. These include (in order of decreasing energy yield): nitrate (NO_3^-), manganese (IV) oxides, iron (III) oxides, and sulfate (SO_4^{2-}). Oxidation of both natural and xenobiotic reduced C compounds can be coupled effectively to the reduction of these redox-active constituents.

In addition to these major alternative electron acceptors, inorganic contaminants can also be reduced in the absence of O_2. As discussed earlier, many inorganic contaminants (e.g., As, Se, Cr, Hg, and Pb) can occur in more than one oxidation state, depending on environmental conditions. Anoxic conditions favor the reduced forms of these elements. For example, the more toxic and mobile aqueous species of selenium is selenate (SeO_4^{2-}), with Se in the +6 oxidation state. Selenate can be reduced to the less toxic and less mobile species, *selenite* (SeO_3^{2-}):

$$2SeO_4^{2-}(aq) + 4H^+(aq) + 4e^-(aq) \leftrightarrow$$
$$2SeO_3^{2-}(aq) + 2H_2O(l); \log K = 60 \qquad (8.22a)$$

This reduction of selenate must be coupled to an oxidation reaction, such as the oxidation of carbohydrate, providing the necessary electrons:

$$CH_2O\,(aq) + H_2O\,(l) \leftrightarrow CO_2\,(g) + 4\,H^+(aq) \\ + 4\,e^-(aq); \log K = 0.8 \qquad (8.22b)$$

Reaction (8.22a) is considered a *reduction half-reaction,* whereas Eq. (8.22b) is an *oxidation half-reaction.* Their sum provides the full, balanced redox reaction:

$$CH_2O\,(aq) + 2\,SeO_4^{2-}\,(aq) \leftrightarrow CO_2\,(g) \\ + 2\,SeO_3^{2-}\,(aq) + H_2O\,(l); \log K = 60.8 \qquad (8.22c)$$

Here, the pairs CH_2O/CO_2 and SeO_4^{2-}/SeO_3^{2-} are real-world examples of molecules A and B in Fig. 8.9, respectively. Note that the oxidation half-reaction (Eq. (8.22b)) could be replaced by one for an organic contaminant, in which case the oxidative transformation of an organic contaminant would be coupled directly to the reductive transformation of an inorganic contaminant.

Whether a chemical species in solution is oxidized or reduced has a profound influence on its biogeochemical cycling and its transport. This is true for metals, nutrients, salts, and organic compounds. For example, redox reactions have an effect on the bioavailability of nutrients. To illustrate, iron exists in solution either as low-solubility oxidized ferric (Fe^{+++}) or as the reduced highly soluble ferrous (Fe^{++}). Phosphorous is an essential nutrient for plant and animal growth and under oxidizing conditions, is bound to ferric iron forming a ferro-phosphate complex that is biologically unavailable. If pollutants enter into a water body, dissolved oxygen may be depleted and reducing conditions prevail. Under these reducing conditions, iron loses its normally close association with phosphorous, with the latter now becoming biologically available. This can lead to massive algal growth and the formation of noxious, and potentially toxic, algal blooms. Reducing conditions often prevail in the bottom of thermally stratified lakes and reservoirs, and phosphorous can accumulate leading to large growths of algae when the lake destratifies.

8.5.3 Photochemical Reactions

Some chemical compounds that are present in the atmosphere, in the top several centimeters of surface waters, or at the land surface can be transformed under the influence of sunlight. Chemical transformations that are induced by light energy are termed *photochemical processes.* Photochemical processes most often result in oxidation or reduction as a result of the absorption of light energy by one or more reactant species. Thus a prerequisite for photochemical transformation is the capability of a molecule to absorb discrete quantities of light energy called *photons.*

The energy (in Joules, J) gained by absorption of a photon is given by:

$$E = h\nu = hc/\lambda \qquad (8.23)$$

where h is Planck's constant (6.626×10^{-34}J s^{-1}), ν is the frequency of light, c is the velocity of light (3×10^8m s^{-1}), and λ is its wavelength (m). Eq. (8.23) indicates that the shorter the wavelength of the light, the greater the energy it transfers to matter when absorbed.

Photochemical transformation of both organic and inorganic pollutants can result from either *direct* or *indirect* photolysis. In *direct photolysis,* the light absorbing substance itself is transformed. An example of direct photolysis is the conversion of a chlorinated, refractory organic contaminant into a hydroxylated organic compound, which is normally less toxic and less refractory. This is very similar to the hydrolysis reaction given in Eq. (8.20), but here, the reaction is being promoted by the presence of light:

$$R-Cl(aq) \overset{h\nu}{\rightarrow} R-Cl^*(aq) \overset{H_2O}{\rightarrow} R-OH + H^+(aq) + Cl^-(aq) \qquad (8.24)$$

The asterisk indicates the *excited state* of the chlorinated pollutant R-Cl that results from photon absorption. This *photoactivation* is characterized by the transition of one electron from its ground state to a higher energy state, which makes the contaminant molecule more susceptible to the subsequent transformation reaction.

Indirect photolysis occurs after sunlight produces highly reactive, transient (short-lived) oxygen species in oxygenated waters. These products, which form from photolysis of dissolved species such as nitrate, nitrite, aqueous iron complexes, and dissolved organic matter, include singlet oxygen ($^1O_2$), superoxide anion (O_2^-), hydroperoxyl ($HO_2\bullet$), hydrogen peroxide (H_2O), ozone (O_3), hydroxyl radical ($OH\bullet$), and organic peroxy radicals ($ROO\bullet$). These reactive photolysis products can then accelerate the oxidation of other compounds that are normally quite refractory to oxidation by the more common oxidizing agents. The reaction of contaminants with these photolysis products is termed *indirect* photolysis because the transformed pollutants are not themselves absorbing the light that induces their transformation.

8.5.4 Radioactive Decay

Radioactive decay is an important transformation reaction for a special class of contaminants—radioactive elements. Radioactive decay is caused by an instability of the nucleus in the atom, whereby either protons and neutrons or electrons are emitted in the form of radiation (see Information Box 8.2). This transformation is spontaneous, but the rate at which it occurs varies widely depending on the element concerned.

There are three main types of ionizing radiation found in both natural and anthropogenic sources: alpha, beta, and gamma/x-ray radiation.

Alpha particles are subatomic fragments consisting of two neutrons and two protons. Alpha radiation occurs when the nucleus of an atom becomes unstable (the ratio of neutrons to protons is too low) and alpha particles are emitted to restore balance. Alpha decay occurs in elements with high atomic numbers, such as uranium, radium, and thorium. The nuclei of these elements are rich in neutrons, which makes alpha particle emission possible. Alpha particles are relatively heavy and slow, and therefore have low penetrating power and can be blocked with a sheet of paper.

Beta radiation occurs when an electron is emitted from the nucleus of a radioactive atom. Beta decay also occurs in elements that are rich in neutrons. Just like electrons found in the orbital of an atom, beta particles have a negative charge and weigh significantly less than a neutron or proton. Beta particles can be blocked by a sheet of metal or plastic and are typically produced in nuclear reactors.

Gamma or x-ray radiation is produced during a nucleus' excited state following a decay reaction. Instead of releasing another alpha or beta particle, it purges the excess energy by emitting a pulse of electromagnetic radiation called a gamma ray. Gamma rays are similar in nature to light and radio waves except that it has very high energy. Gamma rays have no mass or charge. They can travel for long distances and thus pose external and internal hazards for people. An example of a gamma emitter is cesium-137, which is used to calibrate nuclear instruments.

8.5.5 Quantifying Transformation Rates

Many transformation reactions can be described with a first-order equation:

$$\frac{\partial C}{\partial t} = -kC \qquad (8.25)$$

where k is the transformation rate constant ($1/T$). This equation states that the rate of transformation depends on the amount of contaminant present. The minus sign on the right-hand side of the equation denotes that the concentration change is negative (i.e., concentrations of reactants decrease with time). The first-order equation can be used, for example, to represent fixed-pH hydrolysis, radioactive decay, and biodegradation when there is minimal net change in microbial cell numbers.

For transformation reactions, it is useful to define a half-life, which is the time required for half of the original contaminant mass to be transformed. For reactions that follow first-order kinetics, the half-life ($T_{1/2}$) is defined as:

$$T_{1/2} = \frac{0.693}{k} \qquad (8.26)$$

This equation can be used to estimate the time required for a contaminant to be transformed, which is related to its persistence in the environment. For example, less than 1% of the original contaminant mass remains after a time period equal to 7 half-lives. After 10 half-lives, less than 0.1% remains. For a contaminant that has a half-life of 1 day, this would mean that only 0.1% would remain untransformed after 10 days. A comparison of approximate half-lives for five compounds is presented in Table 8.3. Glucose, a labile (readily biodegradable) compound would be expected to be fully degraded in approximately three weeks (7 half-lives × 3 days). Conversely, anthracene is expected to persist for many years.

QUESTIONS AND PROBLEMS

1. List and briefly discuss some of the major differences in the properties and behavior of inorganic and organic compounds.
2. Calculate the time required for 99.9% of the mass to transform for compounds with half-lives of 1 day, 10 days, 100 days, and 1000 days.
3. Which type of radiation is generally the most potentially hazardous to humans, and why?
4. What is hydrolysis?
5. Discuss some examples of Le Chatelier's principle that occur in everyday life.
6. Why are the aqueous solubilities of many organic compounds very low?
7. How does a change in ionic strength of an aqueous solution affect the solubility of (a) inorganic elements? (b) organic compounds? (c) which is generally affected to a greater degree and why?

TABLE 8.3 Half-Lives for Selected Organic Compounds[a]

Compound	Approximate Half-Life in Soil
Glucose	Three days
Benzene	A few days
2,4-D	Several days
Polychlorinated Biphenyl Mixture	Hundreds of days
Anthracene	Hundreds of days

[a]These values are very approximate and are meant for purposes of comparison. Note that all processes contributing to "loss" are incorporated (e.g., biodegradation, abiotic transformation, and volatilization).

Sources: Mackay, D, Shiu, W.Y., Ma, K.C., 1992. Illustrated Handbook of Physical-Chemical Properties and Environmental Fate of Organic Chemicals. Lewis Publ., Boca Raton, FL; Montgomery, J.H., 1996. Groundwater Chemicals, second ed. CRC/Lewis Publ., Boca Raton, FL.

8. Discuss an example of a redox reaction that is important for the fate of a contaminant of your choice in groundwater.

REFERENCES

Langmuir, D., 1996. Aqueous Environmental Geochemistry. Prentice Hall, New York, NY.

Mackay, D., Shiu, W.Y., Ma, K.C., 1992. Illustrated Handbook of Physical-Chemical Properties and Environmental Fate of Organic Chemicals. Lewis Publ, Boca Raton, FL.

Montgomery, J.H., 1996. Groundwater Chemicals, second ed. CRC/Lewis Publ, Boca Raton, FL.

Schwarzenbach, R.P., Gschwend, P.M., Imboden, D.M., 2016. Environmental Organic Chemistry, third ed. John Wiley & Sons, New York, NY.

Verschueren, K., 1983. Handbook of Environmental Data on Organic Chemicals. John Wiley & Sons, New York, NY.

Chapter 9

Biological Processes Affecting Contaminants Transport and Fate

R.M. Maier

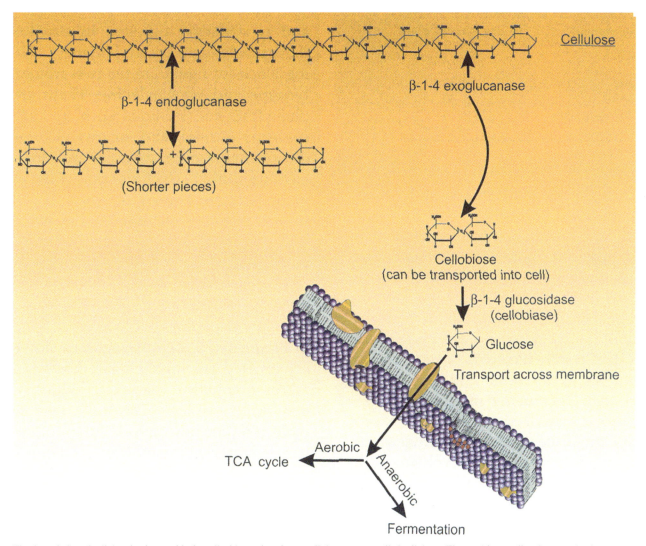

The degradation of cellulose begins outside the cell with a series of extracellular enzymes called cellulases. The resulting smaller glucose subunit structures can be taken up by the cell and metabolized. *From Environmental Microbiology © 2000, Academic Press, San Diego, CA.*

Environmental and Pollution Science. https://doi.org/10.1016/B978-0-12-814719-1.00009-4

9.1 BIOLOGICAL EFFECTS ON POLLUTANTS

Although physical and chemical factors affect the fate of pollutants in soil and water (Chapters 7 and 8), it is apparent that these are not the only critical factors to consider. If they were, the large-scale accumulation of contaminants, including environmental and natural organic substances would dominate the earth, rendering our current pollution problems minuscule by comparison. Thus we must look at a third basic factor—the biological component of soil and water (see Chapter 5). This component, which is responsible for degradation of naturally occurring organic matter, also mitigates the impact of pollutants on the environment. Biological interactions with pollutants are of great interest to scientists and engineers because of the widespread use of biological approaches for the remediation of contaminated sites.

The presence of microorganisms in soil and water can affect the distribution, movement, and concentration of pollutants through a process called biodegradation. Indeed, some pollutants have very short lifetimes under normal environmental conditions because they readily serve as sources of food for actively growing microorganisms. For other pollutants, the effect of microorganisms may be limited for a variety of reasons. Low numbers of degrading microorganisms, microbe-resistant pollutant structures, or adverse environmental conditions can all cause low rates of biodegradation.

In this chapter we will focus on understanding the interaction between microorganisms and pollutants in the environment. We will examine microbial interactions with both organic and inorganic pollutants, as well as the effects that environmental parameters and pollutant structure have on the extent and rate of these biological reactions.

9.2 THE OVERALL PROCESS OF BIODEGRADATION

Biodegradation is the breakdown of organic compounds ("organics") through microbial activity. Biodegradable organic compounds generally serve as a food source, or *substrate*, for microbes. In other words, microorganisms obtain energy and structural components by degrading the organic compound. To be able to degrade a substrate, that substrate must be available to the microbe. This availability of a substrate is termed *bioaccessibility or bioavailability*. Bioaccessibility, which is one important aspect of the biodegradation of any substrate, depends largely on the water phase concentration of the substrate. Microbial cells are 70%–90% water, and the food they obtain comes from the water surrounding the cell. Thus the bioaccessibility of a substrate refers to the amount of substrate in the aqueous solution around the cell. Two important factors

that reduce bioaccessibility are (1) low water solubility (e.g., gasoline—think about how oil and water do not mix) and (2) sorption of substrate by soil (see Chapters 7 and 8).

Biodegradation of organic compounds is really "a series of biological degradation steps or a pathway that ultimately results in the oxidation of the parent compound." Often, the oxidation process generates energy (as described in Chapter 5). Complete biodegradation, or *mineralization*, involves oxidation of the parent compound to form carbon dioxide and water, a process that provides both carbon and energy for growth and reproduction of cells. Fig. 9.1 illustrates the mineralization of any organic compound under aerobic conditions. Mineralization is composed of a series of degradation steps that have much in common, whether the carbon source is a simple sugar such as glucose, a plant polymer such as cellulose, or a pollutant molecule. Each degradation step in the pathway is facilitated by a specific catalyst, or *enzyme*, made by the degrading cell. Enzymes are most often found within a cell, but they are also made and released from the cell to help initiate degradation reactions. Enzymes found external to the cell are known as *exoenzymes*. Exoenzymes are important in the degradation of macromolecules such as the plant polymer cellulose because macromolecules must be broken down into smaller subunits to allow transport into the microbial cell. Both internal enzymes and exoenzymes are essential to the degradation process: degradation will stop at any step if the appropriate enzyme is not present (Fig. 9.2). Lack of appropriate biodegrading enzymes is one common reason for persistence of some pollutants, particularly those with unusual chemical structures that existing enzymes do not recognize.

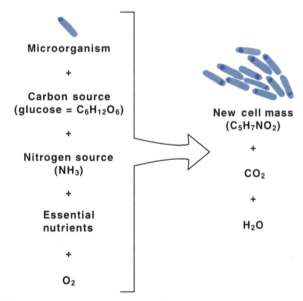

FIG. 9.1 Aerobic mineralization of an organic compound. *(From Pollution Science © 1996, Academic Press, San Diego, CA.)*

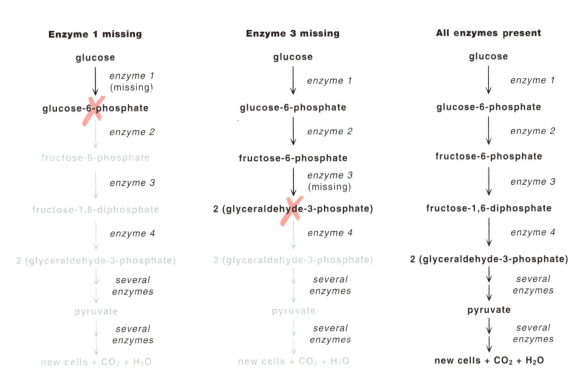

FIG. 9.2 Stepwise degradation of organic compounds. A different enzyme catalyzes each step of the biodegradation pathway. If any one enzyme is missing, the product of the reaction it catalyzes is not formed (denoted with a red cross). The reaction stops at that point and no further product is made (shown in gray). *(From Pollution Science © 1996, Academic Press, San Diego, CA.)*

Thus we can see that degradation depends on chemical structure. Pollutants that are structurally similar to natural substrates usually degrade easily, while pollutants that are dissimilar to natural substrates often degrade slowly or not at all.

Mineralization can also be described as a *mass balance equation* that can be solved to determine the relationship between the amount of substrate consumed; oxygen and nitrogen utilized; and cell mass, carbon dioxide, and water produced. For example, the mass balance equation for glucose (illustrated in Fig. 9.1) can be written as:

$$C_6H_{12}O_6 + NH_3 + O_2 \rightarrow C_5H_7NO_2 + CO_2 + H_2O \quad (9.1)$$

Like glucose, many pollutant molecules—such as most gasoline components and many of the herbicides and pesticides used in agriculture—can be mineralized under the correct conditions (see Example Calculation 9.1).

Example Calculation 9.1 Leaking Underground Storage Tanks

The EPA estimates that more than 1 million underground storage tanks (USTs) have been in service in the United States alone. Over 436,000 of these have had confirmed releases into the environment. Although new regulations now require USTs to be upgraded, leaking USTs continue to be reported at a rate of 20,000/yr and a cleanup backlog of more than 139,000 USTs still exists. In this exercise we will calculate

the amount of oxygen and nitrogen necessary for the biological remediation of a leaking UST site that has released 10,000 gallons of gasoline. To simplify this problem, we will assume that octane (C_8H_{18}) is a good representative of all petroleum constituents found in gasoline. We will use the following mass balance equation to calculate the biological oxygen demand (BOD) and the nitrogen demand:

$$\underset{\text{octane}}{a(C_8H_{18})} + \underset{\text{ammonia}}{b(NH_3)} + \underset{\text{oxygen}}{c(O_2)} \rightarrow \underset{\text{cell mass}}{d(C_5H_7NO_2)} + \underset{\text{carbon dioxide}}{e(CO_2)} + \underset{\text{water}}{f(H_2O)}$$

In this equation, the coefficients a through f indicate the number of moles for each component. To solve the mass balance equation, we must be able to relate the amount of cell mass produced to the amount of substrate (octane) consumed. This is done using the cell yield (Y), where:

$$Y = \frac{\text{Mass of cell mass produced}}{\text{Mass of substrate consumed}}$$

Literature indicates that a reasonable cell yield value for octane is 1.2. Using the cell yield we can now calculate the coefficient d. We will start with 1 mole of substrate ($a = 1$) and use the following equation:

$$d \text{ (MW cell mass)} = a \text{ (MW octane) } (Y)$$
$$d \text{ (113 g/mol)} = 1 \text{ (114 g/mol) (1.2)} \Rightarrow d = 1.2$$

We can then solve for the other coefficients by balancing this equation.

We start with nitrogen. We know that there is one N on the right side of the equation in the biomass term. Examining the left side of the equation, we see that there is similarly one N as ammonia. We can set up a simple relationship for nitrogen and use this to solve for coefficient b:

For b: $b(1\ mol\ nitrogen) = d(1\ mol\ nitrogen)$

$$b(1) = 1.2(1)$$
$$b = 1.2$$

Next we balance carbon and solve for coefficient e:

For e: $e(1\ mol\ carbon) = a(8\ mol\ carbon) - d(5\ mol\ carbon)$

$$e(1) = 1(8) - 1.2(5)$$
$$e = 2.0$$

Next we balance hydrogen and solve for coefficient f:

For f: $f(2\ mol\ hydrogen) = a(18\ mol\ hydrogen) + b(3\ mol\ hydrogen) - d(7\ mol\ hydrogen)$

$$f(2) = 1(18) + 1.2\ (3) - 1.2(7)$$
$$f = 6.6$$

Finally, we balance oxygen and solve for coefficient c:

For c: $c(2\ mol\ oxygen) = d(2\ mol\ oxygen) + e(2\ mol\ oxygen) + f(1\ mol\ oxygen)$

$$c(2) = 1.2(2) + 2(2.0) + 6.6(1)$$
$$c = 6.5$$

Thus the solved mass balance equation is:

$$1(C_8H_{18}) + 1.2(NH_3) + 6.5(O_2) \rightarrow 1.2(C_5H_7NO_2) + 2.0 (CO_2) + 6.6(H_2O)$$

Now we will use this mass balance equation to determine how much nitrogen and oxygen will be needed to remediate the site. First, we will convert gallons of gasoline into mol of octane using the assumption that octane is a good representative of gasoline.

Recall that we started with 10,000 gallons of gasoline:

Convert to liters (L): 10,000 gallons (3.78 L/gallon) $= 3.78 \times 10^4$ L gasoline

convert to grams (g): 3.78×10^4 L gasoline (690 g gasoline/L) $= 2.6 \times 10^7$ g gasoline in the site

Convert to moles: $\dfrac{2.6 \times 107\ g\ gasoline}{114\ g\ octane/mol}$ 2.3×10^5 mol octane in the site

Now we ask—how much nitrogen is needed to remediate this spill? From the mass balance equation we know that we need 1.2 mol NH_3/mol octane (see coefficient b).

$\Rightarrow \dfrac{1.2\ mol\ NH_3}{mol\ octane} \times 2.3 \times 10^5\ mol\ octane\ in\ site$

$\Rightarrow 2.76 \times 10^5\ mol\ NH_3$

$\Rightarrow 2.76 \times 10^5 \times \dfrac{17\ g\ NH_3}{mol}$

$\Rightarrow 4.7 \times 10^6\ g\ NH_3$

$\Rightarrow$ 4.7×10^6 g (1 kg/1000 g) (2.2046 lb/kg) = **This is 10,000 lb or 5 tons of NH_3!!!**

Finally we ask—how much oxygen is needed to remediate this spill? From the mass balance equation we know that we need 6.5 mol O_2/mol octane (see coefficient c).

$\Rightarrow$ 6.5 mol O_2/mol octane $(2.3 \times 10^5$ mol octane in the site) $= 1.5 \times 10^6$ mol O_2

A gas takes up 22.4 L/mol, but remember that air is only 21% oxygen.

$\Rightarrow$ 1.5×10^6 mol O_2 (22.4 L/mol air) (1 mol air/0.21 mol O_2) $= 1.6 \times 10^8$ L air

1 cubic foot of air $= 28.33$ L

$\Rightarrow$ 1.6×10^8 L air (1 cubic foot/28.33 L gas) = **This is 5.5×10^6 cubic feet of air or enough air to fill a football field to a height of 100 ft!!!!**

Some organic compounds are only partially degraded. Incomplete degradation can result from the absence of the appropriate degrading enzyme or it may result from *cometabolism*. In cometabolism, a partial oxidation of the substrate occurs, but the energy derived from the oxidation is not used to support growth of new cells. This phenomenon arises when organisms possess enzymes that coincidentally degrade a particular pollutant; that is, their enzymes are nonspecific. Cometabolism can occur not only during periods of active growth, but also during periods in which resting (nongrowing) cells interact with an organic compound. Although difficult to measure in the environment, cometabolism has been demonstrated for some environmental pollutants. For example, the industrial solvent trichloroethene (TCE) can be oxidized cometabolically by *methanotrophic bacteria* while growing on their normal carbon source, methane. Trichloroethene is of interest for several reasons. It is one of the most frequently reported contaminants at hazardous waste sites, it is classified as a known human carcinogen, and it is generally resistant to biodegradation. As shown in Fig. 9.3, the first step in the methanotrophic oxidation of methane is

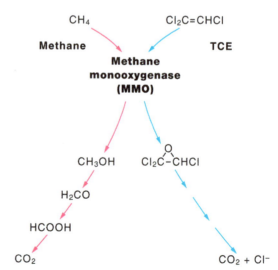

FIG. 9.3 The oxidation of methane by methanotrophic bacteria. This is catalyzed by the enzyme methane monooxygenase. The same enzyme can act nonspecifically on TCE. Subsequent TCE degradation steps may be catalyzed spontaneously, by other bacteria, or in some cases by the methanotroph. *(From Pollution Science © 1996, Academic Press, San Diego, CA.)*

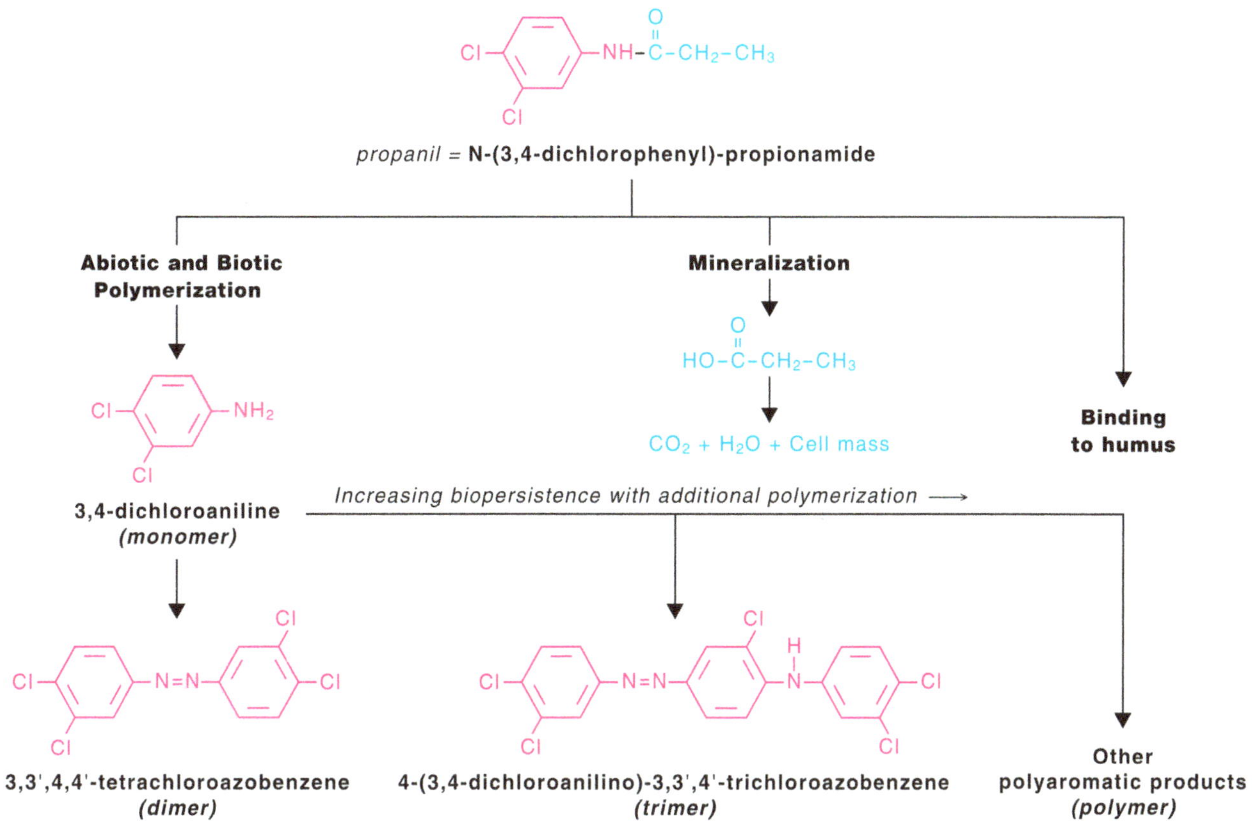

FIG. 9.4 Polymerization reactions that occur with the herbicide propanil during biodegradation. Propanil is a selective postemergence herbicide used in growing rice. It is toxic to many annual and perennial weeds. The environmental fate of propanil is of concern because it, like many other pesticides, is toxic to most noncereal crops. It is also toxic to fish. Care is used in propanil application to avoid contamination of nearby lakes and streams. *(From Pollution Science © 1996, Academic Press, San Diego, CA.)*

catalyzed by the enzyme *methane monooxygenase*. This enzyme is so nonspecific that it can also catalyze the first step in the oxidation of TCE when both methane and TCE are present. However, the methanotrophic bacteria receive no energy benefit from the oxidation of TCE and so it is considered to be a cometabolic reaction. The subsequent degradation steps shown in Fig. 9.3 for TCE may be catalyzed spontaneously by other bacteria, or in some cases by the methanotrophs themselves. This type of cometabolic reaction has great significance in remediation and is not limited to methanotrophs. Other cometabolizing microorganisms that grow on toluene, propane, and even ammonia have also been identified. Currently, scientists are using this technique, the stimulation of cometabolic reactions, for remediation of some TCE-contaminated sites (Chapter 19).

Partial or incomplete degradation can also result in *polymerization*, that is, the synthesis of compounds is more complex and stable than the parent compound. This occurs when initial degradation steps, often catalyzed by exoenzymes, create highly reactive intermediate compounds, which can then combine either with each other or with other organic matter present in the environment. As illustrated in

Fig. 9.4, which shows some possible polymerization reactions that occur with the herbicide propanil during biodegradation, these include formation of stable dimers or larger polymers, both of which are quite stable in the environment. Such stability may be the result of low bioaccessibility (low water solubility, high sorption) or the absence of degrading enzymes.

9.3 MICROBIAL ACTIVITY AND BIODEGRADATION

It is often difficult to predict the fate of a pollutant in the environment because the interactions between the microbial, chemical, and physical components of the environment are still not fully understood. Total microbial activity depends on a variety of factors, such as microbial numbers, available nutrients, environmental conditions (including soil), and pollutant structure. In this section, we will discuss the impact of some of the most important factors affecting microbial activity, with the implicit understanding that microbial activity can be inhibited by any one of these factors even if all other factors are optimal.

9.3.1 Environmental Effects on Biodegradation

The environment around a microbial community—that is, the sum of the physical, chemical, and biological parameters that affect a microorganism—determines whether a particular microorganism will survive and/or metabolize. The occurrence and abundance of microorganisms in an environment are determined by nutrient availability, as well as by various physicochemical factors such as pH, redox potential, temperature, and soil texture and moisture. (These factors are described in detail in Chapters 2 and 5.) Because a limitation imposed by any one of these factors can inhibit biodegradation, the cause of the persistence of a pollutant is sometimes difficult to pinpoint.

Oxygen availability, organic matter content, nitrogen and phosphorus availability, and substrate bioaccessibility are particularly significant in controlling pollutant biodegradation in the environment. Interestingly, the first three of these factors can change considerably, depending on the location of the pollutant. As Fig. 9.5 shows, contamination can occur in terrestrial ecosystems in three major locations: surface soils, the vadose zone, and the saturated zone. The availability of both oxygen and organic matter varies considerably in these zones. In general, oxygen and organic matter both decrease with depth, so biodegradation activity also decreases with depth. There are exceptions to this rule in shallow groundwater regions, which can have relatively high organic matter contents because the rates of groundwater recharge are high.

9.3.1.1 Oxygen and Other Terminal Electron Acceptors

Oxygen is very important in determining the extent and rate of biodegradation of pollutants—it serves as the *terminal electron acceptor (TEA)* in *aerobic* (oxygen is present) biodegradation reactions. As the TEA, oxygen is reduced to water as it accepts electrons from the pollutant substrate or *electron donor* that is being oxidized to carbon dioxide. At lower redox potentials that result in *anaerobic* conditions (oxygen is not available), an alternate TEA must be used. The electron tower shown in Fig. 9.6 illustrates several different TEAs that can be used if available in the environment. Keeping in mind that the higher the place of the TEA on the electron tower, the more energy provided by the reaction, examination of this tower reveals that oxygen provides the most energy for biodegradation. Oxygen is followed closely by iron and nitrate and—at the bottom of the tower—sulfate and carbon dioxide.

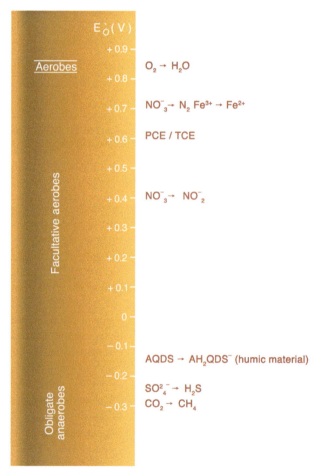

FIG. 9.6 Terminal electron acceptors over a range of redox potentials. The electron tower shows the various terminal electron acceptors (TEAs) used under aerobic (oxygen) and anaerobic (all others) conditions. Note that the higher the place on the electron tower, the more energy that will be produced for metabolism and growth. *(From Pollution Science © 1996, Academic Press, San Diego, CA.)*

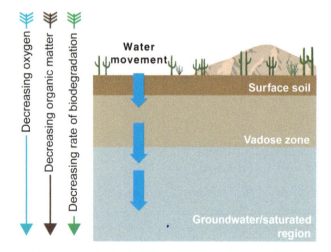

FIG. 9.5 Contamination in different ecosystems. There are three major locations where contamination can occur in terrestrial ecosystems: surface soils, the vadose zone, and the saturated zone. The availability of both oxygen and organic matter varies considerably in these zones. As indicated, oxygen and organic matter both decrease with depth, resulting in a decrease in biodegradation activity with depth. *(From Pollution Science © 1996, Academic Press, San Diego, CA.)*

Because of the energy advantage associated with oxygen, both growth and biodegradation rates are faster under aerobic conditions than under anaerobic conditions.

Some pollutants that degrade aerobically are not degradable anaerobically and vice versa. For example, the highly reduced hydrocarbons found in petroleum, some of which are shown in Fig. 9.7, are readily degraded aerobically, but unless an oxygen atom is present in the structure initially, these compounds are quite stable under anaerobic conditions (Fig. 9.8). This anaerobic stability explains why underground petroleum reservoirs, which contain no oxygen, have remained intact for thousands of years, even though microorganisms are present. In contrast are the highly chlorinated organic compounds which are more stable under aerobic conditions. That is, increasing chlorine content favors anaerobic dehalogenation (removal of chlorines) over aerobic dehalogenation. For instance, tetrachloroethene, a solvent that has been widely used in dry cleaning and as a degreasing agent is not degraded under aerobic conditions but can be sequentially dechlorinated under

	Structure	Name	Physical state
Aliphatic	$CH_3-(CH_2)_n-CH_3$	Propane (n = 1) Octane (n = 8) Hexatriacontane (n = 34)	Gas Liquid Solid
Alicyclic		Cyclopentane	Liquid
		Cyclohexane	Liquid
Aromatic		Benzene	Liquid
		Naphthalene	Solid

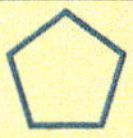

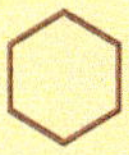

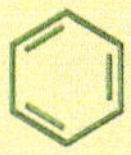

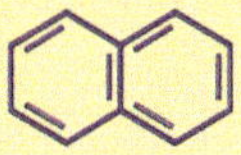

FIG. 9.7 Aliphatic, alicyclic, and aromatic hydrocarbons. Petroleum is usually composed of a mix of these hydrocarbon types. For example, familiar constituents of gasoline are octane and benzene. *(From Pollution Science © 1996, Academic Press, San Diego, CA.)*

		Mineralization	
		Aerobic	Anaerobic
Reduced hydrocarbons	$CH_3-CH_2-CH_2-CH_2-CH_2-CH_3$ Hexane	Yes	No
	or Benzene	Yes	No (or very slow)
Partially oxidized hydrocarbons	$CH_3-CH_2-CH_2-CH_2-CH_2-CH_2-OH$ Hexanol	Yes	Yes
	or OH Phenol	Yes	Yes

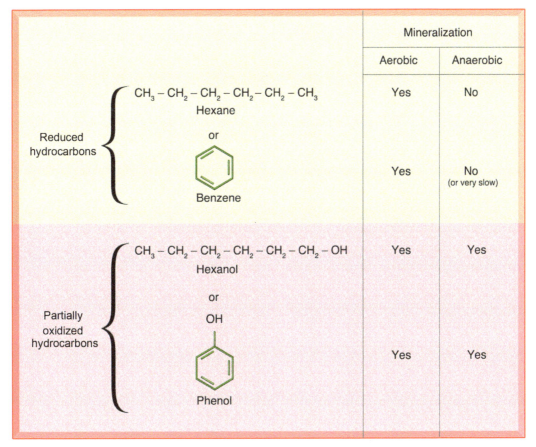

FIG. 9.8 The effects of oxidation on the biodegradability of hydrocarbons. Hydrocarbons with no oxygen, such as hexane or benzene, are only degraded aerobically, while the addition of a single oxygen atom (hexanol) enables both aerobic and anaerobic degradation to occur.

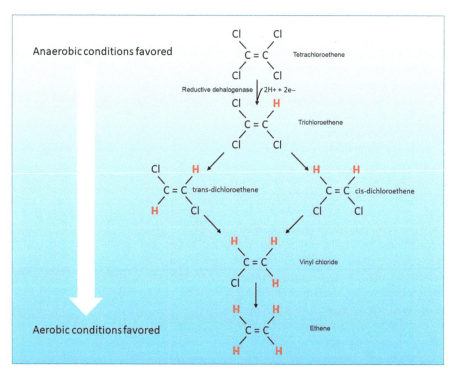

FIG. 9.9 One possible pathway for anaerobic reductive dechlorination of tetrachloroethene. Note that this pollutant is not degraded under aerobic conditions.

anaerobic conditions (Fig. 9.9). As chlorines are removed from tetrachloroethene the degradation products are increasingly amenable to aerobic biodegradation.

In terms of oxygen availability, surface soils and the vadose zone are similar. They both contain significant amounts of air-filled pore spaces and thus tend to favor aerobic degradation of pollutants. However, these regions may contain pockets of anaerobic activity generated by high biodegradative activity or confined zones of water saturation that reduce oxygen levels. In contrast, the oxygen concentrations in the groundwater or water-saturated regions are low. The only oxygen that exists in these regions is that dissolved in water, and the oxygen solubility in water is quite low (~9 mg/L). Therefore, if significant microbial activity occurs, the limited supply of dissolved oxygen is rapidly used up, causing anaerobic conditions to develop. Addition of air or oxygen can often improve biodegradation rates, particularly in subsurface areas that are water saturated (see also Chapter 19).

9.3.1.2 Microbial Populations and Organic Matter Content

Surface soils have large numbers of microorganisms. Culturable bacterial numbers generally range from 10^6 to 10^9 organisms per gram of soil. Fungal numbers are somewhat lower, 10^4 to 10^6 per gram of soil. In contrast, microbial populations in deeper regions, such as the vadose zone and groundwater region, are often lower by two orders

of magnitude or more. This large decrease in microbial numbers with depth is primarily due to differences in organic matter content. Whereas the soil surface may be rich in organic matter, both the vadose zone and the groundwater region often have low amounts of organic matter. One consequence of low total numbers of microorganisms is that the population of pollutant degraders initially present is also low. Thus biodegradation of a particular pollutant may be slow until a sufficient biodegrading population has been built up. The process in which degradation rates are initially low, but then increase over time is sometimes referred to as *adaptation* or *acclimation*. A second reason for slow biodegradation in the vadose zone and groundwater region is that the organisms in this region are often dormant because of the low amount of organic matter present. If microorganisms are dormant, their response to an added carbon source is slow, especially if the carbon source is a pollutant molecule to which they have not previously been exposed.

Given these two factors, oxygen availability and organic matter content, we can make several generalizations about surface soils, the vadose zone, and the groundwater region (see Fig. 9.5):

1. Biodegradation in surface soils is primarily aerobic and rapid.
2. Biodegradation in the vadose zone is also primarily aerobic, but significant acclimation times may be necessary for large biodegrading populations to develop.

3. Biodegradation in the saturated region (groundwater) is initially slow, owing to low numbers, and can rapidly become anaerobic due to lack of available oxygen.
4. Biodegradation in shallow groundwater regions with higher organic matter content may be rapid but conditions will quickly become anaerobic.

9.3.1.3 *Nitrogen*

Nitrogen is another macronutrient that often limits microbial activity because it is an essential part of many key microbial metabolites and building blocks, including proteins and amino acids. As shown by the chemical formula for a cell (see Fig. 9.1), nitrogen is a large component by mass of microorganisms. It is also subject to removal from the soil/water continuum by various processes such as leaching or denitrification (see Chapter 16). Many pollutants are carbon-rich and nitrogen-poor; thus nitrogen limitations can inhibit their biodegradation, whereas the simple addition of nitrogen-rich compounds can often improve it. For example, in the case of petroleum oil spills, where nitrogen shortages can be acute, biodegradation can be significantly accelerated by adding nitrogen fertilizers. In general, microbes have an average C:N ratio within their biomass of about 5:1 to 10:1, depending on the type of microorganism, so the C:N ratio of the material to be biodegraded must be 20:1 or less. The difference in the ratios is due to the fact that approximately 50% of the carbon metabolized is released as carbon dioxide, whereas almost all of the nitrogen metabolized is incorporated into the microbial biomass.

9.3.2 Pollutant Structure

The rate at which a pollutant molecule is degraded in the environment depends largely on its structure. If the molecule is not normally found in the environment—or if its structure does not resemble that of a molecule usually found in the environment—a biodegrading organism may not be present. In this case, chances for biodegradation to occur are low. The bioaccessibility of the pollutant is also extremely important in determining the rate of biodegradation. If the water solubility of the pollutant is extremely low, it will have low bioaccessibility. Many pollutant molecules that are persistent in the environment share the property of low water solubility. Examples include **d**ichlor-o**d**i**p**heny**lt**richloroethane (DDT), a pesticide that is now banned in the United States; **p**oly**c**hlorinated **b**iphenyls (PCBs), similarly banned and have not been manufactured in the United States since 1977, which are used as heat-exchange fluids; and petroleum hydrocarbons. Both PCBs and petroleum hydrocarbons are liquids at room temperature and actually form a phase that separates from water. Although microorganisms are not excluded from this phase,

(A)

(B)

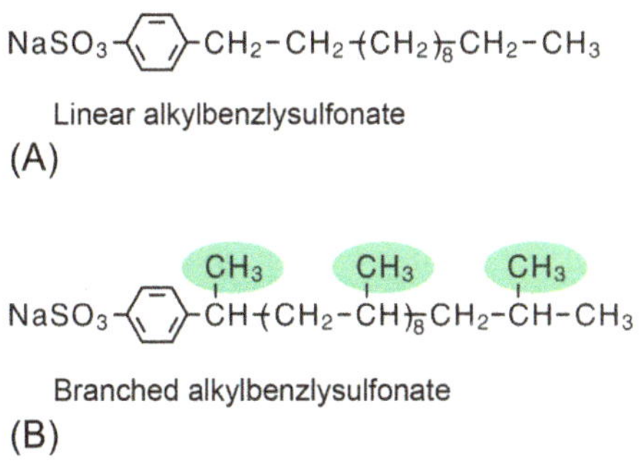

FIG. 9.10 Linear and branched alkylbenzylsulfonates (ABS) are commonly used surfactants. Since the linear variant (A) is readily biodegradable, and the branched form (B) is not, and both work equally well as detergents, the linear ABS has entirely supplanted the branched ABS in environmentally conscious markets. *(From Pollution Science © 1996, Academic Press, San Diego, CA.)*

active metabolism seems to occur only in the aqueous phase or at the oil–water interface. The second factor that reduces bioaccessibility is sorption of the pollutant by soil. Compounds that have low water solubility, such as DDT, PCBs, and petroleum constituents, are also prone to sorption by soil surfaces (see Chapters 7 and 8).

Many pollutants have extensive branching or functional groups that block or sterically hinder the pollutant carbon skeleton at the reactive site, that is, the site at which the substrate and enzyme come into contact during a biodegradation step. For example, we now use biodegradable detergents, namely, the linear alkylbenzylsulfonates (ABSs). The only difference between these readily biodegradable detergents and the slowly biodegradable nonlinear ABS detergents is the absence of branching (Fig. 9.10).

As a result of our increasing knowledge of the effect of pollutant structure on biodegradation in the environment, efforts are focusing on developing and utilizing "environmentally friendly" compounds. For example, slowly biodegradable pesticides are being replaced by rapidly biodegradable ones, which are used in conjunction with integrated pest management approaches (see Chapter 14). This approach means that pesticides are not used on a yearly basis, but instead are rotated. Thus, on the one hand, target insects do not become fully acclimated to these easily degraded pesticides, but on the other hand, soil microorganisms degrade them rapidly so that they are active only during the intended time frame.

9.4 BIODEGRADATION PATHWAYS

The vast majority of the organic carbon available to microorganisms in the environment is material that has been photosynthetically fixed (plant material). Anthropogenic activity has resulted in the addition of many industrial and agricultural chemicals, including petroleum products, chlorinated solvents, and pesticides (see Information Box 9.1 and

Chapter 12). Many of these chemicals are readily degraded in the environment because of their similarity to photosynthetically produced organic material. This allows degrading organisms to utilize preexisting biodegradation pathways. However, some chemical structures are unique or have unique components, which result in slow or little biodegradation. To help understand and predict biodegradation of organic contaminants in the environment, one can classify organic contaminants into one of three basic structural groups: the aliphatics, the alicyclics, and the aromatics (see Fig. 9.7). Constituents of each of these groups can be found in all three physical states—gaseous, solid, and liquid. The general degradation pathways for each of these structural classes have been delineated. These pathways differ for aerobic and anaerobic conditions and can be affected by contaminant structural modifications.

9.4.1 Biodegradation Under Aerobic Conditions

In the presence of oxygen, many heterotrophic microorganisms rapidly mineralize organic compounds (see Chapter 5). During degradation, some of the carbon is completely oxidized to carbon dioxide to provide energy for growth, and some carbon is used as structural material in the formation of new cells (see Fig. 9.1). Energy used for growth is produced through a series of oxidation–reduction (redox) reactions in which oxygen is used as the TEA and reduced to water (see Fig. 9.6).

9.4.1.1 Aliphatic Hydrocarbons

Aliphatic hydrocarbons are straight-chain and branched-chain structures. Most aliphatic hydrocarbons introduced into the environment come from industrial solvent waste and the petroleum industry. Liquid aliphatics readily degrade under aerobic conditions, especially when the number of carbons is between 8 and 16. Longer chain aliphatics are usually waxy substances. Biodegradation of these longer chains is slowed due to limited water solubility, while biodegradation of shorter chains may be impeded by the toxic effects of the short-chain aliphatic on microorganisms. In addition, several common structural modifications can result in severely reduced biodegradation. One of these modifications is extensive branching in the hydrocarbon chain (see Fig. 9.10). Commonly found in petroleum, branched hydrocarbons constitute one of the slowest degraded fractions therein. Another common modification that slows biodegradation is halogen substitution, as seen in TCE and eight other compounds found on the ATSDR top twenty list (Information Box 9.1). The ATSDR list contains a variety of compounds (both aliphatic and nonaliphatic) that not only have limited biodegradability but also pose serious toxicity problems. Thus both the rate of biodegradation and toxicity must be considered in evaluating the potential hazard of pollutants in the environment.

Biodegradation of aliphatic compounds generally occurs by one of the three pathways shown in Fig. 9.11. The most common is a direct enzymatic incorporation of molecular oxygen O_2 (Pathway I). All three of these pathways result in the formation of a common intermediate—a primary fatty acid. The fatty acid formed in the degradation of an alkane is subject to normal cellular fatty acid metabolism. This includes β-oxidation, which cleaves off consecutive two-carbon fragments. Each two-carbon fragment is removed by coenzyme A (CoA) as acetyl-CoA, and shunted to the tricarboxylic acid (TCA) cycle for complete degradation to CO_2 and H_2O. If the alkane has an even number of carbons, acetyl-CoA is the last residue. If the alkane has an odd number of carbons, propionyl-CoA is the last residue, which is also shunted to the TCA cycle after conversion to succinyl-CoA.

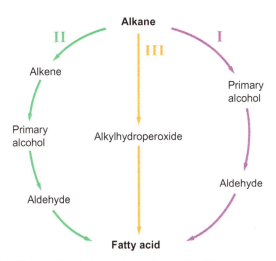

FIG. 9.11 Aerobic biodegradation pathways for aliphatic compounds. *(From Pollution Science © 1996, Academic Press, San Diego, CA.)*

We know that both branching and halogenation slow biodegradation. In the former case, we can see that extensive branching causes interference between the degrading enzyme and the enzyme-binding site. In the latter case, however, we need to know something about the bonds and the reactions involved. For halogenated compounds, the relative strength of the carbon–halogen bond requires two things: (1) an enzyme that can act on the bond and (2) a large input of energy to break the bond. In general, mono-chlorinated alkanes are considered degradable; however, increasing halogen substitution results in increased inhibition of degradation (and a lower amount of net energy gained from degradation). Although increasing halogenation generally slows degradation, aerobic oxidation of highly chlorinated aliphatics can occur cometabolically.

9.4.1.2 Aromatic Hydrocarbons

The *aromatic hydrocarbons* contain at least one unsaturated ring system with the general structure C_6R_6, where R is any functional group (see Fig. 9.7). The parent hydrocarbon of this class of compounds is benzene (C_6H_6), which exhibits the *resonance*, or delocalization of electrons, typical of unsaturated cyclic structures. Owing to its resonance energy, benzene is remarkably inert. (*Note*: As a group, the benzene-like "aromatics" tend to have characteristic aromas—hence the name.)

Aromatic compounds—including *polyaromatic hydrocarbons (PAHs)*, which contain two or more fused benzene rings—are synthesized naturally by plants. For example, they serve as a major component of lignin, a common plant polymer. Release of aromatic compounds into the environment occurs as a result of such natural processes as forest and grass fires. The major anthropogenic sources of aromatic compounds are fossil-fuel processing and utilization (burning). For example, benzene is one component

of gasoline that is often released into the environment; it is of particular concern because it is, like many PAHs, a carcinogen.

Aromatic compounds, especially PAHs, are characterized by low water solubility and are therefore very hydrophobic. As is common with hydrophobic compounds, aromatics are often found sorbed to soil and sediment particles. The combination of low solubility and high sorption results in low substrate bioaccessibility and slow biodegradation rates. This is particularly true for PAHs having three or more rings because water solubility decreases as the number of rings increases. Thus, in general, PAHs having two or three condensed rings are biodegraded rapidly in the environment, often mineralizing completely, whereas PAHs with four or more condensed rings are transformed much more slowly, often as a result of cometabolic attack.

A wide variety of bacteria and fungi can degrade aromatic compounds. Under aerobic conditions, both groups of microbes incorporate oxygen as the first step in biodegradation; however, these two groups of microbes use different pathways, as shown in Fig. 9.12. Bacteria use a *dioxygenase* enzyme that incorporates both atoms of molecular oxygen into the PAH to form a stereo-specific *cis*-dihydrodiol, and then the common intermediate catechol. The ring is then cleaved by a second dioxygenase using either an ortho or a meta pathway. Which of these pathways is used is organism specific. Close examination of the ring cleavage products shows molecules that have a fatty acid character and that can now be degraded using normal cellular fatty acid metabolism, as explained for aliphatics.

In contrast to bacteria, fungi degrade aromatic compounds using a *monooxygenase* enzyme that incorporates only one atom of molecular oxygen into the PAH and reduces the second oxygen to water. The result is the formation of an arene oxide, followed by the enzymatic addition of water to yield a stereo-specific *cis*-dihydrodiol and then the common intermediate, catechol (see Fig. 9.12). This catechol is then mineralized completely as for bacteria. Alternatively, the arene oxide can be isomerized to form a phenol, which can be conjugated with sulfate, glucuronic acid, glucose, or glutathione. These conjugates are similar to those formed in higher organisms, such as humans, and seem to aid in detoxification and elimination of PAH.

9.4.1.3 Alicyclic Hydrocarbons

Alicyclic hydrocarbons are saturated carbon chains that form ring structures (see Fig. 9.7). Naturally occurring alicyclic hydrocarbons are common. For example, alicyclic hydrocarbons are a major component of crude oil, comprising 20%–67% by volume. Other examples of complex, naturally occurring alicyclic hydrocarbons include camphor, which is a plant oil; cyclohexyl fatty acids, which

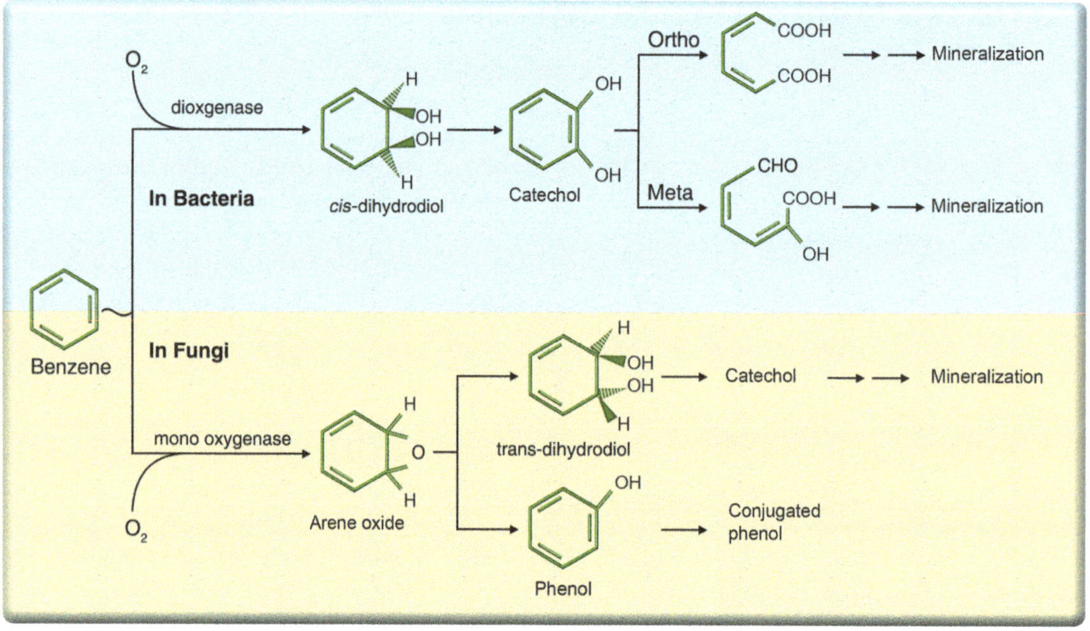

FIG. 9.12 Aerobic biodegradation pathways of aromatic compounds in bacteria and fungi. *(From Pollution Science © 1996, Academic Press, San Diego, CA.)*

are components of microbial lipids; and the paraffins from leaf waxes. Anthropogenic sources of alicyclic hydrocarbons to the environment include fossil-fuel processing and oil spills, as well as the use of such agrochemicals as the pyrethrin insecticides.

It is very difficult to isolate pure cultures of bacteria that can degrade alicyclic hydrocarbons. For this reason, biodegradation of alicyclic hydrocarbon degradation is thought to take place as a result of teamwork among mixed populations of microorganisms. Such a team is commonly referred to as a *microbial consortium*. Another unique aspect of alicyclic degradation is the formation of a lactone ring during one of the biodegradation steps that is one larger than the original ring. For example, in the degradation of cyclohexane, a six-membered ring, we get the formation of ε-caprolactone, a seven-membered ring. Thus, in the degradation of cyclohexane, one population in the microbial consortium performs the first two degradation steps, cyclohexane to cyclohexanone via cyclohexanol, but is unable to lactonize and open the ring. Subsequently, a second population in the consortium, which cannot oxidize cyclohexane to cyclohexanone, performs the lactonization and ring-opening steps, and then degrades the compound completely (Fig. 9.13).

Interestingly, cyclopentane and cyclohexane derivatives, which contain one or two hydroxyl, carbonyl, or carboxyl groups, degrade more readily in the environment than do their parent compounds. In fact, microorganisms capable of degrading cycloalkanols and cycloalkanones are ubiquitous in environmental samples.

9.4.2 Biodegradation Under Anaerobic Conditions

Anaerobic conditions are not uncommon in the environment. Most often, such conditions develop in water or saturated sediment environments. But even in well-aerated soils there are microenvironments with little or no oxygen. In all of these environments, *anaerobiosis* occurs when the rate of oxygen consumption by microorganisms is greater than the rate of oxygen diffusion through either air or water. In the absence of oxygen, organic compounds can be mineralized through *anaerobic respiration*, in which a TEA other than oxygen is used (see Fig. 9.6). The series of alternative TEAs in the environment includes iron, nitrate, manganese, sulfate, and carbonate, which are listed in order from most oxidizing to most reducing conditions. This progression means they are usually utilized in this order because the amount of energy generated for growth depends on the oxidation potential of the TEA. Since none of these TEAs are as oxidizing as oxygen, growth under anaerobic conditions is never as efficient as growth under aerobic conditions although it can come close for TEAs such as iron or nitrate (Fig. 9.6; see also Chapter 5).

Interestingly, many compounds that are easily degraded aerobically, such as saturated aliphatics, are far more difficult to degrade anaerobically. However, as already discussed, in at least one group of compounds—those that are highly chlorinated—the chlorine substituents are removed more rapidly under anaerobic conditions. But once dechlorination has occurred, the remaining molecule

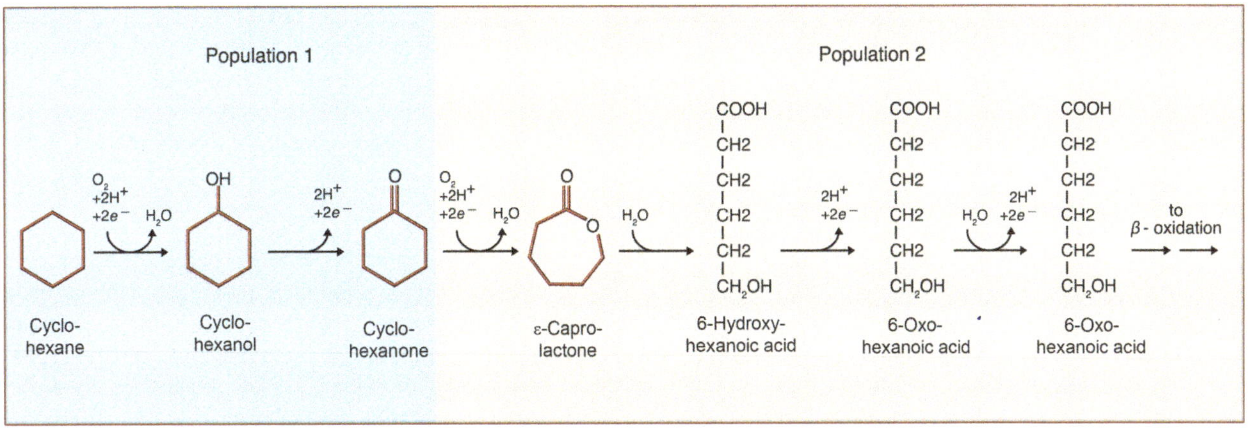

FIG. 9.13 Aerobic biodegradation of cyclohexane by a microbial consortium. *(From Pollution Science © 1996, Academic Press, San Diego, CA.)*

behaves more typically; that is, it is generally degraded more rapidly and extensively aerobically than anaerobically. As a consequence of this sequential process, technologies have been developed that utilize sequential anaerobic–aerobic treatments to optimize degradation of highly chlorinated compounds.

9.4.2.1 Aliphatic Hydrocarbons

Saturated aliphatic hydrocarbons are degraded slowly, if at all, under anaerobic conditions. In general, the longer the hydrocarbon chain, the more likely biodegradation will occur although very slowly. Specifically, methane (CH_4) is not degraded, hexane (C_6H_{14}) may be very slowly or not degraded, and hexadecane ($C_{16}H_{34}$) is slowly degraded. We see evidence of this slow to nonexistent degradation in nature; for example, hydrocarbons in natural underground reservoirs of oil and methane (which are under anaerobic conditions) are not degraded, despite the presence of microorganisms. However, both unsaturated aliphatics and oxygen-containing aliphatics (aliphatic alcohols and ketones) are readily biodegraded anaerobically using a variety of TEAs. The suggested pathway of biodegradation

for a saturated hydrocarbon is the addition of the 4-carbon molecule fumarate, which forms a fatty acid intermediate (Fig. 9.14).

Chlorinated aliphatics can be partially or completely degraded under anaerobic conditions, but the mechanism is very dependent on the actual chlorinated compound in question. For example, C_1 molecules such as chloromethane (CH_3Cl) and dichloromethane (CH_2Cl_2) can support the growth of anaerobic microbes. These microbes first remove the chlorine using a *dehalogenase* enzyme, leaving a methyl group that is oxidized in a complex series of reactions to provide energy for growth. In contrast, C_2 molecules such as chloroethane (CH_3-CH_2Cl) do not support microbial growth under anaerobic conditions. However, for highly chlorinated compounds, *reductive dehalogenation* is used to remove chlorines (Fig. 9.9). In this case, the dechlorination reaction may be cometabolic or linked to respiration, a process called *halorespiration*, in which the chlorinated aliphatic acts as a TEA and the electron donor is either H_2 or a C_1 or C_2 carbon organic compound such as ethanol (CH_3CH_2OH). Halorespiration results in a compound with a reduced number of chlorine atoms that is now more amenable to aerobic

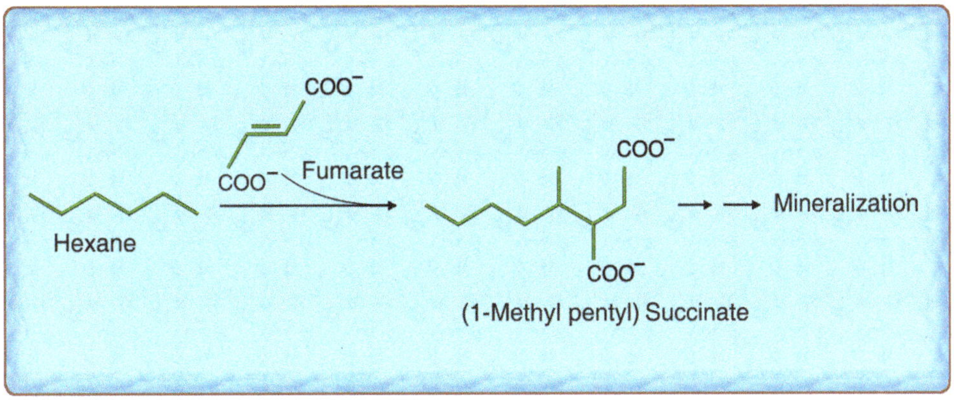

FIG. 9.14 General anaerobic biodegradation pathway of an alkene.

biodegradation. A good example is the common ground-water contaminant perchloroethene (PCE). PCE is not known to degrade at all under aerobic conditions; however, as shown in Fig. 9.9, PCE readily undergoes reductive dehalogenation. In cometabolic reductive dehalogenation, the process may be mediated by reduced transition-metal/metal complexes. The steps in this transformation are shown in Fig. 9.15. In the first step, electrons are transferred from the reduced metal to the halogenated aliphatic, resulting in an alkyl radical and free halogen. Then, the alkyl radical

can either scavenge a hydrogen atom (I) or lose a second halogen to form an alkene (II). In general, anaerobic conditions favor the degradation of highly halogenated compounds, while aerobic conditions favor the degradation of mono- and disubstituted halogenated compounds.

9.4.2.2 Aromatic Hydrocarbons

Recent evidence indicates that nonsubstituted aromatics like benzene can be degraded only very slowly under anaerobic conditions. However, like aliphatic hydrocarbons, substituted aromatic compounds can be rapidly and completely degraded under anaerobic conditions (Fig. 9.16). Anaerobic mineralization of aromatics often requires a mixed microbial community whose populations work together under different redox potentials. For example, mineralization of benzoate can be achieved by growing an anaerobic benzoate degrader in coculture with an aerobic methanogen or sulfate reducer. In this consortium, benzoate is transformed by one or more anaerobes to yield aromatic acids, which in turn are transformed to methanogenic precursors such as acetate, carbon dioxide, or formate. These small molecules can then be utilized by methanogens (Fig. 9.17). This process can be described as an anaerobic food chain because the organisms higher in the food chain cannot utilize acetate or other methanogenic precursors, while the methanogens cannot utilize larger molecules such as benzoate. Methanogens utilize

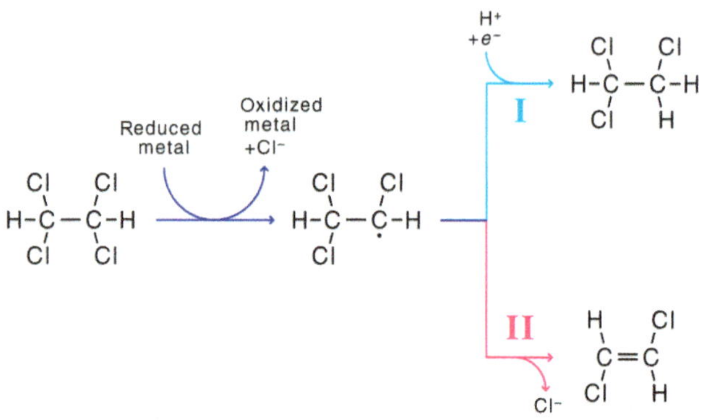

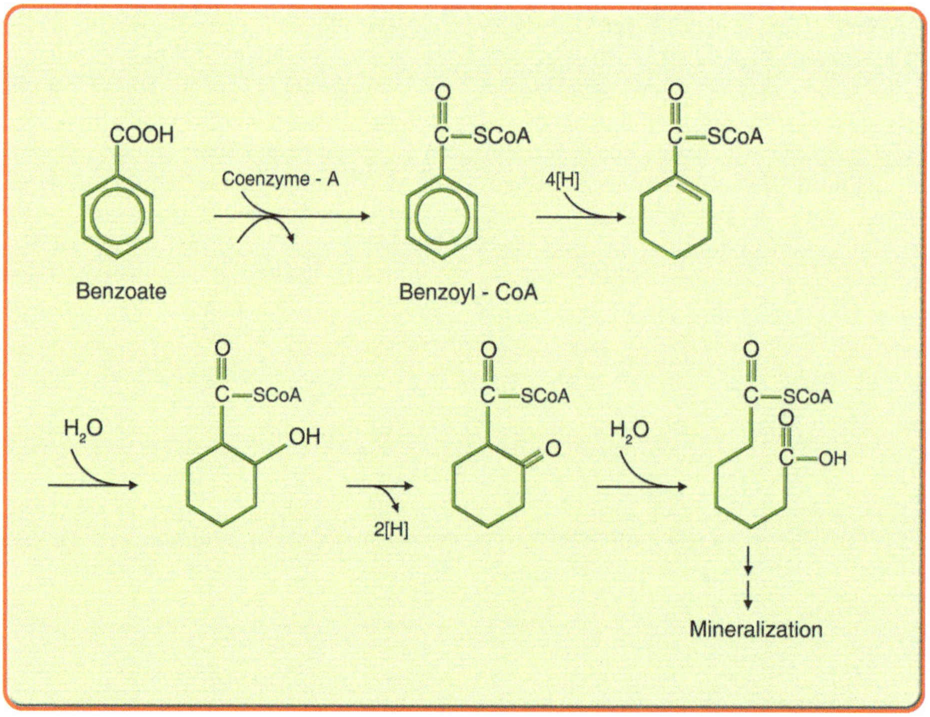

FIG. 9.15 Cometabolic reductive dehalogenation of a chlorinated hydrocarbon in the presence of a metal to form an alkyl radical. (I) The alkyl radical scavenges a hydrogen atom. (II) The alkyl radical loses a second halogen to form an alkene. *(From Pollution Science © 1996, Academic Press, San Diego, CA.)*

FIG. 9.16 Anaerobic biodegradation of benzoate. Note that the intermediate benzoyl-CoA is a common intermediate found in the anaerobic degradation of aromatic compounds. Further note (in contrast to aerobic conditions) that anaerobic microbes completely saturate the ring during biodegradation before it is opened.

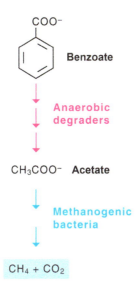

FIG. 9.17 An anaerobic food chain. Shown is the formation of simple compounds from benzoate by a population of anaerobic bacteria and the subsequent utilization of the newly available substrate by a second anaerobic population, the methanogenic bacteria. *(From Pollution Science © 1996, Academic Press, San Diego, CA.)*

carbon dioxide as a terminal electron acceptor, thereby forming methane. (*Note*: Methanogens should not be confused with methanotrophic bacteria, which aerobically oxidize methane to carbon dioxide.)

9.5 TRANSFORMATION OF METAL POLLUTANTS

Metals compose a second important class of pollutants (see Information Box 9.1). However, metals are also essential components of microbial cells. For example, sodium and potassium regulate gradients across the cell membrane, while copper, iron, and manganese provide metalloenzymes for photosynthesis and electron transport. On the other hand, metals can also be extremely toxic to microorganisms. Although the most toxic metals are the nonessential metals such as arsenic, cadmium, lead, and mercury, even essential metals can become toxic in high concentrations.

What is the fate of metals in the environment? Metals and metal-containing contaminants are not degradable in the same sense that carbon-based molecules are for two reasons. First, unlike carbon, the metal atom is not the major building block for new cellular components. Second, while a significant amount of carbon is released to the atmosphere in gaseous form as carbon dioxide, the metals rarely enter the environment in a volatile, or gaseous, phase. (There are some exceptions to this—most notably, mercury and selenium, which can be transformed and volatilized by microorganisms under certain conditions.) In general, however, the nondegradability of metals means that it is difficult to eliminate metal atoms from the environment. Therefore localized, elevated levels of metals are common, especially in industrially developed countries. Consequently, these metals can accumulate in biological systems, where their toxicity poses serious threats to human and ecological health.

Metals and metal-containing molecules can undergo transformation reactions, many of which are mediated by microorganisms. The nature of these reactions is important for consideration of metal toxicity in the environment, because toxic effects more often depend on the form and bioaccessibility of the metal than on the total metal concentration. In general, the most active form of added metals are free metal ions. The metals having the highest toxicity are the cations of mercury (Hg^{2+}) and lead (Pb^{2+}), although other metallic cations (arsenic, beryllium, boron, cadmium, chromium, copper, nickel, manganese, selenium, silver, tin, and zinc) also exhibit significant toxic effects. (The specific toxicities of selected metals to humans are discussed in Chapter 28.)

9.5.1 Effects of Metals on Microbial Metabolism

The nature of the interaction between heavy metals and microorganisms is complex. Metal toxicity requires uptake of the metal by a cell, which is dependent on many factors such as pH, soil type, and temperature. For example, accumulation of metals by cell-surface binding increases with increasing pH. Transport of metals into a cell is also pH dependent, with maximal transport rates in the pH range 6–7. Once a metal is taken up by a cell, toxicity can result. After an initial period in which the toxic effects of the metals are evident, microorganisms often acquire tolerance mechanisms that enable them to repair metal toxicity damage, after which they can to start metabolizing and growing again at a nearly normal rate. The length of time required to develop tolerance mechanisms is influenced by both biotic and abiotic factors. The biotic factors of importance may involve the physiological state of the organism in question, such as the nutritional level, or genetic adaptations that result in metal resistance. Abiotic factors include the physicochemical characteristics of the environment such as pH, temperature, and redox potential, all of which affect the precipitation and complexation of metals.

The specific toxic effects of heavy metals on microorganisms are caused by the binding of the metal to cellular ligands such as proteins or nucleic acids. This metal–ligand binding leads to conformational changes and loss of normal ligand activity. For example, the particularly strong affinity of cationic metals for protein sulfhydryl groups can lead to alterations in protein folding. Both the ligand structure and the size of the metal affect the type binding. Large metal ions such as copper, silver, gold, mercury, and cadmium preferentially form covalent bindings with sulfhydryl groups. In contrast, small, highly electropositive metal ions such as aluminum, chromium, cobalt, iron, titanium, zinc, and tin preferentially complex with carboxyl, hydroxyl, phosphate, and amino groups.

9.5.2   Microbial Transformations of Metals

Microorganisms have developed various resistance mechanisms to prevent metal toxicity: among these mechanisms we will briefly discuss metal oxidation/reduction, metal complexation, and alkylation of metals.

Oxidation/reduction: Metal oxidation enhances metal mobility, stimulating metal movement away from the cell. For example, the Gram-positive bacterium *Bacillus megaterium* can oxidize elemental selenium to selenite, a reaction that increases selenium mobility. Alternatively, some microorganisms reduce metals such as chromium, causing them to precipitate and become immobilized. Harnessing microbial reduction processes is one strategy that is being used to immobilize metals and radionuclides within a site (Chapter 19).

Complexation: Other microorganisms can effectively complex metals to polymeric materials either internal or external to the cell. For example, uranium has been found to accumulate extracellularly as needle-like fibrils in a layer approximately 2 μm thick on the surface of a yeast, *Saccharomyces cerevisiae*. In contrast, uranium accumulates as dense intracellular deposits in *Pseudomonas aeruginosa*.

Alkylation: Some microorganisms can transform metals by alkylation, which involves the transfer of one or more organic ligand groups (e.g., methyl groups) to the metal, thus affording stable organometallic compounds. One such metal is mercury, as shown in Fig. 9.18. The physical and chemical properties of organometals are different from those of pure metals. For example, volatility and solubility are increased, thus facilitating the movement of the organometal through the soil solution and, ultimately, into the atmosphere. In general, the extent of organometallic mobility depends on the nature and number of the organic ligands involved. But the presence of even a single methyl group can significantly increase both the volatility and *lipophilicity* (the affinity for lipids) of a metal. Transformations that increase volatility allow microorganisms to help remove a toxic metal from its environment; however, the concomitant increase in lipophilicity can cause biomagnification of the metal, resulting in toxic effects on other members of the ecosystem.

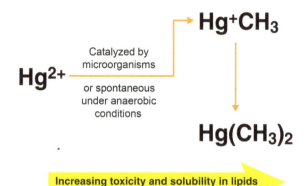

FIG. 9.18   Microbial alkylation of mercury. *(From Pollution Science © 1996, Academic Press, San Diego, CA.)*

QUESTIONS AND PROBLEMS

1. Define biodegradation, making reference to the terms transformation, mineralization, and cometabolism

2. (a) Consider *n*-octane, an eight-carbon straight-chain aliphatic compound. Draw its structure.
   (b) Is this compound biodegradable?
   (c) Beginning with the structure you have drawn, show how you can alter it to make it less biodegradable.
   (d) Compare the structure in part (a) with Eq. (9.1) for biodegradation shown in Section 9.2, and consider the following situation: A site is contaminated by a leaking underground gasoline storage tank. The remediation firm that you work for would like to use bioremediation to clean the site. Given the structure of gasoline components (which typically include simple aliphatic, alicyclic, and aromatic compounds), what other nutrients may be required to complete bioremediation? Explain.

3. List and explain the factors that determine bioaccessibility of an organic compound.

4. Compare the following structures:
   (a) Predict the order of bioaccessibility.
   (b) Predict the order of biodegradability.
   (c) What does a comparison of (a) and (b) tell you about the relationship between bioaccessibility and biodegradability?
   (d) Which is the most likely type of biodegradation for each of the above compounds?

REFERENCE

Atlas, R.M., Bartha, R., 1993. Microbial Ecology, third ed. Benjamin Cummings, Menlo Park, CA.

Part II

Environmental Pollution

The Role of Environmental Monitoring in Pollution Science

J.F. Artiola and M.L. Brusseau

Yellow Stone National Park. *Photo courtesy J.F. Artiola.*

10.1 INTRODUCTION

Environmental Monitoring is based on scientific observations of changes that occur in our environment. Scientists need to observe changes to study the dynamics of not only natural cycles but also anthropogenic-based impacts. The effects of pollution in the environment, as in humans, can be slow (chronic) and may require multiple observations over time, or they can be fast acting (acute), and be assessed with simultaneous observations. The effects of pollution occur at all spatial scales; therefore observations are also made at multiple scales of space. However, since the environment is a continuum, observations must be made using physical, chemical, and biological methods. Only science-based observations and data collection, statistically based data processing, and objective interpretations can produce unbiased information that leads to knowledge and a level of understanding required to evaluate and wisely address environmental challenges from local to global scales. There are numerous examples of knowledge-based regulations that benefit modern society and protect the environment from pollution. These include waste management regulations involving disposal, treatment, or reuse; regulations governing the protection of water resources including natural and public water supplies; air quality regulations; and regulations protecting endangered species.

There are numerous agencies and world institutions involved in environmental monitoring. In the United States, pollution monitoring and prevention is the primary focus of the Environmental Protection Agency. This agency is mandated to develop and enforce laws and regulations that are protective of our health and the environment.

10.2 SAMPLING AND MONITORING

10.2.1 Basics

Developing a program to monitor the extent or effects of pollution in a particular environment requires careful consideration of the following: (i) The *purpose* of the monitoring program—these include determining, for example, pollution-level changes in space and time; (ii) the

Environmental and Pollution Science. https://doi.org/10.1016/B978-0-12-814719-1.00010-0

INFORMATION BOX 10.1 Data Quality Objectives of a Sampling Plan

Data Quality Control and Objectives are needed in a sampling plan. Although the following requirements are reproduced from US EPA pollution monitoring guidelines, these are generic enough that they should be included in any type of environmental sampling plan.

> *Quality*: Discusses statistical measures of:
>
> > *Accuracy (bias)*: Determines how data will be compared to reference values when known. Estimate overall bias of the project based on criteria and assumptions made.
> >
> > *Precision*: Discusses the specific (sampling methods, instruments measurements) variances and overall variances of the data or data sets when possible using relative standard deviations (%CV).
>
> *Defensible*: Insures that sufficient documentation is available after the project is complete to trace the origins of all data.
>
> *Reproducible*: Insures that the data can be duplicated by following accepted sampling protocols, methods of analyses, sound statistical evaluations, and so on.
>
> *Representative*: Discusses the statistical principles used to ensure that the data collected represents the environment targeted in the study.
>
> *Useful*: Insures that the data generated meets regulatory criteria and sound scientific principles.
>
> *Comparable*: Shows similarities or differences between this and other data sets, if any.
>
> *Complete*: Addresses any incomplete data and how this might affect decisions derived from these data.

Adopted from Artiola et al., 2004 (Information Box 2.6).

objectives—which may include specific chemical, physical, or biological analyses; and the *approach*, which will assist in defining the number and types of measurements. To achieve our objectives, we must also consider the environmental characteristics or uniqueness of each environment. Finally, we must consider sampling methods including locations, timing, and type. There are several types of sampling methods; *random* sampling, for example, assumes that all units from an environment have an equal chance of being selected. Another statistically valid approach is *systematic* sampling, which selects sampling locations at predefined intervals in space or time.

Detailed information is needed to ensure that all aspects of sampling plans are thoroughly described. The US EPA defines seven critical elements of a sampling plan to ensure that all the data quality objectives are met and these are described in Information Box 10.1. In the development of sampling plans, environmental scientists must be thoroughly familiar with types of sampling such as destructive and nondestructive methods. They should also know the

accepted methods of analyses by consulting the latest scientific and regulatory reference manuals and books. Environmental scientists should also be familiar with the basic concepts and applications of analysis and measurement principles.

All data must be reported with appropriate degree of precision by considering the sampling methods, instrumental methods, and the number of samples analyzed. Typically, the more uncertainty associated with a data value, the lower the number of digits it has. Thus a number with two significant figures is less precise than a number with four significant figures. Often, data values are derived from various data processing and analysis efforts, including statistical (such as the means of two or more measurements), method dependent (such as dilutions), and field sampling (means of two or more samples). As a rule, precision of the final value is determined by the least precise step. For example, if an instrument measurement detects $235\,\mu g\,L^{-1}$ of total Pb in a water sample, and the sample had been diluted 5 times using a pipette with only two digits of precision, the final result should be reported as $1.2\,mg\,L^{-1}$ total Pb $(0.235 \times 5.0 = 1.2)$.

It is also important to report environmental data using commonly accepted units, preferably conforming to the SI (System International), used worldwide. Unfortunately, in the United States, environmental scientists and engineers often must regularly convert units across the US customary system (USCS) and the SI, a process that often leads to errors in data reporting.

10.3 STATISTICS AND GEOSTATISTICS

Statistical methods are necessary in pollution science because it is impossible to characterize all properties of an environment everywhere all of the time. Statistics are used to select *samples* from a *population* in an unbiased manner. It also helps to interpret the data with the appropriate degree of confidence. *Descriptive statistics* are very useful in environmental science as they provide a summary of the properties or characteristics of an environment. Descriptive statistics include sample means, standard deviations, and confidence intervals (Table 10.1).

Data samples should be collected randomly or systematically from an environment. Biased sampling (usually based on convenience) will produce biased data. The range of values that can be expected from sampling an environment varies randomly, but values have a likelihood of occurrence and this is defined by a probability distribution (Fig. 10.1).

A reoccurring question in pollution science is how many samples (n) are needed to define with some statistical certainty, the extent of pollution. This is a near impossible task since pollution distributions are not usually known beforehand. However, we can estimate the number of

TABLE 10.1 Statistical Symbols and Their Definitions

Term	Symbol	Definition
Sample mean or estimated mean	$\overline{x} = \frac{\sum_{i=1}^{n} x_i}{n} = \frac{x_1 + x_2 + \cdots + x_n}{n}$	Mean estimated from the n values $x_1, x_2, ..., x_n$. (Specific cases are identified by x_{bar}, y_{bar}, y_{bar}.)
Mean	μ	Population or true mean. Limiting value of x_{bar} as n becomes large or if all possible values of x_i are included
Sample variance	$s^2 = \frac{\sum_{i=1}^{n} (x_i - \overline{x})^2}{n-1}$	Estimate of variance based on n values given by $x_1, x_2, ..., x_n$. (If all possible values of n are included, then use n in place of $n-1$ for denominator)
Variance	σ^2	True variance for the population
Standard deviation	s, σ	Defined for s^2 and σ^2 above. (Specific cases are identified by subscripts, such as s_x and s_y)
Coefficient of variation	$CV = \frac{\sigma}{\mu}$ or $\frac{s}{\overline{x}}$	A relative standard deviation. Can also be expressed as a percentage
Correlation coefficient	$r = \frac{\sum_{i=1}^{n} (X_i - \overline{X})(Y_i - \overline{Y})}{(n-1)s_X s_Y} = \frac{b s_X}{s_Y}$	Linear relationship between two variables measured n times. The data pairs are $(X_1, Y_1), (X_2, Y_2), ... (X_n, Y_n)$. Range of r is -1 to $+1$
Coefficient of determination	r^2	Square of r (above) for linear correlation. Range is 0 to 1
Slope	$b = \frac{\sum_{i=1}^{n} (X_i - \overline{X})(Y_i - \overline{Y})}{(n-1)s_X^2}$	Slope for linear regression for n data pairs
Intercept	$a = \overline{Y} - b\overline{X}$	Y-intercept for linear regression for n data pairs
Predicted value	$Y = a + bX$	Predicted value of dependent variable Y

Modified from *Pollution Science*, 1996.

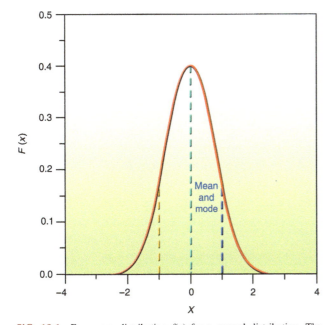

FIG. 10.1 Frequency distribution $f(x)$ for a normal distribution. The dashed lines show the center and $\pm$ standard deviation (σ) from the mean. *(From Environmental Monitoring and Characterization. Elsevier Academic Press, San Diego, CA, 2004.)*

samples needed assuming that the sample standard deviation (s) from previous studies is equivalent to the population standard deviation (σ), and that the sample mean (x) is equivalent to the population mean (μ). Also, we must assume a *normal distribution* (see Fig. 10.1) of values and be willing to accept that our sample mean value will be within a specific confidence interval.

Thus the number of samples (sample size n) needed may be defined as:

$$n = \left[(z_{\alpha/x} \alpha) / d \right]^2$$

where:

$z_{\alpha/x}$ is the probability function value for a given confidence interval (e.g., $z_{\alpha/x} \simeq 2$ for a 95% confidence interval and $n > 30$), and

d is the maximum tolerance or error acceptable of the mean value (x).

See Chapter 3 of Artiola et al., 2004 for a complete description of sample size estimation.

Inferential statistics are also widely used in pollution science to help infer or interpolate information from a partial or incomplete set of data. For example, regression

TABLE 10.2 Calibration of a Spectrophotometer to Measure Aluminum in an Unknown Sample

(AL) in Standard Solution (mg L^{-1})	Spectrophotometer Response (intensity units)
0.100	4142
0.500	17,315
2.00	63,305
5.00	161,486
10.0	320,087
Unknown sample	250,090

Statistical Evaluation

Linear regression equation	$y = a + bx$
y	Instrument response
x	[Al]
a	803.4
b	3.195×10^4
r^2	0.9999
[Al] in the unknown sample	7.80 mg L^{-1}

From *Pollution Science* © 1996, Academic Press, San Diego, CA.

methods can be used to correlate one or more sample attributes with a measurable quantity. This is commonly the case in analytical chemistry where the response of an instrument (measure as light emitted or absorbed) is correlated with a particular parameter (such as concentration). For example, to calibrate a spectrophotometer for aluminum measurement, we would first dilute a certified standard several times to make a series of known solutions, as listed in Table 10.2. When we analyze these solutions on the spectrophotometer, we obtain a reading that corresponds to the concentration of aluminum in each solution. We can then plot the solution concentration (*x*-axis) against the instrument response (*y*-axis) and perform a regression (see Table 10.2). The resulting equation defines the best fit among the (*x*, *y*) points. Thus we have a curve of the true values for each concentration.

We may want to estimate the concentration of a pollutant at a field location that was not sampled. In this case spatial statistics can be used to estimate the concentration of this pollutant. For this application, the interpolation method is used to relate the location (coordinates) of the unknown sample to locations for which concentrations are known. The inverse distance weighting method considers the distances to locations (with unknown values) to be inversely related to the values at those locations. This and other more advanced methods like kriging are often used in the development of maps that show contour lines of pollutant distribution.

10.4 SAMPLING AND MONITORING TOOLS

Modern data collection uses automated data acquisition methods to monitor environmental variables related to pollution. For example, urban air quality is monitored using a network of automated stations that measure pollutants such as carbon monoxide (CO) and ozone (O_3) at close intervals. Coupled with weather stations, these gas monitors produce near-real-time environmental information, which is in turn used to predict future pollution-related events.

The basic components of automated data acquisition systems are shown in Fig. 10.2.

A critical component of the system is the *sensor*, which responds to an environmental stimulus such as temperature. Thermistors, for example, are able to respond to temperature changes by changing their internal resistance. The *data acquisition system (DAS)* usually includes an analog to digital converter used to convert analog (continuous) signals from the sensor to a digital (discrete) value that

FIG. 10.2 Block diagram showing the basics components of a data acquisition system (*DAS*). *ADC*, Analog to digital converter. *(From Environmental Monitoring and Characterization. Elsevier Academic Press, San Diego, CA, 2004.)*

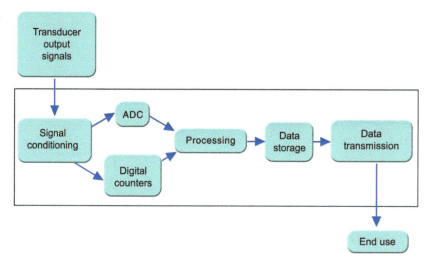

can be stored and manipulated by computer processors. Data storage and transmission systems are also needed in modern DAS to collect raw and processed data, and send it to the user.

Environmental scientists that use modern data collection systems must have a rudimentary working knowledge of electricity, computer processing and programming, and be familiar with basic laws of environmental physics and chemistry.

10.4.1 Maps

Until recently, drawn maps and aerial photographs were the only means at the disposal of environmental scientist to locate places and land features and to navigate (with a magnetic compass). Today, the use of *geographic positioning system (GPS)* of satellites makes it much easier to perform these tasks. However, since maps are small portable representations of portions of our environment, we must be familiar with their principles and types. There are three major types of maps: *planimetric*, which present scaled information in two dimensions; *topographic* maps, which add elevation information; and *thematic* maps discussed in the next section. All maps must have a scale, which is the ratio between the map's distances and the real (earth) distances. Thus most scales are a unit-less ratio or fraction of two distances. For example, if 1 cm on the map represents 1 km in reality, then the map scale is 1/100,000.

The major types of locational maps include the *latitude* and *longitude* system developed in England which divides the earth into parallel (north-south) and meridian (east-west) sections in degrees.

In the United States, the *Public Land survey System* is used exclusively to locate property lines in legal documents. This system is based on the 34 points that defined the origins of principal meridians and baselines coordinates from which townships (6 square miles) and sections (1 square mile) originate (Fig. 10.3).

Example, the filled square (point A) shown in Fig. 10.3— as representing the NW1/4 corner of section 17 of Township 1S and Range 2W (abbreviated: NW1/4, S17, T1S, R2W).

Topographic maps add relief to land maps with the use of contour lines that defined equal elevations drawn at fixed elevation intervals. The US Geologic Survey generates topographic maps which are used in many disciplines including environmental science. Contour lines are very useful in identifying unique land features such as watersheds, rivers, depressions, and mountains.

Soil survey maps combine land features (aerial photographs) with thematic soil data. These maps are very useful in determining the suitability (use) of land for certain activities, such as agriculture, land development, and septic field location. These maps provide a wealth of data on soil characteristics related to pollution (such as infiltration, soil texture, general composition), which makes them very valuable to soil and environmental scientists. An example of a Soil Survey Map is given in Fig. 10.4

Geographic Information Systems (GISs) are very useful in environmental applications because they provide a way to arrange and present layers of data or themes as maps. Thus GIS maps are decision-making tools that provide a visual representation of large amounts and types of data (see Fig. 10.5), which can be easily sorted because all the data and coordinates are in a digital form. For example, a city manager with a GIS map with data overlay of soil properties could ask the question "show which areas in district X are suitable for septic field systems."

Because GIS uses digital data that can be manipulated using Boolean operations (binary 0 and 1 values), data can be reclassified, sorted, and superimposed (data overlay). GIS data is now an accessible community resource available on the Internet for example, to search and locate parcels of land. GIS methods can now visualize environments using digital photography and data, and virtual reality 3-D modeling.

10.4.2 Remote Sensing

Remote sensing is the use of space-based sensors, usually satellite mounted and more recently drone mounted, to observe biophysical and geochemical phenomena in earth's atmosphere and land/ocean surfaces. Most remote sensors measure light from different parts of the electromagnetic spectrum that is reflected from earth's atmosphere and surface. Regions of the electromagnetic spectrum that are commonly used are listed in Table 10.3.

The UV region is commonly used to measure air pollutants (see Chapter 17). But often these gases and others like water vapor, CO, and CH_4 can interfere with land surface radiation measurements. Remote sensing imagery is captured using digital cameras that have a finite optical pixels and therefore spatial resolution in terms of surface area/pixel. Sensors are rated in terms of coarse, medium, or fine resolution. For example, high resolution satellite-based sensors are able to collect data ranging in resolution $\sim$100–<1 m in resolution. But drone-based sensors are now able to resolve <1 cm surface area/pixel over large ground areas thanks to their very low flying capabilities, short focal length lenses, and multimegapixel sensors. These sensors are commonly used to study land cover patterns and changes that result from human activities including agriculture, plant disease, urbanization, or deforestation. Coarse resolution sensors provide data ranging from 100 to <1000 m in resolution and are useful to study cloud cover, dust, and sediment transport over large continent-size areas (Fig. 10.6).

Other remote sensing applications include observing and measuring vegetation, snow cover changes, weather

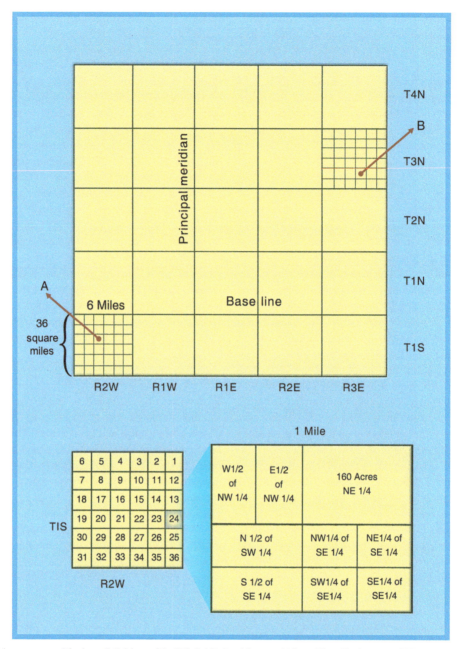

FIG. 10.3 Townships, ranges, and further subdivisions of the U.S. Public Land Survey. *(Adopted from Environmental Monitoring and Characterization. Elsevier Academic Press, San Diego, CA, 2004.)*

changes and forecasting, and physical land degradation including erosion. Satellite and drone imagery can also be used to study pollution migration from waste disposal sites. For example, multispectral data were collected from abandoned mines and other open waste disposal sites using remote sensing. Since minerals have unique spectral signatures such that they emit light in different areas of the spectrum, areas with minerals, such as metal sulfides, that are prone to fast weathering and thus higher potential to release metals into the environment can be identified and mapped.

10.5 SOIL AND VADOSE ZONE SAMPLING AND MONITORING

The soil and vadose zones are defined as regions of porous materials below land surface that are usually not fully saturated with water. Soils, being exposed to the atmosphere, are composed of weathered, heterogeneous, unconsolidated minerals, and biological organic matter including plants, animals, and microorganisms. The heterogeneity of soils, which exists at all scales from landscape to soil particle size, arises from five soil forming processes: parent material,

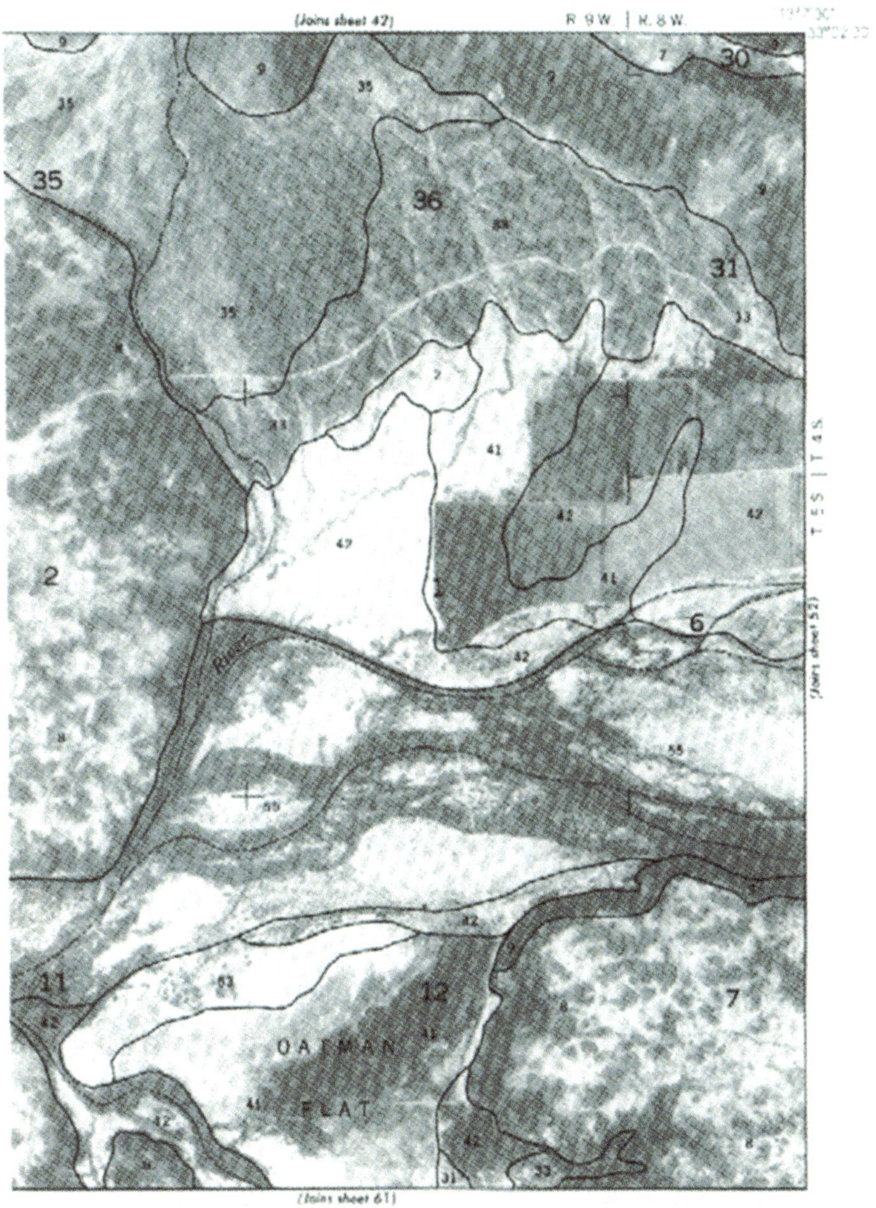

FIG. 10.4   A portion of a detailed soil map showing an area along the Gila River northwest of Gila Bend, Arizona. *(From U.S. Department of Agriculture and Natural Resources Conservation Service, 1997. Soil Survey of Gila Bend-Ajo Area, Arizona, Parts of Maricopa and Pima Counties. Issued May, 1997.)*

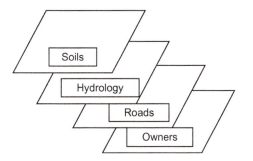

FIG. 10.5   Combining data layers in a geographic information system. *(From Environmental Monitoring and Characterization. Elsevier Academic Press, San Diego, CA, 2004.)*

climate, topography, biological activity, and time (see Chapter 2). Under the influence of all these factors, soils develop unique physical, chemical, and biological characteristics that have been used to classify soils. The USDA Soil Taxonomy (ST) consists of 12 orders according to their degree of weathering and the major diagnostic horizons. (For a detailed discussion on soil classification, see Brady and Weil, (1996).) Horizons in soils appear as the result of two factors, deposition and weathering. For example, clay mineral-rich (argillic) horizons develop as the result of intense chemical weathering conditions (high rainfall and heat) and the translocation of clay particles to

TABLE 10.3 Regions of the Electromagnetic Spectrum Used in Environmental Monitoring

Spectral Region	Wavelengths	Application
Ultraviolet (UV)	0.003–0.4 μm	Air pollutants
Visible (VIS)	0.4–0.7 μm	Pigments, chlorophyll, iron
Near infrared (NIR)	0.7–1.3 μm	Canopy structure, biomass
Middle infrared (MIR)	1.3–3.0 μm	Leaf moisture, wood, litter
Thermal infrared (TIR)	3–14 μm	Drought, plant stress
Microwave	0.3–300 cm	Soil moisture, roughness

From Environmental Monitoring and Characterization. Elsevier Academic Press, San Diego, CA, 2004.

FIG. 10.6 MODIS image showing sediment transport at the mouth of the Yellow River on February 28, 2000. Soil erosion from the Loess Plateau is proceeding at a very high rate. *(Courtesy Jacques Descloitres, MODIS Land Rapid Response Team, NASA/GSFC. From Environmental Monitoring and Characterization. Elsevier Academic Press, San Diego, CA, 2004.)*

subsurface layers. The Unified Soil Classification System (USCS) is used in the engineering and geology disciplines focusing primarily on coarse versus fined grained soils' physical and mechanical properties—information that is often relevant to environmental scientists.

Soil and vadose sampling strategies must consider changes in space at the meso and micro scales, and in time

at geologic as well as daily-hourly time intervals. Changes like physical weathering that occur slowly, for example, can be monitored at long time intervals such as years or decades. But changes such as those induced by improper waste disposal (soil and water contamination) usually require closer sampling intervals such as days to months. Soil sampling is best accomplished by considering (a priori) known soil properties and site-specific past, present, and future uses. Since most counties in the United States have been surveyed, there is information available in Soil Survey reports on the major soil properties and distributions of soil types by county.

Polluted soils should be sampled in an unbiased matter if the source of the pollution is not known. Systematic and directional soil sampling may be more appropriate in cases where the source of pollution is known and may be correlated to spatial location (see Fig. 10.7).

There are numerous types of soil and vadose zone samplers, generally classified as handheld (manually operated) and mechanical (power assisted). Augers and sampling tubes are commonly used to collect soil root zone (~1.5 m depth) samples. Sampling tubes are more appropriate to collect soil samples for pollution-related measurements as they include an inert plastic or metal liner that is used to protect and store the sample (see Fig. 10.8).

Mechanical hollow-stem augers are commonly used to collect samples from the vadose zone. The hollow stem auger is used to drill to the sampling depth by engine power. Then a center rod that plugs the hollow stem during drilling is removed and a clean sampling tube is inserted to collect a soil sample.

Soil pore water is often sampled to monitor movement of water-soluble pollutants in the vadose zone. The use of these devices extends to industrial municipal and agricultural waste treatment and disposal sites (landfills, ponds, septic systems, wetland, irrigation). Porous cups made of ceramic or stainless steel may be inserted into the soil/vadose zone by drilling a hole, as shown in Fig. 10.9.

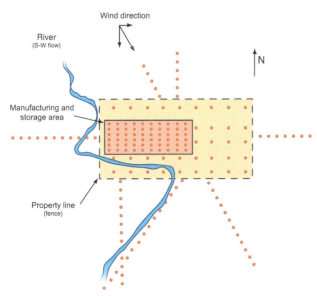

FIG. 10.7 Grid and directional (exploratory) soil sampling patterns assume a pollution point of origin. Note that sampling can be done in the direction of potential pathways of pollution (transport via air and water). Dots represent sampling points. *(From Environmental Monitoring and Characterization. Elsevier Academic Press, San Diego, CA, 2004.)*

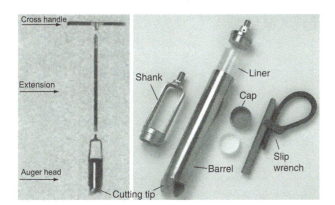

FIG. 10.8 *Left,* basic hand-operated soil auger showing the auger head, an extension rod, and a handle. *Right,* details on the auger head showing cutting tip, barrel, slip wrench, and shank for attachment to extension rods. This illustrated design allows insertion of a replaceable plastic liner and caps for collection and retention of an intact sample suitable for geotechnical studies. Alternative auger head designs are available depending on soil conditions. *(From Ben Meadows company, a division of Lab Safety Supply, Inc. From Environmental Monitoring and Characterization. Elsevier Academic Press, San Diego, CA, 2004.)*

The types of pollutants that can be monitored using these devices include salts, nitrates, and soluble organic carbon compounds.

10.6 GROUNDWATER SAMPLING AND MONITORING

Groundwater sampling is often conducted under the requirements of state or federal regulatory programs. Federal regulatory programs that include requirements or guidelines for groundwater monitoring include: (1) *Comprehensive Environmental Response Compensation and Liability Act (CERCLA)*, commonly known as the "Superfund" program that regulates characterization and cleanup at uncontrolled hazardous waste sites; (2) *Resource Conservation and Recovery Act (RCRA)* that regulates management and remediation of waste storage and disposal sites; (3) *Safe Drinking Water Act (SDWA)* that regulates drinking water quality for public water supplies and also includes the Underground Injection Control (UIC) Program and the Well Head Protection Program; and (4) *Surface Mining Control and Reclamation Act (SMCRA)* that regulates permitting for open pit mining operations. In addition to the federal regulatory programs, many states have regulatory programs that are concerned with groundwater monitoring.

Much of the groundwater sampling conducted during the past four decades has been focused on characterization and cleanup of sites where groundwater has become contaminated through spills, leaks, or land disposal of wastes (see Chapters 14 and 15). In addition to satisfying regulatory requirements, groundwater sampling programs at contaminated sites are usually conducted to obtain data necessary for making decisions on site management or cleanup. The specific objectives of the groundwater sampling program may change as the site becomes better characterized and as site remediation progresses. The frequency of sampling may also be adjusted depending on how rapidly groundwater conditions are changing at the site.

Groundwater sampling is sometimes conducted to establish baseline conditions prior to development of a new facility or to characterize ambient conditions across a large area such as in basin-wide studies. Baseline sampling is usually meant to provide a "snapshot" of groundwater conditions at a particular time. The analytical suite for baseline sampling usually includes common anions and cations and, depending on the objectives of the study, may include trace metals, pesticides, and a range of organic compounds (often those on the primary contaminant list presented in Chapter 12). If baseline sampling is being conducted prior to industrial development of a site, the analytical suite will typically include potential contaminants that may be present on the site once the facility is in operation.

Leak detection monitoring, sometimes referred to as compliance monitoring, is commonly conducted at landfills, hazardous waste storage facilities, and chemical storage and manufacturing facilities. This type of monitoring is designed to provide early warning of contamination associated with releases from these facilities. To be effective, leak detection monitoring must be conducted at wells that are properly located, generally in areas downgradient from the facility of interest. The sampling frequency and list of analytes must be adequate to detect a

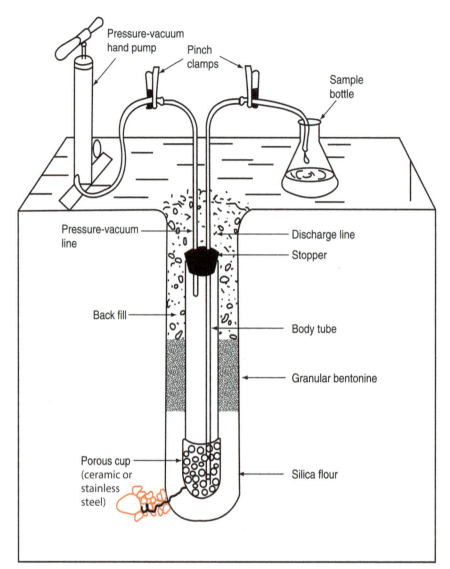

FIG. 10.9   Design and operation of a pressure-vacuum lysimeter used to collect soil pore water samples at shallow depths. The ceramic cup is fragile and can crack or can be crushed under soil pressure. *(After Fetter, C.W., 1999. Contaminant hydrogeology, second ed. Prentice Hall PTR, Upper Saddle River, NJ. From Environmental Monitoring and Characterization. Elsevier Academic Press, San Diego, CA, 2004.)*

release before the contaminants have migrated to any sensitive downgradient receptors. Typically, upgradient monitor wells are also included in leak detection monitoring to provide baseline water quality data for groundwater moving to the site, to allow for comparative evaluations.

Groundwater monitoring is typically conducted by collecting samples from a network of monitor wells. There are many site-specific variables that must be considered in the design of a monitor well network. If the monitor wells are not located or constructed appropriately, the data collected may be misleading. For example, if monitor wells are too widely spaced or improperly screened, the zones of contaminated groundwater may be poorly defined or even completely undetected. The task of designing a monitor well network is generally more difficult for sites with more

complex or heterogeneous subsurface conditions compared to sites with more homogeneous subsurface conditions. More details on this topic may be found in Chapter 8 of Artiola et al. (2004).

Although many site-specific variables must be considered in their design, most monitor wells share some common design elements (Fig. 10.10). Typically, the upper section of the well is protected with a short section of steel casing called surface casing. The surface casing protects the well casing from physical damage by vehicles or from frost heaving in cold climates. Surface casing may also be grouted to provide a seal against surface water infiltration. The well casing, usually comprised of steel or PVC (polyvinyl chloride), extends downward through the surface casing. Grout materials are installed around the outside of

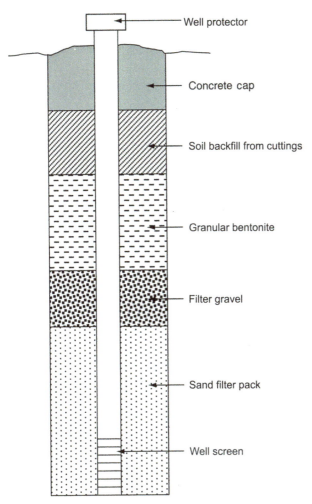

FIG. 10.10 Schematic of generic monitor well design elements. *(From Environmental Monitoring and Characterization. Elsevier Academic Press, San Diego, CA, 2004.)*

The actual process of collecting a groundwater sample, and what is done with it after collection, is an important component of the groundwater sampling program. The methods used for sample collection and processing depend on the type of contaminants present or suspected of being present. The primary concern of sample collection and processing is to maintain the integrity of the samples so that the concentrations and chemical species of analytes are the same as when the samples were collected.

10.7 SURFACE WATER SAMPLING AND MONITORING

Monitoring pollutants in water has become a global concern due to the limited and diminishing water resources and the negative impact of pollutants on ocean and fresh water sources. We have water standards that are protective of our health (National Drinking Water Standards) (see also Chapter 24) and that are protective of the environment by setting pollutant discharge limits from wastewaters into water sources (National Water Quality Criteria). Excessive nutrients in runoff waters, being addressed with new Ecoregional Nutrient Criteria, are also of concern as they can produce a biological and chemical imbalance in natural surface water systems. The quality of water used in agriculture (which accounts for most of the fresh water use in the world) is also important as increased salinity and concentrations of toxic elements reduce plant yields and degrade agricultural soils. Fig. 10.11 presents a list of the major types of pollutants found in surface water (see Chapter 16 for more details).

Sampling the surface water environment may seem trivial at first glance since often we assume that mixing of pollutants is complete. However, surface water bodies often exhibit significant spatial and temporal variability. For example, enclosed bodies of water such as lakes and reservoirs often develop water quality stratification due to temperature changes. These in turn affect oxygen saturation, pH, and total dissolved solids (TDS) by depth.

Field portable devices are commonly used to measure major water quality parameters such as pH, TDS, oxygen saturation, temperature, and turbidity (see Fig. 10.12).

Water sampling followed by analysis in the laboratory is common to monitor specific pollutants such as trace organic and inorganic chemicals (pesticides, hydrocarbons, toxic metals), and pathogens like *Escherichia coli* bacteria. Field portable water quality kits, to measure basic water quality parameters and some types of specific pollutants, are becoming popular. They are convenient and inexpensive, but lack the automation, accuracy, and detection limits of conventional laboratory testing.

the well casing to prevent movement of surface water down the borehole. Well screen is installed in the interval where the groundwater is to be monitored. In monitor wells, the well screen is typically either wire wrapped (also known as continuous slot) or slotted. Surrounding the well screen is a gravel pack, which serves to filter out fine sand and silt particles that would otherwise enter the well through the well screen.

Many groundwater sampling devices are currently commercially available. The use of appropriate sampling equipment is critical for implementation of a successful sampling program. There is no one sampling device that is ideal for all circumstances. The choice of sampling equipment is dependent on several site-specific factors, including the monitor well design and diameter, the depth to groundwater, the constituents being monitored, the monitoring frequency, and the anticipated duration of the sampling program.

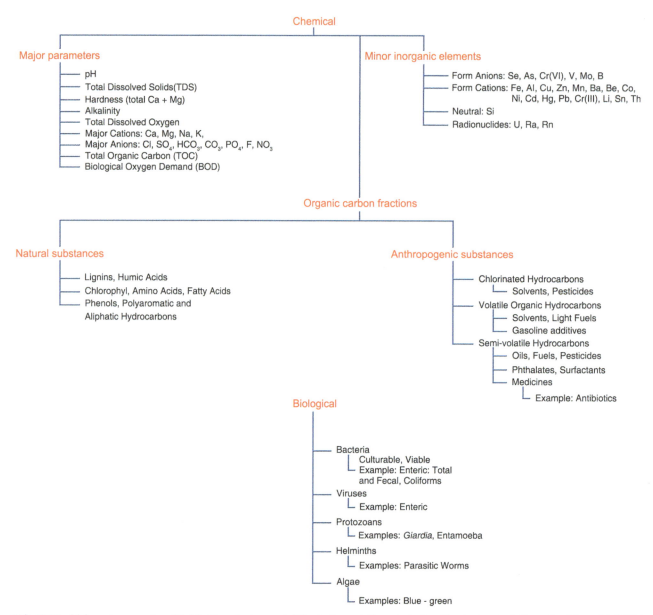

FIG. 10.11   Major water properties. *(Modified from Environmental Monitoring and Characterization. Elsevier Academic Press, San Diego, CA, 2004.)*

10.8   ATMOSPHERE SAMPLING AND MONITORING

Legislative efforts to control air pollution have evolved over a long period of time. Major efforts in the United States to control and prevent air pollution did not begin until the mid-20th century. Monitoring of air pollution is the result of legislation requiring polluters to demonstrate compliance with air quality standards and emissions limits. The legislative framework behind these requirements was, and still is, motivated largely by concerns for human health, as well as concerns about the impacts of air pollution on natural and agricultural ecosystems and global climate. Efforts by the federal government to control air pollution nationwide began in 1955 with enactment and promulgation of the Air Pollution Control Act. This Act permitted federal agencies to aid state and local governments who requested assistance to carry out research of air pollution problems within their jurisdictions. This Act was important in that it established the ongoing principle that state and local governments are ultimately responsible for air quality within their jurisdictions.

The overall objectives of the EPA air quality monitoring program are as follows:

- To judge compliance with and/or progress made toward meeting ambient air quality standards;

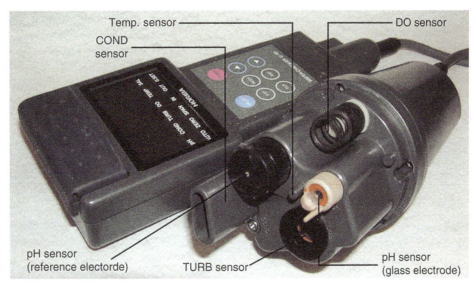

FIG. 10.12 Field portable multiprobe water quality meter (manufacturer: Horiba Ltd.). The delicate multisensor array is covered with a protective sleeve. Each sensor requires a different level of care and periodic calibration. *(From Environmental Monitoring and Characterization. Elsevier Academic Press, San Diego, CA, 2004.)*

- To activate emergency control procedures that prevent or alleviate air pollution episodes;
- To observe pollution trends;
- To provide database for research evaluation of effects (urban, land use, transportation planning);
- The development and evaluation of abatement strategies; and
- The development and validation of diffusion models.

The monitoring is accomplished through the use of networks of air monitoring stations (see Fig. 10.13). How are networks established to meet monitoring requirements? First, the objectives of the network need to be identified. These generally include determining the highest concentrations expected to occur in the area covered by the network, determining representative concentrations in areas of high population density, determining the impact of significant sources or source categories on ambient pollution levels, and determining general background concentration levels. Site selection is obviously an important consideration when attempting to meet these objectives.

General EPA guidelines for selecting monitoring sites include locating sites in areas where pollutant concentrations are expected to be highest; locating sites in areas representative of high population density; locating sites near significant sources; locating sites upwind of the target region to establish background conditions; considering topographic features and meteorological conditions (e.g., valleys, inversion frequency); and considering the effects of nearby obstacles, such as buildings and trees, on air flow patterns. Once the sites are selected, consideration must be given to the control of the sampling environment. An air-conditioned trailer is necessary at each site to properly

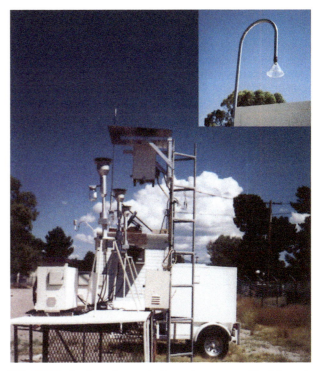

FIG. 10.13 An air quality monitoring station near a residential area in Tucson, Arizona. Meteorological sensors and particulate matter samplers are visible in photograph. Note the air sampling probe *(insert)* through which air is drawn to analytical instruments inside the trailer. *(Photo courtesy A. Matthias. From Environmental Monitoring and Characterization. Elsevier Academic Press, San Diego, CA, 2004.)*

maintain analytic instruments, data recorders, and calibration equipment. This requires access to electrical power, surge protectors, and backup power.

Most monitoring stations include a standard set of meteorological instrumentation, such as a radiation shielded thermometer to measure air temperature, an anemometer and wind vane to record wind speed and direction, and a pyranometer to measure solar radiation (needed for ozone monitoring). In addition, there are other specialized instrumentation used for specific purposes such as PM_{10} and $PM_{2.5}$ air article size samplers, ozone, nitrous oxides, and carbon monoxide analyzers, and volatile organic carbon (VOCs) samplers (see Artiola et al., 2004).

10.9 CONCLUSIONS

Environmental monitoring is critical to the protection of human health and the environment. As the human population continues to increase, as industrial development and energy use continues to expand, and despite advances in pollution control, the continued production of pollution remains inevitable. Thus the need for environmental monitoring is still as great as ever. Continued advances in the development, application, and automation of monitoring devices are needed to enhance the accuracy and cost-effectiveness of monitoring programs. Equally as important is the need to produce more scientists and engineers that have the knowledge and training required to successfully develop and operate monitoring devices and manage monitoring programs.

QUESTIONS AND PROBLEMS

1. The following figure shows an area divided into 20 squares with X and O points marked. Which sampling locations appear to be random and which are systematic? Use the 20 data points lead values ($mg\,kg^{-1}$) to compute the mean, variance, standard deviation, and standard error values of the data with the appropriate number of significant digits. Estimate the 95% Confidence interval of the data (assume that this is ~ sample mean $(x) \pm 2\sigma$), see Fig. 10.1.

2. Using the calibration data shown in Table 10.2. Plot the data points and fit a linear regression using Excel spreadsheet functions. Calculate the acid extractable aluminum concentration (in $mg\,kg^{-1}$) in a soil sample with the following information: (a) the soil acid extract had an instrument response of 110,021; (b) the soil: acid extract ratio was 1:5 on a mass basis, that is, 10 g of soil were extracted with 50 mL of acid.

3. Using U.S. Public Land Survey notation, define the bottom half location of filled square in location B of Fig. 10.2 to the nearest ¼ of ¼ of a section.

4. Which regions of the electromagnetic spectrum are most useful in monitoring vegetation cover? Explain your answer.

5. What is an argillic horizon and what soil forming processes do you think influence most of its formation?

6. Explain why in Fig. 10.7 there are more sampling locations marked below that above the manufacturing plant.

7. Why has ground water sampling and monitoring has become such an important issue in the past two decades? Explain your answer.

8. Why has surface water sampling and monitoring become such an important issue in the past two decades? Explain your answer.

9. What water quality parameters can be routinely monitored directly in the field using portable monitoring devices?

10. What is the Air Pollution Control Act and how does it impact our local air quality?

REFERENCES

Artiola, J.F., Brusseau, M.L., Pepper, I.L. (Eds.), 2004. Environmental Monitoring and Characterization. Elsevier Academic Press, San Diego, CA.

Brady, N.C., Weil, R.R., 1996. The Nature and Properties of Soils, 11th ed. Prentice Hall, Upper Saddle River, NJ.

FURTHER READING

Fetter, C.W., 1999. Contaminant Hydrogeology, second ed. Prentice Hall PTR, Upper Saddle River, NJ.

Pepper, I.L., Gerba, C.P., Brusseau, M.L., 1996. Pollution Science. Academic Press, San Diego, CA.

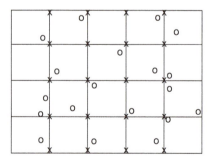

Chapter 11

Physical Contaminants

J. Walworth and I.L. Pepper

Aerosol production from tractor operations. *Photo courtesy J.P. Brooks*

11.1 PARTICLE ORIGINS

Small particles, whether of natural or anthropogenic origin, can pollute air and water resources. These particles pose a hazard to human health and to the environment in a variety of ways. We will explore the properties of particulate contaminants, the health threats they present, where they come from, and how they behave in the environment.

Particulate sources are divided into those arising from a single, well-defined emission source, which is called *point-source pollution*, and *nonpoint-source pollution*, contamination generated from a wide area. Particulate emissions can have natural origins or be of human origin. The threat

from naturally occurring particulates may be exacerbated when mobilized by human activities such as agriculture, logging, and construction, although particulates are also generated by natural processes such as volcanoes or soil erosion. Human-made particulates include those created by industrial processes; combustion in power plants, wood stoves, fireplaces, and internal combustion engines; rubber particles from tire wear; and other sources.

11.2 PARTICLE SIZE

Particle behavior is, to a large degree, determined by size. The diameters of some particles and familiar substances are

Environmental and Pollution Science. https://doi.org/10.1016/B978-0-12-814719-1.00011-2

shown in Fig. 11.1. Many particle properties, as well as their environmental and health impacts, are related to their size. In general, smaller particles pose a greater health threat than do larger particles. Nanoparticles are among the smallest man-made particles.

11.2.1 Nanoparticles

Nanotechnologies refer to "technology of the tiny," with dimensions in the range of 1–100 nanometers (nm) (nano is the SI unit prefix for 10^{-9} or one-billionth, 0.000000001, of a meter). As an illustration of the scale of interest, a molecular chain of 5–10 atoms is about 1 nm long, the helix of DNA has a diameter of about 2 nm, and the average human hair is about 80,000 nm in diameter.

Nanotechnology and nanoscience research represent a key aspect of the development of innovative materials and new productive sectors. Nanomaterials (nanoparticles, nanospheres, nanotubes, and nanostructured surfaces) are used in ceramic, textile, computer, cosmetic, optic, energy, manufacturing, and chemical industries. In addition, they are applied in biomedicine as nanobiomaterials, nanospheres for drug delivery and release, and nanotubes for gene therapy. Most nanoparticles that are currently used today are made from silicon, carbon, and transition metals.

Human exposure to nanoparticles can occur as environmental (e.g., elemental Pt^0 on larger Al_2O_3 carrier particles emitted from automotive catalytic converters, ultrafine TiO_2 or ZnO in cosmetic ingredients such as sunscreen, antimicrobial Ag^0 particles in clothing), occupational (e.g., large-scale preparation of nanoparticles), and biomedical (e.g., ultrafine TiO_2 for tumor tissue targeting and delivery of killing compounds for cancer cells by UV light). At the occupational level, there are four main groups of nanoparticles production processes: gas phase, vapor deposition, colloidal, and attrition, all of which may potentially result in exposure by inhalation, dermal, or ingestion routes. All processes may give rise to exposure to agglomerated nanoparticles during recovery, powder handling, and production processing. In spite of the potential occupational and public exposure to nanomaterials that is dramatically increasing, information on nanoparticle fate in the environment and the human health impact of nanoparticles is severely lacking.

11.3 PARTICLES IN AIR (AEROSOLS)

Particles suspended in air are called *aerosols*. These pose a threat to human health mainly through respiratory intake and deposition in nasal and bronchial airways. Smaller aerosols travel further into the respiratory system and

generally cause more health problems than larger particles. For this reason the United States Environmental Protection Agency (U.S. EPA) has divided airborne particulates into two size categories: PM_{10}, which refers to particle matter (PM) with diameters less than or equal to 10 μm (10,000 nm), and $PM_{2.5}$ which are particles less than or equal to 2.5 μm (2500 nm) in diameter. For this classification, the diameter of aerosols is defined as the *aerodynamic diameter*:

$$d_{pa} = d_{ps} \left(\rho_p / \rho_w \right)^{1/2} \qquad (11.1)$$

where

d_{pa} = aerodynamic particle diameter (μm)
d_{ps} = Stokes' diameter (μm)
ρ_p = particle density ($g\,cm^{-3}$)
ρ_w = density of water ($g\,cm^{-3}$)

Atmospheric particulate concentration is expressed in micrograms of particles per cubic meter of air ($\mu g\,m^{-3}$). The U.S. EPA established a National Ambient Air Quality Standard (NAAQS) for PM_{10} of $150\,\mu g\,m^{-3}$ averaged over a 24-h period, and $50\,\mu g\,m^{-3}$ averaged annually. More recently, separate standards for $PM_{2.5}$ of $65\,\mu g\,m^{-3}$ for 24 h and $15\,\mu g\,m^{-3}$ annually have been introduced.

Symptoms of particulate matter inhalation include decreased pulmonary function, chronic coughs, bronchitis, and asthmatic attacks. The specific causal mechanisms are poorly understood. One well-documented episode occurred in London in 1952, when levels of smoke and sulfur dioxide aerosols, largely associated with coal combustion, reached elevated levels due to local weather conditions. Approximately 4000 deaths over a 10-day period were attributable to cardiovascular and lung disorders brought on or aggravated by these aerosols. Sources of $PM_{2.5}$ and PM_{10} in the US are given in Table 11.1.

Airborne particles can travel great distances. Intense dust storms during 1998 and 2001 in the Gobi desert of Western China and Mongolia (Fig. 11.2) elevated aerosol levels to concentrations near the health standard in Western North America several thousand miles away!

Smaller particles tend to travel greater distances than large particles because they remain airborne longer. Stokes' Law (Eq. (11.2)) is used to describe the fall of particles through a dispersion medium, such as air or water.

$$V = \left[D^2 \times \left(\rho_p - \rho_l \right) \times g \right] / 18\eta \qquad (11.2)$$

where

V = velocity of fall ($cm\,s^{-1}$)
g = acceleration of gravity ($980\,cm\,s^{-2}$)
D = diameter of particle (cm)
ρ_p = density of particle (density of quartz particles is $2.65\,g\,cm^{-3}$)

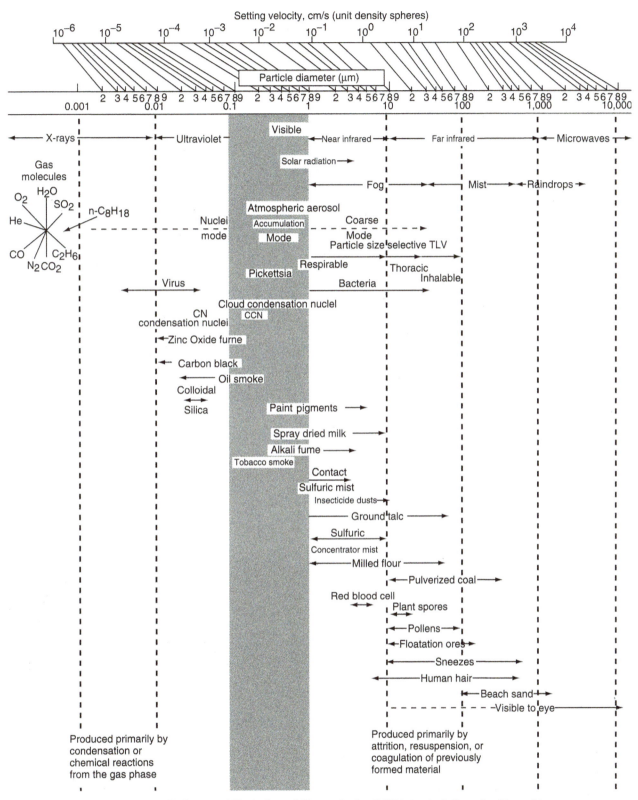

FIG. 11.1 Particle diameters. *From* Environmental Monitoring and Characterization © *2004, Academic Press, San Diego, CA.*

TABLE 11.1 U.S. PM$_{10}$ and PM$_{2.5}$ Production From Various Sources

	Source	PM$_{10}$	PM$_{2.5}$
		(Thousands of tons)	
Industrial processes	Chemical industries	20	14
	Metals processing	58	45
	Petroleum industries	36	32
	Other industrial processes	777	307
	Solvent utilization	4	4
	Storage and transport	67	18
	Waste disposal and recycling	305	253
Fuel Combustion	Electric utilities	229	177
	Industrial	325	240
	Other	368	363
Vehicles	Highway vehicles	287	133
	Off-highway vehicles	163	151
Other	Wildfires	1030	873
	Miscellaneous	21803	4425

USEPA, 2017. Available from: https://www.epa.gov/air-emissions-inventories/air-pollutant-emissions-trends-data

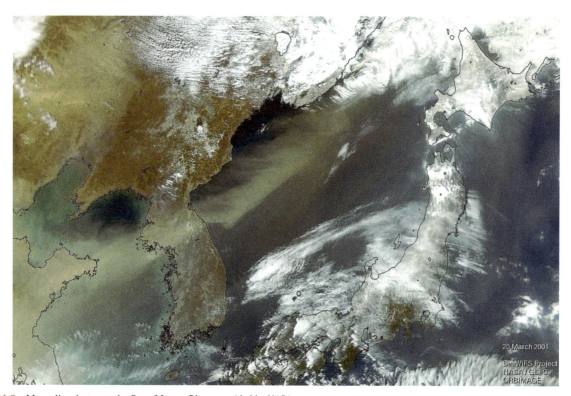

FIG. 11.2   Mongolian dust over the Sea of Japan. *Photo provided by NASA.*

INFORMATION BOX 11.1 Influence of Particle Size on Velocity of Deposition of Particles in Air, Calculated Using Stokes' Law

Particle Diameter (mm)	Particle Type	Rate of Fall in Air (cm s^{-1})
1	Sand	7880
0.1	Silt	79
0.001	Clay	7.9×10^{-5}

$\rho_1 =$ density of dispersion medium (air has a density of about $0.001213 \, g \, cm^{-3}$; water has a density of about $1 \, g \, cm^{-3}$)

$\eta =$ viscosity of the dispersion medium (about 1.83×10^{-4} poise or $g \, cm^{-1} \, s^{-1}$ for air; 1.002×10^{-2} poise for water)

Using Stokes' Law, we can estimate the rate of fall in air (Information Box 11.1). Small particles are a greater concern than larger particles for several reasons. Because small particles stay suspended longer, they can travel greater distances. Risk of exposure to small particles is amplified by their extended suspension times. Small particles also tend to move further into the respiratory system, exacerbating their effects on health. Atmospheric transport of aerosols is discussed further in Chapter 17.

11.3.1 Aerosols of Concern

11.3.1.1 Asbestos

Asbestos particles are a special case of mineral aerosols that are known to lead to debilitation, disease, and death. Asbestos is a group of six naturally occurring fibrous silicate minerals (amosite, anthophyllite asbestos, actinolite asbestos, crocidolite, chrysotile, and tremolite asbestos), which are identified using standard mineralogic analyses employing X-ray diffraction and optical and electron microscopy in transmission, diffraction, and scanning modes. Asbestos, and many other silicate mineral species (e.g., talc, erionite, and vermiculite), may occur in fibrous form (Wilson and Spengler, 1996). The fibrous structure of these minerals permits the formation of sharp aerosols that can be embedded in the lungs. It is likely that no matter what the environment, at home or outdoors, people have been and are continually exposed to particulates and aggregates that are mixtures, and some will contain asbestos or other minerals.

Asbestosis is the disease that results when lungs generate scar tissue (fibrosis) as a result of high exposure. Lung scarring may continue postasbestos exposure, and although affected portions of the lung can be rejected, the quality of life thereafter is diminished. Lung cancer, although associated with asbestos mineral exposure, is more frequently

directly associated with smoking. Many occupations, for example, construction workers or brake repair technicians, often include individuals who smoke, resulting in multiple opportunities for respiratory trauma and therefore multiplying the risks. Another deadly disease associated with asbestos particles is mesothelioma, a cancer of the pleura (tissue surrounding lungs) rather than of the lung tissues.

Although the public health issues related to asbestos have been well documented, the specific mechanisms of fibrogenesis and carcinogenesis related to the exposures, especially at low doses (nonoccupational) are not fully elucidated and remain under discussion and investigation. Asbestos in the built indoor environment is also a potential source of exposure (Skinner et al., 1988). Asbestos removal from buildings is closely regulated to prevent exposure to construction workers. Asbestos issues were raised by the 2001 terrorist attack that destroyed the World Trade Center buildings which contained 400 tons of asbestos and generated widespread dust throughout lower New York City.

11.3.1.2 Silica

One of the most common natural materials and a major component of beach sand, quartz (SiO_2), may become an offending airborne material causing a particulate-based disease. Silicosis is a result of exposure to crystalline silica, and is exclusively occupational, with the size and morphologic characteristics of the particle key to respiratory problems. Construction workers, especially those jack hammering or sand blasting building surfaces using a stream of silica without proper nasal and mouth protection are at great risk. Often biological as well as mineral materials that become airborne in both cases can pose health threats.

Case Study 11.1. Asbestos Exposure in California Near Coalinga From Asbestos-Bearing Serpentinite

The deposit was one of the oldest (1885) and best known mercury mines in the country. Chrysotile, naturally occurring in serpentinite, has been mined at the site since the late 1950s, as it was available in pure mineral form and had many industrial applications. The grain size of the natural "short-fiber" chrysotile was milled, so the product probably approached the respirable size range and was certainly small enough to become airborne. The bagged powdery material was cheap and shipped to markets inside and outside the United States. The site and the large waste dumps created by mining are in the Atlas Asbestos Mine Superfund Site within the Clear Creek Management Area, a popular off-roading site. The climate, elevation, topography, and usage of the site mean that there are increased amounts of dust in the local atmosphere blown by prevailing winds toward populated areas.

There are several crystal forms of SiO_2, including diatoms, the source of diatomaceous earth. The use of these various silica materials is not monitored nor are those at risk necessarily aware of their exposure, but great efforts have been made by some industries and one can anticipate future actions, especially in response to U.S. Occupational Safety and Health Administration (OSHA) regulations.

Silicosis is characterized by nodular lesions, which can be detected radiologically in the upper lung. This expression of silicosis is distinct from that of asbestosis, where fibrosis is usually diffuse and in the lower portions of the lung. Lung function may not be markedly affected initially although under continuing exposure, the nodules coalesce and the fibrosis becomes massive and pervasive for large parts of the lung, even after silica exposure has ended. The formally pliable lung tissues become occluded by scarring and the deposition of the fibrous protein, collagen, often calcify or harden, further compromising respiration and the transmission of the essential gases in these portions of the respiratory system.

11.3.1.3 Human-Made Aerosols

Particulate matter in the atmosphere can be from direct emissions or pollution that enters the atmosphere as previously formed particles. These are called *primary particles*. Alternatively, *secondary particles* are formed in the atmosphere from precursor components, such as ammonia, volatile organics, or oxides of nitrogen (NO_x) and sulfur (SO_x). Primary particles may fall into either the $PM_{2.5}$ or the PM_{10} size ranges, whereas secondary particles fall mainly into the $PM_{2.5}$ category.

Industrially generated primary particles arise largely from incomplete combustion processes and high-temperature metallurgical processes. Secondary particles, on the other hand, are produced from gases emitted from industrial processing and various combustion processes (including automobiles, power plants, wood burning, and incinerators) that undergo gas-to-particle conversion and then growth and coagulation. In the atmosphere, sulfur oxides are oxidized to form sulfuric acid and fine sulfate particles. Gases condense to form ultrafine aerosols (less than $0.01\,\mu m$), either from supersaturated vapor produced in high temperature combustion processes or through photochemical reactions. These particles grow in size through condensation and coagulation to form larger particles ($0.1–2.5\,\mu m$). The principal anthropogenic sources of SO_x in the United States include coal power plants, petroleum refineries, paper mills, and smelters. In contrast, NO_x is largely produced by industrial and automotive combustion processes.

Health threats from $PM_{2.5}$ and PM_{10} generated by human activities are much like those from naturally occurring particulates. Adverse health effects are most severe in senior citizens and those with preexisting heart or lung problems. Recent studies estimate that with each $10\,\mu g\,m^{-3}$ increase in PM_{10} above a base level of $20\,\mu g\,m^{-3}$, daily respiratory

mortality is estimated to increase by 3.4%, cardiac mortality increases by 1.4%, hospitalizations increase by 0.8%, emergency room visits for respiratory illnesses increase by 1.0%, days of restricted activity due to respiratory symptoms increase by 9.5%, and school absenteeism increases by 4.1% (Vedal, 1995). Particles formed from incomplete combustion, such as those formed by wood burning and diesel engines, contain organic substances that may have additional health effects. Diesel exhaust has been shown to increase lung tumors in rats and mice, and long-term human exposure to diesel exhaust may be responsible for a 20%–50% increase in the risk of lung cancer (Koenig, 1999).

In addition to human health concerns, $PM_{2.5}$ associated with wood burning, automobile exhaust, and industrial activities is responsible for much of the atmospheric haze in the United States (Chapter 17). Aerosols (especially $PM_{2.5}$) absorb and scatter light, producing haze and reducing visibility. When severe, this interferes with automobile and aviation navigation, posing safety threats. Atmospheric particulates can also be a nuisance by settling on and in cars and homes and other buildings.

11.3.1.4 Bioaerosols

Biological contaminants include whole entities such as bacterial and viral human pathogens. They also include airborne toxins, which can be parts or components of whole cells. In either case, biological airborne contaminants are known as bioaerosols, which can be ingested or inhaled by humans (see also Chapter 23).

Coccidioidomycosis (also known as *Valley Fever*) is a disease caused by inhalation of spores of the fungus *Coccidioides immitis*, which is indigenous to hot, arid regions, including the Southwestern United States. The fungus can travel from the respiratory tract to the skin, bones, and central nervous system and can result in systemic infection and death.

Endotoxin, also known as lipopolysaccharide, is ubiquitous throughout the environment and may be one of the most important human allergens. Endotoxin is derived from the cell wall of Gram-negative bacteria and is continually released during both active cell growth and cell decay. Hence, endotoxin is found wherever Gram-negative bacteria are found. In soils, bacterial concentrations routinely exceed 10^8 per gram, with a majority of the bacteria being Gram negative. Soil particles containing sorbed microbes can be aerosolized and hence act as a source of endotoxin. Farming operations such as driving a tractor across a field has been shown to result in endotoxin levels of 469 endotoxin units (EU) m^{-3} (Table 11.2). EU units are related to a turbidimetric Limulus Amebocyte Assay. Daily exposures of as little as $10\,EU\,m^{-3}$ from cotton dust can cause asthma and chronic bronchitis. However, dose response is

TABLE 11.2 Aerosolized Endotoxin Concentrations Detected Downwind of Biosolids Operations, a Wastewater Treatment Plant Aeration Basin, and a Tractor Operation[a]

Sample Type	No. of Samples Collected	Distance from Site (m)	Aerosolized Endotoxin			
			Avg	Median	Minimum	Maximum
			EU m[a]			
Controls						
Background	12	NA	2.6	2.49	2.33	3.84
Biosolids Operations						
Loading	39	2–50	343.7	91	5.6	1807.6
Slinging	24	10–200	33.5	6.3	4.9	14.29
Biosolids Pile	6	2	103	85.4	48.9	207.1
Total Operation	33	10–200	133.9	55.6	5.6	623.6
Wastewater Treatment Plant Aeration Basin	6	2	627.3	639	294.4	891.1
Nonbiosolids Field						
Tractor	6	2	469.8	490.9	284.4	659.1

[a]*EU m^{-3}—Endotoxin units per m^3.*

From Brooks, J.P., 2004. Biological aerosols generated from the land application of biosolids: microbial risk, assessment. Ph.D. Dissertation, The University of Arizona, Tucson, Arizona.

dependent on the source of the material, the duration of exposure, and repeated exposures (Brooks, 2004). When inhaled by humans, endotoxin has demonstrated the ability to cause a wide variety of health effects including fever, asthma, and shock.

Data in Table 11.2 demonstrate that endotoxin aerosolization can occur during both wastewater treatment and land application of biosolids. However, the data also show that endotoxin of soil origin resulting from dust generated during tractor operations results in similar amounts of aerosolized endotoxin. In 1997 over 80% (26.3 million tons) of PM_{10} in the United States was attributed to unpaved roads, agriculture and forestry, construction, and wind erosion (Council on Environmental Quality, 1997). Given that these particulates are of soilborne origin, it is possible that endotoxin associated with wind-blown soil particles is a major contributor to respiratory problems.

Mycotoxins are secondary metabolites produced by fungal molds. Fungi such as species of the genera *Aspergillus*, *Alternaria*, *Fusarium*, and *Penicillium* are common soilborne fungi capable of producing mycotoxins. Most notably, aflatoxin is produced by *Aspergillus flavus*, a widespread fungus that grows on grains and other plant materials. Aflatoxin is one of the most potent carcinogens

known and is linked to a variety of health problems, particularly to workers occupationally exposed to grain dust.

11.4 PARTICULATES IN WATER

In water, suspended particulates pose quite different risks than aerosols. Inorganic particles cause an increase in the turbidity of affected water, and the particles themselves can cause problems through sedimentation that can fill lakes, dams, reservoirs, and waterways. In the United States, waterborne soil particles fill over 123 million cubic meters of reservoir capacity each year, reducing water storage capacity and necessitating expensive dredging operations. Suspended particles increase wear on pumps, hydroelectric generators, and related equipment. Also, soils from which suspended particles are derived suffer damage from soil erosion that can reduce agricultural productivity and land values. Severe soil erosion can threaten buildings, roads, and other structures. This issue is discussed in more detail in Chapter 14.

Using Stokes' Law, settling rates can be calculated for various size particles suspended in water (see Information Box 11.2). Sedimentation is not an important factor for particles less than 0.0005 mm in diameter because they act as

INFORMATION BOX 11.2 Influence of Particle Size on Velocity of Deposition of Particles in Water, Calculated Using Stokes' Law

Particle Diameter (mm)	Particle Type	Rate of Fall in Water (cm s^{-1})
1	Sand	90
0.1	Silt	0.9
0.001	Clay	9×10^{-5}

INFORMATION BOX 11.3 Soil Particle Diameter Sizes

Sand:	2–0.05 mm
Silt:	0.05–0.002 mm
Clay:	<0.002 mm (<2 μm)

colloids, which are so small that once suspended in a dispersion medium, they can remain in suspension indefinitely and do not settle out through the force of gravity. Large particles remain suspended for only a short time, and their range of movement is limited. Turbulence, disorderly flow or mixing of the dispersion medium, can increase suspension times considerably, so the calculated settling rates are valid only in the absence of turbulence. Actual settling rates are generally slower.

Decomposition of organic particles such as plant and animal residues, composed of carbohydrates, proteins, and more complex compounds, occurs largely through microbially mediated oxidation processes and can also impact waters. The oxidation of organic substrate via aerobic heterotrophic respiration is described in Chapter 5.

The oxidation of waterborne organic particulates consumes dissolved oxygen and produces CO_2. The consumption of oxygen in biological oxidation reactions is called biological oxygen demand (BOD). Increasing BOD and the resulting oxygen depletion, or *hypoxia*, can have a profound effect on aquatic animals, such as fish, that depend on a consistent supply of dissolved oxygen (see also Chapter 16).

Both organic and inorganic particles (such as soil) can act as carriers of other contaminants that are sorbed to particle surfaces. Nutrients, herbicides, insecticides, fuels, solvents, preservatives, and other industrial and agricultural chemicals can adsorb and desorb from waterborne particulates. The health threats associated with this very diverse group of chemicals are similarly broad. These contaminants are discussed in detail in Chapters 12, 14, and 15.

11.4.1 Soil Particles

Soil particles, which are natural contaminants, are not considered to be pollutants unless they move into the atmosphere or surface waters. They are classified on the basis of size into three categories in the U.S. Department of Agriculture (USDA) classification system (Information Box 11.3).

Sand and silt particles are dominated by primary minerals such as quartz, feldspar, and mica formed directly from molten magma. Clay particles, on the other hand, are composed largely of secondary layered silicate clay minerals (weathering products of primary minerals), hydrous oxides of aluminum and iron, and organic

materials. Most of these particles have negative electric charge, although some can have a positive charge, and others no charge at all. This charge arises from either characteristics of their crystalline structure, in the case of layered silicate clay minerals, or due to surface chemical reactions (see also Chapter 8). Positively charged cations surrounding clay particles balance the negative surface charge of the clay particles. Additionally, certain anions and organic chemicals can be sorbed to clay particles. Migrating soil particles carry associated compounds, so clays can act as carriers in the distribution and transport of molecules attached or attracted to soil particles.

11.4.1.1 Soil Particle Flocculation

As indicated before, the settling rates of particles suspended in water depends, among other things, on their effective diameter. This can be greatly increased when particles adhere to one another or *flocculate* to form aggregates made up of groups or aggregates of soil particles. "Like" charged particles repel each other, so the forces of repulsion must be overcome for particles to aggregate. The negative charge on clay particles is balanced by a cloud of counterions (cations) surrounding the clay particle, called a *diffuse double layer*. In this double layer, the concentration of cations increases as the clay surface is approached, and, similarly, the concentration of anions decreases. The thickness of this diffuse double layer depends on several factors, the most important of which are counterion valence and concentration.

$$1/\chi = \left(K/z^2 n \right)^{1/2} \tag{11.3}$$

where

$1/\chi$ = the effective thickness of the double layer
z = valence of the counterions (cations)
K = a constant dependent on temperature and the dielectric constant of the solvent
n = electrolyte concentration in the equilibrium solution

The thickness of the double layer decreases as the valence of counterions increases, or as the electrolyte concentration in the solvent increases (Fig. 11.3). As the double layer thickness decreases, attractive energy between particles becomes greater than repulsive forces, and the particles flocculate. The electrolyte concentration at which this occurs is called the *critical flocculation concentration* (CFC), which is a

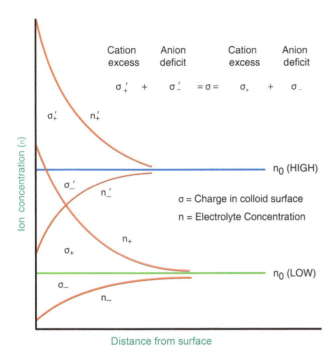

Cation   Anion   Cation   Anion
excess   deficit   excess   deficit

$$\sigma'_+ + \sigma'_- = \sigma = \sigma_+ + \sigma_-$$

$\sigma = $ Charge in colloid surface
$n = $ Electrolyte Concentration

n_0 (HIGH)

n_0 (LOW)

Ion concentration (n)

Distance from surface

FIG. 11.3 Thickness of the electrical double layer and ion distribution at two electrolyte concentrations. *From Sumner, M.E. and Stewart, B.A., eds. (1992) Soil Crusting and Physical Processes. Lewis Publishers, Boca Raton, FL. Reprinted by CRC Press.*

function of $1/z^6$. Although this relationship does not completely explain the varying ability of different cations to flocculate soil particles, we can see that the effect of counterion valence is extremely important.

The relative flocculating power of common soil cations is presented in Table 11.3. In general, counterion valence is the most important factor determining CFC; therefore higher valence cations are more effective flocculators than lower valence cations. The hydrated radius of cations is also important and explains much of the difference between cations of identical valence.

Soil particles suspended in water will tend to be dispersed if the electrolyte concentration is low and the dominant counterions are potassium or sodium, whereas flocculation will occur with a high electrolyte concentration

TABLE 11.3 Relative Flocculating Power of Common Monovalent and Divalent Cations

Ion	Relative Flocculating Power (Relative to Na^+)
Na^+	1.00
K^+	1.70
Mg^{2+}	27.00
Ca^{2+}	43.00

From Rengasamy, P., Sumner, M.E., 1998. Sodic Soils: Distribution, Properties, Management, and Environmental Consequences. Oxford University Press, New York, NY.

or if the dominant counterions are of higher valence. Where rivers flow into the ocean, sediment-laden river water mixes with saline ocean water. Suspended particles flocculate and settle out to form river deltas (Fig. 11.4). To flocculate and settle particles out of wastewater, high-valence flocculating agents such as aluminum, iron, and copper sulfates or chlorides are added as clarifiers. Organic polymers and synthetic polyelectrolytes that have anionic or cationic functional groups are also used as flocculants.

Depending on the properties of the suspended particles and the impacted water, particles may flocculate into aggregates and quickly settle out of suspension, or they may remain dispersed. As indicated before, colloidal particles can remain in suspension indefinitely when in an unflocculated or dispersed state, affording them the opportunity to spread in air and water and increasing environmental impact. Flocculated soil particles, on the other hand, are less mobile and largely benign.

11.5 SUMMARY

Particulate contaminants can be of either natural or human-made origin. These particles can pollute both air and water. Naturally occurring particles include soil or other mineral particles, pollen, ocean spray, bacteria, viruses, and spores. Mineral particles can contaminate water and air via natural processes such as volcanic activity, wave action, and wind and water erosion. Wind and water erosion can be accelerated by human activities that leave soil in a susceptible condition, whereas many of the other processes are beyond human control. Human-made particles include nitrogen and sulfur oxides and a wide range of solid particles formed during combustion, and industrial processes that grind or abrade materials into fine particles.

Airborne particulates are a major health concern because they can penetrate the human respiratory system, causing a wide range of health problems. They can degrade visibility and become a health hazard by impeding air and ground vehicular traffic. Waterborne particulates can be a direct threat by clogging reservoirs and waterways, and by acting as carriers for other contaminants, such as phosphorus and pesticides that contaminate surface waters. Organic particles increase the BOD of contaminated waters, leading to oxygen depletion (hypoxia) that can harm aquatic fauna.

There are many ways of reducing production of particulate contaminants or of removing them from industrial exhaust streams. However, all natural waters contain suspended particles, as does even the cleanest of air. Some of the processes that distribute particulate contaminants to air and water are natural and can be only partially controlled, if at all. On the other hand, human activities that contribute to particulate degradation of air and water often can be modified to minimize their impact on these natural resources (see Chapter 20).

FIG. 11.4   Sediment-laden water from the Ganges and Brahmaputra Rivers flows into the Bay of Bengal. *Photo provided by NASA.*

QUESTIONS AND PROBLEMS

1. Define the terms PM_{10} and $PM_{2.5}$.
2. How do waterborne particles reduce dissolved O_2 in surface water? What are the resulting effects on aquatic fauna?
3. Explain why sediments carried in freshwater rivers quickly settle out when these sediment-laden waters combine with ocean water.
4. Explain the differences between primary and secondary particles in terms of size, formation, and composition.
5. Calculate the rate of deposition of a particle with a diameter of 0.003 mm using Stokes' Law: (a) in air; and (b) in water.

REFERENCES

Brooks J.P. Biological aerosols generated from the land application of biosolids: microbial risk, assessment. 2004, Ph.D. Dissertation, The University of Arizona, Tucson, Arizona.

Council on Environmental Quality, 1997. Environmental Quality: The 1997 Report of the Council on Environmental Quality. U.S. Government Printing Office, Washington, DC.

Koenig, J.Q., 1999. Health Effects of Ambient Air Pollution. Kluwer Academic Publishers, Boston, MA.

Skinner, H.C.W., Ross, M., Frondel, C., 1988. Asbestos and Other Fibrous Materials: Mineralogy, Crystal Chemistry and Health Effects. Oxford University Press, New York, NY.

USEPA, 2017. Air pollutant emissions trends data [online]. Available from: https://www.epa.gov/air-emissions-inventories/air-pollutant-emissions-trends-data. Accessed 19 January 2018.

Vedal, S., 1995. Health Effects of Inhalable Particles: Implications for British Columbia. Air Resources Branch, Ministry of Environment, Lands and Parks.

Wilson, R., Spengler, J. (Eds.), 1996. Particles in Our Air: Concentrations and Health Effects. Harvard University Press, Cambridge, MA.

FURTHER READING

Liu, D.H.F., Liptak, B.G., 2000. Air Pollution. Lewis Publishers, Boca Raton, FL.

Rengasamy, P., Sumner, M.E., 1998. Processes Involved in Sodic Behavior. In: Sumner, M.E., Naidu, R. (Eds.), Sodic Soils: Distribution, Properties, Management, and Environmental Consequences. Oxford University Press, New York, NY.

Sumner, M.E., 1992. The electrical double layer and clay dispersion. In: Sumner, M.E., Stewart, B.A. (Eds.), Soil Crusting: Chemical and Physical Processes. Lewis Publishers, Boca Raton, FL (Chapter 2).

Chapter 12

Chemical Contaminants

M.L. Brusseau and J.F. Artiola

Chemical contamination is a major source of pollution. *Photo courtesy U.S. Environmental Protection Agency, https://www.epa.gov/nj/superior-barrel-drum-photos.*

12.1 INTRODUCTION

It can be argued that all matter in one form or another can become a contaminant when found out of its usual environment or at concentrations above normal. However, chemical contaminants become pollutants when accumulations are sufficient to adversely affect the environment or to pose a risk to living organisms. Today, there are thousands of industrial chemicals that can be dangerous to humans and the environment. Fortunately, many of these chemicals are not produced in large enough quantities to be a human or environmental threat. However, there are many other human-made and natural chemicals that are toxic and are produced in sufficient quantities to be a potential environmental or human health hazard. Thus the production, storage, transport, and disposal of these chemicals are regulated by government agencies. There are numerous sources of chemical contaminants released to the environment, but these generally fall into a few general categories. This chapter will present an overview of the various types of chemical contaminants and their sources.

12.2 TYPES OF CONTAMINANTS

There are three basic categories of chemical contaminants: organic, inorganic, and radioactive. In turn, there are several classes of contaminants within each of these categories. Major classes of contaminants are listed in Table 12.1. Some of these contaminants are considered in greater detail in other Chapters 17–19.

Thousands of chemicals are released into the environment every day. Thus, when conducting site characterization studies, it is important to prioritize the suite of chemicals under investigation. For most sites this is done by focusing on so-called priority pollutants, those that are regulated by federal, state, or local governments. The primary such list of priority pollutants is that governed by the National Primary Drinking Water Regulations, which provide legally enforceable standards that apply to all public water systems. These standards protect public health by limiting the levels of contaminants that are allowed to exist in drinking water. The first page of this list is presented in Table 12.2 as an example. Note that the full list also includes microorganisms, radionuclides, and water disinfection by-products.

The specific contaminants that occur, their frequency of occurrence, and their potential hazard differ greatly for each specific contaminated site. The top 20 contaminants prioritized by frequency of occurrence, human toxicity, and potential for human exposure at U.S. Environmental Protection Agency (EPA) designated Superfund sites are presented in Table 12.3. It is quite likely that one or more of these contaminants will be present at most hazardous waste sites. Inspection of Table 12.3 shows that the contaminants comprise a wide variety of classes or contaminant types, ranging from metals to solvents to pesticides to fuel compounds.

The U.S. EPA has developed special reporting rules for certain chemicals of concern under the Toxic Release Inventory program. These chemicals are classified as **persistent, bioaccumulative, and toxic** (PBT) chemicals. These compounds pose increased risk to human health not only because they are toxic, but also because they

Environmental and Pollution Science. https://doi.org/10.1016/B978-0-12-814719-1.00012-4

TABLE 12.1 Examples of Organic, Inorganic, and Radioactive Chemical Contaminants

Organic Contaminants

Petroleum hydrocarbons (fuels)—Benzene, toluene, xylene, polycyclic aromatics

Chlorinated solvents—Trichloroethene, tetrachloroethene, trichloroethane, carbon tetrachloride

Pesticides—DDT (dichloro-diphenyl-trichloro-ethane), 2,4-D (2,4-Dichlorophenoxyacetic acid), atrazine

Polychlorinated biphenyls (PCBs)—insulating fluids, plasticizers, pigments

Coal tar/creosote—Polycyclic aromatics

Pharmaceuticals/food additives/cosmetics—Drugs, surfactants, dyes

Gaseous compounds—Chlorofluorocarbons (CFCs), hydrochlorofluorocarbons (HCFCs)

Inorganic Contaminants

Inorganic "salts"—Sodium, calcium, nitrate, sulfate

Heavy/trace metals—Lead, zinc, cadmium, mercury, arsenic

Radioactive Contaminants

Solid elements—Uranium, strontium, cobalt, plutonium

Gaseous elements—Radon

remain in the environment for long periods of time, are not readily destroyed, and build up or accumulate in body tissue.

In a related development, an international treaty was enacted to control the future production of a class of chemicals termed **persistent organic pollutants** (POPs). The Stockholm Convention is a global treaty to protect human health and the environment from POPs, which are chemicals that remain intact for long periods, become widely distributed geographically, accumulate in the fatty tissue of living organisms, and are toxic. There are 26 chemicals currently on the Annex A and B of the POP list, including aldrin, chlordane, DDT, dieldrin, dioxins, endrin, furans, heptachlor, hexachlorobenzene, mirex, polychlorinated biphenyls, and toxaphene.

12.3 CHEMICAL CONTAMINANT SOURCES

12.3.1 Agricultural Activities

Agricultural systems consist of highly managed tracts of land that generally receive large inputs of chemical fertilizers and pesticides. The ultimate goal of these chemical additions is to generate optimum amounts of food and fiber. However, fertilizers are often applied in excess of the crop needs or are in chemical forms that make them very mobile in soil and water environments. For example, nitrate pollution of groundwater is often caused by excessive nitrogen fertilizer applications that result in leaching below the root zone. Agricultural activities can cause land, water, and air pollution.

Fertilizers, which are generally inorganic chemicals, are routinely applied at least once a year and include, in order of decreasing amounts, N, P, K, and metals. The annual applications of these chemicals range from 50 to $200 \, kg \, ha^{-1}$, as N, P, or K. Micronutrient (e.g., Fe, Zn, Cu, B, and Mo) fertilizer additions are also applied regularly to agricultural fields but with less frequency because of lower crop requirements. These chemicals are applied to agricultural lands every 2–5 five years at average rates of $0.5–2 \, kg \, ha^{-1}$, in their respective elemental forms. A third group of inorganic chemicals applied to agricultural land consists of soil amendments. These materials are applied to agricultural fields with some frequency for two reasons: (1) to decrease or increase soil pH, decrease soil salinity, and improve soil structure and (2) to replenish macronutrients like Ca^{++}, Mg^{++}, K^+, and $SO_4^=$. To control macronutrient deficiencies, the application rates of these chemicals range from 50 to $500 \, kg \, ha^{-1}$. To control soil pH and salinity, applications typically range from 2000 to $10,000 \, kg \, ha^{-1}$. The common forms of these chemicals are listed in Table 12.4 in order of decreasing probable impact to the environment.

The inorganic chemicals listed in Table 12.4 can act as a nutrient and as a pollutant, depending on the amounts applied, the location of application, and soil-plant-water dynamics. For example, Fig. 12.1 shows the soil nitrogen cycle, which illustrates the transformations, sinks, and sources of this element. Plants and some soil minerals can act as sinks for the two major forms of N. Conversely, some plants, animals, the atmosphere, and humans (fertilizer additions) can contribute to excessive N (NO_3^-) concentrations that lead to groundwater pollution. Groundwater polluted with high levels of nitrate has been shown to cause methemoglobinemia (blue baby syndrome) in infants and some adults. Methemoglobinemia occurs when nitrate is converted to nitrite by the digestive system. Nitrite reacts with oxyhemoglobin (oxygen carrying blood protein), forming methemoglobin. Methemoglobin cannot carry oxygen resulting in a decreased ability of the blood to carry oxygen. Consequently, oxygen deprivation in body tissues can occur. Infants suffering from methemoglobinemia develop a blue coloration of their mucous membranes and possible digestive and respiratory problems.

TABLE 12.2 National Primary Drinking Water Standards

National Primary Drinking Water Regulations

Contaminant	MCL or TT[1] (mg/L)[2]	Potential Health Effects From Long-Term[3] Exposure Above the MCL	Common Sources of Contaminant in Drinking Water	Public Health Goal (mg/L)[2]
Acrylamide	TT[4]	Nervous system or blood problems; increased risk of cancer	Added to water during sewage/wastewater treatment	zero
Alachlor	0.002	Eye, liver, kidney, or spleen problems; anemia; increased risk of cancer	Runoff from herbicide used on row crops	zero
Alpha/photon emitters	15 picocuries per Liter (pCi/L)	Increased risk of cancer	Erosion of natural deposits of certain minerals that are radioactive and may emit a form of radiation known as alpha radiation	zero
Antimony	0.006	Increase in blood cholesterol; decrease in blood sugar	Discharge from petroleum refineries; fire retardants; ceramics; electronics; solder	0.006
Arsenic	0.010	Skin damage or problems with circulatory systems, and may have increased risk of getting cancer	Erosion of natural deposits; runoff from orchards; runoff from glass & electronics production wastes	0
Asbestos (fibers >10 micrometers)	7 million fibers per Liter (MFL)	Increased risk of developing benign intestinal polyps	Decay of asbestos cement in water mains; erosion of natural deposits	7 MFL
Atrazine	0.003	Cardiovascular system or reproductive problems	Runoff from herbicide used on row crops	0.003
Barium	2	Increase in blood pressure	Discharge of drilling wastes; discharge from metal refineries; erosion of natural deposits	2
Benzene	0.005	Anemia; decrease in blood platelets; increased risk of cancer	Discharge from factories; leaching from gas storage tanks and landfills	zero
Benzo(a)pyrene (PAHs)	0.0002	Reproductive difficulties; increased risk of cancer	Leaching from linings of water storage tanks and distribution lines	zero
Beryllium	0.004	Intestinal lesions	Discharge from metal refineries and coal-burning factories; discharge from electrical, aerospace, and defense industries	0.004
Beta photon emitters	4 millirems per year	Increased risk of cancer	Decay of natural and man-made deposits of certain minerals that are radioactive and may emit forms of radiation known as photons and beta radiation	zero
Bromate	0.010	Increased risk of cancer	Byproduct of drinking water disinfection	zero
Cadmium	0.005	Kidney damage	Corrosion of galvanized pipes; erosion of natural deposits; discharge from metal refineries; runoff from waste batteries and paints	0.005
Carbofuran	0.04	Problems with blood, nervous system, or reproductive system	Leaching of soil fumigant used on rice and alfalfa	0.04

Legend

Disinfectant | Disinfectant by product | Inorganic chemical | Microorganism | Organic chemical | Radionuclides

Example first page. [a]MCL = maximum contaminant level, the highest level of a contaminant that is allowed in drinking water; TT = treatment technique level. From: https://www.epa.gov/ground-water-and-drinking-water/national-primary-drinking-water-regulations.

Most pesticides are organic compounds and are often applied in agricultural systems at least once a year, albeit in much smaller quantities than fertilizers. However, synthetic pesticides, designed to be very toxic to plants and pests, may have deleterious effects at very low concentrations. Most synthetic pesticides are broadly classified as insecticides, herbicides, and fungicides. While most pesticides are solids, they are usually dissolved in water or oil to facilitate their handling and application. Fumigants are gaseous pesticides typically used to control

TABLE 12.3 Substance Priority List for Superfund Sites

1	Arsenic
2	Lead
3	Mercury
4	Vinyl Chloride
5	Polychlorinated Biphenyls
6	Benzene
7	Cadmium
8	Benzo(A)Pyrene
9	Polycyclic Aromatic Hydrocarbons
10	Benzo(B)Fluoranthene
11	Chloroform
12	Aroclor 1260
13	DDT[a], P,P'-
14	Aroclor 1254
15	Dibenzo(A,H)Anthracene
16	Trichloroethene
17	Chromium, Hexavalent
18	Dieldrin
19	Phosphorus, White
20	Hexachlorobutadiene

[a]DDT = dichlorodiphenyltrichloroethane.
From: https://www.atsdr.cdc.gov/spl/index.html.

insects. A list of common organic pesticides is presented in Table 12.5. Less common forms of inorganic pesticides are used to control roaches and rats. These chemicals, which have all too often been used in close proximity to humans, have, as their primary acting agent, toxic forms of arsenic (AsO_4^{3-}), boron (H_3BO_3), and S (SO_2).

The chemical structure of organic pesticides controls their water solubility, mobility, environmental persistence, and toxicity. The first generation of organic pesticides had multiple chlorine groups inserted into their structures to give them a broad spectrum of biotoxic effects. However, the chlorine groups also made them very difficult to degrade, making them very persistent (see Chapter 9). The next step in pesticide development sought a compromise between persistence and toxicity, with chemical structures that were moderately soluble in water and with more targeted toxicity effects. The next generation of pesticides again sought to decrease the persistence of these chemicals in the environment by making them even more water soluble and continued to focus their toxic effects. This class of pesticides seldom bioaccumulate in humans or animals and have short life spans (days) in the environment. However, when misused, these chemicals can be found in water sources. For example, today the members of the triazine family are the most commonly found pesticides in surface and groundwater resources. Conversely, chlorinated pesticides are seldom found in water but can still be found in soils and sediments.

Animals generate significant amounts of residues that are benign to the environment in open environments with low concentrations of animals. However, in the last 100 years, large-scale animal production systems have created

TABLE 12.4 Common Fertilizer and Soil Amendments Materials and Potential Contaminant Forms

Fertilizers	Nutrient Form	Pollutant Properties
NH_3(gas), $CO(NH_2)_2$ (urea),	NH_4NO_3, $(NH_4)_2SO_4$, KNO_3	- very mobile, promotes microbial growth
NH_4–PO_4 solutions.	NO_3^-, NH_4^+ PO_4^{3-}	- toxic, volatile as NH_3 - promotes eutrophication
Superphosphate, triple superphosphate, N-P solutions	PO_4^{3-}, Ca^{++}	- variable mobility, promotes microbial growth - increases water hardness
Ammonium phosphate	NO_3^-, NH_4^+, PO_4^{3-}	- see prior
Calcite ($CaCO_3$)	Ca^{++}, CO_3^-	- increases soil water alkalinity
Gypsum ($CaSO_4 \cdot 2H_2O$)	Ca^{++}, $SO_4^=$	- mobile, may pollute water sources
Micronutrients, salt forms, chelates	Fe^{++}, Mn^{++}, Zn^{++}, Cu^{++}, $MoO_4^=$, H_3BO_3, Cl^-	- cations are mobile in acid soils - anions are mobile in alkaline soils

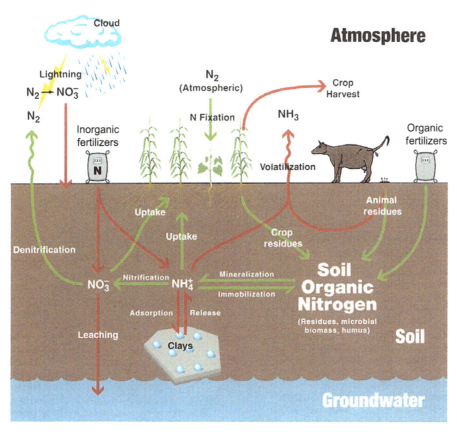

FIG. 12.1   Soil-nitrogen transformations. *(From Environmental Monitoring and Characterization © 2004, Elsevier Academic Press, San Diego, CA.)*

TABLE 12.5 Major Classes of Organic Pesticides and Their Potential Pollutant Properties

Class/Elemental Composition	Common Examples	Pollutant Properties
Organochlorines	DDT	Resistant to degradation (persistent)
Organophosphates	Chlorpyrifos	Mobile in the soil environment
Carbamates	Carbaryl	Very mobile in the soil environment
Triazines	Atrazine	Very mobile in the soil environment
Plant Insecticides	Pyrethroids	Some toxic to fish
Fumigants	Dichloropropene	Toxic to animals, volatile

Note: All of these chemicals have some degree of toxicity (acute and/or chronic) toxicity to humans.

concentrated sources of animal-derived contaminants. Large-scale animal feeding operations include feedlots for beef, swine, and poultry production, dairies, and fish farms. These operations act as point sources for the common chemicals listed in Table 12.5 (see Fig. 12.2). Nitrate-N, ammonium-N, and phosphate-P are the three most common contaminants derived from unregulated animal waste disposal practices. These three chemicals are usually found at concentrations ranging from 1000 to 50,000 mg kg^{-1} (elemental form) in animal wastes.

Small quantities of phosphate (>1 mg L^{-1}) can be extremely deleterious to stagnant water bodies because phosphates can trigger excessive microbial growth that leads to eutrophication. Information Box 12.1 shows a list of contaminants, in addition to N and P, that concentrated animal wastes can introduce in significant amounts into the environment. Pharmacuetical compounds used to treat animals are one set of contaminants of particular potential concern for impacts on human health. These are discussed in greater detail in a following section.

FIG. 12.2 Runoff from feedlots may enter nearby surface water and degrade water quality. *(Photo courtesy USDA National Resources Conservation Service.)*

INFORMATION BOX 12.1

Pollutants Released from Animal Wastes

- Total dissolved solids (TDS) (Na, Cl, Ca, Mg, K, soluble N and P forms): Most animal wastes are high ($>>10,000\,mg\,L^{-1}$) in TDS.
- Organic carbon: Excessive amounts of soluble carbon together with soluble P can quickly reduce O_2 availability in water by raising the biochemical oxygen demand.
- Residual pesticides: Used to control pests in animal facilities.
- Residual metals: Cu, As, from animal diets and pesticides.
- Pharmaceuticals: Antibiotics, growth regulators.
- Gases: From waste storage facilities and waste disposal activities, Greenhouse—(CO_2, N_2O), toxic (NH_3, H_2S), Odors—H_2S, mercaptans, indoles, org-sulfides.

12.3.2 Sources: Industrial and Manufacturing Activities

There are numerous sources of industrial chemical contaminants, the result of controlled or uncontrolled waste disposal and releases into the environment. Industrial wastes may contain contaminants classified by the Federal government as hazardous and nonhazardous. However, this classification primarily separates wastes containing high concentrations of pollutants versus wastes that contain low concentrations. For example, metal-plating industrial

---

INFORMATION BOX 12.2

Industrial Wastes and Sources of Contaminants

Solid, liquid, and slurry wastes with high concentrations of metals, salts, and solvents

Industries: Metal plating, painting.

Types of pollutants: Metals, solvents, toxic aromatic and nonaromatic hydrocarbons.

Liquid wastes with high concentrations of hydrocarbons and solvents

Industries: Chemical manufacturing, electronics manufacturing, plastics manufacturing.

Types of pollutants: Chlorinated solvents, hydrocarbons, plastics, plasticizers, metals, catalysts, cyanides, sulfides.

Wastewaters containing organic chemicals

Industries: Paper processing, tanneries, food processing, industrial wastewater treatment plants, pharmaceuticals.

Types of pollutants: Various organic chemicals.

---

wastes contain high concentrations of toxic metals such as Cr, Ni, and Cd and are usually classified as hazardous. However, municipal wastes, classified as nonhazardous, also contain these metals and many others, but at much lower concentrations.

Most industrial contaminants originate from a few general categories of industrial wastes. These are summarized in Information Box 12.2, with examples of industries and their common classes of contaminants. Industrial and manufacturing activities have produced many pollution problems for soil, surface water, and groundwater resources (see Chapters 14–16).

12.3.3 Sources: Municipal Waste

Municipal solid waste, more commonly known as trash or garbage, is a primary potential source of pollution. Municipal solid waste consists of items such as paper, food scraps, grass clippings, product packaging, bottles, clothes, and furniture. Many households also improperly discard hazardous household waste into their municipal waste receptacles. Hazardous household waste products can be dangerous to human health and the environment, and should be sent to a proper disposal facility. Examples of hazardous household waste include paint, cleaners, oils, pesticides, and batteries. Municipal solid waste is collected and disposed of by landfill or combustion/incineration. Burning municipal solid waste will reduce its volume by up to 90% and its weight by up to 75%. However, air emissions pose an environmental concern. Landfilling municipal solid waste also causes an environmental concern. Landfills produce carbon dioxide and methane, both of which are greenhouse gases. Many landfills capture methane to use as an energy source. Another source of landfill pollution is landfill leachate, which is formed when water percolates through the landfill, dissolving compounds along the way. Landfill leachate may contain heavy metals, ammonia, toxic organic compounds, and pathogens, and is of concern as a groundwater pollutant (see Chapter 15).

Municipal wastewater treatment plants produce wastes that contain many potential contaminants (see Chapter 22). Reclaimed wastewater is usually clean enough to be used for irrigation, but routinely contains higher (~1.5 times) concentrations of dissolved solids than the source water. Also, chlorine-disinfected reclaimed water can contain significant trace amounts of disinfection by-products such as trihalomethanes and haloacetic acids. In addition, an emerging issue for municipal wastewater treatment is pharmaceutical waste. There is growing concern that pharmaceuticals (including hormones from birth control pills and antibiotics) that are excreted in urine and disposed of in wastes may end up in water supply resources. Many of these compounds are not fully treated in current wastewater treatment systems. There is concern about the effects that these compounds may have on humans and wildlife.

The solid residues of wastewater treatment plants, called biosolids, typically contain common inorganic chemicals such as those listed in Table 12.5 and may also contain heavy metals, synthetic organic compounds found in household products, and microbial pathogens. Since biosolids usually contain macro- and micronutrients and

organic carbon, they are routinely applied to agricultural lands as fertilizer and soil amendments (see Chapter 23). Regulations in many states allow for the annual application of up to 8 tons (dry weight) of biosolids on farmland, depending on the metal content of each biosolids source. Land disposal of biosolids completes the natural C and N cycle in the environment. However, repeated application of biosolids often increases the concentrations of metals, P, and some salts in the soil environment. In addition, excessive, concentrated, or uneven applications of biosolids can result in surface and groundwater pollution.

Stormwater is a source of nonpoint-source pollution for both urban and rural communities. Stormwater runoff entrains pollutants as it flows over the ground surface. In urban areas, stormwater runoff will flow over a variety of impervious surfaces, including driveways, parking lots, and streets, acquiring pollutants such as dirt, debris, and hazardous wastes such as insecticides, pesticides, paint, solvents, used motor oil, and other auto fluids (Fig. 12.3). In agricultural areas, stormwater runoff may include dirt, debris, excess nutrients, pesticides, bacteria, and other pathogens. Stormwater will either flow into a sewer system or directly into a lake, stream, river, wetland, or coastal water. In some cities, stormwater runoff flows into a storm sewer system and the collected water is discharged untreated into water bodies. In many areas, stormwater and municipal wastewater enter the same sewer system. During large storm events, wastewater treatment facilities often receive more municipal and stormwater than the facility can handle. When facilities are unable to handle incoming waste, untreated municipal wastewater and stormwater are discharged without treatment.

Septic systems are another repository for municipal waste (Fig. 12.4). Approximately one-fifth of all homes in the United States use a septic system for household wastewater disposal, with several billion gallons of wastewater disposed below the ground surface daily. Septic systems use microbial communities to decompose and digest waste. Most bacteria recover quickly after small amounts of cleaning products have entered the system. However, excess chemical use can cause a septic system to fail. Table 12.6 presents examples of items that can either clog a septic system or kill the microbial populations in the system. To prevent pollutants in household wastewater from entering groundwater, it is extremely important to maintain household septic systems and to make sure they are functioning properly. Typical household wastewater pollutants include nitrogen, phosphorous, and disease-causing bacteria and viruses. To ensure that a septic system is working properly, it should be inspected every three years and pumped every three to five years.

FIG. 12.3 Stormwater runoff will flow over impervious surfaces, acquiring pollutants. *(Photo courtesy USGS. http://www.umesc.usgs.gov/flood_2001/surface.html.)*

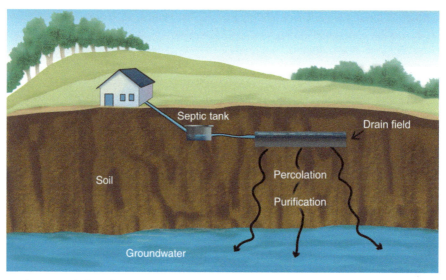

FIG. 12.4 A septic system is composed of a septic tank and a drain field. Wastewater in the drain field will percolate through the subsurface, which acts as a purification system. If the system is working properly, wastewater is free of pollutants before reaching groundwater. *(Image from A Homeowner's Guide to Septic Systems. EPA-832-B-02-005, December 2002. U.S. EPA, Washington, DC.)*

12.3.4 Sources: Service-Related Activities

There are many service activities that produce waste materials that are potential sources of environmental pollution, especially for groundwater (see Chapter 15). The service industries that produce substantial amounts of waste include dry cleaners and laundry plants, automotive service and repair shops, and fuel stations. These facilities are subject to regulation under the Resource Conservation and Recovery Act (RCRA) if they generate wastes that fall under RCRA's definition of a hazardous waste (see Chapter 30).

Dry cleaning, a service industry involved in the cleaning of textiles, uses solvents in the cleaning process that

TABLE 12.6 Items That Can Clog or Damage Septic Systems

Cloggers	Damage Microbial Communities
Diapers	Household chemicals
Cat litter	Gasoline
Cigarette filters	Oil
Coffee grounds	Pesticides
Grease	Antifreeze
Feminine hygiene products	Paint

Adapted from U.S. EPA, *A Homeowner's Guide to Septic Systems.*

are considered as hazardous waste. These solvents include tetrachloroethene, petroleum solvents, and 1,1,1-trichloroethane. Along with spent solvents, other wastes produced are solvent containers, spent filter cartridges, residues from solvent distillation, and solvent-contaminated wastewater.

Underground storage tanks (USTs) are used to hold petroleum products and certain hazardous substances for several service-related activities. Until 1984, many USTs were not equipped with spill, overfill, and corrosion protection. As a result, these USTs have leaked and polluted soil and groundwater (Fig. 12.5). Vapors and odors from leaking underground storage tanks (LUSTs) can collect in basements, utility vaults, and parking garages. Collected vapors can cause explosions, fires, asphyxiation, or other adverse health effects. Petroleum-based fuels, such as gasoline, diesel fuel, and aviation fuels, are ubiquitous sources of contamination at automotive, train, and aviation fuel stations. The lower molecular weight, more soluble constituents, such as

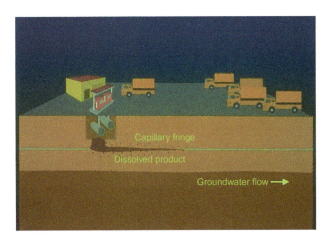

FIG. 12.5 Underground storage tanks can leak causing pollution of soil and groundwater. *(Image courtesy U.S. EPA. http://www.epa.gov/swerust1/graphics/cca017.jpg.)*

benzene and toluene, are of special concern with respect to groundwater contamination potential. In addition, some fuel additives may also be of concern. For example, **methyl-tertiary-butyl ether** (MTBE) is a hydrocarbon derivative that was added to gasoline in the recent past to boost the oxygen content of the fuel. This was done in accordance with federal regulations formulated to improve air quality. However, MTBE is a very soluble compound that is also resistant to biodegradation. It is a very mobile and persistent compound, and this nature has led to widespread groundwater contamination (Chapter 15). Low levels of MTBE can make water supplies undrinkable due to its offensive taste and odor. The use of MTBE in gasoline was phased out as a result of this situation.

Automotive service and repair shops can be a source of numerous contaminants. Various types of solvents are used to degrease and clean engine parts. Metal contaminants can originate from batteries, circuit boards, and other vehicle components. Fuel-based contaminants are also typically present.

12.3.5 Sources: Resource Extraction/Production

Mineral extraction (mining) and petroleum and gas production are major resource extraction activities that provide the raw materials to support our economic infrastructure. An enormous amount of pollution is generated from the extraction and use of natural resources. The Environmental Protection Agency's Toxic Releases Inventory report lists mining as the single largest source of toxic waste of all industries in the United States. Mineral extraction sites, which include strip mines, quarries, and underground mines, contribute to surface water and groundwater pollution, erosion, and sedimentation (see Chapter 14). The mining process involves the excavation of large amounts of waste rock in order to remove the desired mineral ore (Fig. 12.6). The ore is then crushed into finely ground tailings for chemical processing and separation to extract the target minerals. After the minerals are processed, the waste rock and mine tailings are stored in large aboveground piles and containment areas (see also Chapter 14). These waste piles, along with the bedrock walls exposed from mining, pose a huge environmental problem because of the metal pollution associated primarily with acid mine drainage. Acid mine drainage is caused when water draining through surface mines, deep mines, and waste piles comes in contact with exposed rocks containing pyrite, an iron sulfide, causing a chemical reaction. The resulting water is high in sulfuric acid and contains elevated levels of dissolved iron. This acid runoff also dissolves heavy metals such as lead, copper, and mercury, resulting in surface and groundwater contamination. Wind erosion of mine tailings is also a significant problem.

FIG. 12.6   Acid mine drainage has collected at the bottom of this pit mine in Bisbee, Arizona. *(Photo courtesy Alex Merrill.)*

Petroleum and natural gas extraction pose environmental threats such as leaks and spills that occur during drilling and extraction from wells, and air pollution as natural gas is burned off at oil wells (Chapter 14). The petroleum and natural gas extraction process generates production wastes including drilling cuttings and muds, produced water, and drilling fluids. Drilling fluids, which contain many different components, can be oil based, consisting of crude oil or other mixtures of organic substances like diesel oil and paraffin oils, or water based, consisting of freshwater or seawater mixed with bentonite and barite. Each component of a drilling fluid has a different chemical function. For example, barite is used to regulate hydrostatic pressure in drilling wells. As a result of being exposed to these drilling fluids, drilling cuttings and muds contain hundreds of different substances. This waste is usually stored in waste pits, and if the pits are unlined, the toxic chemicals in the spent waste cuttings and muds, such as hydrocarbon-based lubricating fluids, can pollute soil, surface, and groundwater systems. **Produced water** is the wastewater created when water is injected into oil and gas reservoirs to force the oil to the surface, mixing with formation water (the layer of water naturally residing under the

hydrocarbons). At the surface, produced water is treated to remove as much oil as possible before it is reinjected, and eventually when the oil field is depleted, the well fills with the produced water. Even after treatment, produced water can still contain oil, low-molecular-weight hydrocarbons, inorganic salts, and chemicals used to increase hydrocarbon extraction.

Mined and extracted resources can also be potential pollutants once they are used for production. For example, fossil fuels are key resources for energy production. Coal-burning power plants produce nitrogen and sulfur oxides, which are known to be the primary causes of acid rain (see Chapter 17). In addition, fossil fuel combustion produces carbon dioxide, which is a primary culprit in global climate change.

12.4   RADIOACTIVE CONTAMINANTS

Radioactive waste primarily originates from nuclear fuel production and reprocessing, nuclear power generation, military weapons development, and biomedical and industrial activities. The largest quantities of radioactive waste, in terms of both radioactivity and volume, are generated by commercial nuclear power and military nuclear weapons production industries, and by activities that support these industries, such as uranium mining and processing. However, radioactive material can also originate from natural sources. Groundwater contamination by radioactive waste is a major problem at several Department of Energy facilities in the United States. Selected examples of radioisotopes are presented in Table 12.7.

Naturally occurring sources of radioactive materials, including soil, rocks, and minerals that contain radionuclides, can be concentrated and exposed by human industrial activities such as uranium mining, oil and gas production, and phosphate fertilizer production. For example, when uranium is mined using in situ leaching or surface methods, bulk waste material is generated from excavated topsoil, uranium waste rock, and subgrade ores, all of which can contain radionuclides of radium, thorium,

TABLE 12.7 Selected Natural and Anthropogenic Radioisotopes

Element	Radioisotope	Origin	Activity
Uranium	^{238}U	Natural, enriched	Uranium mining
Radium	^{226}Ra	Natural, enriched	Uranium mining
Radon	^{222}Rn	Natural, enriched	Uranium mining, construction
Strontium	^{90}Sr	Fission product	Reactors, weapons
Cesium	^{137}Cs	Natural, fission product	Reactors, weapons

and uranium. Other extraction and processing practices that can generate and accumulate radioactive wastes similar to that of uranium mining are aluminum and copper mining, titanium ore extraction, and petroleum production. According to EPA reports, the total amounts of naturally occurring radioactive waste that are enhanced by industrial practices number in excess of 1 billion tons annually. Sometimes, the levels of radiation are relatively low in comparison to the large volume of material that contains the radioactive waste. This causes a problem because of the high cost of disposing of radioactive waste in comparison with the relatively low value of the product from which the radioactive waste is separated. Additionally, relatively few licensed disposal locations can accept radioactive waste.

Radioactive wastes are classified for disposal according to their physical and chemical properties, along with the source from which the waste originated. The half-life of the radionuclide and the chemical form in which it exists are the most influential of the physical properties that determine waste management. The United States divides its radioactive waste into the following categories: high-level waste, transuranic waste, and low-level waste. High-level waste consists of spent irradiated nuclear fuel from commercial reactors and the liquid waste from solvent extraction cycles along with the solids that liquid wastes have been converted into from reprocessing. Transuranic wastes are alpha-emitting residues that contain elements with atomic numbers greater than 92, which is the atomic number of uranium. Wastes are considered transuranic when the elements have half-lives greater than 20 years and concentrations exceeding $100 \, nCi \, g^{-1}$. Wastes in this category originate primarily from military manufacturing, with plutonium and americium being the principal elements of concern. Low-level waste encompasses the radioactive waste that is not classified under the other two categories. Low-level wastes are separated into subcategories: Classes A, B, and C, with Class A being the least hazardous and C being the most hazardous. Commercial low-level waste is generated by industry, medical facilities, research institutions and universities, and a few government facilities.

In some commercial and military activities, radioactive wastes are mixed with hazardous waste, creating a complex environmental problem. Mixed waste is dually regulated by the EPA and the United States Nuclear Regulatory Commission, and waste handlers must comply with both the Atomic Energy Act and the Resource Conservation and Recovery Act statutes and regulations once a waste is deemed a mixed waste. Military sources are regulated by the Department of Energy and comply with the Atomic Energy Act in regard to radiation safety.

Radon, a naturally occurring radioactive gas that is produced by the radioactive decay of uranium in rock, soil, and water, is of great concern because of the potential for the gas to become concentrated in buildings and homes (see also Chapter 18). The higher the uranium levels in the rocks, the greater the chances that a home or building may have radon gas contamination. Once the parent material decays into radon, it dissolves into the water contained in the pore spaces between soil grains. A fraction of the radon in the pore water volatilizes into the soil atmosphere gas, rendering it more mobile via gas-phase diffusion.

Exposure of humans to radon occurs in several ways. Decay products of radon are electrically charged when formed, so they tend to attach themselves to atmospheric dust particles that are normally present in the air. This dust can be inhaled, and while the inert gases are mostly exhaled immediately, a fraction of the dust particles deposit on the lungs, building up with every breath. Radon dissolved in groundwater is another source of human exposure, mainly because radon gas is released into the home atmosphere from water as it exits the tap. Another source of human exposure in home and building settings is the tendency for radon gas to enter structures via diffusion through their foundations and from certain construction materials. Radon gas availability in structures is mainly associated with the concentration of radon in the rock fractures and soil pores surrounding the structure and the permeability of the ground to gases. Slight pressure differentials between structure and soil foundations, which can be caused by barometric changes, winds, and temperature differentials, create a gradient for radon gas to move from soil gas, through the foundations, and into the internal atmosphere of the structure (indoor air).

12.5 NATURAL SOURCES OF CONTAMINANTS

The contaminant sources presented before are associated with human activities involving the production, use, and disposal of resources, chemicals, and products. It is important to realize that there are also natural sources of contaminants. A major source of such contaminants is drinking water pumped from aquifers composed of sediments and rocks containing naturally occurring elements that dissolve into the groundwater. One example, that of radioactive contaminants such as radon, was discussed in the previous section. Another major example is arsenic, which has become of great concern in recent years (see Case Study 12.1). More discussion of this topic is presented in Chapter 27.

12.6 EMERGING CONTAMINANTS

As mentioned before, some chemicals are regulated to prevent impacts to human health. For example, a number of chemicals are regulated in drinking water through the National Primary Drinking Water Regulations under the

Case Study 12.1 Arsenic Pollution in Bangladesh

Arsenic occurs naturally in aquifers of the country of Bangladesh. As a result, perhaps as many as 50% or more of the 125 million people of this country may be exposed to high (from 50 to $<1000\,\mu g\,L^{-1}$) arsenic (As) concentrations found in their drinking water. Long-term chronic exposure to As promotes several skin diseases (from dermatitis to depigmentation). More advanced stages of As exposure produce gastroenteritis, gangrene, and cancer, among other diseases. More than 2 million people in Bangladesh suffer from one or more of these As-induced diseases. High As concentrations in the groundwater have been associated with As-rich sediments from the Holocene period. These sediments are primarily found in the flood and delta plains of Bangladesh. In these areas, $>60\%$ of the wells have elevated As concentrations.

Arsenic exists in two oxidation states—arsenate, As(III), and arsenite, As(V)—both of which are anions (see also Chapter 8). Although both forms are toxic, arsenite is much more toxic and is also very soluble and mobile in water environments. The exact mechanism of As enrichment in the groundwater of Bangladesh is not known but is likely related to the presence of arsenite-bearing minerals and the reductive dissolution of arsenate to the much more soluble form of arsenite. Iron reacts with As anions and can form insoluble and eventually very stable Fe-As complexes that remove As from water. In fact, amorphous Fe oxide is commonly used by water utilities to decontaminate drinking water. Another possible means of treating As-contaminated water include the use of natural soil material (as filtering devices) that contain high concentrations of iron minerals such as goethite and hematite, which can adsorb As.

No country is immune to the effect of this natural pollutant. In the United States, the drinking water standard is $10\,\mu g\,L^{-1}$ as set by the EPA. The states most likely to have groundwater sources with elevated As concentrations include Arizona, New Mexico, Nevada, Utah, and California. More information about arsenic in groundwater is presented in Chapters 15 and 27.

Safe Drinking Water Act (Chapter 30). However, not all chemicals of concern are currently regulated. These include chemicals that have appeared in the environment more recently and those that have been present for some time but for which new information has indicated greater toxicity than originally thought. These chemicals are called **emerging contaminants** (ECs).

The US EPA has a special program, conducted under the Unregulated Contaminant Monitoring Rule of the Safe Drinking Water Act (SDWA), to manage some emerging contaminants of greatest concern. EPA is required to routinely identify and analyze emerging contaminants and provide guidance to states, local officials, and the public about the potential public health risks and acceptable contamination levels for these materials. As part of this effort,

they periodically publish a Contaminant Candidate List—a list of contaminants that:

- Are not regulated by the National Primary Drinking Water Regulations
- Are known or anticipated to occur in public water systems
- May warrant regulation under the SDWA due to toxicity concerns

Examples of chemicals placed on the Contaminant Candidate List are presented in Table 12.8. Chemicals placed on the CCL undergo review by the EPA for eventual decisions on whether they should become a regulated compound. In the meantime, EPA may issue advisory levels for limits in drinking water. These are not enforceable, but rather serve as guidelines for consideration in water management and site cleanup.

Inspection of Table 12.8 shows that emerging contaminants (ECs) comprise many different types of chemicals. A list of major classes of ECs is provided in Table 12.9. Many of the emerging organic contaminants of concern are endocrine disruptor compounds (EDCs)—chemicals that interfere with endocrine glands, their hormones, or the activities of hormones. A primary source of EDCs is pharmaceuticals and personal care products introduced into soil or surface waters via treated wastewater or biosolids applications. Pharmaceuticals and other personal care products have been reported in the water cycle, including surface waters, wastewater, groundwater and, to a lesser extent, drinking water. The reported levels are typically in the nanograms to low micrograms per liter range (WHO, 2011). Some examples of the hormones and hormone mimics found in U.S. surface waters are presented in Table 12.10.

A reasonable perspective of the potential adverse health risks of EDCs can be obtained by comparing the maximum concentrations found in waters with the medicinal dosage of various pharmaceuticals. For example, drinking water that contained the highest concentration of ibuprofen found in a USGS survey of EDCs in water across the United States (1 μg/L) would require ∼270 years to be equivalent to two Advil tablets (400 mg) assuming consumption of 4 L of water per day. As another example, the maximum concentration of caffeine found in the USGS study was 6 μg/L. Assuming that an 8 oz. cup of coffee contains 135 mg of caffeine, a consumer would need to drink 22,500 L of water to ingest the amount of caffeine equivalent to one cup of coffee.

To date, adverse human health effects of pharmaceuticals in drinking water have not been clearly documented. For example, analysis of human health risk assessments by the World Health Organization indicated that appreciable adverse health impacts to humans are very unlikely from exposure to the trace concentrations of pharmaceuticals that

TABLE 12.8 Example Emerging Contaminants on the EPA Contaminant Candidate List (CCL)

Chemical	CCL	Source/Use	Notes
Alachlor	1-1998	Pesticide	
Methyl-t-butyl ether	1-1998	Gasoline additive	MTBE
Perchlorate	1-1998	Rocket fuel	
RDX	1-1998	Explosive	
Triazines	1-1998	Pesticides	Examples: atrazine, cyanazine
1,1,2,2-Tetrachloroethane	2-2005	Solvent	
1,4-Dioxane	3-2009	Solvent, stabilizer	
Ethylene glycol	3-2009	Antifreeze	
PFOA	3-2009	Textile treatment, other	Perfluorooctanoic acid
PFOS	3-2009	Textile treatment, other	Perfluorooctanesulfonic acid

Selected from full list at https://www.epa.gov/ccl.

TABLE 12.9 EC Groups Detected in Arizona Waters

Constituent Categories	Colorado River	Other Rivers Streams Lakes	Groundwater	Wastewater Reclaimed Water	Drinking Water
Pharmaceuticals	Yes	Yes	Yes	Yes	Yes
Personal Care Products	Yes	Yes	Yes	Yes	Yes
Industrial Chemicals	Yes	Yes	Yes	Yes	Yes
Flame Retardants	Yes	Yes	Yes	Yes	Yes
Pesticides/Herbicides	Yes	Yes	Yes	Yes	Yes
Surfactants	Yes	Yes	Yes	Yes	Yes
Steroids	Yes	No	No	Yes	No
Illicit Drugs	Yes	Yes	No	Yes	No

From: Emerging Contaminants in Arizona Water, 2016, Arizona Department of Environmental Quality.

TABLE 12.10 Examples of Hormones and Pharmaceuticals Found in US Surface Waters

Compound	Description	Compound	Description
17β-Estradiol	Reproductive hormone	Caffeine	Stimulant
Esrone	Reproductive hormone	Ibuprofen	Antiinflammatory
4 Nonylphenol	Detergent metabolite	Erythromycin	Antibiotic
Testosterone	Reproductive hormone	Ciprofloxacin	Antibiotic

Source: New original table.

could potentially be found in drinking water (WHO, 2011). However, concern remains regarding the potential impacts of long-term low-dose exposure to EDCs to human health. A related concern is the potential effects of exposures to mixtures of EDCs including synergistic effects.

In contrast to human health, significant adverse impacts of EDCs to aquatic life have been demonstrated. These include developmental abnormalities in fish and amphibians such as intersex characteristics.

Much of the prior focus on emerging contaminants was directed toward pharmaceuticals and their potential endocrine disrupting effects. However, many emerging contaminants are not pharmaceuticals or personal care products. These chemicals may have other impacts on human health, including carcinogenicity. There is still much work to do to determine the potential human health impacts of exposure to the many different ECs present in the environment.

Emerging contaminants can enter the environment through many different avenues, depending upon their life cycle of production, use, and disposal. As noted before, many ECs are associated with products used routinely by humans, such as pharmaceuticals and personal care products. An illustration of how these types of chemicals enter the environment is shown in Fig. 12.7. More generally, ECs can be introduced into soils via irrigation of crops with treated wastewater or land application of biosolids. ECs can be introduced into surface water or groundwater via disposal of treated wastewater. ECs can also enter groundwater through waste disposal in landfills. Additional discussion of selected ECs in groundwater is presented in Chapter 15.

12.7 IMPACT OF CHEMICAL PROPERTIES ON TRANSPORT IN THE ENVIRONMENT

The physicochemical properties of the contaminant control its transport and fate behavior. For example, as noted in Chapters 7 and 8, chemicals with moderate to large vapor pressures may evaporate or volatilize into the gas phase, thus becoming subject to atmospheric transport and fate processes. Such chemicals can also undergo transport in the gaseous phase in the vadose zone. As another example, chemicals with larger aqueous solubilities will more readily transfer to water, and thus be subject to transport by water flow. Thus the physicochemical properties of contaminants are critical for their migration potential and persistence in the environment, and mediate their overall pollution potential (Chapter 7). The physicochemical properties of contaminants are controlled by their molecular structure (see Chapter 8). The biodegradability of contaminants is also dependent upon their molecular structure (see Chapter 9).

A critical property to consider when evaluating transport and fate behavior is the phase state of the contaminant. Under "natural" conditions (temperature $T = 25°C$, pressure $P = 1$ atm), chemicals in their pure form exist as solids, liquids, or gases (see Table 12.11). Clearly, the mobility of a chemical in the environment will depend in part on the phase in which it occurs, with gases generally being most mobile and solids least mobile.

Many of the organic contaminants of greatest concern happen to exist as liquids in their pure state under natural conditions. These organic compounds are referred to as immiscible or **nonaqueous phase liquids** (NAPLs). Examples of NAPLs include fuels (gasoline, aviation fuel), chlorinated solvents, and polychlorinated biphenyls. The presence of NAPLs in the subsurface at a contaminated site greatly complicates remediation efforts (see Chapter 19). Once released into the subsurface, the NAPL becomes trapped in pore spaces, after which it is very difficult to physically remove. Hence, they serve as long-term sources of contamination as the molecules transfer to other phases (see Chapter 15). An additional complicating factor is that many NAPLs comprise multiple constituents. Examples of such multicomponent NAPLs include fuels (gasoline, diesel fuel, and aviation fuel), coal tar, and creosote, all of which contain hundreds of organic compounds. These multicomponent NAPLs can contain individual compounds, such as naphthalene and anthracene, that normally occur as solids but which are "dissolved" in the organic liquids.

Most inorganic contaminants of concern occur as solids in their elemental state. One notable exception is mercury, which is a liquid under standard conditions. An important factor for inorganic contaminants is their "speciation." For example, many inorganics occur primarily in ionic form in the environment (e.g., Pb^{+2}, Cd^{+2}, NO_3^-). Speciation can greatly influence aqueous solubility and sorption potential. In addition, many inorganics may combine with other inorganics, forming complexes whose transport behavior differs from that of the parent ions. These concepts are discussed further in Chapter 8.

QUESTIONS AND PROBLEMS

1. What are "POPs," and why are they of such great environmental concern?
2. What is a critical factor that controls the transport and fate behavior and pollution potential of contaminants?
3. Describe three concerns associated with disposal of municipal solid waste.
4. What is MTBE, what was it used for, and why is it an environmental concern?
5. What are emerging contaminants?

Drugs in the Water Cycle

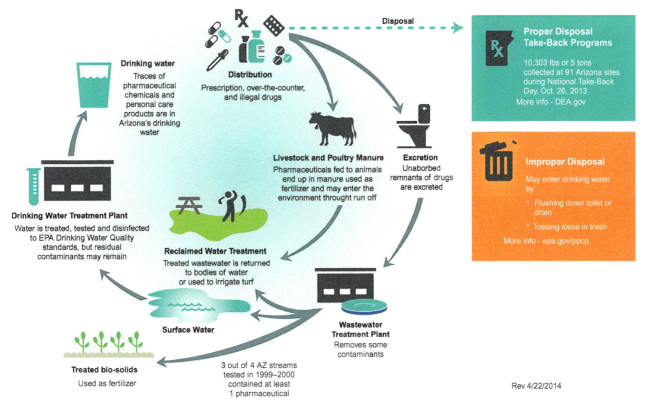

Disposal

Proper Disposal Take-Back Programs

10,303 lbs or 5 tons collected at 91 Arizona sites during National Take-Back Day, Oct. 26, 2013
More info - DEA.gov

Improper Disposal
May enter drinking water by
- Flushing down toilet or drain
- Tossing losse in trash
More info - epa.gov/ppcp

Distribution
Prescription, over-the-counter, and illegal drugs

Drinking water
Traces of pharmaceutical chemicals and personal care products are in Arizona's drinking water

Drinking Water Treatment Plant
Water is treated, tested and disinfected to EPA Drinking Water Quality standards, but residual contaminants may remain

Livestock and Poultry Manure
Pharmaceuticals fed to animals end up in manure used as fertilizer and may enter the environment throught run off

Excretion
Unaborbed remnants of drugs are excreted

Reclaimed Water Treatment
Treated wastewater is returned to bodies of water or used to irrigate turf

Surface Water

Wastewater Treatment Plant
Removes some contaminants

Treated bio-solids
Used as fertilizer

3 out of 4 AZ streams tested in 1999–2000 contained at least 1 pharmaceutical

Rev 4/22/2014

FIG. 12.7 Illustration showing how pharmaceuticals and personal care products enter the water cycle. *(From Emerging Contaminants in Arizona Water, 2016, Arizona Department of Environmental Quality.)*

TABLE 12.11 Properties of Selected Contaminants

	Representative Contaminants	Solubility	Vapor Pressure	Volatility	Sorption Potential	Biodegradation Rate
Solids						
Organic	Naphthalene	Low	Medium	Medium	Medium	Medium
	Pentachlorophenol	Low	Medium	Low	High	Medium
	DDT	Low	Low	Low	High	Low
Inorganic	Lead	Low	Low	Low	Medium	Nondegradable
	Chromium	High	Low	Low	Low	Nondegradable
	Arsenic	Medium	Low	Low	Low	Nondegradable
	Cadmium	Low	Low	Low	Medium	Nondegradable
Liquids						
Organic	Trichloroethene	Medium	High	Medium	Low	Low
	Benzene	Medium	High	Medium	Low	Medium
Inorganic	Mercury	Low	Medium	Low	Medium	Nondegradable
Gases						
Organic	Methane	Medium	Very high	Very high	Low	Low
Inorganic	Carbon dioxide	Medium	Very high	Very high	Low	Nondegradable
	Carbon monoxide	Low	Very high	Very high	Low	Nondegradable
	Sulfur dioxide	Medium	Very high	Very high	Low	Nondegradable

From *Environmental Monitoring and Characterization* © 2004, Elsevier Academic Press, San Diego, CA.

REFERENCE

World Health Organization, 2011. Pharmaceuticals in Drinking-water WHO/HSE/WSH/11.05.

FURTHER READING

Eisenbud, M., Gesell, T.F., 1997. Environmental Radioactivity: From Natural, Industrial, and Military Sources. Academic Press, San Diego, CA.

Kathren, R.L., 1984. Radioactivity in the Environment: Sources Distribution, and Surveillance. Harwood Academic Publishers, New York, NY.

Schwartzenbach, R.P., Gschwend, P.M., Imboden, D.M., 2003. Environmental Organic Chemistry. Wiley, Hoboken, NJ.

Smith, A.H., Lingas, E.O., Rahman, M., 2000. Contamination of drinking-water by arsenic in Bangladesh: a public health emergency. Bull. World Health Organ. 78 (9), 1093–1103.

Chapter 13

Microbial Contaminants

C.P. Gerba and I.L. Pepper

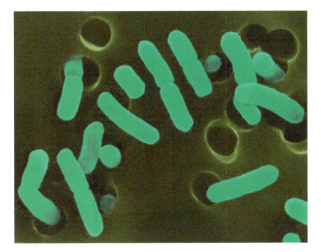

Rod-shaped *Escherichia coli* bacteria 0157:H7 (magnification 22, 245 ×), © Dr. Dennis Kunkel, University of Hawaii. Used with permission.

13.1 WATER-RELATED MICROBIAL DISEASE

London's Dr. John Snow (1813–58) was one of the first to make a connection between certain infectious diseases and drinking water contaminated with sewage. In his famous study of London's Broad Street pump, published in 1854, he noted that people afflicted with cholera were clustered in a single area around the Broad Street pump, which he identified as the source of the infection. When, at his insistence, city officials removed the handle of the pump, Broad Street residents were forced to obtain their water elsewhere. Subsequently, the cholera epidemic in that area subsided. However successful the effect, Snow's explanation of the cause was not generally accepted because disease-causing germs had not been discovered at the time.

In the United States, the concept of *waterborne disease* was equally poorly understood. During the Civil War (1860–65), encamped soldiers often disposed of their waste upriver, but drew drinking water from downriver. This practice resulted in widespread dysentery. In fact, dysentery, together with its sister disease typhoid, was the leading cause of death among the soldiers of all armies until

the 20th century. It was not until the end of the 19th century that this state of affairs began to change. By that time, the germ theory was generally accepted, and steps were taken to properly treat wastes and protect drinking water supplies.

In 1890 more than 30 people out of every 100,000 in the United States died of typhoid. But by 1907 water filtration was becoming common in most U.S. cities, and in 1914 chlorination was introduced. Because of these new practices, the national typhoid death rate in the United States between 1900 and 1928 dropped from 36 to 5 cases per 100,000 people. The lower death toll was largely the result of a reduced number of outbreaks of waterborne diseases. In Cincinnati, for instance, the yearly typhoid rate of 379 per 100,000 people in the years 1905–07 decreased to 60 per 100,000 people between 1908 and 1910 after the inception of sedimentation and filtration treatment. The introduction of chlorination after 1910 decreased this rate even further (Fig. 13.1).

Poor water quality and sanitation account for approximately 1.8 million deaths a year worldwide (Chapter 26), mainly through infections and diarrhea. Nine out of 10 are children and virtually all are from developing countries.

Although many diseases have been eliminated or controlled in the economically developed countries, microorganisms continue to be the major cause of waterborne illness today. Most outbreaks of such diseases are attributable to the use of untreated water, inadequate or faulty treatment (i.e., no filtration or disinfection), or contamination after treatment. In addition, some pathogens, such as *Cryptosporidium*, are very resistant to removal by conventional drinking water treatment and disinfection (Chapter 24).

The true incidence of waterborne disease in the United States is not known because neither investigation nor reporting of waterborne disease outbreaks is required. Investigations are difficult because waterborne disease is not easily recognized in large communities, and epidemiological studies are costly to conduct.

Nevertheless, between 12 and 20 drinking water disease outbreaks per year have been documented in the United States, and the true incidence may be 10–100 times greater (Fig. 13.2).

Environmental and Pollution Science. https://doi.org/10.1016/B978-0-12-814719-1.00013-6

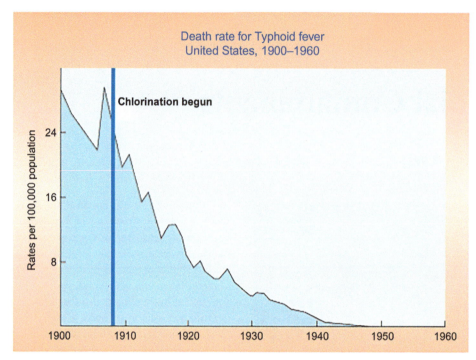

FIG. 13.1 Death rate for typhoid fever. *(Source: U.S. Centers for Disease Control and Prevention, Summary of Notifiable Diseases, 1997.)*

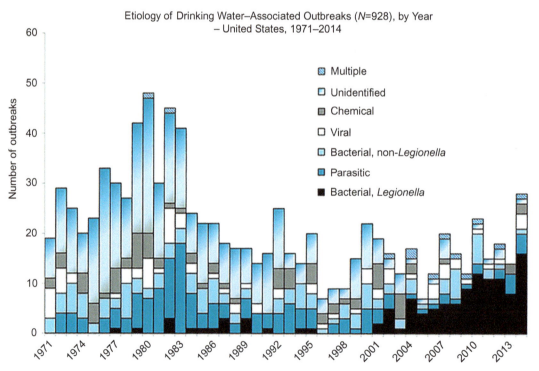

FIG. 13.2 Number of waterborne disease outbreaks associated with drinking water, by year and etiologic agent—United States, 1971–2014 ($n=928$). (*The total from previous reports has been corrected from $n=691$ to $n=688$. [†]Acute gastrointestinal illness of unknown etiology.) *(Source: Centers for Disease Control 2018.)*

13.2 CLASSES OF DISEASES AND TYPES OF PATHOGENS

Disease-causing organisms, or pathogens, that are related to water can be classified into four groups (as shown in Information Box 13.1).

Waterborne diseases (Table 13.1) are those transmitted through the ingestion of contaminated water that serves as the passive carrier of the infectious or chemical agent. The classic waterborne diseases, cholera and typhoid fever, which frequently ravaged densely populated areas throughout human history, have been effectively controlled by the protection of water sources and by treatment of contaminated water supplies. In fact, the control of these classic diseases gave water supply treatment its reputation and played an important role in the reduction of infectious diseases. Other diseases caused by bacteria or by viruses, protozoa, and helminths may also be transmitted by contaminated drinking water. However, it is important to remember that waterborne diseases are transmitted through the fecal-oral route, from human to human or animal to human, so that drinking water is only one of several possible sources of infection.

Water-washed diseases are those closely related to poor hygiene and improper sanitation. In this case, the availability of a sufficient quantity of water is generally considered more important than the quality of the water. The lack of water for washing and bathing contributes to diseases that affect the eye and skin, including infectious conjunctivitis and trachoma, as well as to diarrheal illnesses, which are a major cause of infant mortality and morbidity in the developing countries. The diarrheal diseases may be directly transmitted through person-to-person contact or

indirectly through contact with contaminated foods and utensils used by persons whose hands are fecally contaminated. When enough water is available for hand washing, the incidence of diarrheal diseases has been shown to decrease, as has the prevalence of enteric pathogens such as *Shigella*.

Water-based diseases (Table 13.2) are caused by pathogens that either spend all (or essential parts) of their lives in water or depend upon aquatic organisms for the completion of their life cycles. Examples of such organisms are the parasitic helminth *Schistosoma* and the bacterium *Legionella*, which cause schistosomiasis and Legionnaires' disease, respectively.

The three major schistosome species that develop to maturity in humans are *Schistosoma japonicum*, *S. haematobium*, and *S. mansoni*. Each has a unique snail host and a different geographic distribution. It is estimated that more than 200 million people in Asia, Africa, South America, and the Caribbean are currently infected with one, or perhaps two, of these species of schistosome. Although schistosomiasis is not indigenous to North America, schistosomiasis dermatitis has been documented in the United States, immigrants to the United States have been found to be infected with schistosomiasis, and some 300,000 persons in Puerto Rico are probably infected. The economic effects of schistosomiasis have been estimated at some $642 million annually—a figure that includes only the resource loss attributable to reduced productivity, not the cost of public health programs, medical care, or compensation of illness.

Legionella pneumophila, the cause of *Legionnaires' disease*, was first described in 1976 in Philadelphia, Pennsylvania. This bacterium is ubiquitous in aquatic environments. Capable of growth at temperatures above 40°C, it can proliferate in cooling towers, hot water heaters, and water fountains. If growth occurs at high temperatures, these bacteria become capable of causing pneumonia in humans if they are inhaled as droplets or in an aerosol.

Water-related diseases, such as yellow fever, dengue, filariasis, malaria, onchocerciasis, and sleeping sickness, are transmitted by insects that breed in water (like the mosquitoes that carry malaria) or live near water (like the flies that transmit the filarial infection onchocerciasis). Such insects are known as *vectors*.

INFORMATION BOX 13.1 Classification of Water-Related Illnesses Associated With Microorganisms

Modified from White GF, Bradley DJ: Drawers of water, Chicago, IL, 1972, University of Chicago Press.

Class	Cause	Example
Waterborne	Pathogens that originate in fecal material and are transmitted by ingestion	Cholera, typhoid
Water-washed	Organisms that originate in feces and are transmitted through contact because of inadequate sanitation or hygiene	Trachoma
Water-based	Organisms that originate in the water or spend part of their life cycle in aquatic animals and come in direct contact with humans in water or by inhalation	Schistosomiasis
Water-related	Microorganisms with life cycles associated with insects that live or breed in water	Yellow fever

13.3 TYPES OF PATHOGENIC ORGANISMS

Pathogenic organisms identified as capable of causing illness when present in water include such microorganisms as viruses, bacteria, protozoan parasites, and blue-green algae, as well as some macroorganisms—the helminths, or worms, which can grow to considerable size. Some of

TABLE 13.1 Waterborne and Water-Based Human Pathogens

Group	Pathogen	Disease or Condition
Viruses	Enteroviruses (polio, echo, coxsackie)	Meningitis, paralysis, rash, fever, myocarditis, respiratory disease, diarrhea
	Hepatitis A and E	Hepatitis
	Norovirus	Diarrhea
	Rotavirus	Diarrhea
	Astrovirus	Diarrhea
	Adenovirus	Diarrhea, eye infections, respiratory disease
Bacteria	Salmonella	Typhoid dysentery, diarrhea
	Shigella	Diarrhea
	Campylobacter	Diarrhea
	Vibrio cholerae	Diarrhea, cholera
	Yersinia enterocolitica	Diarrhea
	Escherichia coli (certain strains)	Diarrhea
	Legionella	Pneumonia, other respiratory infection
Protozoa	Naegleria	Meningoencephalitis
	Entamoeba histolytica	Amoebic dysentery
	Giardia lamblia	Diarrhea
	Cryptosporidium	Diarrhea
	Toxoplasma	Mental retardation, loss of vision
	Cyclospora	Diarrhea
Blue-green algae	Microcystis	Diarrhea, possible production of carcinogens
	Anabaena	
	Aphanizomenon	
Helminths	Ascaris lumbricoides	Ascariasis
	Trichuris trichiura	Trichuriasis-whipworm
	Necator americanus	Hookworm
	Taenia saginata	Beef tapeworm
	Schistosoma mansoni	Schistosomiasis (complications affecting the liver, bladder, and large intestines)

TABLE 13.2 Characteristics of Waterborne and Water-Based Pathogens

Organism	Size (μm)	Shape	Environmentally Resistant Stage
Viruses	0.01–0.1	Variable	Virion
Bacteria	0.1–10	Rod, spherical, spiral, comma	Spores or dormant cells
Protozoa	1–100	Variable	Cysts, oocysts
Helminths	$1–10^9$	Variable	Eggs
Blue-green algae	1–100	Coccoid, filamentous	Cysts

the characteristics of these organisms are listed in Table 13.2.

- *Viruses* are organisms that usually consist solely of nucleic acid (which contains the genetic information) surrounded by a protective protein coat or *capsid*. The nucleic acid may be either ribonucleic acid (RNA) or deoxyribonucleic acid (DNA). They are always obligate parasites; as such, they cannot grow outside of the host organism (i.e., bacteria, plants, or animals), but they do not need food for survival. Thus they are potentially capable of surviving for long periods of time in the environment. Viruses that infect bacteria are called *bacteriophages*, and those bacteriophages that infect intestinal, or coliform, bacteria are known as *coliphages*.
- *Bacteria* are prokaryotic single-celled organisms surrounded by a membrane and cell wall. Bacteria that grow in the human intestinal or gastrointestinal (GI) tract are referred to as *enteric bacteria*. Enteric bacterial pathogens usually cannot survive for prolonged periods of time in the environment.
- *Protozoa* are single-celled animals. Protozoan parasites that live in the GI tract are capable of producing environmentally resistant cysts or oocysts. These cysts or oocysts have very thick walls, which make them very resistant to disinfection.
- *Helminths* (literally "worms") are multicellular animals that parasitize humans. They include roundworms, hookworms, tapeworms, and flukes. These organisms usually have both an intermediate and a final host. Once these parasites enter their final human host, they lay eggs that are excreted in the feces of infected persons and spread by wastewater, soil, or food. These eggs are very resistant to environmental stresses and to disinfection.
- *Blue-green algae*, or *cyanobacteria*, are prokaryotic organisms that do not contain an organized nucleus—unlike the green algae. Cyanobacteria, which may occur as unicellular, colonial, or filamentous organisms, are responsible for algal blooms in lakes and other aquatic environments. Some species produce toxins that may kill domestic animals or cause illness in humans.

13.3.1 Viruses

More than 140 different types of viruses are known to infect the human intestinal tract, from which they are subsequently excreted in feces. Viruses that infect and multiply in the intestines are referred to as *enteric viruses*. Some enteric viruses are capable of replication in other organs such as the liver and the heart, as well as in the eye, skin, and nerve tissue. For example, hepatitis A virus infects the liver, causing hepatitis. Enteric viruses are generally

very host specific; therefore human enteric viruses cause disease only in humans and sometimes in other primates.

During infection, large numbers of virus particles, up to 10^8–10^{12}, per gram may be excreted in feces, whence they are borne to sewer systems.

Enteroviruses, which were the first enteric viruses ever isolated from sewage and water, have been the most extensively studied viruses. The more common enteroviruses include the polioviruses (3 types), Coxsackieviruses (30 types), and the echoviruses (29 types). Although these pathogens are capable of causing a wide range of serious illness, most infections are mild. Usually only 50% of the people infected actually develop clinical illness. However, coxsackieviruses can cause a number of life-threatening illnesses, including heart disease, meningitis, and paralysis; they may also play a role in insulin-dependent diabetes.

Infectious viral hepatitis is caused by *hepatitis A* virus (HAV) and *hepatitis E* virus (HEV). These types of viral hepatitis are spread by fecally contaminated water and food, whereas other types of viral hepatitis, such as hepatitis B virus (HBV), are spread by exposure to contaminated blood. Hepatitis A and E virus infections are very common in the developing world, whereas much as 98% of the population may exhibit antibodies against HAV. HAV is not only associated with waterborne outbreaks, but is also commonly associated with foodborne outbreaks, especially shellfish. HEV has been associated with large waterborne outbreaks in Asia and Africa, but no outbreaks have been documented in developed countries. HAV is one of the enteric viruses that is very resistant to inactivation by heat.

Rotaviruses (nine types) have been identified as the major cause of infantile gastroenteritis, that is, acute gastroenteritis in children under 2 years of age. This condition is a leading cause of morbidity and mortality in children and is responsible for hundreds of thousands of childhood deaths per year in Africa, Asia, and Latin America. These viruses are also responsible for outbreaks of gastroenteritis among adult populations, particularly among the elderly, and can cause "traveler's diarrhea" as well. Several waterborne outbreaks have been associated with rotaviruses.

The *Norovirus*, first discovered in 1968 after an outbreak of gastroenteritis in Norwalk, Ohio, causes an illness characterized by vomiting and diarrhea that lasts a few days. This virus is the agent most commonly identified during water- and foodborne outbreaks of viral gastroenteritis in the United States (Fig. 13.3). It has not yet been grown routinely in the laboratory. Norovirus is a genus in the calicivirus family.

The ingestion of just a few viruses is enough to cause infection. But because enteric viruses usually occur in relatively low numbers in the environment, large volumes of environmental samples must usually be collected before the presence of these viruses can be detected. For example, from 10 to 1000 L of water must be collected in order to

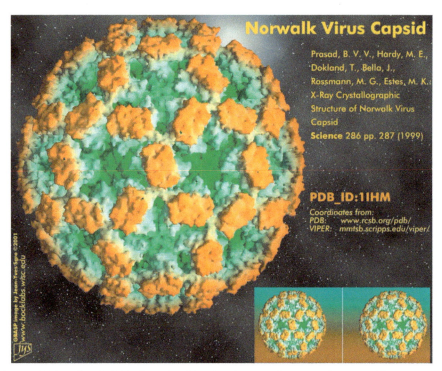

FIG. 13.3   Norwalk virus capsid (Prasad et al., 1999). *Reproduced from http://www.virology.wisc.edu/virusworld/images/r14.jpg, © 2005 Virusworld, Jean-Yves Sgro, Institute for Molecular Virology, University of Wisconsin-Madison.*

assay these pathogens in surface and drinking water. This volume must first be reduced in order to concentrate the viruses. The water sample is thus passed through microporous filters to which the viruses adsorb; then the adsorbed viruses are eluted from the filter. This process is followed by further concentration, down to a few milliliters of sample, leaving a highly concentrated virus population. Next, the concentrate is assayed by using either cell culture or molecular techniques. Cell culture techniques involving animal cells are effective (Fig. 13.4A), but they may require several weeks for results; thus bacteriophages may sometimes be used as timely and cost-effective surrogates (Fig. 13.5). For example, coliphages are commonly used as models to study virus fate during water and wastewater treatment and in natural waters.

13.3.2   Bacteria

13.3.2.1   Enteric Bacteria

The existence of some enteric bacterial pathogens has been known for more than a hundred years. At the beginning of the 20th century, modern conventional drinking water treatment involving filtration and disinfection was shown to be highly effective in the control of such enteric bacterial diseases as typhoid fever and cholera. Today, outbreaks of bacterial waterborne disease in the United States are relatively rare: they tend to occur only when the water treatment process breaks down, when water is contaminated after treatment, or

when nondisinfected drinking water is consumed. The major bacteria of concern are members of the genus *Salmonella, Shigella, Campylobacter, Yersinia, Escherichia,* and *Vibrio.*

Salmonella is a very large group of bacteria comprising more than 2000 known serotypes. All these serotypes are pathogenic to humans and can cause a range of symptoms from mild gastroenteritis to severe illness or even death. *Salmonella* are capable of infecting a large variety of both cold- and warm-blooded animals. Typhoid fever, caused by *S. typhi,* is an enteric fever that occurs only in humans and primates. In the United States, salmonellosis is primarily due to foodborne transmission since the bacteria infect beef and poultry and are capable of growing in foods. The pathogen produces a toxin that causes fever, nausea, and diarrhea and may be fatal if not properly treated.

Shigella spp. infect only human beings, causing gastroenteritis and fever. They do not appear to survive long in the environment, but outbreaks from drinking and swimming in untreated water continue to occur in the United States. *Campylobacter* and *Yersinia* spp. occur in fecally contaminated water and food and are believed to originate primarily from animal feces.

Campylobacter, which infects poultry is often implicated as a source of foodborne outbreaks; it is also associated with the consumption of untreated drinking water in the United States. *Escherichia coli* is found in the gastrointestinal tract of all warm-blooded animals and is usually considered a harmless organism. However, several strains

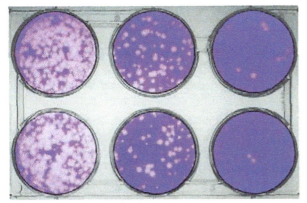

(A)   Enumeration of viruses

(B)   Enumeration of bacteria

FIG. 13.4   Quantitative assays for viruses and bacteria: (A) Viruses can be enumerated by infection of buffalo green monkey kidney cells (BGMK cells). The infections lead to lysis of the host animal cells resulting in a clearing (plaque) on the medium. This yields a measure of the infectious virus in terms of plaque forming units (PFU). (B) Fecal coliforms can be enumerated by growth on a selective medium, such as the above pictured mEndo agar. The resulting bacterial colonies are counted as colony forming units (CFU). Note that in both (A) and (B), more than one original virus or bacterium may have participated in the formation of macroscopic entities. *(Photographs courtesy of (A) C.P. Gerba and (B) I.L. Pepper. From Environmental Microbiology. Academic Press, San Diego, CA, 2000.)*

are capable of causing gastroenteritis; these are referred to as *enterotoxigenic* (ETEC), *enteropathogenic* (EPEC), or *enterohemorrhagic* (EHEC) strains of *E. coli*. Enterotoxigenic *E. coli* causes a gastroenteritis with profuse watery diarrhea accompanied by nausea, abdominal cramps, and vomiting. This bacterium is another common cause of travelers' diarrhea. EPEC strains are similar to ETEC isolate but contain toxins similar to those found in the shigellae. Entero-hemorrhagic *E. coli* almost always belong to the single serological type *0157:H7*. This strain generates a potent group of toxins that produce bloody diarrhea and damage the kidneys. It can be fatal in infants and the elderly. This organism can contaminate both food and water. Cattle are a major source of this organism in the environment (Fig. 13.6).

The genus *Vibrio* comprises a large number of species, but only a few of these species infect human beings. One such is *Vibrio cholerae*, which causes cholera exclusively in humans. Cholera can result in profuse diarrhea with rapid loss of fluid and electrolytes.

Fatalities exceed 60% for untreated cases, but death can be averted by replacement of fluids. Cholera outbreaks did not occur in the Western Hemisphere in the 20th century until 1990, when an outbreak that began in Peru spread through South and Central America. The only cases that occur in the United States are either imported or result from consumption of improperly cooked crabs or shrimp harvested from Gulf of Mexico coastal waters. *V. cholerae* is a native marine microorganism that occurs in low concentrations in warm coastal waters.

Usually, the survival rate of enteric bacterial pathogens in the environment is just a few days, which is less than the survival rates of enteric viruses and protozoan parasites. They are also easily inactivated by disinfectants commonly used in drinking water treatment. Analysis of environmental samples for enteric bacteria is not often performed because they are difficult to isolate. Instead, indicator bacteria are used to indicate their possible presence.

13.3.2.2   Legionella

The pathogen *L. pneumophila* was unknown until 1976, when 34 people died after an outbreak at the annual convention of the Pennsylvania Department of the American Legion in Philadelphia. *Legionellosis*, the acute infection resulting from *L. pneumophila*, is currently associated with two different diseases: Pontiac fever and Legionnaires' disease. Since 1976, numerous deaths from Legionnaires' disease have been reported. Pontiac fever is a milder type of legionellosis. Both these diseases are *noncommunicable*, that is, not transmitted person to person. It has been estimated that more than 100,000 cases of legionellosis occur annually in the United States, an unknown number of which are due to contaminated drinking water.

Scientists, however, point out the error of referring to *L. pneumophila* as a classical contaminant. Although this organism occupies an ecological niche (just as do hundreds of other microorganisms in the water environment), no outbreak of legionellosis has yet been directly associated with a natural waterway such as a lake, stream, or pond. The only scientifically documented habitats for *L. pneumophila* are damp or moist environments. Evidently, it takes human activity—and certain systems like cooling towers, plumbing components, or even dentist water lines—to harbor or grow the organisms. Therefore while *L. pneumophila* may be common to natural water, they can proliferate only when taken into distribution systems where water is allowed to stagnate and temperatures are favorable.

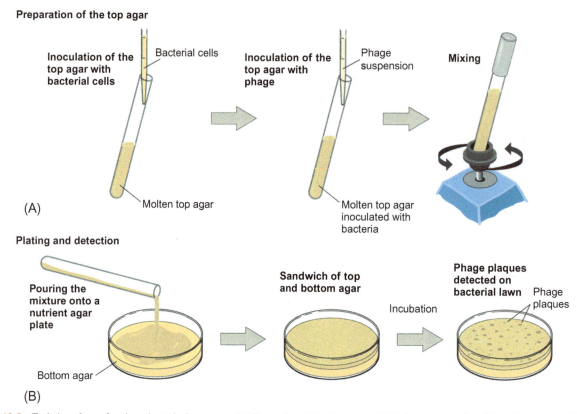

FIG. 13.5 Technique for performing a bacteriophage assay. (A) Preparation of the Top Agar, (B) Plating and detection. *(From Environmental Microbiology. Academic Press, San Diego, CA, 2000.)*

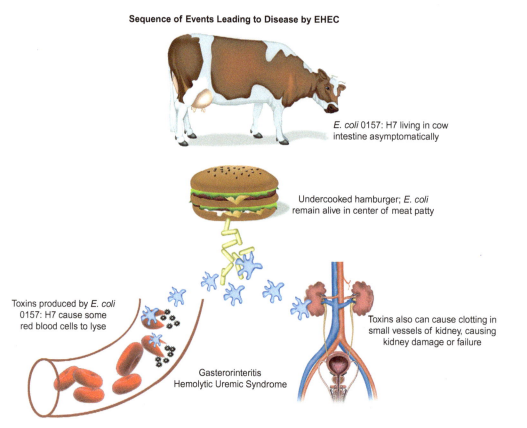

FIG. 13.6 Sequence of events leading to disease by EHEC.

L. pneumophila can grow to a level that can cause disease in areas that restrict water flow and cause buildup of organic matter. Moreover, the optimum temperature for the growth of *L. pneumophila* is 37°C. Thus *L. pneumophila* has been discovered in the hot water tanks of hospitals, hotels, factories, and homes (Fig. 13.7). Ironically, some hospitals and hotels keep their water heater temperatures low to save money and to avert lawsuits, thereby rendering themselves vulnerable to *Legionella* growth. And once established, *Legionella* tends to be persistent. One survey of a hospital water system showed that *L. pneumophila* can exist for long periods under such conditions, collecting in showerheads and faucets in the system. It is believed that showerheads and faucets can emit aerosols composed of very small particles that harbor *L. pneumophila*. Such aerosols, owing simply to their smaller size, can reach the lower respiratory tract of humans.

A link between the presence of *L. pneumophila* in the water system and Legionnaires' disease in susceptible hospital patients has been established by the medical community. It is this abundance of susceptible people, together with the nature of the water system, that has resulted in outbreaks in hospitals. The great majority of people who have contracted Legionnaires' disease were immunosuppressed or compromised because of illness, old age, heavy alcohol consumption, or heavy smoking. Although some healthy people have come down with Legionnaires' disease, outbreaks that included healthy individuals have usually resulted in the milder Pontiac fever. But the fact that *L. pneumophila* exists in a water system does not necessarily mean disease is inevitable. *Legionella* bacteria have been detected in systems where no disease or only a few random cases were found. Therefore the condition or susceptibility of the host or patient is considered to be the single most important factor in whether the infection develops.

Legionella has the ability to survive conventional water treatment. It appears to be considerably more resistant to chlorination than coliform bacteria and can survive for extended periods in water with low chlorine levels.

13.3.2.3 Opportunistic Bacterial Pathogens

Some bacteria common in water and soil are, at times, capable of causing illness. These are referred to as opportunistic pathogens. Segments of the population particularly susceptible to opportunistic pathogens are the newborn, the elderly, and the sick. This group includes heterotrophic Gram-negative bacteria belonging to the following genera: *Pseudomonas*; *Aeromonas*; *Klebsiella*; *Flavobacterium*; *Enterobacter*; *Citrobacter*; *Serratia*; *Acinetobacter*; *Proteus*; and *Providencia*.

These organisms have been reported in high numbers in hospital drinking water, where they may attach to water distribution pipes or grow in treated drinking water.

However, their public health significance with regard to the population at large is not well understood. Other opportunistic pathogens are the nontubercular mycobacteria, which cause pulmonary and other diseases. The most frequently isolated nontubercular mycobacteria belong to the species *Mycobacterium avium-intracellular*. Potable water, particularly that found in hospital water supplies, can support the growth of these bacteria, which may be linked to infections of hospital patients.

13.3.2.4 Indicator Bacteria

The routine examination of water for the presence of intestinal pathogens is currently a tedious, difficult, and time-consuming task. Thus scientists customarily tackle such examinations by looking first for certain indicator bacteria whose presence indicates the possibility that pathogenic bacteria may also be present. Developed at the turn of the 19th century, the indicator concept depends upon the fact that certain nonpathogenic bacteria occur in the feces of all warm-blooded animals. These bacteria can easily be isolated and quantified by simple bacteriological methods.

Detecting these bacteria in water means that fecal contamination has occurred and suggests that enteric pathogens may also be present (see also Section 13.3.2.7).

For example, *coliform bacteria*, which normally occur in the intestines of all warm-blooded animals, are excreted in great numbers in feces. In polluted water, coliform bacteria are found in densities roughly proportional to the degree of fecal pollution. Because coliform bacteria are generally hardier than disease-causing bacteria, their absence from water is an indication that the water is bacteriologically safe for human consumption. Conversely, the presence of the coliform group of bacteria is indicative that other kinds of microorganisms capable of causing disease also may be present, and that the water is unsafe to drink.

The coliform group, which includes *Escherichia*, *Citrobacter*, *Enterobacter*, and *Klebsiella* genus, is relatively easy to detect: specifically, this group includes all aerobic and facultatively anaerobic, Gram-negative, nonspore forming, rod-shaped bacteria that produce gas upon lactose fermentation in prescribed culture media within 48 h at 35°C. In short, they are hard to miss.

Three methods are commonly used to identify total coliforms in water. These are the *most probable number* (MPN), the *membrane filter* (MF), and the *presence-absence* (P-A) tests.

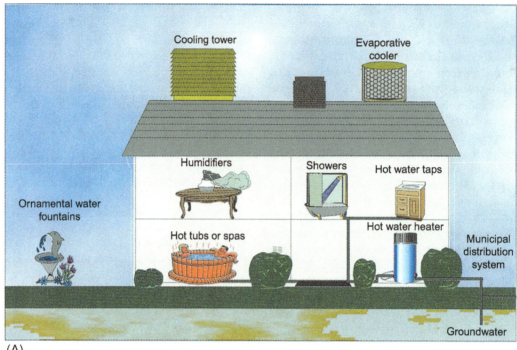

(A)

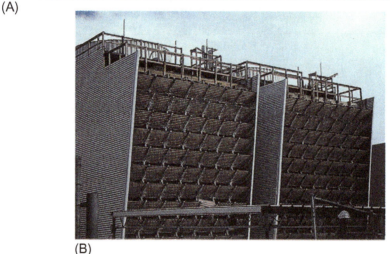

(B)

FIG. 13.7 (A) Sources of *Legionella* in the environment. (B) Cooling towers. Outbreaks of Legionnaires' disease have been commonly traced to cooling towers. *(From Environmental Microbiology. Academic Press, San Diego, CA, 2000.)*

13.3.2.5 The Most Probable Number (MPN) Test

The MPN test allows detection of the presence of coliforms in a sample and an estimate of their numbers. This test consists of three steps: a presumptive test, a confirmed test, and a completed test. In the *presumptive test* (Fig. 13.8), lauryl sulfate tryptose lactose broth is added to a set of test tubes containing different dilutions of the water to be tested. Usually, three to five test tubes are prepared per dilution. These tests tubes are incubated at 35°C for 24–48 h, then examined for the presence of coliforms, which is indicated by gas and acid production. Once the positive tubes have been identified and recorded, it is possible to estimate the total number of coliforms in the original sample by using an MPN table that gives numbers of coliforms per 100 mL (Table 13.3; Information Box 13.2).

In the *confirmation test*, the presence of coliforms is verified by inoculating such selective bacteriological agars as Levine's Eosin Methylene Blue (EMB) agar or Endo agar with a small amount of culture from the positive tubes. Lactose-fermenting bacteria are indicated on the media by the production of colonies with a green sheen or colonies with a dark center. In some cases a *completed test* (not shown in Fig. 13.8) is performed in which colonies from

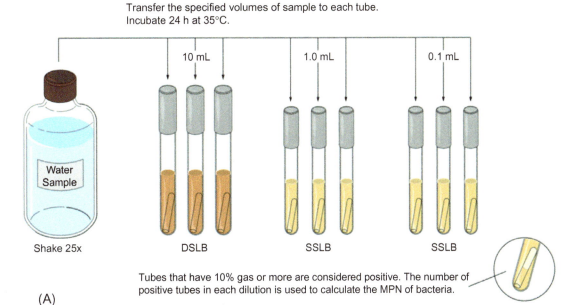

Transfer the specified volumes of sample to each tube.
Incubate 24 h at 35°C.

10 mL   1.0 mL   0.1 mL

Water
Sample

Shake 25x   DSLB   SSLB   SSLB

Tubes that have 10% gas or more are considered positive. The number of
positive tubes in each dilution is used to calculate the MPN of bacteria.

(A)

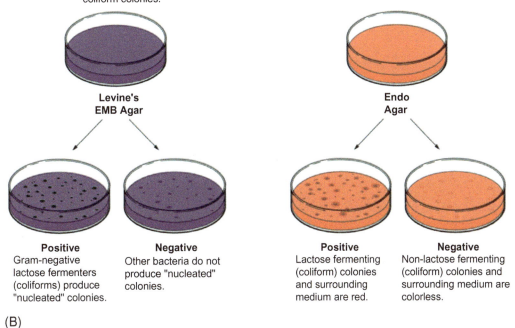

One of the positive tubes is selected, as indicated by the presence of gas trapped
in the inner tube, and used to inoculate a streak plate of Levine's EMB agar and
Endo agar. The plates are incubated 24 h at 35°C and observed for typical
coliform colonies.

Levine's
EMB Agar

Endo
Agar

Positive
Gram-negative
lactose fermenters
(coliforms) produce
"nucleated" colonies.

Negative
Other bacteria do not
produce "nucleated"
colonies.

Positive
Lactose fermenting
(coliform) colonies
and surrounding
medium are red.

Negative
Non-lactose fermenting
(coliform) colonies and
surrounding medium are
colorless.

(B)

FIG. 13.8 Procedure for performing an MPN test for coliforms on water samples: (A) presumptive test and (B) confirmed test. *(From Environmental Microbiology: A Laboratory Manual, second edition. Elsevier Academic Press, San Diego, CA, 2005.)*

the agar are inoculated back into lauryl sulfate tryptose lactose broth to demonstrate the production of acid and gas.

13.3.2.6 The Membrane Filter (MF) Test

The MF test also allows confirmation of the presence and estimate the number of coliforms in a sample, but it is easier

to perform than the MPN test because it requires fewer test tubes and less handling (Fig. 13.9). In this technique, a measured amount of water (usually 100 mL for drinking water) is passed through a membrane filter (pore size 0.45 μm) that traps bacteria on its surface. This membrane is then placed on a thin absorbent pad that has been saturated with a

TABLE 13.3 Most Probable Number (MPN) Table Used for Evaluation of the Data in This Experiment, Using Three Tubes in Each Dilution

Number of Positive Tubes in Dilutions				Number of Positive Tubes in Dilutions			
10 mL	1 mL	0.1 mL	MPN per 100 mL	10 mL	1 mL	0.1 mL	MPN per 100 mL
0	0	0	<3	2	0	0	9.1
0	1	0	3	2	0	1	14
0	0	2	6	2	0	2	20
0	0	3	9	2	0	3	26
0	1	0	3	2	1	0	15
0	1	1	6.1	2	1	1	20
0	1	2	9.2	2	1	2	27
0	1	3	12	2	1	3	34
0	2	0	6.2	2	2	0	21
0	2	1	9.3	2	2	1	28
0	2	2	12	2	2	2	35
0	2	3	16	2	2	3	42
0	3	0	9.4	2	3	0	29
0	3	1	13	2	3	1	36
0	3	2	16	2	3	2	44
0	3	3	19	2	3	3	53
1	0	0	3.6	3	0	0	23
1	0	1	7.2	3	0	1	39
1	0	2	11	3	0	2	64
1	0	3	15	3	0	3	95
1	1	0	7.3	3	1	0	43
1	1	1	11	3	1	1	75
1	1	2	15	3	1	2	120
1	1	3	19	3	1	3	160
1	2	0	11	3	2	0	93
1	2	1	15	3	2	1	150
1	2	2	20	3	2	2	210
1	2	3	24	3	2	3	290
1	3	0	16	3	3	0	240
1	3	1	20	3	3	1	460
1	3	2	24	3	3	2	1100
1	3	3	29	–	–	–	–

specific medium designed to permit growth and differentiation of the organisms being sought. For example, if total coliform organisms are sought, a modified Endo medium is used. For coliform bacteria, the filter is incubated at 35°C for 18–24 h. The success of the method depends on using effective differential or selective media that can facilitate identification of the bacterial colonies growing on the membrane filter surface (Fig. 13.9). To determine the number of coliform bacteria in a water sample, the colonies having a green sheen are enumerated.

13.3.2.7 The Presence-Absence (P-A) Test

Presence-absence tests are not quantitative tests—rather they answer the simple question of whether the target organism is present in a sample or not. The use of a single tube of lauryl sulfate tryptose lactose broth as used in the MPN test, but without dilutions, would be used as a P-A test. Enzymatic assays have been developed that allow for the detection of both total coliform bacteria and *E. coli* in water and wastewater at the same time. These assays can be a simple P-A test or an MPN assay. One commercial P-A test commonly used is the Colilert test, also called the ONPG-MUG (for *O*-nitrophenyl-β-D-galactopyranoside 4-methylumbelliferyl-β-D-glucuronide) test (Fig. 13.10). The test is performed by adding the sample to a single bottle (P-A test) or MPN tubes that contain(s) powdered ingredients consisting of salts and specific enzyme substrates that serve as the only carbon source for the organisms (Fig. 13.11A). The enzyme substrate used for detecting total coliform is ONPG and that used for detecting of *E. coli* is MUG. After 24 h of incubation, samples positive for total coliforms turn yellow (Fig. 13.11B), whereas *E. coli*-positive samples fluoresce under longwave UV illumination (Fig. 13.11C).

Although the total coliform group has served as the main indicator of water pollution for many years, many of the organisms in this group are not limited to fecal sources. Thus methods have been developed to restrict the enumeration to those coliforms that are more clearly of fecal origin, that is, the *fecal coliforms* (Fig. 13.4). These organisms, which include the genera *Escherichia* and *Klebsiella* are differentiated in the laboratory by their ability to ferment lactose with the production of acid and gas at 44.5°C within 24 h. In general, then, this test indicates fecal coliforms; it does not, however, distinguish between human and animal contamination.

Although coliform and fecal coliform bacteria have been successfully used to assess the sanitary quality of drinking water, they have not been shown to be useful indicators of the presence of enteric viruses and protozoa. While outbreaks of enteric bacterial waterborne disease are rare in the United States, outbreaks of waterborne disease have occurred in which coliform bacteria were not found. Enteric viral and protozoan pathogens are more resistant to inactivation by disinfectants than bacteria. Viruses are also more difficult to remove by filtration, due to their smaller size. For this reason, other potential indicators have been investigated. These include bacteriophages (i.e., bacterial viruses) of the coliform bacteria (known as coliphages), fecal streptococcus, enterococci, or *Clostridium perfringens*. The criteria for an ideal indicator organism are shown in Information Box 13.3.

None of these potential indicators has yet been proven ideal.

13.3.2.8 Fecal Streptococci

The fecal streptococci are a group of gram-positive Lancefield group D streptococci. The fecal streptococci belong to the genera *Enterococcus* and *Streptococcus*. The genus *Enterococcus* includes all streptococci that share certain biochemical properties and have a wide range of tolerance of adverse growth conditions. They are differentiated from other streptococci by their ability to grow in 6.5% sodium chloride, pH 9.6, and 45°C and include *Ent. avium*, *Ent. faecium*, *Ent. durans*, *Ent. facculis*, and *Ent. gallinarium*. In the water industry the genus is often given as *Streptococcus* for this group. Of the genus *Streptococcus*, only *S. bovis* and *S. equinus* are considered to be true fecal streptococci. These two species of *Streptococcus* are predominately found in animals; *Ent. faecalis* and *Ent. faecium* are more specific to the human gut. Fecal streptococci are considered to have certain advantages over the coliform and fecal coliform bacteria as indicators.

- They rarely multiply in water.
- They are more resistant to environmental stress and chlorination than coliforms.
- They generally persist longer in the environment

A relationship between the number of enterococci in water and gastroenteritis in bathers has been observed in several studies (National Research Council, 2004) and have been used as standards for recreational waters.

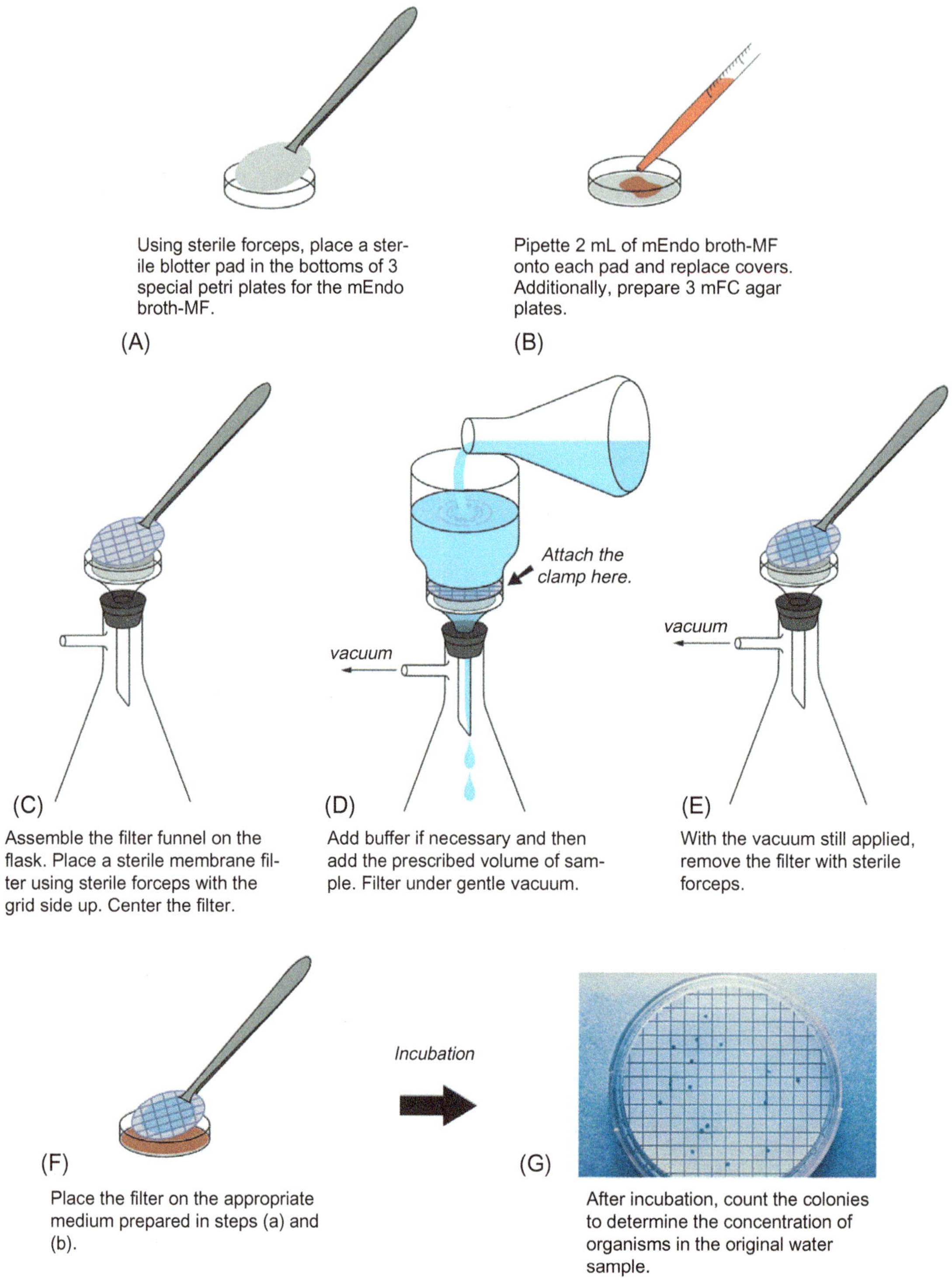

Using sterile forceps, place a sterile blotter pad in the bottoms of 3 special petri plates for the mEndo broth-MF.

(A)

Pipette 2 mL of mEndo broth-MF onto each pad and replace covers. Additionally, prepare 3 mFC agar plates.

(B)

Attach the clamp here.

vacuum

vacuum

(C)

Assemble the filter funnel on the flask. Place a sterile membrane filter using sterile forceps with the grid side up. Center the filter.

(D)

Add buffer if necessary and then add the prescribed volume of sample. Filter under gentle vacuum.

(E)

With the vacuum still applied, remove the filter with sterile forceps.

Incubation

(F)

Place the filter on the appropriate medium prepared in steps (a) and (b).

(G)

After incubation, count the colonies to determine the concentration of organisms in the original water sample.

FIG. 13.9 The membrane filtration method for determining the fecal coliform count in a water sample. *(From Environmental Microbiology. Academic Press, San Diego, CA, 2000.)*

13.3.2.9 Heterotrophic Plate Count

An assessment of the numbers of aerobic and facultatively anaerobic bacteria in water that derive their carbon and energy from organic compounds is conducted via the *heterotrophic plate count* (HPC). This group includes gram-negative bacteria belonging to the following genera: *Pseudomonas*; *Aeromonas*; *Klebsiella*; *Flavobacterium*; *Enterobacter*; *Citrobacter*; *Serratia*; *Acinetobacter*; *Proteus*; *Alcaligenes*; *Enterobacter*; and *Moraxella*. In drinking water, the number of HPC bacteria may vary from less than one to more than 10^4 colony forming units (CFU)/mL and their numbers are influenced mainly by temperature, presence of residual chlorine, and the level of organic matter.

In reality, these counts themselves have no or little health significance. However, there has been concern because the HPC can grow to large numbers in bottled water and charcoal filters on household taps. In response to this concern, studies have been performed to evaluate the impact of HPC on illness. These studies have not demonstrated a conclusive impact on illness in persons who consume water with high HPC. Although the HPC is not a direct indicator of fecal contamination, it does indicate variation in water quality and potential for pathogen survival and regrowth. These bacteria may also interfere with coliform and fecal coliform detection when present in high numbers. It has been recommended that the HPC should not exceed 500/mL in tap water (LeChevallier et al., 1980).

Heterotrophic plate counts are normally done by the spread plate method using yeast extract agar incubated at 35°C for 48h. A low-nutrient medium, R2A (Reasoner and Geldreich, 1985), has seen widespread use and is

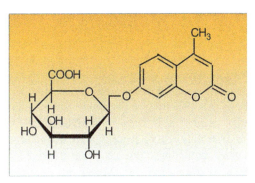

FIG. 13.10 The structure of 4-methylumbelliferyl-β-D-glucuronide (MUG). *(From Environmental Microbiology. Academic Press, San Diego, CA, 2000.)*

recommended for disinfectant-damaged bacteria. This medium is recommended for use with an incubation period of 5–7 days at 28°C. HPC numbers can vary greatly depending on the incubation temperature, growth medium, and length of incubation (Fig. 13.12).

13.3.2.10 Bacteriophage

Because of their constant presence in sewage and polluted waters, the use of bacteriophage (or bacterial viruses) as appropriate indicators of fecal pollution has been proposed. These organisms have also been suggested as indicators of viral pollution. This is because the structure, morphology, and size, as well as the behavior in the aquatic environment of many bacteriophage closely resemble those of enteric viruses. For these reasons, they have also been used extensively to evaluate virus resistance to disinfectants, to evaluate virus fate during water and wastewater treatment, and as surface and groundwater tracers. The use of bacteriophage as indicators of fecal pollution is based on the assumption that their presence in water samples denotes the presence of bacteria capable of supporting the replicator of the phage. Two groups of phage in particular have been studied: the *somatic coliphage*, which infect *E. coli* host strains through cell wall receptors, and the *F-specific RNA coliphage*, which infect strains of *E. coli* and related bacteria through the F+ or sex pili. A significant advantage of using coliphage is that they can be detected by simple and inexpensive techniques that yield results in 8–18h. Both a plating method (the agar overlay method) and the MPN method can be used to detect coliphage (Fig. 13.5) in volumes ranging from 1 to 100mL. The F-specific coliphage (male-specific phage) have received the greatest amount of attention because they are similar in size and shape to many of the pathogenic human enteric viruses. Because F-specific phage are infrequently detected in human fecal matter and show no direct relationship to the

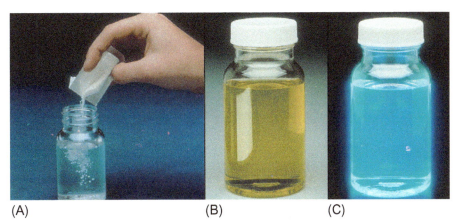

FIG. 13.11 Defection of indicator bacteria with Colilert reagent. (A) Addition of salts and enzyme substrates to water sample; (B) yellow color indicating the presence of coliform bacteria; (C) fluorescence under longwave ultraviolet light indicating the presence of *E. coli. (Photographs used with permission of IDEXX Laboratories, Inc., Westbrook, Maine. From* Pollution Science © *1996, Academic Press, San Diego, CA.)*

Criteria for an Ideal Indicator Organism.

- The organism should be useful for all types of water.
- The organism should be present whenever enteric pathogens are present.
- The organism should have a reasonably longer survival time than the hardiest enteric pathogen.
- The organism should not grow in water.
- The testing method should be easy to perform.
- The density of the indicator organism should have some direct relationship to the degree of fecal pollution.

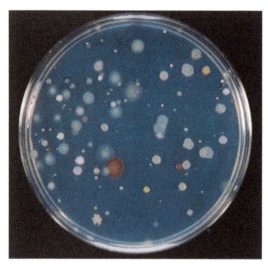

FIG. 13.12 Heterotrophic plate count bacteria. *(Photo courtesy S. Maxwell.)*

fecal pollution level, they cannot be considered indicators of fecal pollution. However, their presence in high numbers in wastewaters and their relatively high resistance to chlorination contribute to their consideration as an index of wastewater contamination as potential indicator of enteric viruses.

13.3.3 Protozoa

13.3.3.1 Giardia

Anton van Leeuwenhoek (1632–1723), the inventor of the microscope, was the first person to identify the protozoan *Giardia* in 1681. However, *Giardia lamblia*, the specific microorganism responsible for giardiasis, was unknown in the United States until 1965, when the first case of giardiasis was reported in Aspen, Colorado. Giardiasis is a particularly nasty disease whose acute symptoms include gas, flatulence, explosive watery foul diarrhea, vomiting, and weight loss. In most people these symptoms last from 1 to 4 weeks, but have been known to last as little as 3 or

4 days or as long as several months. The incubation period ranges from 1 to 3 weeks.

G. lamblia occurs in the environment—usually water—as a cyst, which can survive in cold water for months and is fairly resistant to chlorine disinfection.

Humans become infected with *G. lamblia* by ingesting the environmentally resistant stage, the cyst (Fig. 13.13). Once ingested, it passes through the stomach and into the upper intestine. The increase in acidity via passage through the stomach stimulates the cyst to excyst, which relates two trophozoites into the upper intestine. The trophozoites attach to the epithelial cells of the small intestine (Fig. 13.14). It is believed that the trophozoites use their sucking disks to adhere to epithelial cells. It can cause both acute and chronic diarrhea within 1–4 weeks of ingestion of cysts resulting in foul-smelling, loose, and greasy stools.

The majority of outbreaks are the result of consumption or contact with surface water that is either untreated or treated solely with chlorine. In fact, outbreaks in swimming pools used by small children have also been identified.

Researchers have identified three factors contributing to the greatest risk of *Giardia* infection: drinking untreated water, contact with contaminated surface water, and having children in a day-care center.

Giardia cysts may be constantly present at low concentrations, even in isolated and pristine watersheds. Moreover, wild animals have been implicated as the cause of giardiasis outbreaks. For example, beavers have been blamed as the source that originally transferred the disease to humans. Another study showed that, although giardiasis outbreaks occur most often in surface water systems, *Giardia* cysts may also be present in groundwater supplies. For instance, groundwater sources are usually contaminated springs and wells.

Giardia outbreaks occur most often in the summer months, especially among visitors of recreational areas.

13.3.3.2 Cryptosporidium

Cryptosporidium, an enteric protozoan first described in 1907, has been recognized as a cause of waterborne enteric disease in humans since 1980.

Within 5 years of its recognition as a human pathogen, the first disease outbreak associated with *Cryptosporidium* was described in the United States. Since then, many outbreaks have been reported, and several studies have documented that *Cryptosporidium parvum* is widespread in US surface waters. Moreover, this species is responsible for infection both in human beings and domesticated animals. For example, this species infects cattle, which, in turn, serve as a major source of the organism in surface waters.

The prevalence of *Cryptosporidium* infection is largely attributable to its life cycle (Fig. 13.15). The organism produces an environmentally stable oocyst that is released into

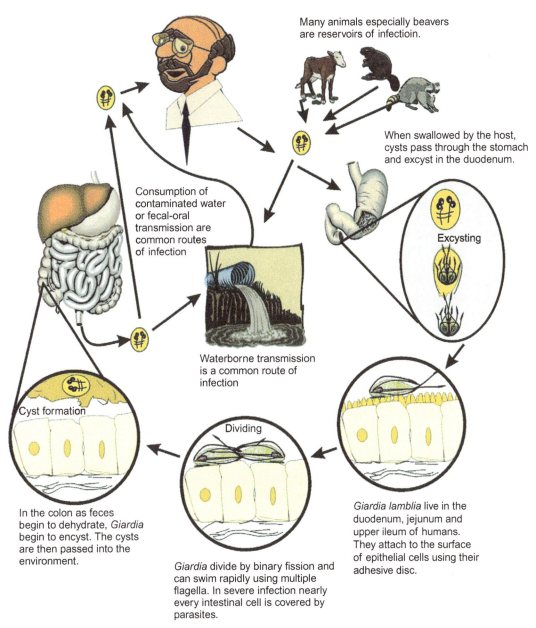

Many animals especially beavers are reservoirs of infectioin.

When swallowed by the host, cysts pass through the stomach and excyst in the duodenum.

Consumption of contaminated water or fecal-oral transmission are common routes of infection

Excysting

Waterborne transmission is a common route of infection

Cyst formation

Dividing

In the colon as feces begin to dehydrate, *Giardia* begin to encyst. The cysts are then passed into the environment.

Giardia divide by binary fission and can swim rapidly using multiple flagella. In severe infection nearly every intestinal cell is covered by parasites.

Giardia lamblia live in the duodenum, jejunum and upper ileum of humans. They attach to the surface of epithelial cells using their adhesive disc.

FIG. 13.13   Life cycle of *Giardia lamblia*. *(From Environmental Microbiology. Academic Press, San Diego, CA, 2000.)*

the environment in the feces of infected individuals. The spherical oocysts of *C. parvum* range in size from 3 to 6 μm in diameter. After ingestion, the oocysts undergo excystation, releasing sporozoite, which then initiate the intracellular infection within the epithelial cells of the gastrointestinal tract (Fig. 13.15). Once in the GI tract, *Cryptosporidium* in humans causes *cryptosporidiosis*, characterized by profuse watery diarrhea, which can result in fluid losses averaging three or more liters a day. Other symptoms of cryptosporidiosis may include abdominal pain, nausea, vomiting, and fever. These symptoms, which usually set in about 3–6 days after exposure, can be severe enough to cause death.

Studies have indicated that *Cryptosporidium* oocysts occur in 55%–87% of the world's surface waters. Thus *Cryptosporidium* oocysts are generally more common in surface water than are the cysts of *Giardia*. But like *Giardia* cysts, *Cryptosporidium* oocysts are very stable in the environment, especially at low temperatures, and may survive for many weeks.

Levels are lowest in pristine waters and protected watersheds, where human activity is minimal and domestic animals scarce.

Occurrence of *Cryptosporidium* outbreaks associated with conventionally treated drinking water suggests that the organism is unusually resistant to removal by this

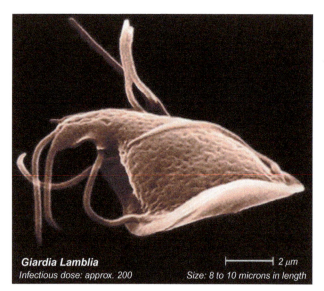

Giardia Lamblia
Infectious dose: approx. 200 |———————| 2 μm
                                       Size: 8 to 10 microns in length

FIG. 13.14 Trophozoites of *Giardia lamblia* the reproductive stage of this waterborne protozoa. *(Photo courtesy F. Myer.)*

process. So far, the oocysts of *Cryptosporidium* have proved to be the most resistant of any known enteric pathogens to inactivation by common water disinfectants. Concentrations of chlorine commonly used in drinking water treatment ($1–2\,\mathrm{mg\,L}^{-1}$) are not sufficient to kill the organism. Water filtration is the primary technique used to protect water supplies from contamination by this organism. However, the oocysts are easily inactivated by ultraviolet light disinfection. Several large outbreaks of waterborne disease in the United States and Europe attest to the fact that conventional drinking water treatments involving filtration and disinfection may not be sufficient to prevent disease outbreaks when large concentrations of this organism occur in surface waters. For example, one U.S. outbreak in Carrollton, Georgia, involved illness in 13,000 people—fully one-fifth of the county's total population. In this case, oocysts were identified in the drinking water and in the stream from which the conventional drinking water plant drew its water. Another outbreak of *Cryptosporidium*, which occurred in Milwaukee, Wisconsin, was the largest outbreak of waterborne disease ever documented in the United States (see Case Study: *Cryptosporidiosis* in Milwaukee). In other cases, *Cryptosporidium* infection has been transmitted by contact with, or swimming in, contaminated water. In fact, *Cryptosporidium* is now the largest cause of recreational outbreaks, almost entirely in chlorinated swimming pools.

Case Study: *Cryptosporidiosis* in Milwaukee

Early in the spring of 1993, heavy rains flooded the rich agricultural areas of Wisconsin. These rains produced an abnormal run-off into a river that drains into Lake Michigan, from which the city of Milwaukee obtains its drinking water.

The city's water treatment plant seemed able to handle the extra load: it had never failed before, and all existing water quality standards for drinking water were properly met. Nevertheless, by April 1, thousands of Milwaukee residents came down with acute watery diarrhea, often accompanied by abdominal cramping, nausea, vomiting, and fever. In a short period of time, more than 400,000 people developed gastroenteritis, and more than 100—mostly elderly and infirm individuals—ultimately died, despite the best efforts of modern medical care. Finally, after much testing, it was discovered that *Cryptosporidium* oocysts were present in the finished drinking water after treatment. These findings pointed to the water supply as the likely source of infection; and on the evening of April 7, the city put out an urgent advisory for residents to boil their water. This measure effectively ended the outbreak. All told, direct costs and loss of life are believed to have exceeded $150 million dollars.

The Milwaukee episode was the largest waterborne outbreak of disease ever documented in the United States. But what happened? How could such a massive outbreak occur in a modern U.S. city in the 1990s? And how could so many people die? Apparently, high concentrations of suspended matter and oocysts in the raw water resulted in failure of the water treatment process—a failure in which *Cryptosporidium* oocysts passed right through the filtration system in one of the city's water treatment plants, thereby affecting a large segment of the population. And among this general population were many whose systems could not withstand the resulting illness. In immunocompetent people, *Cryptosporidiosis* is a self-limiting illness; it is very uncomfortable, but it goes away of its own accord. However, in the immunocompromised, *Cryptosporidiosis* can be unrelenting and fatal.

As in the case with other enteric organisms, only low numbers of *Giardia* cysts and *Cryptosporidium* oocysts need to be ingested to cause infection. Thus large volumes of water are sampled (from 100 to 1,000 L) for analysis. The organisms are entrapped on filters with a pore size smaller than the diameter of the cysts and oocysts. After extraction from the filter, they can be further concentrated and detected by observation with the use of a microscope. Antibodies tagged with a fluorescent compound are used to aid in the identification of the organisms.

13.3.4 Helminths

In addition to the unicellular protozoa, some multicellular animals—the helminths—are capable of parasitizing humans. These include the Nematoda (roundworms) and the Platyhelminthes, which are divided into two subgroups: the Cestoda, or tapeworms, and the Trematoda, or flukes. In these parasitic helminths, the microscopic ova, or eggs, constitute the infectious stage. Excreted in the feces of infected persons and spread by wastewater, soil, or food, these ova are very resistant to environmental stresses and to

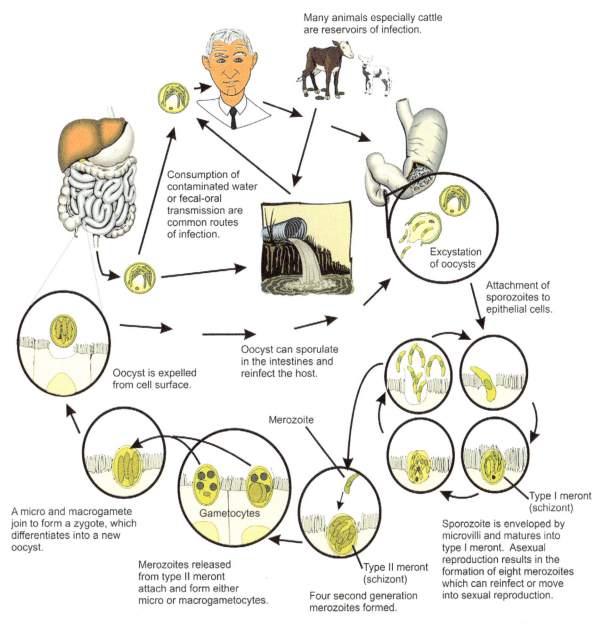

Many animals especially cattle are reservoirs of infection.

Consumption of contaminated water or fecal-oral transmission are common routes of infection.

Excystation of oocysts

Attachment of sporozoites to epithelial cells.

Oocyst is expelled from cell surface.

Oocyst can sporulate in the intestines and reinfect the host.

Merozoite

Type I meront (schizont)

A micro and macrogamete join to form a zygote, which differentiates into a new oocyst.

Gametocytes

Sporozoite is enveloped by microvilli and matures into type I meront. Asexual reproduction results in the formation of eight merozoites which can reinfect or move into sexual reproduction.

Merozoites released from type II meront attach and form either micro or macrogametocytes.

Type II meront (schizont)

Four second generation merozoites formed.

FIG. 13.15 Life cycle of *Cryptosporidium parvum. (From Environmental Microbiology. Academic Press, San Diego, CA, 2000.)*

disinfection. The most important parasitic helminths are listed in Table 13.4.

Ascaris, a large intestinal roundworm, is a major cause of nematode infections in humans. This disease can be acquired through ingestion of just a few infective eggs (Fig. 13.16). Since one female *Ascaris* can produce approximately 200,000 eggs per day and each infected person can excrete a large quantity of eggs, this nematode is very common. Worldwide estimates indicate that between 800 million and 1 billion people are infected, with most infections being in the tropics or subtropics. In the United States, infections tend to occur in the Gulf coast. The life cycle of this parasite includes a phase in which the larvae migrate

through the lungs and cause pneumonitis (known as Loeffler's syndrome). Although the eggs are dense, and hence readily removed by sedimentation in wastewater treatment plants, they are quite resistant to chlorine action. Moreover, they can survive for long periods of time in sewage sludge after land application, unless they were previously removed by sludge treatment.

Although *Taenia saginata* (beef tapeworm) and *Taenia solium* (pig tapeworm) are now relatively rare in the United States, they can still be found in developing countries around the world. These parasites develop in an intermediate animal host, where they reach a larval stage called *cysticercus*. Then these larvae are passed, via meat products, to

TABLE 13.4   Helminth Pathogens in the Environment

Organism	Disease (Main Site Affected)
Nematodes (roundworms)	
Ascaris lumbricoides	Ascariasis: intestinal obstruction in children (small intestine)
Trichuris trichiura	Whipworm (trichuriasis): (intestine)
Hookworms	
Necator americanus	Hookworm disease (intestine)
Ancylostoma duodenale	Hookworm disease (intestine)
Cestodes (tapeworms)	
Taenia saginata	Beef tapeworm: results in abdominal discomfort, hunger pains, chronic indigestion
Taenia solium	Pork tapeworm (intestine)
Trematodes (flukes)	
Schistosoma mansoni	Schistosomiasis (liver [cirrhosis], bladder, and large intestine)

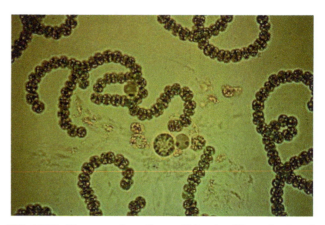

FIG. 13.17   Blue-green algae. *(Source: University of Pennsylvania. cal. vet.upenn.edu/poison/index)*

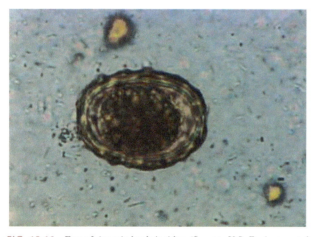

FIG. 13.16   Egg of *Ascaris lumbricoides*. *(Source: U.S. Environmental Protection Agency.)*

humans, who serve as final hosts. For instance, cattle that ingest the infective oval while grazing serve as intermediate hosts for *T. saginata*, while pigs are the intermediate hosts for *T. solium*. The cysticerci invade the muscle, eye, and brain tissue of the intermediate host and can cause severe enteric disturbances, such as abdominal pains and weight loss in their final hosts.

13.3.5   Blue-Green Algae

Blue-green algae, or *cyanobacteria* (Fig. 13.17), occur commonly in all natural waters, where they play an important role in the natural cycling of nutrients in the environment and the food chain.

However, a few species of blue-green algae, such as *Microcystis*, *Aphanizomenon*, and *Anabaena*, produce toxins capable of causing illness in humans and animals. These toxins can cause gastroenteritis, neurological disorders, and possibly cancer. In this case, illness is caused by the ingestion of the toxin produced by the organisms, rather than ingestion of the organism itself, as is the case with helminths. Numerous cases of livestock, pet, and wildlife poisonings by the ingestion of water blooms of cyanobacteria have been reported, and evidence has been mounting that humans are also affected. Heavy blooms of cyanobacteria can occur in surface waters when sufficient nutrients are available resulting in sewage-contaminated water supplies.

13.4   SOURCES OF PATHOGENS IN THE ENVIRONMENT

Waterborne enteric pathogens are excreted, often in large numbers, in the feces of infected animals and humans, whether or not the infected individual exhibits the symptoms of clinical illness. In some cases, infected individuals may excrete pathogens without ever developing symptoms; in other cases, infected individuals may excrete pathogens for many months—long after clinical signs of the illness have passed. Such *asymptomatic* infected individuals are known as carriers, and they may constitute a potential source of infection for the community. Owing to such sources, pathogens are almost always present in the sewage of any community. However, the actual concentration of pathogens in a community sewage depends on many factors: the incidence of enteric disease (i.e., the number of individuals with the disease in a population), the number of carriers in the community, the time of year, sanitary conditions, and per capita water consumption.

The peak incidence of many enteric infections is seasonal in temperate climates. Thus the highest incidence of enterovirus infection is during the late summer and early fall, while rotavirus infections tend to peak in the early winter, and *Cryptosporidium* infections peak in the early spring and fall. The reason for the seasonality of enteric infections is not completely understood, but several factors may play a role. It may be associated with the survival of different agents in the environment during the different seasons: *Giardia*, for example, can survive winter temperatures very well. Alternatively, excretion differences among animal reservoirs may be involved, as is the case with *Cryptosporidium*. Or it may well be that greater exposure to contaminated water, as in swimming, is the explanation for increased incidence in the summer months.

Certain populations and subpopulations are also more susceptible to infection. For example, enteric infection is more common in children because they usually lack previous protective immunity. Thus the incidence of enteric virus and protozoa infections in day-care centers, where young children are in close proximity, is usually much higher than that in the general community (Table 13.5). A greater incidence of enteric infections is also evident in economically disadvantaged groups, particularly where lower standards of sanitary conditions prevail. Concentrations of enteric pathogens are much greater in sewage in the developing world than the industrialized world. For example, the average concentration of enteric viruses in sewage in the United States has been estimated at $10^3 L^{-1}$ (Table 13.6), while concentrations as high as $10^5 L^{-1}$ have been observed in Africa and Asia.

13.4.1 Sludge

During municipal sewage treatment, *biosolids*—(or sludges)—are produced (Chapter 23). Biosolids are a by-product of physical (primary treatment), biological (activated sludge), and physicochemical precipitation of suspended solids (by chemicals) treatment processes. Although treatment by anaerobic or aerobic digestion and/or dewatering reduces the numerical population of disease agents in these biosolids, significant numbers of the pathogens present in raw sewage often remain in biosolids. On a volume basis, the concentration of pathogens in biosolids can be fairly high because of settling (of the large organisms, especially helminths) and adsorption (especially viruses). Moreover, most microbial species found in raw sewage are concentrated in sludge during primary sedimentation. And although enteric viruses are too low in mass to settle alone, they are also concentrated in sludge because of their strong binding affinity to particulates.

The densities of pathogenic and indicator organisms in *primary sludge* presented in Table 13.7 represent typical, average values detected by various investigators. Note that the indicator organisms are normally present in fairly constant amounts. But bear in mind that different sludges may contain significantly greater or fewer numbers of any organism, depending upon the kind of sewage from which the sludge was derived. Similarly, the quantities of pathogenic species are especially variable because these figures depend on which kind are present in a specific community at a particular time. Finally, note that concentrations determined in any study are dependent on assays for each microbial species; thus these concentrations are only as accurate as the assays themselves, which may be compromised by such factors as inefficient recovery of pathogens from environmental samples.

Secondary sludges are produced following the biological treatment of wastewater. Microbial populations in sludges following these treatments depend on the initial concentrations in the wastewater, die-off or growth during treatments, and the association of these organisms with sludge. Some treatment processes, such as the activated sludge process, may limit or destroy certain enteric microbial species. Viral and bacterial pathogens, for example, are reduced in concentration by activated sludge treatment. Even so, the ranges of pathogen concentration in

TABLE 13.5 Incidence and Concentration of Enteric Viruses and Protozoa in Feces in the United States

Pathogen	Incidence (%)	Concentration in Stool (per gram)
Enterovirus	10–40	10^3–10^8
Hepatitis A	0.1	10^8
Rotavirus	10–29	10^{10}–10^{12}
Giardia	3.8	10^6
	18–54[a]	10^6
Cryptosporidium	0.6–20	10^6–10^7
	27–50[a]	10^6–10^7

[a]*Children in day-care centers.*

TABLE 13.6 Estimated Levels of Enteric Organisms in Sewage and Polluted Surface Water in the United States

	Concentration (Per 100 mL)	
Organism	*Raw Sewage*	*Polluted Stream Water*
Coliforms	10^9	10^5
Enteric viruses	10^2	1–10
Giardia	10	0.1–1
Cryptosporidium	0.1–1	$0.1–10^{2a}$

[a]*Greatest numbers in surface waters result from animal (cattle) sources.*

Data from USEPA, 1986. United States environmental protection agency. ambient water quality. Criteria—1986. EPA440/5-84-002, Washington, DC.

TABLE 13.7 Densities of Microbial Pathogens and Indicators in Primary Sludges

Type	Organism	Density (Number Per Gram of Dry Weight)
Virus	Various enteric viruses	$10^2–10^4$
	Bacteriophages	10^5
Bacteria	Total coliforms	$10^8–10^9$
	Fecal coliforms	$10^7–10^8$
	Fecal streptococci	$10^6–10^7$
	Salmonella sp.	$10^2–10^3$
	Clostridium sp.	10^6
	Mycobacterium tuberculosis	10^6
Protozoa	*Giardia* sp.	$10^2–10^3$
Helminths	*Ascaris* sp.	$10^2–10^3$
	Trichuris vulpis	10^2
	Toxocara sp.	$10^1–10^2$

Modified from Straub et al., 1993.

secondary sludges obtained from this and most other secondary treatments are usually not significantly different from those of primary sludges, as presented in Table 13.8.

13.4.2 Solid Waste

Municipal solid waste may contain a variety of pathogens, a source of which is often disposable diapers—it has been found that as many as 10% of the fecally soiled disposable diapers entering landfills contain enteroviruses. Another primary source of pathogens is sewage biosolids, where codisposal is practiced. Pathogens may also be present in domestic pet waste (e.g., cat litter) and food wastes. Municipal solid wastes from households have been found to average 7.7×10^8 coliforms and 4.7×10^8 fecal coliforms

per gram. *Salmonella* have also been detected in domestic solid waste. In unlined landfills, such pathogens may be present in the leachate beneath landfills.

13.5 FATE AND TRANSPORT OF PATHOGENS IN THE ENVIRONMENT

There are many potential routes for the transmission of excreted enteric pathogens. The ability of an enteric pathogen to be transmitted by any of these routes depends largely on its resistance to environmental factors, which control its survival, and its capacity to be carried by water as it moves through the environment.

Some routes can be considered "natural" routes for the transmission of waterborne disease, but others—such as the

TABLE 13.8 Densities of Pathogenic and Indicator Microbial Species in Secondary Sludge Biosolids

Type	Organism	Density (Number Per Gram Dry Weight)
Virus	Various viruses	3×10^2
Bacteria	Total coliforms	7×10^8
Fecal coliforms		8×10^6
Fecal streptococci		2×10^2
	Salmonella sp.	9×10^2
Protozoa	*Giardia* sp.	$10^2 - 10^3$
Helminths	*Ascaris* sp.	1×10^3
	Trichuris vulpis	$<10^2$
	Toxocara sp.	3×10^2

Modified from Straub TM, Pepper IL, Gerba CP: Hazards of pathogenic microorganisms in land-disposed sewage sludge, Rev Environ Contam Toxicol 132:55–91, 1993.

use of domestic wastewater for groundwater recharge, large-scale aquaculture projects, or land disposal of disposable diapers—are actually new routes created by modern human activities.

Human and animal excreta are sources of pathogens. Humans become infected by pathogens through consumption of contaminated foods, such as shellfish from contaminated waters or crops irrigated with wastewater; from drinking contaminated water; and through exposure to contaminated surface waters as may occur during bathing or at recreational sites.

Furthermore, those individuals infected by the earlier processes become sources of infection through their excrement, thereby completing the cycle.

In general, viral and protozoan pathogens survive longer in the environment than enteric bacterial pathogens (Fig. 13.18). How long a pathogen survives in a particular environment depends on a number of complex factors, which are listed in Information Box 13.4. Of all the factors temperature is probably the most important. Temperature is a well-defined factor with a consistently predicable effect on enteric pathogen survival in the environment. Usually, the lower the temperature, the longer the survival time. But freezing temperatures generally result in the death of enteric bacteria and protozoan parasites. Viruses, however, can remain infectious for months or years at freezing temperatures. Moisture—or lack thereof, can cause decreased survival of bacteria. The UV light from the sun is a major factor in the inactivation of indicator bacteria in surface waters; thus die-off in marine waters can be predicted by amount of exposure to daylight.

Viruses are much more resistant to inactivation by UV light.

Many laboratory studies have shown that the microflora of natural waters and sewage are antagonistic to the survival of enteric pathogens. It has been shown, for example, that enteric pathogens survive longer in sterile water than in water from lakes, rivers, and oceans. Bacteria in natural waters can feed upon indicator bacteria. Suspended matter (clays, organic debris, and the like) and fresh or marine sediments has been shown to prolong their survival time (Fig. 13.19).

13.6   STANDARDS AND CRITERIA FOR INDICATORS

Bacterial indicators such as coliforms have been used for the development of *water quality standards*. For example, the U.S. Environmental Protection Agency (U.S. EPA) has set a standard of no detectable coliforms per 100 mL of drinking water. A drinking water standard is legally enforceable in the United States (see Chapter 30). If these standards are violated by water suppliers, they are required to take corrective action or they may be fined by the state or federal government. Authority for setting drinking water standards was given to the U.S. EPA in 1974 when Congress passed the Safe Drinking Water Act (see Chapter 30). Similarly, authority for setting standards for domestic wastewater discharges is given under the Clean Water Act. In contrast, standards for recreational waters and wastewater ruse are determined by the individual states. Microbial standards set by various government bodies in the United States are presented in Table 13.9. Standards used by the European Union are given in Table 13.10.

Criteria and *guidelines* are terms used to describe recommendations for acceptable levels of indicator microorganisms. They are not legally enforceable but serve as guidance indicating that a potential water quality problem exists. Ideally, all standards would indicate that an unacceptable public health threat exists or that some relationship

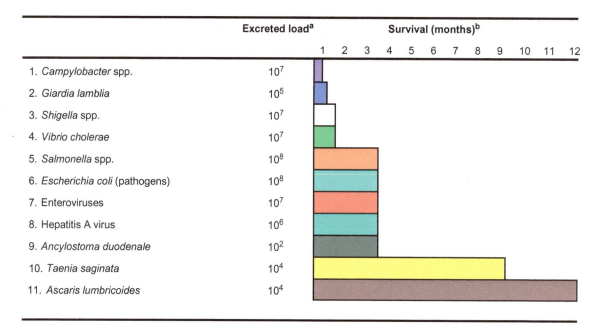

	Excreted load[a]	Survival (months)[b]
1. *Campylobacter* spp.	10^7	
2. *Giardia lamblia*	10^5	
3. *Shigella* spp.	10^7	
4. *Vibrio cholerae*	10^7	
5. *Salmonella* spp.	10^8	
6. *Escherichia coli* (pathogens)	10^8	
7. Enteroviruses	10^7	
8. Hepatitis A virus	10^6	
9. *Ancylostoma duodenale*	10^2	
10. *Taenia saginata*	10^4	
11. *Ascaris lumbricoides*	10^4	

FIG. 13.18 Survival times of enteric pathogens in water, wastewater, soil, and on crops. ([a]Typical average number of organisms/g feces. [b]Estimated average life of infective stage at 20–30°C.) (*Modified from Feachem, R.G., Bradley, D.J., Garelick, H., Mara, D.D., 1983. Sanitation and Disease Health Aspects of Excreta and Wastewater Management, World Bank, Washington, DC.*)

INFORMATION BOX 13.4 Environmental Factors Affecting Enteric Pathogen Survival in Natural Waters

Factor	Remarks
Temperature	Probably the most important factor; longer survival at lower temperatures, freezing kills bacteria and protozoan parasites, but prolongs virus survival.
Moisture	Low moisture content in soil can reduce bacterial populations.
Light	UV in sunlight is harmful.
pH	Most are stable at pH values of natural waters. Enteric bacteria are less stable at pH > 9 and <6.
Salts	Some viruses are protected against heat inactivation by the presence of certain cations.
Organic matter	The presence of sewage usually results in longer survival.
Suspended solids or sediments	Association with solids prolongs survival of enteric bacteria and virus
Biological factors	Naive microflora is usually antagonistic

exists between the amount of illness and the level of indicator organisms. Such information is difficult to acquire because of the involvement of costly epidemiological studies that are often difficult to interpret because of confounding factors. An area where epidemiology has been used to develop criteria is that of recreational swimming. Epidemiological studies in the United States have demonstrated a relationship between swimming-associated gastroenteritis and the densities of enterococci (Fig. 13.20) and *E. coli*. No relationship was found for coliform bacteria

(Cabelli et al., 1982). It was suggested that a standard geometric average of 35 enterococci per 100 mL be used for marine bathing waters. This would mean accepting a risk of 1.9% of the bathers developing gastroenteritis. Numerous other epidemiological studies of bathing-acquired illness have been conducted. These studies have shown slightly different relationships to illness and that other bacterial indicators were more predictive of illness rates. These differences probably arise because of the different sources of contamination (raw versus disinfected wastewater), types of recreational water (marine versus fresh), types of illness (gastroenteritis, eye infections, skin complaints), immune status of the population, length of observation, and so on. Various guidelines for acceptable numbers of indicator organisms have been in use (Table 13.11), but there is no general agreement on standards.

The use of microbial standards also requires the development of standard methods and quality assurance or quality control plans for the laboratories that will do the monitoring. Knowledge of how to sample and how often to sample is also important. All of this information is usually defined in the regulations when a standard is set. For example, frequency of sampling may be determined by the size (number of customers) of the utility providing the water. Sampling must proceed in some random fashion so that the entire system is characterized. For drinking water, no detectable coliforms are allowed in the United States (Table 13.9). However, in other countries some level of coliform bacteria is allowed. Because of the wide

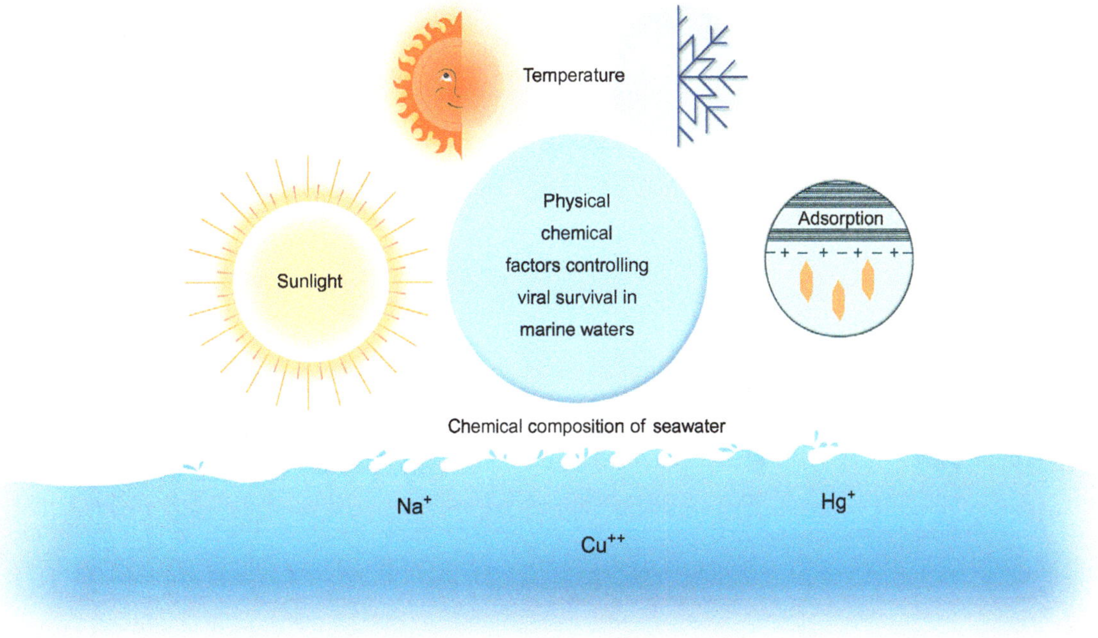

FIG. 13.19   Factors affecting the survival of enteric bacteria and viral pathogens in seawater.

TABLE 13.9 U.S. Federal and State Standards for Microbial Quality of Water

Authority	Standards
U.S. EPA	
Safe Drinking Water Act	0 coliforms/100 mL
Clean Water Act	
Wastewater discharges	200 fecal coliforms/100 mL
Sewage sludge	<1000 fecal coliforms/4 g
	<3 *Salmonella*/4 g
	<1 enteric virus/4 g
	<1 helminth oval/4 g
California	
Wastewater reclamation for irrigation	<2.2 MPN coliforms
Arizona	
Wastewater reclamation for irrigation of golf courses	25 fecal coliforms/100 mL

TABLE 13.10 Drinking Water Criteria of the European Union

Tap Water	
Escherichia coli	0/100 mL
Fecal Streptococci	0/100 mL
Sulfite-reducing clostridia	0/20 mL
Bottled Water	
Escherichia coli	0/250 mL
Fecal streptococci	0/250 mL
Sulfite-reducing clostridia	0/50 mL
Pseudomonas aeruginosa	0/250 mL

variability in numbers of indicators in water, some positive samples may be allowed or tolerance levels or averages may be allowed. Usually, *geometric averages* are used in standard setting because of the often skewed distribution of bacterial numbers. This prevents one or two high values from giving overestimates of high levels of contamination, which would be the case with *arithmetic averages* (Table 13.12).

Geometric averages are determined as follows.

$$\log \overline{X} = \frac{G(\log x)}{N} \qquad (13.1)$$

$$\overline{X} = \text{antilog}\left(\log \overline{X}\right) \qquad (13.2)$$

where N is the number of samples $\overline{X}$ is and the geometric average, and x is the number of organisms per sample volume.

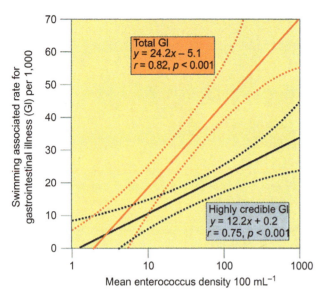

FIG. 13.20 Dose-response relationships produced by the work of Cabelli et al. (1982). *(From Environmental Microbiology. Academic Press, San Diego, CA, 2000.)*

As can be seen, standard setting and the development of criteria is a difficult process and there is no ideal standard. A great deal of judgment by scientists, public health officials, and the regulating agency is required.

QUESTIONS AND PROBLEMS

1. What are pathogens? What is an enteric pathogen?
2. What is the difference between a waterborne and a water-based pathogen?
3. Which group of enteric pathogens survives the longest in the environment and why?
4. Describe some of the methods that can be used to detect indicator bacteria in water.
5. What are some of the criteria for indicator bacteria?
6. Protozoan parasites cause waterborne disease outbreaks in the United States. Why?
7. What are some of the factors that control the survival of enteric pathogens in the environment?
8. Why are geometric means used to report average concentrations of indicator organisms?
9. Calculate the arithmetic and geometric averages for the following data set: Fecal coliforms/100 mL on different days at a bathing beach were reported as 3, 7, 1000, 125, 150, and 3000. Did the bathing beach make the standard of 200 fecal coliforms per 100 mL for bathing?
10. Calculate the most probable number (MPN) for the following data set.
    Sample

TABLE 13.11 Guidelines for Recreational Water Quality Standards

Country or Agency	Regime (Samples/Time)	Criteria or Standard[a]
U.S. EPA[b]	5/30 days	200 fecal coliforms/100 mL
		<10% to exceed 400 mL
		Fresh water[b]
		33 enterococci/100 mL
		126 *E. coli*/100 mL
		Marine waters[b]
		35 enterococci/100 mL
European Economic Community	2/30 days[c]	500 coliforms/100 mL
		100 fecal coliforms/100 mL
		100 fecal streptococci/100 mL
		0 *Salmonella*/L
		0 Enteroviruses/10 L
Ontario, Canada	10/30 days	1000 coliforms/100 mL
		100 fecal coliforms/100 mL

[a]All bacterial numbers in geometric means.
[b]Acceptance is by individual state.
[c]Coliforms and fecal coliforms only.
From U.S. EPA, 1986.

TABLE 13.12 Arithmetic and Geometric Averages of Bacterial Numbers in Water

MPN[a]	Log
2	0.30
110	2.04
4	0.60
150	2.18
1100	3.04
10	1.00
12	1.08
198 = arithmetic average	1.46 = arithmetic average of log values
	antilog = 29
	29 = geometric average

[a]MPN, most probable number.

1 mL	three positive tubes
0.1 mL	two positive one negative tube
10^{-2}	three negative tubes

11. What are enterococci?
12. Why have bacteriophage been suggested as standards for water quality?
13. Why is *Cryptosporidium* so difficult to control in drinking water treatment? What animal is a source of *Cryptosporidium* in water?
14. Why do enteric viruses and protozoa survive longer in the environment than enteric bacteria.
15. What are some of the niches that *Legionella* bacteria can grow to high numbers?

REFERENCES

Cabelli, V.J., Dufour, A.P., McCabe, L.J., Levin, M.A., 1982. Swimming associated gastroenteritis and water quality. Am. Epidemiol. 115, 606–615.

LeChevallier, M.W., Seidler, R.J., Evans, T.M., 1980. Enumeration and characterization of standard plate count bacteria in chlorinated and raw water supplies. Appl. Environ. Microbiol. 40, 922–930.

National Research Council, 2004. Indicators for Waterborne Pathogens. The National Academies Press, Washington, DC.

Prasad, B.V.V., Hardy, M.E., Dokland, T., Bella, J., Rossmann, M.G., Estes, M.K., 1999. X-ray crystallographic structure of Norwalk virus capsid. Science 286, 287.

Reasoner, D.J., Geldreich, E.E., 1985. A new medium for enumeration and subculture of bacteria from potable water. Appl. Environ. Microbiol. 49, 1–7.

Straub, T.M., Pepper, I.L., Gerba, C.P., 1993. Hazards of pathogenic microorganisms in land-disposed sewage sludge. Rev. Environ. Contam. Toxicol. 132, 55–91.

FURTHER READING

Bitton, G., 2011. Wastewater Microbiology, fourth edition Wiley-Liss, New York.

Cortruvo, J.A., Dufour, A., Rees, G., Bartram, J., Carr, R., Cliver, D.O., Craun, G.F., Fayer, R., Gannon, V.P.J., 2004. Waterborne Zoonoses. IWA Publishing, London.

European Union (EU), 1995. Proposed for a council directive concerning the quality of water intended for human consumption. Com (94) 612 Final. Offic. J. Eur. Union 131, 5–24.

Feachem, R.G., Bradley, D.J., Garelick, H., Mara, D.D., 1983. Sanitation and Disease Health Aspects of Excreta and Wastewater Management. World Bank, Washington, DC.

Pepper, I.L., Gerba, C.P., Brendecke, J.W., 2000. Environmental Microbiology: A Laboratory Manual, second edition Academic Press, San Diego, CA.

Pepper, I.L., Gerba, C.P., Gentry, T.J., 2015. Environmental Microbiology. Academic Press, San Diego, CA.

USEPA, 1986. United States environmental protection agency. ambient water quality. Criteria—1986. EPA440/5-84-002, Washington, DC.

USEPA, 1988. Comparative Health Effects Assessment of Drinking Water. United States Environmental Protection Agency, Washington, DC.

White, G.F., Bradley, D.J., 1972. Drawers of Water. University of Chicago Press, Chicago, IL.

Chapter 14

Soil and Land Pollution

J.F. Artiola, J.L. Walworth, S.A. Musil and M.A. Crimmins

Irrigation wheels in Southwest United States. *Photo courtesy J.F. Artiola.*

14.1 INTRODUCTION

Mining, agriculture, and activities that lead to deforestation are important energy-intensive activities that impact economies and at the same time directly and indirectly cause soil, air, and land pollution. Mining produces vast quantities of almost sterile and structureless geologic materials (crushed rock) that particularly in the past contained significant amounts of toxic metals, such as lead, arsenic, and cadmium in the form of primary minerals such as metal sulfides, and

secondary minerals such as calcium, magnesium, and metal sulfates and carbonates. Mine overburden and tailings are often stockpiled next to large open pit excavations. Modern agricultural production requires the use of large quantities of commercial fertilizers and pesticides, and produces animal wastes, all of which can pollute land, air, and water. Land deforestation indirectly affects the quality of land and water by increasing the rates of soil erosion and sediment transport, and accelerating the loss of the nutrient-rich soil surface.

Environmental and Pollution Science. https://doi.org/10.1016/B978-0-12-814719-1.00014-8

Invasive exotic plant species also have a significant impact on the quality of our lands by creating soil conditions that may be toxic to other plants and by increasing fire hazards. All of these activities in turn can affect soil salinity and acidity of surrounding land areas by releasing or concentrating unwanted metals, salts, and acid or acid-forming minerals. These materials can also be released into air or water sources.

Because these activities continue to play a key role in the growth and development of our modern society, their impacts on the environment are now monitored and regulated. It is important to recognize the impacts that these activities have on our environment and there is a need to achieve a balance between their economic and social benefits and the need for the preservation of our environment. The restoration of land adversely affected by these activities is discussed in Chapter 20.

14.2   SURFACE MINING

Mining of coal and metal ores was one of the earliest contributors to the industrial revolution. When transformed, these ores became both the fuel and the building blocks of industrialization. Although carbon-based materials such as pesticides, solvents, fuels, and plastics have dominated industrial production since the early 1950s, metal-based construction materials and goods remain fundamental to modern industry and society. Numerous modern goods—from cars to paints—require the use of such common metals as iron, aluminum, copper, and lead. In addition to these four metals, other less common metals such as lead, cadmium, nickel, and mercury, metalloids like arsenic, and selenium are essential for the manufacture of these and other goods. These elements are therefore commonly found in industrial wastes, where they have complex and often poorly understood effects on the environment. What is known is that uncontrolled and concentrated releases of metals into the environment present both short- and long-term hazards to human health and adversely affect the environment.

Industries that mine and process ores, drill for oil and gas, and/or burn coal also generate large volumes of salt-containing wastes. For these industries, the predominant chemical species include sodium (Na^+), calcium (Ca^{++}), sulfate ($SO_4^=$), and chloride (Cl^-), and carbonate ($CO_3^=$) ions, which are also very abundant in the natural environment. Because these wastes are not intrinsically hazardous or acutely toxic, they do not pose an immediate risk to health and the environment. Nonetheless, the volumes of these wastes that are generated each year are massive enough to be of concern as they can adversely impact local and regional land, air, and water environments.

14.2.1   Mine Tailings

Mining activities and, in particular, strip mining of metal ores produce vast quantities of residues called *mine spoils* and *mine tailings* that may contain significant concentrations of metals (Fig. 14.1). Mine spoils or *overburden* consists of surface materials that do not contain the metal(s) of interest and that are therefore stockpiled at the surface, often resembling large "mesas." Mine tailings, in contrast, are the crushed mineral rock that has been processed to release the metal of interest. These wastes are often pumped as a slurry in "lifts" into valleys or depressions. Mine tailings can be tens of meters deep due to successive depositions of lifts. Thus these residues, which are usually composed of unweathered primary minerals, produce environments that are physically and chemically unstable (fast weathering) and prone to wind and water erosion. Strip mining for copper, for example, produces large quantities of tailings that often contain concentrations of ~100 to <10,000 mg kg^{-1} of such metals as arsenic, cadmium, and lead. Similarly, iron pyrites (FeS_2), which are often associated with copper, silver, and lead ores, can have a devastating impact on the aquatic environment because their oxidation releases sulfuric acid into the environment (Fig. 14.2). The overall reaction is described as follows:

FIG. 14.1   Copper mine tailings and spoils in Southwest United States. *(Photo courtesy J.F. Artiola.)*

FIG. 14.2   Acid mine drainage from gold mine in South Dakota. *(From U.S. Fish and Wildlife Service.)*

$$FeS_2 \text{ (pyrite mineral)} + 3.75O_2 + 3.5H_2O \rightarrow$$
$$Fe(OH)_3 \text{ (solid)} + 2SO_4^{2-} + 4H^+ \quad (14.1)$$

However, in an acid stream (pH < 3), fresh pyrite can react in a cascading effect with soluble ferric iron (Fe^{3+}), creating even more acidity (Stumm and Morgan, 1996). The reaction rate is controlled by the oxidation of Fe^{2+} to Fe^{3+} in the presence of O_2, and results in lowering the pH of the environment. This process can also occur biologically via autotrophic bacteria which thrive at pH 2–3 (see Chapter 5).

Mining operations that treat or leach ores and/or store acid chemicals for the extraction of metals can generate large volumes of acidic metal-containing wastewaters and/or leachates. For example, low-grade Cu ore can be extracted by means of sulfuric acid heap leaching. In this process, crushed Cu ore is continuously leached with sulfuric acid until most of the Cu is solubilized due to both the high acidity and formation of Cu-sulfate complexes. Spent acid solutions, usually contaminated with other metals, must be neutralized and stored in lagoons or impoundments. Gold mining also produces vast quantities of spent ores and liquid process streams that usually contain residual levels of cyanide ion (CN^-) complexes. Metal–cyanide complexes are usually either stable in the soil environment or biologically degraded into nontoxic forms of N (see Chapter 5). However, when released into aquatic systems, unstable complexes of cyanide can be extremely toxic to fish if free cyanide is produced.

14.2.2 Air Emissions

Metal smelting and refining processes generate wastes that may contain multiple hazardous metals, such as lead, zinc, nickel, copper, cadmium, chromium, mercury, selenium, arsenic, and cobalt. These elements may be found in the ores used or they may be added as mixed metals into the melts to produce metal alloys. Thus metal-containing smelter wastes have to be treated and disposed of as hazardous wastes. Smelting and refining require very high temperatures to reduce the metal ores (such as pyrite and bauxite for iron and aluminum production) into pure metal and to refine metals and alloys. For example, iron melts at 1536°C, copper melts at 1083°C, and aluminum melts at 660°C. At these temperatures, many other metals and metal compounds also melt or volatilize; for example, the melting point of lead is 328°C and the boiling points of mercury, cadmium, zinc, and arsenic are 357°C, 765°C, 906°C, and 613°C, respectively. Therefore smelter and metal refining stacks that do not have particulate or gas scrubbers can release into the air significant amounts of toxic metals that are relatively volatile or bound to ultrafine clay-size and silt-size particulates, which eventually deposit onto the land and aquatic environments. The

particulate chemical composition (metals and metalloids) varies greatly with source and particle size (Csavina et al., 2014).

14.3 DEFORESTATION

Deforestation is simply the conversion of forested tracts to barren lands. This is usually done by clear-cutting trees and removing the wood (Fig. 14.3). Forested areas are typically cleared to make room for agricultural operations or to harvest wood as a fuel source or for lumber products. Much of the deforestation occurring globally is due to slash-and-burn operations that make room for crop production or animal pastures.

The process of deforestation results in many undesirable environmental impacts at multiple scales. Local impacts include decreasing soil stability, increasing erosion and sediment transport into streams, reduction in biodiversity through loss of habitat or its fragmentation, and alterations to microclimates that typically increase local temperatures because of loss of vegetation and increased numbers of heat islands. Degradation of air quality is often at the regional scale if deforestation is being driven by burning felled trees. This promotes episodes of high levels of atmospheric particulate matter and carbon monoxide gas that are harmful to the health of both humans and wildlife. Deforestation can also produce impacts on a global scale. These are discussed in Chapter 25.

14.3.1 Local Land Pollution Impacts of Deforestation

Removal of forest vegetation increases the potential of soils to become eroded by wind and/or rainfall. Runoff during precipitation events can promote both the erosion of soils and the transport of sediments into river systems. Although nutrients from sediments are beneficial to aquatic plants and

FIG. 14.3 Example of deforested area with downed slash in foreground. United Nations Environmental Program.

organisms, high sediment loads will degrade water quality with excessive turbidity and levels of total and dissolved nutrients (e.g., phosphorus and nitrates) (see also Chapter 16). Experiments to document the effects of deforestation on watershed dynamics and stream water quality have been conducted in several experimental watersheds throughout the United States. Results from the Hubbard Brook Experimental Forest in New Hampshire show large increases in dissolved nutrient levels and sediment loads instream for a deforested watershed area, as compared to a control forested area watershed (Borman and Likens, 1979; Bonan, 2002).

A more serious form of land-based pollution has been tied to deforestation in areas of South America. High levels of mercury have been found in the blood of people in many rural communities in Brazil where fish is a staple food. The high mercury levels were initially attributed to gold mining operations found throughout these areas. Further study has shown that naturally occurring pools of mercury found in soil and organic matter were being readily transported into streams through runoff following deforestation (Veiga et al., 1994). The removal of forest vegetation allowed mercury that was initially stabilized in soil organic matter to become mobile and be transported into streams.

14.3.2 Regional Air Quality Impacts of Deforestation

Deforestation is often accompanied by the burning of biomass. Sometimes the burning is done to clear slashed vegetation, and at other times harvested forest vegetation is burned as a fuel source or to produce charcoal (biochar) for heating and cooking. In either case, the burning of forest biomass on large scales can cause serious air quality problems. Particulate matter, carbon monoxide, and volatile and condensed liquid organics are all produced when forest biomass is burned, and all pose health risks to humans. Fig. 14.4 shows the impact of large-scale forest biomass burning for heating and cooking on atmospheric carbon dioxide levels across Asia (Heald et al., 2003).

14.4 SOIL ACIDITY—SALINITY

14.4.1 Acid Soils

Acid soils occur naturally or develop as the result of continuous additions of acid-forming fertilizers. Natural acid soils are usually found in the tropics, the result of thousands of years of excessive weathering of soil minerals.

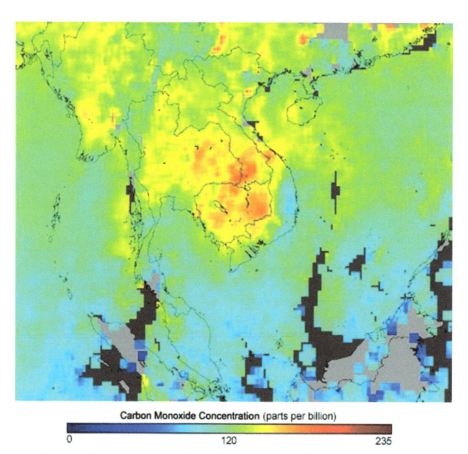

Carbon Monoxide Concentration (parts per billion)

0   120   235

FIG. 14.4 Atmospheric carbon monoxide concentrations over Asia estimated using data collected by the NASA Terra satellite. *(From NASA.)*

TABLE 14.1 Soil Salinity Rankings

Parameter (dS m^{-1})	Nonsaline	Slightly Saline	Moderately Saline	Saline
EC^a	<4	4–8	8–16	>16

^{a}Measured on a water-saturated soil paste extract (USDA, 1954).

TABLE 14.2 Soil Exchangeable Sodium Percentage Rankings

Parameter	Adequate	Border-Line	Inadequate^a	Comments
ESP^b	<10	10–15	>15	Sandy soils may tolerate ESP values up to 15

^{a}Considered a sodic soil (USDA, 1954).
^{b}ESP calculated using mol-equivalent units, using the concentrations of Na$^+$, Ca^{++}, Mg^{++}, and K$^+$ soil water extractable or exchangeable ions.

Year-around high temperatures and high rainfall leaches all basic cations (such as Na, Ca, Mg, and K) and pH buffering minerals (such as carbonates). Also, this climate promotes the transformation and subsequent leaching of Si from Si-based minerals, leaving acidic iron and aluminum oxides minerals. For example, soluble aluminum can release protons into the soil environment by the following general reaction:

$$Al^{3+} + H_2O \rightarrow Al(OH)_3 \text{ (solid)} + 3H^+ \qquad (14.2)$$

Similarly, the presence of pyrite minerals in some soils can lead to the formation of acidic soil conditions, in a reaction similar to Eq. (14.2).

Agricultural soils can also become acidic due to the continuous additions of large amounts of acid-forming fertilizers such as ammonia and urea. For example, one mole-equivalent weight of ammonium (NH_4^+) can produce two mole-equivalent of H^+ after it is fully oxidized to nitrate (NO_3^-) in the soil environment.

Other sources of acid-forming chemicals that impact the soil environment include coal-burning air emissions (SO_2, NO_x) that are hydrolyzed and scrubbed out of the atmosphere by rain (see also Chapter 17).

14.4.2 Salinity

Soil salinity is a measure of the minerals and salts that can be dissolved in water. In most cases, the following mineral ions are found in soil-water extract listed in order of importance:

$$Na^+, Cl^-, Ca^{++}, SO_4^=, HCO_3^-, K^+, Mg^{++}, NO_3^-$$

Increased soil salinity has progressive and often profound effects on the structure, water movement, and microbial and plant diversity of soils. Soil salinity is measured by using electrical conductivity (EC) measurements of a water-saturated soil paste extract (Table 14.1).

An excessive concentration of Na ions in soils produces an imbalance in the ratio of monovalent cations to divalent cations. This is measured by the exchangeable sodium percent (ESP). Salt-affected soils are thus also classified by their ESP as presented in Table 14.2.

There are numerous sources of soil salinity. Natural soil salinity occurs in hot arid and semiarid climates with <27 cm of annual rainfall. Soils and lands that have shallow water tables can develop saline soils due to excessive water evaporation and the concentration of salts. Poor water quality and irrigation practices also contribute to the salinization of thousands of acres of farmland each year around the world. Salt-affected soils occupy, on a global basis, 952.2 million ha of land. These soils constitute nearly 7% of the total land area or nearly 33% of the potential agricultural land area of the world (Gupta and Abrol, 1990).

14.5 SOIL EROSION

Soil particles can act as carriers of other contaminants that are sorbed to particulate surfaces. For example, phosphorus is often associated with soil particles and sediments. When soil particles are eroded and discharged into an aquatic environment, the resulting sediments increase the P nutrient levels of the water and can cause excessive growth of algae and other aquatic plants. This process, together the concomitant reduction in oxygen, is known as eutrophication (see also Chapter 16).

Organic chemicals, including herbicides, insecticides, fuels, solvents, fire retardants, surfactants, and other industrial and agricultural chemicals, can be similarly adsorbed and desorbed from waterborne soil particulates.

Naturally occurring particulate contaminants come from many sources, including agricultural operations, logging, construction-related activities, mining and quarrying, and unpaved roads, and from wind erosion (see this chapter). Soil erosion is a natural process that occurs continuously but is often accelerated by human activities. Several factors are required for soil material to become dislodged and transported into air or water. Soil must be susceptible to erosional processes, which generally requires that the soil be exposed to erosional forces. In the first phase of soil erosion, soil particles become dislodged. Energy inputs must be adequate to dislodge particles. In the second phase, the particles are transported. Various soil properties, which we will examine, determine the susceptibility of soil particles to dislodgement.

14.5.1   Soil Water Erosion

We will first consider particle movement caused by water. As indicated in Chapter 2, soil particles are usually formed into aggregates, which vary considerably in size, shape, and stability. Organic and inorganic materials and certain soil cations are the primary aggregating and interparticle cementing agents. In the detachment phase, individual particles are dispersed or separated from aggregates or cemented particles. The source of the energy responsible for detaching soil particles is either raindrop impact or the flow of runoff water. When raindrops, which travel at approximately $9\,m\,s^{-1}$, hit bare soil, the kinetic energy of the raindrops is transferred to the soil particles, breaking apart aggregates and dislodging particles. Dislodged particles can be moved over 1 m in the splash from raindrop impact. They are moved larger distances by runoff water, which can dislodge additional particles through scouring action. Smaller particles are transported more easily than large ones, and faster flowing water can carry a heavier particulate load than slow-moving water. When uniform shallow layers of soil are eroded off areas of land, this is called *sheet erosion*. Directed water flow cuts channels into the soil. Small channels are called *rills*; large channels are *gullies* (Fig. 14.5).

The process of water erosion has been described by the *United States Department of Agriculture* (USDA) using the *Universal Soil Loss Equation* (USLE), later modified to the *Revised Universal Soil Loss Equation* (RUSLE).

$$A = 2.24R \times K \times LS \times C \times P \qquad (14.3)$$

where:

A = the estimated average annual soil loss (metric tons/hectare)
R = the rainfall and runoff erosivity index. This describes intensity and duration of rainfall in a given geographical area. It is the product of the kinetic energy of raindrops and the maximum 30-min intensity.
K = the soil erodibility factor. K is related to soil physical and chemical properties that determine how

FIG. 14.5   Soil erosion in abandoned farmland in southern Arizona that results in a gully. *(Photo courtesy J.F. Artiola.)*

easily soil particles can be dislodged. It is related to soil texture, aggregate stability, and soil permeability or ability to absorb water. It ranges from 1 (very easily eroded) to 0.01 (very stable soil).
LS = a dimensionless topography factor determined by length and steepness of a slope. The LS factor is related to the velocity of runoff water. Water moves faster on a steep slope than a more level one, and it picks up speed as it moves down a slope. Therefore the steeper and longer the slope, the faster runoff water will flow. The faster water flows, the more kinetic energy it can impart to the soil surface (kinetic energy = mass × velocity²).
C = the cover and management factor. Cover of any kind can help protect the soil surface from raindrop impact and can force runoff water to take a longer, more tortuous path as it moves downslope, slowing the water and reducing its kinetic energy.
P = the factor for supporting practices. This factor takes into account specific erosion control measures. Erosion control practices reduce the P factor.

In the past few decades, farmers have tried to reduce tillage that leaves soil bare and to minimize the amount of time that the soil surface is exposed to raindrops. These new agricultural practices are collectively known as *Reduced Tillage* or *Minimum Tillage* systems (see Fig. 14.6). On highly erodible lands, specific erosion control practices include contour planting strip cropping or terracing, all of which can effectively reduce erosion. Other methods can be used for specific applications, for example, *Gabions* or wire

FIG. 14.7 A rock-filled dam erosion control structure. *(From Environmental Monitoring and Characterization, Elsevier Academic Press, San Diego, CA, 2004.)*

It should be noted that finer textured soils (those with more silt and clay sized particles) are often less erodible than sandy soils. This reflects the ability of fine particles to form soil large aggregates that hold the soil in place during high wind events. On the other hand, air pollution is measured using the PM_{10} standard, which consists largely of silt-sized particles, and the $PM_{2.5}$ standard, which captures mainly clay size particles. Therefore soil wind erosion and soil texture are not always directly related. Methods to control or reduce wind erosion are covered in Chapter 20.

14.6 AGRICULTURAL ACTIVITIES

14.6.1 Fertilizers

Plants need numerous chemicals in order to complete their life cycles. There are at about 18 *essential elements* required for the growth of plants: C, H, O, N, P, K, Ca, Mg, S, Fe, Mn, Zn, Cu, Mo, Co, B, Ni, and Cl in various ionic forms. Other elements deemed beneficial to plant growth include Si, Na, V, Se, and Ti. In undisturbed ecosystems, plants obtain these nutrients from the soil solution via mineral weathering, atmospheric inputs, inputs from stream deposition, and nutrient recycling due to death and decomposition of vegetation. The availability of the nutrients depends on abiotic soil factors (Chapter 2) and chemical and biological properties (Chapters 8 and 9). Agricultural crop production has always relied on soil components for nutrient sources. However, excessive cropping and in particular dense monoculture practices deplete soil plant nutrients, especially N, P, K, Ca, and Mg and lower precultivation organic carbon content significantly. Thus, over years of continuous crop production, large amounts of

FIG. 14.6 No-till cotton planted into crop residue to protect the soil surface. From USDA, image number K8550–8. *(From Environmental Monitoring and Characterization, Elsevier Academic Press, San Diego, CA, 2004.)*

mesh containers filled with rocks can be used to control water erosion in gullies and drainage ways (Fig. 14.7). Methods to control erosion are discussed in more detail in Chapter 20.

14.5.2 Soil Wind Erosion

Like water erosion, wind erosion has two phases: detachment and movement. As the wind blows, soil particles are dislodged and begin to roll or bounce along the soil surface in a process called *saltation*. Larger soil particles can move relatively short distances in this way, but, more importantly, as the large particles bounce and strike smaller particles and aggregates, they provide the energy necessary to break aggregates apart and suspend smaller particles in the air. Smaller particles remain suspended in air for longer periods of time and are therefore more likely to travel much longer distances. As in the case of water erosion, models that examine the factors important in wind erosion are useful in predicting wind erosion (Saxton et al., 2000).

nutrients are removed including carbon, with a concomitant decline in soil productivity. Therefore N, P, K, and other plant nutrients must be periodically augmented by the use of fertilizers, plant residues, and animal (manure) or human wastes (biosolids). Fertilizers may contain any of the essential nutrients, but the majority of fertilizers applied to agricultural soils contain nitrogen (N), phosphorus (P), potassium (K), or some combination thereof. These are the so-called *macronutrients* because plants take them up in larger amounts than the other essential nutrients.

Fig. 14.8 illustrates trends for macronutrient use in the United States. Fertilizer use dramatically increased around the time of World War II, as improved crop varieties and management practices, together with increased mechanization, made fertilizer use both practicable and profitable. In the 1980s, however, fertilizer use began to level off, reflecting both lower agricultural profitability and increased environmental concerns related to fertilizer use. Nonetheless, the combined annual per capita use in the United States of NPK fertilizers is about more than 150 lbs (~70 kg). The aforementioned concerns associated with the use of NPK fertilizers and wastes include excessive surface and groundwater pollution by water-soluble nitrates and colloid-bound phosphates due to poor agricultural fertilizer and waste management practices (see also Chapters 18 and 19).

14.6.2 Pesticides

Extensive use of synthetic pesticides began in the 1940s with dichlorodiphenyldichloroethane (DDT) used to control mosquitoes. This was quickly followed by the adoption of pesticides in large-scale monocultural agricultural production. Initially pesticide use was credited with significant increases in food production. However, the negative aspects of their indiscriminate use also became evident. For example, extensive use of insecticides and herbicides has created new generations of pesticide-resistant insect pests and weeds.

In 1962, Rachel Carson's book *Silent Spring* brought public attention to the fact that chlorinated pesticides were very persistent in the environment. These chemicals can accumulate in animal fatty tissue and produce fish kills when released into waterways. DDT, associated with the rapid decline of some birds of prey, was banned for agricultural use in the United States in 1973. Other chlorinated pesticides were also banned, but have been replaced by much less persistent, but more acutely toxic, pesticides. In addition, in recent years new links have been discovered between some types of cancer and low-level exposure to pesticides like 2,4-dichlorophenoxyacetic acid (2,4-D), 2,4,5-trichlorophenoxyacetic acid (2,4,5-T), and other pesticides. These two plant herbicides, used extensively as

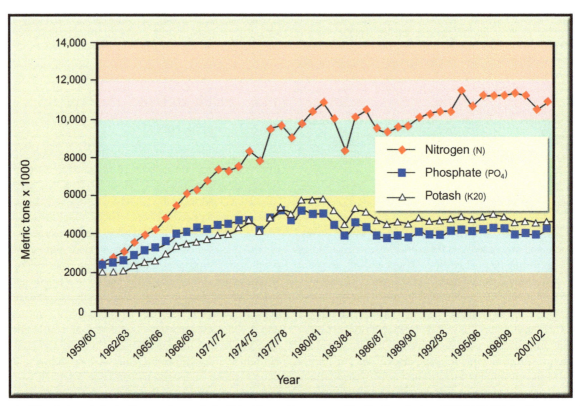

FIG. 14.8 Historical trends for fertilizer nutrient use in the United States. *(Data source: The Fertilizer Institute (TFI, 2005. U.S. fertilizer use statistics. http://www.tfi.org/Statistics/Usfertuse).)*

defoliants in the Vietnam War, were contaminated with dioxins, a family of very toxic polychlorinated dibenzene derivatives and very insoluble and persistent (decades) in the environment.

Less persistent pesticides are usually much more soluble in water than chlorinated hydrocarbons. Unfortunately, these new pesticides are more likely to leach to groundwater or be found in the agricultural runoff if they are not degraded fast enough in the soil environment (see Section 14.6.2.3). Today, pesticides continue to be used extensively in modern farming, urban lawns, parks, and golf courses primarily to control weeds, fungi, and insect infestations. Unfortunately, even less persistent pesticides have their problems. In 2003, the U.S. EPA concluded that atrazine, the second most widely used pesticide (herbicide) in the United States, could cause sexual abnormalities in frogs. In addition, atrazines, the most common family of herbicide chemicals found in groundwater are also potential endocrine disruptors. Other common pesticides have been linked to or are being studied as possible endocrine disruptors. In recent years, the introduction and extensive use of yet another family of chlorinated systemic pesticides called neonicotinoids (nicotine-like) has been linked to the worrisome decline in honey bees and other pollinating insects.

14.6.2.1 Types of Pesticides

The technical definition, stated in the amended *Federal Insecticide, Fungicide, and Rodenticide Act* (FIFRA), is that a pesticide is any substance or mixture of substances intended for destroying, preventing, or mitigating insects, rodents, nematodes, fungi, weeds, or any other undesirable pests. This also includes plant or insect growth regulators as well as defoliants that are used to cause leaves to drop from plants to facilitate harvest and desiccants that dry up unwanted plant tissue. Under this definition, many chemicals, both newly developed and familiar, may be considered as pesticides and be regulated as such. For example, insect pheromones (sex attractants) may be used to attract certain insect populations, to confuse mating patterns, and thereby control insect population. In addition, ordinary dish detergent may be used to kill whiteflies or bees. Common table salt (sodium chloride) is used to control weeds in beet fields in humid regions.

Insecticides are formulated to control particular insects. Two common insecticides are chlorpyrifos and malathion. Herbicides are formulated to control weeds. Glyphosate and atrazine are the two most common herbicides, accounting for 70%–90% of the total herbicide use in the United States (U.S. EPA, 2004). Fungicides are formulated to control fungi including molds and mushrooms. Chloropicrin, metam-sodium, and 1,3-dichloropropene are the three commonly used fumigants applied to soil to control nematodes and soil fungi in the United States (U.S. EPA, 2004).

Pesticides may also be classified according to their mode of entry into the target pest. Contact pesticides enter the target pest upon direct application, while systemic pesticides must pass through a host organism before they enter their targets. For example, a contact insecticide, or its residue, kills target plants or insects on direct application, while a systemic insecticide kills insects only after moving through the system of the plant hosting the target insect. Thus if a particular insect does not feed on the plant, it will not be harmed.

Finally, pesticides can be classified by the forms in which they are used. Fumigants, for example, are pesticides applied as gases. Fumigants may be used selectively to control drywood termites in houses or to control the pest population in stored products such as fruits, vegetables, and grains. They may also be released over large areas to remove many pests from soil.

14.6.2.2 Extent of Pesticide Use

Pesticides are sold or distributed by intra- or interstate commerce in the United States, and they must be registered by the U.S. EPA. The EPA has compiled substantial lists of pesticide ingredients whose applications must be reported. The EPA is also authorized, by the *Federal Food, Drug, and Cosmetic Act* (FFDCS), to establish tolerances for pesticide residues in raw and processed foods. The *Food and Drug Administration* (FDA) of the Department of Health and Human Services monitors and enforces the established tolerances.

In addition, many individual states in the United States have established other regulatory agencies to control pesticide applications in order to protect wildlife and water supplies. For example, Arizona has compiled a list of chemicals—the Groundwater Protection List—whose use must be reported. Similar requirements exist for the sales of these pesticides, so that significant underreporting of applications cannot occur without alerting the regulatory agency.

According to the U.S. EPA, in 2012 the use at the producer level of conventional pesticides in the United States was estimated to be about 1.2 billion lbs (545 million kg), reflecting a flat if slightly declining trend in use since the mid-1980s (Fig. 14.9). These figures place the annual per capita use of pesticides at about 4 lbs (~1.9 kg). The U.S. EPA estimates that in 2001 about 78% of these products were used in agricultural production; 12% in home and garden settings; and the remaining 10% in forestry, industry, and government programs. Therefore most of these chemicals were applied directly onto plants and animals on agricultural lands and water systems. In addition, industry and water utilities also use chemicals with pesticide-like properties. For example, according to U.S. EPA estimates in 2001, about 790 million lbs (360 million kg) of wood preservative chemicals and 2.6 billion

FIG. 14.9 Pesticide usage in the United States by Pesticide Type, 2012 Estimates. *(From U.S. EPA 2012 Pesticides Industry Sales and Usage 2008–2012 Market Estimates. Biological and Economic Analysis Division Office of Pesticide Programs Office of Chemical Safety and Pollution Prevention U.S. Environmental Protection Agency Washington, DC 20460, 2017.)*

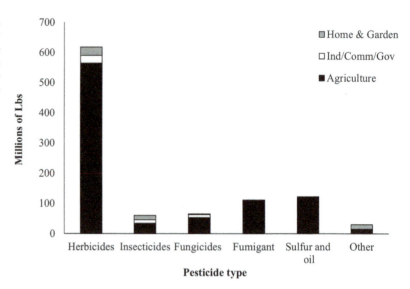

lbs (1.19 billion kg) of chlorine and hypochlorite chemicals were used in the United States. These highly toxic chemicals include creosote, pentachlorophenol, and CCA (chromate copper arsenate). Presently, pesticide product labeling must list their active ingredients and the EPA's registration number, as well as safe use instructions to minimize personal exposure and damage to soil and water environments.

It is interesting to note that despite the public awareness about numerous links between pesticide residues, their adverse health and environmental effects, and the increasing public demands for "organic" pesticide-free food, the largest growth sector for pesticides was in home and garden applications until 2001 (U.S. EPA, 2004). In 2012 pesticide use in homes and gardens was about 5% of the total usage in the United States (U.S. EPA, 2017).

14.6.2.3   Fate of Pesticides

Depending upon their physicochemical properties, patterns of use, and local conditions, some pesticides may leach through the crop root zone and eventually contaminate groundwater at certain locations (see Chapter 15). The two most important properties of a chemical that determine whether a pesticide represents a threat to groundwater are its *persistence* and *mobility* in soil, as discussed in Chapter 7. During the registration of new pesticides, fate and transport models are used to estimate the potential for groundwater to be contaminated by the specific use of a particular chemical at various locations in the United States. Several states, including Arizona and California, consider the capacity of a compound to leach through the soil into groundwater as a criterion for inclusion in their lists of controlled chemicals.

After a pesticide is applied to a field, it may meet a variety of fates, as shown in Fig. 14.10. Some may be lost to the atmosphere through volatilization, carried away to surface waters by runoff and erosion, or photodegraded by sunlight. Pesticides that have entered into soil may be taken up by plants (and subsequently removed), degraded into other chemical forms, or leached downward with water below the crop root zone. The amount of any particular chemical that ends up volatilized, leached, degraded, or in surface runoff depends upon site conditions, weather conditions, management practices, soil properties, and pesticide properties (as described in several previous chapters).

In evaluating the contamination potential of a particular pesticide, it is essential to consider its sorption (retardation) and transformation half-life behavior jointly, as discussed in Chapter 8. For example, a pesticide with low retardation and a long half-life (e.g., more than 100 days) poses a considerable threat to groundwater through leaching, particularly in soils having low organic matter. Conversely, a pesticide with large retardation and a long half-life is more likely to remain on or near the surface of soils with moderate levels of organic carbon content, thereby increasing its chances of being carried to a lake or stream in runoff water. In terms of water quality protection, pesticides with intermediate retardation and short half-lives may be considered the "safest." Although they are not readily leached, they move into the soil with water, thereby reducing their potential for loss from erosion, and they degrade fairly rapidly, thereby reducing the chance for losses below the root zone. Fig. 14.11 provides a schematic representation of the depth of movement of a strongly sorbed (glyphosate), a moderately sorbed (atrazine), and a weakly sorbed (aldicarb) chemical. It was assumed that the rainfall and irrigation

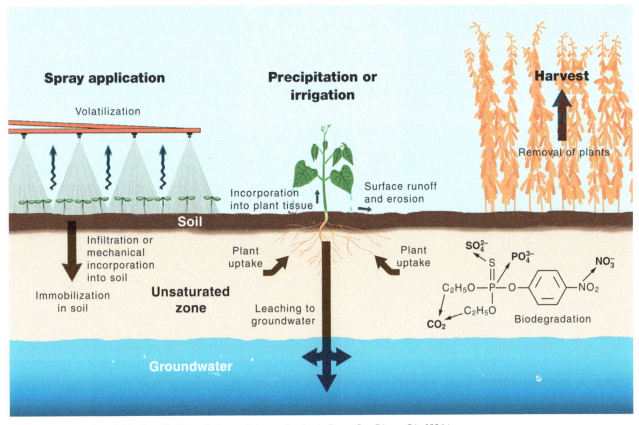

FIG. 14.10 Fate of pesticides in soil. *(From Pollution Science, Academic Press, San Diego, CA, 1996.)*

amounts exceeded the crop water use by twice the amount of water contained in the root zone at an optimum water content that moved the chemicals downward.

As shown in Fig. 14.11, glyphosate would be concentrated in the root zone to a depth of about 25 cm, atrazine would be concentrated near the bottom of the root zone (about 125 cm), and aldicarb would be concentrated at a depth of about 250 cm. A slightly higher percentage of the applied atrazine would exist in the system, compared with the other two pesticides, because it has a slightly larger half-life. For a growing season of about 120 days, about 6% of the applied aldicarb and glyphosate would remain, while about 20% of the atrazine would remain. This example does not account for numerous differences in management practices that would influence the persistence and soil distribution of these pesticides.

14.7 ANIMAL WASTES

Animal wastes contain several types of land pollutants that are of concern both to the public and regulators. Besides traditional pollutants, discussed later, increasing evidence suggests that excessive use of animals waste on land releases measurable amounts of antibiotics, growth hormones, and pesticides containing toxic elements like arsenic. Animal

agricultural wastes can be divided by two production types: range and pasture production, and confined or concentrated animal production (see Chapter 23 for more detail).

In range and pasture systems, the concentration of wastes is generally much more diffuse or dispersed than it is when large numbers of animals are confined to relatively small areas. Range and pasture systems have two principal measurable effects on surface water quality: (1) increased turbidity through the movement of soil particles into streams, rivers, and lakes; and (2) increased fecal coliform and parasite counts in areas of heavy animal use. Although we know that grazing systems may adversely affect some measures of water quality, we will focus here on the highly concentrated animal production units and the methods of preventing and controlling pollution from these concentrated units. Concentrated animal production is very common and is occurring in increasingly controlled environments to raise productivity and diminish climatic, feeding, and mortality variables. Larger numbers of animals are being raised in *concentrated animal feeding operations* or *CAFOs*—principally, feedlots, dairies, swine operations, poultry houses, and intensive aquaculture.

Following World War II, manure was displaced as the primary fertilizer by fossil-fuel-based fertilizers as farms became increasingly specialized.

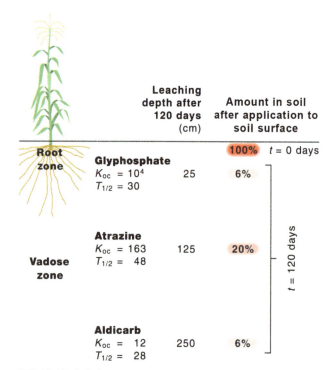

	Leaching depth after 120 days (cm)	Amount in soil after application to soil surface
Root zone		100% $t = 0$ days
Glyphosphate $K_{oc} = 10^4$ $T_{1/2} = 30$	25	6%
Vadose zone **Atrazine** $K_{oc} = 163$ $T_{1/2} = 48$	125	20%
Aldicarb $K_{oc} = 12$ $T_{1/2} = 28$	250	6%

$t = 120$ days

FIG. 14.11 Relative movement and persistence of pesticides in soil. The transport of a particular pesticide is strongly influenced by retardation. Here, the degree of retardation increases in the order aldicarb < atrazine < glyphosate. The amount of pesticide remaining in the soil is illustrated by the intensity of color 120 days after the application. *(From Pollution Science, Academic Press, San Diego, CA, 1996.)*

With the breakdown of the traditional cycle of reincorporation of wastes back in to the land, what was once an essential source of nutrients has now become a potential pollutant. Thus the production of large numbers of animals on a small land base has resulted in the stockpiling of wastes, the construction of large waste-storage ponds, and, oftentimes, waste applications to land in excess of agronomic crop needs. To date, few states regulate the land application of animal wastes to the degree that biosolids are regulated (see Chapter 23).

14.7.1   Nonpoint Versus Point Source Pollution

The term "nonpoint pollution" is misleading and is often misused in the context of animal wastes. In animal agricultural systems, true *nonpoint sources* are those in which potential contaminants are not concentrated during production and do not pass through a single or small number of conduits for disposal. These nonpoint sources include corrals, feedlots, and extensive and intensive pasture systems.

Point sources are those facilities that concentrate pollutants or contaminants to a significant degree and pass these contaminants through a pipe, ditch, or canal for disposal. The

most common point sources are milksheds and barns, dairy and other food-processing plants, intensive indoor swine facilities, anaerobic and aerobic lagoons, and evaporative storage ponds. In addition, certain types of intensive aquaculture may also be point sources of contaminants, with return flows highly nutrient laden with fish excreta.

According to EPA regulations, however, some concentrated animal feeding operations may be designated as point sources requiring an individual National Pollution Discharge Elimination System (NPDES) permit. In this case, a concentrated animal feeding operation is defined as a lot or facility without vegetation where animals are confined for 45 or more days per year. The number of animals needed to meet this definition as a CAFO depends on several factors; the key determinant is whether or not the facility discharges into navigable waters, as determined by the method of discharge. The method of discharge is judged by the 25-year, 24-h storm event, which is the required event that a facility must be designed to meet.

Nonpoint sources, such as nondischarging concentrated animal feeding operations, require a different approach to prevention and mitigation of pollutants than do point source emissions from a pipe or conduit. At present, the nonpoint source approach to mitigation employs *Best Management Practices* (BMPs), as defined by the 1987 Amendments to the Federal Water Pollution Control Act. In contrast, point source methods employ methods termed *Best Available Demonstrated Control Technology* or *Best Available Control Technology*. In 2003, the U.S. EPA (2003) published a Final Rule on CAFOs that is now used to permit animal feeding operations by establishing requirements that are more protective of the environment. Large amounts of animal wastes are land applied (see Chapter 23).

14.7.2   Specific Pollutants

Concentrated animal agriculture produces specific pollutants in the wastes resulting from animal metabolic activity (Information Box 14.1).

14.8   INDUSTRIAL WASTES WITH HIGH SALTS AND ORGANICS

14.8.1   Oil and Gas Drilling

The process of drilling for crude oil requires powerful drill rigs that use large quantities of drilling fluids. These fluids contain high-density weighing agents such as barium sulfate (barite) and scale inhibitors of unknown compositions. Other drilling fluids are composed of sodium chloride solutions, which are used to force crude oil up to the surface. These fluids must be disposed of once they are "spent" or no longer useful. Prior to 1985, these spent fluids were stored in ponds near the drill sites and often simply bulldozed over when the well was completed. Consequently, many older oilfields have large tracts of land contaminated with spent drilling wastes. These wastes are not classified as hazardous because they do not contain significant amounts of metals. Although free barium (Ba^{+2}) is very toxic, the mineral barite ($BaSO_4$) is quite inert in the environment. On the other hand, NaCl is very soluble in water and can increase the salinity of surface waters, rendering them nonpotable. However, with increased gas extraction using newly developed hydraulic fracturing (fracking) of oil/gas-rich sedimentary rocks, the geologic formation water often has higher levels of radioactive elements such as uranium, thorium, and radium. When formation waters comingle with fracking fluids they can make oil and gas production wastewaters (called produced water) and associated sludges more radioactive (U.S. EPA 2018).

14.8.2 Coal-Burning Electric Power Plants

Electric power plants generated about 110 million of tons of coal ash in 2014 (U.S. EPA, 2017b) Coal (or fly) ash may be stored either in ponds or landfills and may be used as fill material. Fly ash is recovered from electrostatic precipitators that scrub out silt-size particulate matter from the flue gases generated from coal combustion. These particles generally arise from the incombustible silt and clay found in coal deposits. Upon exposure to high temperatures, silt and clay (which consists mostly of silica and alumina) combine to yield amorphous Si-Al-based spheres onto which other elements may condense. Typically, fly ash

spheres also include Ca, Na, Fe, Mg, K, and Ti, with small amounts of other elements sorbed onto them, such as As, B, Ba, Cd, Cr, Cu, F, Mo, Ni, Pb, S, and Zn. The concentrations of these elements in fly ash vary widely, depending on the source of the coal and leached metals from fly ash impoundments have been linked to groundwater contamination and increased cancer rates in some parts of the United States (PSR, 2018) and other countries like India (Singh et al., 2016). A typical empirical composition of fly ash is

$$10\,Si + 5\,Al + 0.5\,Ca + 0.5\,Na + 0.4\,Fe + 0.2\,Mg + 0.2\,Mg$$
$$+ 0.2\,K + 0.1\,Ti + 0.05\,S + \text{trace amounts of more than}$$
$$15\,\text{other elements}$$

Given the large volumes of that are generated of this waste, the USEPA Coal Ash Disposal Rule requires stricter disposal guidelines of these wastes in landfills and surface impoundments including groundwater monitoring around coal ash disposal sites.

The removal (scrubbing) of sulfur dioxide (SO_2) gas from flue gases produces large quantities of *flue gas desulfurization wastes*, which consist largely of calcium carbonates, sulfates, and sulfites. These wastes may also contain trace quantities of some of the elements in fly ash, but the concentration of these elements depends on the source of the coal and the type of scrubbing systems used. Because flue gas desulfurization products are usually more than 70% water, these wastes are disposed of in drying ponds and are often treated along with power plant wastewaters. This waste mixing may add significant amounts of soluble salts (e.g., NaCl) that increase the salinity of sludges (see Table 24.2).

Despite new gas scrubbing technology and stricter emission standards, sulfur dioxide and nitrous oxide have been reduced but not completely eliminated. For example, in 2015 sulfur dioxide emissions from coal-fired power plants were ~2 million metric tons (Mt) (U.S. EIA 2018). In the last 10 years significant reductions in SO_2 occurred in response to the 2009 *Clean Air Interstate Rule* (CAIR), which affects 28 eastern U.S. states. In 2015 the CAIR was replaced by the Cross-State Air Pollution Rule (CSAPR), which also addresses ozone-producing nitrogen oxides (NO_x) emissions from power plants (U.S. EPA 2018b).

Mercury metal emissions from coal-burning electric power plants remain a controversial issue. For the first time, reductions in Hg emissions from coal-burning power plants are being mandated under the new March 2005 *Clean Air Mercury Rule*. Under this rule, by 2020, reductions of 70% Hg emissions were expected from coal-burning power plants. However, this rule was replaced in 2011 by Mercury and Air Toxics Standards (MATS) and its effect on reducing mercury emissions from coal-fired power plants is not known.

14.8.3 Industrial Wastes High in Organic Chemicals

Most industrial wastes contain varying amounts of organic chemicals. With few exceptions, carbon-based chemicals, reagents, solvents, feedstocks, and raw materials are extensively used in most phases of industrial processing. Exceptions to this rule may include mine tailings and metal-plating wastes. Wastes containing large quantities of organic chemicals include those originating from oil refineries, as well as petrochemical, chemical, pharmaceutical, and food-processing industries, and paper mills. Hazardous waste sites associated with these types of activities are discussed in Chapter 15. Driven by environmental laws and regulations (Chapter 30), these industries have reduced their polluting waste streams by applying pollution prevention strategies such as wastewater treatment processes before discharge, the implementation of waste reduction techniques that include recycling, and changes in industrial processes with emphasis on waste minimization (see Chapter 21).

14.9 INVASIVE SPECIES

Invasive species are an environmental problem of growing concern worldwide. Invasive species are organisms that have been introduced to a new ecosystem and that have a severe, often irreversible effect on agriculture and natural ecosystems (see Information Box 14.2). Any organism can become an invasive species, including microorganisms, invertebrates, insects, fish, plants, and animals. Invasive organisms often find few enemies (predators and diseases) in their new location, allowing them, at least initially, to grow and reproduce relatively easily.

It is important to note that a key component of this issue is that humans typically introduce the invasive species. Species can gradually spread into new areas as a natural process; this process is usually slow and involves adjustments by all members of the ecosystem. Conversely, human-initiated introductions are usually relatively fast,

---

INFORMATION BOX 14.2 Invasive Species and Associated Problems

Invasive species:	Nonnative species, introduced to a new area primarily by human activity, that can reproduce and spread independently and that cause, or is likely to cause, economic or environmental damage.
Invasive species problems:	Competition with and replacement of native species, damage to wildlife habitat, altered fire ecology, potential to spread disease.

INFORMATION BOX 14.3 Adaptive Traits of Invasive Species

- High reproductive rates
- High dispersal rates
- High genetic variability (allows them to adapt more quickly to a broad range of environmental conditions)
- Broad range of native habitat, that is, adapted to a variety of soil and climatic conditions
- Moved by humans to new locations, whether deliberately (kudzu) or accidentally (tumbleweed)

---

often resulting in large disruptions of the ecosystem. As noted in Information Box 14.2, these disturbances can be obvious or subtle.

In most cases, invasive species take advantage of opportunities in ecosystems that are disturbed by human activity (see Information Box 14.3). Disturbances can be flow control of rivers, disturbed soil along roadways and agricultural areas, human structures (e.g., pigeons and sparrows are better adapted to cities than many native birds), water temperature change due to power plant outflows, and so on.

Some invasive species were introduced to provide erosion control, such as kudzu and salt cedar. Others, such as Lehman lovegrass (*Eragrostis lehmanniana*), buffelgrass (*Cenchrus ciliaris*), red brome (*Bromus madritensis*), were repeatedly introduced over large areas of the western United States as forage grasses for cattle and sheep. These grasses have radically altered the fire ecology of western ranges, which in turn has changed plant populations, wildlife distribution, and nutrient cycling. Buffelgrass in particular, is a very aggressive, drought and fire-resistant plant that has become a major concern (Fig. 14.12). It is quick to ignite and it burns very hot, often killing nearby native vegetation in wildfires. Fountain grasses (*Pennisetum*) related to buffelgrass are dangerous invasive grasses, which escaped from gardens. Another widely known example is purple loosestrife (*Lythrum salicaria*), which is a colorful perennial plant that is prized in gardens. Purple loosestrife has spread widely in the eastern and northern United States, choking waterways and supplanting native riparian species. Animals released as hunting stock have caused problems. Examples include opossums (*Didelphidae*) in the northwestern United States, rabbits (*Leporidae*) in Australia, and red deer (*Cervus elaphus*) in New Zealand. Invasive Zebra mussels (*Danaea polymorpha*) in the Great Lakes have caused millions of dollars worth of damage to water intake systems, while simultaneously decreasing lake biodiversity. Fire ants (*Solenopsis invicta*) decrease biodiversity while causing major economic damage.

FIG. 14.12 Buffelgrass invading parts of the Sonoran desert. The grass is a fire hazard initially favoring expansion along disturbed roadside areas. Inset: Roadside buffelgrass plant with distinct bottlebrush-shaped flower stacks in early spring. This invasive grass also thrives in desert scrup environments of Africa, Australia, and other parts of the world. *(Photos courtesy of J.F. Artiola.)*

14.9.1   Kudzu

Kudzu (*Pueraria montana* var. *lobata*) is a well-known, highly visible example of an invasive plant species. Kudzu is a broad leaved, fast growing perennial vine (Fig. 14.13) from Japan that was deliberately introduced in the early 1900s to control erosion and provide forage in the southern United States. Kudzu grows rapidly and stabilizes loose soil with large, fleshy roots. In the United States, kudzu has few serious checks on its growth by insects or disease, which allows it to grow as much as 20 m (60 ft) per season. This growth tends to completely cover existing vegetation, and can break branches and block sunlight from the native plants, eventually greatly weakening or killing them. Kudzu can also cover cars and entire buildings. Once it is well established, kudzu is difficult to remove. In addition, there is some evidence that kudzu is becoming more cold tolerant, extending its range to the north. It is estimated that kudzu covers about $25,000 \text{km}^2$ (10,000 square miles) in the United States and that it costs somewhere between $100 million to $500 million dollars a year in lost cropland and control costs.

14.9.2   Salt Cedar

Kudzu is quite obvious, even to the untrained eye, and causes visible damage as it smothers other plants and buildings with its extremely rapid growth. Many invasive species do not cause such obvious problems. Salt cedar (*Tamarix* spp.) has rapidly colonized riparian areas throughout the western United States, causing major changes in this habitat. These

FIG. 14.13 Kudzu is an invasive perennial vine that grows over everything in its path, including trees and buildings. (A) A cabin in early spring, with kudzu visibly starting to climb on trees and the abandoned cabin. (B) The same location in late summer, showing kudzu that has covered the cabin and most of the bushes and trees in the foreground. Kudzu blocks light from reaching the plants underneath, greatly weakening them, and can become heavy enough to break branches. *(Photo courtesy Jack Anthony http://www.jjanthony.com/kudzu.)*

changes are not obvious to the casual observer and yet are causing profound changes in riparian ecosystems.

Multiple species of salt cedar were originally introduced more than 100 years ago for erosion control and ornamental use (Fig. 14.14). Salt cedar is tolerant of drought and saline and alkaline soils, grows quickly if there is sufficient water, has high seed production, and resprouts easily after fire. In addition, salt cedar has been implicated in lowering water tables at the expense of native species and also of salinizing soil. It is generally thought that these characteristics have enabled them to supplant native stands of cottonwood (*Populus* spp.) and willow (*Salix* spp.) that provide wildlife habitat, while producing little useful habitat of their own. As a result, land managers consider salt cedar a prime example of a detrimental invasive species.

Salt cedar is now found throughout the western United States, thriving in response to human disturbances related to dams and diversion projects along large drainages. Research indicates that salt cedar changes the species

FIG. 14.14 Salt cedar thickets growing along the edge of a seasonally dry retention pond with high salinity in the southwestern United States. Dark shrubs on the hills above the salt cedar are juniper. The inset shows a flowering salt cedar. *(Photo courtesy J.F. Artiola.)*

composition of riparian communities and reduces their biodiversity. Thickets of salt cedar tend to replace native shrubs and trees such as willow and cottonwood, generally without replacing their usefulness as nesting sites and food sources. This may be due in part to the control of floods by dams, as willows and cottonwoods tend to establish new seedlings after flood events. In addition, regulation of river flow in general is causing many riparian areas to become drier and more saline. Recent research indicates that salt cedar may be better adapted to these new growing conditions, which allows them to outcompete the native species.

Salt cedar is also widely regarded as a cause of salinization of soil, which is thought to prevent or diminish the growth of many native species. Salt cedar is very tolerant of salinity and is able to use low-quality surface and groundwater sources that many natives cannot. The plants store excess salts in salt glands in the leaves and also excrete salt onto the surface of the leaves themselves. Since salt cedar is deciduous, the leaves eventually fall to the soil surface and build up a salt-rich litter. In areas with floods or sufficient rain, this salt is moved out of the root zone, but salt cedar usually grows in arid areas with low rainfall and along regulated waterways (preventing regular floods), thus allowing some salt buildup. Recent research indicates that there may not be salt buildup over a period of years, as rains and occasional floods may leach the salt out of the root zone. Other research indicates that the salt buildup is small enough that it does not affect some of the more salt-tolerant native species.

There is no doubt that salt cedar is an invasive species. Land managers are finding it difficult to remove salt cedar and reestablish native populations, and have blamed salt cedar for drier conditions and more saline soil. Salt cedar may be able to adapt more readily than some native species to human-caused ecosystem disturbances.

Invasive species have caused major ecosystem changes throughout the world, resulting in billions of dollars of damage to agriculture, forestry, power plants, and the like each year. They are almost impossible to eradicate and difficult to control once established (see Chapter 20).

QUESTIONS AND PROBLEMS

1. Explain why mine tailings can be a source of pollutants in the soil and air environments. Give two examples of pollutants associated with mine tailings.
2. Give two examples of land pollution associated with deforestation.
3. Soils can become acidic when (a) basic cations are leached, (b) Al oxides accumulate, (c) carbonate and bicarbonate minerals disappear, or (d) all of the above. Explain your answer.
4. What is the difference between a saline and a sodic soil? Explain the difference, and the EC and ESP values that a soil would have to be called saline-sodic.
5. Which factor would you add to or replace in the water erosion model RUSLE equation presented in Section 14.5.1 to make it a wind erosion model?
6. Explain why aggregated fine particles are less affected by wind erosion than loose coarser soil particulates.
7. Give two examples (each) of chlorinated and non-chlorinated pesticides and their use.
8. Why are chlorinated pesticides more persistent in the environment than nonchlorinated pesticides?
9. Explain how invasive plant species can change soil salinity and water quality.
10. Explain how deforestation affects water quality.

REFERENCES

Bonan, G., 2002. Ecological Climatology. Cambridge University Press, Cambridge, England.

Borman, F.H., Likens, G.E., 1979. Pattern and Process in a Forsted ecosystem. Springer-Verlag, New York.

Csavina, J., Taylor, M.P., Félix, O., Rine, K.P., Eduardo Sáez, A., Betterton, E.A., 2014. Size-resolved dust and aerosol contaminants associated with copper and lead smelting emissions: implications for emissions management and human health. Sci. Total Environ. 493, 750–756.

Gupta, R.K., Abrol, I.P., 1990. Salt-affected soils: their reclamation and management for crop production. In: Lal, R., Stewart, B.A. (Eds.), Soil Degradation: Advances in Soil Science. vol. 11. Springer-Verlag, New York, pp. 223–288.

Heald, C.L., Jacob, D.J., Fiore, A.M., Emmons, L.K., Gille, J.C., et al., 2003. Asian outflow and trans-pacific transport of carbon monoxide and ozone pollution: an integrated satellite, aircraft, and model perspective. J. Geophys. Res. 108 (D24), 4804. https://doi.org/10.1029/2003JD003507.

PSR, 2018. Coal ash: hazardous to your health. http://www.psr.org/assets/pdfs/coal-ash-hazardous-to-human-health.pdf.

Saxton, K., Chandler, D., Stetler, L., Lamb, B., Clairborn, C., Lee, B.H., 2000. Wind erosion and fugitive dust fluxes on agricultural lands in the Pacific Northwest. Trans. ASAE 43, 623–630.

Singh, R.K., Gupta, N.C., Guha, B.K., 2016. Fly ash disposal in ash ponds: a threat to groundwater contamination. J. Inst. Eng. India. Ser. A 97 (3), 255–260.

Stumm, W., Morgan, J.J., 1996. Aquatic Chemistry, third edition. A Wiley-Interscience Series of Texts and Monographs, John Wiley & Sons, New York.

U.S. EPA, 2004. Pesticide Industry Sales and Usage—2000–2001 Market Estimates. In: Biological and economic analysis division—Office of Pesticide Programs. Office of Prevention, Pesticides, and Toxic Substances. U.S. Environmental Protection Agency, Washington, DC, May 2004.

USDA, 1954. Saline and sodic soils. In: Agriculture Handbook No. 60. United States Department of Agriculture.

U.S. EIA, 2018. Sulfur dioxide emissions. https://www.eia.gov/todayin energy/detail.php?id=29812.

U.S. EPA, 2017. Pesticide industry sales and usage 2008–2012 market estimates. https://www.epa.gov/sites/production/files/2014-01/documents/pesticides-industry-sales-usage-2016_0.pdf.

U.S. EPA, 2017b. Frequent questions about the coal ash disposal rule. https://www.epa.gov/coalash/frequent-questions-about-coal-ash-disposal-rule#2.

U.S. EPA, 2018a. TENORM: oil and gas production wastes. https://www.epa.gov/radiation/tenorm-oil-and-gas-production-wastes.

U.S. EPA, 2018b. FACT SHEET. Final cross-state air pollution rule update for 2008. NAAQS. https://www.epa.gov/sites/production/files/2014-06/documents/final_finalcsaprur_factsheet.pdf.

Veiga, M.M., Meech, J.A., Oñate, N., 1994. Mercury pollution from deforestation. Nature 368, 816–817.

FURTHER READING

Anthony, J., 2005. Kudzu, The vine. http://www.jjanthony.com/kudzu.

Blake, J., Donald, J., Magette, W. (Eds.), 1992. National livestock, poultry and aquaculture waste management. Proceedings of the National Workshop, July 1991. American Society of Agricultural Engineers, St. Joseph, Michigan.

Bucklin, R. (Ed.), 1994. Dairy systems for the 21st century. Proceedings of the 3rd International Dairy Housing Conference, 1994. American Society of Agricultural Engineers, St. Joseph, Michigan.

Glenn, E.P., Nagler, P.L., 2005. Comparative ecophysiology of *Tamarix ramosissima* and native trees in western U.S. riparian zones. J. Arid Environ. 61, 419–446.

Hegg, R.O., 2005. Trends in animal manure management research. CRIS database. In: Symposium State of the Science Animal Manure and Waste Management, January 5–7, 2005. San Antonio, Texas.

Luken, J.O., Thieret, J.W., 1997. Assessment and Management of Plant Invasions. Springer-Verlag, New York.

Merkel, J.A., 1981. Managing Livestock Wastes. Avi Publishing Company, Westport, CT.

Texas Environmental Profiles. n.d. http://www.texasep.org.

TFI, 2005. U.S. fertilizer use statistics. http://www.tfi.org/Statistics/Usfertuse.

U.S. EPA-CAFO Final Rule, 2003. http://www.usawaterquality.org/themes/animal/default.html.

U.S. DA-CREES. n.d. http://www.usawaterquality.org

Chapter 15

Subsurface Pollution

M.L. Brusseau

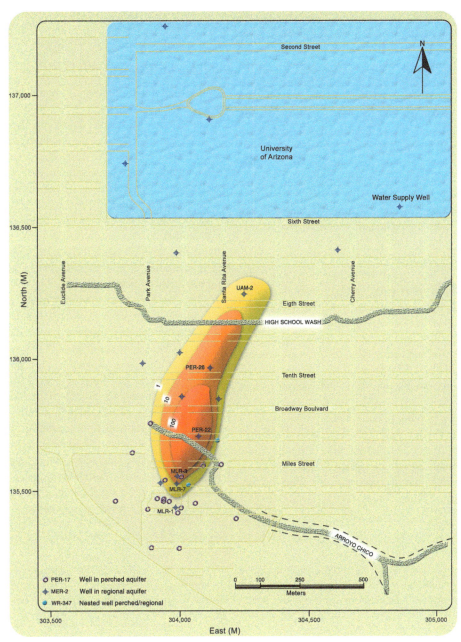

Groundwater contaminant plume (tetrachloroethene) contour map for a site in Tucson, Arizona. *Source: Data provided by Arizona Department of Environmental Quality; drawn by Concepción Carreón Diazconti.*

Environmental and Pollution Science. https://doi.org/10.1016/B978-0-12-814719-1.00015-X

15.1 GROUNDWATER AS A RESOURCE

Freshwater comprises approximately 3% of all water on Earth (see also Chapter 3). Approximately 95% of this small fraction occurs as water in the subsurface, that is, groundwater. Thus groundwater is a critical resource throughout the world. Groundwater is a major source of potable water, supports food and crop production, and is used for myriad industrial activities. As such, the availability, quality, and sustainability of groundwater resources are issues of great significance. We will briefly explore these issues in this chapter.

Groundwater has long been used by humankind. This likely occurred initially through the use of natural springs and later via hand-dug wells. There are numerous references to groundwater in ancient texts, such as the works of Plato and Aristotle (Fetter, 2001). Chinese archaeologists have found wells in the Hunan Province dating back to 2000 BC (Xinhua News Agency, 2002). Evidence of hand-dug wells has been found at archaeological sites thousands of years old (see Information Box 15.1).

Groundwater use has increased greatly in the past several decades. Worldwide, groundwater use tripled during the mid-late 20th century (Fig. 15.1). This is also true for the United States (Fig. 15.2). These increases are a direct result of increases in population and economic development. A primary use of groundwater is to supply potable water for drinking and other domestic uses. Groundwater serves as a significant source of potable water throughout the world, providing approximately half of all drinking water globally and ranging from 15% in Australia to 75% in Europe (Table 15.1). In the United States, groundwater is the source of drinking water for approximately half of the total population and nearly all of the rural population. The percentage of the US population relying on groundwater as their primary source of potable water varies greatly by state (Fig. 15.3).

In addition to supplying potable water, groundwater is also used for many other purposes. Agricultural applications, primarily as irrigation for crops, constitute the single largest use of groundwater in the United States (Fig. 15.4). The majority of irrigation use occurs in the semiarid and arid regions of the United States, as one would expect (Fig. 15.5). Other uses of groundwater include industrial, mining, and power generation. In total, groundwater provides approximately one-fourth of all water use in the United States.

The interconnectedness of water and energy is recognized explicitly in the relatively new concept titled the **Water-Energy Nexus**. This concept formalizes the reality that energy and water systems are interconnected. Energy is required to extract, treat, and deliver water. Conversely, water is used in multiple stages of energy production and electricity generation, from hydraulic fracturing and irrigating crops for biofuels to providing cooling water for power plants.

15.2 GROUNDWATER POLLUTION

Hand in hand with the use of groundwater by humans is the pollution of groundwater by humans. This pollution can occur in many forms, including hazardous industrial organic compounds, fuel components, heavy metals, agrochemicals, pathogenic microorganisms, and salinity. Groundwater can be contaminated through numerous means, as illustrated in Fig. 15.6. Major sources of groundwater contamination in the United States are reported in Fig. 15.7. There are two general categories of groundwater contamination: those produced from point sources and those that develop from diffuse or nonpoint sources. Specific examples of major groundwater pollution issues will be presented in later sections.

The growth of population centers (urbanization), with the attendant increase in population densities and industrial/commercial development, has had an enormous impact on groundwater use and quality. Water resource development during the evolution of an urban center follows a typical pattern, as illustrated in Fig. 15.8. Generally, increasing demand for potable water that occurs as population increases results in overpumping of groundwater from the original well field located in the city center. The amount of water being extracted is greater than the amount recharged, causing a decline in water levels. This eventually requires the city to supplement their potable water supply from other sources. An ancillary effect of the overpumping and falling water levels is subsidence of the land surface.

INFORMATION BOX 15.1 7,000-Year-Old Hand-Dug Wells in New Mexico and Texas

After the last Ice Age (about 11,000 years ago), the region encompassing western Texas and eastern New Mexico was covered by grasslands with numerous playa lakes. This region experienced a large-scale, long-term drought beginning about 7500 years ago, resulting in the loss of the playas. It appears that the humans occupying the region adapted to this change in climate by digging wells to obtain water. Evidence of wells dating to approximately 7000 years ago has been found at sites in New Mexico and Texas. These wells are 1–2 m in depth, with openings of approximately 1 m in diameter. Artifacts found at the sites indicate that sticks and stones were used to excavate the wells.

(**Source:** *Available from: http://www.mnsu.edu/emuseum/archaeology/sites/ northamerica/blackwaterdraw.html; and http://archaeology.about.com/od/ mesolithicarchaic/a/archaicwells.htm*).

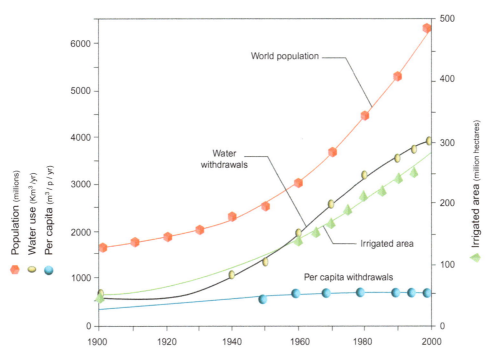

FIG. 15.1 Global trends in water use. *(From Morris, B.L., Lawrence, A.R.L., Chilton, P.J.C., Adams, B., Calow, R.C., Klinck, B.A., 2003. Groundwater and its susceptibility to degradation. A global assessment of the problem and options for management. Early Warning and Assessment Report Series.)*

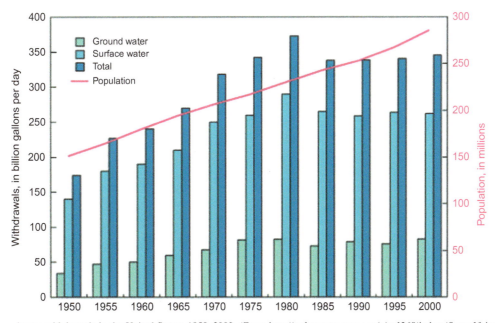

FIG. 15.2 Groundwater withdrawals in the United States, 1950–2000. *(From http://pubs.water.usgs.gov/circ1268/htdocs/figure13.html.)*

Another significant effect of overpumping in coastal areas is seawater intrusion. This will be discussed in detail in a forthcoming section.

The development of high-intensity agriculture and the widespread use of fertilizers and pesticides have led to major groundwater pollution issues for rural areas. This is of particular concern because groundwater generally is the predominant source of potable water for rural areas. For example, more than 95% of the rural population of the United States uses groundwater as their potable water supply. Additionally, most of the potable water supply is obtained from individual or small community wells. Groundwater from these wells does not typically undergo the extensive treatment and monitoring that is prevalent

TABLE 15.1 Estimated Percentage of Potable Water Supply Obtained From Groundwater (Morris et al., 2003)

Region	Percent	Population Served (millions)
Asia-Pacific	32	1000–2000
Europe	75	200–500
Central and South America	29	150
USA	40	135
Australia	15	3
Africa	NA	NA
World	—	1500–2750

FIG. 15.3 Percentage of population relying on groundwater as a drinking-water source. *(From Environmental Protection Agency (U.S. EPA), 1999. Safe Drinking Water Act, Section 1429 Groundwater Report to Congress. https://nepis.epa.gov/Exe/ZyPDF.cgi/20002893.PDF?Dockey=20002893.PDF.)*

for centralized urban water supply systems (see Chapter 24).

A comprehensive monitoring program is instrumental in managing and protecting groundwater resources. The U.S. Geological Survey's National Water-Quality Assessment (NAWQA) Program is a nation-wide program that is the principal source of information on groundwater quality in the United States. Under this program, the U.S. Geological Survey collects water quality data in 60 special study regions of the country, conducts retrospective analyses of existing data (such as state data), and prepares national-scale syntheses of the results. In addition, many state and local governments in the United States carry out groundwater monitoring programs. Specialized monitoring programs are conducted in association with characterization and remediation of contaminated sites (see Chapter 10).

National ground water use

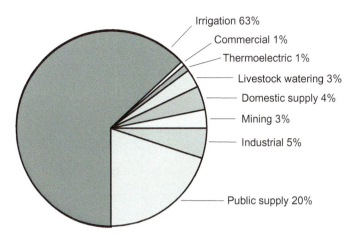

Irrigation 63%
Commercial 1%
Thermoelectric 1%
Livestock watering 3%
Domestic supply 4%
Mining 3%
Industrial 5%
Public supply 20%

Source: *Estimated Use of Water in the United States in 1995.*
U.S. Geological Survey Circular 1200, 1998.

FIG. 15.4 Groundwater use in the United States for 1995. *(From Environmental Protection Agency (U.S. EPA), 1998. National Water Quality Inventory Report to Congress. United States Environmental Protection Agency. https://www.epa. gov/waterdata/1998-national-water-quality-inventory-report-congress.)*

Volume of ground water used
for irrigation in 1995

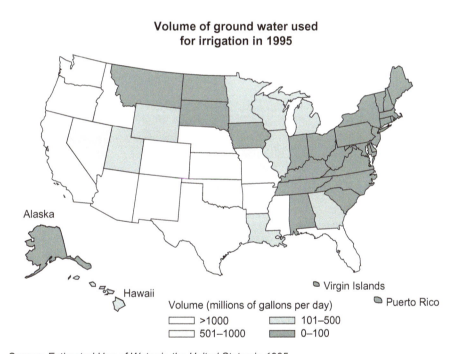

Alaska

Hawaii

Virgin Islands

Puerto Rico

Volume (millions of gallons per day)
>1000   101–500
501–1000   0–100

Source: *Estimated Use of Water in the United States in 1995.*
U.S. Geological Survey Circular 1200, 1998.

FIG. 15.5 Volume of groundwater used for irrigation in the United States for 1995. *(From Environmental Protection Agency (U.S. EPA), 1998. National Water Quality Inventory Report to Congress. United States Environmental Protection Agency. https://www.epa.gov/waterdata/1998-national-water-quality-inventory-report-congress.)*

15.3 GROUNDWATER POLLUTION RISK ASSESSMENT

Physical properties of the subsurface result in significant differences in the behavior of groundwater compared to that of surface water. For example, residence times for groundwater range from years to hundreds of years or more; these times are much greater than those for streams. Dilution effects are much less significant for groundwater compared to surface-water systems. In addition, the absence of light eliminates the possibility of photochemical reactions, which are a major route of transformation for

Sources of ground water contamination

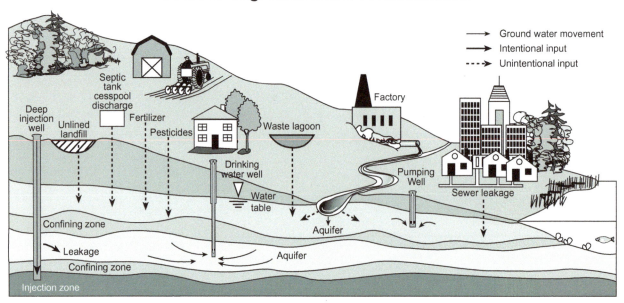

FIG. 15.6   Sources of groundwater contamination. *(From Environmental Protection Agency (U.S. EPA), 1998. National Water Quality Inventory Report to Congress. United States Environmental Protection Agency. https://www.epa.gov/waterdata/1998-national-water-quality-inventory-report-congress.)*

Major sources of ground water contamination

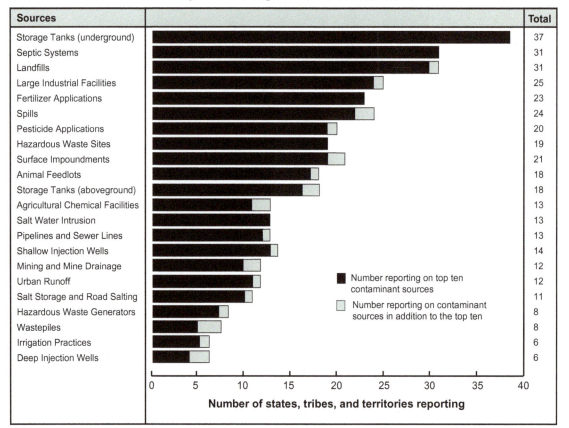

Sources		Total
Storage Tanks (underground)		37
Septic Systems		31
Landfills		31
Large Industrial Facilities		25
Fertilizer Applications		23
Spills		24
Pesticide Applications		20
Hazardous Waste Sites		19
Surface Impoundments		21
Animal Feedlots		18
Storage Tanks (aboveground)		18
Agricultural Chemical Facilities		13
Salt Water Intrusion		13
Pipelines and Sewer Lines		13
Shallow Injection Wells		14
Mining and Mine Drainage		12
Urban Runoff		12
Salt Storage and Road Salting		11
Hazardous Waste Generators		8
Wastepiles		8
Irrigation Practices		6
Deep Injection Wells		6

Number reporting on top ten contaminant sources

Number reporting on contaminant sources in addition to the top ten

Number of states, tribes, and territories reporting

FIG. 15.7   Major sources of groundwater contamination in the United States. *(From Environmental Protection Agency (U.S. EPA), 1998. National Water Quality Inventory Report to Congress. United States Environmental Protection Agency. https://www.epa.gov/waterdata/1998-national-water-quality-inventory-report-congress.)*

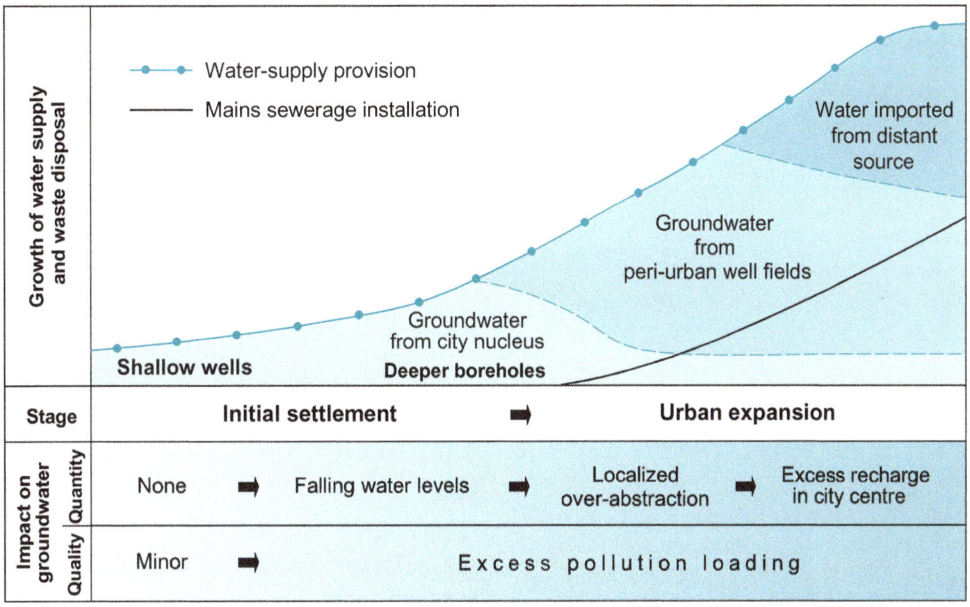

FIG. 15.8 Stages in the development of water resources during evolution of an urban center. *(From Morris, B.L., Lawrence, A.R.L., Chilton, P.J.C., Adams, B., Calow, R.C., Klinck, B.A., 2003. Groundwater and its susceptibility to degradation. A global assessment of the problem and options for management. Early Warning and Assessment Report Series.)*

aboveground environments. The net result is that once groundwater and the subsurface are contaminated, it is very difficult to decontaminate (the remediation of subsurface contamination is discussed in Chapter 19). Thus pollution prevention is critical to maintaining sustainable groundwater resources. Groundwater pollution risk assessment is a key aspect of pollution prevention.

Preventing contamination from entering the environment is the only sure way of preventing pollution. Laws and regulations promulgated during the past few decades have helped to reduce the overall contaminant load to the environment (see Chapter 30). Obviously, however, preventing all contamination from entering the environment is not possible, hence the need for groundwater pollution risk assessment, the goal of which is to evaluate the risk posed by a given activity or event to groundwater resources. Implementing risk assessments enhances the effective management of groundwater resources and helps to minimize potential contamination (see also Chapter 10).

Groundwater pollution risk is composed of two components: groundwater vulnerability and contaminant load. Groundwater vulnerability is the intrinsic susceptibility of the specific aquifer in question to contamination. Several factors affect groundwater vulnerability, as presented in Table 15.2. An aquifer that is close to ground surface, overlain by sandy soil, and located in an area with high precipitation and infiltration rates would clearly be more vulnerable to contamination than an aquifer that is hundreds of meters below ground surface and located in an area of low precipitation and infiltration.

Factors involved in the contaminant load are the type of contaminant, the amount of contaminant released, the timescale of release, and the mode of release. The pollution potential of a contaminant is controlled by its transport and fate behavior, as discussed in Chapters 7 and 8. Contaminants that are transported readily (e.g., those with high aqueous solubility and low sorption) and that are not transformed to any great extent (i.e., are persistent) generally have greater potential to pollute groundwater. For most contamination events, the contamination enters the environment in close proximity to land surface (e.g., surface spills, leaking storage tanks, and landfills). Thus transport of contaminants from the source zone to groundwater necessitates travel through the soil and vadose zone. Attenuation processes such as sorption and biodegradation (see Chapters 7–9) can act to reduce and limit the transport of contaminants to groundwater. For this reason, the soil and vadose zone is often referred to as a "living filter." The degree to which contaminants will be attenuated is a function of the type of contaminant and the nature of the subsurface. General transport and attenuation properties of common subsurface contaminants are presented in Table 15.3. Generally, the greater the amount of contamination released, the greater the pollution potential. The timescale and mode of release can also affect pollution potential. For example, releases from buried storage tanks may be more prone to cause groundwater contamination than releases from tanks stored aboveground on concrete pads.

The risk of groundwater pollution results from the combination of intrinsic vulnerability and contaminant load

TABLE 15.2 Factors Affecting Groundwater Vulnerability to Contamination

Factor	Increased Vulnerability	Decreased Vulnerability
Depth to groundwater	Shallow	Deep
Soil type	Well drained (sandy)	Poorly drained (high clay, organic matter content)
Vadose zone physical properties	Preferential flow channels	Horizontal low-permeability layers
Recharge	High precipitation, high infiltration	Low precipitation, low infiltration
Subsurface attenuation processes	Minimal attenuation	Significant attenuation

factors. Thus an aquifer that is very vulnerable may have little to no risk of pollution if the contaminant load remains negligible. Conversely, an aquifer that has a relatively low degree of vulnerability may have a significant pollution risk if the contaminant load factor is very high. The greatest pollution risk will be associated with locations where the aquifer has a high vulnerability and the contaminant loading is high.

In general, not much can be done to modify or change the inherent vulnerability of an aquifer. That is why it is critical to focus on controlling the contaminant load for managing and preventing groundwater pollution. This can be done through implementing land-use and facility operation regulations. For example, aquifer vulnerability assessments can be used in the siting of new facilities that involve production, storage, or disposal of hazardous materials. Including aquifer vulnerability as a siting factor can prevent building such facilities in locations where groundwater is most susceptible to pollution. For existing facilities, operation procedures can be implemented to minimize the production and disposal of wastes (see Chapter 21).

Special procedures and tools have been developed to assess groundwater pollution risk. This is usually done by employing a geographic information system. The first step typically involves constructing an aquifer vulnerability map (Fig. 15.9). This map incorporates one or more factors that influence aquifer vulnerability, such as those listed in Table 15.2. A commonly used aquifer vulnerability tool is DRASTIC (see Information Box 15.2). An example of an aquifer vulnerability map is shown in Fig. 15.10. Once the vulnerability map has been created, land-use factors can be superimposed to evaluate risk. An example of a groundwater risk map is shown in Fig. 15.11, which shows the nitrate contamination risk for shallow aquifers in the United States. The map was developed using soil drainage as the aquifer vulnerability factor, and fraction of cropland acreage, population density, and nitrogen loading as the loading (land use) factors.

15.4 POINT-SOURCE CONTAMINATION

Point-source systems are characterized by very localized contamination releases. Primary examples include surface spills, leaking storage tanks, disposal pits, and waste injection wells. The distribution of contamination at sites associated with point-source releases follows a general pattern as illustrated in Fig. 15.12. The region of the subsurface where the majority of the original contamination is present is referred to as the **source zone.** It is usually in close proximity to the location of the contaminant release. The source zone generally encompasses a relatively small area, such as the approximate size of a soccer field, and contains the majority of the contaminant mass. Conversely, the region where groundwater is contaminated by dissolved compounds originating from the source zone is referred to as a **groundwater contaminant plume.** The plume is often much larger than the source zone. The size of the plume depends on the migration potential and persistence of the contaminant, as well as groundwater flow behavior. Plumes can extend for several kilometers in some cases.

The configuration or "architecture" of the source zone (e.g., porous-medium heterogeneity, total contaminant mass, contaminant distribution) and source-zone "dynamics" (e.g., mass-transfer processes, transformation processes) is central to the pollution risk posed by the site. For example, the magnitude (size and concentration level) of the groundwater contaminant plume generated from the source zone is clearly dependent on the magnitude and rate of contaminant mass transfer from the source zone to surrounding regions (i.e., the source-zone mass flux). This mass transfer will be influenced by groundwater flow patterns (which are mediated by the physical properties of the porous media), and by the type, amount, and distribution of contaminant. The groundwater contaminant plume is generally the primary source of human health risk posed by these types of contaminated sites, as use of contaminated groundwater is usually the major route of potential exposure to subsurface contamination. Thus it is critical

TABLE 15.3 Transport and Attenuation Properties of Major Subsurface Contaminants

Contaminant	Source	Attenuation Mechanism				Permitted Drinking Water Concentration	Mobility	Persistence	Potential to Develop Extensive Groundwater Plume
		Biochemical degradation	Sorption	Filtration	Precipitation				
Pathogens	Sewage	**	**	***	×	Very low (<1 per 100mL)	Low	Low–moderate	Low
Nitrogen (N)	Agriculture, sewage	*	*	×	×	Moderate (10–20mg N/L)	Very high	Very high	High
Chloride (Cl)	Sewage, industry, road deicer	×	×	×	×	High	Very high	Very high	High
Sulfate (SO$_4$)	Road-runoff, industry	* 2	*	×	*	High	High	High	High
Dissolved organic carbon (DOC)	Sewage, industry (esp. food processing, textiles)	**	**	*	×	Not controlled	Moderate	Low–Moderate	Moderate
Heavy metals	Industry	×	***	* 3	**	Low (variable)	Generally low unless pH low (except CR [VI])	High	Low
Halogenated solvents	Industry, commercial	*	*	×	×	Low (5–30µg/L)	High	High	High
Fuels, lubricants, oils, other hydrocarbons (LNAPLs)	Fuel station, industry	***	**	×	×	Low (10–700µg/L BTEX 4)	Moderate	Low–Moderate	Moderate
Other synthetic organics	Industry, sewage	Variable	Variable	×	×	Low (variable)	Variable	Variable	Variable

KEY=*** highly attenuated, ** significant attenuation, * some attenuation, × no attenuation
1—Ammonia is absorbed, 2—Can be reduced, 3—When it occurs as organic complexes, 4—Aromatic compounds with health guideline limits.
Modified from Morris, B.L., Lawrence, A.R.L., Chilton, P.J.C., Adams, B., Calow, R.C., Klinck, B.A., 2003 Groundwater and its susceptibility to degradation. A global assessment of the problem and options for management. Early Warning and Assessment Report Series.

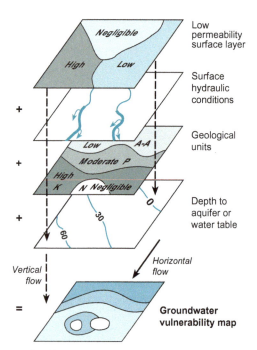

FIG. 15.9 Combining factors to create a groundwater vulnerability map. *(From Morris, B.L., Lawrence, A.R.L., Chilton, P.J.C., Adams, B., Calow, R.C., Klinck, B.A., 2003. Groundwater and its susceptibility to degradation. A global assessment of the problem and options for management. Early Warning and Assessment Report Series.)*

industrialized countries. Major types of organic compounds that are prevalent groundwater contaminants are listed in Table 15.4. It is estimated that there are tens of thousands of sites across the United States where groundwater is contaminated with one or more of the chemical classes listed in Table 15.4 (see Information Box 15.3). Heavy metals such as lead and cadmium are also a major concern at many hazardous waste sites.

This contamination results from the release of the chemicals during their transport, storage, use, and disposal. For example, thousands of underground fuel storage tank releases have been confirmed for every state in the United States (Fig. 15.13). For another example, it has been reported that up to 75% of the 36,000 active dry-cleaning facilities in the United States have experienced releases of tetrachloroethene or other dry-cleaning solvents (EPA, 2003). Many of these compounds are of concern with respect to human health (e.g., carcinogenic), as discussed in Chapter 28.

One of the most critical issues associated with hazardous waste sites is the potential presence of immiscible-liquid contamination in the subsurface. The chemicals listed in Table 15.4 are liquids under natural conditions, and they are immiscible with water, meaning they do not mix. Immiscible liquids released into the subsurface

---

INFORMATION BOX 15.2 Aquifer Vulnerability is DRASTIC

DRASTIC was developed by the EPA as a standardized system for evaluating aquifer vulnerability to pollution. The method is based on assumptions that the contaminant is released at the ground surface and that it enters the subsurface via infiltrating water. Contaminant attenuation is not considered. Thus it may be considered to provide a conservative estimate of vulnerability for those situations where attenuation may be significant. The acronym DRASTIC is derived from the seven hydrogeologic factors considered:

1. **D**epth to groundwater
2. net **R**echarge
3. **A**quifer media
4. **S**oil media
5. **T**opography (slope)
6. **I**mpact of the vadose zone media
7. hydraulic **C**onductivity of the aquifer.

Each of the hydrogeologic factors is assigned a rating from 1 to 10, based on the properties of the specific site. For example, a rating of 10 is given for "depth to groundwater" if the water table is located from 0 to 5 ft from ground surface. A rating of 1 is given if it is more than 100 ft from ground surface. The ratings are then multiplied by a weighting coefficient ranging from 1 to 5. The most significant factors have a weight of 5; the least significant have a weight of 1. The weight factors are D=5, R=4, A=3, S=2, T=1, I=5, and C=3. The products of the rating and weighting coefficient for each factor are summed to produce the final DRASTIC score. The smallest possible DRASTIC score is 23 and the largest is 226.

The DRASTIC score represents a relative measure of aquifer vulnerability. The higher the DRASTIC score, the greater the vulnerability of the aquifer to contamination.

---

to characterize source-zone and contaminant plume properties at a given site (See Chapter 19).

15.4.1 Hazardous Waste Sites

The contamination of groundwater by chemicals at hazardous waste sites and the associated risk to human health and the environment is one of the primary groundwater pollution issues facing the United States and other

become trapped in pore spaces due to capillary forces. Once entrapped, the immiscible liquid is very difficult to remove. Hence, immiscible liquids serve as long-term sources of subsurface contamination as they dissolve in groundwater or soil-pore water. The presence of immiscible liquids in the subsurface of a site can greatly impact the costs and time required for site remediation. For example, for sites contaminated by dense nonaqueous phase liquids (DNAPLs), it is estimated that upwards of hundreds of years may be

Drastic index

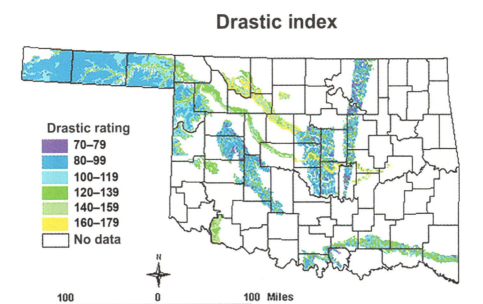

FIG. 15.10   DRASTIC-based vulnerability assessment of major aquifers in Oklahoma. *(From Osborn, N.I., Eckenstein, E., Koon, K.Q., 1998. Vulnerability assessment of twelve major aquifers inOklahoma. Oklahoma Water Resources Board Technical report 98.)*

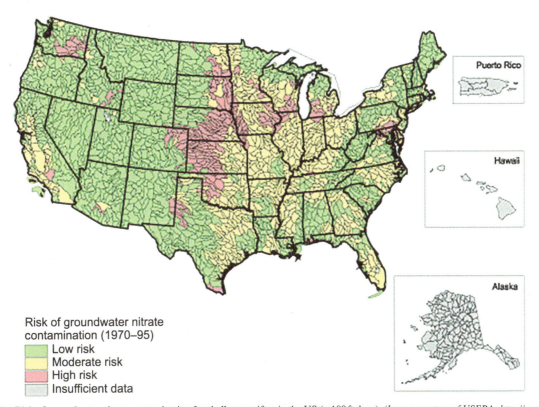

FIG. 15.11   Risk of groundwater nitrate contamination for shallow aquifers in the US (<100 ft deep). *(Image courtesy of USEPA; http://www.epa.gov/iwi/1999sept/iv21_usmap.html.)*

necessary to achieve health-based groundwater cleanup objectives using standard pump-and-treat systems (ITRC, 2002). This clearly illustrates the critical importance of addressing immiscible-liquid contamination when it is present at a site. Unfortunately, as is widely acknowledged, cleaning up sites contaminated by immiscible liquids is one of the greatest challenges in the field of environmental remediation. In fact, according to several reviews

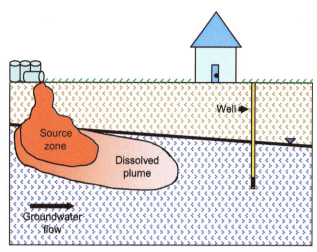

FIG. 15.12 Distribution of contamination at a point-source groundwater contamination site. *(From C.M. McColl.)*

complicates site characterization and risk assessment efforts. The estimated cost to clean up the immiscible-liquid contaminated sites in the United States is $100 billion or more (EPA, 2003).

The distribution of immiscible liquids in the subsurface is controlled by the physical and chemical properties of the porous media and by the properties of the chemical. A primary property of concern is the density of the immiscible liquid in comparison to that of water. In fact, this is such an important property that the immiscible liquids are classified based on whether they are more (DNAPL) or less (LNAPL) dense than water (see Table 15.4), where NAPL=nonaqueous phase liquid. Because DNAPLs are denser than water, they can sink below the water table with sufficient volume of release. A conceptual diagram of the distribution of a DNAPL in the subsurface is presented in Chapter 7 (see Fig. 7.1). In contrast to DNAPLS, LNAPLS

TABLE 15.4 Major Classes of Hazardous Organic Chemicals of Significance for Groundwater Pollution

Class	Example Compounds	Sources/Uses	Properties[a]
Chlorinated solvents	Trichloroethene	Manufacturing	DNAPL
	Tetrachloroethene	Degreasing	
		Dry cleaning	
Coal tar, Creosote	Polynuclear	Coal gasification	DNAPL
	Aromatic		
	Hydrocarbons		
	Phenols	Wood treatment	
Hydrocarbon fuels	Benzene	Fuel	LNAPL
Polychlorinated Biphenyls	Aroclor	Transformer fluid	DNAPL

[a]DNAPL = denser than water; LNAPL = less dense than water.

INFORMATION BOX 15.3 Chlorinated-Solvent Contamination of Groundwater in Arizona

Chlorinated solvents, such as tetrachloroethene, trichloroethene, dichloroethene, carbon tetrachloride, and vinyl chloride, are among the most common groundwater contaminants in the United States due to their widespread use as dry cleaning solvents and as degreasing and cleaning agents for military, industrial, and commercial applications. In Arizona, chlorinated solvents are the primary contaminants at an overwhelming majority of both the state Superfund sites (30 out of 34) and the federal Superfund sites (13 out of 14).

Hundreds of millions of dollars have been spent to date on characterization and remediation of these sites. Furthermore, additional hundreds of millions of dollars have been spent to settle toxic-tort lawsuits. In aggregate, these sites encompass hundreds of km² in land area and contain billions of liters of contaminated groundwater. As such, chlorinated solvents are a major source of groundwater contamination in Arizona and pose an immediate, significant, and continuing threat to the sustainability of the state's potable water supplies.

conducted by expert panels convened by the National Research Council, the presence of immiscible liquids is usually the single most important factor limiting the cleanup of organic-chemical contaminated sites. In addition, the presence of immiscible liquids greatly

float on the water table because they are less dense than water (Fig. 15.14). The distribution and amount of immiscible liquid present in the source zone has a major impact on the nature of the groundwater contaminant plume that forms at the site.

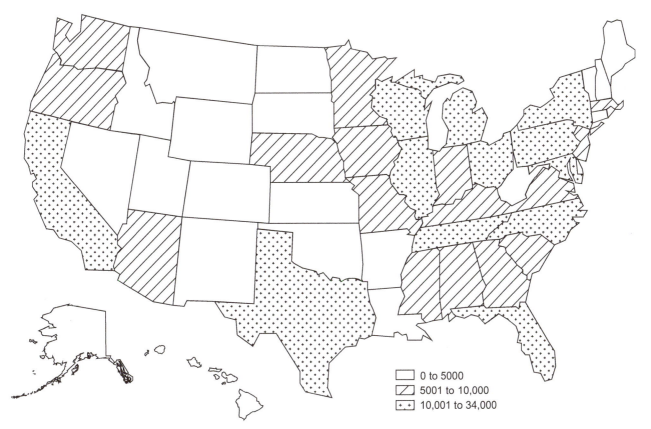

FIG. 15.13 Confirmed underground storage tank releases as of February 1999. *(From EPA, 1999.)*

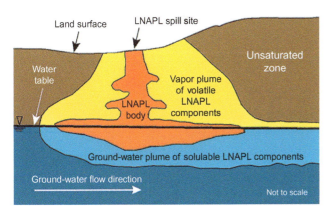

FIG. 15.14 Conceptual diagram illustrating the distribution of an LNAPL in the subsurface. *(Image courtesy of USGS; http://toxics.usgs.gov/definitions/lnapls.html.)*

In addition to source-zone properties, the size of the dissolved contaminant plume will be influenced by the nature of the chemicals and by physical, chemical, and biological properties of the subsurface. Plumes are generally relatively small for chemicals that undergo significant attenuation. Conversely, very large contaminant plumes can form for chemicals that undergo minimal attenuation. The general potential of the major classes of subsurface contaminants to develop extensive groundwater plumes is listed in

Table 15.3. For example, groundwater plumes comprised of chlorinated-solvent constituents are typically hundreds to thousands of meters long because these compounds are difficult to biodegrade and have relatively low sorption. Conversely, plumes generated from fuel hydrocarbons such as gasoline are generally much shorter, primarily because the compounds are more amenable to biodegradation.

15.4.2 Landfills

The contamination of soil and groundwater from landfills is another important problem that continues to threaten groundwater resources, posing risks to human health and the environment. Chemicals, both hazardous and nonhazardous, can be leached from the materials that are disposed of in landfills (see Fig. 21.7). This landfill-derived contamination or **leachate** can enter the soil and migrate through the vadose zone, eventually contaminating groundwater resources. Landfills are used for the disposal of numerous types of wastes, garbage, and materials. Thus contamination emanating from landfills can contain mixtures of numerous types of compounds, increasing the complexity of the pollution problem (Table 15.5).

Contaminated water or leachate is produced when water (e.g., precipitation or irrigation) enters the landfill and

TABLE 15.5 Common Contaminants and Concentrations at Landfills

Parameter	"Typical" Concentration Range	Average
BOD	1000–30,000	10,500
COD	1000–50,000	15,000
TOC	700–10,000	3500
Total volatile acids (as acetic acid)	70–28,000	NA
Total Kjeldahl Nitrogen (as N)	10–500	500
Nitrate (as N)	0.1–10	4
Ammonia (as N)	100–400	300
Total phosphate (PO_4)	0.5–50	30
Orthophosphate (PO_4)	1.0–60	22
Total alkalinity (as $CaCO_3$)	500–10,000	3600
Total hardness (as $CaCO_3$)	500–10,000	4200
Total solids	3000–50,000	16,000
Total dissolved solids	1000–20,000	11,000
Specific conductance (mhos/cm)	2000–8000	6700
pH	5–7.5	6.3
Calcium	100–3000	1000
Magnesium	30–500	700
Sodium	200–1500	700
Chloride	100–2000	980
Sulfate	10–1000	380
Chromium (total)	0.05–1	0.9
Cadmium	0.001–0.1	0.05
Copper	0.02–1	0.5
Lead	0.1–1	0.5
Nickel	0.1–1	1.2
Iron	10–1000	430
Zinc	0.5–30	21
Methane gas	60%	
Carbon dioxide	40%	

All values mg/L except as noted.
NA = not available
From Jones Lee, A., Lee, G.F., 1993. Groundwater pollution by municipal landfills: Leachate composition, detection and water quality significance. In: Proceedings of Sardinia '93 IV International Landfill Symposiums, 11–15 October, Margherita di Pula, Italy. http://www.gfredlee.com/lf-conta.htm.

contacts the waste materials (Fig. 15.15). The infiltrating water can remove hazardous and nonhazardous chemicals, including metals, minerals, salts, organic chemicals (e.g., chlorinated solvents, petroleum hydrocarbons, and pesticides), and various other toxic compounds. Millions of gallons of leachate can percolate through a landfill, depending on the size of the landfill or disposal facility. A well-designed landfill should contain the disposed materials, isolating potentially harmful leachate from the environment and specifically from groundwater and drinking

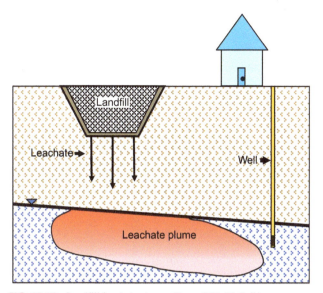

FIG. 15.15 Landfill leachate plume. *(From C.M. McColl.)*

water resources. However, it is impossible to contain and control all wastes produced at landfill and disposal sites. Leachate can enter groundwater systems as a result of poorly designed or improperly constructed landfills, deterioration of landfill liners, and landfills constructed without liners (e.g., typically older designs).

Given that leachate is likely to leak to some degree from all landfills, it is essential to implement a well-designed monitoring scheme to help manage landfill pollution problems. For example, a series of monitoring wells can be constructed around the perimeter of the landfill. The presence of high salt concentrations (e.g., Cl^-) indicates the potential threat of contamination to groundwater. Similarly, total dissolved solids (TDS) can be used to indicate potential contamination of groundwater by landfill leachate. Salts serve as good indicators of leachate contamination because of their high mobility and persistence (see Table 15.3), which means they usually constitute the leading (downgradient) front of the groundwater leachate plume. Monitoring pH is another means to detect potential contamination of groundwater by leachate. Generally, the pH of landfill leachate is lower than that of uncontaminated groundwater. A third method to detect potential leachate contamination of groundwater is by determining the oxidation–reduction potential. Highly reducing conditions typically indicate either low pH or high microbial degradation activity. Waste materials disposed of in landfills are often subject to microbially mediated decay and decomposition. The microorganisms consume oxygen during the degradation process, thus reducing the oxidation-reduction potential. In addition, some microorganisms can degrade waste under anaerobic conditions and in the process release methane and other gases. Thus the presence of such gases in groundwater is another indicator of landfill leachate contamination.

As noted before, landfill leachate comprises many compounds. Thus the transport and fate behavior of leachate is complex and highly variable. Compounds with high mobilities and low degradation potentials (high persistence) will tend to be transported much further than compounds that sorb to soil and/or that are easily degraded. For this reason, it is difficult to predict the transport and fate behavior of leachate in soil and groundwater.

In arid or semiarid environments, groundwater may be many tens or several 100 m below ground surface. In such environments, precipitation and infiltration is low, and as a result there is little to no recharge to groundwater. This means that landfill leachate will likely not migrate to the deep groundwater. However, groundwater below landfills in these environments is often contaminated with chlorinated solvent compounds for example. Migration of these volatile compounds to the deep groundwater occurred via gas-phase diffusion and perhaps in some cases via advective transport by methane generation.

15.5 NONPOINT-SOURCE CONTAMINATION

Contamination resulting from nonpoint (diffuse) sources of pollution refers to those inputs that occur over a wide area and are associated with particular land uses. This is in contrast to point-source discharges, which occur from a specific, very localized source such as a leaking fuel tank or pipe. Nonpoint sources generally encompass much larger scales (regional and even global scales) and as a result can create very large zones of pollution compared to point sources. However, the contaminant concentrations associated with nonpoint source pollution are generally lower than those associated with point sources. This section will focus on three major diffuse source issues: agrochemical contamination, saltwater intrusion, and geologic contaminant sources.

15.5.1 Agrochemical Pollution of the Subsurface

The advent of intensive agricultural practices during the last century and into the 21st century has greatly increased global food production. However, as discussed in Chapter 14, it has also greatly increased the use of fertilizers and pesticides, so-called **agrochemicals** (Fig. 15.16). The increasing use of these agrochemicals has led to extensive pollution of groundwater. For example, nitrate, derived from fertilizers, pesticides, and animal wastes, is one of the most widespread and pervasive groundwater contaminants in the United States and the world. In a survey of almost 200,000 water sampling reports, the EPA found that more than 2 million people were using water from public potable water supply systems for which nitrate standards

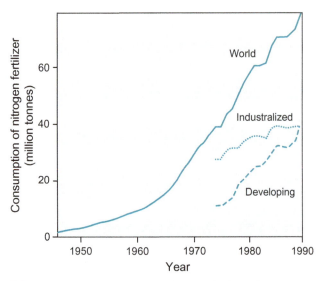

FIG. 15.16 Increases in nitrogen fertilizer use in developing countries. Consumption of nitrogen fertilizer from 1946 to 1989. *(From Morris, B.L., Lawrence, A.R.L., Chilton, P.J.C., Adams, B., Calow, R.C., Klinck, B.A., 2003. Groundwater and its susceptibility to degradation. A global assessment of the problem and options for management. Early Warning and Assessment Report Series.)*

were exceeded at least once between 1986 and 1995 (EWG, 1996). An additional 3.8 million people were using water from private wells that exceed federal drinking water standards. Researchers predict, due to past and current inputs of nitrates into the environment, that the full effect of overapplication of nitrate fertilizers will not be realized for another 30–40 years (Hallberg and Keeney, 1993). The potential impacts of nitrate contamination on human health are discussed in Chapter 28.

As we might expect, agricultural areas generally have the most significant groundwater nitrate contamination problems. For example, major agricultural regions such as the San Joaquin Valley in California and the Ogallala aquifer system extending from Minnesota to Texas are areas with high vulnerability to nitrate groundwater contamination (see Fig. 15.11). In addition, these agricultural regions are also experiencing severe declines in groundwater levels as a result of excessive groundwater extraction for irrigation. It is common in agricultural areas to see the compounding effect of declining groundwater levels coinciding with high levels of nitrate contamination.

Nitrate is generally very mobile in the subsurface. This, in conjunction with the large areal extent of input, results in extensive groundwater plumes of nitrate contamination. It is difficult and expensive to clean up groundwater once it is contaminated by nitrate. For example, the costs associated with managing nitrate contamination problems in California and Iowa are estimated to exceed $200 million per year (EWG, 1996). A standard approach for dealing with shallow groundwater nitrate contamination is to drill

deeper wells. However, this can be done only a limited number of times. Another common practice is to blend contaminated water with uncontaminated water. The objective of this technique is to dilute the nitrate to concentrations below the drinking water standard. This approach increases the overall use of water resources. It is estimated that closing down a well due to nitrate contamination, and drilling another well or blending contaminated water with cleaner supplies, can cost between $200,000 and 500,000 per well (EWG, 1996). Point-of-use treatment for nitrate is also expensive, requiring methods such as reverse osmosis. A key to solving the problem of nitrate groundwater contamination is to prevent future contamination by using best management practices, as discussed in Chapter 14.

Another major class of agrochemicals, also widely used in agricultural practices, are pesticides. Pesticides are typically applied at the land surface, usually as a chemical spray. Once applied to the ground surface, pesticides can migrate downward through the vadose zone with infiltrating water and contaminate groundwater. Pesticides are used throughout the world, primarily for agriculture, to control weeds, insects, and fungal pests (Table 15.6). A study conducted between 1991 and 2001 by the U.S. Geological Survey found that 42% of wells sampled in agricultural regions across the United States contained the common pesticides atrazine and diethylatrazine. Furthermore, it was reported that about 20% of the wells sampled in major aquifer systems throughout the United States contained both atrazine and diethylatrazine. These results illustrate the extent of groundwater pollution by pesticides.

The majority of pesticides are organic compounds. Their transport and fate behavior in the subsurface is a function of their chemical properties. Some pesticides are relatively mobile (e.g., 2,4-D), while others are highly sorbed (e.g., DDT). The classes of pesticides that were initially developed, such as DDT, are very persistent in the environment. Newer pesticides have been designed in part to be less persistent. A complicating factor in the evaluation of pesticide pollution problems is the sheer number of pesticides available for use. It is not customary to analyze for all possible pesticide compounds, their derivatives, and possible degradation products in groundwater monitoring surveys. In addition, analytical limitations have constrained detection capability. However, recent advancements have resulted in more frequent detection of pesticide compounds in groundwater supplies.

15.5.2 Saltwater Intrusion

Salinization of freshwater is one of the most serious and widespread groundwater contamination issues throughout the world. Areas along coasts where seas or oceans meet

TABLE 15.6 Pesticide Use and Occurrence in Groundwater

Region	Dominant Pesticide Use	Typical Compounds Detected
United Kingdom	Pre- and postemergent herbicides on cereals, triazine herbicides on maize and in orchards	Isoproturon, mecoprop, atrazine, simazine
Northern Europe	Cereal herbicides and triazines as before	As before
Southern Europe	Carbamate and chloropropane soil insecticides for soft fruit, triazines for maize	Atrazine, alachlor
Northern United States	Triazines on maize and carbamates on vegetables, e.g., potatoes	Atrazine, aldicarb, metolachlor, alachlor and their metabolites
Southern and Western United States	Carbamates on citrus and horticulture, and fumigants for fruit and crop storage	Aldicarb, alachlor and their metabolites, ethylene dibromide
Central America and Caribbean	Fungicides for bananas, triazines for sugarcane, insecticides for cotton, and other plantation crops	Atrazine
South Asia	Organophosphorous and organochlorine insecticides in wide range of crops	Carbofuran, aldicarb, lindane
Africa	Insect control in houses and for disease vectors	Little monitoring as yet

From Morris, B.L., Lawrence, A.R.L., Chilton, P.J.C, Adams, B., Calow, R.C., Klinck, B.A., 2003. Groundwater and its susceptibility to degradation. A global assessment of the problem and options for management. Early Warning and Assessment Report Series.

continental landmasses and island systems are most vulnerable to salinization of groundwater resources. In most coastal settings, groundwater beneath land surface generally exists as a lens of "freshwater." The freshwater is separated from the denser seawater by a diffuse interface known as the freshwater/saltwater interface or zone of dispersion (Fig. 15.17). The freshwater/saltwater interface typically resides at some point near where the ocean meets the land mass (coastline) and extends vertically in the subsurface, separating freshwater (inland) from high salinity water (seawater). This freshwater lens tends to become thinner as it approaches the shoreline. If recharge into the aquifer equals extraction rates due to pumping, the freshwater/saltwater interface will remain stable (Fig. 15.18).

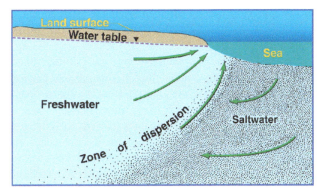

FIG. 15.17 Freshwater and saltwater mix in the zone of dispersion. *(Modified from Cooper, H.H., 1964. A hypothesis concerning the dynamic balance of fresh water and saltwater in a coastal aquifer. U.S. Geological Survey Water-Supply Paper 1613-C, pp. 1–12.)*

However, if groundwater pumping exceeds recharge, the saline water will invade the freshwater aquifer and the freshwater/saltwater interface will progress further inland. As this zone progresses further inland, groundwater supply wells can become contaminated from the invading saltwater. This phenomenon is known as **saltwater intrusion** and it has caused severe degradation and contamination of groundwater.

It docs not take very much saltwater to contaminate a fresh groundwater supply. Only 3%–4% addition of salinity can make a fresh water supply unsuitable for most uses, including drinking water and even irrigation (Morris et al., 2003). As little as 6% addition of saltwater will render a freshwater source (groundwater) unsuitable for any use except cooling or flushing purposes. Once a freshwater resource is degraded by salt contamination, it will take a very long time for that aquifer to recover, and if positive groundwater recharge conditions are not reestablished, it may never recover. Remediation efforts are often cost prohibitive if not impossible due to the technical constraints associated with removing or decreasing the levels of salt concentration in groundwater. The first response is often abandonment of the contaminated wells, accompanied by drilling of new wells further inland. Effectively, the freshwater resource will have been lost, and supplying new water depends on availability of groundwater supplies further inland. Furthermore, if excessive groundwater pumping continues, the saltwater will continue to invade further inland, again contaminating fresh groundwater supplies. The other option currently available is to construct a desalinization plant to treat the contaminated groundwater prior

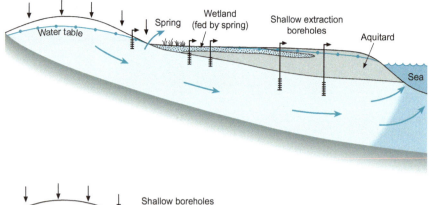

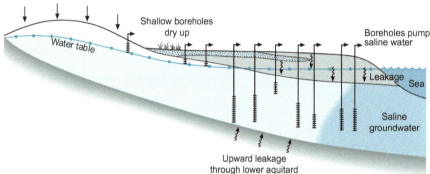

FIG. 15.18 Saltwater intrusion: invading seawater. Top: Groundwater extraction balanced by recharge. Bottom: Groundwater extraction exceeds recharge. *(From Morris, B.L., Lawrence, A.R.L., Chilton, P.J.C., Adams, B., Calow, R.C., Klinck, B.A., 2003. Groundwater and its susceptibility to degradation. A global assessment of the problem and options for management. Early Warning and Assessment Report Series.)*

to use. Such plants are likely to increase in use in the future, as technology improves and the availability of water supplies dwindles.

Although coastal regions may be most susceptible to saltwater intrusion, many noncoastal areas are also being affected by salinization of groundwater. Many natural geologic systems can lead to the salinization of freshwater resources. Areas once occupied by deepwater oceans or seas are now part of continents. These areas, now deep within the subsurface, contain ancient geologic units that contain high concentrations of salt or brine water. These salt-containing geologic units can contaminate freshwater aquifers when over pumping occurs in the region, causing the water table to decline and encroach into the high-salinity geologic units.

Generally there are no harmful health effects associated with low concentrations of chloride in drinking water. In some cases, salt (chloride) can be harmful to people with heart or kidney conditions. The EPA has set unenforceable secondary drinking water guidelines for chloride at 250 mg/L. However, the contamination of water by saltwater intrusion will increase salt concentrations far beyond what can be tolerated by humans. The primary concern with the contamination of drinking water supplies from saltwater intrusion is the large-scale loss of water resources.

15.5.3   Geologic Sources of Contamination

As noted in other chapters, some forms of contamination originate from natural sources. This is a particular issue for groundwater pollution given that groundwater is in intimate contact with geological media, for which the minerals serve as a source of potential contaminants. Arsenic is one major example and is discussed further in a following section. Other examples include iron, fluoride, manganese, selenium, sulfate, and uranium. A complicating factor is that most of these constituents have both natural and anthropogenic sources. A critical step to effective management of groundwater quality affected by these types of contaminants is to delineate the source or sources of the contaminants.

A recent study conducted by the United States Geological Survey examined the different types of contaminants present in California groundwater. The contaminants were organized into three classes: trace elements (such as arsenic, fluoride, uranium), organics (such as solvents, pesticides, fuels), and nitrate. The results indicated that approximately 20% of California's groundwater used for public water supply was contaminated at levels exceeding a human health benchmark. Interestingly, trace elements were the most prevalent contaminants, comprising roughly half of the exceedances based on population affected.

The trace elements originate primarily from natural geologic sources, rather than anthropogenic sources. This study illustrates the potential significance of natural sources of contaminants for groundwater quality. The issue of natural sources of contamination is discussed extensively in Chapter 27.

15.6 OTHER GROUNDWATER CONTAMINATION PROBLEMS

Although we have discussed some of the major groups of contaminants threatening groundwater resources, such as hazardous organic chemicals (e.g., chlorinated solvents and fuel-type hydrocarbons), agrochemical pollutants (e.g., nitrates and pesticides), and salinization (e.g., saltwater intrusion and high-salinity groundwater), it is important to note some other contaminants that also present potential threats to the quality of groundwater supplies. We will briefly discuss some of these groundwater contaminants in the following sections.

15.6.1 Pathogen Contamination of Groundwater

Contamination of groundwater by microbial pathogens, including viruses, bacteria, and protozoa, is of significant concern throughout the world. The types of pathogens and their impact on human health were discussed in Chapter 13. The transport behavior of pathogens in the subsurface environment is discussed in Chapter 23.

Potential sources of pathogens for groundwater contamination include land disposal of sewage treatment by-products (wastewater, biosolids), septic tank systems, and latrines. Risks posed by pathogen-contaminated groundwater are generally believed not to be significant for public supply systems, given the level of treatment applied before use (see Chapter 24). Of much greater concern is potential pathogen contamination of groundwater used for private water supplies, because water from private wells typically undergoes little or no treatment before use. Thus residential areas with septic systems and private wells are particularly susceptible to potential effects of groundwater contamination by pathogens.

Proper siting and construction of septic and well systems is necessary to minimize potential pollution problems. So-called wellhead protection rules have been developed to prevent the siting or application of pathogen sources too close to water supply wells. Several recent surveys of groundwater across the United States have shown that the incidence of human viruses in groundwater is greater than previously believed and may in part be due to septic tank systems (see Information Box 15.4).

15.6.2 Gasoline Additive: Methyl Tertiary-Butyl Ether (MTBE)

In the mid-1990s it was discovered that methyl tertiary-butyl ether (MTBE), an additive in gasoline, had caused extensive contamination of groundwater throughout the United States (see Information Box 15.5). MTBE had been used since the early 1970s as an oxygenate to promote more

---

INFORMATION BOX 15.4 Occurrence of Viruses in U.S. Groundwater

Viruses (10–100 nm) are smaller than bacteria (0.5–3 μm) and protozoa (1–15 μm), and thus viruses are generally more mobile in porous media. For example, viruses have been observed to travel more than 100 m in the subsurface. Accordingly, it would be expected that groundwater is more likely to be contaminated by viruses than by other pathogens. To this end, a large-scale study was instituted to evaluate the occurrence of viruses in groundwater in the U.S. Information pertaining to physical and geological characteristics of wells and associated subsurface environments, along with various microbial and physicochemical water quality parameters, was collected, and possible correlation with the presence of human viruses was investigated. Groundwater samples were collected from 448 sites in 35 states and assayed for microorganisms and chemical contaminants. Infective viruses, viral nucleic acid, bacteriophages, and bacteria were present in approximately 5, 31, 21, and 15% of the samples, respectively.

(**Source:** *Abbaszadegan, M., Lechevallier, M., Gerba, C., 2003. Occurrence of viruses in U.S. groundwaters. J. Am. Water. Works Assoc. 95, 107–120.*)

INFORMATION BOX 15.5 The First Significant Incidence of MTBE Contamination

In 1996, it was discovered that two well fields providing drinking water to the city of Santa Monica, California, were extensively contaminated with MTBE at average levels of 610 and 86 μg L^{-1}. The city of Santa Monica lost 50% of its drinking water supply when these two well fields were shut down. Enormous costs were incurred from aquifer decontamination and remediation efforts, which continue to this day. In addition, the city of Santa Monica has had to supplement its drinking water supply by purchasing replacement water from outside resources.

(**Source:** *Environmental Protection Agency (U.S. EPA), 2003b. Fact Sheet on MTBE, www.epa.gov/mtbe/faq.htm.*)

efficient combustion of fuel in automobiles. Ironically, while the intended use of MTBE has in fact reduced toxic emissions (e.g., carbon monoxide) released to the atmosphere from automobiles, it has contributed to the widespread contamination of groundwater resources from leaking underground fuel storage tanks, posing serious threats to the quality of drinking water supplies. The contamination of groundwater by MTBE is extensive in the United States. The EPA reports, citing a study by Chevron, that MTBE concentrations exceeded $1000 \mu g \, L^{-1}$ in 47% of 251 California sites surveyed, 63% of 153 Texas sites surveyed, and 81% of 41 Maryland sites surveyed (EPA, 2004). Responding to the occurrences of MTBE contamination, California in 1999 became the first state to ban the use of MTBE in gasoline reformulation after 2002. Since then the use of MTBE in gasoline has been phased out nationwide.

The primary concern associated with the release of MTBE to the subsurface environment (from spills or leaking underground fuel storage tanks) is its mobility and persistence. For example, at most fuel station sites affected by leaking storage tanks, it is commonly observed that MTBE has migrated much farther than typical gasoline contaminants (e.g., benzene, toluene). Unlike these contaminants, MTBE has a relatively low biodegradation potential. The health effects associated with MTBE through human ingestion of drinking water are still undergoing review. However, MTBE has been placed on the CCL (Chapter 12), and the EPA has set provisional drinking water health advisory limits at $20–40 \mu g \, L^{-1}$.

15.6.3   Solvent Additives: 1,4-Dioxane

In recent years, 1,4-dioxane (dioxane) has gained considerable attention as a primary emerging contaminant for groundwater resources (Chapter 12). Dioxane, like MTBE, is commonly used as an additive. It is a synthetic organic chemical used as an industrial solvent or solvent "stabilizer" that prevents the breakdown of chlorinated solvents during manufacturing processes. Dioxane is commonly added to chlorinated solvents and other solvents such as tetrachloroethene (PCE), trichloroethene (TCE), 1,1,1-trichloroethane (TCA), and paint thinners. Dioxane is also used as a solvent for the manufacturing of paper, cotton, textiles and various organic products, automotive coolant, shampoos, and cosmetics. It is estimated that

TCE and TCA contain approximately 1% and 2%–8% 1,4-dioxane, respectively (Mohr, 2001). It was estimated that between 10 and 18 million pounds of 1,4-dioxane were produced in the United States in 1990 (Mohr, 2001).

Although 1,4-dioxane has been used as a stabilizer for solvents since the 1940s, the extensive contamination of groundwater by dioxane was not documented until the mid-1990s, when improved analytical methods allowed for the detection of lower concentrations. Dioxane is a very mobile and persistent compound, and is considered by the EPA as a likely human carcinogen. The EPA has not yet defined a national regulatory standard for dioxane. However, dioxane has been placed on the CCL (Chapter 12), and the EPA has set a drinking water health advisory limit of $0.35 \mu g \, L^{-1}$. Dioxane has been found at many hazardous waste sites that are contaminated by chlorinated-solvent compounds.

15.6.4   Perchlorate in Groundwater

Perchlorate, another emerging groundwater contaminant, was first detected in drinking water in 1997 and has since been recognized to pose a significant threat to groundwater resources (EPA, 2005). As mentioned previously, it often requires a significant advancement in analytical capability to first observe an emerging contaminant's presence in the environment and in particular groundwater. The development of an analytical method to detect low concentrations of perchlorate allowed for the recognition of its widespread occurrence in groundwater in the United States (Information Box 15.6). Perchlorate is an inorganic anion and is often present as a salt complex or ammonium salt as ammonium perchlorate. Perchlorate is used for numerous industrial and military purposes. For example, it is a primary constituent in the manufacturing and use of rocket propellants (solid rocket fuel) and other explosives. For this reason, much of the perchlorate contamination of groundwater is derived from military bases and installations. In fact, it is estimated that approximately 90% of perchlorate compounds are produced for use in defense activities and the aerospace industry (EPA, 2005).

Perchlorate, like many salt compounds, is extremely mobile in groundwater. In addition, perchlorate is not readily susceptible to chemical or microbial degradation and is thus persistent in the environment. In studies perchlorate has been shown to interfere with the uptake of

---

INFORMATION BOX 15.6   Perchlorate Contamination in Nevada

The city of Las Vegas gained notoriety as having one of the largest known groundwater perchlorate problems in the United States. The site of the Kerr-McGee Chemical Corporation near Las Vegas produced some of the highest perchlorate concentrations in groundwater ever reported, with 3.7 million $\mu g \, L^{-1}$ in

groundwater and $24 \mu g \, L^{-1}$ in drinking water. Strategies for cleanup and remediation are currently in progress and will likely continue well into the future, with incurred costs estimated into the hundreds of millions if not billions of dollars (Struglinski, 2005).

iodine by the thyroid. This can disrupt thyroid functioning, including hormone regulation, metabolism regulation, fetus development, and child development. To date, the EPA has not established a maximum contaminant level or enforceable regulatory limit for perchlorate in drinking water. However, perchlorate has been placed on the CCL (Chapter 12), and the EPA has set a drinking water health advisory limit of $15 \mu g L^{-1}$.

15.6.5 Arsenic in Groundwater

Contamination of groundwater by arsenic (As) is another important groundwater contaminant problem throughout the world (see Chapter 12). The high toxicity associated with arsenic is of primary concern for human health. As a result of the high associated toxicity of arsenic, regulatory standards were lowered by the EPA, from 50 to $10 \mu g L^{-1}$. The additional water treatment costs associated with meeting the revised standard are projected to be in the billions of dollars.

The contamination of groundwater and drinking water can result from natural or human activities. Arsenic is a naturally occurring metallic element that is found in soil, rocks, air, plants, and animals. Arsenic in soil and rocks can act as sources for groundwater contamination. Through processes such as dissolution, weathering, and erosion, arsenic can be released into the environment, resulting in the contamination of groundwater and drinking water supplies. Arsenic sources associated with human activities include agriculture, use as a wood preservative, the burning of fuels and wastes, smelting and mining, paper production, glass manufacturing, and cement manufacturing. In 1997, almost 8 million pounds of arsenic were released to the environment by human activities (EPA, 2000). Extensive adverse health impacts due to arsenic contamination of groundwater have been recently documented in Bangladesh (see Chapters 12 and 27).

15.6.6 PFAS in Groundwater

A family of emerging contaminants of very recent concern are per- and poly-fluorinated alkyl substances (PFAS). PFAS comprise hundreds of individual compounds. They have been used in a wide variety of consumer products, such as nonstick materials (Teflon), stain-resistant textiles (Scotchgard, Stainmaster), water-resistant textiles (Gore-Tex), cosmetics, cleaning products, paints, food packaging (fast food wrappers, pizza boxes). They were also a primary component of firefighting foams used at many military installations and airports. And, they have been used in other applications such as engineered coatings and medical devices.

The EPA has not established a maximum contaminant level or enforceable regulatory limit for PFAS in drinking water. However, two primary PFAS of concern, perfluorooctanesulfonic acid (PFOS) and perfluorooctanoic acid (PFOA) have been placed on the CCL (Chapter 12). The EPA recently set a drinking water lifetime health advisory limit of $70 \, ngL^{-1}$ for the two combined. Note the extremely low value for this advisory. A survey of US drinking water systems conducted by the EPA indicated that several million people were using drinking water that exceeded the advisory level.

The presence of PFAS at military installations is a major issue driving concern with potential PFAS hazards to human health. Recent surveys indicate that PFAS is present at many military installations particularly due to its use as firefighting foam used at fire-training areas. Because PFAS are extremely resistant to any type of transformation or degradation process, they are very persistent. This has led to issues, for example, with redevelopment of closed military bases (see Information Box 15.7).

15.7 SUSTAINABILITY OF GROUNDWATER RESOURCES

Clearly, the world's population is greatly dependent on groundwater for many uses. Thus managing and protecting our groundwater resources is essential to ensuring that sufficient quantities of quality groundwater will be available for future generations. This concept is referred to as **maintaining long-term sustainability of groundwater resources**.

Ensuring groundwater sustainability requires balancing supply and demand. Hydrologic and geologic factors (climate, topography, subsurface properties) exert primary control on the intrinsic supply of groundwater. The potential impact of global climate change on the hydrologic cycle and groundwater supply is of concern. Groundwater pollution

---

INFORMATION BOX 15.7 PFAS Contamination at Pease AFB in New Hampshire

Pease AFB was closed in 1991. After its closure, it was redeveloped into a commercial trade port. Currently, approximately 250 businesses exist at the park, as well as hotels, restaurants, healthcare facilities, and 2 daycare centers. Several thousand workers and visitors are present each day. Three wells located on site were used to supply potable water to the park. Sampling conducted in 2014 indicated that water from the wells contained several PFAS. Both PFOS and PFOA were present at levels much greater than the EPA health advisory. Efforts are currently in progress to further characterize the contamination and to remediate it.

affects the fraction of the intrinsic supply that is of sufficient quality for use. Thus it imposes a constraint on supply. The demand for groundwater is associated with land use and population density. The significant increase in groundwater use observed over the past few decades (see Fig. 15.1) is a result of population growth and economic expansion.

The primary means by which to ensure long-term sustainability of groundwater resources is to manage supply and demand. The groundwater supply can be extended through moderating demand and can be supplemented with additional sources of water. Water reuse is one primary method being implemented to enhance sustainability. This includes reusing municipal wastewater directly, either as potable water or for secondary uses such as irrigation, or indirectly (e.g., artificial groundwater recharge), as discussed in Chapter 24. Instituting conservation measures to reduce demand is another method. The use of external supplies is another means of supplementation. For example, the **Central Arizona Project** (CAP) provides surface water from the Colorado River to supplement groundwater

resources in Arizona. The CAP canal extends 336 miles from Lake Havasu City to Tucson and cost $3.6 billion to construct. In Tucson, all water provided by the CAP is recharged into groundwater prior to being pumped to the surface and treated for potable use. A potential problem with using external water supplies, versus water reuse and conservation, to manage supply and demand, is that it imposes a demand on water resources at the point of origin. An example of this is, in fact, the Colorado River, which barely exists as a river close to its entry into Mexico because of its great degree of use upstream. Thus the demand has just been shifted in part from one location to another. Changes in land use also exert an impact on supply and demand. For example, groundwater use is now shifting from agriculture to residential as urban centers increase in size and population density.

Balancing supply and demand is encapsulated in the concept of **safe yield.** In essence, the principle of safe yield is that the amount of groundwater extracted should not exceed the amount replenished through recharge. This

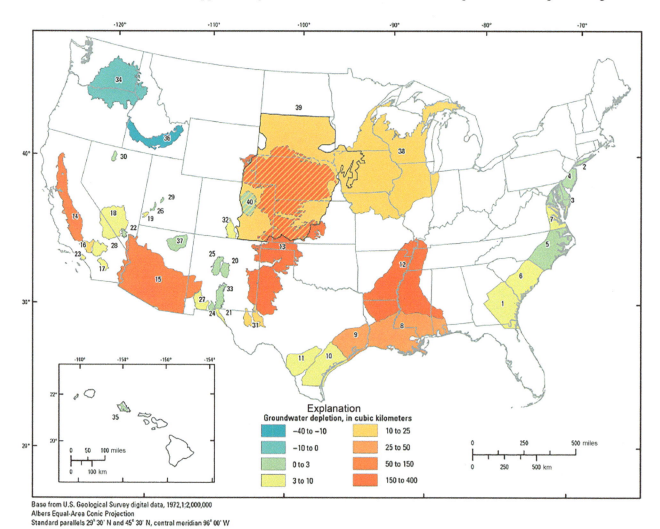

FIG. 15.19 Map of the United States showing cumulative groundwater depletion from 1900 through 2008 in 40 aquifer systems or subareas. *(From USGS.)*

concept is simple in theory. Unfortunately, it is very complex in practice. A primary reason for this is that water resource issues are influenced to a great extent by non-science factors (political, economic, societal, legal). This causes the management of demand among the competing uses prevalent in large urban centers to be a complex and difficult task. In addition, planning for future water resource use is difficult and fraught with uncertainty. For example, there is always uncertainty in estimating the future supply of groundwater available in a region. This uncertainty is compounded by uncertainty in demand, which must be estimated based on future population growth and land-use patterns.

These difficulties are exacerbated by the "invisibility" of groundwater—as opposed to surface water, it cannot be seen. For example, the depletion of a lake or reservoir is readily seen as the water level drops. Conversely, there is no such immediate dramatic visual cue for falling groundwater levels. The prevalent attitude of "out of sight, out of mind" often creates impediments to increasing the awareness of groundwater resource issues. In many regions of the world groundwater is being used at rates greater than it can be replenished (Chapter 25). This is also true for many regions of the United States, as depicted in Fig. 15.19. Thus it is imperative that effective methods that consider all components of the hydrologic cycle, as well as human elements, are employed to manage groundwater in a sustainable manner.

QUESTIONS AND PROBLEMS

1. What is meant by the term "safe yield"?
2. Discuss the factors that affect the supply of groundwater available in a specific location.
3. Discuss the factors that affect the demand for groundwater resources in a specific location.
4. What can be done to modify the supply and demand for groundwater resources?
5. Why is nitrate a widespread groundwater contaminant?
6. Comparing heavy metals and solvents, which would you expect to generate larger groundwater contaminant plumes? Why?
7. What are some examples of contaminants that originate from natural or geologic sources?
8. What is involved in conducting a groundwater pollution risk assessment? What type of information would you need?

REFERENCES

Environmental Protection Agency (U.S. EPA), 2000. Arsenic Occurrence in Public Drinking Water Supplies. EPA 815-R-00-23.

Environmental Protection Agency (U.S. EPA), 2003a. The DNAPL remediation challenge: is there a case for source depletion? EPA/600/R-03/143.

Environmental Protection Agency (U.S. EPA), 2003b. Fact Sheet on MTBE. www.epa.gov/mtbe/faq.htm.

Environmental Protection Agency (U.S. EPA), 2004. MTBE Demonstration Project. www.epa.gov/swerustl/mtbe/mtbedemo.htm.

Environmental Protection Agency (U.S. EPA), 2005. Perchlorate Fact Sheet. www.epa.gov/fedfac/documents/perchlorate.htm.

Environmental Working Group (EWG), 1996. Pouring it on: nitrate contamination of drinking water. www.ewg.org/reports/Nitrate/nitratecontents.html.

Fetter, C.W., 2001. www.appliedhydrogeology.com.

Hallberg, G.R., Keeney, D.R., 1993. Nitrate. In: Alley, W.M. (Ed.), Regional Ground-Water Quality. Van Nostrand Reinhold, New York.

Interstate Technology and Regulatory Council (ITRC), 2002. DNAPL Source Reduction: Facing the Challenge. ITRC, Washington, DC.

Mohr, T., 2001. Solvent Stabilizers White Paper. Santa Clara Valley Water District. San Jose, CA.

Morris, B.L., Lawrence, A.R.L., Chilton, P.J.C., Adams, B., Calow, R.C., Klinck, B.A., 2003. Groundwater and its susceptibility to degradation. In: A global assessment of the problem and options for management. Early Warning and Assessment Report Series.

Struglinski, S., 2005. In: Report Perchlorate Cleanup Needs Tracking. Las Vegas Sun, June 2005 Sun Washington Bureau. www.lasvegassun.com/sunbin/stories/lv-gov/2005/jun/23/518950600.html.

Xinhua News Agency, 2002. Interstate Technology and Regulatory Council, Washington, DC. http://www.china.org.cn/english/24719.htm.

FURTHER READING

Abbaszadegan, M., Lechevallier, M., Gerba, C., 2003. Occurrence of viruses in US groundwaters. J. Am. Water Works Assoc. 95, 107–120.

Cooper H.H. (1964) A hypothesis concerning the dynamic balance of fresh water and salt water in coastal aquifer. United States Geological Survey (USGS) Water-Supply Paper 1613-C, pp. 1–12.

Environmental Protection Agency (U.S. EPA), 1998. National Water Quality Inventory Report to Congress. United States Environmental Protection Agency https://www.epa.gov/waterdata/1998-national-water-quality-inventory-report-congress.

Jones Lee, A., Lee, G.F., 1993. Groundwater pollution by municipal landfills: Leachate composition, detection and water quality significance. In: Proceedings of Sardinia '93 IV International Landfill Symposiums, 11–15 October. Margherita di Pula, Italy. http://www.gfredlee.com/lf-conta.htm.

UNEP/DEWA RS, 2003. United Nations Environment Programme (UNEP). Nairobi, Kenya, ISBN: 92-807-2297-2.

Chapter 16

Surface Water Pollution

D.B. Walker, D.J. Baumgartner, C.P. Gerba and K. Fitzsimmons

Cover art: Water quality sampling from Canyon Lake, Arizona. *Photo courtesy: David Walker.*

16.1 SURFACE FRESHWATER RESOURCES

Freshwater is a scarce and valuable resource—one that can easily be contaminated. Once contaminated to the extent it can be considered "polluted," freshwater quality is difficult and expensive to restore. Thus the study of surface water pollution has focused primarily on streams and lakes, and most of the scientific tools developed by such regulatory agencies as the U.S. Environmental Protection Agency have been applied to protecting water quality in this segment of earth's surface waters. The amount and distribution of fresh surface water was illustrated in Chapter 3.

The water stored in reservoirs and lakes, together with the water that flows perennially in streams, is subject to heavy stress, and because it is used for water supplies, agriculture, industry, and recreation, this water can easily be contaminated.

16.2 MARINE WATER RESOURCES

Oceans contain most of the water of the planet. Yet even with the phenomenal volume of water in which contaminants may be dispersed, marine resources can be polluted. Using various biological and physical parameters, we usually classify the ocean environment as three components: the coastal zone, the upper mixed layer, and the abyssal (deep) ocean. Several regulatory agencies and international organizations share different responsibilities for those components. The coastal zone, which is most susceptible to the day-to-day kinds of contamination found in freshwater lakes and rivers, is often the province of water quality regulatory agencies established by individual nation-states. International organizations have traditionally dealt with pollution concerns of the open ocean and its seabed, which includes the other two components. In addition to these physically described components of the sea, there are legally defined (and disputed) zones, sometimes overlapping, that influence regulatory practices, as indicated in Fig. 16.1.

16.2.1 The Coastal Zone

The *coastal zone* extends from the low-tide line to the 200-m depth contour, tending to match the geophysical demarcation of the continental shelf. The coastal zone can be as wide as 1400 km along some coasts and less than a kilometer along others. The average width of the zone worldwide is about 50 km, comprising about 8% of the surface of the ocean. (The coastal zone of Alaska is larger than that of the rest of the United States.) Within the coastal zone definition, the difference between estuaries and the open coast is important in considering the disposal of wastewater and the potential for pollution problems.

Almost all of the surface-water-carried wastes of a continent enter the coastal zone through an estuary. *Estuaries* are bodies of water with a free connection to the sea whose salinity is measurably diluted with fresh water, as from a river. Because estuaries provide critical and limited habitat for marine organisms to rear and feed their young, water quality is of special concern. Species that inhabit the coastal zone, and especially the estuaries, have to be very resilient to such natural environmental stresses as wide daily variations in salinity, turbidity, temperature, and UV radiation. Owing to this natural resiliency, coastal organisms may be able to tolerate contaminants associated with industrial and municipal wastes better than residents of the

Environmental and Pollution Science. https://doi.org/10.1016/B978-0-12-814719-1.00016-1

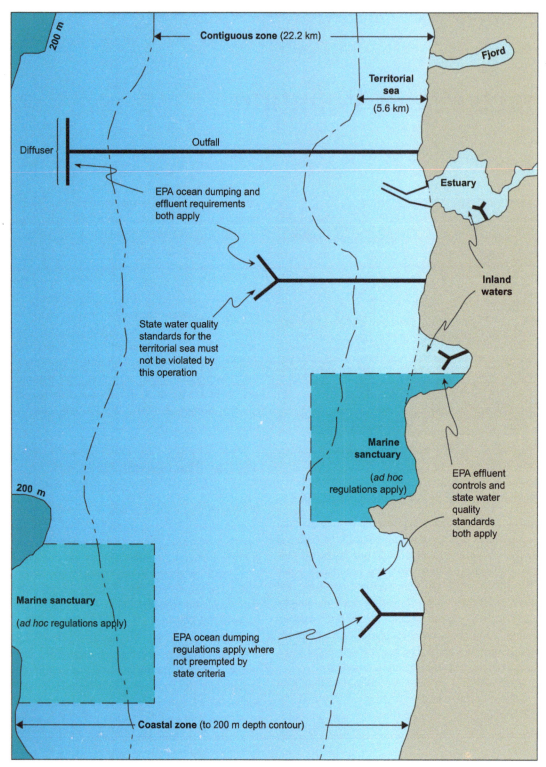

FIG. 16.1 Water pollution regulations in the coastal zone. The outfalls depicted (both T- and Y-shaped) all use diffusers. *(From Pollution Science © 1996, Academic Press, San Diego, CA.)*

continental shelf, where the natural environment is quite stable. The estuarine habitat must be maintained primarily because of its limited extent, as distinguished from the shelf habitat, which is enormous. For this reason, treated wastewater effluents are usually discharged offshore rather than into estuaries in coastal regions. A large pipeline or tunnel, called an *outfall*, is used to transport the effluent to the disposal site.

In disposing of treated wastes offshore, we also need to take the physical features of the coastal zone into account. For example, continental headlands that protrude into the sea can impede both circulation of water and exchange of nearshore water with open ocean water. Outfalls are therefore best located far offshore rather than inside the region of headland influence. Similarly, outfalls should not be located close to shore in the vicinity of estuaries or bays because tidal incursions can carry diluted wastes into the estuary, thereby eroding one of the advantages offered by offshore disposal.

16.2.2 Open Ocean Waters

A variety of the majority of circumstances can contribute to the contamination of the open ocean waters beyond the coastal zone, including atmospheric fallout, oil spills, and dumping of hazardous wastes and sewage sludge as practiced by some countries of the world. Floatable and soluble materials tend to stay in the *upper mixed layer* of the ocean, where they may be decomposed. This upper layer is also the most active photosynthetic zone of the ocean, where the majority of plant—and hence animal—life can thrive. The depth of this layer, which varies between 100 and 1000 m, changes with season and geographic location. Although mixing between the upper and deeper layers of the ocean is impeded by strong density gradients, particles formed in the upper mixed layer, or discharged there, may eventually settle so far that they can no longer be resuspended by surface-generated turbulence and thus become part of the detrital sediment load of the deep ocean waters.

Because the quality of the water in the upper mixed layer can significantly affect all life there, it is important to take precautions with waste disposal operations. When ocean disposal of certain materials is justified, we can use technologies to avoid contamination of the upper mixed layer and facilitate transit and long-term retention of the material in the deep waters of the open ocean, that is, the *abyssal ocean*. For example, containers have been proposed for disposal of such materials as xenobiotic chemicals or radioactive wastes. Pipelines can also be used to carry liquid carbon dioxide to the seabed, where it can be retained for a long time—conceivably long enough to help reduce the rate of global climate change.

16.3 SOURCES OF SURFACE WATER POLLUTION

Water pollution is a qualitative term that describes the situation when the level of contaminants impedes an intended water use (Chapter 30). It takes just a small amount of contaminant to pollute a waterbody intended for a drinking water supply. But the same water might not be considered polluted if the water were to be used, for example, for agriculture. Nor is pollution restricted to chemical contaminants. Physical factors of the environment can also contribute to pollution. For example, heated water discharged from a power plant can change the temperature of an aquatic environment. It might not be a problem in a lake or a river during the winter, but it can certainly be a problem in the summertime. Biological entities such as pathogenic microorganisms can also be pollutants. Moreover, heated water or water containing some contaminant may not be a problem at any time of the year, provided it is rapidly mixed with the surface water, and the diluted material does not accumulate over time. There are also many kinds of contaminants that can usually be accommodated by the natural environment without resulting in pollution, but in many situations, these same contaminants (sometimes in conjunction with other contaminants) can cause pollution even in well-mixed waterbodies.

Major sources of surface water contamination are construction, municipalities, agriculture, resource extraction-production, and industry. In addition, the water delivered to earth in the form of precipitation is not necessarily pure to begin with. Near the coast, it may contain particulate and dissolved sea salts, and farther inland, it may contain organic compounds and acids scrubbed from contaminants added to the atmosphere both by natural processes and by anthropogenic (human) activities. Gases from plant growth and decay, and gases from geological activity are examples of naturally derived atmospheric contaminants that can be returned to earth via precipitation. The acid rain problem of the New England states is a classic example of anthropogenically derived atmospheric contaminants that contribute to surface water pollution (see Chapter 17).

16.4 SEDIMENTS AS SURFACE WATER CONTAMINANTS

The properties of particulates or sediments in water are described in Chapter 11. Soil water erosion and its control are described in Chapters 14 and 20, respectively. The ability of rivers to carry sediment over large distances has resulted in the landscape of continents. Certainly, some background level of sediment load in rivers is considered natural and desirable. Problems ensue when anthropogenic activities in a river's watershed increase, or in some cases decrease, sediment load. Running water, wind, and ice are the major factors responsible for the detachment, entrainment, and transport of particulate matter. Geologic erosion is highest in areas with relatively steep gradients such as low- to intermediate-order streams in mountainous areas. Historically, natural erosion has been the largest source of sediment supplied to rivers. As human land use activities increase in watersheds around the globe, anthropogenic effects now are major contributors to both increased sediment supplied to rivers as well as the blocking and impoundment of this sediment behind dams. Both impoundment and increased erosional processes in watersheds can have profound biological, physical, and chemical impacts on rivers and streams.

Almost any kind of human activity in watersheds can result in an increase of suspended sediment in rivers. A few classic examples of anthropogenic activity known to increase sedimentation are as follows:

- *Logging, deforestation, wildfire*: Specific types of logging activity can increase sediment yield by two orders of magnitude for short periods. Fire suppression and drought can combine to create catastrophic wildfires, which can have devastating impacts on receiving waters from these areas.
- *Overgrazing by domestic animals*: Sedimentation can increase not only due to decreased vegetative interception of precipitation-enhancing erosion, but also through direct trampling of the streambed and channel.
- *Urbanization and road construction*: Road construction commonly results in a 5- to 20-fold increase in suspended sediment yield. Impervious materials such as pavement, parking lots, or rooftops can increase the velocity of stormwater runoff, which will increase erosion once this water comes in contact with soil.
- *Mining operations*: Mines, particularly strip mines, can lead to extraordinarily high levels of erosion and subsequent sedimentation in rivers. An example is the coal strip mines in Kentucky.

On a global scale, rivers discharge roughly $40,000,000\,m^3$ into the world's oceans annually. For every cubic meter of water reaching the ocean, there is (on average) an accompanying $0.5\,kg$ of sediment carried away from the continents.

Suspended sediment is also a major carrier of pollution. While rivers may be transporters of pollution, suspended sediment is the "package" in which these pollutants are often contained. Heavy metals, organic pollutants, pathogens, and nutrients responsible for eutrophication can all be found attached to sediments in flowing water. The "quality" or overall pollutant load of suspended sediment depends on the degree of pollution in the watershed.

Transport of sediments in water is dependent on many factors, including sediment particle size and water flow rate (see Chapter 11). The quantification and predicted rates of transport of sediment are based upon the assumption that for any given flow and sediment, there is a unique transport rate. Estimates of sediment transport rates are based upon measures of water flow (including velocity, depth, shear velocity, viscosity, and fluid density) and both sediment size and density. There are different classification terminologies, based upon the mode of sediment transport in a stream. *Bed load* refers to the sediments moving predominantly in contact with or close to the streambed. In contrast, *suspended load* refers to sediments that move primarily suspended in fluid flow but that may also interact with bed load. Suspended load has a continual exchange between sediment in fluid flow and on the bed as it is constantly being entrained from the bed and suspended, while heavier particles settle out from the flow to the bed. *Solute load* refers to the total amount of dissolved material (ions) carried in suspension and can only be quantified by laboratory analytical techniques. *Total load* is the total amount of sediment in motion and is the sum of bed load plus suspended load. It is important to remember that these classifications are somewhat artificial. The sediment load carrying capacity of a stream or river constantly changes both spatially and temporally as flow changes. Flow in any river or stream is never homogenous, so the resulting sediment movement in any section of stream or river varies greatly.

Particles that are too heavy to be fully suspended may roll or slide along the bed (*traction load*) or hop as they rebound on impact with the bed. In the latter case, ballistic trajectories occur, and the particle is said to move by *saltation*. *Stream competence* refers to the heaviest particles that a stream can carry. Stream competence depends on stream velocity; the faster the current, the heavier the particle that can be carried. *Stream capacity* refers to the maximum amount of total load (bed and suspended) a stream can carry. It depends on both discharge and velocity, since velocity affects the competence and therefore the range of particle sizes that can be transported. Note that as stream volume and discharge increase, so do competence and capacity. This is not a linear relationship, and doubling the discharge and velocity does not automatically double the competence and capacity. Stream competence varies as approximately the sixth power of velocity. For example, doubling velocity usually results in a 64-time increase in competence. For most streams, capacity varies as a range of squared to cubed values. For example, tripling the discharge usually results in a 9–27-time increase in capacity. Most of the work of streams is accomplished during floods, when stream velocity and discharge (and therefore competence and capacity) are many times their level compared to periods of quiescent flow. This work is in the form of bed scouring (erosion), sediment transport (bed and suspended loads), and sediment deposition (Fig. 16.2).

16.4.1 Suspended Solids and Turbidity

It has been stated that *total suspended solids* (TSS) in water are the most important pollutant. Erosion happens constantly around the planet, and some rivers and streams have naturally high TSS levels without any human intervention. The Yangtze River in China and the Colorado and Mississippi Rivers in the United States are examples of

FIG. 16.2 Flooding in the Santa Cruz River, Arizona. Arid regions are especially prone to increases in suspended sediment concentrations during flood events. *(Source: http://az.water.usgs.gov/.)*

rivers that have historically entrained large amounts of sediment due to local topography, geology, and climate.

Total suspended solids are defined as all solids suspended in water that will not pass through a 2.0-μm glass fiber filter (dissolved solids would be the fraction that does pass through the same size filter). The filter is then dried in an oven between 103°C and 105°C, and weighed. The increase in weight of the filter represents the amount of TSS.

Problems with TSS arise when excess erosion occurs in a watershed due to human land use practices. Excess levels of TSS can come from either point (municipal and industrial wastewater) or nonpoint (e.g., agriculture, timber harvesting, mining, and construction) sources. Generally, water with less than 20 mg/L is considered relatively "clear"; levels between 40 and 80 mg/L tend to be "cloudy"; while levels over 150 mg/L would be classified as "dirty" or "muddy." Point sources generally require treatment, usually through settling or flocculation, prior to being released into a river or stream. Nonpoint sources are much more difficult to manage due to several sources acting synergistically. No-till farming, sedimentation basins, and silt fences are common practices to reduce run off from agriculture or construction areas. Stormwater retention ponds and regular street sweeping can reduce the impact of stormwater runoff from urban areas.

Increasing levels of TSS often result in a waterbody being unable to support a diversity of aquatic life. Sedimentation of the stream bed as velocity decreases often results in the suffocation of many aquatic macroinvertebrates and the eggs of fish. Where TSS is deposited results in increased embeddedness (the percentage of any piece of substrate covered in sediment) of cobble, rocks, and boulders within the stream. Several species of macroinvertebrates use the

bottom of rocks as refuge from predators or from fast-flowing water, and as embeddedness increases, this vital habitat is diminished or completely lost. Additionally, TSS absorbs heat and can increase the temperature of a waterbody. In lakes and reservoirs, this can exacerbate thermal stratification as heat accumulates close to the surface. Suspended solids can also decrease the amount of dissolved oxygen due to consumption of organic matter by respiring bacteria. In lakes or reservoirs with a large algal biomass, sudden inputs of water containing suspended sediments have been known to deplete the oxygen of water and cause massive fish kills. The once-photosynthesizing algae switch to respiration as light for photosynthesis was reduced or eliminated.

Besides the relatively direct effects of suspended solids on waterbodies, perhaps the greatest indirect effects are the pollutants that may be attached to suspended sediment. Examples of pollutants known to sorb to sediment particles are nutrients, metals, organic compounds such as polycyclic aromatic hydrocarbons (PAH) and polychlorinated biphenyls (PCBs), and a wide assortment of herbicides and pesticides. All of these contaminants have differing solubilities and therefore differing fates once they enter a river, stream, or lake. Sediment-associated pesticides in water can be a major problem in agricultural areas.

Turbidity is related to, but not a proxy for, suspended sediments. Specifically, turbidity is the quantification of the light that is scattered or absorbed rather than transmitted through a water sample. Turbidity is another measure of water clarity but is not a measure of dissolved substances that can add color to water. Particulates are what add turbidity to water and can include such things as silt, clay, organic matter, algae and other microorganisms, and any other particulate matter that can scatter or absorb light. The amount of light scattered or absorbed is proportional to the concentration of particulates in the sample. The exact amount and wavelength of light scattered by a particle is dependent on the particle's shape, size, and refractive index, which makes any correlation between turbidity and suspended solids difficult and impractical. However, turbidity is directly related to the level of particulates and is an excellent general indicator of water quality in its own right. Units of measure for turbidity are in *nephelometric turbidity units* (NTUs) and are measured on a nephelometer (often called a turbidimeter). Turbidimeters operate by shining an intense beam of light up through the bottom of a glass tube containing the sample. Light scattered by particulates in the sample is detected by a sensitive photomultiplier tube at a 90-degree angle from the incident beam of light. The amount of light reaching the photomultiplier tube is proportional to the level of turbidity in the sample. The photomultiplier tube converts the light energy into an electrical signal, which is amplified and displayed on the instrument meter.

16.5 METALS AS SURFACE WATER CONTAMINANTS

Metals that can be toxic to humans and wildlife are often found in industrial, municipal, and urban runoff and in atmospheric deposition from coal-burning plants and smelters and from natural weathering of rocks and soils. Levels of harmful metals in water have risen globally with increasing urbanization and industrialization. Currently, there are over 50 heavy metals that can be toxic to humans. Of these, 17 are considered very toxic and simultaneously readily accessible.

Common heavy metals known to be toxic to humans include arsenic, cadmium, chromium, copper, lead, mercury, and zinc. Interestingly, chromium, copper, and zinc are essential micronutrients required by the human body for growth, and toxicity depends upon enhanced dose.

Heavy metals are also environmentally persistent, which exacerbates any potentially toxic exposure because these metals often accumulate under certain environmental conditions.

16.5.1 Mercury

Mercury in the environment is one of the most widely recognized and publicized pollutants. Under certain environmental conditions, elemental mercury complexes to form methyl-mercury, which is especially mobile in the environment and toxic to humans and wildlife. Use of mercury in the tanning industry and for making hats was the first time that it was widely recognized as a toxic substance affecting the brain; hence the term "mad as a hatter" (see also Chapters 27 and 28). Today, release of mercury in smokestack emissions from coal-burning power plants is the primary source of contamination. Bioaccumulation of methyl-mercury in long-lived predatory fish from cold freshwater lakes (pikes and walleyes) and in coldwater marine species (swordfish, sharks, and some tunas) has led to public warnings for pregnant and nursing women to limit consumption of shark and swordfish and for all of the public to limit consumption of fish from certain lakes in parts of the United States and Western Europe.

The cycling of mercury through the environment is complex and depends on several physical, chemical, and, most importantly, biological aspects of the system in question. The manner in which mercury cycles through any given area or ecosystem determines its relative toxicity and subsequent bioaccumulation rate upward through the food chain. The term *bioaccumulation* refers to the net accumulation, over time, of pollutants within an organism from both biotic and abiotic factors. The term *biomagnification* refers to the progressive accumulation of persistent toxicants by successive trophic levels. Biomagnification relates to the concentration ratio in a tissue of predator

organisms as compared to that in its prey. Mercury exists in several forms in the environment: elemental mercury (Hg^0), inorganic mercury compounds (Hg^{+1} or Hg^{+2}), and organic mercury compounds ($HgCH_3$ or $Hg(CH_3)_2$), which include both methyl- and dimethyl-mercury (Fig. 16.3).

There are numerous pathways by which mercury can make its way into waterbodies. Inorganic and methyl-mercury can enter water directly from atmospheric deposition, methyl-mercury and Hg^{+2} can be bound to organic substances in runoff, and surface water flow in upper soil layers can transport Hg^{+2} and methyl-mercury to waterbodies. There has been a global increase of mercury released into the atmosphere since the beginning of the Industrial Age, so that atmospheric deposition onto watersheds and surface water often plays as large a role as runoff from natural sources.

Once in an aquatic ecosystem, mercury goes through several complexation and transformation processes. While most forms of mercury are bioavailable, methyl-mercury (MeHg) is the form most readily absorbed and bioaccumulated. The methylation of mercury in aquatic systems not only requires a certain range of physicochemical factors, but also the presence of a group of bacteria known as *sulfate-reducing bacteria* (SRBs). There are several species of SRBs, but some of the most common include strains of *Desulfovibrio* and *Desulfobacter*. The majority of methylation that occurs in lakes and reservoirs is within anaerobic sediments. Factors affecting the methylation of mercury are outlined in Table 16.1, while common transformations of mercury are given in Fig. 16.4.

Mercury in all forms is a potent toxin that can cause developmental effects in the fetus as well as toxic effects on the liver and kidneys of adults and children. Sublethal effects of mercury toxicity can affect the ability to learn, speak, feel, see, taste, and move. Children under the age of 15 are most vulnerable, because their central nervous system is still developing. Mercury is easily passed from pregnant mother to fetus, and even extremely small trace amounts of mercury can have devastating effects to a developing central nervous system. Mercury toxicity can occur through skin contact, inhalation, or ingestion. Due to biomagnification, the most common route of exposure to humans is through consumption of contaminated fish. Hence, women are advised to refrain from eating fish and other seafood during pregnancy.

The methylation, biomagnification, bioaccumulation, and toxicity of mercury are often linked to the problem of increasing eutrophication. Increases in most of the factors that cause eutrophication within a waterbody, including increased sulfate content, also increase the rate of methylation and subsequent toxicity to humans and wildlife. Humans have not only increased the availability of mercury deposited into aquatic systems, but also, through

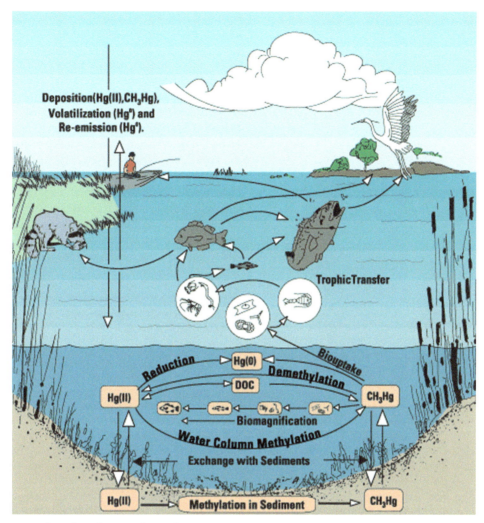

FIG. 16.3 Common transformations of mercury in aquatic ecosystems.

cultural eutrophication, increased the bioavailability and subsequent toxicity of mercury in these systems.

16.5.2 Arsenic

Arsenic is an element widely distributed throughout the earth's crust. As such, it is often introduced into water through the dissolution of minerals and ores and may concentrate in groundwater. Arsenic is also used in industry and agriculture and is a by-product of copper smelting, mining, and coal burning. One form of arsenic, chromated copper arsenate, is the most common wood preservative in the United States and contains 22% arsenic.

Inorganic arsenic occurs in several different forms in the environment, but in natural waters it is most commonly found as trivalent arsenite [As(III)] or pentavalent arsenate [As(V)]. Most of the organic species of arsenic, usually at very high levels in seafood, are less toxic and are readily eliminated by normal body functions.

Symptoms from arsenic exposure include vomiting, esophageal and abdominal pain, and bloody diarrhea. Long-term exposure can cause cancers of the skin, lungs, urinary bladder, and kidney as well as other skin changes, such as pigmentation changes and thickening.

One of the largest mass poisonings in the world occurred in Bangladesh, where 53 out of a total of 64 districts had groundwater contaminated with arsenic. The cause of arsenic contamination was related to the onset of intense agriculture in the region where irrigation resulted in large-scale withdrawal of groundwater via wells (see Chapter 12).

16.5.3 Chromium

Chromium is found in natural deposits as ores containing other elements. Additionally, chromium is an important industrial metal, where it is used in alloys such as stainless steel, protective coatings on other metals and magnetic

TABLE 16.1 Factors Influencing the Methylation of Mercury in Aquatic Ecosystems

Physical or Chemical Condition	Influence on Methylation
Low dissolved oxygen	Enhanced methylation
Decreased pH	Enhanced methylation within the water column
Decreased pH	Decreased methylation in sediment
Increased dissolved organic carbon	Enhanced methylation within sediment
Increased dissolved organic carbon	Decreased methylation within water column
Increased salinity	Decreased methylation
Increased nutrient concentrations	Enhanced methylation
Increased temperature	Enhanced methylation
Increased sulfate concentrations	Enhanced methylation

$$Hg^0 \underset{\text{Reduction}}{\overset{\text{Oxidation}}{\rightleftharpoons}} Hg^{+1} \text{ or } Hg^{+2} \underset{\text{Demethylation}}{\overset{\text{Methylation}}{\rightleftharpoons}} HgCH_3$$

FIG. 16.4 Common transformations of mercury.

tapes, pigments for paints, cement, paper, rubber, and floor coverings. Chromium has several oxidation states, but the most common are $^{+2}$, $^{+3}$, and $^{+6}$, with $^{+3}$ being the most stable. Oxidation states of $^{+4}$ and $^{+5}$ are relatively rare. Toxicity of chromium depends on oxidation state. Chromium (III) is an essential nutrient, while the hexavalent form, chromium(VI), is listed as a known human carcinogen.

16.5.4 Selenium

Selenium occurs naturally in the environment as selenide and is often combined with sulfide, copper, lead, nickel, or silver. Like chromium, selenium is a micronutrient needed in very small quantities in humans and wildlife to produce the amino acid selenocysteine. However, it can be toxic at higher doses. The relatively narrow range between selenium acting as a beneficial nutrient (50 μg/day) and the initiation of toxicity (400 μg/day) in humans means that it needs to be closely monitored in the environment, especially in areas with alkaline soils, because this is where selenium is often found in its most oxidized and toxic form. As with several other naturally occurring metals, problems may arise due to increased availability of selenium in aquatic systems primarily due to irrigation and farming practices. Symptoms of short-term

selenium toxicity include hair and fingernail changes, damage to the peripheral nervous system, and irritability. Long-term symptoms include damage to liver and kidney tissue and nervous and circulatory systems.

Selenium is a bioaccumulative pollutant; however, unlike mercury, selenium concentrations do not increase upward through the food chain, that is, it does not biomagnify. Selenium toxicity can have devastating effects on both terrestrial and aquatic wildlife. Selenium can affect the growth and survival of juvenile fish as well as the offspring of adult fish exposed to sublethal levels. Birds that have eaten fish suffering from selenium toxicity either succumb to the acutely toxic effects of selenium or produce offspring, often stillborn, with gross skeletal deformities. Due to selenium uptake in terrestrial plants, both domestic and wildlife species foraging on these plants can be affected (see Case Study 16.1).

Case Study 16.1. Selenium Toxicity in Kesterson Reservoir, California

California's Kesterson Reservoir in the San Joaquin Valley is a classic example of one of the most dramatic cases of heavy metal toxicity known to date. Kesterson Reservoir was built in the late 1960s to address the issue of California's decreasing wetland habitat by using agricultural drainage for the creation of wetlands solely for the purpose of attracting and harboring native wildlife species. Due to its perceived benefit to wildlife, Kesterson was made into a National Wildlife Refuge under the auspices of the U.S. Fish and Wildlife Service.

Mountains forming the western boundary of the San Joaquin Valley consist of shale enriched with selenium. The San Joaquin Valley is an area of poorly drained soils where intensely irrigated agriculture in the otherwise arid valley resulted in selenium becoming highly concentrated in agricultural drainage. By 1981, almost all of the water entering Kesterson Reservoir was agricultural drainage from poorly drained soils.

Mosquitofish collected by the U.S. Fish and Wildlife Service in the early 1980s from Kesterson Reservoir contained levels of selenium approximately 100 times higher than mosquitofish found in neighboring wetlands not receiving agricultural effluent. Several studies were implemented during the 1980s to determine whether selenium or other toxicants were present at levels that could harm wildlife.

Agricultural drainage entering Kesterson Reservoir had an average selenium concentration of 0.3 ppm, seemingly low levels at first glance. The greatest damage, however, rested in the bioaccumulative nature of selenium. Algae had average selenium concentrations of 69 ppm, aquatic plants had 73 ppm, aquatic insects more than 100 ppm, and mosquitofish 170 ppm, which was more than 500 times the concentration of the aquatic habitat in which these mosquitofish lived. All of these levels were much higher than those found in neighboring wetlands not receiving agricultural drainage.

Numerous birds feeding on aquatic organisms in Kesterson Reservoir suffered and died due to selenium toxicity. Symptoms included emaciation, feather loss, degeneration of live tissue, and muscle atrophy. Adult birds that did not immediately succumb to the toxic effects of selenium produced offspring, usually stillborn, that had abnormal or missing eyes, beaks, legs, wings, and feet. The area created to enhance and preserve wildlife, especially migrating and native waterfowl, now appeared to be a death trap to their survival.

Millions of dollars have been spent studying the effects of selenium toxicity at Kesterson Reservoir and millions more have been spent on cleanup efforts. The circumstances that caused the devastating effects to wildlife at Kesterson Reservoir emphasize the need to find viable solutions to disposing of contaminant-laden waters emanating from agricultural drainage in arid regions.

16.6 NUTRIENTS AND EUTROPHICATION OF SURFACE WATERS

On a global scale, eutrophication has often been cited as the number one cause of impairment to surface water resources. Eutrophication is the gradual accumulation of nutrients, and organic material subsequently utilizing these nutrients as an energy source, within a body of water. While eutrophication is often cited as an example of anthropogenic pollution of inland waters such as lakes and streams, coastal areas, estuaries, and salt marshes are also commonly affected. Eutrophication often results in increases in algal biomass, and therefore some discussion of what these nutrients are, and what specific ratios cause eutrophication, are in order.

Justus Von Liebig, a German analytical chemist and professor of chemistry at the University of Giessen, made great contributions to the science of plant nutrition and soil fertility in the mid-1800s. Liebig's *Law of the Minimum* states that yield is proportional to the amount of the most limiting nutrient, whichever nutrient it may be. From this, it may be inferred that if the deficient nutrient is supplied, yields may be improved to the point that some other nutrient is needed in greater quantity than the soil can provide, and the Law of the Minimum would apply in turn to that nutrient. This same law can be applied to aquatic systems. The nutrients that most often limit primary production in aquatic systems are forms of carbon, nitrogen, and phosphorous. The specific ratio of limitation (on a molar basis) is 106C:16N:1P. *Carbon* is ubiquitous in the environment and atmosphere, but can become limiting during intense photosynthesis by algae or aquatic plants. Since carbon dioxide is used during photosynthesis, it is possible for this carbon nutrient to be temporarily depleted during daylight hours. This situation would be reversed during the evening, when respiration would exceed photosynthesis, and the carbon dioxide used during the day would be released back into the water. Generally, it is uncommon for carbon to be a limiting nutrient.

The idea of nutrient limitation based upon the ratio between C:N:P in aquatic systems only works when one of these essential nutrients is, indeed, "limiting" the growth of primary producers such as algae. In eutrophic or hypereutrophic systems, ratios may indicate that a nutrient is "limiting" in the traditional sense, however, if *all* nutrients are orders of magnitude higher than what it takes to limit primary production then, in this case, ratios can be misleading and nothing is truly limiting growth. Often, in hypereutrophic systems, algal biomass can become so large that the only limiting factor is available light for photosynthesis, as algal cells near the surface shade those at depth.

Nitrogen is an essential plant nutrient used in the synthesis of organic molecules such as amino acids, proteins, and nucleic acids. Nitrogen (mostly as N_2 or "dinitrogen" gas) comprises 78% of the Earth's atmosphere. Most of the abundant nitrogen found in the atmosphere is not yet bioavailable. Nitrogen must be "fixed" into nitrate (NO_3^-), ammonia (NH_3), or ammonium (NH_4) before it can be used by organisms incapable of fixation. Organisms capable of nitrogen fixation include certain species of bacteria, actinomycetes, and cyanobacteria. In aquatic systems, cyanobacteria perform the majority of nitrogen fixation. In aquatic systems, the forms of nitrogen of greatest interest are (in order of decreasing oxidation state) as follows:

- Nitrate (NO_3^-)
- Nitrite (NO_2^-)
- Ammonia (NH_3)
- Ammonium (NH_4^+)
- Organic-N (amino groups)

Total oxidized nitrogen is the sum of $NO_3^- + NO_2^-$. Organic nitrogen is the organically bound fraction and includes such natural materials as proteins and peptides, nucleic acids and urea, and numerous synthetic organic materials. Analytically, organic nitrogen and ammonia can be determined together and referred to as "Kjeldahl nitrogen," a term that reflects the technique used in their determination. Total Kjeldahl nitrogen (TKN) is not synonymous with total nitrogen. If TKN and NH_3 are determined individually, "organic nitrogen" can be estimated by the difference.

$$TKN + NO_3^- + NO_2^- = \text{Total Nitrogen}$$

All forms of nitrogen (organic and inorganic) are interconvertible. The nitrogen cycle is an important component of overall biogeochemical cycling in aquatic systems (Fig. 16.5).

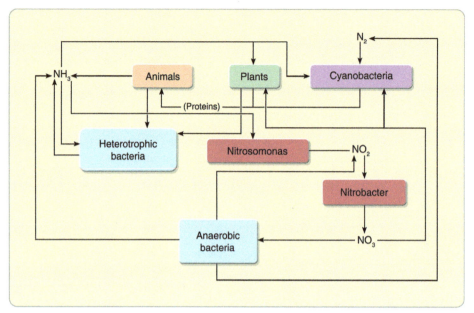

FIG. 16.5   The nitrogen cycle in aquatic systems. *(Courtesy of Dave McShaffrey, Marietta College.)*

Ammonification is an important process in the nitrogen cycle and is, basically, the process of decomposition with production of ammonia or ammonium compounds, especially by the action of bacteria on organic matter. Aquatic animals commonly excrete NH_3 as a waste product of metabolism. The excreted or mineralized NH_3/NH_4 is then available for direct uptake and utilization by other organisms, or it may be converted to more oxidized forms of nitrogen for incorporation into cells. In some nitrogen-poor lakes or reservoirs, the excretory contribution (e.g., ammonification) from zooplankton can provide up to 90% of the nitrogen required by primary producers. Ammonification is difficult to quantify because of the rapid uptake of NH_3 and NH_4 by primary producers. Ammonification is the opposite of assimilation and protein synthesis. Both aerobic and anaerobic bacteria play vital roles in ammonification.

Nitrification is the biological oxidation of NH_4^+ and NH_3 to NO_2^- and then NO_3. Nitrification is important because NH_4^+ and NH_3 are toxic to species of aquatic vertebrates. Nitrification is performed by bacteria that gain energy from oxidizing reduced forms of nitrogen. The aerobic chemoautotrophs involved in nitrification are species of *Nitrosomonas* and *Nitrobacter*. Nitrification consumes and simultaneously requires oxygen, and is a two-part process.

Ammonia

$$NH_2{}^+ + 1/2\ O_2 \xrightarrow{\text{mono} - \text{oxygenase}} NH_2OH + H^+$$

$$NH_2OH + O_2 \rightarrow O_2{}^- + HOH + H^+$$

This process requires 66 Kcal of energy/gram atom of ammonium oxidized.

Under anaerobic conditions:

$$NH_2OH \rightarrow NOH \rightarrow N_2O$$

Ammonium oxidation has important ecological significance in aquatic systems. The microbes that perform nitrification are relatively inefficient autotrophs that use the energy gained from oxidizing ammonia to fix carbon. Thus these bacteria have a dual ecological role: they are involved in recycling nitrogen and in fixing carbon into organics. The microbes that perform nitrification are fragile. These organisms are acid sensitive even though they produce acid. If a large source of nitrogen is added into the environment, these organisms can potentially kill themselves by metabolizing it to nitric acid. Since they are also strict aerobes, they can be killed if introduction of wastes leads to excessive growth of other species that deplete oxygen (i.e., eutrophication).

Denitrification is the reduction of nitrate (NO_3) to nitrogen gas or to organic nitrogen compounds and can be a significant pathway for the loss of nitrogen from aquatic systems. There are two types of denitrification, assimilatory and dissimilatory.

Assimilatory nitrate reduction: Many organisms can only acquire nitrogen in the form of nitrate and must reduce nitrate to form the amino groups needed for metabolism.

$$NO_3{}^- + Energy \rightarrow Amino\ groups$$

The "energy" in the above equation is usually supplied by enzymatic activity (nitrogenase).

Dissimilatory nitrate reduction: Dissimilatory nitrate reduction is performed by anaerobic bacteria that use nitrate as the terminal electron acceptor in the absence of oxygen. The overall equation is:

$$NO_3^- \rightarrow NO_2^- \rightarrow NO \rightarrow N_2O \rightarrow N_2 \text{ gas}$$

The individual steps of dissimilatory nitrate reduction are as follows:

1. Reduction of nitrate to nitrite

$$2HNO_3^- \rightarrow 2HNO_2^- + 4e$$

Enzyme: dissimilatory nitrate reductase

2. Reduction of nitrite to nitric oxide

$$2HNO_2^- \rightarrow 2NO + 2e$$

Enzyme: dissimilatory nitrite reductase

3. Reduction of nitric oxide to nitrous oxide

$$2NO \rightarrow N_2O + 2e$$

Enzyme: dissimilatory nitric oxide reductase

4. Reduction of nitrous oxide to dinitrogen

$$N_2O \rightarrow N_2 + 2e$$

Enzyme: dissimilatory nitrous oxide reductase.

Since reductions are energy yielding, 24 ATPs are generated per mole of nitrate reduced.

Although denitrification requires anoxic conditions, it has been observed in aerated lake sediments and can form relatively thin biofilms on rocks in streams. Evidently, denitrification can occur in microzones of anoxia within sediments and biofilms. Oxygen produced through photosynthesis by benthic algae may inhibit denitrification. Denitrification requires an organic carbon source and proceeds faster where more carbon is available in the water and sediments. Denitrification may contribute a significant portion of the oxidative metabolism in waterbodies where nitrate levels are high. Within any given waterbody, denitrification can occur simultaneously with nitrification. Denitrification occurs due to microorganisms, usually facultative anaerobes and predominantly two genera: *Pseudomonas* and *Bacillus*. Dissimilatory denitrification is used in sewage treatment and bioremediation where denitrifying bacteria aid in converting organic nitrogen to nitrogen gas that escapes to the atmosphere.

As explained by the previous processes, all forms of nitrogen are interconvertible. While there are losses of nitrogen within any aquatic system, there are simultaneous gains from the atmosphere and from recycling within any given region. Problems with eutrophication arise when humans contribute to loading of nitrogen to a waterbody from either point or nonpoint sources of pollution.

Since the 1940s, the amount of nitrogen available for uptake in aquatic systems at any given time has more than doubled (see also Chapter 14). Human activities now contribute more to the global supply of fixed nitrogen each year than natural processes. Anthropogenic nitrogen totals about 210 million metric tons per year, while natural processes contribute about 140 million metric tons. This influx of extra nitrogen has caused serious distortion of natural nutrient cycling in aquatic systems. Excess nitrogen can wreak havoc with aquatic ecosystem structure affecting the number and kind of species found.

Phosphorous, like nitrogen, is essential to all life. Phosphorous functions in the storage and transfer of a cell's energy and in genetic systems. Cells use adenosine triphosphate (ATP) as an energy carrier that drives a number of biological processes, including photosynthesis, muscle contraction, and the synthesis of proteins. Phosphate groups are also found in nucleotides and therefore nucleic acids. Phosphorous is usually more scarce environmentally than other principal atoms of living organisms including carbon, hydrogen, oxygen, nitrogen, and sulfur.

Phosphorous occurs naturally in rocks and other mineral deposits. During weathering, the rocks gradually release the phosphorus as phosphate ions, which are soluble in water, and the mineralized phosphate compounds breakdown. The phosphorus cycle in aquatic systems is shown in Fig. 16.6. Phosphorous exists primarily as phosphates in two forms: orthophosphate and organically bound phosphate. These forms of phosphate occur in living and decaying plant and animal remains as free ions, chemically bonded, or mineralized and chemically bonded in sediments. Analytically, phosphorous in water is usually categorized as being either dissolved or particulate, depending on whether or not it can pass through a 0.45-μm filter. The "dissolved" fraction can have a substantial colloidal component. Within the dissolved fraction, inorganic P (dissolved inorganic phosphorus) occurs as orthophosphate (PO_4). Dissolved inorganic phosphorous is sometimes referred to as *soluble reactive phosphorous* (SRP). *Total phosphorous* (TP) is determined on a nonfiltered sample by heat and acid digestion, which converts the sample to SRP for measurement.

In unpolluted rivers, SRP averages about 0.01 mg/L on a worldwide basis and total phosphorous averages about 0.025 mg/L (Maybeck, 1982). Agricultural activities may increase SRP levels to 0.05–0.1 mg/L, and municipal effluents may increase SRP concentrations to 1.0 mg/L or much higher. Particulate phosphorous includes P incorporated into mineral structures, adsorbed onto clays, and incorporated into organic matter. Worldwide averages of particulate phosphorous concentrations are about 0.5 mg/L. This level can be much higher depending upon land use and erodibility of the watershed.

Phosphorous is often the limiting macronutrient with regard to primary production in aquatic systems. Because of this, and its relative scarcity, it is quickly removed from

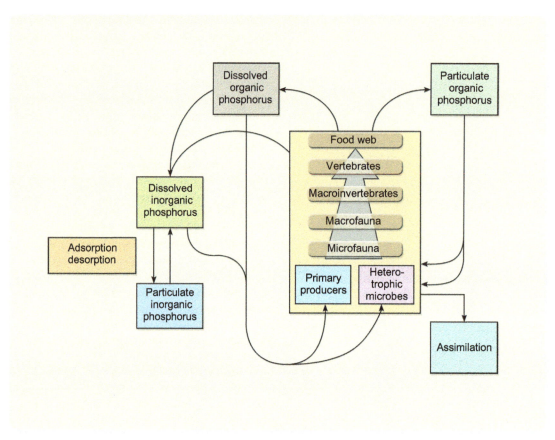

FIG. 16.6 The phosphorus cycle in aquatic systems. *(From Calow, P., Petts, G.E., 1992. The Rivers Handbook, vol. 1. Reprinted with permission of Blackwell Publishing.)*

its dissolved state and incorporated into living biomass. Bacteria and algae are both responsible for turnover rates as fast as 1–8 min (Rigler, 1973). Turnover rates usually follow the order of (in order of decreasing turnover times): Bacteria → algae → zooplankton → vertebrates.

It has been estimated that in freshwater lakes, zooplankton excrete about 20% of the phosphorous required by phytoplankton, whereas bacteria can excrete up to 80%. Therefore food web dynamics play a large role in either the sequestration or the recycling of phosphorous in aquatic systems. The speed at which phosphorous is moved between biotic and abiotic compartments makes interpretation of different forms difficult. It is impossible to distinguish between zooplankton-P, bacterial-P, algae-P, and sometimes even in-organic-P. The best way to quantify phosphorous in a body of water is by analysis of total phosphorous.

Eutrophication, besides increasing algal biomass, often results in depletion of dissolved oxygen, increases in pathogenic bacteria and viruses, increases in potentially toxic species of algae, fish kills, and loss of biodiversity. Remediation efforts of even a small lake are usually cost prohibitive, and it is very difficult, if not often impossible, to return a lake or reservoir back to an earlier trophic state.

The best approach is a proactive, watershed-based one that attempts to protect waterbodies from cultural eutrophication. This usually requires collaboration among several resource agencies, in addition to municipalities and individual landowners in the watershed.

16.6.1 Harmful Algal Blooms

Planktonic (i.e., free-floating algae) are vitally important components of all marine and freshwater systems on the planet. They form the base of the food chain in all aquatic systems. Of the thousands of known species, a few hundred have the potential to produce a wide variety of toxins under certain environmental conditions. Eutrophication greatly exacerbates the growth and prevalence of potentially toxic species.

16.6.1.1 Harmful Algal Blooms in Marine Systems

Most *harmful algal blooms* (HABs) occur in coastal areas where terrestrial runoff of nutrients causes the growth and proliferation of sometimes monospecific blooms of toxic algae. Dinoflagellates (Division Dinoflagellata) are marine phytoplankton often associated with toxic blooms.

Dinoflagellates have the potential to produce a variety of toxins that can be harmful to humans and wildlife. Some affect humans following the ingestion of shellfish or fish that have consumed toxic dinoflagellate species (Fig. 16.7).

Ciguatera poisoning is the most commonly reported disease associated with consumption of seafood. Ciguatera is a lipid-soluble toxin that can affect a variety of fish species and can be very toxic to humans after ingestion of these fish. Tropical and subtropical fish species, including barracuda, grouper, and snapper, are commonly affected. The dinoflagellate species most often associated with ciguatera poisoning is *Gambierdiscus toxicus* but other species including *Proro-centrum mexicanum*, *P. concavum*, *P. lima*, and *Ostreopsis lenticularis* have also been implicated (Fig. 16.8). Ciguatera exhibits both gastrointestinal and neurological symptoms, with the time to onset usually less than 24 h. Gastrointestinal symptoms include diarrhea, abdominal pain, nausea, and vomiting. The most common neurological symptoms include abnormal or impaired skin sensations, vertigo, lack of muscle coordination, cold-to-hot sensory reversal, myalgia (muscular pain), and itching. Neurological symptoms may recur intermittently, with gradually diminishing severity for a long as 6 months. No deaths have been reported from ciguatera in the United States, although worldwide, the mortality rate is 7%–20% of people infected.

Brevetoxin is a large, lipophilic, polyether toxin primarily produced by the dinoflagellate *Karenia brevis* (Figs. 16.9 and 16.10). Brevetoxin poisoning occurs with the most frequency in the Gulf of Mexico and has caused

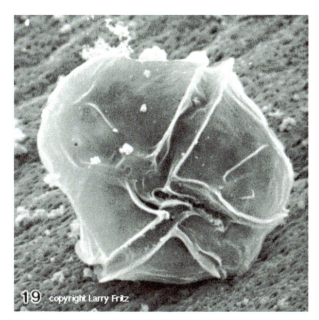

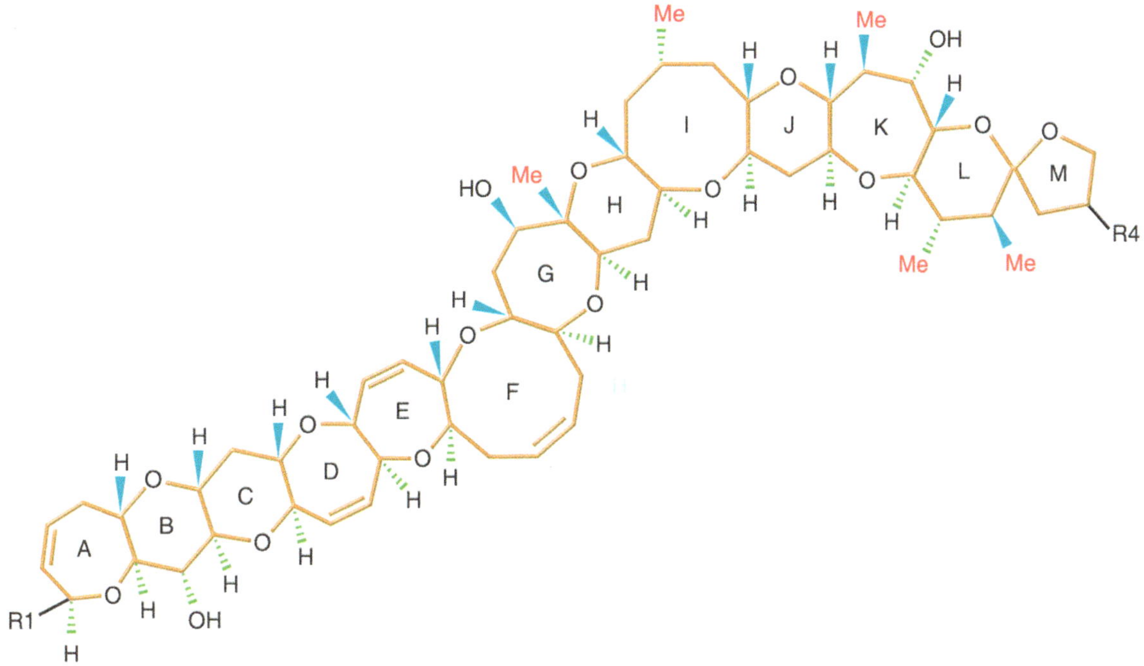

FIG. 16.7 Scanning electron micrograph of *Alexandrium tamarense*, a typical dinoflagellate associated with paralytic shellfish poisoning. *(Source: http://www.whoi.edu/redtide/species/species.html.)*

Dinoflagellates are microscopic, unicellular, flagellated protists that can be either autotrophic (photosynthetic) or heterotrophic (consuming other organisms). Heterotrophic forms often have life history patterns more akin to an animal than a plant. Additionally, dinoflagellates are routinely found in freshwater and often produce some of the same toxins found in marine systems.

FIG. 16.8   Structure of ciguatoxin Type I. *(Source: http://www.aims.gov.au/arnat/arnat-0004.htm.)*

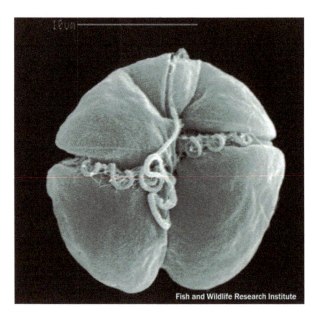

FIG. 16.9 Scanning electron micrograph of *Karenia brevis. (Source: Florida Fish and Wildlife Conservation Commission. http://www.csc. noaa.gov/crs/habf/proceedings/intro.html.)*

sporadic fish kills for decades. These toxins also affect shellfish, which can in turn poison humans who ingest contaminated shellfish. This syndrome is referred to as *neurotoxic shellfish poisoning* (NSP) and produces similar symptoms similar to ciguatera poisoning. There have been

no reported fatalities from NSP, although it has been known to kill laboratory mammals.

Under certain environmental conditions, dinoflagellates can rapidly multiply in numbers and form "tides." Usually, tides are identified by the color of the dinoflagellate causing the bloom. "Red" tides (Fig. 16.11) are often associated with species of *Alexandrium* and "brown" tides with species of *Aureococcus*.

The initiation of either red or brown tides is complex due to the complex life cycles of dinoflagellates but usually involves warm, nutrient-enriched water.

16.6.1.2 Harmful Algal Blooms in Freshwater Systems

While freshwaters often contain many of the same dinoflagellate species found in marine systems, and sometimes the same toxins, the majority of freshwater toxins are caused by several different species of cyanobacteria. Cyanobacteria, like dinoflagellates in marine systems, greatly increase in number in eutrophic waters and occur on a global scale (Fig. 16.12). Cyanobacterial toxins ("cyanotoxins") can affect both humans and wildlife. Humans are usually affected from ingesting water containing cyanotoxins, and the disease is categorized based upon the type of toxin in the water. Toxins produced by cyanobacteria can be either hepatotoxic or neurotoxic.

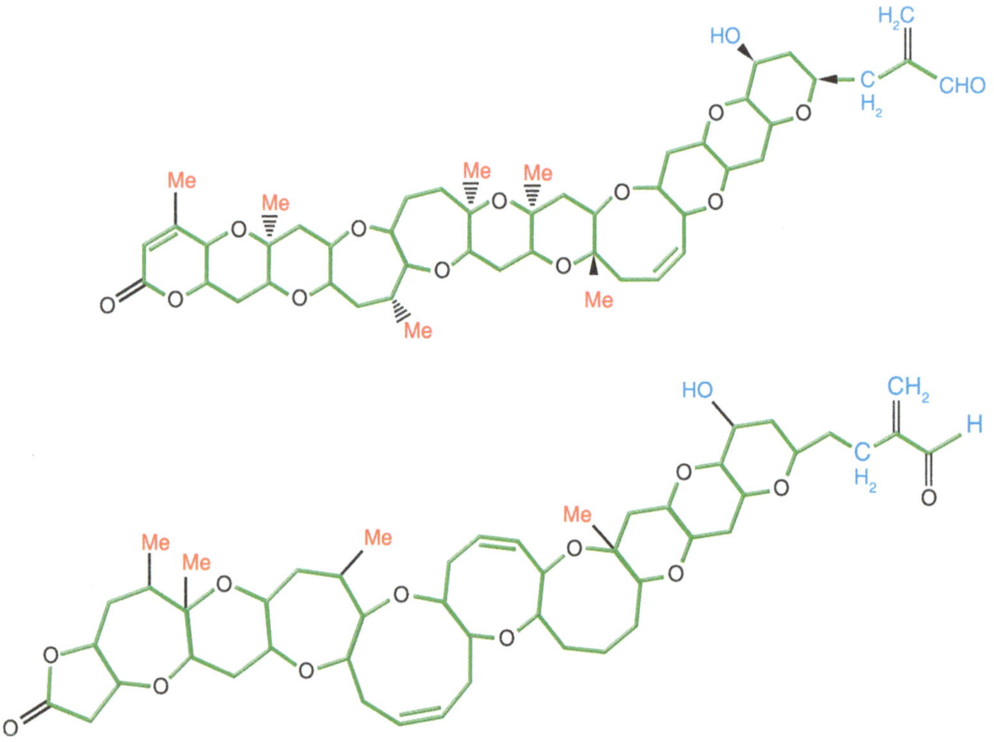

FIG. 16.10 Structure of brevetoxin-A and brevetoxin-B respectively. *(Source: http://www.aims.gov.au/arnat/arnat-0003.htm.)*

How a toxic algal bloom occurs
The life cycle of one cell

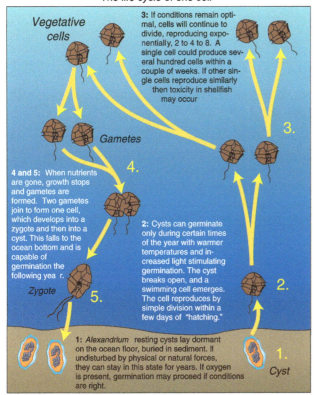

Vegetative cells

3: If conditions remain optimal, cells will continue to divide, reproducing exponentially, 2 to 4 to 8. A single cell could produce several hundred cells within a couple of weeks. If other single cells reproduce similarly then toxicity in shellfish may occur

3.

Gametes

4 and 5: When nutrients are gone, growth stops and gametes are formed. Two gametes join to form one cell, which develops into a zygote and then into a cyst. This falls to the ocean bottom and is capable of germination the following yea r.

4.

2: Cysts can germinate only during certain times of the year with warmer temperatures and increased light stimulating germination. The cyst breaks open, and a swimming cell emerges. The cell reproduces by simple division within a few days of "hatching."

2.

Zygote

5.

1: *Alexandrium* resting cysts lay dormant on the ocean floor, buried in sediment. If undisturbed by physical or natural forces, they can stay in this state for years. If oxygen is present, germination may proceed if conditions are right.

1.

Cyst

FIG. 16.11   Red tide formation. *(Source: Jack Cook, Woods Hole Oceanographic Institution. http://www.whoi.edu/redtide/whathabs/whathabs. html.)*

FIG. 16.12   The curtain divides the two halves of the lake. The area of the lake in the lower half of the picture had phosphorus, a limiting nutrient, experimentally added to it and now contains a massive bloom of cyanobacteria. *(Source: http://www.umanitoba.ca/institutes/fisheries/eutro.html.)*

One of the ubiquitous cyanotoxins is microcystin, which can be produced by species of *Anabaena* (Fig. 16.13), *Nodularia*, *Nostoc*, *Oscillatoria*, and *Microcystis* (Fig. 16.14). There are over 50 different analogs of microcystin (Fig. 16.15). These toxins mediate toxicity by inhibiting liver function (i.e., hepatotoxic) and can often be found at high levels in drinking water reservoirs.

Anatoxin-a is a small, low-molecular-weight neurotoxic alkaloid produced by species of *Anabaena*, *Aphanizomenon*, *Cylindrospermum*, *Microcystis*, and *Oscillatoria*. Anatoxin-a is a powerful, depolarizing, neuromuscular blocking agent that strongly binds to the nicotinic acetylcholine receptor (Fig. 16.16). This is a potent neurotoxin that can cause rapid death in mammals through respiratory arrest.

Cylindrospermopsin, while having many of the properties of hepatotoxin, also resembles neurotoxin. It is produced primarily by *Cylindrospermopsis raciborskii*, but has also been found in *Umezakia natans* and *Aphanizomenon ovalisporum*. Cylindrospermopsin has poisoned at least 149 people, many of them children requiring hospitalization, in Palm Island, Queensland, Australia. At one time believed to be strictly a tropical to subtropical species, *C. raciborskii* has been found in waters in the north

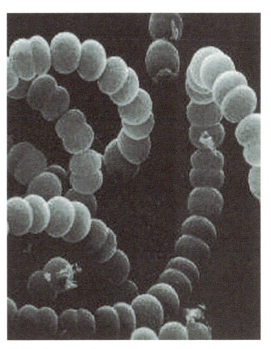

FIG. 16.13   Scanning electron micrograph of *Anabaena flosaquae*. *(Courtesy: Dr. Wayne Carmichael. http://www.nps.gov/romo/resources/ plantsandanimals/names/checklists/other_algae/bluegreens/anabaena_f-a.html.)*

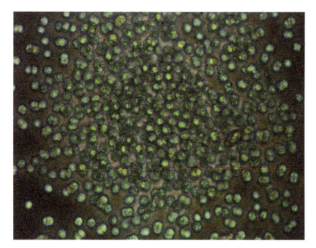

FIG. 16.14 Image of *Microcystis* sp. *(Photo courtesy: David Walker.)*

temperate United States recently. Like most cyanotoxins, cylindrospermopsin can often be found in drinking water reservoirs.

16.7 ORGANIC COMPOUNDS IN WATER

16.7.1 Persistent, Bioaccumulative Organic Compounds

Certain organic compounds, due to their physicochemical properties (Chapter 8), are very persistent in the environment. These compounds are referred to as persistent, bioaccumulative, toxic (PBTs) contaminants (Chapter 12). These compounds are the most important organic contaminants in aquatic systems. These compounds bioaccumulate

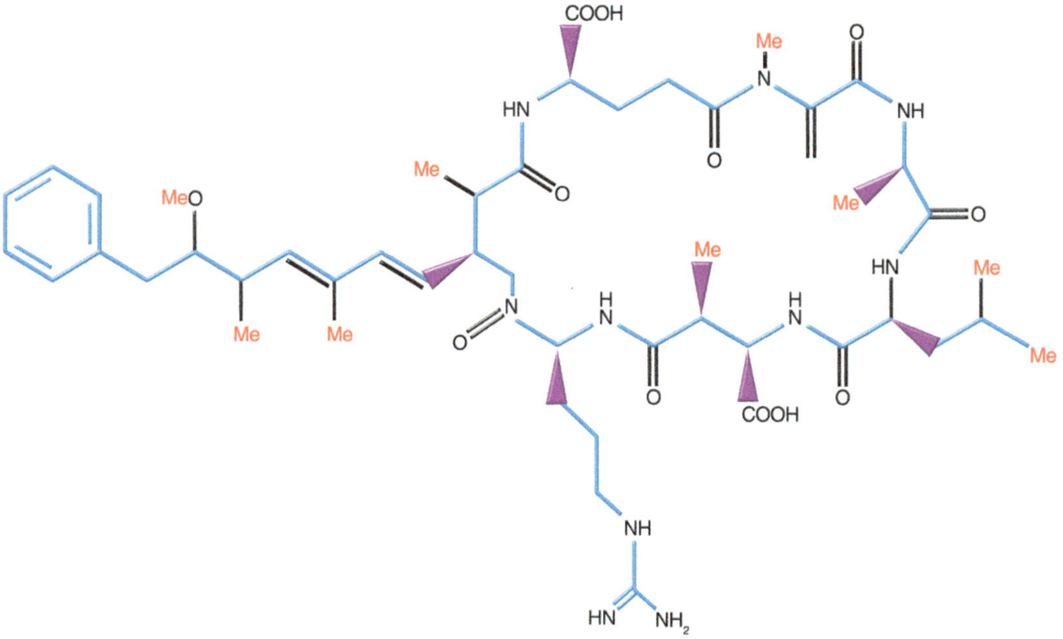

FIG. 16.15 Structure of microcystin. *(Source: http://www.aims.gov.au/arnat/arnat-0002.htm.)*

Anatoxin-a

Homoanatoxin-a

Anatoxin-a(S)

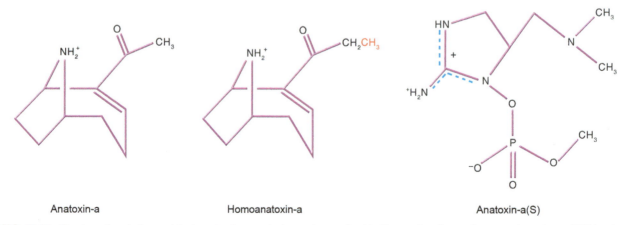

FIG. 16.16 Structure of anatoxin-a and the homolog homoanatoxin-a and anatoxin-a(s). *(Source: http://www.aims.gov.au/arnat/arnat-0002.htm.)*

and biomagnify within the aquatic ecosystem, and they often accumulate in the sediments of surface waterbodies.

Dichlorodiphenyltrichloroethane (DDT) is an organochlorine pesticide that was in widespread use from the 1940s to the early 1970s, when it was used primarily for agricultural crops or vector-pest control (the control of insects known to carry malaria and typhus).

When DDT is released into the environment, it begins to degrade into several different metabolites. Once bound to sediment particles, DDT and its degradation products can persist for many years depending upon environmental conditions. Ingestion is the main route of exposure of DDT and its metabolites to humans and wildlife. Ingestion of foodstuffs and in particular consumption of fish is how humans ingest the largest amounts of DDT, primarily due to bioaccumulation. In fish and other wildlife, especially predatory birds feeding on fish, even if acute toxicity and death does not occur, reproductive failure often results.

The use of DDT in the United States has been banned since 1972. However, the need to protect agricultural crops and humans from insect-borne vectors of disease still exists. Most organochlorine pesticides, including DDT, have been replaced with less environmentally persistent compounds such as organophosphate, carbamate, and synthetic pyrethroid pesticides. While these compounds are degraded in the environment at a much faster rate than DDT, they are also more acutely toxic. Even though DDT was banned over 30 years ago, due to its persistence, we still feel its toxic effects in the United States. Twenty years after the ban of DDT, the U.S. EPA reported that out of 388 sites throughout the nation sampled between 1986 and 1989, total DDT and PCBs (discussed later) were detected at 98% and 90% of all sites, respectively. Fish still remain vulnerable to the effects of DDT. A study by Munn and Gruber in 1997 showed total DDT was detected in 94% of whole-fish samples collected in streams of eastern Washington State.

Polychlorinated biphenyls (PCBs) are a group of organic compounds with similar physical structure and chemistry, ranging from oily liquids to waxy solids (Fig. 16.17). All PCBs are formed from the addition of chlorine (Cl_2) to biphenyl (Cl_2H_{10}), which is a dual-ring structure consisting

of two 6-carbon benzene rings linked by a single carbon-carbon bond. The presence of a benzene ring allows a single attachment to each carbon, meaning that there are 10 possible positions for chlorine to replace the hydrogens in the original biphenyl.

Each unique compound in the PCB category is referred to as a "congener" whose individual name is dependent upon the total number and position of each chlorine substitute. There are 209 PCB congeners.

Due to the chemical stability and high boiling point of PCBs, they were used in hundreds of industrial applications, including electrical insulation, hydraulic equipment, and plasticizers in paints, plastics, and rubber products. Prior to their ban in 1977, total production of PCBs in the United States was more than 1.5 billion pounds. Because of the vast amount of possible congeners, PCBs were sold as many different trade names but one of the most prevalent was Arachlor®.

Similar to DDT, PCBs are environmentally persistent and adhere strongly to particulates in water, meaning that they can remain intact in sediments of lakes and rivers for extended periods. Because of PCBs' strong adherence to sediments and suspended particles in water, the contamination level of a waterbody may be several times higher than the aqueous solubility of a particular PCB. Also like DDT, all PCBs are extremely lipophilic, meaning that they can bioaccumulate and biomagnify in aquatic environments. First detected in the 1960s, PCBs were found to be contaminants on a global scale, occupying virtually every component of the environment including air, water, soil, fish, wildlife, and human blood.

Uptake of PCBs by microorganisms is very rapid and extremely high bioconcentration factors are often observed. Uptake by microorganisms is by true absorption into cells rather than adsorption onto the cell.

Fish are especially susceptible to the accumulation and concentration of PCBs, and all life stages of almost every species readily absorb PCBs from the water. PCB congeners with higher chlorination levels are taken up most rapidly by fish. Due to the fact that PCBs are usually at higher levels in sediments, fish such as bottom feeders are most susceptible; however, route of exposure in fish can occur through water, sediment, or prey. Due to rapid uptake of PCBs in fish tissue, birds, especially those that eat fish, are also vulnerable. Egg-laying females can transfer substantial amounts of PCB to eggs with subsequent reproductive failure.

Route of exposure to humans, like DDT, is generally much greater for aqueous environments (through either direct ingestion of water or through eating contaminated fish) than terrestrial. PCBs are probable carcinogens in humans and are known to be carcinogenic to laboratory animals. The risks associated with consuming fish contaminated with PCBs are more than 1000 times greater than the 1-in-a-million cancer risk used to regulate most hazardous wastes.

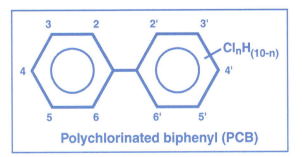

Polychlorinated biphenyl (PCB)

FIG. 16.17   Basic structure of a polychlorinated biphenyl. (*Source:* http://www.epa.gov/toxteam/pcbid/defs.htm.)

16.8 ENTERIC PATHOGENS AS SURFACE WATER CONTAMINANTS

Almost all animals are capable of excreting disease-causing intestinal microorganisms (enteric pathogens) in their feces (see Chapter 13). Sources of pathogens into surface waters include:

- Urban stormwater
- Combined sewer and sanitary sewer overflows
- Animal feeding operations
- Sewage treatment plants
- Septic tanks (onsite systems)

Pathogens can remain infectious for prolonged periods of times in surface waters presenting health risks to recreational users shellfish harvesting, and drinking water treatment plants. While drinking water treatment plants are required to treat water from surface sources, the more pathogens that are present in the raw water, the more treatment is required. Also, after periods of heavy rains, when the amount of suspended matter and pathogens often increases, it is difficult to remove all pathogens. For example, it has been shown that waterborne disease outbreaks in the United States are related to the intensity of rainfall events (Curriero et al., 2001).

Forty percent of rivers and estuaries that fail to meet ambient water quality standards fail because of pathogens, usually measured by fecal coliform bacteria (see Chapter 13) (Smith and Perdek, 2004). See Information Box 16.1.

Stormwater can contain a wide variety of pathogens that originate from the feces of wild and domestic animals. Besides pets, other animal sources in urban areas include pigeons, geese, rats, and raccoons. Animal feces accumulate on the ground, and following a storm event are flushed into nearby streams and lakes. This results in a rapid increase in the concentration of enteric organisms, sometimes exceeding that found in raw sewage. In some cities, sewers that collect domestic sewage are combined with stormwater drains or collection systems. These flows are then transported to a sewage treatment plant for treatment. Unfortunately, after periods of heavy rainfall, this combined flow is greater than the sewage plant can treat, requiring the sewage plant to discharge untreated combined sewage and stormwater. These events are referred to as *combined sewer overflows* (CSOs). CSOs generally occur in older parts of the country, involving approximately 900 cities. To reduce the impacts of CSOs, cities may blend the untreated wastewater with treated wastewater (Fig. 16.18) or may construct large holding reservoirs where the combined flows can be stored until they can be treated later (see Case Study 16.2).

In the United States, there are 212,000 animal feeding operations, or AFOs that produce 350 million tons of manure annually. This figure does not include manure from

INFORMATION BOX 16.1 Microbial Source Tracking

Fecal contamination of surface waters can result from numerous sources, including human sewage, manure from livestock operations, indigenous wildlife, and urban runoff. Effective watershed management requires identification of, and targeting mitigative action towards, the dominant source of fecal contamination in the watershed. Several *microbiological source tracking* (MST) methods have been developed to fill this need. MST methods are intended to discriminate between human and nonhuman sources of fecal contamination, and some methods are designed to differentiate between fecal contaminations originating from individual animal species.

MST methods involve the isolation of fecal bacteria (*Escherichia coli*, enterococci) or viruses (human or bacterial) from a watershed. Subsequently molecular analyses of DNA or RNA, or patterns of sensitivity to different antibiotics, are performed to "fingerprint" the organisms. Samples of potential sources of the fecal bacteria are then collected in the watershed (e.g., from cattle, ducks, pigeons, dogs, sewage treatment plants, and urban runoff), and a "fingerprint" of these bacteria is obtained. By matching the fingerprints, the major sources of fecal bacteria in a watershed or waterbody can be identified. Identification of human enteric viruses (see Chapter 11) is an indication of human sewage as a source. Certain bacteriophages are only found in humans and others only in animals; hence, this is another approach that can be used.

grazing animals. These operations generate approximately 100 times as much manure as municipal wastewater treatment plants produce sewage sludge (biosolids) in the United States. The Clean Water Act (see Chapter 30) requires operations having more than 1000 animals to have a discharge permit. These are defined by the United States Environmental Protection Agency as concentrated animal feeding operations or CAFOs, of which there are currently almost 20,000. Runoff from AFOs and farmland can contribute significant levels of pathogens that can infect humans, such as *Cryptosporidium* and *Escherichia coli* O157:H7 (see Chapter 13).

Septic tanks (also referred to as *decentralized* or *onsite wastewater treatment systems*) collect, treat, and release about 4 billion gallons of treated effluent per day from an estimated 26 million homes, business, and recreational facilities in the United States (see Chapter 22). Poorly treated wastewater from improperly operating or overloaded systems can contain enteric pathogens that by transport through the soil can make their way to nearby streams and lakes. Overflows from failing onsite systems can result in the sewage reaching the surface. The discharge of partially treated sewage from malfunctioning onsite systems was identified as a principal or contributing source

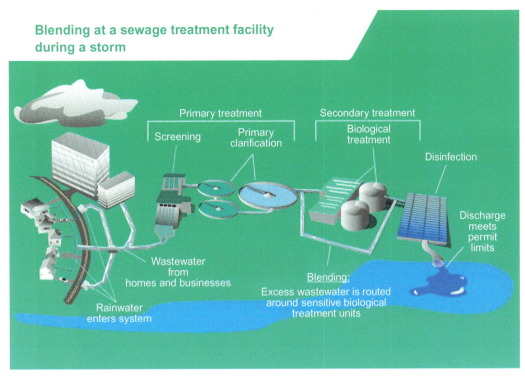

Blending at a sewage treatment facility during a storm

FIG. 16.18   Blending at a sewage treatment facility. *(Source: www.epa.gov.)*

Case Study 16.2. Tunnel and Reservoir Plan (Tarp) to Control Excess Stormwater

Chicago and 51 older municipalities in Cook County, Illinois, have combined sewer systems. This means when rain falls, stormwater runoff drains into a combined sewer, where it mixes with the sewage flow from homes and industry. The net result is one massive quantity of dirty water! A system that was designed to treat 2 billion gallons of wastewater per day may be inundated with more than 5 billion gallons of stormwater runoff (about 1″ of rain) during a single rainstorm.

When the urban area grew and treatment plants were at capacity, there was no alternative but to allow the excess mixture of raw sewage and stormwater to spill directly into the rivers and canals as combined sewer overflow or CSO. This meant that much untreated sewage, diluted with storm runoff, was bypassing treatment plants and polluting lakes, rivers, and streams, and also causing street and basement flooding. A better solution had to be found.

In the 1970s, a team of engineers from the City of Chicago, Cook County, and state agencies considered various plans to solve the problem of flooding and water pollution. The hybrid plan selected as best and most cost effective was the *Tunnel and Reservoir Plan* (TARP) (Fig. 16.19). Under this plan, 109 miles of huge underground tunnels would be burrowed under the city to intercept combined sewer overflow and convey it to large storage reservoirs. After the storm had subsided, the overflow could then be conveyed to treatment plants for cleaning before going to a waterway.

The Mainstream tunnel is 35 ft in diameter, bored in limestone rock 240–350 ft below ground, and holds 1 billion gallons of water (Fig. 16.20). Mainstream is one of the largest rock tunnel bores on record. Since tunnel contractors would be working beneath homes, businesses, and streets, excavation by extensive blasting was ruled out. Boring by huge tunnel-boring machines (TBMs) was selected instead, to cause less rock disturbance, noise, and vibration.

The success of this project is evident by the dramatic improvements in the water quality of the Chicago River, the Calumet River, and other waterways. Game fish have returned, marinas and riverside restaurants abound, and river recreation and tourism has improved. http://www.mwrdgc.dst.il.us/plants/tarp.htm

of degradation in 32% of all harvest-limited shellfish growing areas in the United States. Problems with surface water contamination by onsite systems are most likely to occur in areas with shallow groundwater tables (e.g., within a few feet of the surface).

In the United States it is required that sewage discharges be disinfected to reduce the level of pathogens (see Chapter 22). While very effective in eliminating most enteric bacterial pathogens, significant levels of enteric virus and protozoan parasites may remain.

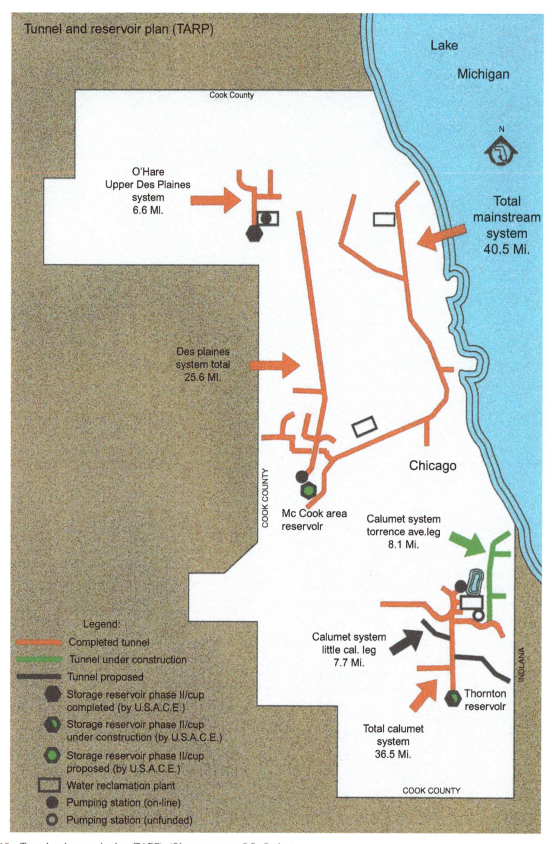

FIG. 16.19   Tunnel and reservoir plan (TARP). *(Photo courtesy: C.P. Gerba.)*

16.9 TOTAL MAXIMUM DAILY LOADS (TMDL)

A *total maximum daily load* (TMDL) is the maximum amount of pollution that a waterbody can assimilate without violating water quality standards. A TMDL is the sum of the allowable loads of a single pollutant from all contributing point and nonpoint sources, so that a waterbody can meet a designated use, such as swimming or fishing. Under the Clean Water Act of 1972, states are required to identify surface waters not meeting water quality standards and develop a TMDL for each pollutant for each listed waterbody. The processes of TMDL development and implementation are shown in Fig. 16.21. Once the impaired body of water is identified, a study is usually conducted to identify the sources and concentration of pollutants. From this, an informational plan is developed to reduce the most significant source(s) so that water quality standards can be met.

$$\begin{aligned} \text{TMDL} = \; & \text{point sources of waste allocations} \\ & (\text{e.g., sewage treatment plant discharge}) \\ & + \text{nonpoint source load allocations (urban runoff)} \\ & + \text{natural sources (mineral deposits, wild animals)} \\ & + \text{growth factor (growth of enteric bacteria)} \\ & + \text{a margin of safety (to compensate for uncertainties} \\ & \text{about the link between pollutant loads and impairments)} \end{aligned}$$

A load allocation is the part of a TMDL/water quality restoration plan that assigns reductions to meet identified water quality targets. For example, in a given watershed it was found that cattle were the major source of fecal coliform bacteria in the streams. To reduce fecal coliform loading, fences were placed to limit direct access of the cattle to the stream.

Fecal coliform bacteria and temperature (largely thermal waters from power plants) are the major contaminants that cause most waterbodies to not meet water quality standards for intended uses in the United States (Fig. 16.22). See Case Study 16.3 for an example.

16.10 QUANTIFICATION OF SURFACE WATER POLLUTION

Even as treatment methods were being developed, methods were simultaneously developed to quantify and assess both the disease threat and the dissolved oxygen problem posed by the discharge of municipal wastes into waterbodies. Quantitative methods are also used to calculate the degree of treatment needed. Such methods are based on an understanding of the physical and biochemical processes controlling the decay of microbes and chemicals over time.

FIG. 16.20 The Mainstream tunnel. *(Photo courtesy: C.P. Gerba.)*

16.10.1 Die-Off of Indicator Organisms

Early in the development of quantitative assessment methods, scientists realized that the many different microorganisms that exist in human wastes could not be effectively cultured and counted. Consequently, they settled on the coliform group of organisms to serve as an indicator of fecal pollution because they could be cultured and counted easily. Known to exist in large numbers in the gut of all warm-blooded animals, the coliform group provides a good indication of fecal pollution; however, it is not very specific, so other indicators such as fecal coliforms and streptococci may also serve as indicators of the sanitary quality of water.

Tests have shown that 99.99% of the indicator bacteria can be removed by wastewater treatment. But the residual 0.01% remains a problem raising concerns about the quality of water for recreational use, for example. The fact is that the number of bacteria in sewage is tremendous (500 million to 2 billion per 100 mL), so that, in assessing quality, the percentage of removal is not as useful as is the actual remaining concentration. The microorganism concentrations allowed for various uses of water are relatively small. For drinking purposes, the concentration of fecal coliforms should, of course, be 0, but a concentration

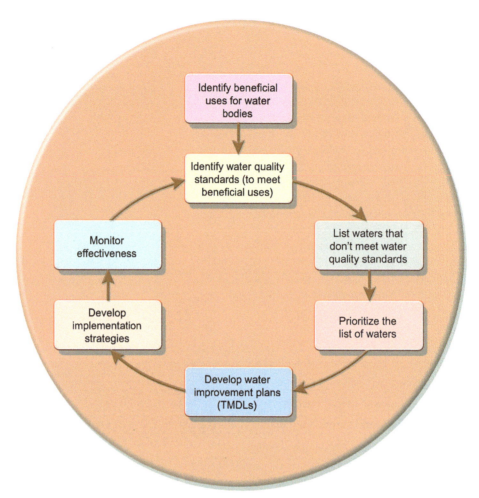

FIG. 16.21   TMDL development and implementation.

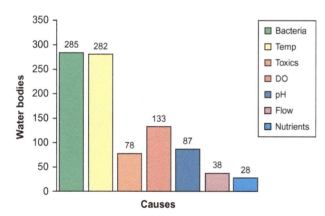

FIG. 16.22   Major causes for waterbodies that fail to meet water quality standards. *(Source: www.epa.gov.)*

of less than 1 per 100 mL may be allowed; for bathing, a concentration of 1000 per 100 mL is frequently accepted. If, for purposes of demonstration, we assume that raw sewage has a population of 1 billion (10^9) per 100 mL, a removal efficiency of 99.99% would still leave 100,000 per 100 mL. The good news is that the concentration of organisms decreases with time and distance downstream of the discharge point owing to natural processes. Depending on the location and proximity of uses, the point of discharge and method of discharge can be designed to optimize the rate of natural purification, further reducing downstream pollution problems. The bad news is that we have only a partial understanding of all the processes that affect die-off.

The concentration C of bacterial indicators of fecal contamination has been observed to decrease with time t according to a first-order reaction, the equation of which is

$$\frac{dC}{dt} = -KC \qquad (16.1)$$

where K is the *die-off rate constant*. One parameter frequently employed in water pollution analyses obtained by solution of the first-order reaction equation is t_{90}, which is the time required for 90% die-off of the bacteria. This parameter is analogous to the half-life (t_{50}) used in radioactivity studies and is calculated in the same fashion. Thus given the value of K

Case Study 16.3. Total Maximum Daily Load (TMDL) for Fecal Coliform Bacteria in the Waters of Duck Creek in Mendenhall Valley, Alaska

Duck Creek is on the list of impaired waters in Alaska because of fecal coliform bacteria. The primary sources of fecal coliform bacteria in the creek were found to be urban runoff and animal waste. As the watershed became more developed, urban runoff and pet populations increased. This increased the level of fecal coliforms entering Duck Creek. Duck Creek is used as a source of drinking water, and the State of Alaska standard for such waterbodies is a geometric average of 20 fecal coliforms per 100 mL. This standard was usually exceeded several times per year (Fig. 16.23). Based on the water quality standards for fecal coliform bacteria and the hydrologic conditions of Duck Creek, the loading capacity for fecal coliform bacteria was established at 2.23×10^{11} fecal coliforms per year. To meet these objectives it was recommended that wetlands be constructed to retain stormwater flows, greenbelts be developed to serve as buffers to overland flow, and eroding banks be stabilized. In addition, pet owners were encouraged to clean up and properly dispose of pet waste.

$$t_{90} = \frac{2.3}{K} \qquad (16.2)$$

The general solution of Eq. (16.1) is used to find the concentration at any time C_t, after the initial concentration C_i is determined:

$$C_t = C_i e^{-Kt} \qquad (16.3)$$

This information can be used to calculate freshwater or marine die-off. To find the concentration of bacteria after effluent has traveled in the river for, say, 8 h, it is necessary to determine the value of K for the specific situation being studied. Results of many studies in lakes and streams have shown that K varies widely, depending on the temperature of the water, the amount of sunlight, and the depth at which the plume travels. An average value for fresh water is about $K = 0.038$ per hour; however, values from 0.02 to 0.12 per hour have been measured. Using Eq. (16.3) and $K = 0.038$ per hour, the initial concentration $C_0 = 100,000$ per 100 mL would be reduced to about 74,000 per 100 mL in $t = 8$ h due to die-off alone. In many cases, it is preferable to predict the concentration at a given distance downstream, rather than at time increments. Most U.S. rivers have a remarkably uniform low-flow current of 1.5–2 km per hour, which can be used to convert low-flow travel time to distance. In our example, this travel distance would be between 12 and 16 km. Travel times for other flow conditions vary from river to river, so field measurements may be needed to relate bacterial concentrations to specific locations downstream.

In analyzing coliform die-off cases in the marine environment, we often use an average value of 1.2 per hour for K, which is nearly 30 times greater than that of freshwater. This rapid die-off rate is usually attributed to the salinity of the marine environment, although it may also be related to a greater concentration of predatory animals. In addition, the natural flocculation and sedimentation of particles that occurs in estuaries could account for removal of bacteria from the water column. There are, however, other factors

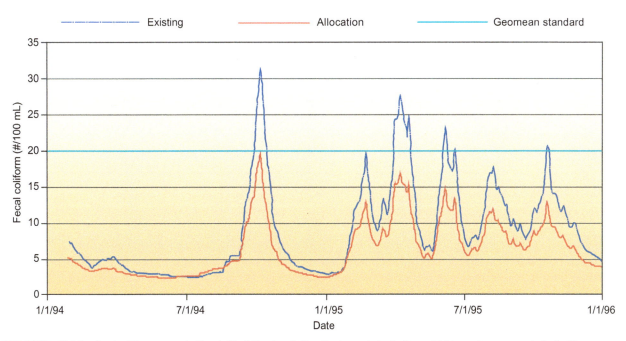

FIG. 16.23 Existing fecal coliform concentrations in Duck Creek and allocation to meet standards as a drinking water source waterbody. *(Source: www. epa.gov./owow/tmdl.)*

that can reduce the die-off rate. For example, when an effluent is discharged at a great depth, the die-off can slow down considerably because sunlight cannot penetrate deeply enough.

Calculating seawater die-off is similar to freshwater die-off. The value of K for marine waters usually ranges from 0.3 to 3.8 per hour. Recently, however, K values as low as 0.02 per hour have been found where an effluent plume is transported in a layer far below the surface, say, 40 m, suggesting that die-off is reduced because of the low penetration of UV radiation to that depth. What difference would this low K value make in the concentration?

Using the average value of $K = 1.2$ per hour, the original concentration, $C_0 = 100,000$ per 100 mL, would die off to $C_1 = 6.7$ per 100 mL in $t = 8$ h using Eq. (16.3). But using $K = 0.02$ per hour instead of 1.2 per hour, we get 85,000 per 100 mL. Thus for a 60-fold reduction in K, the concentration is increased by a factor of $85,000/6.7 = 13,000$! It is evident from this example how important it is to have accurate values for K and how widely the results can vary with equally good, but different, estimates for K.

Variations in the rates of indicator bacterial die-off are not the only problem we have to contend with. Noncoliform pathogenic microorganisms may decay at rates different from those of our coliform indicators. Therefore as water analysis techniques become more sophisticated, we will need to conduct many field observations to establish values that can be used to predict die-off rates for specific pathogens.

In addition to the decrease of bacterial concentrations due to die-off in either fresh or marine surface waters, bacterial concentrations are decreased as the water is diluted with upstream ambient water at the point of effluent discharge and further diluted as it flows downstream. The effect of dilution may or may not be important in meeting water quality criteria, depending on the initial mixing, the nature of the subsequent flow patterns, and the distance to water use areas. But before discussing the mechanics of the dilution process, observe how the same first-order decay process used for assessing indicator bacteria die-off can be applied to the analysis of the fate of biodegradable organics.

16.10.2   Organic Matter and Dissolved Oxygen

Biodegradable organic compounds are decomposed by bacteria and other organisms that live in surface waters. While some organics are mineralized to carbon dioxide and oxides of nitrogen, others are synthesized into more microbial biomass, most of which is subsequently decomposed as well. All this decomposition consumes dissolved oxygen (DO), upon which many desirable species of fish, other aquatic organisms, and wildlife depend. Thus depressed dissolved oxygen concentrations adversely affect these life forms. For example, some fish can survive at concentrations near 1 mg L^{-1}, most are adversely affected at DO concentrations below 4 mg L^{-1}. The maximum amount of oxygen that pure surface water can hold is a function of salinity, temperature, and atmospheric pressure; compared to the maxima of many other substances, however, it is remarkably low. Note that solubility of O_2 *decreases* with increasing temperature, which is the opposite of the temperature-solubility relationship observed for most substances in water.

Domestic sewage can contain about 300–400 mg L^{-1} of organic compounds, 60% of which is readily degradable by bacteria commonly found in nature. Readily degradable implies that most of the material will be decomposed within about a week in a stream or other body of water that is sufficiently large. The change in the concentration of the organic matter with time is conveniently described by the first-order decay equation used to describe bacterial die-off. However, the value of this K depends on the specific organic compounds in the sewage. For domestic sewage, an average value is about $K = 0.4$ per day, ranging from 0.1 to 0.7 per day. As more industrial wastes are contributed to the sewer system, the rate constant may increase or decrease. The amount of organic material discharged to a surface water depends on the population served by the municipal sewer system and treatment technology employed. Each person contributes about 90 g per day of organics; thus if the population is 100,000 people, the mass emission rate is 9 metric tons per day. The concentration of organics in the sewage depends on the amount of water added by the individual households and that added by other water uses in the community. If the average water use is 300 L per person per day, the concentration is 300 mg L^{-1}.

16.10.3   Measurement of Potential Oxygen Demand of Organics in Sewage

16.10.3.1   COD

A parameter frequently used for industrial wastes, particularly where industrial wastes contribute heavily to the sewer system, is the *chemical oxygen demand* (COD), which is a measure of the amount of oxygen required to oxidize the organic matter—and possibly some inorganic materials—in a sample. Note that the method employed to obtain this parameter, which involves reflux of a sample in a strong acid with an excess of potassium dichromate, does not specifically measure the organic content in the sample, but rather the amount of oxygen required for oxidation. This approach therefore provides a direct measure of the potential impact of oxygen consumption on the oxygen content of the waterbody.

16.10.3.2 BOD

The *biochemical oxygen demand* (BOD) is the most commonly used parameter in the analysis of oxygen resources in water. The BOD is the amount of oxygen consumed over time, usually 5–20 days, as the organic matter is oxidized both microbially and chemically.

16.11 DETERMINING BOD

The laboratory method used to determine BOD has changed very little since it was initiated in the 1930s. We begin by setting up many sample bottles to contain a sample of waste, mixed in water either from the disposal site or from a standard laboratory supply. Then we use standard methods to find the initial concentration of DO, after which the bottles are incubated in a dark water bath at a given temperature, usually 20°C. Every day for 5 or more days, we open a few of the bottles and measure the remaining DO. The difference between the initial value and the value at each time period, that is, the *demand*, is plotted as the BOD for the series of days.

From the data obtained in the laboratory, we construct a smooth curve that lets us calculate the reaction rate coefficient K by graphical or analytical methods. The curve we construct, whose equation is

$$BOD_t = BOD_L\left(1 - e^{-Kt}\right) \qquad (16.4)$$

becomes more and more horizontal as time progresses (Fig. 16.24). By extrapolating the curve to horizontal, we can make an estimate of the ultimate value, called the *limiting value of BOD*, or BOD_L.

We can use the curve in this example to compute the value of K for the wastewater sample in this laboratory test.

First, by extrapolating this curve to horizontal, we see that the estimated value of BOD_L would be approximately $7.6\,\text{mg}\,\text{L}^{-1}$. Next, by substituting the value read from the curve for day 5 ($6.1\,\text{mg}\,\text{L}^{-1}$), so that $BOD_t = BOD_5 = 6.1$ and $t = 5$, we can find K from Eq. (16.4):

$$BOD_5 = BOD_L\left(1 - e^{-Kt}\right)$$

$$6.1 = 7.6\left(1 - e^{-5K}\right)$$

so that

$$1 - \frac{6.1}{7.6} = e^{-5K} \text{ or } \ln 0.197 = -5K$$

Thus

$$K = \frac{-1.62}{-5} = 0.32\,\text{per day}$$

16.11.1 Impact of BOD on Dissolved Oxygen of Receiving Waters

Typically, municipal sewage is treated to some degree to remove organics, and hence to reduce BOD, before discharge to surface waters in developed countries of the world (Chapter 22). The amount of BOD remaining after treatment can range from 10% to 70% of the amount originally in the sewage. The impact on receiving waters depends on many environmental and waste characteristics, most of which are briefly mentioned later in this chapter. After the BOD parameters already explained, the next most important considerations are the amount of DO in the waterbody before the sewage is added, called the *initial DO* (DO_i), and the rate at which additional oxygen is transferred from the atmosphere to the receiving water.

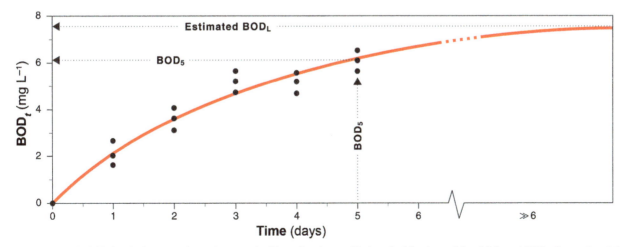

FIG. 16.24 Typical biochemical oxygen demand test results illustrating the graphical method for determining BOC_L and BOD_5 for use in solving Eq. (16.4) to find K as described in the text. Note that, in actuality, the extrapolated curve is asymptotic to $y = BOD_L$. *(From Pollution Science ©1996, Academic Press, San Diego, CA.)*

Many rivers and streams have depressed DO concentrations, that is, a *DO deficit*, because of wastewater added by cities upstream. The saturation value of DO is 100% with the quantification in mg/L depending upon temperature, salinity, atmospheric pressure. The local deficit—the difference between the saturation value and the observed initial value at the location of waste discharge—must be included in the computation of the downstream DO deficit caused by a new effluent discharged to the stream. Before presenting an example calculation, let us first examine how nature deals with the DO deficit.

Deficits tend to be redressed by oxygen gas derived from the atmosphere. Such replenishment occurs by a process of gas-liquid mass transfer at the surface and subsequent mixing throughout the depth of water. This overall process can be described in an approximate manner by a first order-equation, the solution of which is

$$D_t = D_i e^{-Rt} \qquad (16.5)$$

where

D_t is the deficit at time t;
D_i is the initial deficit (i.e., $DO_s - DO_t$);
R is the reaeration coefficient.

Note that the larger the magnitude of R, the quicker a given deficit is removed. The value of R depends on the degree of vertical mixing in a waterbody, as well as its overall depth. It varies from 0.1 per day in small ponds to above 1 in rapidly moving streams.

Most DO problem situations require that we determine the oxygen deficit resulting from the simultaneous effects of the oxygen demand of a waste and the competing restoration of oxygen from atmospheric reaeration. The combined result is termed the *self-purification capacity* of the waterbody. The deficit shown graphically as a function of time (or distance) is known as the *oxygen sag curve* because of its characteristic spoon shape. The equation for the curve is

$$D_t = \left[\frac{K(BOD_L)}{R - K}\right]\left(e^{-Kt} - e^{-Rt}\right) + D_t e^{-} \qquad (16.6)$$

where the terms are as defined before.

An example of the curve produced by the equation is shown in Fig. 16.25. The values of K and BOD_L are taken from Fig. 16.24; the value for D_i, the initial deficit, was $3\,mg\,L^{-1}$ due to upstream discharges; and a reaeration coefficient R of 0.5 per day was chosen for a large, slow-flowing stream. This example might represent a case of highly treated sewage, a typical effluent plume after an initial dilution of 4, or poorly treated sewage dispersed from a diffuser that provides an initial dilution factor of 25.

One interesting result from the solution of the oxygen sag equation is that the deficit, irrespective of the saturation

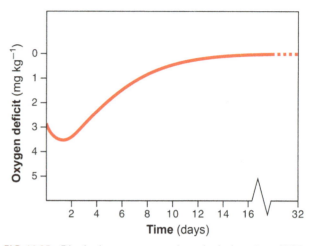

FIG. 16.25 Dissolved oxygen sag curve determined using values of BOD_L and K from Fig. 16.24. The equation for this curve is given in Eq. (16.6). *(From* Pollution Science *©1996, Academic Press, San Diego, CA.)*

value of DO, is the same as long as the initial deficit is the same. Sometimes, water quality standards impose limits both on the amount of deficit per se and on the resulting DO value itself. For example, a regulation could require that a waste discharge must neither increase the deficit by more than 10% nor depress resulting DO concentration below $5\,mg\,L^{-1}$.

When water systems are heavily used or highly valued water uses are threatened, additional factors require consideration:

1. The diurnal demands and supplies of oxygen from photosynthesizing organisms.
2. The oxygen demand of organic materials deposited in the sediment layer.
3. The oxygen demand of nitrogen compounds discharged in the effluent.
4. Variations in the reaeration coefficient R with travel time due to flow conditions in the waterbody.
5. The wide range of sewage flows encountered over the lifetime of a river.

Comprehensive computer programs are available to describe the concentration of oxygen in large watersheds consisting of dozens of interconnecting streams and dozens of wastewater inputs; however, data on plume travel, K values, and R values still have to be obtained with time-consuming laboratory studies and physically demanding field studies. Dilution of wastes is one of the most important factors to consider in assessing impacts. The methods used to assess effects on the dissolved oxygen resource of a waterbody and bacterial contamination are applicable to a wide variety of toxic chemical problems.

16.12 DILUTION OF EFFLUENTS

Dilution can be—but is not necessarily—an effective way to prevent pollution of surface waters. Environmental scientists do not categorically embrace the old saying, "Dilution is the solution to pollution." Instead, we recognize through the analytical process of risk assessment that certain principles underlie the utility of dilution in managing waste discharges. The first principle concerns the concentration dependence of the pollutant response mechanism. Here, we want to know if the effect of the contaminant is directly related to its concentration. That is, we ask if the concentration is reduced sufficiently, will the degree of effect be directly reduced?

Further, once the contaminant concentration is reduced, can it subsequently become more concentrated? Reconcentration is a phenomenon often associated with sediments and persistent organic chemicals such as PCBs. Dilution of waste streams containing high concentrations of suspended solids may prevent significant pollution near the discharge site, but such sediments may eventually settle out of the water column and concentrate in depressions in the streambed, where they can cause a variety of problems. Even in the ocean, waste disposal can lead to accumulation of sediments in the seabed. With persistent organic chemicals, whose solubility in water is low and whose affinity for sorption to animal tissues is high, adverse bioaccumulation can occur even from highly diluted mixtures. Thus dilution may or may not solve a potential concentration-related problem.

16.12.1 Dilution in Streams and Rivers

Aside from concentration-toxicity considerations, social and economic considerations enter into the decision to use dilution. In addition, its use is also dependent on the availability of a sufficient quantity of dilution water. In the arid southwestern United States, for example, many streams are ephemeral; that is, they contain water only after major rainfall events. In other streams in arid climates, the effluent from municipal treatment plants is the predominant flow for more than 50% of the year; that is, they are effluent-dominated streams. In such situations, dilution is somewhere between small and nil, except during storm runoff. For example, if the stream flow is 10 million liters per day and the effluent flow is 30 million liters per day, the contaminants dissolved in the effluent will be reduced in concentration by just 25%, assuming the concentration of each contaminant upstream is zero (generally, the effective dilution will be even less than that because there is almost always some measurable concentration of contaminants upstream).

The general equation used in determining the concentration after dilution is

$$c_f = \frac{c_e v_e + c_a v_a}{v_e + v_a} \qquad (16.7)$$

where

c_f = cross-sectional average final concentration in the stream

c_e = concentration in the effluent

v_e = volume flux of the effluent

c_a = concentration in the ambient dilution water upstream

v_a = volume flux of the ambient dilution water.

In large rivers, the amount of dilution achieved depends on the method used to discharge the effluent into the river. To maximize the dilution, it is necessary to employ a diffuser, which consists of a pipeline with many exit orifices across the width of the river. Thus if the river flow is 120 MLD, an effluent discharge of 30 MLD yields an 80% reduction in concentration of contaminants if the ambient concentration is zero. That is, the final concentration would be one-fifth the effluent concentration. Of course, the ambient concentration is almost always greater than zero, so Eq. (16.7) must be used to estimate the final concentration accurately.

In many streams, construction of diffusers may be either inappropriate or prohibited. Consequently, the amount of dilution depends on natural mixing processes that occur during stream flow. But in large streams, the effluent plume may hug the bank for many miles, so it is not actively diluted with the main flow. In cases like this, it is not possible to estimate a range of values for the dilution rate. It is reasonable to assume that without a physical structure, like a diffuser, to mix the effluent into the river, we may consider only the natural die-off process in assessing the impact of bacterial contamination on downstream water uses. Similarly, if there is no dilution of the BOD, we can count only on the decomposition of organics and reaeration to restore or maintain the DO resource of the river.

16.12.2 Dilution in Large Bodies of Water

Large bodies of water, particularly open coastal waters, offer much greater opportunity for effective dilution of waste streams. Effective initial dilution is achieved by a multiport diffuser on the end of the outfall discharge pipe. Because the density of most wastewaters is very close to that of freshwater, the discharge of effluent to deep marine waters creates a strong buoyant force. Thus the effluent, no matter how deep the discharge, will rise to the surface of the sea if it is not trapped by density gradients below the surface. As it rises toward the surface, the effluent effectively mixes with the surrounding ambient water, resulting in more and more dilution. A good analogy is the increasing width of a smoke plume as it rises in the atmosphere. The dilution is proportional to the square of the plume width.

In the effective placement of ocean outfalls, however, depth is not the only determining factor. All other things being equal, the greater the extent of vertical travel, the greater the amount of initial dilution. But those "other things" must be truly equal. If, for example, a location chosen for its great depth has poor circulation, the net result may be less effective dilution of wastes than that offered by placement in shallower, but more open, water. Such considerations are a major concern in the placement of outfalls in fjords, bays, and, sometimes, estuaries.

Depth does not always provide the same opportunity for greater initial dilution in lakes and reservoirs because the difference in density between wastewater and receiving waters may be very small. In fact, it is not uncommon for industrial wastes to have a density greater than that of lake water, so these wastes tend to settle along the bottom rather than rise to the surface. However, because of their high temperatures, the cooling waters from large thermal-electric power stations have a density less than that of most lake waters. Thus a deep discharge site can be advantageous for achieving effective reduction of thermal effects.

Many countries still allow the practice of dumping partially treated municipal sewage sludge into the ocean. For many years, before being banned in the United States in the 1980s, such sludge from New York City and Philadelphia was dumped in the Atlantic Ocean. Typically a portion of the sludge is particulate matter possessing a sufficiently high density to settle to the seabed, although currents and turbulence spread the material throughout the disposal zone. Other materials disposed of in the ocean, such as dredged sediments from harbors and waterways and some industrial wastes, behave similarly to sewage sludge.

16.12.2.1 Initial Dilution and Transport

The term *initial dilution* specifically identifies the amount of dilution achieved in a plume owing to the combined effects of the momentum and buoyancy-induced mixing of the fluid discharged from the orifice. This term is used both in regulatory practice and in plume hydrodynamics. The rate of dilution caused by these forces is quite rapid in the first few minutes after exiting the orifice and then decreases markedly after the momentum and buoyancy are dissipated. Ambient currents also influence the rate of dilution during the buoyant rise of the plume irrespective of momentum and buoyancy. As current speed increases, so does initial dilution. In many cases, an initial dilution of 100 to 1, commonly sought in design of outfalls, is sufficient to reduce the toxicity of chemical contaminants to an acceptable level. (Note: When bacterial die-off is an important consideration, the distance from a designated use area is usually a more important factor than initial

dilution per se. In this case, the time it takes microbes to travel a long distance increases the likelihood that they will be inactivated.)

Following initial dilution, waste streams undergo additional dilution, or dispersion, as they are transported by ambient currents and mixed with the surrounding water by turbulence. The process is analogous to the dispersion that takes place when smoke plumes dissipate in the atmosphere after the smoke has risen to an equilibrium level. In some aquatic systems, however, the effluent plume is not as easily observed.

Most modern coastal cities employ multiport ocean outfalls far offshore to protect beaches and nearshore recreational areas from the effects of bacterial contamination. These outfalls are frequently designed to maintain a diluted waste stream below the surface of the sea. Such systems are especially useful during the summer recreational season because they keep the immediate area of discharge free of unsightly messes. Moreover, they reduce landward transport of the diluted waste by onshore wind currents toward peak beach activities. The disadvantage of subsurface trapping lies in the fact that initial dilution is reduced compared with plumes rising to the surface, but this disadvantage is offset by the reduced risk of onshore transport.

16.12.2.2 Measurements and Calculations

The dilution achieved in ambient transport of waste stream plumes in large waterbodies can be described by physical laws—essentially the same laws used for describing aqueous flow in groundwater and gaseous flow in the atmosphere. However, we cannot solve the equations completely for the general case, because we lack existing data. That is, data obtained from field studies or from reports of previous studies are needed for empirical coefficients in the equations. However, many computer programs are available to obtain *approximate* solutions of the equations for complex cases involving multiple waste inputs and variable current speeds. In addition, we can sometimes use simplifications of the governing equations for many pollution assessment problems to obtain satisfactory estimates of contaminant concentration as a function of travel time or distance.

One simplified equation that has been used successfully over the past 30 years for large bodies of water gives us the maximum concentration at a distance X:

$$C_{max} = C_{pi} \, \mathrm{erf} \sqrt{\frac{Ub^2}{16\varepsilon_o X}} \qquad (16.8)$$

where

c_{max} = centerline (maximum) concentration at distance X
c_{pi} = plume concentration at the end of initial dilution

erf (#) = standard error function of (#)

U = current velocity in the X direction

b = width in the Y direction (orthogonal to X) at the end of initial dilution

ϵ_o = constant horizontal (Y direction) eddy diffusivity

X = travel distance [note that U/X can be replaced by $1/t$ (time)].

In using this equation, it is important to use values that are expressed in consistent units. For example, the parameters U, b, and ϵ_o all contain a time unit. Another way to appreciate this requirement is to recognize that the argument *arg* of the error function must be dimensionless. The standard error function (erf) serves here as a mathematical representation of the way contaminants are observed to vary laterally (Y-direction) as the plume is transported in the X-direction. It describes the normal distribution curve used in evaluating variance around a mean value. Values can be found from a tabular listing in a handbook or by using the standard error function in a spreadsheet program. The *transport dilution factor* is equal to the reciprocal of the value of erf(*arg*).

In a typical problem involving finding the transport dilution factor, we might estimate the highest concentration that would occur near the beach from an outfall 8 km (i.e., 8×10^5 cm) offshore when the onshore current velocity U is 15 cm s^{-1}. The width b of the plume after initial dilution, which would be obtained from local observations, would likely be about 1000 m (i.e., 10^5 cm). We can use a commonly cited value for the eddy diffusivity of $\epsilon_o = 10^4$ cm^2 s^{-1}. Now substituting these values into Eq. (16.9), we get:

$$c_{max} = c_{pi} \operatorname{erf} \sqrt{1.17} = 0.87 \, c_{pi} \qquad (16.9)$$

The result is typically surprising: It shows that even with a travel distance of 8 km in a large, open body of water, the concentration is diluted to only 87%, resulting in a transport dilution factor of 1.2. (Contrast this factor of 1.2 to an initial dilution factor of about 100, which is what we expect for an ocean outfall.)

This example shows us that initial dilution is often more important in reducing harmful levels than is dilution due to transport. This is true not only for BOD and DO problems, but also for most contaminants, including metals, ammonia, and toxic organics, that are found in partially treated effluents. However, in the case of indicator bacteria, the value of transport distance is realized, as demonstrated in the following example.

Assume an outfall whose degree of treatment is only 90% effective, so that instead of the 100,000 coliforms per 100 mL used in the example on the freshwater die-off of indicator organisms, the initial count of coliforms is 100,000,000 per 100 mL. The initial dilution of 100:1 would reduce the concentration to 1% of the initial count, or

1,000,000. Then, using the reciprocal of the transport dilution factor found before (0.87), we can calculate that transport to the beach would further reduce the count to $C_0 = 870,000$ per 100 mL. The 8-km distance would be covered in $t = 14.8$ h at a velocity of 15 cm s^{-1}. If the die-off rate constant K is found to be above 0.46 per hour, near the usually expected lower range of values, the beach concentration of bacteria ($C_t = 14.8$), after substituting the values for C_0, K, and t into Eq. (16.3), would be approximately 960 per 100 mL.

Even with some form of secondary treatment, regulatory agencies in the United States usually require disinfection to reduce bacterial concentrations to bathing water standards.

16.13 DYE TRACING OF PLUMES

Frequently, an oceanographic study is necessary to measure the bacterial concentration in the drifting plume as the current carries the water toward shore. Such studies usually employ a tracer, or dye, which is added to the sewage so that the plume can be followed for several kilometers. The data obtained can then be used to calculate a die-off rate constant, which may be useful for predicting the bacterial concentrations at different distances under a variety of current conditions.

Dye tracing (Fig. 16.26A) is a well-known technique commonly used in hydraulic models and prototype field settings, although in deep outfall situations, tracers can be quite costly because of the large volumetric flow rates and large dilutions usually achieved within a short time frame. The rate of dye addition Q_d to the effluent flow V_e needed to provide a dye concentration of C_d following dilution of S_a is

$$Q_d = \frac{V_e \, C_d \, \alpha_a \, S_a}{W \alpha_d} \qquad (16.10)$$

where

α_a = specific gravity of the diluted plume

α_d = specific gravity of the dye solution

W = weight fraction of dye in stock solution.

Fig. 16.26B shows the required dye rate in liters per hour for various dilution factors, and effluent flows in million liters per day, to achieve an ambient dye concentration of 1 µg L^{-1} in seawater. Rhodamine WT, typically used in dye studies, is available as a 20% solution ($\infty_d = 1.19$) in small (57-L) drums. Fluorometers used in field sampling can easily detect this dye at concentrations of 0.5–1 µg L^{-1}.

We can use Fig. 16.26B to estimate the amount of dye needed to trace an effluent flow in a waterbody or any similar aquatic mixing question. If the flows marked on the x-axis of the graph and the slanted lines representing the dilution factors do not match exactly with the problems we have, Eq. (16.10) may be used to refine the estimate.

(A)

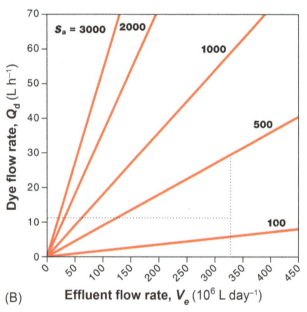

(B)

FIG. 16.26 (A) A surfacing plume dyed with rhodamine WT of partially treated sewage offshore from San Francisco, California. The dye serves as a tracer for monitoring bacterial counts in drifting sewage plumes. Two monitoring vessels are visible. Photo courtesy W. Smith. (B) A graph used to determine the dye required to provide $1 \mu g L^{-1}$ in diluted effluent as used in the study in (A). The dotted lines illustrate using the graph to solve the problem in the text. *(From* Pollution Science *©1996, Academic Press, San Diego, CA.)*

Suppose we have an effluent of 330 million liters (86.7 MGD) per day. A regulatory permit requires the effluent to be diluted by a factor of 200 at the end of a mixing zone. Suppose we set up our sampling boat at the mixing zone boundary and hope to measure $1 \mu g L^{-1}$ of dye as the plume passes under our boat. How much dye do we need to add to the effluent? Using the graph, estimate 330 along the abscissa and draw a line up to the dilution factor line $S_a = 500$. Now estimate where $S_a = 200$ would lie on that line. From that point draw a line horizontally to the ordinate and estimate the dye requirement as $12 L h^{-1}$.

For a more precise estimate, we will use Eq. (16.10). First convert $V_e = 330$ million liters per day (i.e., $330 \times 10^6 L\,day^{-1}$) to 13.7×10^6 liters per hour ($L\,h^{-1}$). For C_d, use 10^{-9} g dye per gram of seawater (this is approximately equal to $1 \mu g L^{-1}$, assuming a specific gravity of 1). The specific gravity of sewage effluent diluted 100:1 with seawater is $\infty_a = 1.023$. (If this were a discharge to fresh water, the specific gravity would be 1.0.) Using $S_a = 200$, $W = 0.2$, and $\infty_d = 1.19$ for rhodamine WT, as cited before, and substituting into Eq. (16.10), we obtain $Q_d = 12 L h^{-1}$ (rounded from 11.8), verifying our estimate from Fig. 16.26B.

16.14 SPATIAL AND TEMPORAL VARIATION OF PLUME CONCENTRATIONS

The concentrations of water quality indicators are neither uniform nor steady with respect to the space and time scales involved in regulating the concentrations at the end of the mixing zone. In general, we assume that the concentrations of constituents in the horizontal extent of a plume from an outfall diffuser are uniform. But we can make no such assumption about the vertical direction. Vertical nonuniformity is commonly encountered in design, performance analysis, and compliance monitoring, although in rivers it is not nearly the problem it is in estuaries, coastal water, and some lakes and reservoirs. Generally associated with density stratification in the receiving water, vertical nonuniformity is also associated with transport of a plume in a relatively thin lens as compared to the depth of the water column. For instance, if the plume is traveling on the surface, its constituents will be dispersed downward, and as these constituents disperse into the water column, the concentration of pollutants near the bottom edge of the plume gradually becomes less than that at the surface. (Thus if a permit condition requires that a maximum value be reported, sampling should be done at the surface, not at mid-depth.) Similarly, the dilution water mixed with the effluent being discharged is also vertically variable due to physical processes influencing the advection of ambient water into the region of the discharge. Dissolved oxygen (DO) is an example of one water quality indicator that exhibits vertical nonuniformity in many riverine impoundments (reservoirs), lake, estuarine, and coastal situations.

Some transport and dispersion models produce estimates in terms of the *centerline concentration*, which is the maximum concentration for the cross section of the plume at a given distance downstream from the orifice. As the plume width expands with increasing distance, the maximum concentration progressively decreases. For example, the centerline (maximum) concentration at a distance of 60 m from the diffuser may be $100 mg L^{-1}$, while at 120 m from the orifice, the maximum concentration would be closer to $70\,mg\,L^{-1}$. Other models calculate an average concentration for the cross section of the plume, and this of course also decreases downstream: the average concentration is always smaller than the maximum concentration. Both values need to be considered in field or lab verification studies, and both values may be useful for regulatory purposes.

16.15   COMPLIANCE MONITORING

Water pollution regulatory practice in the United States is founded on a system of discharge permits—known as the National Pollution Discharge Elimination System (NPDES) permits (see Chapter 30). Holders of these permits, for example, municipal sewage treatment authorities and industries, must comply with the restrictions and requirements of their particular permit, such as limits on concentrations and mass emission rates of specific constituents. They also have to meet water quality standards established for the waterbody into which they discharge their effluents. Some permits, especially for coastal water discharges, require elaborate environmental monitoring projects. The permit holder is required to conduct monitoring activities and report the results to demonstrate compliance with permit conditions. On occasion, regulatory agencies conduct studies to verify and revise ongoing programs. Monitoring data reflect the wide variations of conditions found in the natural environment, and dischargers and regulators are often challenged to rationalize monitoring results with predictions used in setting permit conditions.

16.15.1   Mixing Zones

Permit conditions of regulatory agencies usually allow exceptions to one or more of the water quality criteria within a mixing zone adjacent to the point of discharge. A *mixing zone* might be established by purely arbitrary considerations or by use of data and simulations with mathematical models. Many mixing zone determinations are made on the basis of the expected dilution rate that will be provided by efficient diffuser designs intended to optimize initial dilution. For large bodies of water, a common approach is to describe the width of the zone as the depth of water at the disposal site and the length as the length of the diffuser. For large rivers, it is common to restrict a mixing zone so it does not extend completely across the river, thereby leaving a "safe passage" that lets aquatic species avoid high concentrations of wastewater constituents. But sometimes the shape of a mixing zone is entirely arbitrary, say, a rectangular zone downstream from a discharge pipe equal to one-fourth the width of the stream and extending downstream for one kilometer. Frequently, there are two or three mixing zones for different groups of contaminants and degrees of toxicity.

16.15.2   Regulatory Use

Regulatory interest may be appropriately directly toward both discrete and average values of contaminants. For example, the state of California and the U.S. EPA specify maximum allowable instantaneous values for some parameters as well as several temporal average values (e.g., 30-day and 6-month arithmetic means). In some cases, these regulations are based on knowledge of the effects on aquatic organisms. In other cases, these values are specified to acquire statistics on the performance of the wastewater treatment plant.

Criteria that are expressed in terms of temporal averages (daily to semiannual) suggest that plume concentrations be assessed extensively in three dimensions, both at the boundary of the mixing zone and, in some cases, at sensitive biological resource locations down current. Current speed and direction play significant roles when assessing the concentrations at the boundary. By incorporating data on the cyclical variation of effluent composition, density profiles, and current direction, it is possible to construct a running 6-month average (or median) for a number of points on the mixing zone boundary. The 6-month average is expected to be quite variable at these points, and the point with the highest exposure frequency may not have the highest average concentration.

Beyond the mixing zone, there may be regions where current streams of diluted effluent, each leaving the zone at a different time in different direction, would converge over a reef, a kelp forest, or a swimming area. In this case, the frequency and duration of exposure may be more important than the highest observed concentration in assessing the overall impact on these resources.

16.15.3   Verification Sampling

Aside from the question of whether discrete values or cross-sectional averages are used to test compliance with criteria, the way in which field samples are used to verify or compare with model results is an important consideration.

In laboratory or field verification studies of plume performance, the average value is measured or captured in a sample bottle only by chance. Characteristically, the field value measured is from a very small spatial region and represents a signal over a certain time span. Many samples are sought from the same cross section in order to arithmetically compute an average. In the laboratory, using hydraulic models, this is relatively easy to do. But in the field, where multiple plumes are usually involved, sampling is more complicated. We are usually trying to take samples from a moving flow field too deep below the surface to see, using a moving sampler mounted on a moving boat. It is therefore reasonable to assume some uncertainty as to what portion of the cross section the value represents.

For these reasons, field verification studies of submerged plumes in deep rivers, lakes, and coastal waters are best attempted for a cross section as far from the outfall as practical, as long as the region is still within the range where the plume is continuous. Nearer to the outfall, the values are changing more readily and the dimensions of the plume are much smaller, making it much harder to get the sampler in the right place or even in the plume. In addition, it is best to conduct the study when currents are low, so that the plume rises nearest to the surface. Placement of the sampling device may be improved because it may even be possible to see the plume. Aside from the ease of sampling, samples taken during low currents may be especially useful for verification of regulatory compliance.

QUESTIONS AND PROBLEMS

1. Calculate the time required for the coliform count to diminish to 1000 per 100 mL (the bathing water criterion) from a sewage discharge containing 10^7 per 100 mL. The die-off coefficient, K, is expected to be 1.5 per day.

2. Suppose currents in the receiving body of water average 0.8 km per hour. How far away from the bathing area must the discharge be located?

3. The example BOD curve in the chapter shows a BOD value of 6.1 mg L^{-1} at day 5. It looks like the ultimate BOD (BOD$_L$) is about 7.6 mg L^{-1}. We solved the BOD equation using these values to find that the value of K was about 0.32 per day. Suppose the ultimate value was 15. What would be the value at day 5?

4. Suppose another waste has an ultimate value of 15, but the waste is more readily degradable. Would the value at day 5 be lower or higher? Why?

5. Calculate the oxygen concentration in the receiving body of water at 1, 2, 4, and 8 days caused by discharge of partially treated sewage with a BOD of 30 mg L^{-1} and a decay coefficient of 0.1 per day. The reaeration coefficient is 0.3 per day. This is a pristine body of water with no initial deficit in the region of discharge. Consequently, the ambient water is essentially saturated at 10 mg L^{-1} of dissolved oxygen.

REFERENCES

Curriero, F.C., Patz, J.A., Rose, J.B., Lele, S., 2001. The association between extreme precipitation and waterborne disease outbreaks in the United States, 1948–1994. Am. J. Public Health 91, 1194–1199.

Maybeck, M., 1982. Carbon, nitrogen and phosphorus transport by world rivers. Amer. J. Sci. 282, 401–450.

Rigler, F.H., 1973. A dynamic view of the phosphorus cycle in lakes. In: Griffith, E.J., Beeton, A., Spencer, J.M., Mitchell, D.T. (Eds.), Environmental Phosphorus Handbook. John Wiley & Sons, New York.

Smith, J.E., Perdek, J.M., 2004. Assessment and management of watershed microbial contaminants. Crit. Rev. Environ. Sci. Technol. 34, 109–139.

FURTHER READING

Calow, P., Petts, G.E., 1992. The Rivers Handbook. vol. 1. Blackwell Scientific Publishers, Oxford, United Kingdom.

Harwood, V.J., Staley, C., Badgley, B.D., Borges, K., Korajkic, A.A., 2014. Microbial source tracking markers for detection of fecal contamination in environmental waters: relationships between pathogens and human health outcomes. FEMS Microbiol. Rev. 38, 1–40.

Scott, T.M., Rose, J.B., Jenkins, T.M., Farrah, S.R., Lukasik, J., 2002. Microbial source tracking: current methods and future directions. Appl. Environ. Microbiol. 68, 5796–5803.

Chapter 17

Atmospheric Pollution

M.L. Brusseau, A.D. Matthias, A.C. Comrie and S.A. Musil

Smelter-produced air pollutants trapped by an inversion layer. *Photo courtesy J.F. Artiola.*

17.1   AIR POLLUTION CONCEPTS

In this chapter, we discuss air pollutants, including their sources and effects on human activity, as well as their transport to, and fate in, the atmosphere. We also describe the role of air pollution in such major environmental issues as global climate change and stratospheric ozone depletion that are covered in Chapter 25.

An *air pollutant* is any gas or particulate that, at high enough concentration, may be harmful to life, the environment, and/or property. A pollutant may originate from natural or anthropogenic sources, or both. Pollutants occur throughout much of the troposphere (see Chapter 4); however, pollution close to the earth's surface within the boundary layer is of most concern because of the relatively high concentrations resulting from sources at the surface.

Atmospheric pollutant concentrations depend mainly on the total mass of pollution emitted into the atmosphere, together with the atmospheric conditions that affect its transport and fate. Obviously, air pollution has many and varied sources, including cars, smokestacks, and other industrial inputs into the atmosphere as well as wind erosion of soil. Large emissions from both anthropogenic and natural sources over long periods enhance concentrations, as do the chemical and physical properties of these pollutants. For example, when nitrogen oxides and hydrocarbons in car exhaust are emitted into warm, sunlit air,

they readily form ozone molecules (O_3). Similarly, the solubility of a pollutant affects how efficiently it is removed by rainfall.

Atmospheric conditions have a major effect upon pollutants once these pollutants are emitted into (e.g., nitrogen oxides from car exhaust) or formed within (e.g., O_3) the atmosphere. Pollution dispersal is controlled by atmospheric motion, which is affected by wind, stability, and the vertical temperature variation within the boundary layer. Stability, in turn, influences both air turbulence and the depth at which mixing of polluted air takes place.

Wind determines the horizontal movement of pollution in the atmosphere. Pollution emitted from a point source, such as a smokestack, is generally dispersed downwind in the form of a *plume* (see Chapter 7). Wind speed establishes the rate at which the plume contents are transported. Strong winds flowing over rough land surfaces enhance mixing of air by producing shear stress (mechanical mixing) much like that created when an electric fan circulates air in a room. Also, wind direction establishes the path followed by the pollution.

Once present in the atmospheric boundary layer, a pollutant may undergo a series of complex transformations leading to new pollutants, such as O_3. Also, the removal of pollutants from air by rain and snow, by gravity, or by surface deposition is influenced by boundary-layer conditions. These removal processes, in turn, are also affected by the type and roughness of the underlying ground surface.

Environmental and Pollution Science. https://doi.org/10.1016/B978-0-12-814719-1.00017-3
Copyright © 2019 Elsevier Inc. All rights reserved.

Even when emissions are relatively constant, pollutant concentrations can quickly change, owing to variations in atmospheric conditions. When atmospheric conditions are stable, relatively low emissions can cause a buildup of pollution to hazardous levels. This situation can occur during radiation inversions at night (see Section 4.2.2). In contrast, unstable conditions may effectively dilute pollution to relatively "safe" concentrations despite a fairly high rate of emissions.

Air pollution, which is of major public concern (Chapter 26), is currently the object of extensive scientific research. Its effects on life, including human health, productivity, and property, are not yet fully understood, even though exposure to high levels of pollution is a daily experience for many people. The cost of such pollution, whether expressed in terms of direct biological consequences or in terms of economic impact, is enormous. Worldwide, urban air pollution affects nearly a billion people, exposing them to possible health hazards. In the United States alone, billions of dollars are spent annually to prevent, control, and clean up air pollution; other developed nations are incurring similar costs. The United Nations considers air pollution to be a major global problem.

Most commonly, air pollution poses a health risk that can and does harm life. It harms the human respiratory and pulmonary systems. Emphysema, asthma, and other respiratory illnesses may result from or be aggravated by chronic exposure to certain pollutants, such as O_3 or particulates. Research conducted in Southern California indicates that breathing polluted air slows lung development in children as much as having a parent who smokes tobacco. Teenagers who are chronically exposed to polluted air are five times as likely to have reduced lung function as are teens who breathe clean air. Children and adults with chronic exposure to elevated ozone levels are more likely to develop asthma than individuals in cleaner air (see Information Box 17.1). Breathing polluted air can also thicken human artery walls, which is an important risk factor for heart failure and strokes. In 1991, researchers at the U.S. Environmental Protection Agency (EPA) concluded that about 60,000 U.S. residents die each year due to heart attacks and respiratory illness caused by breathing particulates (dust) at concentrations within the federal PM_{10} (see Section 17.2.1.3) air quality standard.

Air pollution also poses an ecological risk. Vegetation can be harmed by uptake of pollutants through the leaf stomata or by deposition of pollutants on the leaf surfaces. Sufficiently high concentrations of sulfur dioxide or ozone, for example, may cause leaf lesions in susceptible plants. Chronic exposure to relatively low levels of pollution can harm plants by reducing their resistance to disease and insect predators. Crop yields can be lowered by air pollution. The presence of certain air pollutants, such as ozone

INFORMATION BOX 17.1 Asthma and Air Pollution

There is much concern about the possible adverse health effects of air pollution on respiratory health, especially of children within inner-city areas. The U.S. EPA is currently conducting studies of the potential role that several air pollutants may have in inducing and/or exacerbating asthma or asthma symptoms.

The incidence of asthma has increased dramatically in the United States during recent decades, the causes of which are not well understood. Increased air pollution in the form of complex mixtures of particulate matter, metals, and tobacco smoke is thought to play an important role in the increase. Indoor air quality is a particular concern (see Chapter 18). Another class of hazardous chemicals that may have an important role is the carbonyl compounds, such as aldehydes and ketones. They can originate from industrial sources and from the burning of diesel fuel. They also form within the troposphere during complex photochemical processes involving organic compounds.

(formed from nitrogen oxides and hydrocarbons), enhances the earth's natural greenhouse effect within the troposphere. This enhancement warms the earth and may change rainfall patterns, which could markedly alter the distribution of life on earth (see Chapter 25).

Air pollution can also damage property. It can erode the exterior surfaces of buildings, particularly those constructed of limestone materials that react with acids in precipitation (see Chapter 4). Further evidence of the deleterious effects of air pollution on property can be seen in the damaged paint finishes on cars regularly parked downwind from ore smelters.

17.2 SOURCES, TYPES, AND EFFECTS OF AIR POLLUTION

Air pollution is not a new problem. Lead in Swedish lake sediments indicates that air pollution produced from lead mining and silver production in ancient Greece and Rome affected air quality throughout Europe. Early written accounts of air pollution refer mainly to smoke from burning wood and coal. For example, in the 13th century, King Edward I of England prohibited the use of sea coal, the burning of which produced large amounts of soot and sulfur dioxide (SO_2) in the atmosphere over London. The Industrial Revolution increased pollution so markedly that air quality deteriorated significantly in Europe and North America. By the mid-19th century, many cities in the United States and Europe were experiencing the consequences of air pollution. By the early 20th century, the term "smog" was coined to describe the adverse combination of smoke and fog in London. In Los Angeles, photochemical

INFORMATION BOX 17.2 Air Pollution in China

China is the second largest energy user in the world, due in part to its rapid industrialization. In 2002, about two-thirds of China's energy came from burning coal, which tends to release large amounts of soot and SO_2, depending on the quality of the coal. This is particularly troublesome, since much of the coal emissions are from older industrial plants with poor pollution controls, as well as from domestic use (individuals using coal to cook and heat homes). The high levels of domestic use create numerous small point sources, low to the ground, that are difficult to control or abate (see Section 17.3 for more information on air pollution dispersion). As a result, some cities are establishing "coal-free zones" and are attempting to increase the use of cleaner burning natural gas.

The World Health Organization noted in a 1998 report that seven of the ten most polluted cities in the world are in China. The government has begun to implement pollution reduction programs, but they face enormous problems, including acid rain, which is estimated to fall on about 30% of China's territory. The government is trying to improve enforcement of existing laws, as well as instituting fines on polluters, designing systems for emissions trading (already in use in North America and Europe), and focusing on new technology to reduce energy use and pollution.

FIG. 17.1   A smokestack contributing to smog in Tianjin, China. *(Photo courtesy J. Walworth.)*

smog alerts became common by the mid-1940s. The first major air pollution disaster in the United States occurred in 1948, when approximately 20 lives were lost as a result of industrial pollutants trapped in very stable air over Donora, Pennsylvania, in the Monongahela River Valley. During one week in December 1952, stagnant air and coal burning caused severe smog conditions in London that ultimately took the lives of nearly 12,000 people over a three-month period.

Virtually all metropolitan areas are affected by air pollution, especially those situated in valleys surrounded by mountains (e.g., Mexico City) or along coastal mountain ranges (e.g., Los Angeles). Air pollution has recently become a critical issue for cities in countries experiencing rapid economic development, such as China and India (see Information Box 17.2 and Fig. 17.1 for an example from China). But even unpopulated areas far from cities may be affected by long-range transport of pollution, either from urban areas or from such rural sources as ore smelters or coal-burning power plants. For example, pollution from a coal-burning power plant in northern Arizona reduces visibility in Grand Canyon National Park, located 400 km west of the plant.

Most of the air we breathe is elemental oxygen (O_2) and nitrogen (N_2) (see Chapter 4). About 1% is composed of naturally occurring trace constituents such as carbon dioxide (CO_2) and water vapor. A small part of this 1%

may, however, be air pollutants, including gases and *particulate matter* suspended as aerosols. Anthropogenic air pollution enters the atmosphere from both fixed and mobile sources. Fixed sources include factories, electrical power plants, ore smelters, and farms, while mobile sources include all forms of transportation that burn fossil fuels. Mobile sources account for 56% of the pollutants emitted to the atmosphere in the United States (Fig. 17.2A). Fuel combustion from stationary sources accounts for nearly 15%, and industrial processes account for about 7% of emissions in the United States. Natural sources of air pollution include winds eroding dust from cultivated farm fields, smoke from forest fires (Fig. 17.3), and volcanic ash that is emitted into the troposphere and stratosphere (see Information Box 4.1).

There are many types of air pollutants. Some gases, such as CO_2, produced by burning fossil fuels, are central to the issue of global climate change but are also essential to plant life. Many pollutants, such as dust particles, exist naturally in the atmosphere and become hazardous only when their concentrations exceed air-quality standards set by such regulatory agencies as the U.S. EPA. The EPA classifies air pollutants according to two broad categories: primary and secondary air pollutants.

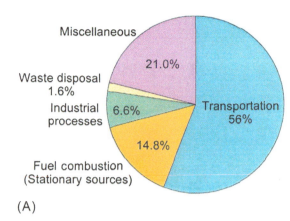

(A)

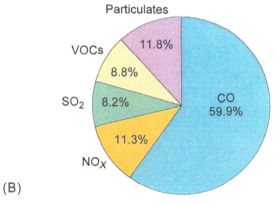

(B)

FIG. 17.2 (A) Source contributions of primary air pollutants in the United States in 2002 and (B) the percentage of total primary air pollutants by type. *(Data: U.S. Environmental Protection Agency.)*

FIG. 17.3 A natural source of atmospheric pollution. The "Aspen Fire" burned for weeks in the Catalina Mountains near Tucson, Arizona, releasing large amounts of particulates and other air pollutants. *(Photo courtesy J.F. Artiola.)*

17.2.1 Primary Pollutants

Primary air pollutants enter the atmosphere directly from various sources. The EPA designates six types of primary air pollutants for regulatory purposes:

- carbon monoxide
- hydrocarbons
- particulate matter
- sulfur dioxide
- nitrogen oxides
- lead

17.2.1.1 Carbon Monoxide

Carbon monoxide (CO), which is the major pollutant in urban air, is a product of incomplete combustion of fossil fuels. Carbon monoxide has relatively few natural sources. It is a part of cigarette smoke, but the internal combustion engine is the major source, with about 50% of all CO emissions in the United States originating from cars and trucks. Emissions, therefore, are highest along heavily traveled highways and streets (Fig. 17.4). Of the EPA-designated primary pollutants in the United States, CO emissions currently contribute about 60% of the total emissions (see Fig. 17.2B). Fortunately, CO concentrations are decreasing in the United States, because newer cars have higher fuel efficiencies.

Carbon monoxide is highly poisonous to most animals. The EPA standards currently limit human exposure to a 24-h average of $9\,nL\,L^{-1}$ or a 1-h average of $35\,nL\,L^{-1}$. (Note: The alternative unit of parts per billion (ppb) is also commonly used. See Chapter 4 for more information on units for describing gas concentrations.) When inhaled, CO reduces the ability of blood hemoglobin to attach oxygen. Although relatively stable, it is short-lived in the atmosphere because it is quickly oxidized to CO_2 by reaction with hydroxide radicals. Some atmospheric CO may be removed by soil microbes. In order to increase the oxidation of CO to CO_2 during fuel combustion, some cities require the use of oxygenated gasoline containing ethanol or other additives during winter months.

FIG. 17.4 Emissions of CO and nitrogen oxides are highest along heavily traveled highways and streets. *(Photo courtesy S.A Musil.)*

17.2.1.2 Hydrocarbons

Hydrocarbons (HCs), or *volatile organic compounds* (VOCs), are compounds composed of hydrogen and carbon. *Methane* (CH_4), the most abundant hydrocarbon in the atmosphere, is an active greenhouse gas. Volatile organics include the *nonmethane hydrocarbons* (NMHCs), such as benzene, and their derivatives, such as formaldehyde. Some of these compounds (e.g.,benzene) are carcinogenic, and some relatively reactive HCs contribute to ozone production in photochemical smog.

Hydrocarbons are produced naturally from decomposition of organic matter and by certain types of plants (e.g., pine trees, creosote bushes). In fact, HCs emitted from vegetation may be a major factor in smog formation in some cities, particularly those near forested areas of the southeastern United States. A large proportion of HCs and NMHCs are generated by human activity. Some NMHCs, including formaldehyde, are readily emitted from indoor sources, such as newly manufactured carpeting. Hydrocarbons are also emitted into the atmosphere by fossil fuel combustion and by evaporation of gasoline during fueling of cars. To mitigate this latter source, some municipalities require that service-station gasoline pumps be fitted with a special trap to collect HC vapors emitted during fueling of vehicles. Because transportation is the primary source of HCs, concentrations tend to be highest near heavily traveled roadways.

17.2.1.3 Particulate Matter

The category of particulate matter comprises solid particles or liquid droplets (aerosols) small enough to remain suspended in air. Such particles have no general chemical composition and may, in fact, be very complex. Examples include soot, smoke, dust, asbestos fibers, and pesticides, as well as some metals (including Hg, Fe, Cu, and Pb). We can characterize particulate matter by size. Particles whose diameters are $10 \mu m$ or larger generally settle out of the atmosphere in less than a day, whereas particles whose diameters are $1 \mu m$ or less can remain suspended in air for weeks. Smaller particulate matter, whose particles are $10 \mu m$ or less, have come to be known as PM_{10}. The very small particles with diameters $2.5 \mu m$ or less are known as $PM_{2.5}$ (see Information Box 17.3 and Chapter 11).

INFORMATION BOX 17.3 Aerosols and Visibility (See also Chapter 11)

Aerosols are solid or liquid particles suspended in a gas. In relation to environmental pollution, our major concern is with microscopic aerosol particles produced primarily from combustion or windblown dust, or secondarily from gas-to-particle conversion. Aerosols in the atmosphere typically fall into two distributions by mass or size, with coarse particles in the range 2.5–10 μm and fine particles from about 0.1 to 2.5 μm. The coarse fraction is usually composed of soil dust including minerals and organic particles. Fine particulate matter <2.5 μm ($PM_{2.5}$) comprises a range of combustion products such as elemental carbon, sulfates, and nitrates. Aerosols are important constituents of acid rain, and they can also alter the atmospheric radiation balance. Because their small size allows them to penetrate deep into the lungs, aerosols have important health effects, and therefore the EPA has established health-based standards for $PM_{2.5}$ and PM_{10}.

Visibility impairment, especially at the regional scale, is typically caused by fine aerosols, especially <1 μm, as this size range tends to remain suspended longest in the atmosphere (coarser particles may settle out and ultrafine particles can be removed as condensation nuclei in rain-out) (see Fig. 17.5). These fine particles also scatter relatively large amounts of light, leading to greater haziness and decreased visibility. Visibility impairment has become a problem in many areas, not only in the eastern United States, but also in relatively remote places such as the Grand Canyon. The Western Regional Air Partnership is a coordinated effort by the Western Governors' Association and the National Tribal Environmental Council to develop data, tools, and policies needed by states and tribes to improve visibility in parks and wilderness areas across the western United States.

(A)   (B)

FIG. 17.5 Photos taken of Mount Trumbull in the Grand Canyon National Park on a "clear" (A) day and a "hazy" (B) day at noon. The figure to the right clearly shows the impairment of visibility due to aerosols, some of which originate in California urban areas. The National Park Service has a visibilitymonitoring program that systematically collects photos at numerous parks across the United States. (*Photos courtesy National Park Service.*)

The effects of particulates in the air are various. Some particulates, especially those containing sulfur compounds, are emitted by volcanoes. These particulates can reach the stratosphere, where they may significantly alter the radiation and thermal budgets of the atmosphere and thus produce cooler temperatures at the earth's surface (see Chapter 4). Tropospheric particulates may cause or exacerbate human respiratory illnesses. Especially harmful to the human respiratory system is the fraction of mid-sized particles, PM_{10} and $PM_{2.5}$. In large cities, particulates also reduce visibility. In the United States in 2002, about 62% of particulates came from roads and transportation, with another 26% contributed by agriculture, forestry, and fires (Fig. 17.6). Construction is now considered to make a large contribution to PM_{10} levels.

17.2.1.4 Sulfur dioxide

About 90% of *sulfur dioxide* (SO_2) emissions come from burning sulfur-containing fossil fuels, such as coal, which may contain up to 6% sulfur. Ore smelters and oil refineries also emit significant amounts of SO_2. At relatively high concentrations, SO_2 causes severe respiratory problems. Sulfur dioxide is also a source of acid rain, which is produced when SO_2 combines with water droplets to form sulfuric acid (H_2SO_4). At sufficiently high concentrations, SO_2 exposure is harmful to susceptible plant tissue. Sulfur dioxide and other tropospheric aerosols containing sulfur are believed to affect the radiation balance of the atmosphere, which may cause cooling in certain regions. See Chapter 4 for information about the contribution of SO_2 to acid rain.

17.2.1.5 Nitrogen Oxides (NO_x)

Nitrogen oxides (NO_x stands for an indeterminate mixture of NO and NO_2) are formed mainly from N_2 and O_2 during high-temperature combustion of fuel in cars. Catalytic converters are used to reduce emissions. Nevertheless, NO

FIG. 17.6 Agricultural crops and livestock contributed about 19% of particulate matter in 2002 in the United States. *(Photo: College of Agriculture and Life Sciences, The University of Arizona.)*

causes a reddish-brown haze in city air that contributes to heart and lung problems and may be carcinogenic. Nitrogen oxides also contribute to acid rain because they combine with water to produce nitric acid (HNO_3) and other acids. Natural sources of nitrogen oxides include those produced during the metabolism of certain soil bacteria.

17.2.1.6 Lead

Lead is highly toxic, and its effects on humans have been recognized since the Roman Empire era. Lead can produce chronic impairment of the formation of blood and it affects infant neurological development (see Chapter 26). Concentrations of lead in the environment in the United States are no longer as high as they were prior to the introduction of nonleaded gasoline in the 1970s, but it is still a concern in many localities. Lead from human sources may be present in soil and it is often found in particulate matter in older urban environments. Lead-based paints continue to be a source of concern in situations where children are exposed to paints that have peeled from building surfaces. Lead-based paint was banned in the United States in 1978. Homes built before 1978 may have lead-based paint that can be a health hazard when sanded, chipped, or removed.

17.2.2 Secondary Pollutants

Secondary air pollutants are formed during chemical reactions between primary air pollutants and other atmospheric constituents, such as water vapor. Generally, these reactions must occur in sunlight; thus they ultimately produce *photochemical smog* (see Information Box 17.4). Photochemical smog is most common in the urban areas where solar radiation is very intense.

A simplified set of some of the reactions involved in photochemical smog formation is given as follows:

$$
\begin{aligned}
N_2 + O_2 &\rightarrow 2NO && \text{(inside an engine)} \\
2NO + O_2 &\rightarrow 2NO_2 && \text{(in the atmosphere)} \\
NO_2 + h\nu &\rightarrow NO + O && \text{(in the atmosphere)} \\
O + O_2 &\rightarrow O_3 && \text{(in the atmosphere)} \\
NO + O_3 &\rightarrow NO_2 + O_2 && \text{(in the atmosphere)} \\
HC + NO + O_2 &\rightarrow NO_2 + PAN && \text{(in the atmosphere)}
\end{aligned}
$$

$$(17.1)$$

As indicated by the reactions in Eq. (17.1), photochemical smog is composed mainly of O_3, *peroxyacetyl nitrate* (PAN), and other oxidants. Ozone formation is closely tied to weather conditions. Favorable conditions for O_3 formation include:

- air temperatures exceeding 32°C
- low winds
- intense radiation
- low precipitation

INFORMATION BOX 17.4 Photochemical Smog in Los Angeles

Photochemical smog can be severe in the Los Angeles basin of the California coast (see Fig. 17.7). Commuting in Los Angeles requires many cars, which produce high emissions of NO_x and hydrocarbons. At certain times of the year, particularly spring and fall, weather conditions in this area are dominated by subtropical high pressure with clear, calm air conditions that exacerbate air stagnation. The factors influencing smog formation in the Los Angeles basin can be summarized as follows:

- Numerous sources of primary pollutants.
- Inversions that inhibit turbulent mixing of air.
- Few clouds, which result in higher UV intensity.
- Light winds that are unable to disperse pollutants.
- Complex coastal mountain terrain that slows pollutant dispersal.

FIG. 17.7 Hydrocarbons interact with nitrogen oxides under the influence of ultraviolet light, resulting in photochemical smog. In urban centers such as Los Angeles, pictured here, atmospheric pollutants can concentrate and pose severe health hazards. *(Photo: U.S. Environmental Protection Agency.)*

Unfortunately, many major U.S. cities exceed the federal air-quality standard for O_3 (an average O_3 concentration $>80 nL L^{-1}$ for 1h 1 day per year averaged over a 3-year period).

As the reactions indicate, HCs are necessary for ozone buildup in the atmosphere. In the absence of HCs, solar UV breaks down the NO_2 into NO and O. Next, the O atom combines with O_2 to form O_3, which then combines with the NO to reform NO_2 and O_2. Ozone would not accumulate in the atmosphere if it was not for the fact that HCs disrupt the reaction cycle by reacting with NO to form $PAN+NO_2$. Hydrocarbons from car emissions and other sources, therefore, play an important part in O_3 formation in urban environments. However, not all O_3 in the lower atmosphere results from human activities; natural sources include lightning and the diffusion of some O_3 downward from the stratosphere.

In most western U.S. cities, photochemical smog is often referred to as *brown cloud* $(O_3+PAN+NO_x)$. Industrial eastern and midwestern U.S. cities also have photochemical smog, but they generally receive less intense sunlight than western cities; thus smog in those cities is sometimes referred to as *gray air* because of particles (especially $PM_{2.5}$) and SO_2 emanating from burning fossil fuel.

Ozone may be either hazardous or beneficial, depending largely on where it is. For example, it is hazardous as an oxidant in smog (ground-level ozone), but in the O_3 layer in the stratosphere, it is beneficial because it absorbs UV radiation. Smog ozone reduces the normal functioning of lungs because it inflames the cells that line the respiratory tract. Other health effects include increased incidence of asthma attacks, increased risk of infection, and reduced heart and circulatory functions.

Smog O_3 can also damage plant life. In vegetation, the main damage occurs in foliage, with smaller effects on growth and yield. In the United States, it has been implicated in the loss of conifer trees near Los Angeles and is suspected of doing similar damage to trees in the Appalachian Mountains. Some plant species are also very susceptible to PAN in smog, and this is known to affect plants in the Los Angeles area.

Ozone and NO_x pollution in the troposphere is not limited to urban areas. In the mid-1990s, the EPA reported that O_3 and NO_x concentrations were increasing in rural areas in the southeastern and midwestern United States (see Information Box 17.5). Most of this increase is probably attributable to upwind urban sources; however, in some rural areas, soil bacteria may be a greater NO_x source than is fossil fuel combustion. In fact, some estimates indicate that soil bacteria may emit as much as 40% of the total amount of NO_x emitted into the atmosphere. This percentage is very uncertain, since data on measurements of NO_x fluxes from soils are scanty. Soil NO_x fluxes may also be highly spatially and temporally variable.

Tropospheric NO_x strongly controls the concentrations of oxidants such as OH and O_3, which may affect the health of about one-quarter of the U.S. population. As in urban areas, O_3 in the rural atmosphere is controlled by reactions involving NO_x, HCs, OH, and other tropospheric species. The study of the production and destruction processes of O_3 in the rural troposphere is currently an area of active research by the EPA and other government agencies. Many questions remain unanswered concerning the reasons why O_3 concentrations continue to be high in both urban and rural areas despite recent efforts to curb emissions of the O_3 precursors, NO_x, and HCs.

17.2.3 Toxic and Hazardous Air Pollutants

In addition to primary and secondary pollutants, the EPA has identified 188 chemicals (or classes of chemicals) that

INFORMATION BOX 17.5 Regional Ozone Transport

Ground-level ozone ("smog") problems are not just limited to cities. Several times each summer, the central and eastern United States comes under the influence of stagnating high-pressure weather systems. Sunshine, high temperatures, and stagnant nonmixing conditions exacerbate the buildup of ozone precursor pollutants and the production of ozone over areas with high pollutant emissions, such as the industrialized Midwest. The slow flow of the polluted air mass traverses multiple states between the Midwest and the Atlantic Coast and can raise concentrations of ozone to unhealthy levels in rural as well as urban areas (Fig. 17.8).

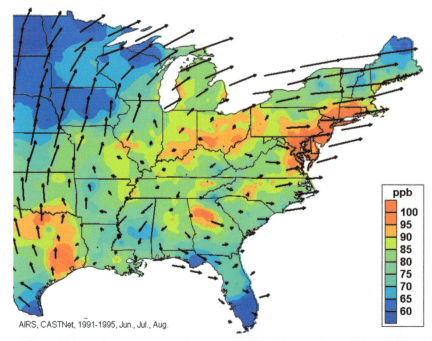

AIRS, CASTNet, 1991-1995, Jun., Jul., Aug.

FIG. 17.8 Wind can move polluted air across long distances. As a result, rural areas may experience polluted air that originated in metropolitan areas. The black arrows indicate transport winds, where longer arrows indicate faster movement. The color contours indicate ozone concentration. *(Source: Ozone Transport Assessment Group, 1997.)*

are considered to be *hazardous air pollutants* (HAPs) or *urban air toxics* (UATs). Many of these are volatile organic chemicals, such as benzene found in gasoline and used as a solvent, and trichloroethene, which is used as a solvent/degreaser (see Chapter 12). Mercury is an example of a hazardous inorganic compound (see Information Box 17.6).

17.2.4 Pollutants With Radiative Effects

Some air pollutants greatly influence the interactions between radiation of various wavelengths and the atmosphere. Some radiatively active pollutants contribute strongly to the natural greenhouse effect, while others impact the amount of ozone present within the stratosphere. This section describes these pollutants and their radiative effects.

17.2.4.1 Greenhouse Gases

Carbon dioxide is sometimes not considered to be an air pollutant because it is not hazardous to human health at ambient atmospheric concentrations; moreover, it is essential for carbon fixation by plants. It is, however, an important greenhouse gas and a major by-product of fossil-fuel burning, which steadily increases the atmospheric concentration of carbon dioxide. It therefore plays a central role in global climate change. This increase is discussed in detail in Chapter 25.

Carbon dioxide is by far the most abundant and important atmospheric trace gas contributing to the natural *greenhouse effect* (see also Chapter 4). It is released to the atmosphere by various processes including deforestation and land clearing, fossil-fuel combustion, and respiration from living organisms. Carbon dioxide is readily absorbed by water, with warm water absorbing more than cold water, so it is removed from the atmosphere by the oceans and other bodies of water. Photosynthesis by land and water plants (phytoplankton) also removes significant amounts of CO_2 from the atmosphere. Removal by plants is particularly apparent during the summer, when average CO_2 concentrations decrease. In addition, large amounts of CO_2

INFORMATION BOX 17.6 How Does Mercury Get in My Food?

More than 40% of all mercury emissions in the United Statesare from power plants burning coal (about 50 tons $Hg\,yr^{-1}$ in 2005). The mercury is released as a gas, which eventually is deposited into water or soil. Microbes can then convert inorganic mercury to methyl-mercury in a process called methylation. Small organisms can take up small amounts of methyl-mercury as they feed, storing it in their tissues. As these organisms are eaten by animals higher in the food chain, the methyl-mercury continues to accumulate in body tissues, until they may reach relatively high levels in predators at the top of the food chain, such as swordfish and sharks. The methyl-mercury then becomes part of our food as we consume shellfish or fish from saltwater and freshwater sources.

Mercury tends to accumulate in tissue, so toxic levels can build up slowly. High levels of exposure to methyl-mercury can damage the human brain, resulting in neurological problems, such as increased irritability, shyness, tremors, vision and/or hearing loss, mental retardation, and memory loss. It is also harmful to the kidneys. It is especially hazardous for the fetus, infants, and young children. As many as 4.9 million women of childbearing age in the United Statesmay have unsafe levels of mercury, and as many as 60,000 newborns per year are at risk due to dietary exposure.

Fish can be eaten regularly, but some care should be taken in terms of the quantity, frequency, and type of fish eaten. Some fish accumulate much higher levels of mercury than others. For instance, studies show that swordfish tend to have much higher levels of mercury than salmon, so it might be wise to eat swordfish only occasionally (see also Chapter 26).

may eventually be fixed as limestone by deposition of the skeletons of some marine invertebrates in oceans. The atmosphere currently contains about 750 billion metric tons (BMT) of carbon in the form of CO_2, and a 3-BMT excess enters the atmosphere each year. Research indicates that this excess gives rise to a mean annual increase of about $1.5\,\mu L\,L^{-1}$ in the global concentration of CO_2.

In addition to CO_2, the other main greenhouse gases are CH_4, N_2O, *chlorofluorocarbons*(CFCs), and O_3. Water vapor is also an important, but variable, greenhouse gas. All of these gases are, as the term "greenhouse" implies, efficient absorbers of longwave radiation. This absorption helps maintain a relatively warm climate on earth. However, because greenhouse gas concentrations continue to increase, the earth's climate is undergoing rapid alteration and global temperatures are rising. Therefore much scientific research is currently being directed toward improving our understanding of the atmospheric budgets of these trace gases and their role in the greenhouse effect.

Atmospheric CH_4 concentrations have also steadily increased at a rate of about 1% per year in recent decades.

This rapid increase is commonly attributed to increased worldwide rice and livestock production. Increased mining of natural gas resources for energy production may also be an important factor.

Synthetic CFCs are also significant contributors to the greenhouse effect. Chlorofluorocarbons are used in refrigerators, air conditioners, foam insulation, and industrial processes. In addition to being very efficient longwave absorbers, CFCs are also involved in depleting stratospheric O_3. Fortunately, because of concerted international effort resulting in the *1987 Montreal Protocol on Substances that Deplete the Ozone Layer*, CFC emissions to the atmosphere have decreased substantially in recent years (see Chapter 32 for more details on the Montreal Protocol).

Nitrous oxide (N_2O) is an especially good absorber of longwave radiation; 1 molecule of N_2O is equivalent to about 200 CO_2 molecules in terms of its ability to absorb longwave radiation. Currently, atmospheric N_2O accounts for only about 5% of the greenhouse effect, but this percentage is expected to increase in coming years. A 25% N_2O increase in atmospheric concentration may, according to numerical model predictions, increase global mean temperature by about 0.1 K. Worse, N_2O has a very long atmospheric lifetime, estimated to be about 150 years, which is far longer than the atmospheric lifetime of any other nitrogen oxide. Thus the current buildup of N_2O could affect the earth's climate well beyond the 21st century.

17.2.4.2 Stratospheric Pollution

Stratospheric O_3 depletion is another global environmental concern related to pollution. Concern about O_3 first emerged in the early 1970s, when modeling studies indicated that a proposed fleet of supersonic transport (SST) aircraft could emit enough NO_x to damage the O_3 layer. The results from the modeling studies helped put an end to plans to build the fleet, but such considerations remain major factors in plans for aircraft development.In the mid-1970s, concern shifted to the possible O_3-depleting effects of manufactured CFCs used as refrigerants, propellants, cleaning compounds, and foam insulation. Intensive study of the effects of CFCs on stratospheric O_3 led to a 1979 U.S. ban on the use of CFC propellants in aerosol spray cans.

Stratospheric O_3 absorbs UV light, decreasing the amount of UV striking living organisms on the earth's surface. Satellite and ground-based measurements have shown that there is a temporary decrease in O_3 concentrations (50%–75% of total) over Antarctica each year. This has come to be known as the ozone hole, which is defined as the geographic area above the Antarctic where the total ozone is less than 220 Dobson units between 1 October and 30 November (Fig. 17.9). Fig. 17.10 shows

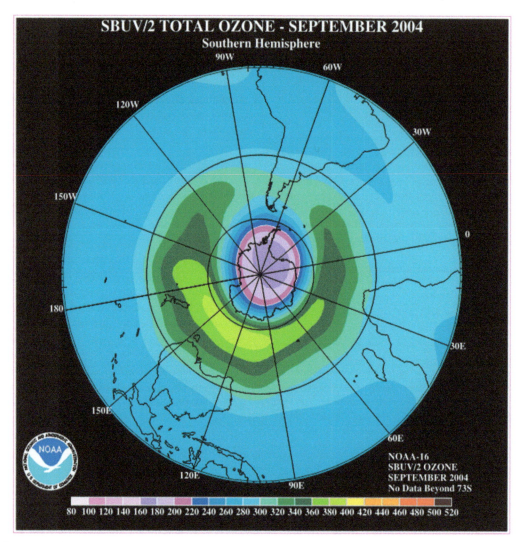

FIG. 17.9   Southern Hemisphere map of total ozone for September 2004. The ozone hole, with total ozone lower than 220 Dobson Units, is shown in purple and magenta. *(Image: U.S. National Oceanic and Atmospheric Administration.)*

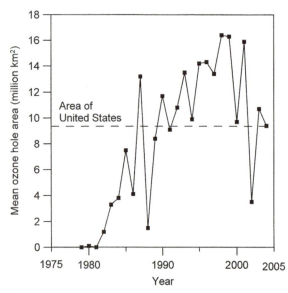

FIG. 17.10   Year-to-year variations in the average size of the Antarctic ozone hole between 1 October and 30 November. The size of the ozone hole appears to be decreasing since the phaseout of CFCs. *(Data: U.S. National Oceanic and Atmospheric Administration.)*

the average area of the Antarctic ozone hole in recent years. Stratospheric O_3 depletion has engendered serious concerns about the causes and possible ecological and human health consequences if this trend continues. In humans, increased UV would probably increase the incidence of skin cancer, including melanoma. Other organisms are also vulnerable to UV; phytoplankton, for example, has declined by 6%–12% in areas near Antarctica. The decline in this one-celled organism is thought to be due to increased amounts of UV that are reaching surface waters.

The chemical pathways leading to the formation of stratospheric O_3 start with the photodissociation of molecular oxygen (O_2) by solar UV radiation (photons of energy hv, where h is Planck's constant and v is the frequency). The UV photon splits O_2 into two oxygen atoms (O), each of which recombines with undissociated O_2 (in the presence of another chemical species, M) to form two O_3 molecules. These two reactions, which result in a net formation of O_3 are given as follows:

$$O_2 + h\nu \quad\quad \rightarrow O + O$$
$$\underline{2O_2 + 2O + M \rightarrow 2O_3 + M} \quad (17.2)$$
Net Reaction: $\quad 3O_2 + h\nu \quad\quad \rightarrow 2O_3$

The two O_3 molecules quickly convert back to molecular oxygen via

$$O_3 + h\nu \rightarrow O + O_2$$
$$\underline{O_3 + O \rightarrow 2O} \quad (17.3)$$
Net Reaction: $\quad 2O_3 + h\nu \rightarrow 3O_3$

The process of production and loss of the O_3 molecules by photodissociation is very important because, overall, it helps prevent harmful UV from reaching the earth's surface. These production/destruction schemes indicate that the chemistry of stratospheric O_3 would be straightforward if there were no other reactive chemical species in the stratosphere. However, other chemicals, such as CFCs and NO_x, are present and play an important destructive role. This is indicated by the following general catalytic cycle:

$$X + O_3 \rightarrow XO + O_2$$
$$O_3 + h\nu \rightarrow O + O_2$$
$$\underline{XO + O \rightarrow X + O_2} \quad (17.4)$$
Net Reaction: $\quad 2O_3 + h\nu \rightarrow 3O_3$

where X and XO represent the compounds or free radicals that catalyze the destruction of O_3 molecules. Mainly NO_x, water vapor, and CFCs, these species are summarized in Table 17.1.

Of these catalysts, CFCs are entirely anthropogenic, whereas nitrogen oxides come from both natural and synthetic sources. Stratospheric water vapor comes mainly from natural processes. In addition to the three main catalysts, other chemicals may play a role in controlling stratospheric O_3 levels. For example, recent evidence indicates that methyl bromide, which is used as a soil fumigant, may reach the stratosphere, where it can undergo a catalytic reaction sequence with O_3 similar to those of the three main chemical species.

The catalytic reactions do not destroy all the O_3 present in the stratosphere. The reason they donot is that reactions also occur between the catalysts, and these reactions result in chemicals that do not deplete O_3. Some of the chemicals

TABLE 17.1 Chemical Species that are Believed to Catalyze the Destruction of O_3 Molecules in the Atmosphere

Cycle	X	XO
NO_x	NO	NO_2
Water	HO•	HO_2•
CFC	Cl•	ClO•

eventually return to the earth's surface (e.g., HNO_3 in rain).

Further aspects of each of these catalytic cycles are described in Graedel and Crutzen (1993) and are briefly discussed in the following paragraphs.

NO_x/O_3 Destruction Cycle

The nitrogen oxides in the stratosphere come mainly from photodissociation of nitrous oxide, which originates mostly from microbial processes at the earth's surface. Nitrous oxide is also a major greenhouse gas. It is produced mainly within moist soils by microbial denitrification of nitrate fertilizer, but it can also be biologically produced in oceans. Since the 1970s there has been concern that increased agricultural use of nitrogen fertilizers could increase the amount of nitrous oxide reaching the stratosphere, ultimately depleting O_3. Measurements of atmospheric N_2O indicate that its concentration is increasing by about 0.25% per year. A 25% increase in N_2O by the late 21st century could reduce total stratospheric O_3 by 3%–4%, which could increase the incidence of skin cancer by 2%–10%.

Nitrous oxide is not known to be lost within the troposphere; however, it is converted to NO in the stratosphere mainly by the following reactions:

$$N_2O + h\nu \rightarrow N_2O(^1D)$$
$$N_2O + O(^1D) \rightarrow 2NO \quad (17.5)$$

The two NO molecules formed initiate the O_3 destruction reactions described previously. Note that $O(^1D)$ in Eq.(17.5) denotes atomic oxygen in an electronically excited state.

H_2O/O_3 Destruction Cycle

The stratosphere is generally very dry. However, enough water vapor is present to react with electronically excited atomic oxygen to produce the free radical HO• via

$$H_2O + O(^1D) \rightarrow 2HO• \quad (17.6)$$

The catalytic water cycle has less influence upon O_3 concentrations than do the other reaction cycles, but it can be significant when sufficient water vapor is present.

CFC/O_3 Destruction Cycle

Chlorofluorocarbons (e.g., $CFCl_3$ and CF_2Cl_2) are relatively stable in the troposphere, but once in the stratosphere, they are photodissociated by UV. This photodissociation produces the catalysts Cl and ClO, both of which deplete O_3. There is evidence that links the O_3 depletion in the Antarctic region to CFCs and other pollutants that carry chlorine and bromine into the stratosphere. Chlorine monoxide (ClO) has been identified as the chief cause of O_3

depletion in polar regions. Weather patterns and volcanic eruptions may also play a part.

Chlorofluorocarbons are also implicated in possible global climate change. Because many are extremely efficient absorbers of longwave radiation, they contribute to the earth's greenhouse effect.

17.3 WEATHER AND POLLUTANTS

What happens to pollutants in the atmosphere? The answer depends on several factors. Pollutants are transported by wind and turbulence, and they may undergo chemical transformations before being deposited on the earth's surface. Thus weather conditions strongly affect the fate of air pollutants.

17.3.1 Stability and Inversions

The stability of boundary-layer air (see Chapter 4) largely determines how quickly pollutants are moved upward from their ground sources. Stability is primarily a function of the vertical air temperature gradient relative to the adiabatic lapse rate. Strong instability associated with buoyancy causes efficient air mixing and pollution dispersal over a large mixing depth of the boundary layer (from 100 to 1000 m). Good mixing often occurs on warm days when the ground is heated by sunlight. In contrast, pollution is poorly dispersed on days or nights when the atmosphere is stable. At those times, turbulent movement of pollution upward is slow or nearly nonexistent.

We know from Chapter 4 that temperature inversions influence atmospheric stability; thus they play an important role in determining the concentrations of air pollutants. The effects of inversions are intensified by limited air drainage out of enclosed valleys, as is the case in Los Angeles (Fig. 17.7) and Mexico City. Various processes may generate inversions, including surface cooling caused by loss of infrared radiation or by evaporation, atmospheric subsidence, and topographic effects.

Ground-surface cooling is caused mainly by infrared radiation emission from the surface to the sky. It generally occurs during clear, calm nights, with inversion heights extending about 100 m above the ground. Such inversions commonly occur throughout the western United States during fall, winter, and spring, when the air is relatively dry and skies are clear. Such conditions readily permit cooling by longwave loss of energy from the ground. Tucson, Arizona, for example, often experiences radiation inversions in the cooler months, so that wintertime pollution problems are exacerbated in the area. Cooling of the ground by evaporation of water from soil and plants may also establish inversions. Evaporative cooling can occur during the day or night, particularly over irrigated fields. This type of inversion may be important in relation to certain agricultural activities, such as the aerial application of pesticides

over large irrigated fields. The depth of an evaporatively cooled inversion layer is usually just a few meters.

Warming of the atmosphere by subsidence causes inversions over regions that have semipermanent high pressure (anticyclonic flow), such as the southwestern United States. As air subsides (sinks), it encounters higher pressure and thus is warmed. Within regions of high pressure, an inversion height may be several hundred meters above ground; thus the air may be very stable over a large depth of the atmosphere. Because subsidence inversions may last from several days to weeks, the result is highly polluted conditions at ground level. For example, subsidence inversions are a major factor in reducing air quality in the Los Angeles basin.

Inversions associated with topography result from adiabatic warming of air as it flows downslope over mountainous terrain. These inversions may exacerbate air pollution problems in populated, mountainous areas such as Denver, Colorado, or Salt Lake City, Utah.

17.3.2 Wind and Turbulence in Relation to Air Pollution

Wind affects turbulence near the ground, thus affecting the dispersion of pollutants released into the air. Turbulence (largely fine scale vertical and horizontal motion of air) is generated in part by air flow over rough ground. The greater the wind speed, the greater the turbulence, and hence the greater the dispersion of pollutants that are near the ground (Fig. 17.11).

We can visualize the dispersion of pollutants in air by looking at the familiar cloud or "plume" of pollution emitted continuously by a smokestack (Fig. 17.12). As the plume contents are carried away from the stack by the wind, the size of the plume increases owing to dispersion. Because of dispersion, the pollutant concentration within the plume decreases with increasing distance from the source.

Dispersion of pollution downwind from a smokestack is affected by the roughness of the ground surface. Because of friction between the atmosphere and the ground, wind speed is slowed markedly near the ground. If the surface is relatively rough, as it is when trees and buildings are present, the air flow tends to be turbulent and the increase of wind speed with increasing height is relatively small. Greater surface roughness increases turbulence, which helps disperse pollutants. Air flow over a smooth surface, such as a large mowed lawn, tends to be less turbulent and the decrease in wind speed near the ground is relatively small (see Fig. 17.11).

The cone-shaped plume in Fig. 17.12 illustrates the pattern of pollutant dispersal downwind of a point source. Several factors affect the plume, including the effective height (H) of emission, which is a measure of how high the pollutants are emitted into the atmosphere directly

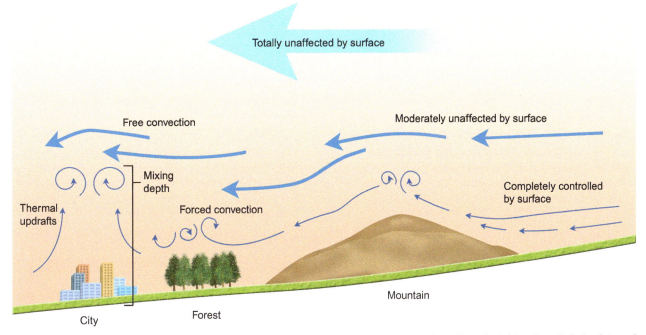

FIG. 17.11 Diagrammatic representation of air flow, mixing, and relative velocity over varying terrain as affected by height. *(From Pollution Science ©* *1996, Academic Press, San Diego, CA.)*

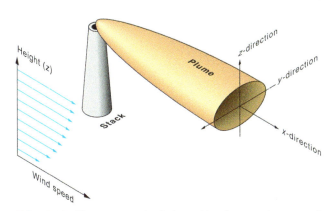

FIG. 17.12 Plume pattern (coning) resulting from continuous stack emission into a near-neutral stable boundary layer under moderate winds. *(From Pollution Science © 1996, Academic Press, San Diego, CA.)*

above the source. The height is dependent upon source characteristics and atmospheric conditions. Generally, a tall smokestack produces relatively low ground-level pollutant concentrations, because turbulence tends to dilute the pollution before reaching the ground. Driven by buoyancy, fast-moving pollutants are initially transported high up into the atmospheric boundary layer because they are warmer than the surrounding air. But as the pollutants cool and merge with the ambient air, the plume begins to move sideways with the wind. Then turbulence caused by the air flow over the surface and by possible instability governs the diffusion of the plume contents.

Usually, turbulence helps mix plume contents uniformly in such a way that the concentration follows a Gaussian distribution about the plume's central axis. Mathematically, pollutant concentration $\chi_{(x,y,z,H)}$ (kg m^{-3}) at any point in the plume is described by

$$\chi_{(x, y, z, H)} = \frac{Q}{2\pi\sigma_y\sigma_z\overline{u}}$$

$$\exp\left(-\frac{y^2}{2\sigma_y^2}\right) \tag{17.7}$$

$$\left[\exp\left(-\frac{(z-H)^2}{2\sigma_z^2}\right) + \exp\left(-\frac{(z-H)^2}{2\sigma_z^2}\right)\right]$$

where:

Q is the rate of emission of pollution from the source (kg s^{-1})

σ_y and σ_z are the horizontal and vertical standard deviations of the pollutant concentration distributions in the y and z directions

$\overline{u}$ is the mean horizontal wind speed within the plume (m s^{-1}).

This model, which is applicable to continuous sources of gases and particulates less than about 10 μm in diameter (larger particles quickly settle to the ground), can be used to model plume concentrations over horizontal distances of 10^2–10^4 m. With this Gaussian plume model, it is assumed that no deposition of plume contents to the ground surface takes place. In fact, it is assumed that plume contents are "reflected" from the ground back to the air. The values of σ in the equation are estimated from any one of several empirical formulas that relate σ to downwind distance (x) and stability conditions. These formulas include the following equations, which were developed by the *Brookhaven National Laboratory* (BNL).

$$\sigma_y = ax^b \text{ and } \sigma_z = cx^d \qquad (17.8)$$

where $a, b, c,$ and d are parameters dependent upon stability. (See Hanna et al., 1982 for a summary discussion of BNL equations as well as other approaches.) At ground level, $z = 0$, and along the plume centerline, $y = 0$. Thus, from Eq. (17.7), the concentration can be calculated by

$$\chi_{(x, H)} = \frac{Q}{\pi \sigma_y \sigma_z \overline{u}} \exp\left(-\frac{H^2}{2\sigma_z^2}\right)$$

$$= \frac{Q}{\pi ax^b cx^d \overline{u}} \exp\left(-\frac{H^2}{2\left(cd^d\right)^2}\right) \qquad (17.9)$$

One type of plume (shown in Fig. 17.12) typically occurs under windy conditions with stability conditions at or near neutral. Within such a plume, mixing occurs mainly by frictionally generated turbulence, and pollutant diffusion is nearly equal in all directions (i.e., the σ values are nearly equal and the plume spreads out in the familiar cone pattern, known as *coning*). Coning can occur day or night, and is often seen during cloudy and windy conditions. Depending upon effective source height and atmospheric conditions, the plume may reach the ground close to the source. Using Eqs. (17.8), (17.9), we can estimate the ground level ($z = 0$) concentration of the plume composed of a pollutant, say, SO_2, emitted into the atmosphere at a known effective height. Suppose we have the following: $Q = 0.5 \text{ kg s}^{-1}$, $H = 25 \text{ m}; \overline{u} = 2 \text{ m s}^{-1}$; near neutral stability, and BNL parameters $a = 0.32$, $b = 0.78$, $c = 0.22$, and $d = 0.78$. Then the ground-level SO_2 concentration along the plume centerline at an arbitrary distance of $x = 500 \text{ m}$ from the source will be $4.7 \times 10^{-5} \text{ kg m}^{-3}$ (or 47 mg m^{-3}).

Plumes may change because of changes in the wind velocity and boundary-layer stability. When the atmospheric boundary layer is strongly stable, such as during radiation inversions at night or during subsidence inversions, a *fanning* pattern may be evident in the plume, as illustrated in Fig. 17.13A. Under these conditions, there is almost no vertical motion and the BNL parameters are $a = 0.31$, $b = 0.71$, $c = 0.06$, and $d = 0.71$. Lack of vertical motion thus effectively forces the plume into a relatively narrow layer, while changes in wind direction may spread the plume out laterally, resulting in a V- or fan pattern; hence, the term. A constant wind direction, however, forces the plume into a tightly closed fan pattern, which follows a relatively straight and narrow path. Over flat terrain the plume in Fig. 17.13A

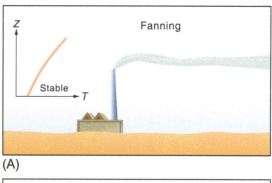

(A)

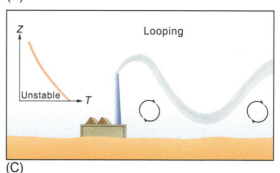

(B)

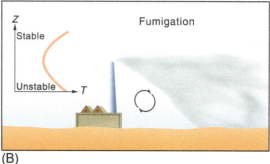

(C)

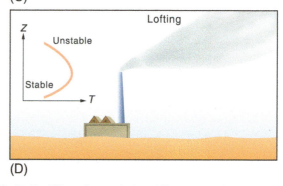

(D)

FIG. 17.13 Effect of atmospheric stability upon resultant stack plume pattern during (A) inversion (fanning pattern), (B) dissipation of inversion near ground (fumigation pattern), (C) lapse conditions (looping pattern), and (D) lofting pattern.

may be unchanged for very long distances. If there is no vertical air movement, ground-level concentrations downwind of a tall smokestack can be nearly zero. However, if the source is close to the ground (i.e., H is small), or if changes in topography cause the plume to intercept the ground, the ground-level concentrations can be very large.

By midmorning, surface heating by solar radiation typically begins to break down the inversion developed during the previous night, as illustrated in Fig. 17.13B. Unstable conditions develop near the ground, resulting in vertical mixing of the air. With moderately unstable conditions, $a = 0.36$, $b = 0.86$, $c = 0.22$, and $d = 0.86$. In this situation, pollution is transported downward toward ground level. Stable conditions above, however, limit dispersion of pollutants upward. Thus the remaining inversion effectively puts a "lid" over the ground-level pollution. This situation is known as *fumigation* and generally lasts for periods of an hour or less. Fumigation is highly conducive to enhanced ground-level pollutant concentrations.

By early afternoon, lapse conditions (i.e., negative vertical temperature gradient) generally become fully established within the boundary layer due to strong surface heating by the sun. During much of the afternoon, air motion mainly exhibits the large turbulent eddies associated with buoyancy. These eddies are generally larger than the plume diameter and thus transport the plume upward and downward in a sinusoidal path or *looping* pattern, as illustrated in Fig. 17.13C. The loops are carried with the overall wind pattern and generally increase in size with increasing distance downwind from the source. The motion may bring the plume contents to ground level quite close to the source. Because of turbulence, however, the plume eventually becomes dispersed at relatively large distances.

By early evening, a radiation inversion often rebuilds from ground level upward. Stable conditions near the ground inhibit transport of plume contents downward, but unstable air aloft (above the inversion height) allows dispersal upward. This upward transport, known as *lofting*, is highly favorable for dispersing pollutants, as shown in Fig. 17.13D. Lofting is only effective when the source is above the inversion height. Plume contents emitted below the inversion height are essentially trapped in a fan-type plume configuration.

Topography downwind from pollution sources affects air quality, especially in mountainous areas. For example, air drainage into relatively enclosed valleys during winter and/or inversion conditions can cause accumulation of pollutants within the valleys. Thus urban areas in valleys with restricted air flow are particularly prone to high pollution levels. In addition, in coastal areas, air flow from the ocean (sea breezes) can be blocked or channeled by mountain ranges. This situation is common in the Los Angeles basin, which is surrounded by mountains that restrict air flow from the Pacific Ocean. Thus dispersal of pollutants from sources in the basin is inhibited.

17.3.2.1 Pollutant Transformation and Removal

As pollutants move with the wind, chemical reactions often occur between the pollutants and other atmospheric chemical species. Although the pathways and rates of many of these chemical reactions are poorly understood, they are an important factor affecting the fates of many air pollutants.

Most pollutants, such as CO, remain in the atmosphere for relatively brief periods, lasting only a few days or weeks. Thus, if emissions were completely curtailed, the lower atmosphere would quickly lose nearly all of its pollutants. However, some pollutants—volcanic ash and sulfur-containing aerosols, for example—emitted high into the stratosphere can remain there for months before settling back to the surface. These long-lasting upper-atmospheric pollutants can alter the earth's climate, as evidenced by lower air temperatures resulting from volcanic eruptions (see Chapter 4). In addition, synthetic *chlorofluorocarbon* (CFC) compounds can remain in the atmosphere for many years before they break down.

Pollution can be removed from the atmosphere by gravitational settling, dry deposition, condensation, and wet deposition.

Gravitational settling. Gravitational settling removes most particles whose diameters are greater than about $10 \mu m$. Particles less than $10 \mu m$ in diameter are often small enough to stay suspended in the atmosphere for long periods. Particles greater than about $10 \mu m$ in diameter quickly settle out.

Dry deposition. Dry deposition is a mass-transfer process that results in adsorption of gaseous pollutants by plants and soil. Dry deposition to plants is dependent upon uptake of the pollution through stomatal openings in plant leaves and upon turbulent transport in the air. Dry deposition to bare soil involves not only turbulent transport of pollutants in air above the soil, but also soil microorganisms that take up such pollutants such as CO.

Condensation. Volatile organic compounds can condense on cold surfaces during winter in temperate and polar regions. The process of evaporation, transport, and condensation of toxic compounds, such as dioxins and the pesticide, dichlorodiphenyltrichloroethane (DDT), may be responsible for causing high levels of toxic organic pollutants in the Arctic (see Chapter 25).

Wet deposition. Rain is very effective at removing gases and small particulates. Raindrops increase in size as they fall toward the ground, and thus they increasingly capture more pollutants. Raindrops, in effect, "sweep up" pollution as it falls through the air. The ability of the rain to remove pollutants depends upon the rainfall intensity, the size and electrical properties of the drops, and the solubility of the polluting species.

17.4 POLLUTION TRENDS IN THE UNITED STATES

Emissions of nearly all types of primary air pollutants have generally declined or held steady in the United States since about 1970 (Fig. 17.14). This decline is mostly attributable

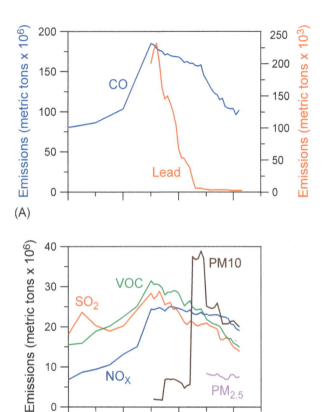

FIG. 17.14 Air pollution emissions have generally decreased since the 1970s with the exception of PM_{10}. PM_{10} emissions rose sharply in 1985, largely due to the adoption of reporting PM_{10} emissions in the miscellaneous category, which includes roads, construction, and agriculture. (A) CO and lead, (B) Other pollutants. *(Data: U.S. Environmental Protection Agency.)*

to general compliance with the federal air quality regulations set forth in response to the U.S. Clean Air Act of 1970. Although air quality is improving overall, many specific urban areas fail to meet the air quality standards set for some pollutants. Poor air quality is estimated to affect the lives of about 100 million people in the United States alone.

Despite the fact that transportation continues to be a major source of pollution in the United States, the proportion of its contribution is diminishing. While the number of cars is increasing in most urban areas, fuel efficiencies have increased and pollutant emissions per vehicle are declining owing to improvements in technology such as catalytic converters and other pollution control devices. Evidence of improved air quality is shown by the marked decline of atmospheric lead (Pb) concentrations since 1970. Atmospheric lead comes mainly from the burning of lead-containing gasoline in cars and trucks. Thus the introduction of unleaded gasoline was a significant factor in this decline. Now required for cars in the United States because of environmental health concerns, unleaded

gasoline is also used because leaded fuels deactivate catalytic converters.

It is generally recognized that reducing air pollution through control of emissions at the source is the best approach, which is the goal of the EPA and other regulatory agencies. Total control of pollutant emissions is certainly not feasible for various economic and technological reasons, but efforts at reducing emissions are helping to improve air quality in most locations. There are various physical and chemical precipitators/concentrators/burners that can be used to control emissions. Consult an environmental engineering handbook for more details.

QUESTIONS AND PROBLEMS

1. Describe how surface air temperature inversions form. Why are airtemperature inversions important relative to air pollution in urban areas?
2. What factors affect atmospheric stability? Explain.
3. Based upon the Brookhaven National Laboratory Eq. (17.8), how do numerical values of σ_z values compare at $x = 100\,m$ for pollution plumes during stable and unstable atmospheric conditions? How do the σ_y values compare? During which condition (stable or unstable) would you expect the plume to intercept the ground closer to the source? Explain.
4. Describe the processes that remove air pollution.
5. What is the difference between EPA-designated primary and secondary air pollution? Give an example of each type of pollutant.
6. What is photochemical smog? Explain how it is formed.
7. Explain how O_3 in the stratosphere is beneficial, whereas O_3 in the troposphere is harmful.
8. Explain how anthropogenic chlorofluorocarbons (CFCs) destroy stratospheric O_3.

REFERENCES

Graedel, T.E., Crutzen, P.J., 1993. Atmospheric Change: An Earth System Perspective. W.H. Freeman, New York.
Hanna, S.R., Briggs, G.A., Hosker Jr., R.P., 1982. Handbook on Atmospheric Diffusion. U.S. Department of Energy, Washington, DC.

FURTHER READING

Ahrens, C.D., 2003. Meteorology Today: An Introduction to Weather, Climate and the Environment, seventh ed. Brooks/Cole-Thomson Learning Inc., Pacific Grove, CA.
Albritton, D.L., Monastersky, R., Eddy, J.A., Hall, J.M., Shea, E., 1992. Our Ozone Shield: Reports to the Nation on Our Changing Planet. Fall 1992. University Cooperation for Atmospheric Research, Office for Interdisciplinary Studies, Boulder, CO.
Dickinson, R., Monastersky, R., Eddy, J., Bryan, K., Matthews, S., 1991. The Climate System: Reports to the Nation on Our Changing Planet.

Winter 1991. University Cooperation for Atmospheric Research, Office for Interdisciplinary Studies, Boulder, CO.

EPA. (2005) Asthma research results highlights. EPA 600/R-04/161. Available from:http://www.epa.gov/ord/articles/2005/asthma_fact_sheet.htm.

Mitchell, J.F.B., 1989. The "greenhouse" effect and climate change. Rev. Geophys. 27, 115–139.

Oke, T.R., 1987. Boundary Layer Climates. Routledge, New York.

Ozone Transport Assessment Group, 1997. Telling the OTAGozone story with data. In: Final Report, Vol. I: Executive Summary. OTAG Air Quality Analysis Workgroup. Co-chairs, D. Guinnup and B. Collom. Available from: http://capita.wustl.edu/otag/reports/aqafinvol_I/animations/v1_exsumanimb.html.

Schlesinger, W.H., 1991. Biogeochemistry: An Analysis of Global Change. Academic Press, San Diego, CA.

Chapter 18

Urban and Household Pollution

J.F. Artiola, K.A. Reynolds and M.L. Brusseau

Chapter cover art: urban pollution issues. *From: United Nations Environment Programme; https://environmentlive.unep.org/pollution.*

18.1   INTRODUCTION

The majority of people in the world live in cities and smaller urban areas. Urban areas include densely populated clusters of 2500 or more people that are generally smaller than cities. The United Nations estimates that 55% of the World's population lives in urban areas (United Nations, 2018). The urban–rural distribution varies from country to country. The most urbanized regions include Northern

America (82% living in urban areas), Latin America and the Caribbean (81%), Europe (74%), and Oceania (68%). The level of urbanization in Asia is ∼50%, while Africa remains the least urbanized, with 43% of its population living in urban areas. The UN projects that the proportion of the World's population living in urban areas will increase to 68% by 2050.

In 1990, there were ten "mega-cities" with 10 million inhabitants or more, which were home to 153 million

Environmental and Pollution Science. https://doi.org/10.1016/B978-0-12-814719-1.00018-5

people or slightly less than 7% of the global urban population at that time. Currently, there are 33 mega-cities worldwide, home to ~12% of the world's urban dwellers. By 2030, the world is projected to have 43 mega-cities with 10 million inhabitants or more.

According to the U.S. Census Bureau, approximately 63% or the U.S. population lives in cities and 80% live in urban areas. Thus only 20% of the population lives in rural areas. The urban areas hosting the 80% of the population represent only ~3% of the nation's land area. Overall, the U.S. population density is ~90 people per square mile (ppsm). However, the population density of U.S. cities is much greater than that of rural areas, 1600 ppsm versus 35 ppsm.

As we see from the preceding, the majority of people live in high-density urban areas, and the proportion of people who do so is projected to increase in the future. The high population densities and the infrastructure required to support these high populations generate unique environmental and human health issues. Managing urban development in a sustainable manner will be key to improving and maintaining human well-being while preserving environmental quality. This issue of sustainable urban development is one of the most critical issues of the 21st century.

This chapter will present some of the major environment-related issues associated with urban areas. Topics related to waste management, treatment, and disposal are covered in several other chapters in this text. This chapter will focus on topics not covered elsewhere, specifically sensory pollutants derived from sources of heat, light, noise, and odorous air contaminants, as well as issues of indoor air quality. It is important to recognize that some of the issues discussed herein may also be relevant for rural areas, particularly, for example, the issue of indoor air quality.

18.2 HEAT

18.2.1 Heat Islands

The term "heat island" is used to describe the increase in urban surface and air temperatures above those observed in rural areas. The United States Environmental Protection Agency (U.S. EPA) has reported that downtown urban and suburban air temperatures may be up to 10°F (5.6°C) warmer than temperatures in surrounding areas with natural land cover (Fig. 18.1). Heat islands form when natural land cover such as grasses, shrubs, and trees is replaced with urban infrastructure that includes pavement and buildings. An important surface property that influences heat island formation is *albedo*. Albedo is a measurement of a surface's ability to reflect incoming solar radiation and is reported as the ratio of reflected light to incident light. Thus albedo is

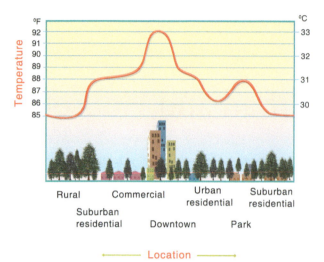

FIG. 18.1 Urban heat island late-afternoon temperature profile. *(From U.S. Environmental Protection Agency. Available from: http://www.epa.gov/heatisland/about/index.html)*

measured on a scale from 0 to 1, where a low albedo value indicates high absorbance, and a high albedo indicates high reflectivity of a material (Table 18.1). In general, the lower the albedo value, the higher the surface temperature within a given environment. Albedo accounts for the reflectivity of visible, infrared, and ultraviolet wavelengths, and is also called shortwave reflectivity. In general, city structures have a lower albedo than natural cover, yielding an increase in surface and ambient temperatures. Although some flora also have a relatively low albedo, plants provide natural cooling through shading and evapotranspiration. As such, when an area is developed, lower albedo surfaces replace natural land cover, and the natural cooling effect provided by native vegetation is removed. In addition, traffic and urban congestion cause city temperatures to rise as a result of increased waste heat released from vehicles and many other electromechanical devices such as air conditioners. Industrial activity and smokestacks can also increase the magnitude of heat islands.

18.2.2 Effects of Heat Islands

Heat islands can adversely affect human health and the environment. In general, higher rates of heat-related illness, including heat exhaustion and heat stroke, and even death are observed in cities. Populations in cities particularly vulnerable to high temperatures include the elderly and outdoor workers. For example, approximately 700 deaths were attributed to a heat wave in Chicago, Illinois, in 1995, and thousands of deaths in France resulted from a massive heat wave in 2003. In 2005, in Phoenix, Arizona, dozens of people died from heat-related causes. Global climate change will exacerbate the urban heat island effect,

TABLE 18.1 Shortwave Reflectivity (Albedo) of Soils and Vegetation Canopies

Surface	Reflectivity	Surface	Reflectivity
Grass	0.24–0.26	Snow, fresh	0.75–0.95
Wheat	0.16–0.26	Snow, old	0.40–0.70
Maize	0.18–0.22	Soil, wet dark	0.08
Beets	0.18	Soil, dry dark	0.13
Potato	0.19	Soil, wet light	0.10
Deciduous forest	0.10–0.20	Soil, dry light	0.18
Coniferous forest	0.05–0.15	Sand, dry white	0.35
Tundra	0.15–0.20	Road, blacktop	0.14
Steppe	0.20	Urban area (average)	0.15

From Campbell, G.S., Norman, J.M., 1998. Introduction to Environmental Biophysics, 2nd ed. New York, Springer-Verlag. Reprinted with kind permission of Springer Science and Business Media.

but the degree of impact will depend on local urban development strategies and conditions (McCarthy et al., 2010).

Heat islands can also benefit society, particularly in cold-climate cities in the winter. Warmer winter temperatures reduce energy demand for heating and help to melt snow and ice on city streets. However, the consequences of the heat island effect in the summer are significant, especially in areas with a warmer climate. Higher summer temperatures increase air conditioning and energy demand, which increases air pollution. In addition, the financial cost associated with cooling warmer cities is substantial.

Los Angles, California, is a prime example of an urban heat island. According to the Heat Island group at the Lawrence Berkeley National Laboratory, the Los Angeles basin was primarily comprised of irrigated orchards in 1934, and the high summer temperature was 97°F. As urbanization of the Los Angeles basin occurred, the high temperature steadily rose to 105°F, and to even higher values in the 21st century. This increase in summer temperatures caused by the heat island effect in Los Angeles has been estimated by Heat Island Group researchers to cost ratepayers approximately $100 million a year. Global climate change is expected to aggravate the heat island effect, adding 0.5% to 8.5% more electricity usage per degree (°C) increase (Santamouris et al., 2015).

18.2.3 Controlling Heat Islands

A variety of measures can be taken to mitigate the heat island effect in urban areas, including the application of cool roofs, cool pavements, green roofs, and urban forestry. *Cool roofs* are composed of materials that have a high albedo and therefore high reflectivity. Many cool roof materials also have a high emittance, which refers to the ability of a material

to release thermal radiation. Cool roofs can significantly reduce roof temperatures and therefore reduce heat transfer into a building. More specifically, the U.S. EPA (http://www.epa. gov/heatisland) reports that summertime temperatures on a traditional roof may reach as high as 190 °F (88°C). Cool roofs, by comparison, will only reach 120°F (49°C). *Cool pavements* are generally composed of light-colored material with a high permeability. Lighter colored pavements reflect more light and absorb less heat. Pavements with a high permeability allow water to percolate and evaporate, thereby cooling the pavement and surrounding air. *Green roofs* are roofs that have been planted with vegetation to reduce rooftop temperature and cool the surrounding air. A green roof is comprised of a waterproof membrane, a drainage system, a growing medium, and plants. The benefits of green roofs are shown in Information Box 18.1.

Urban forestry is the process of incorporating vegetation into an urban area to increase cooling through shading and evapotranspiration. The United States Department of Agriculture Forest Service has reported that urban forestry can decrease midday maximum air temperatures by 0.07°F (0.04°C) to 0.36°F (0.2°C) for every 1% increase in canopy cover. Similarly, the incorporation of urban agriculture can also help reduce the heat island effect.

INFORMATION BOX 18.1 Benefits of Green Roofs

1. Reduce roof top temperatures.
2. Reduce temperature of surrounding air.
3. Reduce rainwater runoff to sewer systems.
4. Work as filtration systems for pollutants such as heavy metals and excess nutrients.
5. Provide habitat for birds and other small animals.
6. Esthetic green space for building residents.

18.3   LIGHT

The part of the electromagnetic radiation spectrum (see Chapter 4) known as light is of vital importance not only to society, but also to plants and animal life. However, excess light at night induced by human activities can result in light pollution, which has become an issue of environmental concern in recent years. *Light pollution* can be described as the illumination of the night sky by artificial sources of light. *Illuminance*, or *luminous flux density*, is a measure of the total amount of visible light falling on a surface area. Illuminance is measured in either the older unit of foot-candle (fc) or the newer SI unit lux (lx) (1 fc = 1 lumen ft^{-2} and 1 lux = 1 lumen m^{-2}, where lumen [lm] is a unit of measurement of light). A lumen, also defined as luminous flux, is the SI unit of measurement for the amount of brightness that comes from a light source. For example, a 60-W bulb may generate 700 lumens and is not as bright as a 100-W bulb, which can generate 1200 lumens. Light pollution often occurs when night lights are directed upward or outward or fail to deliver all of their light downward.

Key terms used to describe different aspects of light pollution include light trespass, glare, and sky glow. *Light trespass* is defined as intrusive or objectionable light that shines onto neighboring properties. An example of light trespass would be a floodlight shining into a neighbor's bedroom window. *Light glare* is direct light shining into the eye from a bright surface. Glare can come directly from a light fixture or it can be reflected off a surface. Glare can harm vision, cause discomfort, and reduce visibility. *Sky glow* is the yellowish-orange glow seen over many cities and towns and is the composite excess illumination being released from many electrical lighting fixtures.

Measuring urban sky glow is not an easy task and depends on a number of variables. Merle Walker (IDA, 1996) measured sky glow for a number of cities in California in the 1970s and developed a formula to estimate sky glow from different distances and populations. The following equation is a simplified form of *Walker's Law*, proposed by the *International Dark-Sky Association* (IDA), and can be used to estimate the typical level of urban sky glow:

$$I = 0.01 \, P d^{-2.5} \qquad (18.1)$$

where

    I is the increase in sky glow above natural background light (illuminance) level
    P is the population of the city
    d is the distance to the city center in km (IDA, 1996).

18.3.1   Sources of Light Pollution

The main source of light pollution is inefficient outdoor lighting. Examples of inefficient outdoor lighting include street lights with inappropriate lamps and fixtures, spot lights pointed toward the sky to mark an event or illuminate an object such as a billboard, yard lights used to illuminate landscaping, and floodlights used for security. Another source of light pollution is light escaping from the inside of buildings at night. Light emitting diode (LED) lights are 8–10 times more energy efficient than incandescent lights and are more compact than filament of fluorescent light sources. However, their high lumens output is likely to make them new sources of light pollution.

18.3.2   Effects of Light

Light pollution first attracted attention when the general population, and astronomers in particular, were increasingly unable to view the night sky due to excess visible and UV light from cities. In recent years, however, it has become apparent that artificial night lighting is not only a concern for astronomers. Research has shown that artificial night lighting yields negative consequences for a variety of animals including aquatic and terrestrial invertebrates, amphibians, sea turtles and other reptiles, fish, birds, and mammals. For example, light pollution has been shown to affect the reproductive patterns of sea turtles. In Florida, female sea turtles come ashore to dig nests and lay eggs between May 1 and October 31. Bright artificial light on coastlines may deter females from coming ashore to lay their eggs. Further, if female turtles do come to shore, artificial light can disorient them, causing the turtles to wander onto roads, where they may be hit by a vehicle. In addition, sea turtle hatchlings generally hatch at night and instinctively head toward light. Before artificial light was installed along the coastline, the moon reflecting off of the ocean surface was brighter than the inland area. Light pollution can cause hatchlings to wander the shoreline or to head inland instead of into the sea. Therefore hatchlings are exposed to a higher risk of predation or can be run over by vehicles on nearby roads.

Light pollution can also affect migrating bird navigation. Many bird species use the light from constellations to guide them during their migration. Artificial light can confuse migrating birds, causing them to fly off course. Bird kills are common near floodlit smokestacks and radio transmission towers, where birds become confused by the artificial light and fly into these structures. It is hypothesized that light pollution may affect human health by disrupting hormone regulation. Light pollution also contributes to an increase in other environmental pollutants. The excess energy generated to power unnecessary artificial night lighting increases the volume of pollutants released from power plants.

Presently, it is estimated that about two-thirds of the world's population live in areas where the night sky brightness is considered polluted (Cinzano et al., 2001).

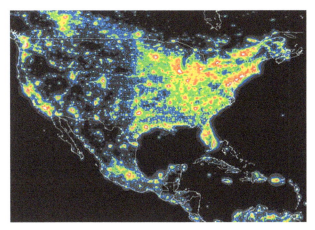

FIG. 18.2   North American map of artificial night sky brightness. *(Courtesy Cinzano, P., et al., 2001. The First World Atlas of the artificial night sky brightness, Mon Not R Astron Soc, 328, 689–707. Reprinted with permission of Blackwell Publishing.)*

In the United States, the same source estimates that 99% of the population lives in areas polluted with excessive night sky brightness. Of those, about 40% can no longer see stars with their eyes adjusted to night vision (Fig. 18.2).

18.3.3   Controlling Light Pollution

There are currently no federal or state regulations for controlling light pollution. However, several cities in the United States have developed guidelines for light pollution. For example, along the Florida coastline, cities have established ordinances that regulate artificial light at night during the nesting season of sea turtles. In addition, the city of Tucson, Arizona, initiated lighting ordinances to control outdoor night lighting to reduce interference with astronomical observations at the nearby Kitt Peak National Observatory (Fig. 18.3A and B).

Light pollution is expensive and contributes to other forms of environmental pollution. The International Dark-Sky Association has estimated that light pollution costs the United States an estimated $2 billion per year in wasted energy. Controlling light pollution conserves energy therefore decreasing energy costs and the pollution associated with energy production.

Light pollution can be controlled by simply turning off unnecessary lights and using appropriate lamps, fixtures/casings, and designs for lighting efficiency. Street lights commonly contain either *high-pressure sodium (HPS) lamps* or *low-pressure sodium (LPS) lamps,* and produce yellowish colored lighting. LPS lamps are more efficient ($125 \, \text{lumens} \, \text{W}^{-1}$) (W = watt) than the more common street light HPS lamps ($83 \, \text{lumens} \, \text{W}^{-1}$). However, LPS lamps do not give as true a color to surfaces as HPS lamps do. Other examples of types of lamps include mercury vapor lamps ($42 \, \text{lumens} \, \text{W}^{-1}$), which are not only less efficient than sodium lamps, but also produce bluish light, glare, and UV light pollution; metal halide lamps, a newer and more efficient but expensive source of white light ($59 \, \text{lumens} \, \text{W}^{-1}$); the durable compact mercury-containing fluorescent lamps also produce UV light (~ 60–$70 \, \text{lumens} \, \text{W}^{-1}$); and the most common, less durable and inefficient incandescent (halogen and tungsten) lamps ($15 \, \text{lumens} \, \text{W}^{-1}$) (Calgary Centre, 2005). The recent development of low cost LED (up to $200 \, \text{lumens} \, \text{W}^{-1}$) technology is now ushering a new era in home and also urban lighting with much welcomed reduction is energy use without UV radiation.

Full-cut-off (FCO) and *semi-cut-off* (SCO) lamps are important lamp features designed to control light pollution. FCO lamps, when mounted correctly, emit no light above the horizontal. SCO lamps, when mounted correctly, emit little to no light upward. There are also a variety of fixtures and casings that can be used for "shielding" lamps and controlling light pollution. When night lighting is designed and

(A)

(B)

FIG. 18.3   (A) Car lot lights in Tucson before light pollution control City ordinance. (B) Car lot lights in Tucson after light pollution control City ordinance. *(Photo courtesy: International Dark-Sky Association(IDA), Tucson, AZ. Available from: http://www.darksky.org)*

installed correctly, the lamp wattage will be selected to appropriately light the task without overlighting, and light will only be directed to the area intended to be illuminated. In addition, the light should evenly spread across the surface area intended.

18.4 NOISE POLLUTION

Noise is defined as unwanted or unwelcome sound that produces annoyance or physiological stress. If noise is loud enough and continuous, it can produce temporary and even permanent damage to our hearing system. Loud noises can also place people and animals in danger, because they may prevent the hearing of potential or impending dangers. In the United States, the *Noise Control Act of 1972* gave the U.S. EPA powers to set noise emission standards for major transportation and industrial sources of noise, to protect public health and welfare. In 1990 an amendment to the Clean Air Act EPA added title IV, which deals with noise abatement. Since the 1970s the Occupational Safety and Health Administration (OSHA) has implemented standards for noise exposure in the workplace for worker protection. However, to date the EPA has set only transportation-related noise regulations. In addition, EPA does not have the funding to enforce or revise these regulations. Thus there are no explicit national, state, or local laws that protect the public against noise pollution. Existing local noise ordinances vary widely and often they are disregarded or poorly enforced.

This section presents a short summary of the physics of noise, followed by a discussion of the sources of activities that produce noise and the increasing impacts of noise on modern life. Noise is an insidious form of pollution that is increasing in modern day life.

18.4.1 The Physics of Sound

Sound is energy in the form of airborne vibrations or pressure waves that can be sensed or heard through our hearing system. The vibrations of audible sounds range from about 16 to 20,000 Hz (oscillations per second). All media, including air, liquids, and solids, transmit sound waves, but the speed of sound movement varies dramatically through each medium. For example, sound travels about $346 \, \mathrm{m \, s^{-1}}$ through air, and about 4.2 and 14.8 times faster through water and iron metal, respectively.

Since sound is felt as pressure (P_r), its energy is usually given per unit area and in relation to its source or point of origin. Under ideal conditions of a homogeneous medium with no boundary surfaces, the pressure felt from a single point source (P_o) of sound felt at a distance (r) is defined by the following formula:

$$P_r = P_o / r$$

This assumes that under ideal conditions the spherical wave transmits its expanding power to a portion of an area ($\mathrm{cm^2}$) of a sphere. Note that the pressure (P_r) is in dynes $\mathrm{cm^{-2}}$, but the source pressure (P_o) is in dynes $\mathrm{cm^{-1}}$ (pressures are the root-mean-squares values) (Liu et al., 1997).

Since the range of sound pressures that the human ear can detect is at least eight orders of magnitude, a logarithmic scale is used to report sound pressure, with a base reference of 0.0002 dynes $\mathrm{cm^{-2}}$ being equal to 0 decibels (dB). Note that each order of magnitude in pressure change corresponds to a log scale change of 20 dB.

The sound intensity (I_r) is defined by the following equation:

$$I_r = W / 4\pi r^2$$

where

W = power of the sound per unit area (watts $\mathrm{cm^{-2}}$)
r = distance from the source (cm)

The sound pressure (P_o), distance from the source (r), medium density (ρ), and wave velocity through the medium (c) are related as follows:

$$I_r = P_o{}^2 / r^2 \rho c$$

Table 18.2 presents a range of sound pressures, levels, and power sources. Note that while the sound pressures vary by about eight orders of magnitude, the power sources and therefore the sound intensities vary by about 18 orders. Table 18.3 presents the scales of common sound sources and their effects on humans and community responses.

Air sound can be transmitted, reflected, absorbed, and distorted by solid objects. For example, a sound wave may pass through a window, transferring some of its energy into the glass in the form of mechanical energy. The remainder of the sound wave is likely to be distorted and much lower in intensity. These and many other complex sound phenomena are studied in detail in the advanced environmental physics field of acoustics (Liu et al., 1997 and Boeker and van Grondelle, 1999).

18.4.2 How We Hear Noise

Noise is felt by our hearing system and converted into electrical impulses that travel to our brain where they are processed. Briefly, sound waves enter our ear canal (outer ear) and travel through our eardrum. On the other side of the eardrum (middle and inner ear), the cochlear organ converts the sound signal into electrical impulses that are sent to the brain. The sensitivity of the human ear to sound depends on the sound frequency and its intensity, which in turn are related to the distance from the source. Humans have different hearing sensitivities to sound. For example, children can hear high-frequency sounds (above 150 Hz) much better

TABLE 18.2 Representative Sound Pressures and Power of Sources

Source and Distance	Sound Pressure (dynes/cm^{-2})	Sound Level (dB)
Saturn rocket motor, close by	1,100,000	195
Military rifle, peak level at ear	20,000	160
Jet aircraft takeoff; artillery, 2500′	2,000	140
Planing mill, interior	630	130
Textile mill	63	110
Diesel truck, 60′	6	90
Cooling tower, 60′	2	80
Private business office	0.06	50
Source	**Acoustic Power of Source**	
Saturn rocket motor	30,000,000 W	
Turbojet engine	10,000 W	
Pipe organ	10 W	
Conversational voice	10 μW	
Soft whisper	1 millimicrowatt	

From Liu, H.F., et al., 1997. Environmental Engineer's Handbook, second ed. Lewis Publishers, Boca Raton, FL. Reprinted by permission of Taylor and Francis Group, LLC, a division of Informa plc.

than low-frequency sounds (below 125 Hz, bass) (Boeker and van Grondelle, 1999). However, hearing begins to decline significantly after the age of 30. This change is most pronounced in men and for high-frequency sounds (at or above 4000 Hz) (Liu et al., 1997). Hearing loss, due to repeated exposure to loud noise, can also occur prematurely from injury to the hair cells (sound receptors) found inside the cochlear. There is evidence that exposure to noises from modern society accelerates hearing losses (Liu et al., 1997).

18.4.3 Sources of Noise

Modern life has brought about many types of sources of noise generated from industry, construction, transportation, and community and household activities. With few exceptions, these sources of noise are artificial by-products of industrialization, mechanization, crowded urban living, mechanized transport, and electronically reproduced and broadcasted sound. Table 18.3 presents the typical noise levels of common noise sources. Noises can be continuous or intermittent and sporadic, depending on the source. In general, industrial noises tend to be continuous repetitive or momentary. Transportation noises can be continuous, random, or intermittent. In addition, they can have increasing or decreasing sound intensities. Urban environments produce all types of noises that tend to vary significantly by location and time of day. The most disturbing urban noises are sporadic noises produced by sirens, trucks, motorcycles,

and car alarms. In suburbia, background noise associated with urban environments is minimal, making sporadic noises associated with motorcycles, car and truck engines, and dogs barking, much more noticeable. Household noises may be internal or include significant outside noise encroachment. Modern homes are better sound insulated, but still have near-constant sources of low-level noise, including air conditioning or heating units, refrigerators, and personal computers. Sporadic sources of noise in household include telephones, music, dogs, and assorted mechanical noises from doors, chairs, and dishes.

18.4.4 Effects of Noise

Noise affects us in different ways and to varying degrees, depending on our age, habits, health, and mood. Very loud sustained, repetitive, and even sporadic noises such as gunshots can damage our hearing progressively or traumatically (see Table 18.3). Progressive loss of hearing starts with sustained exposure to noise around 75 dB or above (Liu et al., 1997). Eardrum rupture can occur when exposed to very loud sharp noises, such as firecrackers or rifle shots. Other factors that affect the degree of hearing loss include the duration of exposure, including constant versus short duration noises. While the eyelid can protect the eye from bright light, the ear cannot close itself to sound or protect itself well against loud instantaneous noises of frequencies above 2000 Hz (Liu et al., 1997).

TABLE 18.3 Sound Intensity Factors, Sound Levels of Common Sound Sources and Their Effects

Sound Intensity Factor	Sound Level (dB)	Sound Sources	Effects (as Ranges About Each Designation Listed)		
			Perceived Loudness	Damage to Hearing	Community Reaction to Outdoor Noise
1×10^{18}	180	Rocket engine		Traumatic injury	
1×10^{17}	170				
1×10^{16}	160			Injurious range; irreversible damage	
1×10^{15}	150	Jet plane at takeoff	Painful		
1×10^{14}	140				
1×10^{13}	130	Maximum recorded rock music			
1×10^{12}	120	Thunderclap			
		Textile loom		Danger zone; progressive loss of hearing	
		Auto horn, 1 m (3.3 ft) away	Uncomfortably loud		
1×10^{11}	110	Riveter			
		Jet flyover at 300 m (985 ft)			
1×10^{10}	100	Newspaper press			
1×10^{9}	90	Motorcycle, 8 m (26 ft) away			Vigorous action
		Food blender	Very loud		Threats
1×10^{8}	80	Diesel truck, 0 km/h (50 mph), 15 m (50 ft) away		Damage begins after long exposure	
		Garbage disposal			Widespread complaints Occasional complaints
1×10^{7}	70	Vacuum cleaner	Moderately loud		
1×10^{6}	60	Ordinary conversation			
		Air-conditioning unit, 6 m (20 ft) away			
1×10^{5}	50				
		Light traffic noise, 30 m (100 ft) away			No action
1×10^{4}	40	Average living room			
1×10^{3}	30	Bedroom			
		Library	Quiet		
1×10^{2}	20	Soft whisper			
1×10^{1}	10	Broadcasting studio	Very quiet		
1×10^{0}	0	Rustling leaf Threshold of hearing	Barely audible		

From Turk, J., Turk, A., 1977. Physical Science with Environmental and Other Practical Applications. W. B. Saunders Company, Philadelphia.

Noises can have varying psychological impacts, depending on the level, duration, location, and time of occurrence, and our mental state. In general, noises can be distracting or annoying and degrade the quality of life. More serious effects include interference with verbal communication, reduced work efficiency, and the production of fatigue. Being awoken from sleep by barking dogs, a passing motorcycle, or car alarm is unpleasant and disruptive of sleep patterns. Repeated sleep disruptions can lead to poor concentration, mood changes, and stress during the day. The physiological effects of noise to humans are well documented and include changes in heartbeat, blood pressure, and increased respiration and pupil dilation (Welch and Welch, 1970; Vijayalakshimi and Phil, 2003; Muzet, 2007).

18.5 ODOR AS A SENSORY POLLUTANT

The sensation of odor is due to the stimulation of the olfactory organ in response to an exposure to volatile chemical odorant. Located in the nose, this organ likely evolved as a sensor capable of remotely detecting physical dangers and the presence of other animals at a distance without visual cues. Thus odors provide an early warning system for natural dangers and modern forms of air pollution. Although low-level exposure to some odorous chemicals may or may not be detrimental to one's health, the sensations they produce can be annoying. Studies have evaluated the complex relationships between odors and health effects, such as psychosomatic disease (Schiffman et al., 2000; Bulsing et al., 2008). In this section, we will discuss some common odors and the responses associated with them.

18.5.1 Odor Response

The response to a particular odor has a detection threshold that varies significantly from person to person. Thus odor thresholds are determined by a panel in which 50% of the individuals respond to the odor and 50% do not (Altwicker et al. in Liu et al., 1997). Environmental odors are usually complex mixtures of chemicals whose components are difficult to identify and quantify. In this case, an odor recognition threshold can be determined as described before, but using air/odor dilution ratios. This is commonly done to determine the potential for adverse effects of background or sporadic environmental odors, associated for example, with municipal wastewater treatment plants or industrial air emissions. This odor threshold unit is defined as:

Odor Unit = Volume of sample diluted to threshold response/volume of original sample

See Liu et al. (1997) for an extended description of this and other types of odor thresholds.

18.5.2 Odor Perceptions

The human response to odor types and concentrations varies widely, but has a normal distribution with threshold values that can vary over several orders of magnitude (Altwicker et al. in Liu et al., 1997). The acceptance of an odor is referred to as *hedonic tone*, which determines how pleasant or unpleasant a particular odor is perceived (Table 18.4). Repeated or prolonged exposure to a particular odor can become unpleasant or we may become desensitized to it. For example, persons living near natural sources of hydrogen sulfide gas such as swamps and thermal waters, may no longer be bothered by the rotten eggs smell sensed by newcomers. Tolerance to odors deemed otherwise unpleasant is common in industrial workers and chemists, for example, who work with the same chemical sources daily. However, nuisance complaints from communities about odors related to industrial and agricultural activities are more common, due in part to the growth of suburbia encroaching into industrial and agricultural environments. Also, there is increased public awareness about the potential health effects of vapor-phase chemicals originating from indoor sources, such as emissions (degassing) from construction materials such as carpets and insulation and furniture such as foam mattresses, and from the use of cleaning and other products (see Section 18.6).

18.5.3 Sources of Odor

Odors associated with raw sewage or municipal wastewater treatment are easily recognized by these two characteristics: the odor of rotten eggs and the odor rotten cabbage. There are numerous chemicals associated with these odors, including sulfur-based compounds like hydrogen sulfide and mercaptans, nitrogenous compounds like amines and scatole, aldehydes like butyraldehyde, and acids like butyric acid (Stuetz and Frechen, 2001). Individually and collectively, these chemicals produce odor responses with very low thresholds and are commonly associated with neighborhood nuisance complaints.

There is increasing evidence that repeated exposure to these odors produces long-lasting adverse effects including psychological effects, nausea, and stress (Liu et al., 1997). Although chemicals like hydrogen sulfide are very toxic, most odors act as warning signs that often become a nuisance at levels well below what are considered toxic by industrial and ambient air quality standards. For example, in the case of hydrogen sulfide, the human threshold of detection is ~0.15–0.5 ppb–v, but OSHA has set an occupational air safety exposure of 10 ppm–v. To avoid nuisance complaints, the World Health Organization (WHO) recommends that ambient air hydrogen sulfide concentrations do not exceed 0.5 ppb–v.

TABLE 18.4 Common Odors

Odor Character Descriptor	Potential Sources
Nail polish	Painting, varnishing, coating
Fishy	Fish operation, rendering, tannin
Asphalt	Asphalt plant
Plastic	Plastics plant
Damp earth	Sewerage
Garbage	Landfill, resource recovery facility
Weed killer	Pesticide, chemical manufacturer
Gasoline	Refinery
Airplane glue	Chemical manufacturer
Household gas	Gas leak
Rotten egg	Sewerage, refinery
Rotten cabbage	Pulp mill, sewage sludge
Cat urine	Vegetation

From Liu, H.F., et al., 1997. Environmental Engineer's Handbook, second ed. Lewis Publishers, Boca Raton, FL. Reprinted by permission of Taylor and Francis Group, LLC, a division of Informa plc.

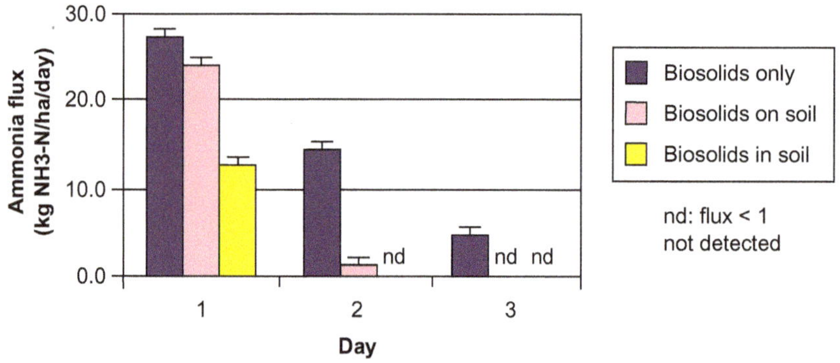

FIG. 18.4 Ammonia fluxes were measured using a chamber placed over samples of pure biosolids (~8% biosolids), biosolids applied to the surface of dry soil, and biosolids incorporated within dry soil. The figure illustrates the effectiveness of reducing ammonia emissions by incorporation of the biosolids into the soil. Ammonium N applied within the biosolids was ~364 kg NH_4-N ha^{-1}. (From Matthias, A.D., Artiola, J.F., unpublished data.)

Ammonia gas is a large component of odors derived from animal wastes and biosolids. Ammonia gas has a very sharp, pungent odor, which is generally only clearly recognized by humans at concentrations above 50 ppm–v (Merck Index, 1996). Common sources of this odor include dairy corrals and animal waste lagoons. Ammonia and other odorous gases emissions are also of concern in the land treatment of biosolids. Fig. 18.4 shows the fluxes of ammonia from biosolids over a 3-day period under laboratory conditions.

There are numerous types of odorous chemicals associated with industrial and transportation activities. A readily recognized odor is that of methyl methacrylate, which is used in the manufacture of plastics and resins. This synthetic toxic chemical is an irritant that like ammonia affects the mucous membranes and that, like hydrogen sulfide, has a very low odor threshold of ~0.1–100 ppb–v. Odors associated with transportation include distinctive gasoline fumes that are composed primarily of chemicals such as benzene, xylenes, and toluene that have relatively high odor thresholds (~0.05–4 ppm–v) and that are known to be carcinogenic and flammable.

18.6 INDOOR AIR QUALITY

The U.S. Environmental Protection Agency (U.S. EPA) ranks indoor air pollution among the top five risks to public health. Potential hazards that may be associated with indoor air include particulates, microbes, and chemicals. Defining the relative impact of specific indoor air pollutants is difficult because of individual genetic susceptibility and exposure to multiple hazards. Harmful exposure levels are defined for many chemical air pollutants. However, for other indoor pollutants, such as mold, little data is available with respect to acceptable exposure levels. Frequently, the consequences of exposure are evaluated based on retrospective, or even anecdotal, evidence.

In industrialized countries, about 90% of an individual's time is spent indoors, where air quality may be two to five times worse than outdoor air (U.S. EPA, 2001). Approximately one-third of all buildings are expected to have *indoor air quality* problems at some point during their operational lifespan. Health effects of poor indoor air quality range from mild and acute (cold and flu-like symptoms, headaches, and nausea), to severe and chronic (allergies, asthma, developmental disorders, cancer, and death). While a number of these building-related illnesses have been traced to specific building problems, conditions of complex symptomology related to chemical and/or biological indoor air quality problems are often vaguely diagnosed as *sick building syndrome* (SBS) and likely involve multiple pollutants acting collectively or synergistically. Excessive complaints, related to indoor air quality, are generated within 30% of new and remodeled buildings, worldwide.

Human activities and climate control efforts that were originally designed to make our lives more comfortable can exacerbate pollutant levels. For example, the production of energy-efficient homes has resulted in a decrease of indoor–outdoor air exchange, leading to decreased dilution of indoor air pollutants. Elevated humidity levels and temperature throughout buildings or in specific microzones, such as behind bathroom showers, or within cooking areas, also play a role in the increase of certain pollutants. Fuel-burning appliances, humidifiers, pesticides, cleaning products, and construction and textile materials certainly have beneficial applications but are also associated with toxic emissions and adverse health effects.

Hazards of the work environment are not necessarily different than those of the home, but exposures may be much greater. Longer periods of time may be spent in hazardous work environments amid larger concentrations of the pollutant. As discussed in Chapter 26, the effects of human exposure in the workplace are often our initial indicator of adverse health impacts of a particular constituent.

18.6.1 Sources of Indoor Air Pollutants

Indoor environments harbor a multitude of pollutant sources, making it difficult to identify the specific source of adverse health effects or the potential synergistic effects of multiple toxic exposures (Fig. 18.5). In the past, indoor air quality problems or SBS were primarily blamed on volatile organic compounds originating from cleaning solutions, paints, or chemicals in building materials and carpets. Biological contaminants (such as mold, dust mites,

and cockroach allergens) are now known to be significantly responsible. Other bioaerosols, such as viruses and bacteria, are microscopic and easily spread from person to person, where air and inanimate objects serve as intermediary transmission routes of infection. For many disease-causing microbes, infection rates are highest in winter months, when people spend more time indoors. Increases in indoor temperature and humidity, and reduced ventilation, have a substantial impact on the exposure levels of many biological and chemical pollutants. In fact, one in three buildings has damp conditions conducive to mold and bacterial growth (U.S. EPA, 2001).

Volatile organic compounds (VOCs) are low molecular weight chemicals that contain carbon and a variety of other elements, such as hydrogen, oxygen, fluorine, chlorine, bromine, sulfur, or nitrogen. They readily vaporize at room temperature, releasing noxious vapors into the air. Common household products, such as cleaning agents, disinfectants, and pesticides often contain high levels of VOCs. Some of these are capable of persisting in the indoor air space for a long time. Even seemingly innocuous materials, like cosmetics, air fresheners, or dry-cleaned fabrics can emit harmful compounds, such as diethanolamine, paradichlorobenzene, and tetrachloroethene, respectively.

VOCs present in indoor air may also originate from outside. For example, buildings located near hazardous waste sites may be affected by *vapor intrusion*, which is the movement of VOCs from the subsurface (contaminated soil and groundwater) into the building. Vapor intrusion is a potentially important exposure route that is evaluated for sites that are contaminated by high levels of VOCs. Vehicle exhaust is another source of VOCs, particularly for buildings located near major thoroughfares. It is often difficult to delineate the impacts of vapor intrusion because of other potential sources of VOCs within the building.

Construction materials and furnishings used in home and building construction and indoor furnishings frequently contain and emit hazardous compounds. Historically, asbestos was used in a variety of construction materials for insulation and as a fire retardant. Many asbestos products have been removed from buildings and continued use has been banned, but older homes may still contain potentially harmful materials. If in good condition and left undisturbed, asbestos products are generally not a risk; however, aerosolized fibers can be respired, damaging the lungs and abdominal lining and leading to irreversible scarring and cancer.

Carpeting and installation materials, such as adhesives and padding, are known to emit volatile organic compounds. Eye, nose, and throat irritation; headaches; rashes; coughing; fatigue; and shortness of breath have all been reported following new carpet installations. Carpeting may also act as a sink for a multitude of chemical and biological pollutants, including pesticides, dust mites, and molds that

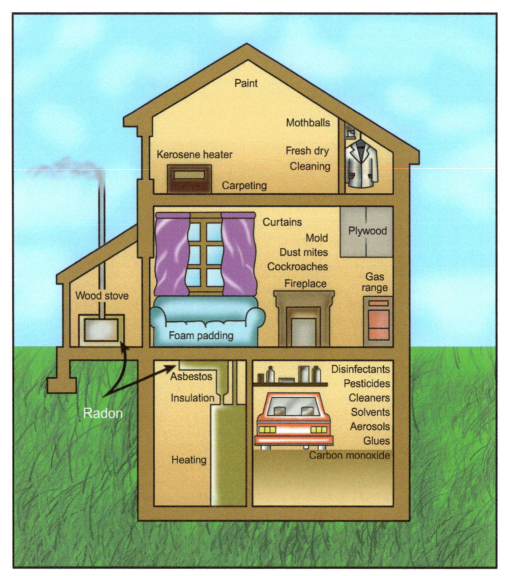

FIG. 18.5   Common sources of indoor air pollutants.

may collect in carpet fibers and remain protected from cleaning and vacuuming.

Formaldehyde is a known human carcinogen but is still widely used in the manufacture of household fabrics, paints, and furniture. Pressed wood products, such as particleboard, are generally produced with formaldehyde-containing resins that are released over time into the air. Those with urea–formaldehyde resins are associated with the highest pollutant emissions. Exposure to formaldehyde toxin is generally via the nasal passages or adsorption through the skin. Symptoms of allergies, asthma, throat and nose irritation, headaches, and nausea are well documented at indoor levels above 0.1 ppm, and some states recommend target threshold levels of 0.05 ppm or lower (Liteplo et al., 2002). Government agencies have set limits on allowable formaldehyde emissions and the use of

certain construction materials, such as pressed wood and insulation.

Combustion products such as oil, gas, kerosene, wood, and coal are common to indoor environments due to the use of fuel-burning appliances, space heaters, fireplaces, and gas or wood stoves. If not properly vented, harmful pollutants, such as CO, NO_2, and particle or chemical irritants may be released into the air. Improperly installed or maintained chimneys or other ventilation outlets can cause a backdraft of pollutants into the home. *Environmental tobacco smoke* is also considered a combustion product, and exposure via second-hand smoke is a major health concern.

Lead is considered to be one of the greatest environmental threats to children's health in the United States (Chapter 26). Prior to recognizing the health risks associated with lead, it was used widely in plumbing materials, gasoline, and paints. Atmospheric pollutants from combustion of leaded gasoline, ore smelting, and the burning of fossil fuels are sources of lead from the outdoor environment that may be tracked indoors via dust, shoes, clothing, or pets. Although soils near roads are contaminated from years of leaded gasoline use, and exposures still occur from hobbies using lead solder, the greatest exposure to lead indoors is from peeling, chipped, or sanded paint. Although banned from use in 1960, older homes may still contain heavily leaded paint.

Lead particulates settle on surfaces of indoor environments and are readily redispersed into the air via air currents common to indoor climates. Contaminated lead particles may be inhaled and ingested. Both result in absorption into the blood, where it is then distributed to soft tissues and bone. As lead accumulates over time, it can eventually affect nearly every system in the body. At levels above $80 \mu g \, dL^{-1}$ of blood, convulsions, coma, and even death can occur. Levels as low as $10 \mu g \, dL^{-1}$ can impair mental and physical development, resulting in lower IQ levels, shortened attention spans, and increased behavioral problems (Lin-Fu, 1992). This is particularly true for actively growing children.

Radon is a radioactive gas released during the natural decay of uranium, a common element of rocks, soil, and water. Radon enters homes through cracks, drains, and even wells, similar to vapor intrusion discussed earlier for VOCs. Once inside, the colorless, odorless gas can become trapped in living spaces, increasing in concentration and remaining undetected. Extended exposure via breathing the radioactive gas into the lungs can result in lung cancer. Radon gas is listed as the second leading cause of lung cancer in the United States. An estimated 21,000 deaths per year could be prevented by addressing radon gas exposures, particularly among smokers, who are known to be at increased risk due to synergistic interactions between radon and smoking (U.S. EPA, 2004).

The only way to be sure that your home, school, or workplace is free of radon is to conduct individual testing. High indoor levels of radon have been documented in every state. Although long-term exposure to any level of radon may be harmful, indoor environments with radon levels <4 picoCuries per liter ($pCi \, L^{-1}$) should be treated. Radon reduction systems are available that can decrease radon levels by 99% in the home.

The term *bioaerosol* encompasses any biological agent transmitted by the airborne route, that is, bacteria, viruses, mold, mites, cockroach particles, pollen, and animal dander and saliva. All of these agents have been associated with adverse health effects, including allergies and asthma, and often coexist in common environments. Allergic diseases have significantly increased worldwide over the last 30 years. More than 50% of all allergic diseases are caused

FIG. 18.6 Bathroom flooring colonized with mold. *(Photo courtesy K.A. Reynolds.)*

by allergens out of the indoor environment, and nearly one in six persons in the United States is effected by hypersensitivity reactions. Indoor molds, in particular, are a rising concern with ambient air contamination, ranking among the most important allergens of indoor environments.

Biological agents can persist in dust particles and animal droppings, or proliferate in humid microzones until they become aerosolized. Natural breezes, air-conditioning systems, humidifiers, and active movement all create eddies that aid in the aerosolization of spores, microbes, and other toxins. In moist environments, mold and bacteria grow in less than 72 h, colonizing solid surfaces and subsequently releasing toxins, particulates, and allergens into air spaces (Fig. 18.6).

18.6.2 Factors Influencing Exposure to Indoor Air Pollution

Identification of specific hazards and determining exposure levels for pollutants are critical steps in assessing an overall health risk (see Chapter 29). In most instances, minimizing exposure to specific hazards will reduce the risk of adverse health effects. In the case of building-related illnesses, symptoms may be apparent only when a particular environment (work, home, or school) is occupied, and they may be chronic and continue until appropriate treatment is sought. Exposure may be continuous or intermittent. Symptoms may be acute or develop over time, and manifest as large outbreaks or only among individuals that are particularly susceptible.

Indoor pollutant exposures are generally controlled by targeting source prevalence, poorly maintained environments or operated systems, and improper building design, or a combination of these factors. For example, an analysis of multiple effects on the prevalence of mucosal irritation and other general adverse symptoms among office workers showed that the concentration of visible floor dust, the type of floor covering, the number of workplaces in the office, the age of the building, and the type of ventilation were all associated with the prevalence of symptoms (Skov et al., 1990). To minimize poor indoor air quality, it is imperative that proactive efforts in building design, construction, operation, and maintenance be implemented.

Removing the pollutant source is one way to reduce indoor air quality exposures. However, this may not always be possible. Active educational campaigns on how individuals can reduce indoor exposures to contaminants in addition to efforts to reduce the sources, such as lead and asbestos, have led to dramatic reductions in reported disease related to these hazards. Similarly, smoking cessation education is important, particularly in the prenatal care medical sector. Following label information for product use regarding exposure precautions and ventilation suggestions, as well as minimizing use of products with known carcinogens, are two primary steps that allow source reduction when use cannot be completely avoided. This is particularly important for common household cleaners and pesticides.

Everyone is exposed to some level of indoor chemical and biological pollution, but the health effects of those pollutants is highly individualized. Symptom manifestation and severity depends largely on host response/susceptibility, dose, exposure route, the time and frequency of the exposure, and the nature and toxicity of the pollutant. Health effects resulting from exposures are dependent upon the health status of an individual, as well as the dosage and route of exposure. As a group, immunocompromised populations are more likely to experience increased negative outcomes following exposure to indoor air pollutants. Sensitive populations constitute 20%–25% of the population, and include the very young, the elderly, pregnant women, and persons with diminished immunity either due to medical intervention (organ transplants), previous illnesses (cancer, liver disease), preexisting respiratory conditions (allergies, chemical sensitivity, asthma), or infections (AIDS patients).

18.6.3 Monitoring Indoor Air Quality

Overall, many indoor air quality problems are due to inadequate ventilation. Therefore the American Society of Heating, Refrigerating, and Air-Conditioning Engineers (ASHRAE) established recommended ventilation rates for indoor air environments in 1973 and currently recommend a standard of 20 cubic feet per minute of outdoor air per person for general office space. In the occupational arena, airborne, contaminant standards are set for specific hazards (Information Box 18.2). For example, the OSHA PEL (permissible exposure limit) for carbon monoxide is 50 parts per million (ppm) averaged over an 8-h workday.

INFORMATION BOX 18.2 Exposure Standards for Airborne Contaminants

Airborne contaminant standards are set, for individual hazards, by various public health, occupational health, and regulatory agencies to guard against exposure levels that could cause adverse health effects.

Exposure Standards are guideline exposure values given below for contaminants that produce irritating or hazardous effects. They may consider ceiling values of acceptable acute exposures, or averages of chronic exposures, where contaminants can accumulate over time. The American Conference of Governmental Industrial Hygienists (ACGIH) sets *Threshold Limit Values* (TLVs) for protecting workers from occupational exposures. These are values thought to be safe for all workers to be exposed to daily without harmful effects. Samples are collected in consideration of

8-h Time-Weighted Averages (TWAs), reflecting the average exposure level over an 8-h shift, 5 days per week. TWA evaluate long-term chronic exposures.

Short-term Exposure Limits (STELs), evaluating 15- to 30-min peak exposure levels. STEL should not exceed three times the TWA more than 30 min in an 8-h workday and never exceed 5 times the TWA.

Permissible Exposure Limits (PELs) are enforceable standards set by Occupational Safety and Health Agency (OSHA) based on 8-h TWAs.

Monitoring for biological pollutants can be more problematic, since molds initiate allergic reactions in both the viable and nonviable state. Thus a variety of monitoring methods must be used, including cultural analysis for viable spore counts and microscopic methods for inactive allergenic particles. Additionally, chemical monitoring methods of mold metabolites, such as mycotoxins, are necessary to fully evaluate the contaminated environment.

Currently, there are no EPA regulations or standards for airborne mold contaminants; therefore sampling of airborne microorganisms is often inconclusive. In addition, there are no absolute monitoring standards or Threshold Limit Values (TLVs) for airborne concentrations of mold or mold spores. Therefore decisions on whether a building has a mold problem are often made arbitrarily. Different methods of collection can give divergent results and adverse outcomes may be heavily subject to interpretation.

Improving indoor air quality requires a multidisciplinary approach utilizing educational entities, research programs, public health administrators, architects, engineers, industrial hygienists, and environmental scientists. Benefits are realized by pollution prevention approaches as well as treatment measures following exposure. With ample unknowns and growing concerns, the health benefit of improved indoor air quality is expected to be a rapidly expanding and highly integrated field of study.

18.7 OTHER ISSUES

18.7.1 Radiation

Radiation is an invisible form of pollution. There are two types of radiation—ionizing and nonionizing (Table 18.5). Ionizing radiation contains sufficient energy in the waves (photons) to break chemical bonds and force electron transitions in atoms. It is well established that that exposure to ionizing radiation can be harmful to living organisms.

Radiation generated from low-frequency electromagnetic fields (EMF), such as microwaves and radio waves, is nonionizing. We are constantly exposed to the earth's permanent magnetic field. However, much stronger EMFs are present around us as the result of extensive use of electrical appliances and electrical power lines.

There are concerns about the long-term exposure effects of ever more common electrical and magnetic fields (EMF) generated by power lines, radio and TV, portable telephone transmitters, and other modern wireless communication devices. Numerous studies have thus far failed to show conclusively any adverse health effects associated with exposures to EMF sources associated with modern occupational and general public activities. The World Health Organization states the following: "current evidence does not confirm the existence of any health consequences from exposure to low level electromagnetic fields. However, some gaps in knowledge about biological effects exist and need further research." (WHO, 2018a).

18.7.2 Electronic Waste

Electronic waste, or e-waste, represents waste generated by the disposal of electronic products such as cell phones, computers, and televisions. It is a growing global problem, with an estimated 41.8 million metric tons produced globally in 2014 (UNEP, 2018). Open burning of e-waste can cause release of harmful substances (including dioxins, polycyclic aromatic hydrocarbons, and heavy metals) into the atmosphere. In addition, hazardous substances such as lead and mercury may leak from E-waste and contaminate nearby soils, surface water, and groundwater. People living near E-waste disposal sites may be exposed to these released contaminants.

The intrinsic material value of global e-waste was estimated to be $58 billion in 2014 (UNEP, 2018). Recycling of valuable elements contained in e-waste such as copper and gold has become a source of income mostly in the informal sector of developing or emerging industrialized countries (WHO, 2018b). However, poor recycling techniques such as burning cables for retaining the inherent copper expose both adult and child workers as well as their families to a range of hazardous substances.

The issue of e-waste and human health is particularly important in urban areas of developing countries due to the exporting of e-waste from developed to developing countries, and poor regulation of e-waste processing and disposal. Efforts are currently underway to improve e-waste management and reduce potential human health impacts.

QUESTIONS AND PROBLEMS

1. (a) Identify and describe two factors that influence heat island formation.
   (b) Why are heat islands a concern for human health?
   (c) Describe two measures that can be implemented to help mitigate heat islands.

TABLE 18.5 Types and Sources of Radiation

Radiation Type	Definition	Forms of Radiation	Source Examples
Nonionizing	Low to mid-frequency radiation which is generally perceived as harmless due to its lack of potency.	• Extremely Low Frequency (ELF) • Radiofrequency (RF) • Microwaves • Visual Light	• Microwave ovens • Computers • House energy smart meters • Wireless (wifi) networks • Cell Phones • Bluetooth devices • Power lines • MRIs
Ionizing	Mid to high-frequency radiation which can, under certain circumstances, lead to cellular and or DNA damage with prolonged exposure	• Ultraviolet (UV) • X-Rays • Gamma	• Ultraviolet light • X-Rays ranging from 30×10^{16} Hz to 30×10^{19} Hz • Some gamma rays

From: the National Institutes of Environmental Health Sciences, https://www.niehs.nih.gov/health/topics/agents/emf/index.cfm

2. Why is light pollution an environmental concern? Include examples of the effects of light pollution in your answer.
3. How can light pollution be reduced or prevented? Explain.
4. Can loud and repeated noise damage our hearing permanently? At what sound levels does damage to our ears begin? Explain.
5. Describe the characteristics of sick building syndrome. Discuss how you would differentiate potential causative sources of a case of SBS in your workplace.
6. Tour your home or workplace and identify the five top priority contaminants of your indoor environment. Discuss likely approaches to minimize your risk of exposure.

REFERENCES

Boeker, E., van Grondelle, R., 1999. Environmental Physics. John Wiley & Sons, New York (Chapter 6).

Bulsing, P., Smeets, M.A.M., Van den Hout, M.A., 2008. The implicit association between odors and illness. Chem. Senses 34 (2), 111–119.

Calgary Centre. (2005) Light pollution abatement site—Calgary Centre. Available from: http://calgary.rasc.ca/lp/greengas.html.

Cinzano, P., Falchi, F., Elvidge, C.D., 2001. The First World Atlas of the artificial night sky brightness. Mon. Not. R. Astron. Soc. 328, 689–707.

Lin-Fu, J.S., 1992. Modern history of lead poisoning: a century of discovery and rediscovery. In: Needleman, H.L. (Ed.), Human Lead Exposure. CRC Press, Boca Raton, FL, pp. 23–43.

Liteplo, R.G., Beauchamp, R., Meek, M.E., 2002. Formaldehyde. Concise International Chemical Assessment Document 40. World Health Organization, Geneva, Switzerland.

Liu H.F., Liptak B.G. and Bouis P.A. (1997) Environmental Engineer's Handbook. 2nd ed. Lewis Publishers, Boca Raton, FL (Chapters 5 and 6).

Merck Index, 1996. Merck Research Laboratories Division of, 12th ed. Merck & Co., Inc., Whitehouse Station, NJ.

McCarthy, M.P., Best, M.J., Betts, R.A., 2010. Climate change in cities due to global warming and urban effects. Geophys. Res. Lett. 37,https://doi.org/10.1029/2010GL042845, L09705.

Muzet, A., 2007. Environmental noise, sleep and health. Sleep Med. Rev. 11 (2), 135–142.

Santamouris, M., Cartalis, C., Synnefa, A., Kiloktsa, D., 2015. On the impact of urban heat island and global warming on the power demand and electricity consumption of buildings—a review. Energy Build. 98, 119–124.

Schiffman, S.S., Walker, J.M., Dalton, P., Lorig, T.S., Raymer, J.H., Shusterman, D., Williams, C.M., 2000. Potential health effects of odor from animal operations, wastewater treatment and recycling of byproducts. J. Agromed. 7, 7–81.

Skov, P., Valbjoern, O., Pedersen, B.V., 1990. Influence of indoor climate on the sick building syndrome in an office environment. Scand. J. Work Environ. Health, 16, 363–371.

Stuetz, R., Frechen, F.-B., 2001. Odours in Wastewater Treatment; Measurement, Modelling and Control. IWA Publishing, London England.

United Nations, 2018. World urbanization prospects 2018. Available from: https://esa.un.org/unpd/wup/.

United Nations Environment Programme, UNEP, (2018). E-waste. Available from: http://web.unep.org/ietc/what-we-do/e-waste. Accessed July 2018.

United States Environmental Protection Agency (USEPA), (2001). Healthy buildings, healthy people: a vision for the 21st century. Office of Air and Radiation 402–K–01–003.

United States Environmental Protection Agency (USEPA), (2004). A citizen's guide to radon: the guide to protecting yourself and your family from radon. Indoor Environments Division (6609J), Washington, DC. U.S. EPA 402–K–02–006.

Vijayalakshimi, K.S., Phil, M., 2003. Noise pollution. In: Bunch, M.J., Madha Suresh, V., Vasantha Kumaran, T. (Eds.), Proceedings of the Third Environmental Conference on Environmental Health. Chennai, India. 15–17 December 2003, pp. 597–603.

Welch, L., Welch, S., 1970. Physiological Effects of Noise. Plenum Press, New York and London.

World Health Organization, WHO, (2018a). Electromagnetic fields. Available from: http://www.who.int/peh-emf/about/WhatisEMF/en/index1.html. Accessed July 2018.

World Health Organization, WHO (2018b). Electronic waste. Available from: http://www.who.int/ceh/risks/ewaste/en/. Accessed July 2018.

FURTHER READING

Campbell, G.S., Norman, J.M., 1998. Introduction to Environmental Biophysics, 2nd Ed. Springer-Verlag, New York.

Chiras, D.D., 2001. Environmental Science, Creating a Sustainable Future, sixth ed. Jones and Bartlett Publishers, Sudbury, MA.

International Dark-Sky Association, IDA, 1996. Estimating the level of sky glow due to cities. Information Sheet 11. Available from: http://www.darksky.org/infoshts/is011.html.

International Dark-Sky Association, IDA. (2018). Light pollution wastes energy and money. Available from: http://www.darksky.org/light-pollution/energy-waste/.

Lawrence Berkeley National Laboratory, Available from: http://eetd.lbl.gov/heatisland.

Mizon, B., 2002. Light Pollution: Responses and Remedies. Springer-Verlag London Limited, Singapore.

Stern, D.P., 2001. The Exploration of the Earth's Magnetosphere. (Chapter 5: Magnetic field lines). Available from: http://www-istp.gsfc.nasa.gov/Education/Intro.html.

United States Environmental Protection Agency (U.S. EPA), 2005. Heat island effect. Available from: http://www.epa.gov/heatisland/about/healthenv.html.

Part III

Remediation, Restoration, Treatment, and Reuse

Chapter 19

Soil and Groundwater Remediation

M.L. Brusseau

Field-scale demonstration project for source-zone remediation at a site in Tucson, Arizona. *Photo courtesy William Blanford.*

19.1 INTRODUCTION

Public concern with polluted soil and groundwater encouraged the development of government programs designed to control and remediate this contamination, as well as to prevent further contamination. These programs were covered in Chapter 30. The institution of Superfund and other regulatory programs generated an entire industry focused on the characterization and remediation of hazardous waste sites. This industry is composed of regulators working for environmental agencies at all levels of government, research scientists and engineers developing and testing new technologies, scientists and engineers working for consulting firms contracted to characterize and remediate specific sites, and attorneys involved in myriad legal activities. In the following sections, basic concepts associated with site characterization and remediation will be presented.

19.2 SUPERFUND PROCESS

We will briefly discuss the major components of Superfund, given its central importance to characterization and remediation activities. There are two types of responses available within Superfund: (1) removal actions, which are responses to immediate threats, such as leaking drums, and (2) remedial actions, which involve cleanup of hazardous waste sites. The Superfund provisions can be used either when a hazardous substance is actually released or when the possibility of such a release is substantial. They may also be used when the release of a contaminant or possibility thereof poses imminent and substantial endangerment to public health and welfare. The process by which Superfund is applied to a site is illustrated in Fig. 19.1.

The first step is to place the potential site in the **Superfund Site Inventory**, which is a list of sites that are candidates for investigation. The site is then subjected to a preliminary assessment and site inspection, which may be performed by a variety of local, state, federal, or even private agencies. The site then undergoes a Hazard Ranking System (HRS) analysis, which is a numerically based screening system that uses information from the initial investigations—the preliminary assessment and the site inspection—to assess the relative potential of sites to pose a threat to human health or the environment. This approach assigns numerical values to factors that relate to risk based on conditions at the site. The factors are grouped into three categories:

- likelihood that a site has released or has the potential to release hazardous substances into the environment;
- characteristics of the waste (e.g., toxicity and waste quantity); and
- people or sensitive environments (targets) affected by the release.

Four pathways can be scored under the HRS—groundwater, surface water, soil, and air. The results of this analysis determine whether or not the site qualifies for the **National Priorities List** (NPL), which is a list of sites deemed by the U.S. Environmental Protection Agency (EPA) to require remedial action. Note that non-NPL sites may also need to be cleaned up, but their remediation is frequently handled by other federal or nonfederal agencies, with or without the help of the EPA.

The two-component **remedial investigation/feasibility study** (RI/FS) is the next step in the process. The purpose of the RI/FS is to characterize the nature and extent of risk posed by the contamination and to evaluate potential

Environmental and Pollution Science. https://doi.org/10.1016/B978-0-12-814719-1.00019-7

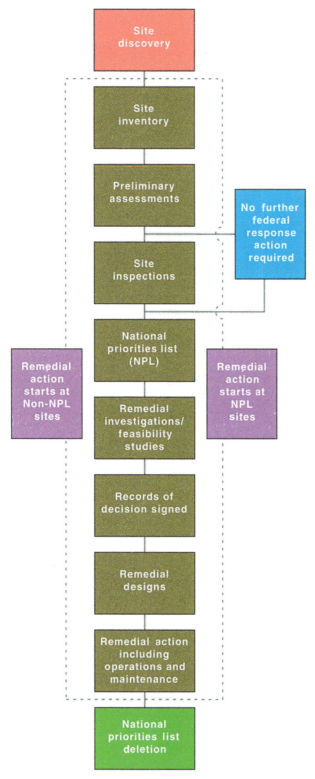

FIG. 19.1   The Superfund process for evaluating and remediating a hazardous waste site. *From Environmental Monitoring and Characterization. San Diego, CA, 2004, Elsevier Academic Press.*

remediation options. The investigation and feasibility study components of the RI/FS are performed concurrently, using a "phased" approach that allows feedback between the two components. A diagram of the RI/FS procedure is shown in Fig. 19.2. Implementing the RI/FS process at a site can take many years, with multiple cycles between the RI and FS components. In addition, large, complex sites in particular are often divided into individual "operable units" (OUs) that reflect a specific region of the site or a specific set of contaminants present at the site. For example, the source zone at the site may be designated as one OU, while the groundwater contaminant plume is designated as another. Individual RI/FS studies may be completed for individual OUs.

The selection of the specific remedial action to be used at a particular site is a very complex process. The goals of the remedial action are to protect human health and the environment, to maintain protection over time, and to maximize waste treatment (as opposed to waste containment or removal). Section 121 of CERCLA mandates a set of three categories of criteria to be used for evaluating and selecting the preferred alternative: threshold criteria, primary balancing criteria, and modifying criteria. These criteria are detailed in Information Box 19.1. The **threshold criteria** ensure that the remedy protects human health and the environment and is in compliance with applicable or relevant requirements (ARARs, such as local regulations), the **balancing criteria** ensure that trade-off factors such as cost and feasibility are considered, and the **modifying criteria** ensure that the remedy meets state and community expectations. An example of how remediation action alternatives are assessed is presented in Case Study 19.1 for a Superfund site in Arizona.

After a remedial action has been selected, it is designed and put into action. Some sites may require relatively simple actions, such as removal of waste storage drums and surrounding soil. However, the sites placed on the NPL generally have complex contamination problems and are therefore much more difficult to clean up. Because of this, it may take a long time to completely clean up a site and to have the site removed from the NPL. In fact, according to a recent study conducted by the National Research Council, cleanup of complex groundwater contamination sites is likely to take several to many decades.

An important component of Superfund and all other cleanup programs is the issue of deciding the target level of cleanup. Associated with this issue is the question "How clean is clean?" If, for example, the goal is to lower contamination concentrations, how low does the concentration level have to be before it is "acceptable?" As can be imagined, the stricter the cleanup provision, the greater

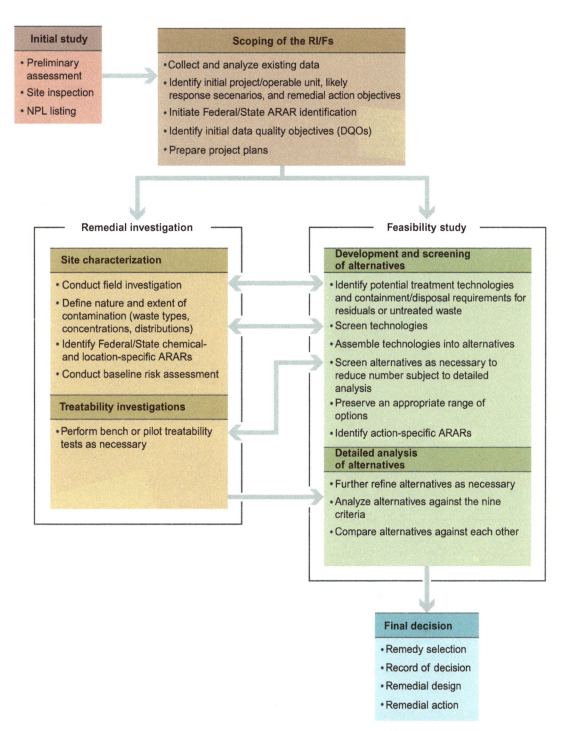

FIG. 19.2 The remedial investigation/feasibility study process (RI/FS). *Adapted from U.S. EPA, 1998. From Environmental Monitoring and Characterization. San Diego, CA, 2004, Elsevier Academic Press.*

will be the attendant cleanup costs. It may require tens to hundreds of millions of dollars to return large, complex hazardous waste sites to their original conditions. In fact, it may be impossible to completely clean up many sites. However, a site need not be totally clean to be usable for some purposes. It is very important, therefore, that the technical feasibility of cleanup and the degree of potential risk posed by

the contamination be weighed against the economic impact and the future planned use of the site. Consideration of the risk posed by the contamination and the future use of the site allows scarce resources to be allocated to those sites that pose the greatest current and future risk. The great difficulty of completely remediating contaminated sites highlights the importance of pollution prevention.

INFORMATION BOX 19.1 Criteria for Evaluating Remedial Action Alternatives

The EPA has developed criteria for evaluating remedial alternatives to ensure that all important considerations are factored into remedy selection decisions. These criteria are derived from the statutory requirements of CERCLA Section 121, particularly the long-term effectiveness and related considerations specified in Section 121 (bX1), as well as other technical and policy considerations that have proved important for selecting among remedial alternatives.

Threshold Criteria

The two most important criteria are statutory requirements that any alternative must meet before it is eligible for selection.

1. *Overall protection of human health and the environment* addresses whether or not a remedy provides adequate protection. It describes how risks posed through each exposure pathway (assuming a reasonable maximum exposure) are eliminated, reduced, or controlled through treatment, engineering controls, or institutional controls.

2. *Compliance with applicable or relevant and appropriate requirements (ARARs)* addresses whether a remedy meets all of the applicable or relevant and appropriate requirements of other federal and state environmental laws or whether a waiver can be justified.

Primary Balancing Criteria

Five primary balancing criteria are used to identify major trade-offs between remedial alternatives. These trade-offs are ultimately balanced to identify the preferred alternative and to select the final remedy.

1. *Long-term effectiveness and permanence* addresses the ability of a remedy to maintain reliable protection of human

health and the environment over time once cleanup goals have been met.

2. *Reduction of toxicity, mobility, or volume through treatment* addresses the anticipated performance of the treatment technologies employed by the remedy.

3. *Short-term effectiveness* addresses the period of time needed to achieve protection. It also assesses any adverse impacts on human health and the environment that may be posed during the construction and implementation period, until cleanup goals are achieved.

4. *Implementability* addresses the technical and administrative feasibility of a remedy, including the availability of materials and services needed to implement a particular option.

5. *Cost* addresses the estimated capital and operation and maintenance costs, and net present worth costs.

Modifying Criteria

These criteria may not be considered fully until after the formal public comment period on the Proposed Plan and RI/FS report is complete, although EPA works with the state and community throughout the project.

1. *State acceptance* addresses the support agency's comments. Where states or other federal agencies are the lead agencies, EPA's acceptance of the selected remedy should be addressed under this criterion. State views on compliance with state ARARs are especially important.

2. *Community acceptance* addresses the public's general response to the alternatives described in the Proposed Plan and the RI/FS report.

Case Study 19.1 Remediation Action Selection at a Superfund Site

This information was extracted from a January 2014 EPA report for the Phoenix-Goodyear Airport Superfund Site: North Area. The PGA Superfund Site was originally listed on the National Priorities List (NPL) in September 1983 as the Litchfield Airport Area Superfund Site. After the airport was transferred to the ownership of the City of Phoenix, the Site was renamed the PGA Area Superfund Site. Groundwater investigations later identified two different sources of contamination and the Site was divided into two areas, PGA-North and PGA-South. The focus of this case study for the proposed improved cleanup plan pertains to PGA-North only.

The PGA-North Source Area is located at a facility that operated as a research, design, development, testing, assembly, and manufacturing plant of ordnance components and related electromechanical devices from 1963 to 1993. Site contamination resulted from past disposal of waste materials from facility operations into a series of drywells. Contaminants from these wells travelled down through the soil to the shallow and deep groundwater aquifers of the Upper Alluvial Unit and have spread over time through the aquifers.

EPA determined that the primary contaminant of concern for remediation is the chlorinated volatile organic compound trichloroethene (TCE). To date, soil vapor extraction has removed most of the TCE in soils (more than 5400 kg), mainly near the Source Area. Based on later investigations, EPA added the inorganic compound perchlorate as a site contaminant of concern.

A groundwater contaminant plume approximately 5 km long is present at the site. The current PGA-North cleanup consists of extraction of contaminated groundwater and reinjection of treated water. TCE and perchlorate in the groundwater are treated by air stripping for TCE and ion exchange for perchlorate. This system of extraction and injection prevents groundwater contamination from spreading, thereby protecting the public water supply wells in the area.

The goal of the cleanup plan selected in the 1989 Record of Decision is restoration of the aquifer to drinking water standards. The Proposed Plan presented in this case study seeks to accelerate treatment of both contaminants in the Source Area in order to shorten the time needed to reach this goal of aquifer restoration.

Remedial Action Objectives are specific goals at each Superfund Site that EPA establishes to protect human health and the environment. These goals also assist EPA in measuring the effectiveness of remedial actions in achieving Superfund cleanups. The Remedial Action Objectives established for groundwater and soil in the 1989 Record of Decision for PGA-North are as follows:

- Restoration of Subunits A and C of the aquifer by reduction of groundwater contamination equal to or less than Applicable or Relevant and Appropriate Requirements (ARARS); ARARS are any state or federal environmental laws which apply to on-site remedial actions.

- Reduction of soil contamination in the source area where soil gas samples show VOCs greater than 1 ug/L, an area which may be expanded or reduced to include removal of 99% of the contamination;
- For soils, prevent migration of TCE into Subunit A and preserve uses of Subunit C groundwater;
- For groundwater, preserve the current use of Subunit C groundwater and protect future uses.

Overall, EPA's goals with the Proposed Plan are to improve and accelerate cleanup of the Source Area TCE and perchlorate in order to reach the Site Remedial Action Objectives established in 1989.

The improved cleanup plan also has Remedial Action Objectives, which are used to evaluate the effectiveness of the cleanup alternatives analyzed in the Focused Feasibility Study:

- Achieve permanent mass reduction within the Source Area of at least 80% for TCE and perchlorate in Subunit A
- Achieve permanent TCE and perchlorate concentration reduction of at least 80% within the Source Area

These RAOs were selected because an 80% reduction inTCE concentrations in the Source Area was projected to result in a major decrease of TCE concentrations and plume size.

Remedial alternatives that were considered in the Focused Feasibility Study are summarized in the table below. The cost information provided is based on preliminary estimates and, in accordance with EPA guidance, the cost estimates have an accuracy of plus 50% to −30%. The estimated time frames for each option to achieve the remedial action objectives are also included.

Inspection of the table reveals that the costs and time frames vary greatly among the options considered. Note that the Superfund process requires that a "no action" alternative be considered in each evaluation as a baseline to compare the remaining alternatives. Because a pump-and-treat system is currently in operation at this site to treat the plume, the "no action" alternative includes continuation of the existing pump-and-treat system but adds no improved remedial measures to speed the cleanup of the Source Area. The term "hydraulic barrier" in the table refers to the operating pump-and-treat system. The degree to which each alternative meets each of the evaluation criteria is presented in the table by the degree of shading within the circles, with an unshaded circle representing "Low" and a completely filled circle representing "High."

Comparative analysis of alternatives

EPA's preferred alternative

Evaluation criteria	Alternative 1 No action	Alternative 2 In-well air stripping + hydraulic barrier	Alternative 3 ARD+Hydraulic barrier	Alternative 4 nZVI + ZVI + ARD + Hydraulic barrier	Alternative 5 ZVI + ARD + Hydraulic barrier	Alternative 6 ISCO (permanganate) + Hydraulic barrier	Alternative 7 ERH + Steam + Hydraulic control
Protection of human health & the environment	Low-moderate	Low-moderate	Moderate	High	High	High	High
Compliance with ARARs	Low	High	High	High	High	High	High
Long-term effectiveness & permanence	Low	Low-moderate	Moderate	Moderate	Moderate	Moderate	High
Reduction of toxicity, mobility, or volume	Low	Low-moderate	Moderate	Moderate	Moderate	Moderate	High
Short-team effectiveness	Low	Low-moderate	Low	High	Moderate	High	High
Implementability	High	Moderate	Moderate	Moderate	Moderate	High	Low-moderate
Cost	High	Moderate	Moderate	Low-moderate	Low-moderate	Moderate	Low
State acceptance	Expected (EPA has worked closely with the state of arizona on this plan)						
Community acceptance	Community acceptance of preferred alternative will be evaluated after the public comment period						

Low Low-moderate Moderate Moderate-High High

Cost:	0	7M	9M	11.6M	13M	7.5M	31M
Time:	Decades	20Y	8Y	8Y	11Y	8Y	1Y

ARD = anaerobic reductive dechlorination

ERH = electrical resistance heating

ISCO = in situ chemical oxidation

ZVI = zero valent iron

n-ZVI = nano zero valent iron

19.3 SITE CHARACTERIZATION

Site characterization is a critical component of hazardous waste site remediation. Site characterization provides information that is required for conducting risk assessments and for designing and implementing remediation systems. A critical element of site characterization is the **Conceptual Site Model** (CSM). The CSM is used to organize and communicate information about site characteristics. It should reflect the best interpretation of available information at any point in time. The CSM is a living entity that needs to be revised and updated as new information is gathered. Among other things, the CSM provides a summary of how and where contaminants are expected to move and what impacts such movement may have. It is useful for supporting risk assessments; identifying additional site characterization needs; and supporting the selection, design, and performance assessments of remedial actions. Best management practices require the development and maintenance of a CSM for each site. The US EPA provides guidance on the development of CSMs.

The primary components of a typical hazardous waste site are the source zone and the groundwater contaminant plume. The source zone represents the location where contamination entered the environment, such as disposal pits, leaking tanks, spill sites. These zones are generally relatively small, in the range of a soccer field in area. Conversely, the groundwater contaminant plume can range in size from a few 10's of meters to many kilometers in length, depending upon the nature of the contamination and site conditions. In some cases extensive vapor contamination plumes may exist in the vadose zones of sites.

The primary objectives of site characterization are to identify the nature and extent of contamination. This includes identifying the types of contaminants present, the amount and location of contamination, and the phases in which it is occurring (e.g., does it occur only as dissolved contamination in groundwater or is NAPL also present?). Generally, the first step in identifying the contaminants potentially present at the site involves determination of prior land uses and operations at the site. This is followed by inspection of material records, if available. Useful information regarding the types of contaminants potentially present at the site can be obtained from chemical stock purchasing records, delivery records, storage records, and other types of documents. Important information may also be obtained from records of specific chemical spills or leaks, as well as waste disposal records. Unfortunately, such information is incomplete for many sites, especially for older or defunct sites.

After examining available records, site inspections are conducted to identify potential sources of contamination. This includes searching the site for drums, leaking storage tanks, and abandoned disposal pits or injection wells. Once located, actions are taken if necessary to prevent further release of contamination. For example, leaking tanks, along with the surrounding porous media, are excavated and removed.

The next step in site characterization generally involves a survey of groundwater contamination. This is accomplished through collecting groundwater samples from wells at and adjacent to the site. This component is often implemented in a phased approach. For example, in the initial characterization phase, when little is known about the type or extent of groundwater contaminants present, it is often necessary to analyze samples for a broad suite of possible contaminants. As noted in Chapter 12, this is typically done using the regulated contaminants list associated with the National Primary Drinking Water Standards (see Table 12.2). After several rounds of sampling, the list of analytes is often shortened to those that are most commonly found at the site or that pose the greatest risk. In addition, the initial phase typically makes use of existing wells and a few new wells placed to obtain a complete (but sparse) coverage of the site. Once the specific zones of contamination are identified, additional wells are often drilled in those areas to increase the density of sampling locations. Site characterization programs are generally focused on obtaining information on the areal (*x-y* plane) distribution of contamination. However, once this information is available, it is important to characterize the vertical distribution of contamination if possible. Specific methods for developing and implementing a groundwater sampling program are discussed in Chapter 10.

Groundwater sampling provides information about the types and concentrations of contaminants present in groundwater. This is a major focus of characterization at most sites because groundwater contamination is often the primary risk driver—meaning that use of contaminated groundwater represents the primary route of potential exposure. However, sampling programs can also be conducted to characterize contamination distribution for other phases. For example, as discussed in Chapter 10, sediment core material can be collected and analyzed to measure the amount of contamination associated with the porous media grains. In addition, various methods can be used to collect gas-phase samples in the vadose zone to evaluate vapor-phase contamination. This is of particular note for sites at which vapor intrusion may be an issue.

One component of site characterization that is often critical for sites contaminated by organic chemicals is determining if immiscible liquids are present in the subsurface. As discussed in Chapter 12, many organic contaminants are liquids, such as chlorinated solvents and petroleum derivatives. Their presence in the subsurface serves as a significant source of long-term contamination and greatly complicates remediation (see Chapter 15). In some cases, it is relatively straightforward to determine the presence of immiscible-liquid contamination at a site. For example, when large quantities of gasoline or other fuels are present, a layer of "free product" may form on top of the water table because the fuels are generally less dense than water (Fig. 15.14). This floating free product can be readily observed by collecting samples from a well intersecting the contamination.

Conversely, in other cases it may be almost impossible to directly observe immiscible-liquid contamination. For example, chlorinated solvents such as trichloroethene are denser than water. Thus large bodies of floating free product are not formed as they are for fuels (see Fig. 7.1). Therefore evidence of immiscible-liquid contamination must be obtained through the examination of sediment core samples and not from well sampling. Unfortunately, it is very difficult to identify immiscible-liquid contamination with the use of core samples. When these denser-than-water liquids enter the subsurface, localized zones of contamination will usually form due to the presence of subsurface heterogeneities (e.g., permeability variability). Within these zones, the immiscible liquid is trapped within pores and may form small pools above capillary barriers. Given the small diameter of cores (5–20cm), the chances of a borehole intersecting a localized zone of contamination are relatively small. Thus it may require a large, cost-prohibitive number of boreholes to characterize immiscible-liquid distribution at a site. Recognizing these constraints, alternative methods for characterizing subsurface immiscible-liquid contamination are being developed and tested. These include methods based on geophysical and tracer test techniques, some of which provide a larger scale of measurement.

While identifying the nature and extent of contamination is the primary site-characterization objective, other objectives exist. A common objective involves determining the physical properties of the subsurface environment. For example, pumping tests are routinely conducted to determine the hydraulic conductivity distribution for the site (see Chapter 3). This information is useful for determining the rates and direction of contaminant movement, and for evaluating the feasibility and effectiveness of proposed remediation systems. In some cases, sediment core samples will be analyzed to characterize geochemical and biological properties and processes pertinent to contaminant transport and remediation. For example, assessing the biodegradation potential of the microbial community associated with the porous medium is critical for evaluating the feasibility of employing bioremediation at the site.

In summary, site characterization is an involved, complex process composed of many components and activities. Generally, the more information available, the better informed are the site evaluations and thus the greater chance

of success for the planned remediation system. However, site characterization activities are generally expensive and time consuming. Thus different levels of site characterization may be carried out at a particular site depending on goals and available resources. Examples of three levels of site characterization are given in Information Box 19.2. Clearly, the basic approach is least costly, while the state-of-the-science approach is much more costly. However, the state-of-the-science approach provides significantly more information about the site compared to the basic approach.

The use of mathematical modeling has become an increasingly important component of site characterization and remediation activities. Mathematical models can be used to characterize site-specific contaminant transport and fate processes, to predict the potential spread of contaminants, to help conduct risk assessments, and to assist in the design and evaluation of remediation activities. As such, it is critical that models are developed and applied in such a manner to provide an accurate and site-specific representation of contaminant transport and fate. Employing advanced characterization and modeling efforts that are integrative and iterative in nature enhances the success of remediation projects.

Numerous approaches are available for modeling transport and fate of contaminants at the field scale. These can be grouped into three general types of approaches, differentiated primarily by the level of complexity (see Information Box 19.3). The standard approach is just that, it is the standard approach used almost exclusively in the analysis of Superfund and other hazardous wastes sites. The primary advantage of this approach is the relative simplicity of the model and the relatively minimal data requirements. Unfortunately, the usefulness of such modeling is very limited. Except for the simplest of systems (e.g., a nonreactive, conservative solute), this type of modeling cannot be used to characterize the contribution of specific processes or factors influencing the transport and fate of contaminants. In addition, because there is no real mechanistic basis to the model, its use for generating predictions is severely constrained.

The state-of-the-science approach is based on implementing cutting-edge understanding of transport and fate processes into fully three-dimensional models that incorporate spatial distributions of all pertinent properties. Clearly, this type of modeling, while desirable, is ultimately impractical for all but the most highly characterized research site.

The state-of-the-art approach is an intermediate-level approach that is process based, but also is developed with recognition of the data availability limitations associated with most sites. Because models based on this approach are process based, this type of modeling can be used to

INFORMATION BOX 19.2 Various Approaches to Site Characterization

Basic Approach

Activities

- Use existing wells; install several fully screened monitoring wells
- Sample and analyze for priority pollutants
- Construct geologic cross sections by using driller's log and cuttings
- Develop water level contour maps
  Advantages
- Rapid screening of site
- Moderate costs involved
- Standardized techniques
  Disadvantages
- True nature and extent of problem not identified

State-of-the-Art Approach

Activities

- Conduct basic site characterization activities
- Install depth-specific monitoring-well clusters
- Refine geologic cross sections using cores
- Conduct pumping tests
- Measure redox potential, pH, dissolved oxygen of groundwater
- Conduct solvent extraction of core samples to evaluate sorbed- and NAPL-phase contamination
- Conduct soil gas surveys
- Conduct limited geophysical surveys (resistivity soundings)
- Conduct limited mathematical modeling

Advantages

- Better understanding of the nature/extent of contamination
- Improved conceptual understanding of the problem
  Disadvantages
- Greater costs
- Detailed understanding of the problem still limited
- Demand for specialists increased

State-of-the-Science Approach

Activities

- Employ state-of-the-art approach
- Conduct tracer tests and borehole geophysical surveys
- Characterize geochemical properties of porous media (organic carbon, oxides, mineralogy)
- Evaluate sorption–desorption and NAPL dissolution behavior using selected cores
- Assess potential for bio- and abiotic transformation using selected cores
- Assess biotransformation potential by PCR analysis and application of "omics"
- Conduct advanced mathematical modeling
  Advantages
- Information set as complete as generally possible
- Enhanced conceptual understanding of the problem
  Disadvantages
- Characterization cost significantly higher
- Field and laboratory techniques not yet fully standardized
- Demand for specialists greatly increased

INFORMATION BOX 19.3 Various Approaches for Mathematical Modeling

Standard Approach

- "One-layer" model with areal hydraulic-conductivity distribution
- No spatial variability of chemical/biological properties
- Lumped macrodispersion term (lumps all contributions to spreading, mixing, and dilution)
- Dissolution of immiscible liquid simulated using a source term function
- Lumped retardation, with mass-transfer processes treated as linear and instantaneous
- Lumped dispersion term for systems with gas-phase transport
- Lumped first-order transformation term

State-of-the-Art Approach

- "Multilayer" model, with vertical and areal hydraulic conductivity distribution
- Layer/areal distribution of relevant chemical and biological properties
- Layer/areal distribution of immiscible liquid
- Rate-limited immiscible liquid dissolution
- Multiple-component immiscible liquid partitioning (using Raoult's Law)
- Rate-limited, nonlinear sorption/desorption
- Gas-phase transport of volatile organic contaminants (gas-phase advection, diffusion)

- Biodegradation processes coupled to mass-transfer processes
- Geochemical zone-dependent (layer/areal distributed) first-order rate coefficients

State-of-the-Science Approach

- Three-dimensional distributed hydraulic conductivity field
- Three-dimensional distribution of relevant chemical and biological properties
- Three-dimensional distribution of immiscible liquid
- Rate-limited immiscible liquid dissolution
- Immiscible liquid composition effects (multiple-component behavior)
- Rate-limited, nonlinear sorption/desorption
- Gas-phase transport of volatile organic contaminants (gas-phase advection, diffusion)
- Gas-phase retention/mass transfer (air–water mass transfer, adsorption at air–water interface, vapor-phase adsorption, immiscible liquid evaporation)
- Biodegradation coupled to mass-transfer processes
- Microbial dynamics: population growth, death, and cell transport
- Biogeochemical properties: multiple electron acceptors

characterize the contribution of specific processes or factors influencing the transport and fate of contaminants. Thus this modeling approach can be used to effectively evaluate the impact of the key factors influencing the contamination potential of waste sites.

19.4 REMEDIATION TECHNOLOGIES

There are three major categories or types of remedial actions: (1) **containment**, where the contaminant is restricted to a specified domain to prevent further spreading; (2) **removal**, where the contaminant is transferred from an open to a controlled environment; and (3) **treatment**, where the contaminant is transformed into a nonhazardous substance. Since the inherent toxicity of a contaminant is eliminated only by treatment, this is the preferred approach of the three. Containment and removal techniques are very important, however, when it is not feasible to treat the contaminant. Although we will focus on each of the three types of remedial actions in turn, it is important to understand that remedial actions often consist of a combination of containment, removal, and treatment.

So-called **Institutional Controls** are another alternative approach that can be considered. These include, for example, the use of fencing to restrict access to a contaminated site and the use of advisories to limit use of contaminated resources (e.g., "do not drink the water" advisories). These approaches typically do not provide treatment or cleanup of the contamination but are used to reduce risk by limiting exposure. They may often be used temporarily while more permanent remediation methods are considered or implemented.

19.4.1 Containment Technologies

Containment can be accomplished by controlling the flow of the fluid that carries the contaminant or by directly immobilizing the contaminant. We will briefly discuss the use of physical and hydraulic barriers for containing contaminated water, which are the two primary containment methods. We will also briefly mention other containment methods.

19.4.1.1 Physical Barriers

The purpose of a physical barrier is to control the flow of water to prevent the spread of contamination. Usually, the barrier is installed in front (downgradient) of the contaminated zone (Fig. 19.3); however, barriers can also be placed upgradient or both up- and downgradient of the contamination. Physical barriers are primarily used in unconsolidated materials such as soil or sand, but they may also be used in consolidated media such as rock if special techniques are employed. In general, physical barriers may be placed to depths of about 15–25 m, with depths

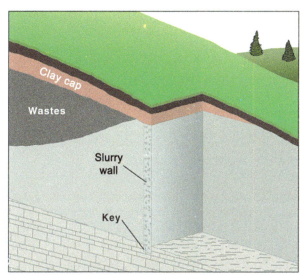

FIG. 19.3 Physical containment of a groundwater contaminant plume with the use of a slurry wall. *From Environmental Monitoring and Characterization. San Diego, CA, 2004, Elsevier Academic Press.*

of up to ~50 m achievable using specialized techniques. The horizontal extent of the barriers can vary widely, depending on the size of the site, from tens to hundreds of meters.

One important consideration in the employment of physical barriers is the presence of an extensive horizontal layer of low permeability beneath the site, into which the physical barrier can be seated. Without such a seating into a low permeability zone (a "key"), the contaminated water could flow underneath the barrier. Another criterion for physical barriers is the permeability of the barrier itself. Since the goal of a physical barrier is to minimize fluid flow through the target zone, the permeability of the barrier should be as low as practically possible. Another factor to consider is the potential of the contaminants to interact with the components of the barrier and degrade its performance. The properties of the barrier material should be matched to the properties of the contaminant to minimize failure of the barrier.

There are three major types of physical barriers: slurry walls, grout curtains, and sheet piling. **Slurry walls** are trenches filled with clay or mixtures of clay and soil. **Grout curtains** are hardened matrices formed by cement-like chemicals that are injected into the ground. **Sheet piling** consists of large sheets of iron that are driven into the ground. Slurry walls are the least expensive and most widely used type of physical barrier, and they are the simplest to install. Application of slurry walls is limited to sites with unconsolidated media (sand, gravels, and clay). Grout curtains, which can be fairly expensive, are used primarily for sites having consolidated subsurface environments. Sheet piles have essentially zero permeability and are generally of low reactivity. They can leak, however, because it is difficult to obtain perfect seals between individual sheets. Moreover, sheet piling is generally more expensive than slurry walls, and it is difficult to drive sheet piles into rocky ground. Thus they are used primarily for smaller scale applications at sites composed of unconsolidated materials.

19.4.1.2 Hydraulic Barriers

The principle behind hydraulic barriers is similar to that behind physical barriers—to manipulate and control water flow. But unlike physical barriers, which are composed of solid material, hydraulic barriers are based on the manipulation of water pressures. They are generated by the pressure differentials arising from the extraction or injection of water using wells or drains. The key performance factor of this approach is its capacity to capture the **contaminant plume**, that is, to limit the spread of the contamination. Plume capture is a function of the number and placement of the wells or drains, as well as the rate of water flow through the wells or drains. Often, attempts are made to optimize the design and operation of the containment system so that plume capture is maximized while the volume of contaminated water removed is minimized.

The simplest hydraulic barrier is that established by a drain system. Such a system is constructed by installing a perforated pipe horizontally in a trench dug in the subsurface and placed to allow maximum capture of the contaminated water. Water can then be collected and removed by using gravity or active pumping. The use of drain systems is generally limited to relatively small, shallow contaminated zones.

Well-field systems are more complicated—and more versatile—than drain systems. Both extraction and injection wells can be used in a containment system, as illustrated in Fig. 19.4. An extraction well removes the water entering the zone of influence of the well, creating a cone of depression. Conversely, an injection well creates a pressure ridge, or mound of water under higher pressure than the surrounding water, which prevents flow past the mound. One major advantage of using wells to control contaminant movement is that this is essentially the only containment technique that can be used for deep (>50m) systems. In addition, well-fields can be used for contaminant zones of any size; the number of wells is simply increased to handle larger problems. For example, some large sites have contaminant plumes that are several kilometers long. For these and other reasons, hydraulic barriers are the most widely used method for containment, despite disadvantages that include the cost of long-term operation and maintenance and the need to store, treat, and dispose of the large quantities of contaminated water pumped to the surface.

19.4.1.3 Other Containment Methods

A variety of techniques have been used to attempt to immobilize subsurface contaminants by fixing them in an impermeable, immobile solid matrix. These techniques are referred to alternatively as solidification, stabilization, encapsulation, and immobilization. They are generally based on injecting a solution containing a compound that will cause immobilization or encapsulation of the contamination. For example, cement or a polymer solution can be added, which converts the contaminated zone into a relatively impermeable mass encapsulating the contaminant.

In another approach, a reagent can be injected to alter the pH or redox conditions of the subsurface, thus causing the target contaminant to "solidify" in situ. For example, promoting reducing conditions will induce chromium to change its predominant speciation from hexavalent, which is water soluble and thus "bioavailable," to trivalent, which has low solubility and thus precipitates on the porous media grains (and is therefore no longer readily available).

In general, it is difficult to obtain uniform immobilization due to the natural heterogeneity of the subsurface. Thus containment may not be completely effective. In addition, a major factor to consider for these techniques is the long-term durability of the solid matrix and the potential for leaching of contaminants from the matrix.

Vitrification is another type of containment, in which the contaminated matrix is heated to high temperatures to "melt" the porous media (i.e., the silica components), which subsequently cools to form a glassy, impermeable block. Vitrification may be applied either in situ or ex situ. For treatment of typical hazardous waste sites, potential applications would primarily be in situ. Aboveground applications are being investigated for use in dealing with radioactive waste. Vitrification is an energy-intensive, disruptive, and relatively expensive technology, and its use would generally be reserved for smaller scale sites for which other methods are not viable.

19.4.2 Removal

19.4.2.1 Excavation

A very common, widely used method for removing contaminants is **excavation** of the porous media in which the contaminants reside. This technique has been used at many sites and is highly successful. There are, however, some disadvantages associated with excavation. First, excavation can expose site workers to hazardous compounds. Second, the contaminated media requires treatment and/or disposal, which can be expensive. Third, excavation is usually feasible only for relatively small, shallow areas. Excavation is most often used to remediate shallow, localized, highly contaminated source zones.

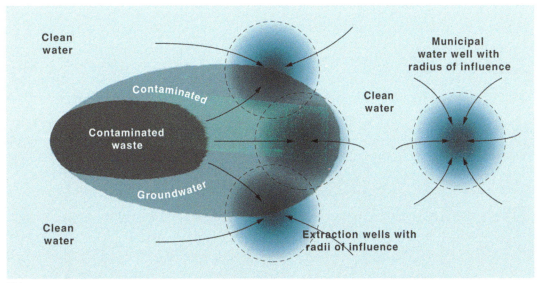

(A) Overhead view of the remediation site showing groundwater surfaces, flow directions, and contaminated waste (soil material removed).

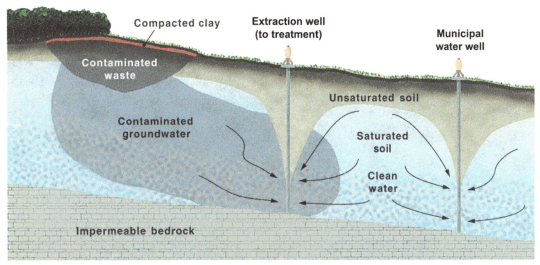

(B) Cut-away view of the above site showing lateral movement of leached contaminants and groundwater.

FIG. 19.4 Containment of a groundwater contaminant plume with the use of a hydraulic barrier. (A) Overhead view of the site, showing the source zone, the contaminant plume, and the direction of groundwater flow. (B) Cross section view. *Adapted from U.S. EPA, 1985. From Environmental Monitoring and Characterization. San Diego, CA, 2004, Elsevier Academic Press.*

19.4.2.2 Pump-and-Treat

Pump-and-treat is the most widely used remediation technique for contaminated groundwater. For this method, one or more extraction wells are used to remove contaminated water from the subsurface. Furthermore, clean water brought into the contaminated region by the flow associated with pumping removes, or "flushes," additional contamination by inducing desorption from the porous media grains and dissolution of NAPL. The contaminated water pumped from the subsurface is directed to some type of treatment operation, which may consist of air stripping, carbon adsorption, or perhaps an aboveground biological treatment system. An illustration of a pump-and-treat system is provided in Fig. 19.5. Pump-and-treat and hydraulic control using wells are essentially the same technology; they just have different objectives. For containment, we usually want to minimize water and contaminant extracted, whereas for pump-and-treat, we want to maximize contamination extracted.

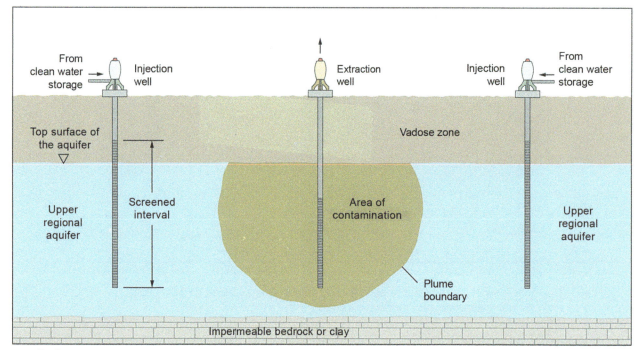

(A) Before treatment

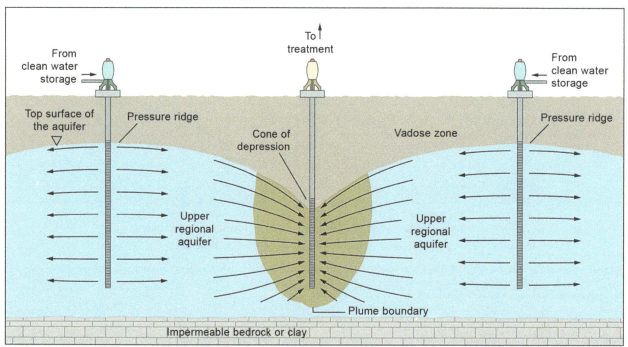

(B) After flow initiation

FIG. 19.5 Remediation of groundwater contamination by the pump-and-treat process. (A) Before treatment. (B) After start of remediation. *From Pollution Science. San Diego, CA, 1996, Academic Press.*

Usually discussed in terms of its use for such saturated subsurface systems as aquifers, the pump-and-treat method can also be used to remove contaminants from the vadose zone. In this case, it is generally referred to as **in situ soil washing.**

For this application, infiltration galleries, in addition to wells, can be used to introduce water to the contaminated zone.

When using water flushing for contaminant removal (as in pump-and-treat), contaminant plume capture and the effectiveness of contaminant removal are the major performance criteria. Studies of operating pump-and-treat systems have shown that the technique is very successful at containing contaminant plumes and, in some cases, shrinking them. However, pump-and-treat is generally ineffective for completely removing contaminants from the subsurface. There are many factors that can limit the effectiveness of water flushing for contaminant removal including:

1. **Presence of low-permeability zones.** When low-permeability zones (e.g., silt/clay lenses) are present within a sandy subsurface, they create domains through which advective flow and transport are minimal in comparison to the surrounding sand (see Chapter 7). The groundwater flows preferentially around the clay/silt lenses, rather than through them. Thus contaminant located within the clay/silt lenses is released to the flowing water primarily by diffusion, which is a relatively slow process. This limits the amount of contaminant present in the flushing water, thereby increasing the time required to completely remove the contamination.

2. **Rate-limited, nonlinear desorption.** Research has revealed that adsorption/desorption of many solutes by porous media can be significantly rate limited. When the rate of desorption is slow enough, the concentration of contaminant in the groundwater is lower than the concentrations obtained under conditions of rapid desorption. Thus less contaminant is removed per volume of water, and removal by flushing will therefore take longer. In addition, many contaminants have nonlinear sorption (see Chapter 8). When this occurs, it becomes more difficult to remove contamination as the concentration decreases, because the proportion sorbed increases.

3. **Presence of immiscible liquid.** In many cases, immiscible organic liquid contaminants may be trapped in portions of the contaminated subsurface. Since it is very difficult to displace or push out this trapped contamination with water, the primary means of removal will be dissolution into water. It can take a very long time to completely dissolve immiscible liquid, thus greatly delaying removal. The immiscible liquid, therefore, serves as a long-term source of contaminant.

Because pump-and-treat is a major remedial action technique, methods are being tested to enhance its effectiveness. One way to improve the effectiveness of pump-and-treat is to contain or remove the **contaminant source zone**, that is, the area in which contaminants were disposed or spilled. If the source zone remains untreated or uncontrolled, it will serve as a continual source of contaminant requiring removal. Thus failure to control or treat the source zone can greatly extend the time required to achieve site cleanup. It is important, therefore, that the source zone at a site be delineated and addressed in the early stages of a remedial action response. This might be done by using a physical or hydraulic barrier to confine the source zone. Other methods of treating the source zone involve enhancing the rate of contaminant removal, as detailed in the following sections.

19.4.2.3 Enhanced Flushing

Contaminant removal can be difficult because of such factors as low solubility, high degree of sorption, and the presence of immiscible-liquid phases, all of which limit the amount of contaminant that can be flushed by a given volume of water. Approaches have been developed to enhance the removal of low-solubility, high-sorption contaminants. One such approach is to inject a reagent solution into the source zone, such as a **surfactant** (e.g., detergent molecule), that will promote dissolution and desorption of the contaminant, thus enhancing removal effectiveness. Such surfactants work like industrial and household detergents, which are used to remove oily residues from machinery, clothing, or dishes: individual contaminant molecules are "solubilized" inside of surfactant micelles, which are groups of individual surfactant molecules ranging from 5 to 10 nm in diameter. Alternatively, surfactant molecules can coat oil droplets and emulsify them into solution. Other enhanced-removal reagents exist, such as alcohols and sugar compounds.

A key factor controlling the success of this approach in the field is the ability to deliver the reagent solution to the places that contain the contaminant. This would depend, in part, on potential interactions between the reagent and the soil (e.g., sorption) and on properties of water flow in the subsurface. An important factor concerning regulatory and community acceptance of this approach is that the reagent should be of low toxicity. It would clearly be undesirable to replace one contaminant with another. Another important factor, especially with regard to the cost effectiveness of this approach, is the potential for recovery and reuse of the reagent.

19.4.2.4 Soil Vapor Extraction

The principle of **soil vapor extraction**, or **soil venting**, is very similar to that of pump-and-treat: a fluid is pumped through a contaminated domain to enhance contaminant removal. In the case of soil venting, however, the fluid is air rather than water. Because air is much less viscous than water, much less energy is required to pump air. Thus to remove volatile contaminants from the vadose zone, it is usually cheaper and more effective to use soil venting rather than soil washing. There are two key conditions for using soil venting. First, the subsurface must contain a gas phase through which the contaminated air can travel. This condition generally limits the use of soil venting to the vadose zone. In some cases, groundwater is pumped to lower the water table, thus allowing the use of soil venting for zones that were formerly water saturated. This is termed dual-phase extraction. Second, contaminants must be capable of transfer from other phases (solid, water, immiscible liquid) to the gas phase. This requirement limits soil venting to volatile and semivolatile contaminants. Fortunately, many of the organic contaminants of greatest concern, such as chlorinated solvents (trichloroethene, tetrachloroethene) and certain components of fuels (benzene, toluene), are volatile or semivolatile. Soil venting is the most widely used method for removing volatile contaminants from contaminated vadose-zone systems.

A blower is generally used to extract contaminated air from the subsurface (Fig. 19.6). In some cases, passive or active air injection is used to increase air circulation. Passive air injection simply involves drilling boreholes through which air can then move, as opposed to the use of a blower for actively injecting air. A cap made of plastic or asphalt is often placed on the ground surface to increase venting effectiveness and to prevent water from infiltrating into the subsurface. Preventing waste infiltration is important, because the presence of greater amounts of water reduces the ability of air to flow through the system.

Once the contaminant is removed from the vadose zone, it is placed into a treatment system. The major performance criteria for soil venting are the effectiveness of capturing and removing the contaminant. The effectiveness of contaminant removal by soil venting can be limited by many of the same factors that limit removal by water flushing.

A closely associated technology is that of **two-phase or dual-phase extraction**. For this technique, both contaminated air and water are extracted simultaneously from the same well (Fig. 19.7). This method is useful for sites at which both groundwater and the vadose zone are contaminated by volatile organic compounds. This approach would be used to target groundwater contamination in the vicinity of the vadose zone (near the water table). Deeper contamination would be targeted with conventional pump-and-treat or other appropriate methods.

19.4.2.5 Air Sparging

In situ air sparging is a means by which to enhance the rate of mass removal from contaminated saturated-zone systems. Air sparging involves injecting air into the target contaminated zone, with the expectation that volatile and semivolatile contaminants will undergo mass transfer (volatilization) from the groundwater to the air bubbles (Fig. 19.8). Due to buoyancy (air is less dense than water), the air bubbles generally move upward toward the vadose

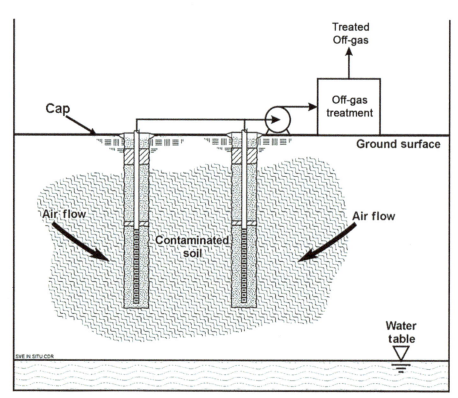

FIG. 19.6 Schematic of a soil-venting system for remediation of volatile organic contamination in the vadose zone. *Image courtesy of U.S. Navy. http://enviro.nfesc.navy.mil/erb/restoration/technologies/remed/phys_chem/phc-26.asp*

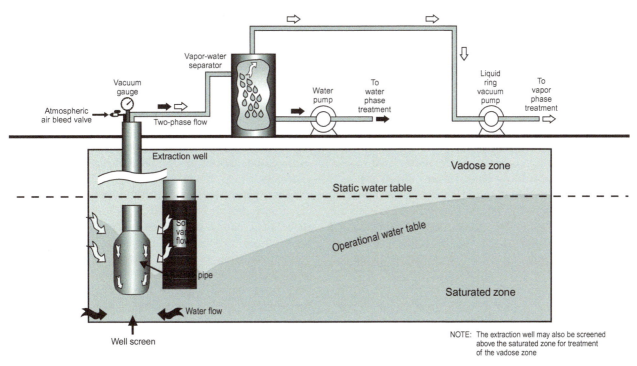

FIG. 19.7 Schematic of a two-phase system for extraction of contaminated air from the vadose zone and contaminated water from the saturated zone. *Image courtesy of the USEPA. (EPA 540-F-97-004 PB 97-963501 April 1997).*

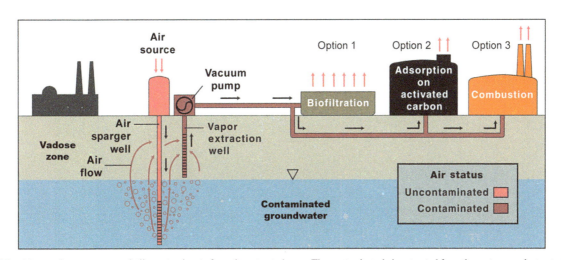

FIG. 19.8 Air sparging to remove volatile contaminants from the saturated zone. The contaminated air extracted from the system can be treated aboveground using a number of different methods. *Adapted from National Research Council: In situ bioremediation, when does it work?, Washington, DC, 1993, National Academy Press. From Environmental Monitoring and Characterization. San Diego, CA, 2004, Elsevier Academic Press.*

zone, where a soil-venting system is usually employed to capture the contaminated air.

Laboratory and pilot-scale research has shown that the effectiveness of air sparging is often limited by a number of factors in practice. One major constraint is the impact of "channeling" on air movement during sparging. Studies have shown that air injected into water-saturated porous media often moves in discrete channels that constitute only a fraction of the entire cross section of the zone, rather than passing through the entire medium as bubbles (as proposed in theory). This channeling phenomenon greatly reduces the "stripping efficiency" of air sparging.

Another significant limitation to air sparging applications is the presence of low-permeability zones overlying the target zone. As noted before, air-sparging systems are designed to operate in tandem with a soil-venting system, so that the contaminated air can be collected and treated. The presence of a low-permeability zone overlying the target zone can prevent the air from passing into the vadose zone, preventing capture by the soil-venting system. In such cases, the air-sparging operation may act to spread the contaminant. While limited by such constraints, air sparging may be of potential use for specific conditions, such as for targeting localized zones of contamination.

A related technology to air sparging is in-well air stripping or in-well aeration. With this method, the air stripping is performed within the water column of a well rather than within the subsurface formation itself (as it is for in situ air sparging). A special well design is used for in-well stripping, that is, constructed with two well screens that are separated by several meters. Groundwater is extracted from the bottom screen for example and pumped to the top of the well, where it is released to free fall down the well casing. This process aerates the water and enhances the release of the volatile contaminants from the water to the air. Vapors are recovered and treated at the surface.

19.4.2.6 Thermal Methods

As noted previously, contaminant removal can be difficult because of such factors as low solubility, large magnitude of sorption, and the presence of immiscible-liquid phases, all of which limit the amount of contaminant that can be flushed by a given volume of water or air. One means by which to enhance desorption, volatilization, dissolution, and evaporation, and thus improve the rate of contaminant removal is to raise the temperature in the contaminated zone.

This can be done in several ways. One method is to inject hot air or water during standard soil-venting or pump-and-treat operations. This approach may increase temperatures by several degrees. Another approach involves injecting steam into the subsurface, which can enhance contaminant removal through a number of complex processes. A third approach is based on the in situ generation of high temperatures using various heating methods. These methods can generate much higher temperatures than the hot water, hot air, and steam methods, but they are much more energy intensive and thus expensive.

Three general methods are used to generate heat in situ: electrical resistance heating (ERH), radio frequency heating (RFH), and thermal conduction heating (TCH). ERH uses arrays of electrodes installed in the target zone to create a concentrated flow of electric current. Resistance to flow of electric current in the soil generates heat greater than 100°C, helping to vaporize contaminants that are then recovered via vacuum extraction and treated at the surface. RFH uses electromagnetic energy (radiowaves or microwaves) to heat soil and enhance soil vapor extraction. The technique can heat soils to over 300°C. TCH supplies heat to the soil through heater wells or with a heater blanket that covers the ground surface. As the target zone is heated, the contaminants are destroyed or evaporated. Heater wells are used for deep contamination (Fig. 19.9) while a blanket is used for shallow contamination. This technique can heat soils to ~900°C.

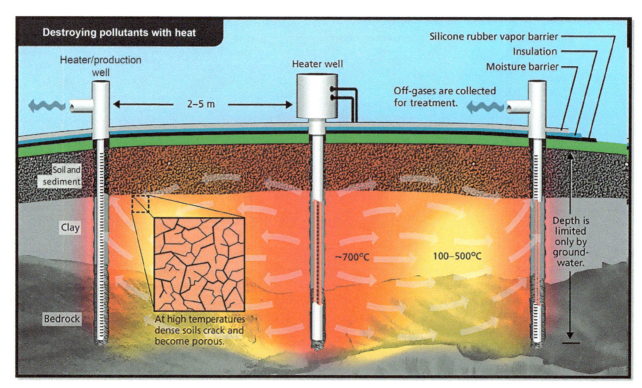

FIG. 19.9 The use of thermal conduction heating for subsurface remediation.

These high-temperature methods can provide rapid remediation of a site, with case studies showing effective treatment of soil and groundwater in less than 40 days. The costs of employing these methods are typically significantly higher than for other methods (see Case Study 19.1). Thus they are typically reserved for special cases. These methods can be particularly useful for remediating sites with extensive low-permeability zones or fractured media, which are difficult to address with most other methods.

19.4.2.7   Electrokinetic Methods

All electrokinetic processes depend on creation of an electrical field in the subsurface. This is done by placement of a pair or series of electrodes within the area to be treated to which direct current (typically between about 50 and 150 V) is applied. Electrokinetic treatment encompasses several different processes that individually or in combination can act to enhance the transport of contaminants, depending on the particular characteristics of the contaminants present.

Electromigration, electrophoresis, and electroosmosis all enhance mobilization of the target contaminant to a location for extraction or treatment (Fig. 19.10). The first two mobilization processes act on contaminants that are charged (ionic) or highly polar. Thus these processes work for metals, radionuclides, and selected ionic or ionizable organic contaminants. Electroosmosis, the movement of water in response to an electrical gradient, can work for charged and uncharged organic contaminants, since the dissolved molecules will be carried along

with the moving water. The presence of water is required for these methods to work.

One potentially promising application for electrokinetic methods is for removing contaminants from low-permeability zones. As noted before, such zones are difficult to remediate with many existing methods, most of which are based on water or air flow. By placing the electrodes directly into the low-permeability zone, contaminant movement can be generated within that zone.

19.4.3   In situ Treatment

In situ treatment technologies are methods that allow in-place cleanup of contaminated field sites. There is great interest in these technologies because they are in some cases cheaper than other methods. Second, in situ treatment can, in some cases, eliminate the risk associated with the contaminant by promoting destruction of the contaminant via transformation reactions. The two major types of in situ treatment are biological (in situ bioremediation) and chemical. The majority of these methods are based on injecting a reagent of some type into the subsurface to promote a transformation reaction (Fig. 19.11).

19.4.3.1   Bioremediation

Bioremediation makes use of the activity of naturally occurring microorganisms to clean up contaminated sites—specifically their ability to biodegrade contaminants (see Chapter 9). There are two types of bioremediation: **in situ bioremediation** is the in-place treatment of a contaminated site, and **ex situ bioremediation** is the above-ground treatment of contaminated soil or water that is

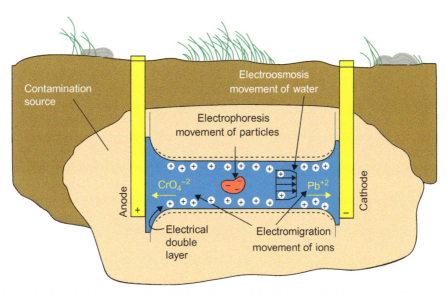

FIG. 19.10   Application of the electrokinetic method of remediation. *Courtesy of Sandia National Laboratories. http://www.sandia.gov/subsurface/factshts/ert/ek.pdf.*

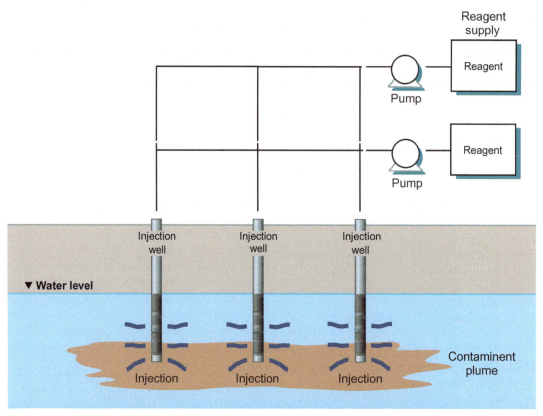

FIG. 19.11   In situ remediation of a contaminated saturated zone using biological or chemical methods. *Image courtesy of the U.S. Navy.*

removed from a contaminated site. Biological methods have been used for many years for aboveground treatment of municipal sewage waste (see Chapter 22) and industrial waste (see Chapter 21). This section will focus on in situ bioremediation.

In situ bioremediation is used primarily for organic contaminants. It has become a proven, dependable remediation method for contaminants such as hydrocarbons (fuels). It can also be effective for other types of organics such as chlorinated solvents, depending upon site conditions. For example, biodegradation is clearly dependent on the structure of the contaminant (see Chapter 9). It should be noted that bioremediation approaches can also be used for inorganic contaminants. For example, in situ bioremediation can be very effective for treating groundwater contaminated by nitrate and sulfate. In addition, microbial processes are central to in situ biosequestration approaches to treating metals in groundwater (see Information Box 19.4). The status of in situ bioremediation feasibility for a range of contaminants is presented in Table 19.1.

Bioremediation, while based on a simple concept, incorporates a complex system of biological processes. Thus it is sometimes difficult to accurately predict the performance of an in situ bioremediation system, which complicates planning and design efforts. Also, in situ bioremediation is a relatively slow process, especially compared to some

of the in situ chemical treatment methods discussed in the next section.

Several key factors are critical to successful application of in situ bioremediation: the presence of microorganisms capable of degrading the contaminant, availability of the contaminant to the microorganisms, environmental conditions, and nutrient availability. Thus when planning a bioremediation effort, it is important to characterize the site—and to identify possible limiting factors. Soil or water samples collected from the site can be tested to determine the presence of microorganisms capable of degrading the contaminant. Monitoring of groundwater at the site can be done to characterize environmental factors that affect biodegradation, such as oxygen concentrations, nutrient concentrations, temperature, and pH.

Not all potentially limiting factors are easy to identify. For example, contaminants are often present as mixtures, and one component of the mixture may inhibit the activity of the microorganisms. Similarly, low bioavailability, which is another factor that can limit bioremediation, can be very difficult to evaluate in the environment. To be biodegraded, the contaminant must be available to the microorganisms. Contaminants that are present as NAPL, or that are sorbed, are generally not directly available to microbial cells. The rate and magnitude of biodegradation of such contaminants is therefore reduced.

INFORMATION BOX 19.4 In Situ Biosequestration for Uranium

Natural restoration, the process of contaminant attenuation under the action of naturally occurring in situ processes, is not always effective in meeting regulatory levels for various groundwater contaminants. Very few alternatives to natural restoration (or monitored natural attenuation) are available for the remediation of large groundwater contaminant plumes containing uranium or similar constituents. Methods to enhance the rates of natural attenuation and improve the feasibility of monitored natural attenuation for inorganic contaminants are a current focus of research. One promising alternative, in situ biosequestration, has potential for the remediation of large, deep contaminant plumes containing uranium.

In situ biosequestration entails the injection of a reagent solution containing an electron donor into a target subsurface zone to stimulate microbial activity and generate reducing conditions, which induces sequestration (via bioprecipitation and/or enhanced adsorption) of targeted contaminants. This sequestration in turn decreases the aqueous concentration and therefore the bioavailability of the contaminant. Three critical questions of concern for this approach are (1) can the sequestration decrease groundwater contaminant concentrations to below the target level, (2) what is the most effective electron donor, and (3) what is the long-term stability of the sequestered phases.

Pilot projects have been conducted by UA personnel to assess the efficacy of in situ biosequestration for the treatment of uranium-contaminated groundwater at the Monument Valley, Arizona, Uranium Mill Tailings Radiation Control Act (UMTRCA) site. Ethanol at 0.5% and 5% concentrations was used as the electron donor to stimulate microbial activity. Analysis of groundwater and sediment samples collected before, during, and after the injection demonstrated the generation of reducing conditions. Uranium concentrations in groundwater decreased by more than a factor of 10 at pilot-test site 1 and by ~25% at pilot-test site 2, which has 30 times higher concentrations.

The sequestration was maintained for approximately 2.5 years for the first pilot test, and approximately 1 year for the second pilot test. It is noteworthy that reducing conditions and uranium sequestration were maintained for a notable period even though uranium-contaminated groundwater was continually flowing into the treatment zone from upgradient. Overall, the results of the pilot tests indicate that ethanol injection successfully stimulated microbial activity and decreased groundwater concentrations of uranium at the site. Based on the results of the study, successful application of this biosequestration approach would likely require periodic electron-donor injections to maintain reducing conditions and associated sequestration effects.

TABLE 19.1 Current Feasibility of Bioremediation

Chemical Class	Frequency of Occurrence	Status of Bioremediation	Evidence of Future Success	Limitations
Hydrocarbons and derivatives				
Gasoline, fuel oil	Very frequent	Established		Nonaqueous-phase liquid
Polycyclic aromatic hydrocarbons	Common	Possible	Aerobically biodegradable under a narrow range of conditions	Sorbs strongly to subsurface solids
Creosote	Infrequent	Possible	Readily biodegradable under aerobic conditions	Sorbs strongly to subsurface solids; forms nonaqueous-phase liquid
Alcohols, ketones, esters	Common	Established		
Ethers	Common	Possible	Biodegradable under a narrow range of conditions using aerobic or nitrate-reducing microbes	
Halogenated aliphatics***	Very frequent	Established	Aerobically biodegradable under a narrow range of conditions; cometabolized by anaerobic microbes; cometabolized by aerobes in special cases	Nonaqueous-phase liquid Production of toxic by-products for anaerobic pathways

Continued

TABLE 19.1 Current Feasibility of Bioremediation—cont'd

Chemical Class	Frequency of Occurrence	Status of Bioremediation	Evidence of Future Success	Limitations
Halogenated aromatics	Common	Possible	Aerobically biodegradable under a narrow range of conditions; cometabolized by anaerobic microbes	Sorbs strongly to subsurface solids; forms nonaqueous phase—solid or liquid
Polychlorinated biphenyls	Infrequent	Possible	Cometabolized by anaerobic microbes	Sorbs strongly to subsurface solids
Nitroaromatics	Common	Emerging	Aerobically biodegradable; converted to innocuous volatile organic acids under anaerobic conditions	
Cr, Cu, Ni, Pb, Hg, Cd, Zn, etc.	Common	Possible	Solubility and reactivity can be changed by a variety of microbial processes	Availability highly variable—controlled by solution and solid chemistry

Revised from National Research Council. In Situ Bioremediation. Washington, DC, 1993, National Academy Press.

One of the most common limiting factors for in situ bioremediation in the saturated zone is the availability of oxygen. Oxygen is required for aerobic biodegradation. However, oxygen has limited solubility in water, and once it is used up, it must be replenished via aqueous diffusion (which is slow). The combination of these factors is the reason that oxygen often limits the use of bioremediation for groundwater systems. Several methods have been developed to supply oxygen to the saturated zone. The simplest method is to inject air directly into groundwater, which is done similarly as the air sparging method discussed before. Other methods involve injecting compounds into the subsurface that release oxygen.

Nutrients, nitrogen and phosphorus in particular, are common additives used to enhance the rate of biodegradation in bioremediation applications. Many contaminated sites contain organic wastes that are rich in carbon but poor in nitrogen and phosphorus. Nutrient solutions are usually injected from an aboveground tank. The goal of nutrient injection is to optimize the carbon/nitrogen/phosphorus ratio (C:N:P) in the subsurface to approximately 100:10:1 (see Chapter 9).

While most of the past in situ bioremediation applications were based on aerobic biodegradation, anaerobic-based processes have also become of interest. There are two potential reasons for considering anaerobic-based bioremediation. First, it is sometimes difficult to establish and maintain aerobic conditions in saturated subsurface systems. Second, some organic contaminants are resistant to aerobic biodegradation. Several alternative electron acceptors have been proposed for use in anaerobic degradation, including nitrate, sulfate, and iron (Fe^{+3}) ions, as well as carbon dioxide. In some cases, labile organic compounds (such as

molasses) are added to the subsurface to promote anaerobic conditions—microorganisms aerobically biodegrade the added carbon supply, thereby using available oxygen. Another approach under investigation is the use of sequential anaerobic/aerobic biodegradation processes, wherein biodegradation is first enhanced under anaerobic conditions, followed by aerobic-based bioremediation. This approach may be useful for mixed-waste sites.

Cometabolic biodegradation, in which microorganisms use one compound for energy (its food source), but in the process also biodegrade other compounds (contaminants) present, is another variation of bioremediation (see also Chapter 9). This approach is of particular interest for the bioremediation of chlorinated solvents. For example, methanotrophic organisms produce the enzyme methane monooxygenase to degrade methane, and this enzyme also happens to cometabolically degrade several chlorinated solvents, such as trichloroethene. Thus methane could be injected into the subsurface to promote biodegradation of the trichloroethene if the appropriate microorganisms were present at the site.

If microorganisms capable of biodegrading the target contaminants are not present in the subsurface at the site, specific microorganisms can be injected into the subsurface. This process is known as **bioaugmentation.** This process can be difficult to accomplish successfully. First, the introduced microbes often cannot establish a niche in the environment and thus may not survive. Second, microorganisms can be strongly sorbed by solid surfaces and trapped in small pores; so there are difficulties in delivering the introduced organisms to the site of contamination. However, commercial mixtures of bacteria are now available for bioaugmentation to promote bioremediation at sites contaminated by chlorinated-solvent compounds.

Bioventing is a technique used to promote biodegradation in the vadose zone. Bioventing is a combination of soil venting technology and bioremediation. Air flow through the contaminated zone is initiated with a blower or vacuum system, which increases the supply of oxygen throughout the zone and hence the rate of contaminant biodegradation. The rate of air flow is significantly lower for bioventing applications compared to soil venting. In the case of volatile contaminants, remediation will result from a combination of biodegradation and removal in the extracted air (i.e., soil venting).

19.4.3.2 In situ Chemical Treatment

In situ chemical remediation is a process in which the contaminant is degraded by promoting an abiotic transformation reaction, such as hydrolysis, oxidation, or reduction, within the subsurface. Although historically this approach has been used much less frequently than in situ bioremediation, it has become a very popular method over the past decade or so. It can be accomplished by using active methods, such as injecting a reagent into the contaminated zone of the subsurface, or by using passive methods, such as placing a permeable treatment barrier downgradient of the contamination. An advantage of these methods is that they promote in situ destruction of the contamination, thus reducing or eliminating the associated hazard potential.

As an alternative to in situ bioremediation, in situ chemical oxidation (ISCO) has recently become a widely used method. The objective of this method is to inject into the contaminated zone an oxidizing compound that will react with organic contaminants, oxidizing them to carbon dioxide and other oxidant-specific by-products. Four oxidizing reagents, in particular, permanganate (MnO_4), hydrogen peroxide (H_2O_2), persulfate (S_2O_8), and ozone (O_3), have been used. Both ozone and permanganate have been used for decades as oxidants for aboveground disinfection and purification of drinking water. As noted previously, chlorinated compounds are in general relatively resistant to biodegradation, which complicates or limits the use of bioremediation methods. Thus ISCO has become a go-to method to remediate sites contaminated by chlorinated compounds. See the Case Study 19.2 for more detailed information on the use of ISCO. All of the oxidizing reagents used for treatment of organic subsurface contamination are relatively nonselective and will thus react with most organics present in the subsurface. Therefore soils and groundwater containing high natural organic carbon concentrations can exhibit a high oxidant demand, competing with the demand from the actual organic contaminant.

In situ chemical reduction (ISCR) involves the introduction of a reductant or reductant-generating material in the subsurface for the purpose of degrading organic compounds to potentially nontoxic or less toxic compounds, and immobilizing metals such as Cr(VI) by adsorption or precipitation. The most common reductant is zero valent iron (ZVI). For their use, microscopic-sized particles of ZVI are suspended in solution and injected into the target zone.

Given that standard ISCO and ISCR methods are generally based on injection (flow) of the reagent solution, the presence of low permeability zones can reduce the overall effectiveness of these techniques. Treatment depends on the reagent being in contact with the contamination; significant heterogeneity will cause preferential channeling of the reagent solution to the high conductivity zones, with poor transport to the low conductivity zones, thus limiting contact with the contamination. Solid forms of the reagent can also be used, such as mixing solid reagent into shallow contaminated soil.

In situ treatment walls or barriers, called **permeable reactive barriers** (PRB), are of particular interest for chlorinated solvents such as trichloroethene. This approach generally involves placing a trench downgradient of the contaminant plume and filling that trench with a wall of permeable, reactive material that can degrade the contaminant to nontoxic by-products (Fig. 19.12). The wall is permeable so that the water from which the contaminant has been removed can pass through and continue to flow downgradient. For example, iron filings can degrade compounds such as trichloroethene via reduction reactions while permitting water to pass through unimpeded. One effective application of PRBs is to use them to prevent the downgradient spread of contamination. For example, the barrier can be emplaced at the boundary of a site property, on the downgradient side of the contaminant zone, with the objective

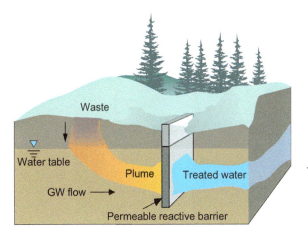

FIG. 19.12 In situ treatment of a groundwater contaminant plume using a permeable reactive wall. *Courtesy of US EPA. (EPA/600/R-98/125 September 1998; http://www.clu-in.org/download/rtdf/prb/reactbar.pdf.)*

being that the barrier would prevent movement of contamination off the site property. One limitation to PRBs is that they are practical only for relatively shallow depths (<50 m). Localized treatment zones can be developed by using reagent injection for deeper systems. In situ treatment barriers are generally used to control the spread of a contaminant plume or to decrease contaminant delivery from a source zone, whereas in situ chemical treatment based on injecting a reagent (ISCO and ISCR) is generally used to remediate source zones.

19.4.4   Other Technologies: Monitored Natural Attenuation

Monitored natural attenuation (MNA) has attracted great interest as a low-cost, low-tech approach for site remediation. Monitored natural attenuation is based upon natural transformation and retention processes reducing contaminant concentrations in situ, thereby containing and shrinking groundwater contaminant plumes (Fig. 19.13). The application of MNA requires strict monitoring to ensure that it is working; thus the name "monitored" natural attenuation. Extensive information has been reported concerning the role of natural attenuation processes for remediation of contaminated sites. This information has been summarized in a report released by the National Research Council (2000). Two of the major conclusions reported in

this document are that: (1) MNA has the potential to be used successfully at many sites, and (2) MNA should be accepted as a formal remedy only when the attenuation processes are documented to be working and that they are sustainable.

Several processes, such as biodegradation, hydrolysis, sorption, and dilution, may contribute to attenuation of the contaminant plume. However, biodegradation or biotransformation is generally the predominant process involved and usually the key factor for successful use of MNA. MNA has proven to be a success for sites contaminated by petroleum hydrocarbons (e.g., gasoline). Conversely, its use for chlorinated-solvent contaminated sites is more difficult because of the greater resistance of chlorinated compounds to biodegradation (see Chapter 9).

There are two major questions to address when evaluating the feasibility and viability of applying MNA to a particular site. These are: (1) Are natural attenuation processes occurring at the site; and (2) Is the magnitude and rate of natural attenuation sufficient to accomplish the remediation goal (e.g., plume containment, plume reduction) and be protective of human health and the environment? Successful implementation of MNA requires an accurate assessment of these two questions.

As noted before, the first step in evaluating the feasibility and viability of applying MNA to a particular site is to determine whether or not natural attenuation processes are occurring at the site. Evaluating and documenting the

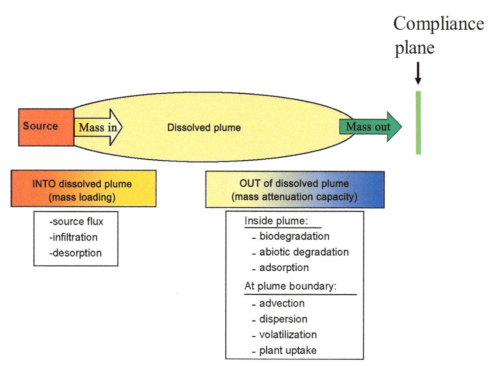

FIG. 19.13 Concept of Monitored Natural Attenuation. *Modified from The Interstate Technology & Regulatory Council, Enhanced Attenuation: Chlorinated Organics, 2008 with permission from ITRC.*

potential occurrence of natural attenuation involves the following steps:

1. Developing a general conceptual model of the site that is used to provide a framework for evaluating the predominant transport and transformation processes at the site.
2. Evaluating contaminant and geochemical data for the presence of known "fingerprints" indicative of active transformation processes. This includes analyzing contaminant concentration histories, the temporal and spatial variability of transformation products and reactants (e.g., O_2, CO_2, and so on), and isotope ratios for relevant products and reactants.
3. Screening groundwater and soil samples for microbial populations capable of degrading the target contaminants.

The second question to address for MNA application is the magnitude and rate of natural attenuation at the site. Characterizing the magnitudes and rates of attenuation at the field scale is a complex task. For example, the initial mass of contaminant released into the subsurface is not known at most sites. This means it is difficult to accurately quantify the magnitude and rate of attenuation. Thus we can generally only obtain estimates of such rates at best. One method is based on characterizing temporal changes in contaminant concentration profiles, electron acceptor concentrations (such as O_2, NO_3^-, SO_4^{2-}, Fe^{3+}), or contaminant transformation products. However, the complexity of typical field sites makes it difficult to accurately determine attenuation rates from temporal changes in these parameters. Another approach often used to characterize potential attenuation processes is to conduct bench-scale studies in the laboratory using core samples collected from the field. However, results obtained from laboratory tests may not accurately represent field-scale behavior. In situ microcosms, in which tests are conducted in a small volume of the aquifer, can be used to minimize this problem. Given the complexity of most field sites, including spatial variability of physical and chemical properties and of microbial populations, a prohibitive number of sampling points may often be required to fully characterize a field site using either of these methods. Biotracer tests have been proposed as an alternative method for field-scale characterization of the in situ attenuation potential associated with subsurface environments.

The rates of natural attenuation processes are not sufficient to maintain plume containment at many sites that have large source zones. The flux of contamination emanating from the source zone is greater than the attenuation capacity, so the plume continues to grow. In such cases, there is great interest in combining MNA with an aggressive source-zone remediation action (such as ISCO). The idea is that the source-zone remedial action will remove sufficient

contamination to reduce the contaminant flux to levels that are below the attenuation capacity of the system. One possible concern for this approach would be that the source-zone remedial action does not impede the attenuation processes of the system (particularly that it does not harm the microbial community responsible for biodegradation).

19.4.5 Other Technologies: Phytoremediation

Another popular method being investigated for the remediation of hazardous waste sites is the application of plants, a process called **phytoremediation.** Phytoremediation can be used to treat both organic and inorganic contaminants (Fig. 19.14). Plants can interact with contaminants in several ways (see Table 19.2).

Accumulation is the uptake of contaminants into plant tissues. For example, some plants, referred to as **hyperaccumulators**, are extremely efficient at metal accumulation. The advantage of this approach is that contaminants are actually removed from the site by harvesting the plants after the contaminants are accumulated. A potential disadvantage of this approach is that wildlife may be exposed to the contaminants if they forage at the site before the plants are harvested.

Stabilization is a second phytoremediation approach and involves stabilization of contaminants in the root zone through complexation with plant root exudates and the plant root surface. This approach is attractive from two perspectives. First, contaminants are not accumulated in aboveground plant tissues, thus avoiding any potential risk of wildlife exposure. Second, this approach does not require plant harvesting.

A third approach is the use of plants as a form of a hydraulic barrier. For example, poplar trees have been used to remove organic contaminants such as trichloroethene

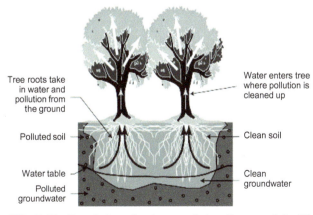

FIG. 19.14 Use of plants for phytoremediation. *Courtesy of the US EPA. http://clu-in.org/download/citizens/citphyto.pdf; EPA 542-F-01-002 April 2001.*

TABLE 19.2 Types of Phytoremediation Processes

Treatment Method	Mechanism	Media
Rhizofiltration	Uptake of metals in plant roots	Surface water and water pumped through troughs
Phytotransformation	Plant uptake and degradation of organics	Surface water, groundwater
Plant-assisted bioremediation	Enhanced microbial degradation in the rhizosphere	Soils, groundwater within the rhizosphere
Phytoextraction	Uptake and concentration of metals via direct uptake into plant tissue with subsequent removal of the plants	Soils
Phytostabilization	Root exudates cause metals to precipitate and become less bioavailable	Soils, groundwater, mine tailings
Phytovolatilization	Plant evapotranspirates selenium, mercury, and volatile organics	Soils, groundwater
Removal of organics from the air	Leaves take up volatile organics	Air
Vegetative caps	Rainwater is evapotranspirated by plants to prevent leaching contaminants from disposal sites	Soils

From Chappell J. Phytoremediation of TCE using Populus. Status report for the U.S. EPA Technology Innovation Office, 1997.

from groundwater. Poplar trees are a type of plant whose root system can extend to shallow water tables. Thus they can draw water directly from the saturated zone, thereby causing a lowering of the water table. In this regard, they can be used to create a hydraulic barrier to contain a groundwater contaminant plume. The contaminated water is drawn up into the trees and expelled via evapotranspiration. It is also possible that some contaminants such as trichloroethene may undergo transformation reactions induced by the tree, either within the plant tissues or within the rhizosphere.

Phytoremediation has become a popular approach, partly because it is relatively inexpensive. However, its use is limited to relatively shallow applications (i.e., the depth of the root system). In addition, it is relatively slow in comparison to some other methods discussed before.

19.4.6 Summary of Remediation Methods

The primary components of a typical hazardous waste site are the source zone, the vapor plume, and the groundwater contaminant plume. It is important to recognize that certain remedial actions are applicable for only one or two of these components. For example, very few options exist to deal with large groundwater contaminant plumes. These are pump and treat, monitored natural attenuation, and permeable reactive barriers. Generally, the use of methods such as in situ thermal, in situ enhanced flushing, in situ electrokinetic, in situ chemical treatment, and excavation would not be considered for treating an entire plume due to the exhorbitant associated costs. These methods are

generally used for small areas such as source zones or other localized hot spots of elevated concentrations. Conversely, pump and treat (e.g., hydraulic containment) and PRBs can both be used to decrease the contaminant load delivered from the source zone to the plume, and thus qualify as source-zone treatment methods. It is critical to keep these distinctions in mind when designing remediation projects.

QUESTIONS AND PROBLEMS

1. What are the essential components of a typical hazardous waste site characterization program? Why are these the essential components?
2. How do so-called LNAPLs and DNAPLs behave when spilled into the ground? How do these behaviors influence our ability to find them during site characterization projects?
3. Compare and contrast the use of physical barriers versus wellfields to contain a groundwater contaminant plume.
4. What is a major limitation of the pump-and-treat method? What are some of the methods that can be used to enhance the performance of pump-and-treat?
5. For which of the following contaminants would you propose to use air sparging as a possible remediation method, and why/why not: trichloroethene, phenol, nitrate, benzene, pyrene, cadmium?
6. What are some of the potential benefits of using in situ bioremediation? What are some of the major potential limitations to its use?

7. How is potassium permanganate injection used for remediation?
8. What are the criteria used to evaluate proposed remedial actions?
9. What are the major advantages and disadvantages of using excavation?
10. What are the differences in remediating a source zone versus a large groundwater contaminant plume?

Detailed information for site characterization and remediation methods is available from two websites:

EPA: Contaminated Site Clean-Up Information. https://clu-in.org/

ITRC: Advancing Environmental Solutions. https://www.itrcweb.org/

Case Study 19.2 Use of ISCO Remediation at a Superfund Site

Pilot-scale tests were performed as part of a collaborative project between academia, industry, consulting firms, and government agencies to assess the efficacy of potassium permanganate for in situ chemical oxidation of trichloroethene (TCE) contamination at a Superfund site in Tucson, Arizona. From the early 1950s and 1960s to approximately 1978, industrial solvents including TCE, trichloroethane (TCA), and methylene chloride; machine lubricants and coolants; paint thinners and sludges; and other chemicals were disposed of in open unlined pits and channels at the site. In addition, wastewater containing chromium from plating operations flowed through a drainage ditch that crosses part of the site. A 1981–1982 investigation determined that TCE had moved down through overlying sediments into the regional aquifer. In 1983, the site was added to the national priorities list as a federal Superfund site.

A groundwater pump-and-treat remediation system was started in 1987 and a soil-venting system was initiated in 1996. The pump-and-treat system has successfully controlled and reduced the extent of contamination, while the soil-venting system removed large quantities of contaminant mass from the vadose zone. Unfortunately, the cleanup target has not yet been attained. The results of advanced characterization studies indicated that contaminant removal is constrained by site heterogeneities, physical and chemical mass transfer limitations, and the presence of immiscible-liquid contamination within the source zones. Remediation of the site is constrained by the difficulty of removing contamination from the source zones.

In situ chemical oxidation (ISCO) is a technique that shows promise for directly treating source-zone contaminants. The simplified proposed reaction mechanism for potassium permanganate ($KMnO_4$) with chlorinated compounds is as follows:

C_2Cl_3H (TCE) $+ 2MnO_4^-$ (Permanganate) $\rightarrow 2CO_2 +$
$2\ MnO_2$ (s) $+ 3\ Cl^- + H^+$

with 0.81 kilogram (kg) of chloride (Cl^-) produced and 2.38 kg manganese oxide (MnO_2) reduced per kg of TCE

oxidized. While this method promotes destruction of TCE and other contaminants, there are some possible negative impacts. Production of manganese oxide (MnO_2) precipitate has the potential to clog the aquifer, and production of hydrochloric acid (HCl) has the potential to dissolve aquifer material, especially carbonates. Additionally, interaction with cocontaminants is possible. For example, chromium is a fairly common contaminant in the industrial waste sites for which ISCO is a potential remediation strategy. When oxidized, chromium converts to the hexavalent form, which is more toxic as well as more mobile, and is therefore a health concern. In situ oxidation for remediation of chlorinated compounds temporarily creates an oxidizing environment in the subsurface. Thus the potential for the mobilization of chromium and other metals, and the subsequent potential for reduction and stabilization (natural attenuation), is an important consideration for this technology.

ISCO pilot tests with potassium permanganate were conducted at two source zones at the site. These tests were designed to affect a relatively small area of the subsurface and provide useful information for the design of a full-scale application. The ISCO pilot-scale test was performed with the collaboration of Raytheon Systems Environmental Division, the University of Arizona, Errol L. Montgomery and Associates, Inc., and IT Corporation.

Groundwater samples from all wells in which permanganate was observed showed a decrease in TCE concentrations coincident with the arrival of the potassium permanganate in concentrations above approximately 50 mg/L. During the operational period, when permanganate solution was active in the treatment area, the average TCE concentration reduction was 84% for Site 3 and 64% for Site 2. As permanganate dissipated within the treatment area, TCE concentrations in all wells increased to some extent. Possible causes of rebounding include movement of contaminated groundwater into the treatment zone from upgradient, and mass transfer of untreated TCE from the aquifer material by desorption from aquifer material, diffusion from areas of lower permeability, and dissolution from nonaqueous phase liquid (NAPL) contamination. Performance monitoring tests conducted at the site indicated that the ISCO project resulted in an overall decrease of ~75% in the contaminant mass discharge for the site. These results indicate that in situ oxidation using potassium permanganate was successful in removing TCE associated with the two source zones.

The presence of chromium as a cocontaminant at the site was an important aspect of this pilot project. Chromium concentrations increased as expected. However, concentrations quickly returned to low levels. Further attenuation is expected with time based on the history of the site and review of other ISCO projects. The observed increase in soluble chromium was anticipated and may be due more to the presence of chromium as an impurity in the injection solution than the mobilization of chromium associated with the aquifer solids.

Potassium permanganate is fairly inexpensive; for example, the cost for the industrial grade potassium permanganate purchased for the field project was approximately $4.40/kg.

Since extraction of groundwater is not required, disposal costs are minimized. Mobilization, demobilization, and costs associated with the delivery system are the primary costs for ISCO treatment. Overall, ISCO is a relatively inexpensive technology compared to many other methods. Since oxidation occurs rapidly with adequate delivery, the time frame of ISCO projects may be significantly shorter in comparison to other methods such as pump-and-treat and bioremediation.

REFERENCES

Boulding, J.R., 1995. Practical Handbook of Soil, Vadose Zone, and Ground-Water Contamination. Lewis Publishers, Boca Raton, FL.

National Research Council, 1993. Situ Bioremediation, When Does It Work? National Academy Press, Washington, DC.

National Research Council, 2000. Natural Attenuation for Groundwater Remediation. National Academy Press, Washington, DC.

National Research Council, 2013. Alternatives for Managing the Nation's Complex Contaminated Groundwater Sites. National Academy Press, Washington, DC.

Natural Research Council, 1994. Alternatives for Groundwater Cleanup. National Academy Press, Washington, DC.

Natural Research Council, 1999. Groundwater and Soil Cleanup. National Academy Press, Washington, DC.

Chapter 20

Reclamation and Restoration of Disturbed Systems

M.L. Brusseau, E.P. Glenn and I.L. Pepper

Cover Art: The Cienega de Santa Clara, a human-made wetland in the delta of the Colorado River where it enters the Gulf of California in Mexico. This wetland is formed by the discharge of agricultural wastewater onto salt flats (visible in background) in the intertidal zone. It supports tens of thousands of resident and migratory water birds and endangered fish and bird species. *Photo courtesy E.P. Glenn.*

20.1 INTRODUCTION

Many different types of human activities can lead to disturbances that impair ecosystems and their functions, as covered in Chapters 14–17, and 25. Major impacts include degraded soil quality, water and air pollution, scenic impairment, and habitat destruction. These disturbances occur over a vast range of scales, from a single site such as a manufacturing plant, mine, or agricultural field, to watershed systems, to entire regions (such as deforestation of the Amazon rainforest). Different approaches are used to address these different scales of problems. This chapter will focus primarily on disturbances associated with single sites, or what may be referred to as local-scale disturbances. Methods for dealing with regional and global-scale issues are discussed in Chapter 32.

The first step in addressing disturbed systems is to understand the "four Rs"—reclamation, remediation, rehabilitation, and restoration (Information Box 20.1). The objective of *Reclamation* is to return the site to a productive use. It is focused primarily on conditions of the land surface and the ability to use the site for a particular land use activity (redevelopment). *Remediation* is focused on removing or treating soil, air, and water contamination to reduce or eliminate risks to the health of humans and other lifeforms. Remediation methods are covered in Chapter 19. Reclamation and Remediation do not focus on the broader aspects of ecological conditions and ecosystem functions. However, reclamation and/or remediation actions may be implemented in support of larger scale ecosystem restoration efforts. *Rehabilitation* is focused on improving site conditions so that an impaired ecosystem service can be recovered. For example, an impaired wetland may be rehabilitated to recover its ability to naturally filter and treat water pollution. *Restoration* is a general term that is often used to represent any type of activity conducted to improve site conditions. Thus Reclamation, Remediation, and Rehabilitation are subsets of Restoration.

Ecological or *Ecosystem restoration* is formally defined by The Society for Ecological Restoration as "assisting the recovery of an ecosystem that has been degraded, damaged or destroyed." This comprises a range of activities designed to initiate or accelerate the recovery of an ecosystem with respect to its health, integrity, and sustainability. Ecological restoration is focused on the entire ecosystem, including flora, fauna, and overall ecosystem functioning, and not just on soil, air, and water quality. Generally, the goal of ecological restoration is to support recovery to a natural state.

Remediation and Reclamation are primarily concerned with improving site conditions to reduce human health risk and to support future human activities. In addition, Rehabilitation is often focused on improving ecosystem services for human well-being. In other words, they are human-centric focused. Conversely, ecological restoration is focused on

Environmental and Pollution Science. https://doi.org/10.1016/B978-0-12-814719-1.00020-3

INFORMATION BOX 20.1 Restoration Terminology

Reclamation: repairing some or most of the damage done to land so that it can serve some productive function; returning it to a useful state. For example, salinized farmland, unable to support native plants, can be planted with salt-tolerant plants (*halophytes*) to prevent erosion and provide wildlife habitat.

Rehabilitation: reinstating a level of ecosystem functionality where complete ecological restoration is not sought, but rather with the objective of recovering ecosystem services. For example, adding organic amendments to a plot of degraded soil to support planting of crops and resultant food production.

Remediation: removing or treating site contamination (site cleanup) in order to protect human health and the environment. This has a specific meaning for hazardous waste sites (see Chapter 19). Examples include excavation of contaminated soils, pump and treat of contaminated groundwater, and soil vapor extraction for volatile organic contaminants in the vadose zone.

Restoration: A general term for any and all activities conducted at the site to improve conditions. Reclamation and Remediation are subsets of Restoration.

Ecological Restoration (**also called ecosystem restoration**): is formally defined by The Society for Ecological Restoration as assisting the recovery of an ecosystem that has been degraded, damaged, or destroyed; an intentional activity that initiates or accelerates the recovery of an ecosystem with respect to its health, integrity, and sustainability.

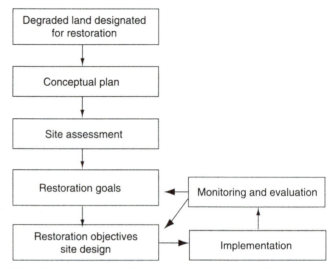

FIG. 20.1 Components of a restoration project. *(From Environmental Monitoring and Characterization, Elsevier, 2004.)*

the ecosystem itself, its functioning and overall health. Keep in mind however that reclamation, remediation, and rehabilitation efforts may be implemented as part of an ecosystem restoration project. This chapter will focus on reclamation and ecological restoration separately. As previously noted, remediation is covered in Chapter 19.

20.2   PROJECT DESIGN

20.2.1   Planning and Site Assessment

The components and overall restoration process is shown in Fig. 20.1. A conceptual plan for the restoration project begins with a detailed history of the site, where information is collected to determine the disturbance and associated changes that have led to the current state of degradation. This information is cross-referenced against historical and current topographic, geologic, and vegetation maps to determine what changes have occurred spatially and temporally. Information on the environmental and ecological properties of the site prior to the disturbance is collected to determine the natural-condition state and evaluate the restoration potential of the area. If this information is not available for the site in question, nearby natural reference sites are used.

Once a complete history of the site has been compiled and it has been determined that the restoration project is feasible, a thorough site assessment of abiotic and biotic conditions takes place. This step is one of the most important in any restoration project, since it not only provides baseline measurements on such parameters as hydrologic features, soil and water conditions, and biological information, but it also serves as the benchmark upon which to evaluate project performance through time. This step first involves placing the site in the context of the regional ecosystem type (or ecoregion) prevalent for that area. This is critical to make sure that the objectives and outcomes selected for the planned restoration effort will be compatible with the prevalent ecological conditions of the region.

After the ecoregion evaluation, site investigations are conducted to characterize specific conditions such as habitat fragmentation, disconnected surface and subsurface hydrologic flows, water quality issues, physical and chemical properties of the soil, and plant and animal inventories. In assessing a site, it is important to determine which basic ecological functions are damaged or fragmented, since this often sets the restoration priorities and ultimately the success of the project over time. Specific site characterization activities are conducted to determine the presence or absence of toxic constituents such as heavy metals, organic compounds, and radioactive waste in soil, water, and air. These are done as described in Chapter 10. It is worth noting that the soil conditions and water quality in a given area generally dictate the type of vegetation cover and thus the biodiversity of biotic components. Therefore extra care should be taken in analyzing and describing these two elements both horizontally and vertically across the landscape.

Plants provide valuable information about site conditions. The occurrence and relative abundance of certain plant species and their physiological and ecological tolerances provide evidence of environmental conditions that are of importance for understanding the nature of the site, potential human health and ecological risks, and the feasibility of different restoration alternatives. Typically, plant ecology investigations include four types of studies: (1) *plant species survey*; (2) estimates of the *percent cover* and age structure of dominant, perennial plant species; (3) evaluation of the composition, relative abundance, and distribution of *plant associations*; and (4) *vegetation mapping*.

The plant species survey is conducted by traversing a site, usually on foot, and noting each species present. The percent cover study attempts to quantify the percent of the site that is covered by bare soil or individual plant species utilizing a *line intercept* method. First, the plant community to be described is delineated on a map and then 30-m transect lines are randomly chosen where actual plant counts will take place. In the field, a 30-m tape is stretched out and the total distance intercepted by each plant species is recorded and used to calculate the percent cover of each species. For example, a transect might consist of 12% fourwing saltbush, 10% black greasewood, and 78% bare soil. Plant associations and ultimately vegetation mapping are used to delineate land management units for future revegetation efforts.

Recently, surveys of microbial communities associated with the soil have become more common for site characterization. They can provide information regarding, for example, overall soil health, the ability to support plant growth, predominant biogeochemical nutrient cycles, and the potential for natural attenuation of contaminants. See Chapters 5 and 9 for more information on this topic. Surveys of animal species are also conducted. This information is used to determine the trophic structure of the system and to identify if certain species are threatened.

20.2.2 Objectives and Approaches

A key component for any successful restoration effort is the identification of clear objectives. There are several reasons for defining objectives for restoration projects, including:

- Focus and sharpen thinking about the desired state or condition of the restored site;
- Describe to others the desired condition of the restored site;
- Provide a desired outcome that serves as a metric for assessing restoration success; and
- Provide direction for the appropriate type of performance monitoring.

Restoration objectives that are inadequately defined lead to ambiguity as to whether a project was successful. The *SMART* acronym can be used to help design effective

objectives: Specific, Measurable, Achievable, Relevant, and Time bound (USGS, 2011).

As noted previously, restoration projects range from limited activities that restore some beneficial land use, to contamination cleanup, to focused efforts to restore a particular ecosystem service, to full-scale ecosystem restoration efforts. The objectives selected to guide a restoration project need to be consistent with the degree of restoration that is feasible for the site. The restoration feasibility as well as the methods chosen for the restoration of a particular site are determined by:

- the nature of the site and surrounding environment,
- the nature and magnitude of disturbance and degradation,
- the desired outcome, and
- the planned future land use.

For example, the site may be contaminated or disturbed to such an extent that it would be cost prohibitive to attempt to fully restore the site (complete ecological restoration). In this case, remediation and reclamation activities would be conducted to return the site to a state that would support alternative land uses. In addition, the extent of reclamation would depend upon the intended land use, with lesser degrees applied for an industrial use versus a site planned for a housing development for example.

It is important to recognize that ecosystems are dynamic, such that they change and evolve over time through the course of natural events. Thus it may not be feasible to restore the site to the original natural state, as it would not be consistent with the current natural status of the region. Therefore the focus of the project should be on the *desired* characteristics of the ecosystem in the future, rather than on the characteristics that were previously there.

20.2.3 Project Completion and Performance Monitoring

A critical component of all restoration projects is monitoring site conditions over time to evaluate success or failure of the project, which is termed performance monitoring or post-project assessment. Effective performance monitoring efforts require development of clearly articulated objectives. In addition, it requires identification of informative indicators or metrics. Examples of metrics would be a reduction in suspended sediment load in lake water, a reduction in erosion of a slope, or an increase in plant density.

Most restoration performance assessments are based on short-term monitoring, perhaps completed several months or a few years after project completion. This is due to a number of reasons—project funding sources typically require a specific project completion time frame, sponsors want to see the results of their investments, and the public expects rapid results. Many reclamation projects can be successfully completed in the span of months to a few years, which is compatible with the short-term focus. Conversely, many remediation and ecological restoration projects can take many years or decades to complete, a time span that is not compatible with the previously stated desires.

Monitoring programs are based on the periodic collection of data. This includes collecting soil, plant, surface water, and groundwater samples and conducting plant and animal inventories. Until recently, these data were collected using standard labor-intensive field sampling techniques. However, with increased computing power and sophisticated data collecting techniques, real-time data can more easily be obtained without additional personnel. This can provide larger, higher resolution data sets that can improve performance assessments. Site monitoring methods are discussed in Chapter 10.

20.3   LAND RECLAMATION METHODS

Major methods used to reclaim disturbed land surfaces and soil quality will be covered in this section. These include:

- Surface grading and contouring,
- Surface covers and caps,
- Revegetation, and
- Soil treatment for various purposes.

20.3.1   Surface Grading and Contouring

A primary approach used in the initial stages of most reclamation projects is grading and contouring of the land surface. This is done for a number of reasons, including improving slope stability, controlling water flow, reducing wind and water erosion, preparing the surface for planned infrastructure, and to enhance aesthetics. The nature and extent of this work will depend on the project objectives.

FIG. 20.2   Slumping of an unstable slope. *(From Unsafe Ground: Landslides and Other Mass Movements; Available from: http://slideplayer.com/slide/4565885/.)*

Slope stability design is based on the *angle of repose*. Slopes with angles larger than the angle of repose are susceptible to slumping or mass wasting (see Fig. 20.2). The angle of repose is a function of the material properties and can be influenced by environmental conditions (water infiltration, erosion, external disturbances). Generally, slopes are designed to less than 3:1 or 33% (a change of three horizontal length units for each vertical unit), which is equivalent to an angle of 18 degrees. Infiltration and erosion control measures can be implemented to help maintain slope integrity. These include application of drainage systems, surface covers, and revegetation, all of which are discussed in forthcoming sections.

Surface contouring involves reshaping of the land surface. It is conducted following these general principles:

- Create a geomorphically mature landscape.
- Create a robust landscape that improves in stability with time.
- Create a landscape that mimics natural systems in the area.
- Use best available demonstrated control technology.

Generally, it is most desirable to produce land forms that mimic natural conditions present in the region. This concept is illustrated in Fig. 20.3. An example of employing natural contouring for reclamation of a mining site in Pennsylvania is given in Fig. 20.4.

20.3.2   Surface Covers and Capping

Different types of covers or caps can be applied to the land surface to achieve a range of objectives. The objectives and types of covers are presented in Information Box 20.2. An example of geomembrane application is presented in Fig. 20.5. The type of cover to use depends on several factors:

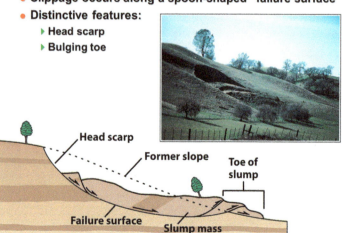

Slumping - Sliding of regolith as coherent blocks
- **Slippage occurs along a spoon-shaped "failure surface"**
- **Distinctive features:**
  - ▸ **Head scarp**
  - ▸ **Bulging toe**

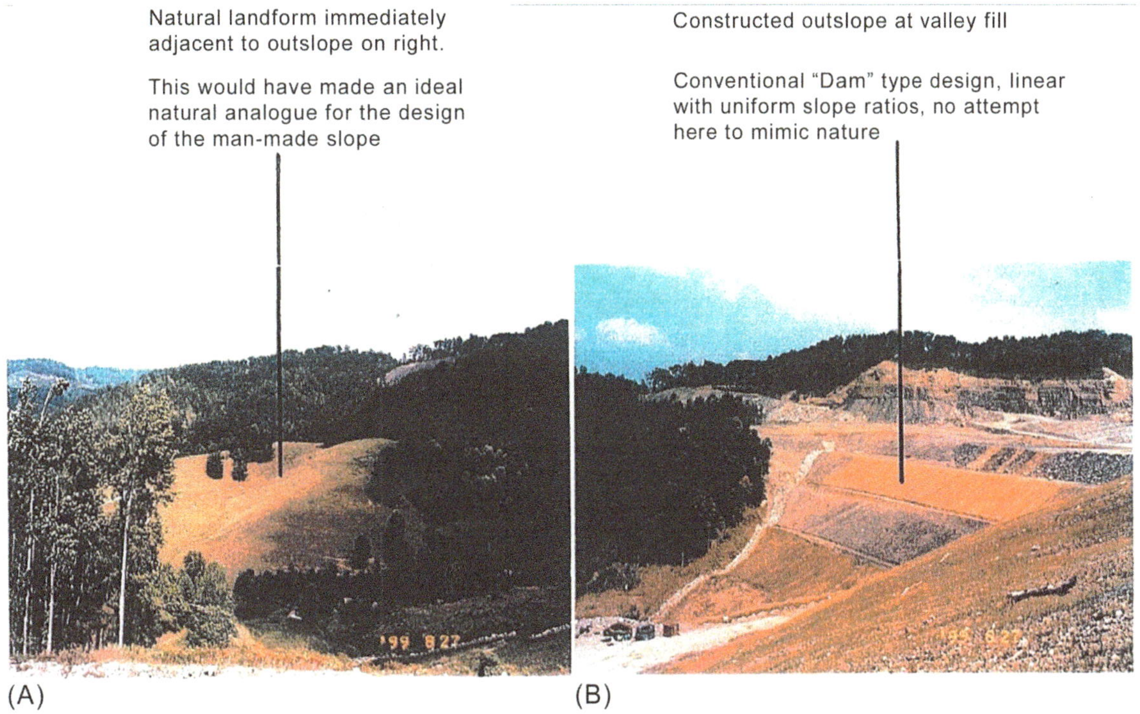

Natural landform immediately adjacent to outslope on right.

This would have made an ideal natural analogue for the design of the man-made slope

Constructed outslope at valley fill

Conventional "Dam" type design, linear with uniform slope ratios, no attempt here to mimic nature

(A)   (B)

FIG. 20.3   Concept of contouring land surfaces to natural forms. *(From Landforming: An Environmental Approach to Hillside Development, Mine Reclamation and Watershed Restoration. Horst J. Schor, Donald H. Gray, John Wiley & Sons, 2007.)*

FIG. 20.4   Example of natural contouring for mine site reclamation. *(From https://en.wikipedia.org/wiki/Mine_reclamation#/media/File: Northeast_from_Providence_Quaker_Chapel.jpg.)*

- *Site conditions*—material properties, climate (rainfall, wind)
- *Source of borrow material* (is there a source of soil nearby)
- *Planned land use*
- *Site location* (nearby human populations?)

Several factors need to be considered when designing the cover system. These include:

- *Drainage layers or devices*: layers (sand and gravel) or devices (drainage pipe) to manage seepage flow
- *Stormwater control*: structures such as drainage channels and retention ponds to manage runoff

- *Institutional controls*: fencing, driving restrictions, no-access signage
- *Long-term monitoring and maintenance*

Examples of surface water control are presented in Fig. 20.6A and B.

In addition to large-scale covers, additional measures are available to control *water erosion*. Permeable barriers made of straw bales or woven fabrics can be used to slow water and reduce its ability to carry sediments. Runoff water can be trapped in settling ponds in which water velocity is eliminated or greatly reduced, allowing suspended particles to settle out, and reducing sediment loads

The objectives of covers include:

- Limit water infiltration
- Control air entry (oxygen)
- Reduce erosion by wind and water
- Support vegetation
- Contain contamination

General types of covers:

Soil caps: soil spread on surface.

Geotextiles: Geotextiles are permeable fabrics that allow passage of water through the material but prevent soil from moving through to lower layers.

Impermeable caps: Engineered, impermeable barrier caps, to prevent percolation into underlying layers. These layers are typically composed of clay or geosynthetic membranes.

Hardened cover: Typically composed of rock, screened to an appropriate size, to reduce erosion and improve stability.

Vegetative cover: Vegetation planted on surface.

before overflow water is released. If suspended colloids are in a dispersed condition, flocculating agents may be added to aggregate particles into larger assemblages that rapidly settle out of suspension. *Gabions* or wire mesh containers filled with rocks can also be used to control water erosion in gullies and drainage ways (Fig. 14.7).

Wind velocity, a major factor in soil *wind erosion*, can be decreased with windbreaks. These may be living windbreaks of planted trees, shrubs, or grasses, or they can be constructed material such as fences or screens. Windbreaks are most effective when placed perpendicular to the direction of the prevailing wind. Effects of windbreaks extend to as much as 40–50 times the height of the windbreak; however, the area adequately protected by the windbreak is usually smaller. Effective control is usually considered to extend to about 10 times the height of the windbreak.

Wind erosion is reduced by a rough soil surface. Surface roughness can be controlled by creating ridges or a rough surface with tillage implements. Ridges 5–10 cm in height are most effective for controlling wind erosion. Soil surface can also be protected by providing vegetative or other surface cover, such as straw, hay, waste residues such as manure, and gravel or rocks. The soil water erosion control measures discussed earlier also provide effective wind erosion control.

Various amendments that help bind soil particles together, including calcium chloride ($CaCl_2$), soybean feedstock processing by-products, calcium lignosulfate, polyvinyl acrylic polymer emulsion, polyacrylamide, and emulsified petroleum resin are applied to unpaved roads to reduce particulate emissions. Unpaved roads can also be covered in gravel or similar nonerodible surfacing materials. However, most of these treatments generally offer only temporary dust control and must be periodically repeated.

20.3.3 Revegetation

Many reclamation and restoration projects involve some type of revegetation of the land surface. Revegetation serves to accomplish multiple objectives, including:

- Limit water infiltration (evapotranspiration barrier)
- Reduce erosion by wind and water
- Reduce contaminant migration
- Restore habitat
- Enhance aesthetics

The types of plants to employ for revegetation depends on the site conditions and intended land use (Table 20.1, Heikkinen et al., 2008).

Special consideration needs to be given to assessing the potential ability of the plant to prosper under the site conditions. This can often be enhanced by selecting species that are native to the region. Another consideration is the potential impact of soil contamination on the

FIG. 20.5 Example of applying a geomembrane cap. *(From: https://commons.wikimedia.org/wiki/File:Geomembrane_installation.jpg.)*

(A)

(B)

FIG. 20.6   Surface water control. (A) Drains, from mining waste treatment technology selection, interstate technology and regulatory council (www. itrcweb.org), (B) trench, *(From Heikkinen, P.M., Noras, P., Salminen, R. (Eds.), 2008. Mine Closure Handbook. Vammalan Kirjapaino Oy, Helsinki. ISBN 978.952-217-055-2.)*

vegetation. Some plants sequester contaminants such as metals and organic compounds into their leaves and stems. This is termed phytoextraction. If this occurs, there is potential for foraging wildlife to become exposed to the contaminants.

Soil quality is another primary factor to consider for revegetation. The soil quality at many disturbed sites may be impaired to such a degree that it cannot support plant growth. In these cases, soil treatment methods need to be employed prior to revegetation. Soli treatments typically involve addition of one or more amendments such as lime for pH modification, organic matter, and nutrients. Soil

remediation may be required if contaminants are present. Soil treatment methods are covered in the next sections.

In semiarid and arid areas of the world, where water is scarce, it is often problematic to include long-term irrigation in the revegetation project design. Hence, it is best to design the project so that the naturally occurring precipitation is sufficient to support vegetation and therefore irrigation is not required. In some cases, irrigation may be needed for the initial application stage, but not over the long term. Low-cost techniques can be used to capture and retain precipitation where it falls to support the revegetation effort. The simplest and cheapest methods involve placing logs, rocks, or mulch

TABLE 20.1 Revegetation Strategy as a Function of Intended Land Use

Type of Land Use	Revegetation Strategy
Plantation forestry	Recommended, if intention is economically viable forest production Fast growing plants with high yield
Recreation	Plants resistant to physical impact, such as turf for sportsgrounds and playing fields Slow-growing ground cover species
Agriculture or grazing	Requires careful risk assessment and monitoring during and after rehabilitation Plants susceptible to uptake of toxic compounds
Natural landscape with no specific commercial or conservation objectives	Facilitate rapid colonization of vegetation by endemic species
No specifically defined land use	Species that rapidly stabilize the site, including commercial grass varieties
Nature conservation	Maximize floral diversity in accordance with original ecosystem

From Heikkinen, P.M., Noras, P., Salminen, R. (Eds.), 2008. Mine Closure Handbook. Vammalan Kirjapaino Oy, Helsinki. ISBN 978.952-217-055-2.

on bare ground to capture moisture, nutrients, seeds, and soil from the surrounding area. Contour furrows, pits, and small depressions in surface soils play the same role in capturing essential elements for plant establishment.

It is important to recognize that most reclamation projects will incorporate more than one method. Thus surface grading and contouring, capping, and revegetation may all be applied for one project. An example of integrated reclamation for mine sites is presented in Fig. 20.7.

Reforestation is a specific type of revegetation strategy for returning trees to a deforested ecosystem. Trees are removed for a variety of reasons. Tropical rainforests are cleared for agriculture, but often the farms fail due to lack of soil nutrients. Temperate zone forests are often clear-cut for timber. Dryland forests in the subtropics are often denuded due to overgrazing of livestock and harvesting of trees for charcoal and firewood. Forests prevent soil erosion, protect against floods, store carbon, and are sites of biodiversity. Hence, efforts are underway to restore forests on a global scale. A wide variety of techniques have

been developed for reforestation. The least expensive method is direct sowing of tree seeds, but the success rate is low. Tree nurseries set up near the site of disturbance can be used to grow juvenile trees for transplanting, a more successful method than direct seeding. Reforestation can use either native or exotic tree species. Fast-growing Eucalyptus trees from Australia have been planted over vast areas of the world to replace native forests that were cleared by humans. Native trees may be more difficult to grow than exotics. However, there is a trend to use native species where possible, because they usually provide greater ecological benefits than exotic species.

20.3.4   Soil Treatment for Acidic Soils

The optimum pH of soils for plant growth is generally between pH 6 and 8. However, agricultural crop production over a period of several years can result in a decreased soil pH or the production of acidic soils (pH < 6).

FIG. 20.7 Integrated mine site reclamation mine tailings in Globe AZ. *(Courtesy of J. Fehmi, Univ of Arizona.)*

In many areas of the world, as for example in the U.S. Midwest and several parts of Europe, agricultural crop production is carried out in environments where rainfall exceeds the amount of water used by the plants. This excess water results in the leaching of basic cations through the soil profile. In essence, these basic cations are replaced from cation exchange sites (see Chapter 2) by H^+ ions that are produced during the degradation of plant residues and organic matter. In addition, applications of ammonium-based fertilizers can also result in increased soil acidity following nitrification. The decreased soil pH can adversely influence plant growth in a number of ways (Information Box 20.3).

Although some crops such as pineapples (grown in Hawaii) or azaleas can tolerate and even grow well at soil pH values of 4.5–5.0, most crops do not grow well. Therefore soil treatment must be conducted to allow viable crop production.

To correct soil acidity, lime amendments are recommended. The liming materials are frequently *calcific limestone*, which consists of calcium carbonate, or *dolomitic limestone*, which is a mixture of calcium and magnesium carbonates. There are many factors that influence the amount of lime to be added which include initial soil pH, final target pH, cation exchange capacity, and other soil characteristics that are site specific. Many states in the United States have specific recommendations for lime requirements issued through the land grant universities or the United States Department of Agriculture (USDA). As an example, however, approximately $10 \, tonne \, ha^{-1}$ of ground limestone might be necessary to raise the pH of a Midwest agricultural soil from pH 6.2 to pH 7. In contrast, once the soil pH falls below 5.0, the amount of lime required increases dramatically. Hence, the same soil at a pH of 5.6 might need $20 \, tonne \, ha^{-1}$ to bring it back up to 7.0. In general, smaller amounts of lime are added more frequently to agricultural land to maintain the pH between 6.0 and 7.0.

20.3.5 Soil Treatment for Saline and Sodic Soils

In contrast to soil acidity problems, agricultural crop production can also result in the build up of salts in high pH

soils. The original source of salts in soils is derived from the primary minerals found in soils including various proportions of the cations—sodium, calcium, and magnesium, and the anions—chloride and sulfate. In humid regions with adequate rainfall, these salts are transported or leached downwards into groundwater or streams, and carried to the oceans. Therefore salt accumulations in soils of the humid regions are rare. In arid areas where rainfall is limited, there is no such mechanism to remove salts, and salt concentrations can increase. However, for salt concentrations to increase to problematic levels, additional sources of salts are necessary.

Irrigation water used for crop production in arid regions can be a source of such salts. In hot arid areas, evaporation of applied irrigation water from soils, and evapotranspiration of water from plants, removes water from the land, but retains the salts. Therefore over time, salt concentrations will increase. This situation can be compounded in low-permeability soils with poor drainage characteristics. In addition, if sodium is the dominant cation, it will replace calcium and magnesium, and result in cation exchange sites dominated by sodium. Therefore salt-affected soils can be separated into three groups: saline, saline-alkali, and nonsaline alkali or sodic soils.

The characterization of these three groups is based on: (1) the electrical conductivity in mmhos (mho is a conductance unit that is the reciprocal of resistance measured in ohms), (2) the percent of cation exchange sites occupied by sodium, and (3) the soil pH (Information Box 20.4).

Saline soils are generally flocculated and have reasonable drainage characteristics. These soils can adversely affect plant growth due to high and often toxic amounts of salts. In contrast, *sodic* soils are frequently dispersed and have poor drainage characteristics. Plant growth in sodic soils may be restricted by both high pH and toxic amounts of sodium. *Saline-alkali* soils can have the adverse characteristics of both saline and sodic soils.

Soil treatment procedures have been developed to remedy the conditions that produce the poor plant growth and decreased yields associated with the three groups of impacted soils. First, large amounts of irrigation water can be used to leach excess salts through the soil profile

INFORMATION BOX 20.4 Characteristics of Salt Affected Soils

Soil	pH	Electrical Conductivity (mmhos cm^{-1})	Exchangeable Sodium Percentage
Saline	<8.5	>4	<15
Saline-Alkali	8.5	>4	>15
Sodic	>8.5	<4	>15

and out of the root zone. This is appropriate for reclamation of saline soils. This can only be accomplished if the soil has adequate drainage characteristics. For sodic soils, drainage can in turn be improved by amending the soil with calcium amendments, which replace sodium on the exchange sites and cause flocculation of soil particles with increased pore space (see Chapter 2). The most frequently used amendments are limestone ($CaCO_3$) and gypsum ($CaSO_4 \bullet 2H_2O$). Sodic or saline alkali soils may require up to $15\text{--}20\,tonne\,ha^{-1}$ to reduce the exchangeable sodium content to acceptable levels.

Finally, elemental sulfur is also used as a soil amendment to reduce soil pH when values are initially in excess of 8.5. Elemental sulfur is oxidized by autotrophic bacteria such as *Thiobacillus thiooxidans*, resulting in the production of sulfuric acid and a concomitant soil pH decrease (see also Chapter 5). Sulfur amendments can be in the range of $1\text{--}3\,tonne\,ha^{-1}$. In the desert southwest, commercial nitrogen fertilizers frequently have sulfur included to maintain soil pH values within the neutral zone. Overall, the strategies employed to reclaim saline and/or sodic soils can be summarized as: (1) improving drainage by replacing sodium on exchange sites with calcium, (2) leaching soluble salts via applying large volumes of irrigation water, and (3) decreasing soil pH via sulfur amendments.

20.3.6 Soil Treatment Using Organic Amendment

Highly disturbed sites often result in surface soils being devoid of organic matter. This can occur from a variety of human activities including strip mining, where the surface top soil is removed and mine tailings are deposited over existing topsoil, soil erosion, and overuse. In all cases, organic matter is sparse or entirely absent, which reduces the ability of the soil to support plants. It is common for these sites to have extremely low microbial populations, extreme pH, low soil permeabilities, and high soluble metal concentrations. These conditions are not suitable for sustainable plant growth, and they generally require some form of organic amendment to jump-start the revegetation process. Problems that occur due to low organic materials are presented in Table 20.2, and common sources of organic materials used to enhance ecosystem restoration are illustrated in Information Box 20.5.

The concept of organic amendments to enhance plant growth has been used for centuries as in the application of "night soil" (human feces and urine) to agricultural land. The use of raw waste material can spread disease, but in the United States, the solid material (*biosolids*) left after treatment of municipal sewage is further refined to largely eliminate potential pathogens before being applied to agricultural land (see Chapter 23). Biosolids have been successfully applied to mine tailings or smelters that contain high

TABLE 20.2 Problems Related to Low Organic Matter in Soil and Other Surface Material

Parameter	Problem
Poor aggregation of primary particles	Compaction Low infiltration rates Low water holding capacity Limited aeration
Low nutrient status	Infertile soils Low microbial populations
Extreme pH	Affect chemical and biological properties
High metal concentrations	Toxicity

From Environmental Monitoring and Characterization, Elsevier Academic Press, San Diego, CA, © 2004.

INFORMATION BOX 20.5 Sources of Organic Materials for Ecosystem Restoration

Human and Animal Waste	Industrial Waste	Agricultural Waste
Animal manures	Paper mill sludges	Mulch
Biosolids	Sawdust	
Composted wastes	Wood chips	

or even phytotoxic levels of heavy metals. Biosolids that have undergone lime stabilization are particularly useful for such restoration, since the increase in pH reduces the bioavailability of metals to plants. Biosolids and composts have also been used to restore diverse ecosystems such as mountain slopes in the Washington Cascades or stabilize sand dunes in southeastern Colorado.

In addition to pathogens, there is potential concern over other contaminants that may be present in the organic material. These include constituents such as salts, metals, nanoparticles, and hazardous organic compounds. The presence of physical, chemical, and biological contaminants needs to be assessed prior to use of the organic material as an amendment.

In all cases of organic amendment added to restore soils, the critical parameter appears to be the magnitude of organic material applied. This is particularly important in desert ecosystems, where high temperatures result in rapid decomposition and mineralization of organic materials. If insufficient organic matter is added to a disturbed site, the beneficial effect is not maintained for a sufficiently long time to allow stable revegetation to occur. An example of organic amendment application for revegetation of mine tailings is presented in Case Study 20.1.

Case Study 20.1 Mission Copper Mine, Arizona: Reclamation and Revegetation of Mine Tailings Using Biosolids Amendment

In the United States, mining is a large industry that provides valuable raw material and creates economic benefit for local communities. However, the environmental damage incurred from this industry range from unsightly mine tailings to the generation of dust, to the leaching of toxic elements into nearby waterways and aquifers. Mine tailings are formed during the extraction and processing of the ore. Typically these materials are disposed of into large tailings piles that are 30–40 m high.

The physiochemical characteristics of mine tailings are totally unlike the displaced topsoil that once supported vegetation in any given area. By removing and crushing bedrock from the mines and placing it on the surface, minerals will oxidize when exposed to the atmosphere. For example, pyrite (FeS_2) is a common mineral at many mines, and it oxidizes to form acid mine drainage. Leachate from tailings can then contaminate surface and groundwaters in addition to increasing the solubility of toxic metals. Mine tailings are a very inhospitable medium in which to grow plants. The material has poor structure; there is no organic material; the cation exchange capacity (CEC) is very low; the water holding capacity of the material is poor to nonexistent; and there are few macronutrients (NPK) available for the plants. Soil biota, in the form of bacteria and fungi, are present only in low numbers, and finally, the pH is usually low, which increases the likelihood that toxic metals could be taken up by the plants. Therefore the material needs to be treated to provide a conducive environment for plant growth. This is typically done by the addition of amendments, one major type of which is organic amendments (Information Box 20.5).

In 1994, the Arizona Mined Land Reclamation Act was passed that required reclamation of all mining disturbances on private land to a predetermined postmining use, and in 1996 the Arizona Department of Environmental Quality (ADEQ) adopted new rules allowing for the use of biosolids to support this reclamation. The Arizona Mining Association (AMA) has estimated that there are 13,360 ha of active mine sites in Arizona that can be reclaimed with the use of biosolids. Current ADEQ regulations limit the lifetime loading rate of biosolid applications to mine tailings to 333 metric tons/ha (dry weight).

Here we describe a case study illustrating the use of biosolids to restore and stabilize mine tailings derived from copper mining. The *Overall Objective of the project was to evaluate the efficacy of dried biosolids as a mine tailing amendment to enhance site stabilization and revegetation.*

Experimental Plan
Study Site
A 5-acre copper mine tailing plot located near Mission Mine, south of Tucson, Arizona was designated for this study. Biosolids were applied at a rate of 2″ to 3″ (70,100 dry tons/acre) across the site in December 1998. Supplemental irrigation was not used in this experiment. This is an important component, as it is not desirable to design projects that require irrigation for sites in arid environments.

Soil Microbial Response to Biosolids
Pure mine tailings contain virtually no organic matter and very low bacterial populations of approximately 10^3 per gram. A large population of heterotrophic bacteria is essential for plant growth and revegetation, and therefore, monitoring soil microbial populations gives an insight into the probability of revegetation prior to plant growth. Biosolids routinely contain very high concentrations of organic matter, including the macroelements carbon and nitrogen, which are essential for promoting microbial growth and metabolism. Following biosolid amendment of the mine tailings, heterotrophic bacterial populations increased at the surface to around 10^7 per gram. Bacteria decreased with increased depths from the surface, indicating the influence of the biosolid surface amendment on bacterial growth. Overall, the microbial monitoring data showed the success of biosolid amendment in changing mine tailings into a true soil-like material.

Physical Stabilization
One of the main objectives in reclaiming mine tailings is *erosion control* through wind and water erosion. This is generally best accomplished through a revegetation program, since the root structures of the plants help to hold soil particles in place. In this experiment, the application of biosolids and the subsequent growth of plants promoted site stabilization. In the desert Southwest, high summer temperatures and limited rainfall are normal; however, despite these extreme conditions, grasses have become established on these tailings. Vegetation transects surveys were conducted on this site 14 months, 21 months, and 33 months after initial seeding. The vegetation cover increased from 18% at the 14-month survey to 78% after 33 months. At 14 months, the predominant plant species were bermuda grass (*Cynodon dactylon*), and the invasive weed, Russian thistle or tumbleweed (*Salsola tragus*), but by the 33rd month, buffelgrass (*Pennisetum ciliare*), and Lehmann lovegrass (*Eragrostis lehmanniana*) had replaced the Russian thistle. Fig. 20.8 A–C show the progressive increase of vegetation on this site over time. In this case, the use of biosolids for enhanced revegetation and stabilization of mine tailings would be considered a success.

Soil Nitrate and Metal Concentrations
At Site 1, soil nitrate and total organic carbon (TOC) are very high at the surface, but decrease to the levels found in pure mine tailings at lower depths. The fact that nitrate and TOC concentrations are correlated is important, since it creates substrate and terminal electron acceptor concentrations suitable for denitrification (see also Chapter 5). Nitrate concentrations increased during the monsoon rainy season of 2000, most likely due to enhanced ammonification and subsequent nitrification. However, within the soil profile, nitrate concentrations decreased with depth. By the winter and spring of 2001, the nitrate concentrations at all soil depths

had decreased. There was no evidence of the leaching of nitrate since concentrations at the 3′—4′ depth were always minimal. Therefore the most likely explanation for decreased nitrates within the soil profile is the process of denitrification. Soil nitrate concentrations became extremely high at both sites in the summer of 2001, again most likely due to nitrogen mineralization and seasonal nitrogen cycling. Specifically during the warmer summer months, rainfall events appear to have triggered microbial mineralization of nitrogen as nitrate.

The application of biosolids to a project site brings some concern about the introduction of heavy metals to the environment. However, data from this study showed that metal concentrations were fairly consistent with soil depth, indicating that the tailings were the major source of metals, not the biosolids. Further evidence of this is shown by the high molybdenum and copper values typical of mine tailings. At this site there was little evidence of metals leaching through the soil profile.

Summary

This study on the application of biosolids to mining tailing at the Mission Mine in Arizona shows that soil stabilization has been encouraged through revegetation techniques and that the leaching of nitrate and heavy metals to important water resources has not been observed. This case study gives an indication of the extensive monitoring that is necessary to understand and assess the performance of the restoration process, and the necessary duration of the monitoring effort.

20.3.7 Soil Treatment for Metals and Organic Contaminants

Special efforts are needed to treat soils contaminated by metals or organic compounds. A standard approach is excavation and replacement. This is an effective method, particularly for relatively small areas with shallow (<3 m deep) contamination. Soil washing can also be used, wherein for example a surfactant solution (detergent) is applied to the contaminated soil to enhance removal of the contaminants. This can be accomplished either in aboveground reactors, by excavating the soil, or in situ by injecting the solution followed by extraction back to the surface. Biological or chemical treatment methods can also be used to degrade organic contaminants. Phytoremediation can be used to remove metals and organics. These and other methods are covered in Chapter 19.

20.3.8 Rangeland Restoration

Roughly a third or more of the earth's land surface can be classified as rangelands, which includes such fragile ecosystems as grasslands, shrublands, wetlands, and deserts. These environments are particularly vulnerable to the impacts of global climate change. Rangelands in many regions of the world are often used to support pastoral farming—animal production conducted to provide food and fiber for humans. Hence, they are critical for human well-being. Rangelands also provide critical habitat for many plant and animal species, supporting healthy wildlife populations.

Rangelands suffer from many issues that can impair their functioning. Loss of soil quality through pollution or overuse, loss of soil quantity through erosion or desertification, and the advent of invasive species are primary examples of issues that can degrade rangeland quality. The restoration and management of rangelands is very complex due to the nature and size of rangelands. A single rangeland "site" may comprise 100's to 1000's of acres for example. Methods used for reclamation and restoration therefore need to be cost effective over those large scales. Because of the cost and complexity of restoring degraded rangeland, sustainable management of rangelands is particularly critical to reduce the advent of pollution and other disturbances.

20.4 AQUATIC RECLAMATION METHODS

Major methods used to reclaim disturbed aquatic systems and improve water and sediment quality will be covered in this section. These include:

- Physical actions (shoreline stabilization, stream channel modification, dredging)
- Water treatment,
- Sediment treatment,
- Aquatic plant management, and
- Biomanipulation.

Aquatic systems span a wide variety of environments:

- Streams
- Lakes
- Wetlands
- Estuaries
- Marine (coastal and open ocean)

The objectives of and methods used for restoration may vary for these different types of aquatic systems.

20.4.1 Physical Actions

The *stabilization* of shorelines—the banks of streams and lakes—is accomplished using grading and contouring, capping, and revegetation. General aspects of these methods were covered in Section 20.3. There are special types of actions used for aquatic systems. For example,

FIG. 20.8 Mine tailings plot during a revegetation project. (A) prior to biosolid amendment; (B) 2 years after biosolid application; (C) 3 years after biosolid application. *(From Environmental Monitoring and Characterization, Elsevier Academic Press, San Diego, CA, 2004.)*

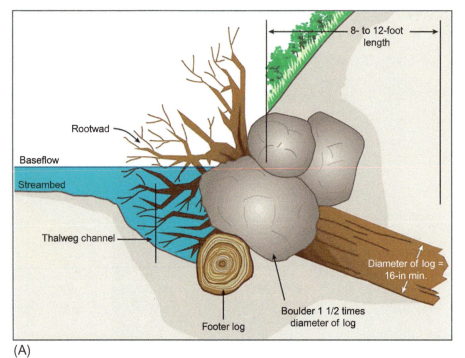

(A)

(B)

FIG. 20.9 Root wads for shoreline stabilization, (A) diagram illustrating implementation, (B) photograph of root wad deployment. *(Available from: USDA, https://www.nrcs.usda.gov/wps/portal/nrcs/detail/national/water/manage/restoration/?cid=stelprdb1044678.)*

one special type of surface cover is called *Root Wads*, which are the stumps of large dead trees. These are placed on the edge of the shoreline along with boulders, as shown in Fig. 20.9. Another type of cover method is *Staking*, which involves the use of wooden stakes driven into the ground to help anchor live vegetation (Fig. 20.10). A third approach is to use *Fascines*, which are rough bundles of

brushwood or other material (Fig. 20.11). *Rock riprap* is a fourth type of cover used for bank stabilization (Fig. 20.12).

A special consideration for streams is the properties and condition of their channel. This influences many aspects of the ecosystem, including stream water velocity, erosion potential and sediment load, flooding potential, and the

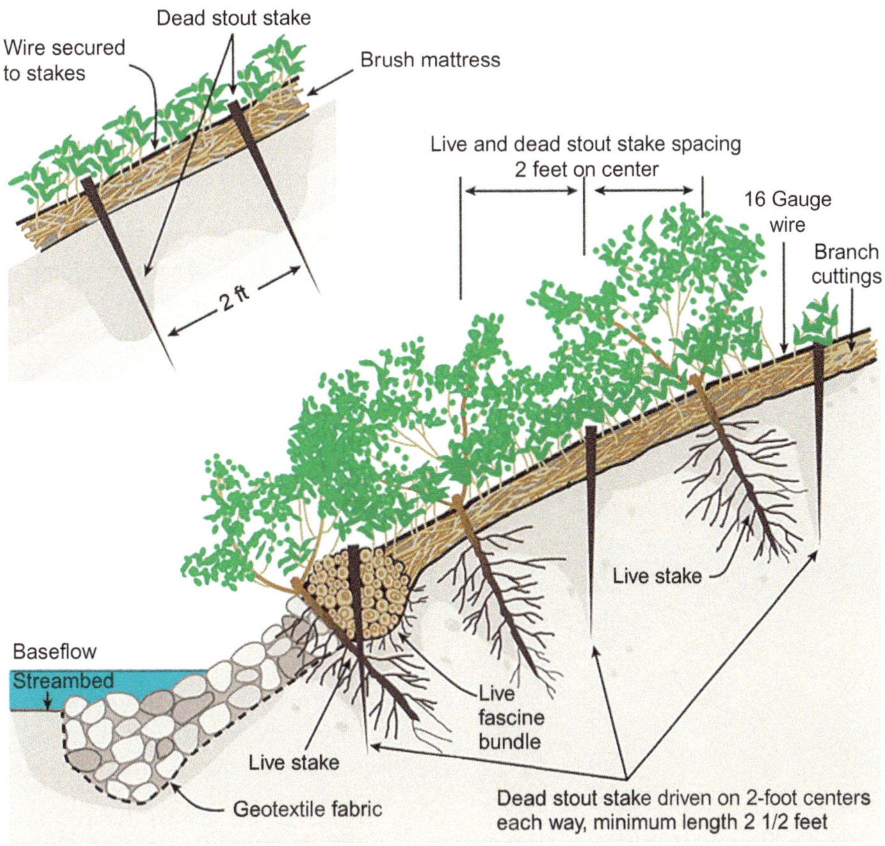

FIG. 20.10 Schematic illustrating the principles of the staking method for shoreline stabilization. *(Available from: USDA, https://www.nrcs.usda.gov/wps/portal/nrcs/detail/national/water/manage/restoration/?cid=stelprdb1044678.)*

FIG. 20.11 Using fascines for shoreline stabilization. A single fascine is shown on the left side of the figure. *(Source: Available from: https://www.ernstseed.com/products/bioengineering-materials/).* Rows of fascines deployed on a shoreline are shown on the right side *(Source: Available from: USDA, https://www.nrcs.usda.gov/wps/portal/nrcs/detail/national/water/manage/restoration/?cid=stelprdb1044678.)*

types of habitats and associated wildlife present. Therefore *stream channel modification* can be a significant component of restoration activities. The first step in channel modification efforts to repair a disturbance and restore function is to determine the optimal conditions to which to restore the channel. This depends on the ecoregion in which the sites resides and the type of stream. For example,

the properties and behavior of a stream located in the headwaters of a watershed are different from those of a stream located in the discharge or outflow zone of the watershed (Fig. 20.13). Thus the types of landforms and habitats present will be different.

A large-scale factor to consider is the *Sinuosity or Meander Ratio*, which is the length of the stream divided

FIG. 20.12 Rock riprap for shoreline stabilization. *(Available from: USDA, https://www.fs.usda.gov/detail/hoosier/about-forest/?cid= fsbdev3_017633.)*

by the length of the valley (Fig. 20.14). For example, a linear stream has a MR of 1. Ideally, one would restore the stream channel pattern to a condition that is compatible with current environmental conditions so that it will remain stable. Large-scale construction work is implemented to make alterations in stream channel patterns. This may include creating artificial meanders and shifting the location of sections of the stream. At smaller scales, *flow changing devices* are a broad category of physical devices used to divert stream flow away from eroding banks. These include devices known as deflectors, bendway weirs, vanes, spurs, kickers, and barbs. In essence, they are structures that project from a streambank and oriented upstream to redirect streamflow away from an eroding bank and to alter secondary currents. These treatments are typically constructed of large boulders and stone, but timber and brush have also been used.

Dredging of sediment is a standard method used for restoration of aquatic systems. Dredging may be implemented for several reasons, including:

- Deepen and increase lake/river volume
- Control excessive plant growth
- Remove excess nutrient-rich muck
- Remove resting cysts of noxious algae
- Remove contaminants

Dredging is conducted under either dry or wet conditions. The water body must first be drained for dry dredging.

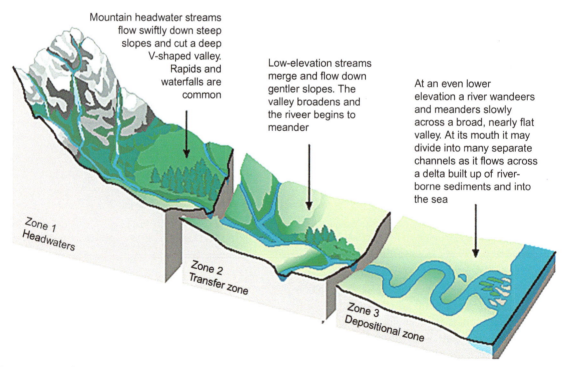

Mountain headwater streams flow swiftly down steep slopes and cut a deep V-shaped valley. Rapids and waterfalls are common

Low-elevation streams merge and flow down gentler slopes. The valley broadens and the riveer begins to meander

At an even lower elevation a river wandeers and meanders slowly across a broad, nearly flat valley. At its mouth it may divide into many separate channels as it flows across a delta built up of river-borne sediments and into the sea

Zone 1 Headwaters

Zone 2 Transfer zone

Zone 3 Depositional zone

FIG. 20.13 Zones of a stream. *(Available from: USDA, https://www.nrcs.usda.gov/wps/portal/nrcs/detail/national/water/manage/restoration/? cid=stelprdb1044678.)*

FIG. 20.14 Example of a meandering stream. *(Available from: USDA, https://www.nrcs.usda.gov/wps/portal/nrcs/detail/national/water/manage/ restoration/?cid=stelprdb1044678.)*

Standard excavation equipment are used for this purpose. Dry dredging may be more disruptive and expensive due to the need to drain the water body, but it provides direct access to the sediment, often resulting in highly effective results.

Wet dredging is conducted while the water remains in the water body. In this case, the dredging equipment are placed on the shoreline or on a barge or ship. There are two general types of wet dredging—mechanical and hydraulic/pneumatic. Mechanical dredging uses a bucket or clamshell device to physically remove sediment (Fig. 20.15). Conversely, hydraulic dredging uses strong suction to remove water and sediment (Fig. 20.16). Pneumatic dredging is similar to hydraulic, except that air rather than water is used.

While dredging can be an effective method, there are several potential concerns. One is the temporary loss of water clarity due to suspension of sediment. Another concern is the mobilization of contaminants and nutrients stored in the sediment. Lastly, a major longer term concern is the potential impact on the benthic habitat and associated wildlife. Another factor is that the sediment will require treatment if it is contaminated. Dredging is typically relatively expensive compared to other restoration methods.

20.4.2 Water and Sediment Treatment

Various methods are available to treat surface water, depending upon the issue. A *chemical coagulant* can be added, for example, to reduce phosphorus concentrations. Aluminum, iron, or calcium salts are added to form floc (small aggregated particles) that sequester P. The floc settles out of of the water column, reducing the P load. *Algicides* can be added to the water to eradicate algal buildups. Aeration devices can be used to provide oxygen to the water column.

Biomanipulation is another option for treating water quality issues. This is the process of purposefully adding or removing an organism from the system. For example, zooplankton such as water fleas can be added to a lake to feed on algae, and thereby reduce algal blooms. Implementing a biomanipulation action requires consideration of how the planned addition or subtraction will affect the overall food web of the system.

As noted before, dredging is a standard method used to remove contaminated sediments. One alternative to dredging is to cap the sediment with clean sediment and/ or a synthetic geomembrane. The cap can stabilize the surface, isolate the contamination, and separate the benthic community from the contamination. Another alternative is to treat the sediment in place. One example currently being investigated is mixing activated carbon into the surface layer of the sediment. The objective is for organic contaminants present in the sediment to adsorb to the carbon, which removes them from the sediment. Reactive media such as oxidants or reductants could be added to destroy the contaminants. Alternatively, methods could be implemented to promote biotransformation of the contaminants. These latter approaches were discussed in Chapter 19.

One common issue for many aquatic systems is excessive vegetation, particularly weeds. Mechanical methods can be used to remove unwanted vegetation. The water level can be manipulated to remove the water source for the vegetation. Herbicides can be added to eradicate the vegetation. The use of herbicides has several associated concerns that must be considered before use.

20.5 ECOLOGICAL RESTORATION

20.5.1 Natural Versus Active Ecological Restoration

Natural restoration is essentially the process of allowing the ecosystem to heal itself without active management or human interference. Depending on site-specific characteristics, natural restoration may not be a viable alternative. Natural or passive ecological restoration projects primarily apply to systems that are still functionally intact, but that have lost vegetative cover and biodiversity from such activities as overgrazing or habitat fragmentation. The implicit goals of these projects are to reduce or eliminate the causes of degradation, often through institutional actions, while encouraging the growth of indigenous plants to increase the productivity of the area in a sustainable way. In general, minimal soil preparation is needed, soil amendments and irrigation are not required, and seeds are simply broadcast in a designated area. Restoration under these conditions is usually coupled with land conservation objectives.

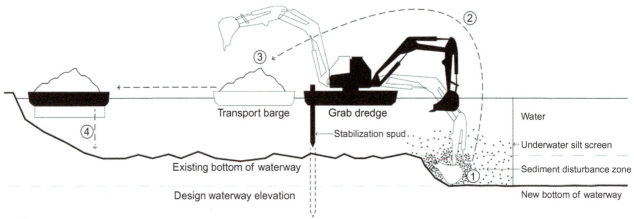

① Dislodging of in situ sediment

② Raising of dredged material to the surface

③ Horizontal transport

④ Placement or further treatment

(A)

(B)

FIG. 20.15 Mechanical wet dredging, (A) Schematic (Available from: https://instrumentalism.wordpress.com/2011/11/06/dredge-landscape-taxon omies/), (B) example rig (Available from: https://www.fema.gov/media-library/assets/images/71444).

Active ecological restoration is necessary for sites where the abiotic and biotic functions of an ecosystem have been impaired to an extent beyond their ability to recover naturally. Active restoration is obviously much more expensive than passive. The cost of restoration will generally rise in proportion to the damage incurred. The restoration goals and objectives developed for a project compared to the existing site conditions will specify the degree of work necessary to achieve the specified goals.

Active ecological restoration employs any and all of the methods discussed in the preceding sections for land and aquatic reclamation, and also remediation methods discussed in Chapter 19. These methods are employed to make the environment suitable to support flora and fauna and to

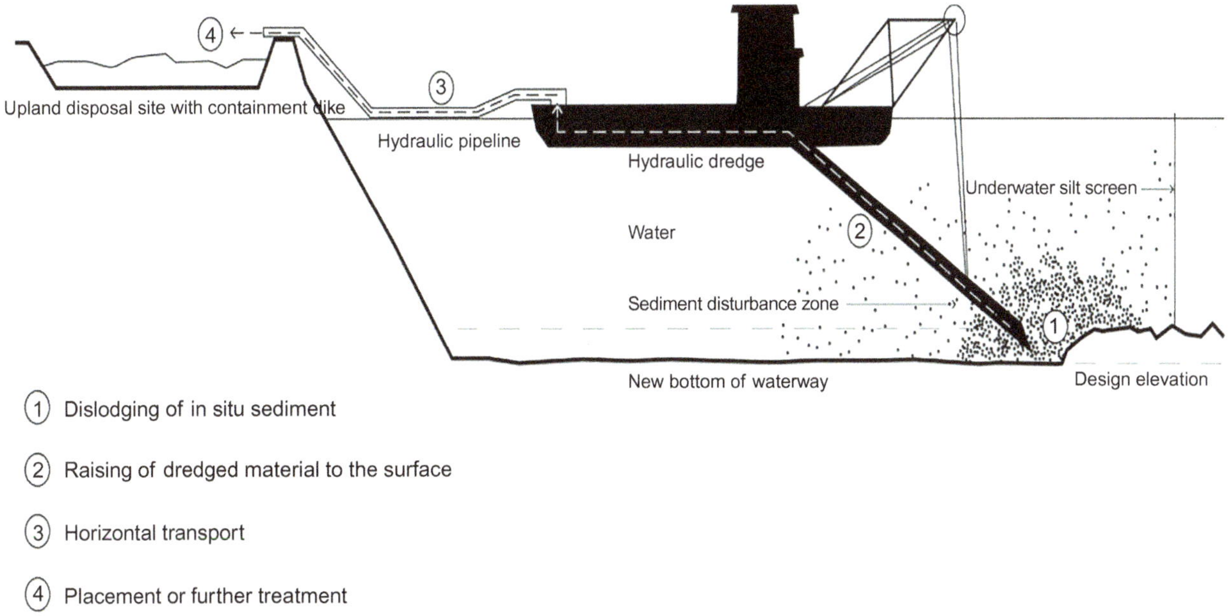

FIG. 20.16 Hydraulic wet dredging schematic. *(Available from: https://instrumentalism.wordpress.com/2011/11/06/dredge-landscape-taxonomies/.)*

① Dislodging of in situ sediment

② Raising of dredged material to the surface

③ Horizontal transport

④ Placement or further treatment

support proper ecosystem functioning. In addition to reclamation and remediation methods, efforts are often made to restore threatened or displaced plant and animal species.

20.5.2 Best Practices

Guiding principles and best practices have been developed for ecological restoration by the Society for Ecological Restoration. These are summarized in Information Box 20.6.

One key concept is to first remove the source of the impairment or disturbance. If this step is not carried out, it is likely that the restoration project will fail because the disturbance will occur again. Another key concept is the use of natural (native) reference ecosystems. A reference system represents what the disturbed site's ecosystem would be like had degradation not occurred, while incorporating capacity for the ecosystem to adapt to existing and anticipated environmental change. It is used to inform the objectives and target outcomes for the restoration project. A third concept

---

INFORMATION BOX 20.6 Values and Principles That Underpin Effective Ecological Restoration

Restoration Should Be Effective, Efficient, and Engaging
Effective ecological restoration reestablishes and maintains values.

Efficient ecological restoration maximizes beneficial outcomes while minimizing costs in time, resources and effort.

Engaging ecological restoration collaborates with partners and stakeholders, promotes participation and enhances experience.

Effective Ecological Restoration
Supports and is modeled on existing native ecosystems and does not cause further harm. Examples of relatively intact land and water ecosystems remain across the globe, which represent an invaluable natural heritage. Appreciation of the long history of evolution of organisms interacting with their natural environments underlies the ethic of ecological restoration.

Is aspirational. The ethic of ecological restoration is to seek the highest and best conservation outcomes. Even if it takes long time frames, full recovery should be the goal wherever it may be ultimately attainable and desirable. Where full recovery is

clearly not attainable or desirable, at least partial recovery and continuous improvement in the condition of ecosystems to provide substantial expansion of the area available to nature conservation is encouraged. This ethic informs and drives high quality restoration.

Is universally applicable and practiced locally with positive regional and global implications. It is inclusive of aquatic and terrestrial ecosystems, with local actions having regional and global benefits for nature and people.

Reflects human values but also recognizes nature's intrinsic values. Ecological restoration is undertaken for many reasons including our economic, ecological, cultural, and spiritual values. Our values also drive us to seek to repair and manage ecosystems for their intrinsic value, rather than for the benefit of humans alone. In practicing ecological restoration, we seek a more ethical and satisfying relationship between humans and the rest of nature.

Is not a substitute for sustainably managing and protecting ecosystems. The promise of restoration cannot be invoked as

Continued

a justification for destroying or damaging existing ecosystems because functional natural ecosystems are not transportable or easily rebuilt once damaged, and the success of ecological restoration cannot be assured.

Efficient Ecological Restoration Depends Upon
Addressing causes at multiple scales to the extent possible. Degradation will continue to undermine restoration inputs unless the causes of degradation are addressed or mitigated.

Recognizing that restoration facilitates a process of recovery carried out by the organisms themselves. Reassembling species and habitat features on a site invariably provides just the starting point for ecological recovery; the longer term process is performed by the organisms themselves.

Taking account of the landscape/aquatic context and prioritizing resilient areas. Sites must be assessed in their broader context to adequately assess complex threats and opportunities.

Applying approaches best suited to the degree of impairment. Many areas may still have some capacity to naturally regenerate, at least given appropriate interventions; while highly damaged areas might need rebuilding "from scratch." It is critical to consider the inherent resilience of a site (and trial interventions that trigger and harness this resilience) prior to assuming full reconstruction is needed.

Addressing all biotic components. Terrestrial restoration commonly starts with reestablishing plant communities but must integrate all important groups of biota including plants and animals (particularly those that are habitat forming) and other biota at all levels from micro- to macro-organisms.

Drawing rigorous, relevant, and applicable knowledge from a dynamic interaction between science and practice. All forms of knowledge, including knowledge gained from science, nature-based cultures, and restoration practice are important for designing, implementing, and monitoring restoration projects and programs.

Knowing your ecosystems and being aware of past mistakes. Success can increase with increased working knowledge of the ecosystem and of appropriate restoration methods.

Taking an adaptive (management) approach. Ecosystems are often highly dynamic, particularly at the early stages of recovery and each site is different. This not only means that specific solutions will be necessary for specific ecosystems and sites; but also that solutions may need to be arrived at after trial and error. It is therefore useful to plan and undertake restoration in a series of focused and monitored steps, guided by initial prescriptions that can be modified as the project develops.

Identifying clear and measurable targets, goals, and objectives. In order to measure progress, it is necessary to identify at the outset how restoration outcomes will be assessed. This will not only ensure that a project collects the right information but that it can also better attune the planning process to devise strategies and actions more likely to end in success.

Adequate resourcing and long-term management arrangements.

Engaging Ecological Restoration Depends Upon
Establishing effective communication and outreach to and with stakeholders. Successful restoration projects have strong engagement with stakeholders including local communities, particularly traditional communities and Indigenous peoples who retain traditional ecological knowledge. This communication and outreach is best achieved if the involvement commences at the planning stage and continues throughout the project and after restoration works are completed.

Involving stakeholders in the development of solutions for improved management and restoration of sites. Ecological restoration outcomes are often more effective and efficient if stakeholders are engaged in assessing problems and devising solutions. The outcome of restoration is also more secure when there are appreciable benefits or incentives available to the stakeholders; and where stakeholders are themselves engaged in the restoration effort, building "ownership" into local cultures.

(**Source:** From Society for Ecological Restoration, 2016.)

is to evaluate the site according to the six key attributes listed in Information Box 20.7. Another key concept is to recognize that full restoration may not always be achievable, and that large magnitudes of recovery may take many years to decades. This concept is captured by the five-star recovery ranking system (Information Box 20.8).

20.5.3   The Invasive Species Problem

As discussed in Chapter 14, due to increases in global trade and travel, alien plant and animal species have been introduced into native ecosystems around the world. Most of these species quickly die out in the new environment, while others become permanent, but minor components of the ecosystem. A few (less than 1%) may become

invasive, spreading over large areas and disrupting ecosystem functions. They may also become pests in human-managed systems such as farms and rangelands. Examples of invasive plants are purple loosestrife and salt cedar, two introductions that have come to dominate many wetland and riparian ecosystems in North America. An example of an invasive animal is the zebra mussel, which clogs the intakes of power plants and depletes the water of phytoplankton that form the base of the native food chain. Over a hundred invasive species have been recognized as causing damage on an ecosystem-wide scale.

Invasive species are a special problem in restoration work, because human-disturbed ecosystems are especially prone to invasion. For example, abandoned farmland in the western United States is often saline and high in

INFORMATION BOX 20.7 Key Ecosystem Attribute Categories and Examples of Broad Goals for Each Attribute Category in a Restoration Project

Attribute	Examples of broad goals—for which more specific goals and objectives appropriate to the project would be developed
Absence of threats	Cessation of threats such as overutilization and contamination; elimination or control of invasive species
Physical conditions	Reinstatement of hydrological and substrate conditions
Species composition	Presence of desirable plant and animal species and absence of undesirable species
Structural diversity	Reinstatement of layers, faunal food webs, and spatial habitat diversity
Ecosystem functionality	Appropriate levels of growth and productivity, reinstatement of nutrient cycling, decomposition, habitat elements, plant-animal interactions, normal stressors, on-going reproduction and regeneration of the ecosystem's species
External exchanges	Reinstatement of linkages and connectivity for migration and gene flow; and for flows including hydrology, fire, or other landscape-scale processes

(**Source:** *From Society for Ecological Restoration, 2016.*)

INFORMATION BOX 20.8 Five-Star Recovery Rating System

Number of stars	Summary of Recovery Outcome (*Note: Modelled on an appropriate local native reference ecosystem*)
★	Ongoing deterioration prevented. Substrates remediated (physically and chemically). Some level of native biota present; future recruitment niches not negated by biotic or abiotic characteristics. Future improvements for all attributes planned and future site management secured.
★★	Threats from adjacent areas starting to be managed or mitigated. Site has a small subset of characteristic native species and low threat from undesirable species onsite. Improved connectivity arranged with adjacent property holders.
★★★	Adjacent threats being managed or mitigated and very low threat from undesirable species onsite. A moderate subset of characteristic native species are established and some evidence of ecosystem functionality commencing. Improved connectivity in evidence.
★★★★	A substantial subset of characteristic biota present (representing all species groupings), providing evidence of a developing community structure and commencement of ecosystem processes. Improved connectivity established and surrounding threats being managed or mitigated.
★★★★★	Establishment of a characteristic assemblage of biota to a point where structural and trophic complexity is likely to develop without further intervention. Appropriate cross boundary flows are enabled and commencing and high levels of resilience is likely with return of appropriate disturbance regimes. Long term management arrangements in place.

(**Source:** *From Society for Ecological Restoration, 2016.*)

nutrients. These conditions favor the establishment of Russian thistle (tumbleweed), an annual plant that completes its life cycle on very little water and releases large amounts of seeds into the soil to germinate in subsequent years. Projects that attempt to reintroduce native plants on abandoned farmland often just stimulate the growth of even more tumbleweed. As another example, western U.S. rivers have been dammed and their flow regulated to prevent the occurrence of normal spring floods. Their floodplains have been invaded by salt cedar, which thrives on the saline soil that develops when the floods are disrupted. By contrast, native cottonwood and willow trees are not salt tolerant, and they have become rare on western rivers.

There is no simple answer to the control of invasive plant species. However, some guiding principles are emerging from restoration studies. First, invasive species tend to be early successional plants, dominating an ecosystem just after it is disturbed. For example, tumbleweed often dominates for five years after a soil is disturbed, but during that time it adds organic matter to the soil, and conditions the soil to eventually become habitable by native species. Hence, patience is necessary in restoration work.

Second, eradication of invasive species by physical or chemical means is often ineffective. Unless the physical environment is restored, the invasive species simply returns. A more successful approach is to attempt to restore the environment to one favoring establishment of native species. For example, native cottonwood and willow trees along western U.S. rivers can be restored even in the presence of invasive salt cedar if a pulse flood regime is returned to the river channel. The floods wash salts from the floodplain soil and germinate native trees, which then overtop the salt cedar and shade them out. Third, invasive species may come to play a positive role in the ecosystem.

Animals that formerly depended on native species may adapt to use the invasive species as a source of food or shelter. Therefore a viable restoration goal might be to aim for a mix of native and introduced species that still can fulfill ecosystem functions.

Invasive species can also be attacked by using biological control agents. Typically, these are insects from the native range of the invasive plant that are introduced into its new habitat to control its spread. For example, Asian beetles that feed on salt cedar are being tested as possible control agents for salt cedar on western U.S. rivers. Extensive research is needed before the release of a biocontrol agent to ensure that it is effective and will not become a pest for native plants as well as the introduced species.

20.6 REDEVELOPMENT AND REVITALIZATION

The presence of an abandoned or otherwise impaired property is often a blight for the local community. Thus there is often great interest in redeveloping such properties for a number of reasons. These include increasing the local tax base, facilitating job growth, lessening development pressures on undeveloped, open land (urban infill), improving environmental quality, and invigorating the local community. The redevelopment of impaired properties is often constrained by the presence of contamination or other risk issues. We learned about the four "Rs" used to repair such sites—Reclamation, Remediation, Rehabilitation, and Restoration. A fifth R, *Revitalization*, connotes the sustainable redevelopment of impaired sites.

Many abandoned properties are classified as *brownfield sites*, which is defined as a property whose expansion, redevelopment, or reuse may be complicated by the presence or potential presence of a hazardous substance. The EPA estimates that there are approximately 450,000 brownfields in the United States. Specific legislation was passed in the United States in 2002 to address these sites (see Chapter 30). Additional types of sites exist that may not be classified as a brownfield site but whose redevelopment is constrained by pollution or related issues. Examples of brownfield and other such sites include abandoned industrial or commercial properties (manufacturing plants, service stations, dry cleaners), unused agricultural fields, closed landfills, closed golf courses, and closed military bases.

Sites may be redeveloped to many different uses, depending upon site conditions, surrounding land uses, and community wishes. Examples of common redevelopment uses include open green spaces (parks), community centers, and commercial uses (hotels, business parks). Successful revitalization efforts require cooperation among all stakeholders in the process.

QUESTIONS AND PROBLEMS

1. Differentiate among: (a) rehabilitation; (b) revegetation; and (c) reclamation.
2. Discuss the different types of surface covers and identify under what conditions each type would be preferred.
3. Discuss why it is important to have clearly defined objectives for a restoration project.
4. What factors primarily determine whether active or passive ecological restoration should be undertaken?
5. What is the influence of organic amendments on soil physical and chemical properties?
6. Discuss the issues that need to be overcome to support a successful revegetation project.
7. What are major issues to deal with for restoration of streams?

REFERENCES

Heikkinen, P.M., Noras, P., Salminen, R., 2008. Mine Closure Handbook. Vammalan Kirjapaino Oy, Helsinki. ISBN 9789522170552.

United States Geological Survey (USGS), 2011. Monitoring habitat restoration projects: U.S. fish and wildlife Pacific region partners for fish and wildlife program and coastal program protocol. In: Chapter 11 of Section A, Biological Science, Book 2, Collection of Environmental Data. Available from: https://pubs.usgs.gov/tm/tm2a11/pdf/tm2a11.pdf.

FURTHER READING

Artiola, J.F., Brusseau, M.L., Pepper, I.L., 2004. Environmental Monitoring and Characterization. Academic Press, San Diego, CA.

Hobbs, R.J., Harris, J.A., 2001. Restoration ecology: repairing the earth's ecosystems in the new millennium. Restor. Ecol. 9, 239–246.

McDonald, T., Gann, G.D., Jonson, J., Dixon, K.W., 2016. International Standards for the Practice of Ecological Restoration—Including Principles and Key Concepts. Society for Ecological Restoration, Washington, DC.

Middleton, N., Thomas, D. (Eds.), 1997. World Atlas of Desertification. Arnold Press, London, England.

Pfadenhauer, J., Grootjans, A., 1999. Wetland restoration in Central Europe: aims and methods. Appl. Veg. Sci. 2, 95–106.

United Nations FAO, Aquatic restoration website document. Available from: http://www.fao.org/docrep/008/a0039e/a0039e00.htm#Contents.

Chapter 21

Industrial Waste and Municipal Solid Waste Treatment and Disposal

J.F. Artiola

Cover art: Solid waste incinerator and ash residues that can have hazardous waste characteristics, due to their metal content. *(Source: U.S. Environmental Protection Agency).*

21.1 INTRODUCTION

Industrial and municipal solid wastes are created by modern societies as unavoidable by-products of mining, industrial production, and the requirements of today's modern consumers. In the United States, the 19th century was an era of rapid growth during which many new and developing industries were established to systematically process raw materials into finished goods. During the 19th and 20th centuries, and now the 21st century, these industries concentrated on the quality of the final product, while discarding the residues and waste products generated during increasingly complicated manufacturing processes.

At the beginning of the Industrial Revolution, mining of coal and metal ores became both the fuel and the building blocks of Western society's industrialization. With the discovery of petroleum and natural gas deposits, chemical-processing industries quickly followed. From the early 1900s to the late 1960s, hundreds of thousands of petroleum-derived and synthetic chemicals were used in the production of goods.

Although carbon-based plastic materials, together with such organic chemicals as pesticides and solvents, have dominated industrial production since the early 1950s, metal-based goods remain fundamental to modern industry. Numerous modern goods—from cars to paints—require the use of common metals such as iron, aluminum, and copper. In addition, less abundant but more toxic metals such as lead, cadmium, nickel, mercury, arsenic, and selenium are also used by industry. Metallic elements are therefore commonly found in industrial wastes, where they have complex and incompletely understood effects on the environment (Fig. 21.1). Consumers also discard massive amounts of hazardous wastes. The advent of extensive industrial development of countries such as China and India has exacerbated the production of industrial wastes. We buy, use, and dispose of increasing quantities of goods that have hazardous characteristics including paints, solvents, pesticides, batteries, light bulbs, and electronic goods.

The disposal of wastes can result in adverse effects not only on the environment but also on human health and well-being (Chapter 26). Early in the 20th century, industries, cities, and towns began to use abandoned or natural depressions to dispose of the ever-growing amounts of municipal waste residues. In essence these were the earliest forms of landfills. Until the 1970s in the United States, it was common practice to dump waste organics on land with little or no regard for the soil and water pollution it produced. Today, there are an increasing number of waste minimization programs, recycle and reuse programs, and numerous solid waste disposal programs, which not only regulate hazardous industrial wastes, but also provide guidelines for the safe recycling and disposal of hazardous household wastes.

21.2 RELEVANT REGULATIONS FOR INDUSTRIAL WASTE AND MUNICIPAL SOLID WASTES

Before 1970, industrial untreated wastes were disposed of in unlined landfills or lagoons or discharged into surface waters. Other wastes were burned in open pits or in

Environmental and Pollution Science. https://doi.org/10.1016/B978-0-12-814719-1.00021-5

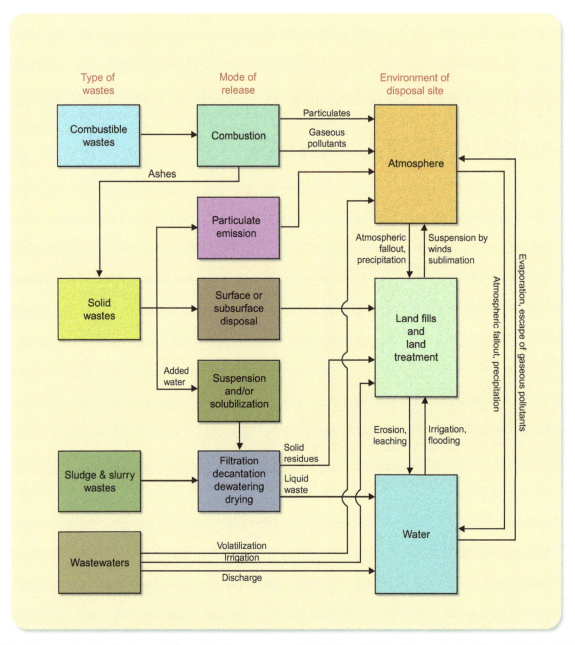

FIG. 21.1   Sources and modes of releases of pollutants into the environment. *(Modified from Pollution Science, Academic Press, San Diego, CA, 1996.)*

incinerators with no pollution controls. As a result, ground-water and surface water resources, such as lakes and rivers, were polluted. Besides the human toll associated with water and air pollution, animals and plants also experienced adverse effects due to industrial pollution associated with the accumulation of pollutants in the environment.

Since the late 1970s, several laws have been implemented to control the disposal of hazardous wastes to protect our health and the environment (Information Box 21.1) (see also Chapter 30).

INFORMATION BOX 21.1 Laws Implemented to Control the Disposal of Hazardous Waste

The Resource Conservation and Recovery Act (RCRA); the Comprehensive Emergency Response, Compensation and Liability Act; the Safe Drinking Water Act; the Federal Clean Air Act; the Clean Water Act; and the Toxics Control and Safety Act. Industrial wastes are classified as *hazardous* and *nonhazardous* by the *Code of Federal Regulations* (CFR 40, Part 261). See Chapter 30.

Most of the materials classified as nonhazardous are regulated under such specific industry categories as mining and oilfield wastes. Of all the wastes generated in the United States, more than 95% are classified as nonhazardous municipal. Although industrial, mining, oil- and gasfield wastes are regulated separately, in this chapter we will treat all of them as "industrial wastes."

Also since the late 1970s, industrial wastes classified as hazardous have received considerable attention due to their obvious potential for deleterious impacts on human health and the environment. Consequently, government scrutiny has focused on new regulations for storage and disposal of these wastes (see Information Box 21.1). Nonetheless, less regulated, nonhazardous wastes constitute the bulk of the waste generated by industry. The largest source of nonhazardous wastes is human sewage, which is treated within wastewater treatment plants prior to land application, incineration, or landfilling (see also Chapters 22 and 23). Nonhazardous wastes often contain the same potentially polluting components found in hazardous wastes but at much lower concentrations. To prevent the disposal of household goods with hazardous properties into *municipal solid waste* (MSW) landfills, new guidelines have been implemented that control the disposal of items such as spent crankcase oil, car batteries, tires, paint thinners, or refrigerants. These are implemented through household hazardous waste collection programs (U.S. EPA, 2005b).

21.3 MAJOR FORMS OF INDUSTRIAL WASTES

Regardless of how wastes are technically categorized by regulatory agencies, their pollutant-releasing capacities depend on their physical characteristics, which determine their mode of transport into the environment (Chapters 7 and 12). The four major waste types based on physical characteristics—combustible wastes, solid wastes, sludge and slurry wastes, and wastewaters—are presented in Fig. 21.1, which shows how each of these types can eventually release pollutants into the atmospheric, terrestrial, and aquatic environments. *Combustible wastes* yield by-products that can be released directly into the atmosphere as gases or particulate pollutants when they are not properly filtered. *Solid wastes* can release pollutants into the atmosphere via dust or particular transport, or when these wastes come into contact with water, their soluble constituents can be leached out into the soil surface or below. *Sludge and slurry wastes* can release pollutants into the soil and groundwater from both their solid and liquid phases. *Wastewaters*, owing to their liquid state, are always potential sources of pollution if discharged directly into the aquatic or terrestrial environment without proper treatment. In general, we can see that the pollutants whose impact on the environment is most severe are most usually found in liquid or gas phases. By their fluid nature, these pollutants can be transported large distances and thus affect large segments of the environment. (The various physical methods by which pollutants are translocated within and among environments are discussed in Chapter 7.)

Except for a few isotopic forms, none of the 40 elements that are economically important to modern industrial society can be made synthetically (Table 21.1). Thus these elements have to be mined, extracted from the natural state, and subsequently purified. Because of such processes, as well as the processes used in manufacturing, these elements—including many precious and strategic metals such as gold (Au), platinum (Pt), cobalt (Co), antimony (Sb), and tungsten (W)—can be found in significant

TABLE 21.1 Elements of Economic Importance

Element	Chemical Symbol	Common Forms in Wastes
Aluminum	Al	Al^{3+} oxides and hydroxides, Al metal, Al-silicates
Arsenic	As	As^{n+} oxides (arsenate, arsenate)
Barium	Ba	$BaSO_4$ (barite)
Boron	B	B^{3+} hydroxides (borate)
Bromine	Br	Br^-
Cadmium	Cd	Cd^{2+} ion and Cd halides, oxides and hydroxides
Calcium	Ca	$CaCO_3$, $CaSO_4$, Ca oxides and hydroxides
Carbon	C	Inorganic forms: CO_2, CO_3^{2-}, and HCO_3^-
		Organic forms include: C—C—, C—H, C—O—, C—, C—S—
Chlorine	Cl	Cl^- ion

Continued

TABLE 21.1 Elements of Economic Importance—cont'd

Element	Chemical Symbol	Common Forms in Wastes
Chromium	Cr	Cr^{3+} hydroxides, Cr^{6+} oxides, Cr metal
Copper	Cu	Cu^{2+} oxides and hydroxides, Cu metal
Fluorine	F	F^- ion, CaF (apatite)
Iron	Fe	Fe^{n+} oxides and hydroxides, Fe metal
Lead	Pb	Pb^{2+} oxides, hydroxides and carbonates, Pb metal
Magnesium	Mg	$MgCO_3$, $MgSO_4$, Mg oxides and hydroxides, Mg Silicate (asbestos)
Mercury	Hg	Hg^{2+} oxides and halides organo-Hg complexes
Molybdenum	Mo	Mo^{n+} oxides (molybdate)
Nickel	Ni	Ni^{2+} ion, Ni^{2+} amines, Ni metal
Nitrogen	N	NO_3^- (nitrate) and NH_4^+ (ammonium) ions
		—C—N— (organic chemicals)
Phosphorus	P	PO_4^{3+} ion, Ca phosphates
Potassium	K	K^+ ion
Selenium	Se	Se^{n+} oxides
Silicon	Si	Si^{4+} oxides and hydroxides, Al Silicates
Silver	Ag	Ag^+ as AgCl, Ag metal
Sodium	Na	Na^+ ion
Sulfur	S	SO_4^{2-} (sulfate) ion, SO_2, H_2S, FeS_2 (pyrite)
		—C—S— (organic chemicals)
Titanium	Ti	Ti^{4+} oxides
Uranium	U	U^{n+} oxides and halides, U metal
Vanadium	V	V^{n+}-oxides
Zinc	Zn	Zn^{2+} oxides and hydroxides, Zn metal
Gold	Au	Au^{+3} halides, oxides, $AuCN^-$, Hg–Au amalgam
Platinum	Pt	Pb^{+2} sulfates, carbonates, halides
Cobalt	Co	Co^{+2} as amines and oxides
Tungsten	W	W^{+6} oxides
Antimony	Sb	Sb^{+3}, Sb^{+5} oxides, halides

amounts in some industrial wastes. A few elements, such as mercury (Hg) and arsenic (As), can be found in gaseous, liquid, or solid wastes. Yet others, such as lead (Pb) and chromium (Cr), can be discharged into the atmosphere as particulate matter associated with other elements such as sulfur (S) and carbon (C).

However, most of the elements listed in Table 21.1, including metals, metalloids, and salts, can be dissolved in the liquids (usually the aqueous phase) found in wastewaters, sludges, and solid industrial wastes. That is, with

few exceptions, most industrial liquid, slurry, and sludge wastes have a water phase that can range from 99% to less than 10% by weight. Therefore the process of dewatering can be used to reduce much of the mass of these wastes. Moreover, many forms of organic pollutants are also water soluble to various degrees, and are therefore found in the water phase of these wastes. These aqueous phases, once separated from solids, must be processed as wastewaters prior to discharge into the open environment.

21.4 TREATMENT AND DISPOSAL OF INDUSTRIAL WASTES

Technologies and practices used for the treatment and disposal of metal- and salt-containing wastes vary widely, but ultimately, the by-products of these technologies end up being released into the air, land, or water environments (see Fig. 21.1). Methods that separate metals from the other waste constituents are driven by the need to reduce waste disposal costs and ultimately by the potential costs associated with liability. Organic wastes that contain low residual concentrations of metals and salts can often be degraded by using thermal or biological destruction processes that completely transform the waste into carbon dioxide and water. Conversely, wastes that contain significant amounts of metals and salts always leave indestructible residues that may or may not be able to be recycled economically. Therefore these wastes that cannot be eliminated must be disposed of in a manner that minimizes their impact on the environment.

21.4.1 Gas and Particulate Emissions

Gases and dust particulates that are generated in the thermal destruction of wastes and smelting can be prevented from escaping into the air using one or more of the following processes: electrostatic precipitators, baghouse and cyclone separators, and wet scrubbers. These technologies are expensive and difficult to operate efficiently. However, without this equipment, smelters and incinerators would discharge large quantities of toxic metals into the atmosphere, thereby contaminating large tracts of land. In the United States, the Clean Air Act requires the control of hazardous emissions by industries. Originally passed in 1970, by 1990 this act was extended to cover 189 industrial chemicals, requiring the installation of air pollution control equipment on all major industrial sources. These regulations also include requirements for the removal of metal-containing particulate matter (PM_{10}) from air emissions (see also Chapters 17 and 30).

A very controversial issue is the reduction or trapping of greenhouse gases like carbon dioxide (CO_2) that are released by power-generating plants that use fossil fuels (coal, oil, and natural gas). These gases are known to be linked to global warming (see also Chapter 25). To date, the U.S. government does not have or participate in any program to control CO_2 emissions. While mandated in Europe and other countries around the world that agreed to the *Kyoto Treaty*, only a few U.S. industries have reduced CO_2 emissions. However, future options to aggressively reduce these emissions will have to include federal- and/or state-imposed limits of emissions. In addition, trading programs to buy and sell CO_2 emissions may be implemented, and the expanded use of carbon sequestration

techniques is likely. Technologies being evaluated include trapping and liquefying CO_2 gas from smokestacks, followed by storage by injection into depleted deep oil or geologic gas reservoirs (see Chapter 32).

21.4.2 Chemical Precipitation

Wastewater and slurried sludges can be treated with chemical agents to precipitate and remove metals from the rest of the waste components. For example, solutions containing alkaline materials such as calcium carbonate, sodium hydroxide, aluminum oxide, or sodium sulfide are commonly used to precipitate metals from waste streams. Fig. 21.2 shows acidic metal sludges neutralized with lime. These chemicals help form insoluble metal hydroxides, carbonates, and sulfides. Similarly, aluminum and iron oxides sorb metals such as cadmium and metalloids such as arsenic, thus removing them from solution. Precipitation and adsorption of inorganics is discussed in Chapter 8.

FIG. 21.2 Unlined pond filled with acid and metals (Fe, Zn, Ni, Cr) wastes. The acid liquids and muds (insert) were treated with hydrated lime to neutralize the acidity and precipitate all the metals. *(Photo courtesy: J.F. Artiola.)*

General precipitation reactions can be represented by the following:

$$M(free) + CaCO_3(slurry) \rightarrow \underset{(precipitates)}{MCO_3} + \underset{(precipitates)}{M(OH)_2}$$

$$M(free) + Na_2S\,(pH > 8) \rightarrow MS$$

where $M = Cd^{2+}$ or Zn^{2+} or Cu^{2+} or Pb^{2+}.

21.4.3 Flocculation, Coagulation, Dewatering-Filtration-Decanting-Drying

Particulate pollutants in wastewaters can be made to settle out quickly by using chemicals known as flocculants and coagulants. *Flocculants*, which react with dissolved chemicals, facilitate the formation of aggregates or clumps, which can be decanted or filtered out of solution. Conversely, *coagulants* destabilize colloids, thereby permitting suspended particles to form aggregates that can settle out of solution. These chemicals can be useful in removing all kinds of pollutants from wastewaters, including metals and organic constituents. Flocculants and coagulants include iron and copper sulfates and chlorides, as well as complex synthetic organic polymers.

Dewatering and drying is an acceptable option for waste reduction where the liquid components of a liquid waste, such as a slurry or wastewater, have very low levels of pollutants, and these can be treated using conventional wastewater treatment methods (see Chapter 22) or even discharged directly into the environment. In this case, liquids are separated from solids via several dewatering options that include air drying, particle filtration, sedimentation, and decanting. The separated solids are then treated and disposed of as solid wastes. Fig. 21.3 shows a pond used to store and dewater flue gas desulfurization waste, a byproduct of sulfur dioxide gas scrubbing generated during coal combustion.

21.4.4 Stabilization (Neutralization) and Solidification

Sludges and slurries containing metals must be chemically and physically stabilized prior to final disposal. Since metals are more soluble in water at low pH, acidic wastes must be neutralized with such basic compounds as calcite $(CaCO_3)$ and hydrated lime $(Ca[OH]_2)$ to form low water solubility metal-carbonates and metal-hydroxides complexes (see Chapter 8).

The process of waste solidification usually involves trapping or encapsulating the waste into a physically stable matrix. For example, when wet cement is mixed in with sludges, it forms a stable block after a few days of curing (drying). This solidification method encapsulates wastes in a matrix that is relatively low in porosity and cannot easily be deformed or cracked under typical landfill overburden pressures. Consequently, percolating water does not readily infiltrate into the matrix, and metals or salts are less likely to leach out. In some cases, waste products like fly ash, collected from electrostatic precipitators during coal burning, can also be used to neutralize and encapsulate other wastes like metal waste streams, due to their strong cementing (pozzolanic) properties.

21.4.5 Oxidation

Carbon-based waste streams can be oxidized to detoxify and destroy organic pollutants. These are two major types of treatment processes: thermal and chemical. The overall goals are the same in both processes. Thermal oxidation reactions can be described as follows:

$$RCCNOCl + O_2 + heat \rightarrow CO_2 + H_2O + N_2$$
$$+ NO_x + SO_x + HCl + intense\ heat$$

Thermal oxidation processes include incineration, which use conventional fuel-driven burners. Solid or liquid organic

FIG. 21.3   Saline flue gas desulfurization (FGD) waste drying impoundment—playa view of dried FGD wastes with sparse vegetation. Slurried saline FGD waste discharge pipe (insert). *(Photos courtesy: C. Salo and J.F. Artiola.)*

wastes having low water content but high heat values are good candidates for thermal oxidation because they burn hot enough to sustain the energy these processes require. The major disadvantages of this process include high costs (usually more than $200 a barrel), limited reliability, and a negative public perception of their safety. In addition, large emissions of pollutants can result when the systems are not operated and maintained properly. Incinerators operate most efficiently when designs are tailored to specific waste stream characteristics. Thus incinerators are not suitable for the treatment of mixed waste streams.

Chemical oxidation generally proceeds via the following reaction pathway:

$$RCCNOCl + (O_2, Cl_2, or\ O_3) \rightarrow R'CO(detoxified)$$
$$+ CO_2 + H_2O + N_x + Cl^- + intense\ heat$$

where R or R' is the remaining part of the organic molecule. For example, chemical oxidation is used in the destruction of cyanide-containing wastewaters. This process involves chlorination, which can be summarized as follows:

$$2CN^-(liquid) + 3Cl_2(gas) + 6NaOH$$
$$\rightarrow N_2\ gas\ (pH > 8-9) + 6Cl^-\ 2HCO_3^- + 2H_2O$$

Many forms and combinations of chemical oxidation processes use chlorine (Cl_2), chlorine dioxide (ClO_2), ozone (O_3), or ultraviolet (UV) radiation to eliminate low concentrations of organics from industrial wastewaters. The chemicals oxidized using these processes include acids, alcohols, and aldehydes (such as oxalic acids and phenols), as well as more stable chlorinated pesticides and some petroleum-derived solvents such as DDT or xylenes.

21.4.6 Landfilling

Industrial wastes that contain significant concentrations of metals that cannot be recycled or recovered economically are usually good candidates for landfill disposal, which is the disposal of waste materials within soil or the vadose zone. However, sludges high in metals must be neutralized and solidified prior to landfill disposal (see Fig. 21.2), and only landfills approved for the disposal of solidified hazardous wastes may be used for the disposal of these wastes. Similarly, wastes high in salts may also be buried in special landfills or used as fill materials for road and dam construction. Although such wastes are not considered toxic, they are soluble in water. Therefore it is important to minimize water contact to prevent the leaching of salts into the environment.

21.4.7 Landfarming of Refinery Sludges and Oilfield Wastes

Landfarming techniques have been used to treat hazardous wastes, particularly oily sludges, whose major constituents

are oil, sediments, and water. In the early 1980s, land treatment of hazardous oily wastes came under intense scrutiny by the U.S. Environmental Protection Agency (EPA). Land disposal restrictions, begun in early 1992, now prohibit land treatment of hazardous oily wastes. These restrictions have compelled the petroleum industry to look at alternative disposal methods, including landfilling and incineration. On the positive side, petroleum refineries have also been spurred to develop waste minimization strategies. Oilfield wastes generated during well drilling activities are treated in treatment facilities that have to be permitted. These facilities dewater and lower the salinity and oil content of these wastes until they meet state-imposed criteria for their safe reuse as fill materials. In addition to residual amounts of oil ($\sim$1%–5%), these wastes often contain significant amounts of metals such as zinc, and minerals like barite, and sodium chloride (Table 21.2).

21.4.8 Deep-Well Injection of Liquid Wastes

Deep-well injection of liquid wastes into the subsurface is another waste disposal method. This method greatly reduces the potential hazard posed by wastes through disposing of them in the deep subsurface. According to the U.S. EPA (2001), there are about 470 Class I wells located mostly in the oil-producing states in the central part of the United States and in heavily industrialized states such as California, Michigan, and New York. Examples of deep-well injection of liquid wastes include oilfield wastes (brines), metal-containing wastewaters, slurries, wastes with toxic organic chemicals such as chlorinated hydrocarbons, and radioactive wastes. According to U.S. EPA (2001), most wells (350 out of 473) inject wastes classified as nonhazardous (mostly oilfield wastes (brines)).

Deep-well injection is most suitable for handling large volumes of liquid or slurry wastes that have a water-like consistency (low viscosity). Watery liquids are usually pumped down into confined, aquifer-like zones composed of highly water-permeable material, such as sandstone or limestone. Similarly, oily wastes can be deep well injected

TABLE 21.2 Examples of Water Solubilities of Minerals Found in Wastes

Mineral Names and Chemical Compositions	Water Solubility[a] ($mg\,L^{-1}$)
Barite ($BaSO_4$)	2
Calcite ($CaCO_3$)	14
Gypsum ($CaSO_4 \bullet 2H_2O$)	2400
Salt (NaCl)	370,000

[a]*Varies, depending on temperature and the presence of other minerals in water.*

Class I Wells

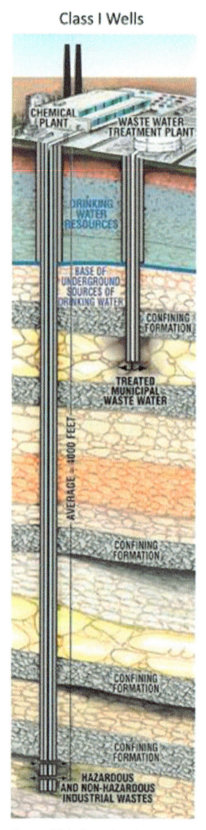

FIG. 21.4   Deep-well injection design. *(Source: https://www.epa.gov/ uic/class-i-industrial-and-municipal-waste-disposal-wells#haz_well)*

into high-permeability subsurface zones. The depth of the injection zones, usually hydrologically confined, ranges from about 200 to 4500 m (600–13,000 ft), and most zones are located between 700 and 2500 m (2000 and 7000 ft) (Fig. 21.4).

Drilling and constructing wells for these depths can be very expensive; therefore deep-well injection is primarily used by the petroleum industry, which already possesses inservice oilfield wells. Once constructed, these systems are comparatively cheap to operate and maintain.

The practice of deep-well injection of wastes still poses some potential problems. Possible clogging of the injection zone due to solid particles or bacterial growth may occur, and the injected waste may contaminate resident groundwater. This possibility is of major concern if the groundwater is a current or potential potable water source. Waste injection can cause groundwater contamination in several ways: (1) direct injection into an aquifer used for drinking water, (2) leaking wells (waste leaks from the well bore into an aquifer used for drinking water), and (3) movement of waste to a zone that supplies drinking water. A more recent concern has been the significant increase in earthquakes recorded in certain regions of the U.S. where deep-well injection of oil and gas production wastewaters is prevalent. For example in Oklahoma, the earthquake rate increased from a many decade average of 21 events/year to more than 100 per year over the past decade and a half.

21.4.9 Incineration

The process of incineration that destroys highly toxic and hazardous organic wastes differs from MSW incineration, where energy is often produced. In general, low-temperature (up to 850°C) and high-temperature (~1200°C) incinerations use energy to oxidize carbon- and water-containing wastes to CO_2, H_2O vapor, and HCl and NO_x gases. However, some incinerators can serve as heat energy sources, especially when oily wastes, such as spent oils, are used as fuel. But these incinerators must meet stringent emissions levels. Incineration efficiencies are also closely regulated, and destruction of all organic compounds must exceed 99.99%.

Incineration cannot be used with wastes that have high concentrations of water and noncombustible solids, nor can it be used for radioactive materials. Moreover, incinerators are very expensive to build and maintain, and waste incineration expenses often exceed $500 a barrel. Thus this technology is mostly limited to low-volume wastes such as medical wastes. Finally, besides the gases previously listed, incinerators emit small but significant amounts of numerous toxic chemicals including dioxins, furans, polynuclear aromatic hydrocarbons, and metals such as mercury, beryllium, and cadmium. Often incinerators produce ash residues (bottom and fly ash) that have hazardous characteristics (particularly fly ash) and must be buried in landfills approved to handle such wastes. Thus incinerators do not totally eliminate the problems of environmental contamination, and this has resulted in the technology being condemned by the public. Grassroots organizations are currently fighting for the elimination of this technology as a waste treatment option, even as a renewable energy recovery source. Because of the public outcry and health-related concerns raised by the scientific community (NRC, 1999), incineration has undergone a large decline in the United States. The number of municipal solid waste (MSW) incinerators and medical waste incinerators in the United States has decreased from a peak of 186 in 1990 to 112 in 2003 to 86 currently.

21.4.10 Stockpiling, Tailings, and Muds

Mining activities produce vast quantities of tailings, which are usually stockpiled in the form of terraces on or near the ore-processing mills. The environmental impacts of mining activities, which vary widely from site to site, are usually associated with runoff or percolation of waters contaminated with sediments (see Chapter 14). They are also related to pH, metal solubility, salt concentrations, and the quantities of wind-blown particulates that are contaminated with metals. Oil- and gas-fields produce large quantities of well cuttings (muds). These well cuttings, which contain barite, salts (see Table 21.2), and crude oil, are usually stockpiled or treated until they can be used as fill materials. The potential for off-site releases of pollutants from these sites is normally associated with water and wastewater releases that contain varying concentrations of these chemicals. Mine tailing stabilization and revegetation is discussed in Chapter 20.

21.4.11 Air Stripping and Carbon Adsorption

Air stripping is a standard method of treatment for wastewater contaminated by volatile organic compounds. Air stripping is, for example, widely used to treat groundwater contaminated by hydrocarbons and chlorinated solvents at hazardous waste sites (Chapter 19). Typically, large towers are used, which are filled with plastic media to create large amounts of air-water interfacial area. The water is released into the top of the tower, and air is blown upwards from the bottom of the tower. This countercurrent flow causes mixing of the air and water, and as a result the volatile contaminants in the water partition to the air (volatilization)

which is termed air stripping. Because of this transfer, the air is now contaminated and must be treated.

Another standard method used to remove low concentrations of organic contaminants from wastewaters and from gases (such as those produced from air stripping) is adsorption by granular activated carbon (GAC). GAC is a highly processed form of charcoal that has very high solid surface area. The GAC is packed into large vessels, and the contaminated wastewater or vapor is passed through the vessel. As the fluid migrates through the packed vessel, the contaminants adsorb to the GAC and are thereby removed from the water or air. At many hazardous waste sites contaminated by hydrocarbons or chlorinated solvents, contaminated groundwater is first passed through an air stripping tower, and then the contaminated air is passed through a GAC vessel.

21.5 REUSE OF INDUSTRIAL WASTES

Because metal- and salt-containing wastes cannot be destroyed, when improperly disposed they can be a hazard to humans and the environment. At the same time, these wastes have a potential economic value that is becoming more evident as the quality of their natural sources diminishes. Precious and strategic metals, such as gold, platinum, cobalt, antimony, and tungsten, are routinely mined out of wastes that contain significant concentrations of these elements (see Table 21.1). However, wastes that contain less valuable metals, such as aluminum and iron, are far less likely to be treated for the removal of these elements. Even more improbable is the extraction of highly reactive metals, nonmetallic elements, and their soluble salts, such as magnesium and calcium sulfates and carbonates, which are found in most industrial wastes.

21.5.1 Metals Recovery

Economically valuable elements such as precious metals can be recovered from waste streams by using complex chemical reactions, including chemical separation or precipitation. Silver, for example, can be recovered from photographic wastes by acidifying the liquid wastes and separating the Ag sludge that precipitates. Subsequently, the supernatant can be neutralized and disposed of safely, while the Ag sludge is sent to a smelter for purification. Metals can also be made to react selectively with synthetic organic chemicals known as chelates (literally, "claws") and synthetic zeolites, which are porous aluminosilicate minerals. In such reactions, metals are grabbed or trapped by the chelates or sequestered in the internal structure of the zeolites, while other materials pass by. Once reacted, the metal of interest is either precipitated or extracted out of the waste solution or is removed with a sorbent. The recovered metal sludge can be smeltered and refined into pure solid metal. Metal-containing residues resulting from the smelting and refining of ores can be further refined by using a combination of chemical and physical separation techniques and high-temperature furnaces.

21.5.2 Energy Recovery

The majority of wastes containing organic carbon forms have large quantities of stored energy. Thus wastes containing high concentrations of reduced carbon (usually organic carbon), such as organic liquids, woody materials, oils, resins, or asphalts, can be used in incinerators and electrical generators as energy sources. For example, kerosene and natural gas have about 44×10^6 to 49×10^6 Joules of energy kg^{-1}. Many solvents such as hexane, xylenes, and paraffins, which are found in oils and alcohols, have similar stored energies (42×10^6 to 58×10^6 Joules kg^{-1}). Therefore many waste mixtures of organic chemicals have energy values approaching those of commercial fuels. However, the limitations of this technology, when applied to wastes, are regulated by very strict air emission standards promulgated under RCRA.

21.5.3 Industrial Waste Solvents

Industrial wastes high in solvents are being successfully recycled using recovery systems that include distillation techniques and chemical and physical fractionation processes. For example, spent solvents used to clean and paint metal parts can be redistilled as a means to separate the heavy impurities from the volatile solvents. Some examples of solvents that can then be reused via solvent recovery stills include chloroform, acetone, benzene, xylene, hexane, and methylene chloride.

21.5.4 Industrial Waste Reuse

Some mine tailings can be used as fills for earthworks. However, economic and potential liability issues related to transportation costs and potential releases of contaminants usually limit their reuse. Consequently, mine tailings are most often stockpiled in place. Oilfield wastes that are high in salts may be leached with water to remove the soluble salts. The salt-free muds are then dried and stockpiled for use as fill material. These treated solid oilfield wastes have been shown to be safe in earthworks and release no residual salts into the surrounding environment. Coal-burning wastes such as fly ash and flue gas desulfurization sludges have been used successfully as soil amendments. Coal-burning fly ash has also been used as a fill material for earthworks. These materials can be good sources of gypsum and calcite, which can neutralize acidity

in soils and replenish macronutrients such as Ca, Mg, and S, and even trace elements like Zn, Cu, Fe, and Mn. Fly ash additions to acidic agricultural soils have also been shown to improve the soil pH, porosity, and overall structure.

21.6 TREATMENT AND DISPOSAL OF MUNICIPAL SOLID WASTE

21.6.1 Municipal Solid Waste

Municipal solid waste (MSW) is the solid waste material commonly called "trash" or "garbage" that is generated by homeowners and businesses. Most state and/or federal rules permit landfills to contain either hazardous or nonhazardous wastes, but more than 95% of landfilling involves MSW. According to the U.S. EPA (2003), municipal solid waste consists primarily of paper, yard waste, food scraps, and plastic (Fig. 21.5).

In 2003, EPA estimated that the United States was producing more than 236 million metric tons of MSW each year, or about 2 kg of trash per person per day, or nearly 1.7 times more waste per capita than in 1960 (U.S. EPA 2003). Developed nations such as the United States and Canada lead the world in the production of trash, producing much more discardable material than developing countries. More encouraging is the fact that trends in MSW recycling have been increasing steeply since the 1980s, now exceeding 30% in the United States. Fig. 21.6 shows the recycling rates of various MSW materials. The widespread increase in recycling in the U.S. has helped cause the per capita production of trash to decrease slightly since 2005, and the rate of increase in annual total MSW production to decline.

Although it was hoped that much of the landfilled material would biodegrade, in reality, most of the land filled waste from the past 40$^+$ years is still present at landfill sites. Surveys of landfills have shown that conditions are often not favorable for biodegradation. Such conditions include low moisture, low oxygen concentration, and high heterogeneity of materials, many of which are nondegradable or very slow to degrade. Thus old landfills serve as "chemical repositories," releasing pollutants to the groundwater and the atmosphere (see Chapter 15).

The number of landfills in the United States has been decreasing, from 18,500 in 1979 to less than 8000 in 1988 and 1762 in 2002, due mostly to the closure of small landfills (U.S. EPA, 2005a,b). However, it appears that the dire predictions of landfill shortages by the turn of the century have not materialized. This is due in part to increases in recycling and the ongoing construction of much larger modern landfills. This is despite new regulations and landfill designs that have increased construction and permitting costs substantially.

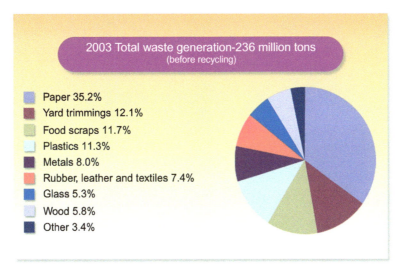

FIG. 21.5   Municipal solid waste composition. *(Source U.S. EPA: Website on MSW basic facts, 2003. Available from: http://www.epa.gov/epaoswer/ non-hw/muncpl/facts.htm.)*

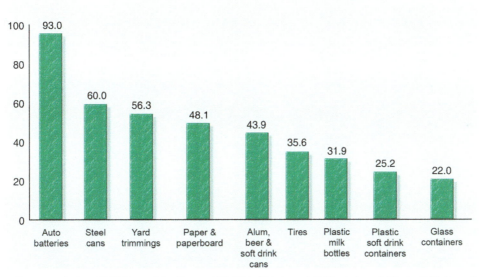

FIG. 21.6   Recycling rates of selected MSW materials. *(Source: U.S. EPA: Website on MSW basic facts, 2003. Available from: http://www.epa.gov/ epaoswer/non-hw/muncpl/facts.htm.)*

21.6.2 Modern Sanitary Landfills

Until the late 1970s, waste materials were buried or dumped haphazardly within soil or the vadose zone, then simply covered with a layer of soil as a cap. There was no regulation of these landfills nor were protective mechanisms in place to prevent or minimize releases of contaminants from the landfills. Old landfills were usually located in old quarries, mines, natural depressions, or excavated holes in abandoned land (Fig. 21.7A). Conditions in these landfills are conducive to pollution, as these sites are often hydrologically connected to surface streams or groundwater sources. Water pollution is caused by the formation and movement of *leachate*, which is produced by the infiltration of water through the waste material. As the water moves through the waste, it dissolves components of the waste; thus landfill leachate consists of water containing dissolved chemicals such as salts, heavy metals, and often synthetic organic compounds (see Chapter 15). In addition to contaminating water, landfills release pollutants into the air. Anaerobic microbial processes generate greenhouse gases, such as nitrous oxide (N_2O), methane (CH_4), and carbon dioxide (CO_2).

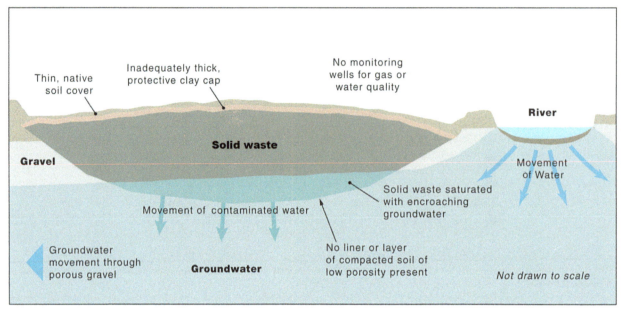

(A) *Old-style sanitary landfill*

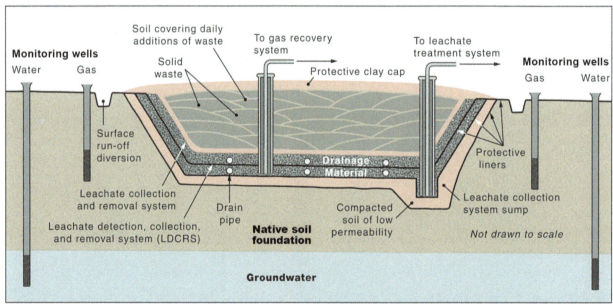

(B) *Modern sanitary landfill*

FIG. 21.7 Landfills: (A) Old-style sanitary landfill where location was chosen more out of convenience or budgetary concerns than out of any environmental considerations. Here, an abandoned gravel pit located near a river exemplifies an all-too-common siting arrangement. (B) Cross section of a modern sanitary landfill showing pollutant monitoring wells, a leachate management system, and leachate barriers. In contrast to the old-style landfill in (A), the modern sanitary landfill emphasizes long-term environmental protection. Additionally, whereas landfills have formerly been abandoned when full, a modern landfill is monitored long after closure. *(From Pollution Science, Academic Press, San Diego, CA, 1996.)*

A modern *sanitary landfill* is designed to meet exacting standards with respect to containment of all materials, including leachates and gases (Fig. 21.7B). New design specifications are predicated on minimal impact to the environment, both short- and long-term, with particular emphasis on groundwater protection. Landfill site selection is based on geology and soil type, together with such groundwater considerations as depth of water and use (*i.e.*, aquifer vulnerability, as discussed in Chapter 15). Fig. 21.7B shows a schematic drawing of a typical modern landfill. In this design, the new landfill is located in an excavated depression, and fresh garbage is covered daily with a layer of soil. The bottom of the landfill is lined with low-permeability liners made out of high-density plastic or clay. In addition, provisions are made to collect and analyze leachate and gases that emanate from the MSW.

New landfills can cost up to $1 million per hectare to construct. Additionally, they require costly permanent monitoring for potential contaminant releases to the surrounding environment. Thus many communities are faced with difficult decisions with respect to disposal of MWS—despite the fact that landfill design has improved efficiency with respect to pollution prevention, and increased recycling rates have slowed down MSW production.

Perhaps the most difficult problem associated with construction of new landfills is that of locating or siting a landfill. Few communities want landfills located nearby. Real or perceived local community concerns about potential health effects and property devaluation often delay the location of a landfill for months or years. Thus the costs of MSW disposal will continue to rise as new landfills are located in increasingly remote areas. However, increasing costs may force communities to look for alternative forms of MSW disposal, such as the development of new, more cost-effective recycling and waste minimization strategies. Currently, in the United States, 30% of MSW is recovered, recycled, or composted; 14% is burned at combustion facilities; and the remaining 56% is disposed of in landfills.

21.6.3 Reduction of MSW

The primary methods used to reduce MSW are recycling or composting (30%) and combustion (14%) (U.S. EPA, 2003). These approaches are widely used in densely populated areas where land scarcity limits the use of landfills. In Europe and Japan, for example, less than 15% of MSW is sent to landfills, in contrast to the United States, where more than 50% is sent to landfills.

Combustion or incineration allows the heat derived from burning to be converted to other forms of energy. *Incineration* usually involves combustion of unprocessed solid waste; however, in some instances nearly 25% of the waste is processed into pellets, termed *refuse-derived fuel*, prior to burning. A new technique called *mass burning* can burn MSW at temperatures up to 1130°C, trapping the resultant heat to generate steam and electricity. Incineration is a relatively efficient method of reducing MSW, often reducing it as much as 90% by volume and 75% by weight. However, this technique is not without drawbacks. The primary problem associated with incineration is air pollution, due to the release into the air of toxic chemicals like dioxin and metals like mercury. This has led to a growing public resistance to this technology. Additionally, incinerator ash contains concentrated toxic metals and refractory organics that often require their disposal as hazardous waste.

In contrast to incineration, *source reduction* is aimed at prevention; thus it is a fundamental way of reducing MSW, because it tends to eliminate the need for landfills. Source reduction can be accomplished by a variety of methods, including the use of less material for packaging and the practice of municipal composting. For example, plastic containers may be reduced in size or eliminated altogether; and yard waste may be separated from other sources of MSW and used as compost. Municipalities often sell the composted material for use as a soil amendment or fertilizer.

Recycling, whose trends and growth rate have been previously discussed (see Fig. 21.6), involves the collection of certain types of trash, separating them into their components, and subsequently using those components to make new products. Materials that can be recycled include aluminum cans, plastics, glass, paper, cardboard, and metal. Also, as previously indicated, MSW may contain significant amounts of hazardous waste, which to date is not regulated under RCRA. However, in many communities, household hazardous waste programs have been implemented to recycle these wastes and prevent them from entering MSW landfills. Among the hazardous wastes or wastes with hazardous properties that have been banned and thus nearly eliminated from landfills include lead car batteries, tires, and spent motor oil.

Municipal sewage is treated within wastewater treatment plants (see Chapter 22) and over 60% of the biosolids that are produced from treatments are land applied (see Chapter 23). The remainder of biosolids are either incinerated or landfilled.

21.7 POLLUTION PREVENTION

Nations like the United States and many European countries that have a long history of industrial development and associated pollution issues are controlling sources and releases of pollution into the environment through implementation of the concept of *pollution prevention*. Pollution can be reduced or in some cases eliminated either by reducing the demand for products or by improving the treatment and manufacturing of raw materials and finished goods. Demand-driven pollution can be mitigated by the discovery and use of alternate but equally acceptable less polluting products. Source reduction implies the modification of chemical processes or the development of alternate processes that ultimately produce less waste or less toxic waste.

In recent years, development of pollution prevention strategies has been driven by stricter environmental regulation and economic factors. For example, new industrial processes that emphasize waste reduction are not only economically desirable, but also increasingly required by government regulations. Thus pollution prevention strategies are usually process driven. Fig. 21.8 shows that, for example, recycling is favored over waste disposal such as landfilling of wastes, which leads to waste minimization and consequently less pollution.

QUESTIONS AND PROBLEMS

1. Using Table 21.1, select the 10 most prevalent elements found in industrial wastes. Give examples of the common mineral compounds of these elements. Example: Oxygen, not listed in Table 21.1, is found in the aqueous phase of wastes as H_2O.

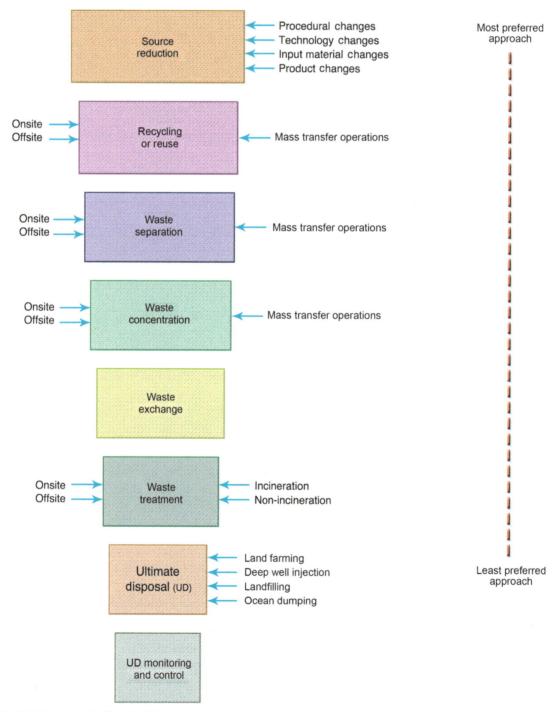

FIG. 21.8 Pollution prevention hierarchy. Note that reducing wastes (at the source) is the most desirable form of pollution prevention. *(From Liu, H.F., et al., 1997. Environmental Engineer's Handbook, second ed. Lewis Publishers, Boca Raton, FL. Reprinted by permission of Taylor and Francis Group, LLC, a division of Informa plc.)*

2. What is the most important treatment option difference between organic and inorganic wastes?
3. Two hundred gallons of an industrial wastewater contains the following materials:
   (a) 10 lbs of table salt
   (b) 5 kg of gypsum
   (c) 1 kg of iron rust [Fe(OH)$_3$] (assume solubility = 0.1 mg L^{-1}
   (d) 0.5 g of the pesticide DDT (assume solubility = 1 mg L^{-1}). Using the additional mineral solubilities data from Table 21.2, estimate amounts of each of these four materials that are dissolved in the wastewater, and what amounts of these four materials are in solid forms. (*Hint*: Assume that 1 L of water weighs 1 kg or 2.2 lbs.)
4. Which elements and their chemical forms would be found in solution in the wastewater of Problem 3? List four, in order of decreasing concentrations.
5. Would you consider incineration as a final disposal method for the wastewater in Problem 3? Explain your answer. Can you suggest alternative treatment and disposal methods?
6. Modern landfills have leachate collection systems. Explain why.
7. What concerns are driving the recycling of municipal solid waste in the United States?

REFERENCES

National Research Council, 1999. Waste Incineration and Public Health. National Academy Press. National Research Council. The National Academies Press, Washington, DC.

U.S. EPA. 2001. Class I Underground injection control program: study of the risks associated with Class I underground injection wells. U.S. Environmental Protection Agency. Office of Water, Washington, DC. EPA 816-R-01-007 March 2001.

U.S. EPA, 2003. Website on MSW basic facts. Available from:http://www.epa.gov/epaoswer/non-hw/muncpl/facts.htm.

U.S. EPA, 2005a. Website on mercury emissions. Available from:http://www.epa.gov/mercury/control_emissions/.

U.S. EPA, (2005b). Website on wastes. Available from: http://www.epa.gov/osw/.

FURTHER READING

Brown, K.W., Carlile, B.L., Miller, R.H., Rutledge, E.M., Runge, E.C.A., 1986. Utilization, Treatment, and Disposal of Waste on Land. Soil Science Society of America, Madison, WI.

Bunce, N., 1991. Environmental Chemistry. Wuerz Publishing Ltd., Winnipeg, Canada.

Common Dreams Progressive Newswire, (2003. Global protests against incineration signal death knell for deadly technology new report shows dramatic decline in U.S. incinerator industry. Available from: http://www.commondreams.org/news2003/0714-06.htm.

Cope, C.B., Fuller, W.H., Willets, S.L., 1983. The Scientific Management of Hazardous Wastes. Cambridge University Press, Cambridge, England.

Cote, P., Gilliam, M., 1989. Environmental aspects of stabilization and solidification of hazardous and radioactive wastes, ASTM STP 1033. American Society of Testing and Materials, Philadelphia, Pennsylvania.

Dohlido, J.R., Best, G.A., 1993. Chemistry of Water and Water Pollution. Ellis Horwood, New York.

ERI (1984) Land Treatability of Appendix VIII Constituents Present in Petroleum Industry Wastes. Environmental Research and Technology, Inc., Houston, TX. Document B-974-220.

Fuller, H.W., Warrick, A.W., 1985. Soils in Waste Treatment and Utilization. 2 CRC Press, Boca Raton, FL.

Hesketh, H.E., Cross, F.L., Tessitore, J.L., 1990. Incineration for Site Cleanup and Destruction of Hazardous Wastes. Technomic Publishing, Lancaster, PA.

Jackman, A.P., Powell, R.L., 1991. Hazardous Waste Treatment Technologies: Biological Treatment, Wet Air Oxidation, Chemical Fixation, Chemical Oxidation. Noyes Data Corporation, Park Ridge, NJ.

LaGrega, M.D., Buckingham, P.L., Evans, J.C., 1994. Hazardous Waste Management. McGraw-Hill, New York.

Liu, D.H.F., Liptak, B.G., 1997. Environmental Engineer's Handbook, 2nd ed. Lewis Publishers, Boca Raton, FL.

Miller, G.T., 1992. Environmental Science, fourth ed. Wadsworth, Belmont, CA.

National Research Council, 2002. Biosolids Applied to Land: Advancing Standards and Practices. National Academy Press. National Research Council. The National Academies Press, Washington, DC.

Pepper, I.L., 1991. Agricultural sludge utilization. In: Annual Report to Pima County Wastewater Management Division. Department of Soil, Water and Environmental Science, University of Arizona, Tucson, Arizona.

U.S. EPA, 1980. Hazardous Waste Land Treatment. United States Environmental Protection Agency. Office of Water and Waste Management, Washington, DC.

U.S. EPA, (1983) Land Application of Municipal Sludge: Process Design Manual. United States Environmental Protection Agency. Municipal Environmental Research Laboratory, Cincinnati, OH. EPA-625/1-83-016.

Wang, L.K., Wang, M.H.S., 1992. Handbook of Industrial Waste Treatment. Marcel Dekker, Inc., New York.

Warner, D.L., Lehr, J.H., 1981. Subsurface Wastewater Injection: The Technology of Injecting Wastewater into Deep Wells for Disposal. Premier Press, Berkeley, CA.

Chapter 22

Municipal Wastewater Treatment

C.P. Gerba and I.L. Pepper

A clarifier at a modern wastewater treatment plant. *(Photo courtesy of K.L. Josephson)*

22.1   THE NATURE OF WASTEWATER (SEWAGE)

The cloaca maxima, the "biggest sewer" in Rome, had enough capacity to serve a city of 1 million people. This sewer, and others like it, simply collected wastes and discharged them into the nearest lake, river, or ocean. This expedient made cities more habitable, but its success depended on transferring the pollution problem from one place to another. Although this worked reasonably well for the Romans, it does not work well today. Current population densities are too high to permit a simple dependence on transference. Thus modern-day sewage is treated before it is discharged into the environment. In the latter part of the 19th century, the design of sewage systems allowed collection and treatment to lessen the impact on natural waters. Today, more than 15,000 wastewater treatment plants treat approximately 150 billion liters of wastewater per day in the United States alone. In addition, septic tanks, which were also introduced at the end of the 19th century, serve approximately 25% of the U.S. population, largely in rural areas.

Domestic wastewater is primarily a combination of human feces, urine, and "graywater." *Graywater* results from washing, bathing, and meal preparation. Water from various industries and businesses may also enter the system. People excrete 100–500 g wet weight of feces and 1–1.3 L of urine per person per day (Bitton, 2011). Major organic and inorganic constituents of untreated domestic sewage are presented in Table 22.1.

The amount of organic matter in domestic wastes determines the degree of biological treatment required. Three tests are used to assess the amount of organic matter: *biochemical oxygen demand* (BOD), *chemical oxygen demand* (COD), and *total organic carbon* (TOC);

The major objective of domestic waste treatment is the reduction of BOD, which may be either in the form of solids (suspended matter) or soluble. BOD is the amount of dissolved oxygen consumed by microorganisms during the biochemical oxidation of organic (carbonaceous BOD) and inorganic (ammonia) matter. The methodology for measuring BOD has changed little since it was developed in the 1930s.

The 5-day BOD test (written BOD_5) is a measure of the amount of oxygen consumed by a mixed population of heterotrophic bacteria in the dark at 20°C over a period of 5 days. In this test, aliquots of wastewater are placed in a 300-mL BOD bottle (Fig. 22.1) and diluted in phosphate buffer (pH 7.2) containing other inorganic elements (N, Ca, Mg, Fe) and saturated with oxygen. Sometimes acclimated microorganisms or dehydrated cultures of microorganisms, sold in capsule form, are added to municipal and industrial wastewaters, which may not have a sufficient microflora to carry out the BOD test. In some cases a nitrification inhibitor is added to the sample to determine only the carbonaceous BOD.

Dissolved oxygen concentration is determined at time 0, and after a 5-day incubation by means of an oxygen electrode, chemical procedures (e.g., Winkler test), or a manometric BOD apparatus. The BOD test is carried out on a series of dilutions of the sample, the dilution depending on the source of the sample. When dilution water is not seeded, the BOD value is expressed in milligrams per liter, according to the following equation (APHA, 1998).

$$BOD(mg/L) = \frac{D_1 - D_5}{P} \qquad (22.1)$$

Environmental and Pollution Science. https://doi.org/10.1016/B978-0-12-814719-1.00022-7

TABLE 22.1 Typical Composition of Untreated Domestic Wastewater

Contaminants	Concentration (mg/L)		
	Low	Moderate	High
Solids, total	350	720	1200
Dissolved, total	250	500	850
Volatile	105	200	325
Suspended solids	100	220	350
Volatile	80	164	275
Settleable solids	5	10	20
Biochemical oxygen demand[a]	110	220	400
Total organic carbon	80	160	290
Chemical oxygen demand	250	500	1000
Nitrogen (total as N)	20	40	85
Organic	8	15	35
Free ammonia	12	25	50
Nitrites	0	0	0
Nitrates	0	0	0
Phosphorous (total as P)	4	8	15
Organic	1	3	5
Inorganic	3	5	10

[a]*Five-day test (BOD$_5$, 20°C).*

From Pepper, I.L., Brooks, J.P., Gerba, C.P., 2006. Pathogens in biosolids. Adv. Agron. 90, 1–41. Environmental Microbiology, 2nd Ed., Table 24.1, p. 504.

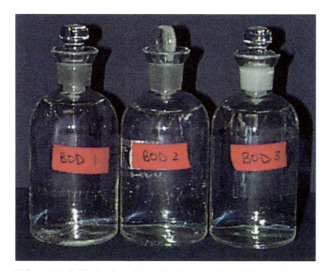

FIG. 22.1 BOD bottle. *(From Environmental & Pollution Science, 2nd Ed.)*

where D_1 = initial dissolved oxygen (DO), D_5 = DO at day 5, and P = decimal volumetric fraction of wastewater utilized.

If the dilution water is seeded,

$$\text{BOD(mg/L)} = \frac{(D_1 - D_5) - (B_1 - B_5)f}{P} \qquad (22.2)$$

where

D_1 = initial DO of the sample dilution (mg/L),
D_5 = final DO of the sample dilution (mg/L),
P = decimal volumetric fraction of sample used,
B_1 = initial DO of seed control (mg/L),
B_5 = final DO of seed control (mg/L), and
f = ratio of seed in sample to seed in control = (% seed in D_1)/(% seed in B_1)

Because of depletion of the carbon source, the carbonaceous BOD reaches a plateau called the ultimate

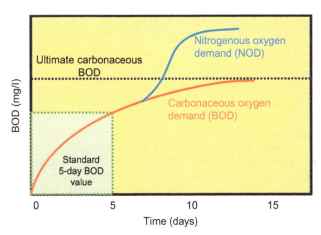

FIG. 22.2 Carbonaceous and nitrogenous BOD. *(From Environmental & Pollution Science, 2nd Ed.)*

carbonaceous BOD (Fig. 22.2). The BOD_5 test is commonly used for several reasons:

- To determine the amount of oxygen that will be required for biological treatment of the organic matter present in a wastewater
- To determine the size of the waste treatment facility needed
- To assess the efficiency of treatment processes
- To determine compliance with wastewater discharge permits

The typical BOD_5 of raw sewage ranges from 110 to 440 mg/L (see Example Calculation 22.1). Conventional sewage treatment will reduce this by 95%.

Chemical oxygen demand (COD) is the amount of oxygen necessary to oxidize all of the organic carbon completely to CO_2 and H_2O. COD is measured by oxidation

with potassium dichromate ($K_2Cr_2O_7$) in the presence of sulfuric acid and silver and is expressed in milligrams per liter. In general, 1 g of carbohydrate or 1 g of protein is approximately equivalent to 1 g of COD. Normally, the ratio BOD/COD is approximately 0.5. When this ratio falls below 0.3, it means that the sample contains large amounts of organic compounds that are not easily biodegraded.

Another method of measuring organic matter in water is the *TOC* or *total organic carbon* test. TOC is determined by oxidation of the organic matter with heat and oxygen, followed by measurement of the CO_2 liberated with an infrared analyzer. Both TOC and COD represent the concentration of both biodegradable and nonbiodegradable organics in water.

Pathogenic microorganisms are almost always present in domestic wastewater (Table 22.2). This is because large numbers of pathogenic microorganisms may be excreted by infected individuals. Both symptomatic and asymptomatic individuals may excrete pathogens. For example, the concentration of rotavirus may be as high as 10^{12} virions per gram of stool or 10^{14} in 100 g of stool (Table 22.3). Infected individuals may excrete enteric pathogens for several days or as long as a few months. The concentration of enteric pathogens in raw wastewater varies depending on the following:

- The incidence of the infection in the community
- The socioeconomic status of the population
- The time of year
- The per-capita water consumption

The peak incidence of many enteric infections is seasonal in temperate climates. Thus the highest incidence of

---

Example Calculation 22.1 Calculation of BOD

Determine the 5-day BOD (BOD_5) for a wastewater sample when a 15-mL sample of the wastewater is added to a BOD bottle containing 300 mL of dilution water, and the dissolved oxygen is 8 mg/L. Five days later the dissolved oxygen concentration is 2 mg/L.

Using Eq. (22.1):

$$BOD(mg/L) = \frac{D_1 - D_5}{P}$$

$$D_1 = 8 \, mg/L$$

$$D_5 = 2 \, mg/L$$

$$p = \frac{15 \, ml}{300 \, ml} = 5\% = 0.05$$

$$BOD_5 = \frac{8 - 2}{0.05} = 120 \, mg/L$$

(From Environmental Microbiology, 2nd Ed., Example Calculation 24.1, p. 505)

TABLE 22.2 Types and Numbers of Microorganisms Typically Found in Untreated Domestic Wastewater

Organism	Concentration (Per mL)
Total coliform	$10^5 - 10^6$
Fecal coliform	$10^4 - 10^5$
Fecal streptococci	$10^3 - 10^4$
Enterococci	$10^2 - 10^3$
Shigella	Present
Salmonella	$10^0 - 10^2$
Clostridium perfringens	$10^1 - 10^3$
Giardia cysts	$10^{-1} - 10^2$
Cryptosporidium cysts	$10^{-1} - 10^1$
Helminth ova	$10^{-2} - 10^1$
Enteric virus	$10^1 - 10^2$

From Pepper, I.L., Brooks, J.P., Gerba, C.P., 2006. Pathogens in biosolids. Adv. Agron. 90, 1–41; Environmental Microbiology, 2nd Ed., Table 24.2, p. 505.

TABLE 22.3 Incidence and Concentration of Enteric Viruses and Protozoa in Feces in the United States

Pathogen	Incidence (%)	Concentration in Stool (Per Gram)
Enteroviruses	10–40	10^3–10^8
Hepatitis A	0.1	10^8
Rotavirus	10–29	10^{10}–10^{12}
Noroviruses	6	10^{10}–10^{12}
Giardia	3.8	10^6
	18–54[a]	10^6
Cryptosporidium	0.6–20	10^6–10^7
	27–50[a]	10^6–10^7

[a]Children in day care centers.
From Environmental Microbiology, 2nd Ed., Table 24.3, p. 506.

TABLE 22.4 Estimated Levels of Enteric Organisms in Sewage and Polluted Surface Water in the United States

Organism	Concentration (# Per 100 mL)	
	Raw Sewage	Polluted Stream Water
Coliforms	10^9	10^5
Enteric viruses	10^2	1–10
Giardia	10–10^2	0.1–1
Cryptosporidium	1–10	0.1–10^2

Source: EPA (U S. Environmental Protection Agency), 1988. Comparative health risks effects assessment of drinking water. Washington, DC; From Environmental Microbiology, 2nd Ed., Table 24.4, p. 506.

enterovirus infection is during the late summer and early fall. Rotavirus infections tend to peak in the early winter, and *Cryptosporidium* infections peak in the early spring and fall. The reason for the seasonality of enteric infections is not completely understood, but several factors may play a role. It may be associated with the survival of different agents in the environment during the different seasons. *Giardia*, for example, can survive winter temperatures very well. Alternatively, excretion differences among animal reservoirs may be involved, as is the case with *Cryptosporidium*. Finally, it may well be that greater exposure to contaminated water, as in swimming, is the explanation for increased incidence in the summer months.

Concentrations of enteric pathogens are much greater in sewage in the developing world than the industrialized world. For example, the average concentration of enteric viruses in sewage in the United States has been estimated to be 10^3 per liter (Table 22.4), whereas concentrations as high as 10^5 per liter have been observed in Africa and Asia.

22.2 CONVENTIONAL WASTEWATER TREATMENT

The primary goal of wastewater treatment is the removal and degradation of organic matter under controlled conditions. Complete sewage treatment comprises three major steps, primary, secondary, and tertiary treatment, as shown in Fig. 22.3.

22.2.1 Primary Treatment

Primary treatment is the first step in municipal sewage treatment and it involves physically separating large solids from the waste stream. As raw sewage enters the treatment plant, it passes through a metal grating that removes large

debris, such as branches and tires (Fig. 22.4). A moving screen then filters out smaller items such as diapers and bottles (Fig. 22.5), after which a brief residence in a grit tank allows sand and gravel to settle out. The waste stream is then pumped into the primary settling tank (also known as a sedimentation tank or clarifier), where about half the suspended organic solids settle to the bottom as sludge (Fig. 22.6). The resulting sludge is referred to as primary sludge. Microbial pathogens are not effectively removed from the effluent in the primary process, although some removal occurs.

Dissolved air floatation (DAF) is a more recent innovation for removing suspended solids from sewage, which is now being introduced into new wastewater treatment plants as an alternative to conventional primary sedimentation processes. DAF clarification is achieved by dissolving air in the wastewater under pressure, and then releasing the air at atmospheric pressure in a floatation tank or basin. This occurs in the front end of the DAF tank known as the "contact zone." The resulting air bubbles that form attach to floc particles and suspended solids. Frequently, coagulants are added to the wastewater prior to the DAF tank, to produce the flocs. The floc-bubble aggregates are then carried by water into the second DAF zone known as the "separation zone." Here, free bubbles and floc-bubble-aggregates rise to the surface of the tank forming a concentrated sludge blanket that can be removed by skimming devices (Edzwald, 2010). DAF clarifiers remove suspended solids more rapidly than conventional primary sedimentation and are cost effective from an engineering standpoint.

22.2.2 Secondary Treatment

Secondary treatment consists of biological degradation, in which the remaining suspended solids are decomposed by microorganisms, and the number of pathogens is reduced. In this stage, the effluent from primary treatment usually undergoes biological treatment in a trickling filter bed

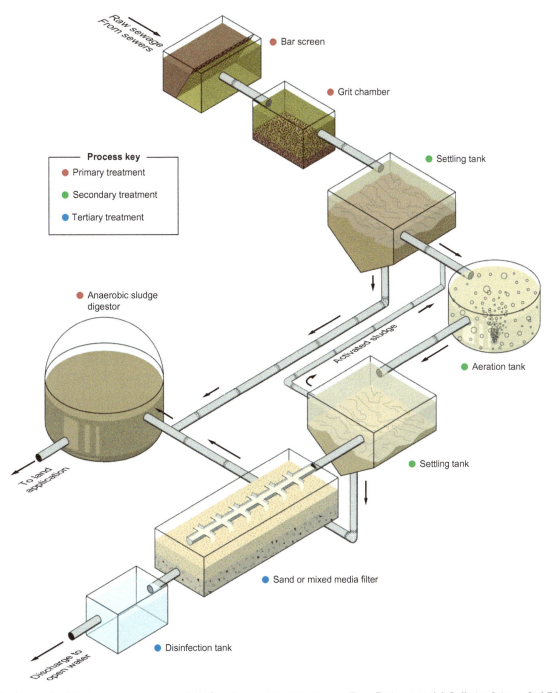

FIG. 22.3 Schematic of the treatment processes typical of modern wastewater treatment. *(From Environmental & Pollution Science, 2nd Ed.)*

(Fig. 22.7), an aeration tank (Fig. 22.8), or a sewage lagoon (see Section 22.3). A disinfection step is generally included at the end of the treatment.

22.2.2.1 Trickling Filters

In modern wastewater treatment plants, the *trickling filter* is composed of plastic units (Figs. 22.9 and 22.10). In older plants, or developing countries, the filter is simply a bed of stones or corrugated plastic sheets through which

wastewater drips (see Fig. 22.7). This is one of the earliest systems introduced for biological waste treatment. The effluent is pumped through an overhead sprayer onto the filter bed, where bacteria and other microorganisms have formed a biofilm on the filter surfaces. These microorganisms intercept the organic material as it trickles past and decompose it aerobically.

The media used in trickling filters may be stones, ceramic material, hard coal, or plastic media. Plastic media of polyvinyl chloride (PVC) or polypropylene are used

FIG. 22.4 Removal of large debris from sewage via a "bar screen." *(From Environmental & Pollution Science, 2nd Ed.)*

today in high-rate trickling filters. As the organic matter passes through the trickling filter, it is converted to microbial biomass which forms a thick biofilm on the filter medium. The biofilm that forms on the surface of the filter medium is called a *zoogleal film.* It is composed of bacteria, fungi, algae, and protozoa. Over time, the increase in biofilm thickness leads to limited oxygen diffusion to the deeper layers of the biofilm, creating an anaerobic environment near the filter medium surface. As a result, the organisms eventually slough from the surface and a new biofilm is formed. BOD removal by trickling filters is approximately 85% for low-rate filters (U.S. EPA, 1977). Effluent from the trickling filter usually passes into a final clarifier to further separate solids from effluent.

FIG. 22.5 Removal of small debris via a "moving screen." *(From Environmental & Pollution Science, 2nd Ed.)*

FIG. 22.7 A trickling filter bed. Here, rocks provide a matrix supporting the growth of a microbial biofilm that actively degrades the organic material in the wastewater under aerobic conditions. *(From Environmental & Pollution Science, 2nd Ed.)*

FIG. 22.6 Primary treatment of sewage with a clarifier, where suspended organic solids settle out as primary sludge. *(From Environmental & Pollution Science, 2nd Ed.)*

FIG. 22.8 Secondary treatment: an aeration basin. *(From Environmental & Pollution Science, 2nd Ed.)*

FIG. 22.9   A unit of plastic material used to create a biofilter (trickling filter). The diameter of each hold is approximately 5 cm. From Pepper, Gerba, and Brusseau, 2006. *(From Environmental & Pollution Science, 2nd Ed.)*

FIG. 22.10   A trickling biofilter or biotower. This is composed of many plastic units stacked upon each other. Dimensions of the biofilter may be 20 m diameter by 10–30 m depth. *(From Pepper, I.L., Brooks, J.P., Gerba, C.P., 2006. Pathogens in biosolids. Adv. Agron 90, 1–41; Environmental & Pollution Science, 2nd Ed.)*

22.2.2.2   Conventional Activated Sludge

Aeration-tank digestion is also known as the activated sludge process. In the United States, wastewater is most commonly treated by this process. Effluent from primary treatment is pumped into a tank and mixed with a bacteria-rich slurry known as *activated sludge*. Air or pure oxygen pumped through the mixture encourages bacterial growth and

decomposition of the organic material. It then goes to a secondary settling tank, where water is siphoned off the top of the tank and sludge is removed from the bottom. Some of the sludge is used as an inoculum for primary effluent. The remainder of the sludge, known as *secondary sludge*, is removed. This secondary sludge is added to primary sludge from primary treatment and is subsequently anaerobically digested to produce biosolids (Chapter 23). The concentration of pathogens is reduced in the activated sludge process by antagonistic microorganisms as well as adsorption to or incorporation in the secondary sludge.

An important characteristic of the activated sludge process is the recycling of a large proportion of the biomass. This results in a large number of microorganisms that oxidize organic matter in a relatively short time (Bitton, 2011). The detention time in the aeration basin varies from 4 to 8 h. The content of the aeration tank is referred to as the *mixed-liquor suspended solids* (MLSS). The organic part of the MLSS is called the *mixed-liquor volatile suspended solids* (MLVSS), which is the nonmicrobial organic matter as well as dead and living microorganisms and cell debris. The activated sludge process must be controlled to maintain a proper ratio of substrate (organic load) to microorganisms or *food-to-microorganism ratio* (F/M) (Bitton, 2011). This is expressed as BOD per kilogram per day. It is expressed as:

$$\frac{F}{M} = \frac{Q \cdot BOD}{MLSS \cdot V} \qquad (22.3)$$

where

Q = flow rate of sewage in million gallons per day (MGD),
BOD_5 = 5-day biochemical oxygen demand (mg/L)
$MLSS$ = mixed-liquor suspended solids (mg/L)
V = volume of aeration tank (gallons)

F/M is controlled by the rate of activated sludge wasting. The higher the wasting rate, the higher the F/M ratio. For conventional aeration tanks the F/M ratio is 0.2–0.5 lb BOD_5/day/lb MLSS, but it can be higher (up to 1.5) for activated sludge when high-purity oxygen is used (Hammer, 1986). A low F/M ratio means that the microorganisms in the aeration tank are starved, leading to more efficient wastewater treatment.

The important parameters controlling the operation of an activated sludge process are organic loading rates, oxygen supply, and control and operation of the final settling tank. This tank has two functions: clarification and thickening. For routine operation, sludge settle ability is determined by use of the *sludge volume index* (SVI) (Bitton, 2011).

SVI is determined by measuring the sludge volume index, which is given by the following formula:

$$SVI = \frac{V \cdot 1000}{MLSS} \qquad (22.4)$$

where V = volume of settled sludge after 30 min (mL/L).

The microbial biomass produced in the aeration tank must settle properly from suspension so that it may be wasted or returned to the aeration tank. Good settling occurs when the sludge microorganisms are in the endogenous phase, which occurs when carbon and energy sources are limited, and the microbial specific growth rate is local (Bitton, 2011). A mean cell residence time of 3–4 days is necessary for effective settling (Metcalf and Eddy, 2003). Poor settling may also be caused by sudden changes in temperature, pH, absence of nutrients, and presence of toxic metals and organics. A common problem in the activated sludge process is *filamentous bulking*, which consists of slow settling and poor compaction of solids in the clarifier. Filamentous bulking is usually caused by the excessive growth of filamentous microorganisms. The filaments produced by these bacteria interfere with sludge settling and compaction. A high SVI (>150 mL/g) indicates bulking conditions. Filamentous bacteria are able to predominate under conditions of low dissolved oxygen, low F/M, low nutrient, and high sulfide levels. Filamentous bacteria can be controlled by treating the return sludge with chlorine or hydrogen peroxide to kill filamentous microorganisms selectively.

22.2.2.3 Nitrogen Removal by the Activated Sludge Process

Activated sludge processes can be modified for nitrogen removal to encourage nitrification followed by denitrification. The establishment of a nitrifying population in activated sludge depends on the wastage rate of the sludge, and therefore on the BOD load, MLSS, and retention time. The growth rate of nitrifying bacteria (μ_n) must be higher than the growth rate

(μ_h) of heterotrophs in the system. In reality, the growth rate of nitrifiers is lower than that of heterotrophs in sewage; therefore a long sludge age is necessary for the conversion of ammonia to nitrate. Nitrification is expected at a sludge age greater than 4 days (Bitton, 2011).

Nitrification must be followed by denitrification to remove nitrogen from wastewater. The conventional activated sludge system can be modified to encourage denitrification. Two such processes are as follows:

- Single sludge system (Fig. 22.11A). This system comprises a series of aerobic and anaerobic tanks in lieu of a single aeration tank.
- Multisludge system (Fig. 22.11B). Carbonaceous oxidation, nitrification, and denitrification are carried out in three separate systems. Added methanol or settled sewage serves as the source of carbon for denitrifiers.

22.2.2.4 Phosphorus Removal by Activated Sludge Process

Phosphorus can also be reduced by the activity of microorganisms in modified activated sludge processes. The process depends on the uptake of phosphorus by the microbes during the aerobic stage and subsequent release during the anaerobic stage. Two of several systems in use are as follows:

- A/O (anaerobic/oxic) process. The A/O process consists of a modified activated sludge system that includes an anaerobic zone (detention time 0.5–1 h) upstream of the conventional aeration tank (detention time 1–3 h). Fig. 22.12 illustrates the microbiology of the A/O

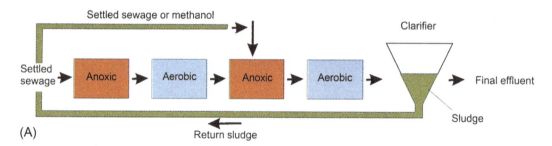

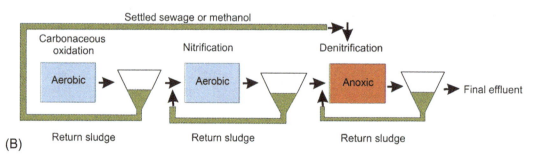

FIG. 22.11 Denitrification systems: (A) Single-sludge system, (B) multisludge system. *(Modified from Curds, C.R., Hawk, H.A. (Eds.), 1983. Ecological Aspects of Used-Water Treatment, vol. 2. Academic Press, London.)*

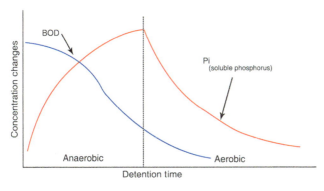

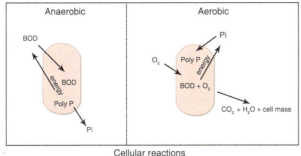

FIG. 22.12 Microbiology of the A/O process. *(From Environmental Microbiology, 2nd Ed., Fig. 24.12, p. 511.)*

process: During the anaerobic phase, inorganic phosphorus is released from the cells as a result of polyphosphate hydrolysis. The energy liberated is used for the uptake of BOD from wastewater. Removal efficiency is high when the BOD/phosphorus ratio exceeds 10 (Metcalf and Eddy, 2003). During the aerobic phase, soluble phosphorus is taken up by bacteria, which synthesize polyphosphates, using the energy released from BOD oxidation.

22.2.2.5 The Bardenpho Process

The Bardenpho process is an advanced modification of the activated sludge process, which results in nutrient removal of N and P via microbial processes that occur in a multistage biological reactor (Fig. 22.13). This reactor removes high levels of BOD, suspended solids, nitrogen, and phosphorus.

- *Fermentation Stage*: Activated sludge is returned from the clarifier and undergoes microbial fermentation and phosphate is released.
- *First Anoxic Stage*: Mixed liquor containing nitrates from the third stage is recycled here, and mixed with conditioned sludge from the fermentation stage, in the absence of oxygen. Heterotrophic denitrifying bacteria reduce BOD by utilizing carbonaceous substrate while using nitrate as a terminal electron acceptor, which is reduced to gaseous nitrogen.
- *Nitrification Stage*: Oxygen is introduced allowing for heterotrophic aerobic respiration, which further oxidizes BOD. At the same time ammonia is aerobically nitrified to nitrate, and phosphate is taken up and utilized by microbes. Mixed liquor containing the nitrates is recycled back to the first anoxic stage.
- *Second Anoxic Stage*: The remaining liquor from the nitrification stage is passed into this second anoxic stage, where nitrate (in the absence of oxygen) is again reduced to nitrogen gas. This results in low effluent nitrate concentrations.
- *Reaeration Stage*: This is an aerobic environment that ensures that phosphate taken up microbially is not released in the final clarifier.

Overall the Bardenpho process results in an effluent that is low in nitrates and phosphates. Bardenpho processes utilizing autotrophic denitrification have also been evaluated utilizing "spent caustic" as an electron donor. Spent caustic is produced in petrochemical processes and contains adsorbed hydrogen sulfide which acts as substrate for autotrophic denitrifiers (Park et al., 2010).

22.2.2.6 Membrane Bioreactors

Membrane bioreactors (MBR) are a combination of biological treatment with membrane separation by microporous or ultrafiltration membranes. The process consists of a tank and a membrane unit either located external to the bioreactor

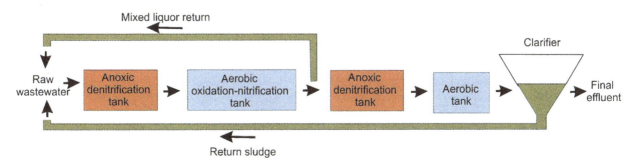

FIG. 22.13 The five stage Bardenpho process for microbial nutrient removal.

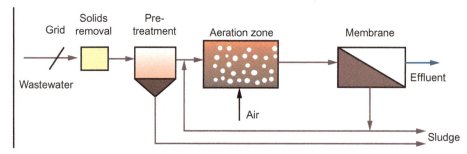

FIG. 22.14 Membrane bioreactor treatment train showing an external membrane unit.

or submerged directly within it (Fig. 22.14). The membranes act to retain suspended solids and maintain a high biomass concentration within the bioreactor thereby functioning as a replacement for sedimentation. Membranes come with various pore sizes and can be dense or porous. Separation by dense membranes relies on physicochemical interactions between the permeating components and the membrane material and is known as reverse osmosis or nanofiltration. Porous membranes have larger pore size, separate particles mechanically, and are referred to as ultrafiltration or microfiltration. The microbial bioreactor is normally maintained aerobically, but can be operated anaerobically, or with alternating aerobic/anaerobic phases to enhance microbial nitrification followed by denitrification.

Membrane bioreactors have several advantages including a much smaller area needed than conventional activated sludge and a high quality effluent. A membrane bioreactor effectively displaces three individual process steps in a conventional treatment plant (primary settling, activated sludge, and reduces the need for disinfection). The major advantages of MBRs include good quality effluent and reduced reactor volume and net sludge production. The major disadvantages include high operating costs and membrane fouling (Chang et al., 2002). Operating costs can be reduced by integrating microbial fuel cells and membrane bioreactors (Wang et al., 2012).

22.2.2.7 Nitrogen Removal by Anammox Process

Conventional removal of nitrogen from wastewater relies on two processes, nitrification which converts ammonium substrate to nitrate utilizing aerobic autotrophic bacteria (*Nitrosomonas* and *Nitrobacter*). In a separate process, nitrate is used as a terminal electron acceptor by anaerobic heterotrophic bacteria, a process known as denitrification. For nitrification, oxygen must be pumped into the wastewater to promote aerobic conditions. The addition of oxygen adds significantly to the cost of wastewater treatment. A useful discovery is that ammonium oxidation can occur under anaerobic conditions using nitrite as the terminal electron acceptor.

$$NH^+_4 + NO^-_2 \rightarrow N_2 + NO_3 \quad (22.5)$$

Anaerobic ammonium oxidation also known as *anammox* has led to a new technology being developed for removal of nitrogen during wastewater treatment. The anammox process is undertaken by anammox bacteria, which require ammonium as substrate and nitrite as a terminal electron acceptor. Thus the overall process is accomplished by two different autotrophic bacteria.

(1) *Nitrosomonas* which perform the first step of nitrification converting ammonium to nitrite under aerobic conditions.

(2) *Anammox bacteria that oxidize the ammonium anaerobically using the nitrite as the terminal electron acceptor.*

The characteristics of anammox bacteria are shown in Information Box 22.1. The overall pathways of nitrogen removal by nitrification/denitrification and anammox are shown in Fig. 22.15. Because oxygen is not required by the anammox process, it results in significant cost savings relative to conventional nitrification/denitrification. The potential advantages and disadvantages of anammox are outlined in Information Box 22.2. Because anammox requires a consortium of aerobic and anaerobic bacteria in close contact, technologies have been developed that utilize biofilms on plastic discs (Fig. 22.16). Aerobic *Nitrosomonas* thrive on the outer layer of the biofilm whereas the anaerobic anammox bacteria occupy the inner portions of the biofilms which are more anaerobic.

Overall, there is tremendous interest in the anammox technology as a novel inexpensive way to remove N from wastewater during wastewater treatment.

22.2.3 Tertiary Treatment

Tertiary treatment of effluent involves a series of additional steps after secondary treatment to further reduce organics, turbidity, nitrogen, phosphorus, metals, and pathogens. Most processes involve some type of physicochemical treatment such as coagulation, filtration, activated carbon adsorption of organics, reverse osmosis, and additional

- Members of the phylum Planctomycetes
- Anaerobic autotrophic bacteria
- Extremely slow growth rates
- Although a prokaryote, anammox bacteria have internal membrane-bound structures where the ammonium oxidation takes place

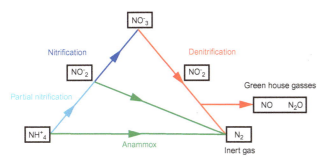

FIG. 22.15 Pathways of N removal.

FIG. 22.16 An anammox bioreactor with biofilms developed on plastic discs.

Advantages
- Reduced aeration
  - 63% reduction in oxygen demand
- No supplemental carbon required
  - 100% reduction
- Loss biomass (biosolids) produced
  - 80% reduction
- Technology has a small footprint

Disadvantages
- Slow doubling time for anammox bacteria, if culture is lost it can take months to regrow
- Sensitive to environmental changes especially pH
- Technology is new and has been infrequently used in municipal wastewater streams

disinfection. Tertiary treatment of wastewater is practiced for additional protection of wildlife after discharge into rivers or lakes. Even more commonly, it is performed when the wastewater is to be reused for irrigation (e.g., food crops, golf courses), for recreational purposes (e.g., lakes, estuaries), or for drinking water.

22.2.4 Removal of Pathogens by Sewage Treatment Processes

There have been a number of reviews on the removal of pathogenic microorganisms by activated sludge and other

wastewater treatment processes (Leong, 1983). This information suggests that significant removal especially of enteric bacterial pathogens can be achieved by these processes (Table 22.5). However, disinfection and/or advanced tertiary treatment is necessary for many reuse applications to ensure pathogen reduction. Current issues related to pathogen reduction are treatment plant reliability, removal of new and emerging enteric pathogens of concern, and the ability of new technologies to effect pathogen reduction. Wide variation in pathogen removal can result in significant numbers of pathogens passing through a process for various time periods. The issue of reliability is of major importance if the reclaimed water is intended for recreational or potable reuse, where short-term exposures to high levels of pathogens could result in significant risk to the exposed population.

Compared with other biological treatment methods (i.e., trickling filters), activated sludge is relatively efficient in reducing the numbers of pathogens in raw wastewater. Both sedimentation and aeration play a role in pathogen reduction. Primary sedimentation is more effective for the removal of the larger pathogens such as helminth eggs, but solid-associated bacteria and even viruses are also removed. During aeration, pathogens are inactivated by antagonistic microorganisms and by environmental factors such as temperature. The greatest removal probably occurs

TABLE 22.5 Pathogen Removal During Sewage Treatment

	Enteric Viruses	Salmonella	Giardia	Cryptosporidium
Concentration in raw sewage (number per liter)	10^5–10^6	5000–80,000	9000–200,000	1–3960
Removal during				
Primary treatment[a]				
% Removal	50–98.3	95.8–99.8	27–64	0.7
Number remaining L^{-1}	1,700–500,000	160–3360	72,000–146,000	
Secondary treatment[b]				
% Removal	53–99.92	98.65–99.996	45–96.7	
Number remaining L^{-1}	80–470,000	3–1075	6480–109,500	
Secondary treatment[c]				
% Removal	99.983–99.9999998	99.99–99.999999995	98.5–99.99995	2.7[d]
Number remaining L^{-1}	0.007–170	0.000004–7	0.099–2.951	

[a]Primary sedimentation and disinfection.
[b]Primary sedimentation, trickling filter or activated sludge, and disinfection.
[c]Primary sedimentation, trickling filter or activated sludge, disinfection, coagulation, filtration, and disinfection.
[d]Filtration only.
From Environmental Microbiology, 2nd Ed., Table 24.5, p. 512.

by adsorption or entrapment of the organisms within the biological floc that forms. The ability of activated sludge to remove viruses is related to the ability to remove solids. This is because viruses tend to be solid associated and are removal along with the floc. Activated sludge typically removes 90% of the enteric bacteria and 80% to 90%–99% of the enteroviruses and rotaviruses (Rao et al., 1986). Ninety percent of *Giardia* and *Cryptosporidium* can also be removed (Rose and Carnahan, 1992), being largely concentrated in the sludge. Because of their large size, helminth eggs are effectively removed by sedimentation, and are rarely found in sewage effluent in the United States, although they may be detected in the sludge. However, although the removal of the enteric pathogens may seem large, it is important to remember that initial concentrations are also large (i.e., the concentration of all enteric viruses in 1 L of raw sewage may be as high as 100,000 in some parts of the world).

Tertiary treatment processes involving physical-chemical processes can be effective in further reducing the concentration of pathogens and enhancing the effectiveness of disinfection processes by the removal of soluble and particulate organic matter (Table 22.6). Filtration is probably the most common tertiary treatment process. Mixed-media (sand, gravel, coal) filtration is most effective in the reduction of protozoan parasites. Usually, greater removal of *Giardia* cysts occurs than of *Cryptosporidium* oocysts because of the larger size of the cysts (Rose and Carnahan, 1992). Removal of enteroviruses and indicator bacteria is usually 90% or less. Addition of coagulant can increase the removal of poliovirus to 99% (EPA, 1992).

Coagulation, particularly with lime, can result in significant reductions of pathogens. The high-pH conditions (pH 11–12) that can be achieved with lime can result in significant inactivation of enteric viruses. To achieve removals of 90% or greater, the pH should be maintained above 11 for at least an hour (Leong, 1983). Inactivation of the viruses occurs by denaturation of the viral protein coat. The use of iron and aluminum salts for coagulation can also result in 90% or greater reductions in enteric viruses. The degree of effectiveness of these processes, as in other solids separating processes, is highly dependent on the hydraulic design and in particular, coagulation and flocculation. The degree of removal observed in bench-scale tests may not approach those seen in full-scale plants, where the process is more dynamic.

Reverse osmosis and ultrafiltration are also believed to result in significant reductions in enteric pathogens. Removal occurs by size exclusion. Removal of enteric viruses in excess of 99.9% can be achieved (Leong, 1983).

22.2.5 Removal of Organics and Inorganics by Sewage Treatment Processes

In addition to nutrients such as nitrogen and phosphorus, and microbial pathogens, there are other constituents within sewage that need to be kept at low concentrations. These

TABLE 22.6 Average Removal of Pathogen and Indicator Microorganisms in a Wastewater Treatment Plant, St. Petersburg, Florida

	Raw Wastewater to Secondary Wastewater		Secondary Wastewater to Postfiltration		Postfiltration to Postdisinfection		Postdisinfection to Poststorage		Raw Wastewater to Poststorage	
	Percentage	*log_{10}*	*Percentage*	*log_{10}*	*Percentage*	*log_{10}*	*Percentage*	*log_{10}*	*Percentage*	*log_{10}*
Total coliforms	98.3	1.75	69.3	0.51	99.99	4.23	75.4	0.61	99.999992	7.1
Fecal coliforms	99.1	2.06	10.5	0.05	99.998	4.95	56.8	0.36	99.999996	7.4
Coliphage[a]	82.1	0.75	99.98	3.81	90.05	1.03	90.3	1.03	99.999997	6.6
Enterovirus	98.0	1.71	84.0	0.81	96.5	1.45	90.9	1.04	99.999	5.0
Giardia	93.0	1.19	99.0	2.00	78.0	0.65	49.5	0.30	99.993	4.1
Cryptosporidium	92.8	1.14	97.9	1.68	61.1	0.41	8.5	0.04	99.95	3.2

[a]*Escherichia coli host ATCC 15597.*

From Environmental Microbiology, 2nd Ed., Table 24.6, p. 512.

include inorganics exemplified by metals and organic priority pollutants. Metals and organics are normally associated with the solid fraction of sewage, and neither are significantly removed by sewage treatment. However, when point source control mechanisms are implemented to prevent industrial discharges, the concentration of metals and organics within sewage can be significantly reduced. In particular, over the past few decades in the United States this has resulted in decreased metal concentrations. More recently there has been concern over the presence of emerging contaminants such as pharmaceuticals and perfluorinated alkyl compounds in sewage.

22.3 OXIDATION PONDS

The next two sections discuss several alternatives to large-scale modern wastewater treatment process discussed in Section 22.2. The first of these are *sewage lagoons* and are often referred to as *oxidation* or *stabilization ponds*. These are the oldest of the managed wastewater treatment systems. Usually no more than a hectare in area and just a few meters deep, oxidation ponds are natural "stewpots," where wastewater is detained while organic matter is degraded (Fig. 22.17). A period of time ranging from 1 to 4 weeks (and sometimes longer) is necessary to complete the decomposition of organic matter. Light, heat, and settling of the solids can also effectively reduce the number of pathogens present in the wastewater.

The following four categories of oxidation ponds are often used in series:

- *Aerobic ponds* (Fig. 22.18A), which are naturally mixed, must be shallow (up to 1.5 m) because they depend on penetration of light to stimulate algal growth

FIG. 22.17 An oxidation pond. Typically these are only 1–2 meters deep, and small in area. *(From Environmental Microbiology, 2nd Ed., Fig. 24.13, p. 513.)*

that promotes subsequent oxygen generation. The detention time of wastewater is generally 3–5 days.

- *Anaerobic ponds* (Fig. 22.18B) may be 1–10 m deep, and require a relatively long detention time of 20–50 days. These ponds, which do not require expensive mechanical aeration, generate small amounts of sludge. Often, anaerobic ponds serve as a pretreatment step for high-BOD organic wastes rich in protein and fat (e.g., meat wastes) with a heavy concentration of suspended solids.

- *Facultative ponds* (Fig. 22.19) are most common for domestic waste treatment. Waste treatment is provided by both aerobic and anaerobic processes. These ponds range in depth from 1 to 2.5 m and are subdivided in three layers: an upper aerated zone, a middle facultative

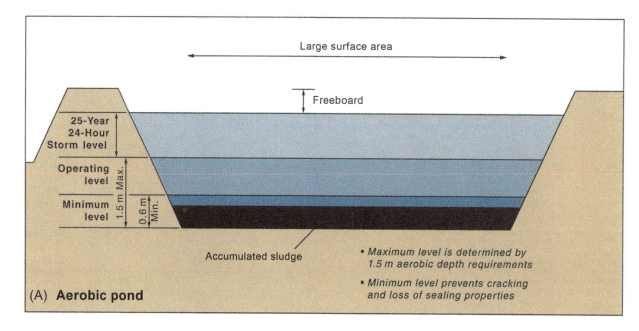

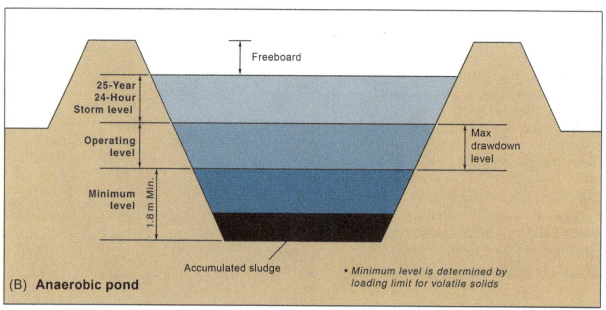

FIG. 22.18 Pond profiles: (A) Aerobic waste pond profile, and (B) Anaerobic waste pond profile. *(From Environmental Microbiology, 2nd Ed., Fig. 24.14, p. 514.)*

zone, and a lower anaerobic zone. The detention time varies between 5 and 30 days.

- *Aerated lagoons* or *ponds* (Fig. 22.20) which are mechanically aerated may be 1–2 m deep and have a detention time of less than 10 days. In general, treatment depends on the aeration time and temperature, as well as the type of wastewater. For example, at 20°C an aeration period of 5 days results in 85% BOD removal.

Because sewage lagoons require a minimum of technology and are relatively low in cost, they are most common in developing countries. However, biodegradable organic

matter and turbidity are not as effectively reduced as during activated sludge treatment.

Given sufficient retention times, oxidation ponds can cause significant reductions in the concentrations of enteric pathogens, especially helminth eggs. For this reason, they have been promoted widely in the developing world as a low-cost method of pathogen reduction for wastewater reuse for irrigation. However, a major drawback of ponds is the potential for short-circuiting because of thermal gradients even in multipond systems designed for long retention times (i.e., 90 days). Even though the amount of short-circuiting may be small, detectable levels of

FIG. 22.19 Microbiology of facultative ponds. *(Modified from Bitton, 2011; From Environmental Microbiology, 4th Ed., Fig. 24.15, p. 514.)*

Domestic wastes and waste treatment

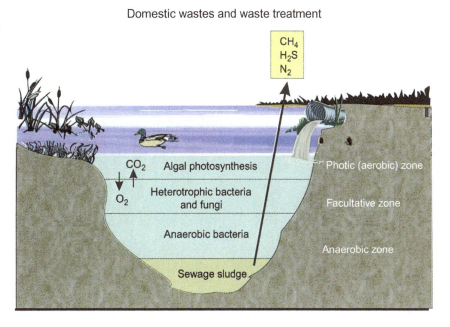

FIG. 22.20 An Aerated lagoon. *(From Environmental Microbiology, 2nd Ed., Fig. 24.16, p. 515.)*

pathogens can often be found in the effluent from oxidation ponds.

Inactivation and/or removal of pathogens in oxidation ponds is controlled by a number of factors including temperature, sunlight, pH, bacteriophage, predation by other microorganisms, and adsorption to or entrapment by settleable solids. Indicator bacteria and pathogenic bacteria may be reduced by 90%–99% or more, depending on retention times.

22.4   SEPTIC TANKS

Septic systems are the second alternative to large-scale centralized wastewater treatment systems. Until the middle of the 20th century in the United States, many rural

families and quite a few residents of towns and small cities depended on pit toilets or "outhouses" for waste disposal. In rural areas of developing countries these are still used. These pit toilets, however, often allowed untreated wastes to seep into the groundwater, allowing pathogens to contaminate drinking water supplies. This risk to public health led to the development of septic tanks and properly constructed drain fields. Primarily, septic tanks serve as repositories where solids are separated from incoming wastewater, and biological digestion of the waste organic matter can take place under anaerobic conditions. In 2007, 20% (26.1 million) of the homes in the United States depended on septic tanks. Approximately 20% of all new homes constructed use septic tanks. Most septic tanks are located in the Eastern United States (Fig. 22.21).

In a typical septic tank system (Fig. 22.22), the wastewater and sewage enter a tank made of concrete, metal, or fiberglass. There, grease and oils rise to the top as scum, and solids settle to the bottom. The wastewater and sewage then undergo anaerobic bacterial decomposition, resulting in the production of a sludge. The wastewater usually remains in the septic tank for just 24–72 h, after which it is channeled out to a drain field. This drain field or leach field is composed of small perforated pipes that are embedded in gravel below the surface of the soil. Periodically, the residual sludge in the septic tank known as septage is pumped out into a tank truck and taken to a treatment plant for disposal.

Although the concentration of contaminants in septic tank separate is typically much greater than that found in domestic wastewater (Table 22.7), septic tanks can be an

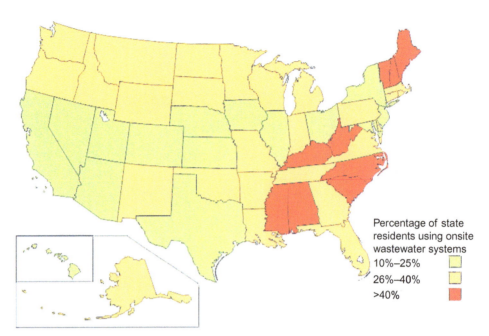

FIG. 22.21 Percentage of U.S. residents utilizing septic tanks for on-site wastewater treatment. Source: U.S. Census Bureau, 1990. *(From Environmental Microbiology, 2nd Ed., Fig. 24.17, p. 515.)*

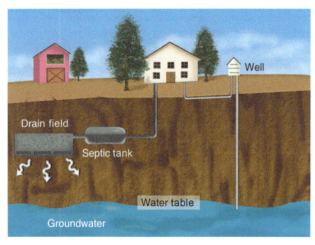

FIG. 22.22 Septic tank (On-site treatment system). *(From Environmental Microbiology, 2nd Ed., Fig. 24.18, p. 515.)*

effective method of waste disposal where land is available and population densities are not too high. Thus they are widely used in rural and suburban areas. But as suburban population densities increase, groundwater and surface water pollution may arise, indicating a need to shift to a commercial municipal sewage system. In fact, private septic systems are sometimes banned in many suburban areas. Moreover, septic tanks are not appropriate for every area of the country. They do not work well, for example, in cold, rainy climates, where the drain field may be too wet for proper evaporation, or in areas where the water table is shallow. High densities of septic tanks can also be responsible for nitrate contamination of groundwater. Finally, most of the waterborne disease outbreaks associated with

groundwater in the United States are thought to result from contamination by septic tanks.

22.5 LAND APPLICATION OF WASTEWATER

Although treated domestic wastewater is usually discharged into bodies of water, it may also be disposed of via land application for crop irrigation, or as a means of additional treatment and disposal. The three basic methods used in the application of sewage effluents to land include: low-rate irrigation, overland flow, and high-rate infiltration. Characteristics of each of these are listed in Table 22.8. The choice of a given method depends on the conditions prevailing at the site under consideration (loading rates, methods of irrigation, crops, and expected treatment).

With low-rate irrigation (Fig. 22.23A), sewage effluents are applied by sprinkling or by surface application at a rate of 1.5–10 cm per week. Two-thirds of the water is taken up by crops or lost by evaporation, and the remainder percolates through the soil matrix. The system must be designed to maximize denitrification in order to avoid pollution of groundwater by nitrates. Phosphorus is immobilized within the soil matrix by fixation or precipitation. The irrigation method is used primarily by small communities and requires large areas, generally on the order of 5–6 ha per 1000 people.

In the overland flow method (Fig. 22.23B), wastewater effluents are allowed to flow for a distance of 50–100 m along a 2%–8% vegetated slope and are collected in a ditch. The loading rate of wastewater ranges from 5 to 14 cm a week. Only about 10% of the water percolates through

TABLE 22.7 Typical Characteristics of Septage

Constituent	Concentration (mg/L)	
	Range	Typical Value
Total solids	5000–100,000	40,000
Suspended solids	4000–100,000	15,000
Volatile suspended solids	1200–14,000	7000
5-day, 20°C BOD	2000–30,000	6000
Chemical oxygen demand	5000–80,000	30,000
Total Kjeldahl nitrogen (as N)	100–1600	700
Ammonia, NH_3 as N	100–800	400
Total phosphorus as P	50–800	250
Heavy metals[a]	100–1000	300

[a]*Primarily iron (Fe), zinc (Zn), and aluminum (Al).*

Pepper, I.L., Brooks, J.P., Gerba, C.P., 2006. Pathogens in biosolids. Adv. Agron. 90, 1–41; From Environmental Microbiology, 2nd Ed., Table 24.7, p. 516.

TABLE 22.8 General Characteristics of the Three Methods Used for Land Application of Sewage Effluent

Factor	Application Method		
	Low-Rate Irrigation	Overland Flow	High-Rate Infiltration
Main objectives	Reuse of nutrients and water Wastewater treatment	Wastewater treatment	Wastewater treatment Groundwater recharge
Soil permeability	Moderate (sandy to clay soils)	Slow (clay soils)	Rapid (sandy soils)
Need for vegetation	Required	Required	Optional
Loading rate	1.3–10 cm/week	5–14 cm/week	>50 cm
Application technique	Spray, surface	Usually spray	Surface flooding
Land required for flow of 10^6 L/day	8–66 ha	5–16 ha	0.25–7 ha
Needed depth to groundwater	About 2 cm	Undetermined	5 m or more
BOD and suspended solid removal	90%–99%	90%–99%	90%–99%
N removal	85%–90%	70%–90%	0%–80%
P removal	80%–90%	50%–60%	75%–90%

From Pepper, I.L., Brooks, J.P., Gerba, C.P., 2006. Pathogens in biosolids. Adv. Agron. 90, 1–41. Environmental Microbiology, 2nd Ed., Table 24.8, p. 516.

the soil, compared with 60% that runs off into the ditch. The remainder is lost as evapotranspiration. This system requires clay soils with low permeability and infiltration.

High-rate infiltration treatment is also known as soil aquifer treatment (SAT) or rapid infiltration extraction (RIX) (Fig. 22.23C). The primary objective of SAT is the treatment of wastewater at loading rates exceeding 50 cm per week. The treated water, most of which has percolated through coarse-textured soil, is used for groundwater recharge, or may be recovered for irrigation. This system requires less land than irrigation or overland flow methods.

Drying periods are often necessary to aerate the soil system and avoid problems due to clogging. The selection of a site for land application is based on many factors including soil type (e.g., texture, permeability), distance to groundwater, groundwater movement, slope, and degree of isolation of the site from the public.

Inherent in land application of wastewater are the risks of transmission of enteric waterborne pathogens. The degree of risk is associated with the concentration of pathogens in the wastewater and the degree of contact with humans. Land application of wastewater is usually

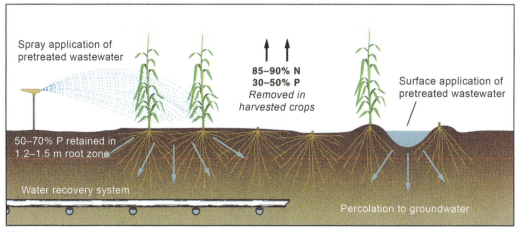

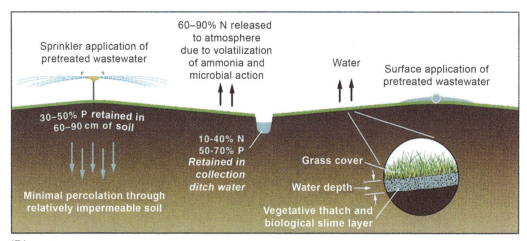

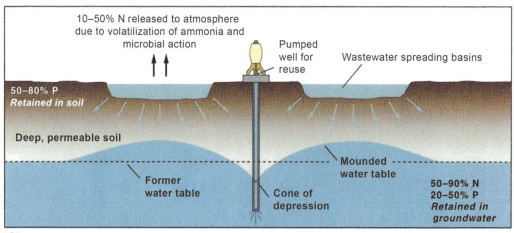

FIG. 22.23   Three basic methods of land application of wastewater. *(From Environmental Microbiology, 2nd Ed., Fig. 24.19, p. 517.)*

considered an intentional form of reuse and is regulated by most states. Because of limited water resources in the western United States, reuse is considered essential. Usually, stricter treatment and microbial standards must be met before land application. The highest degree of treatment is required when wastewater will be used for food crop irrigation, with lesser treatment for landscape irrigation or fiber crops. For example, the State of California requires no disinfection of wastewater for irrigation and no limits on coliform bacteria. However, if the reclaimed wastewater is used for surface irrigation of food crops and open landscaped areas, chemical coagulation (to precipitate suspended matter), followed by filtration and disinfection to reduce the coliform concentration to 2.2/100 mL is required. In some cities excess effluent is disposed of in river beds that are normally dry. Such disposal can create riparian areas (Fig. 22.24).

Because high-rate infiltration may be practiced to recharge aquifers, additional treatments of secondary wastewater may be required. However, as some removal of pathogens can be expected, treatment requirements may be less. The degree of treatment needed may be influenced by the amount or time it takes the reclaimed water to travel from the infiltration site to the point of extraction, and the depth of the unsaturated zone. The greatest concern has been with the transport of viruses, which, because of their small size, have the greatest chance of traveling large distances within the subsurface. Generally, several meters of moderately fine-textured, continuous soil layer are necessary for virus reductions of 99.9% or more (Yates, 1994).

FIG. 22.24 Effluent outfall of the Roger Road Wastewater Treatment Plant in Tucson, Arizona. Here, extensive growth of vegetation due to the effluent produces a riparian habitat. *(From Environmental Microbiology, 2nd Ed., Fig. 24.20, p. 518.)*

22.6 WETLANDS SYSTEMS

Wetlands, which are typically less than 1 m in depth, are areas that support aquatic vegetation and foster the growth of emergent plants such as cattails, bulrushes, reeds, sedges, and trees. They also provide important wetland habitat for many animal species. Recently, wetland areas have been receiving increasing attention as a means of additional treatment for secondary effluents. The vegetation provides surfaces for the attachment of bacteria and aids in the filtration and removal of such wastewater contaminants as biological oxygen and excess carbon. Factors involved in the reduction of wastewater contaminants are presented in Table 22.9. Although both natural and constructed wetlands have been used for wastewater treatment, recent work has focused on constructed wetlands because of regulatory requirements. Two types of constructed wetland systems are in general use: (1) *free water surface* (FWS) systems and (2) *subsurface flow systems* (SFS). An FWS wetland is similar to a natural marsh because the water surface is exposed to the atmosphere. Floating and submerged plants, such as those shown in Fig. 22.25A, may be present. SFS consist of channels or trenches with relatively impermeable bottoms filled with sand or rock media to support emergent vegetation.

During wetland treatment, the wastewater is usable. It can, for instance, be used to grow aquatic plants such as water hyacinths (Fig. 22.25B) and/or to raise fish for human consumption. The growth of such aquatic plants provides not only additional treatment for the water but also a food source for fish and other animals. Such aquaculture systems, however, tend to require a great deal of land area. Moreover, the health risk associated with the production of aquatic animals for human consumption in this manner must be better defined (Fig. 22.26).

There has been increasing interest in the use of natural systems for the treatment of municipal wastewater as a form of tertiary treatment (Kadlec and Wallace, 2008). Artificial or constructed wetlands have a higher degree of biological activity than most ecosystems; thus transformation of pollutants into harmless by-products or essential nutrients for plant growth can take place at a rate that is useful for the treatment of municipal wastewater (Case Study 22.1). Most artificial wetlands in the United States use reeds or bull rushes, although floating aquatic plants such as water hyacinths and duckweed have also been used. To reduce potential problems with flying insects, subsurface flow wetlands have also been built (Fig. 22.27). In these types of wetlands all of the flow of the wastewater is below the surface of a gravel bed containing plants tolerant of water-saturated soils. Most of the existing information on the performance of these wetlands concerns coliform and fecal coliform bacteria. Kadlec and Wallace (2008) have summarized the existing literature on this topic. They point

TABLE 22.9 Principal Removal and Transformation Mechanisms in Constructed Wetlands Involved in Contaminant Reduction

Constituent	Free Water System	Subsurface Flow	Floating Aquatics
Biodegradable organics	Bioconversion by aerobic, facultative, and anaerobic bacteria on plant and debris surfaces of soluble BOD, adsorption, filtration	Bioconversion by facultative and anaerobic bacteria on plant and debris surfaces	Bioconversion by aerobic facultative, and anaerobic bacteria on plant and debris surfaces
Suspended solids	Sedimentation, filtration	Filtration, sedimentation	Sedimentation, filtration
Nitrogen	Nitrification/denitrification, plant uptake, volatilization	Nitrification/denitrification, plant uptake, volatilization	Nitrification/denitrification plant update, volatilization
Phosphorus	Sedimentation, plant uptake	Filtration, sedimentation, plant uptake	Sedimentation, plant uptake
Heavy metals	Adsorption of plant and debris surfaces	Adsorption of plant roots and debris surfaces, sedimentation	Absorption by plants sedimentation
Trace organics	Volatilization, adsorption, biodegradation	Adsorption, biodegradation	Volatilization, adsorption biodegradation
Pathogens	Natural decay, predation, UV irradiation, sedimentation, excretion of antimicrobials from rots of plants	Natural decay, predation, sedimentation, excretion of anti-microbials from roots of plants	Natural decay, predation sedimentation

From Pepper IL, Brooks JP, Gerba CP: Pathogens in biosolids, Adv Agron 90:1–41, 2006; Environmental Microbiology, 2nd Ed., Table 24.9, p. 519.

out that natural sources of indicators in treatment wetlands never reach zero because wetlands are open to wildlife. Reductions in fecal coliforms are generally greater than 99%, but there is a great deal of variation, probably depending on the season, type of wetland, numbers and type of wildlife, and retention time in the wetland. Volume- and area-based bacterial die-off models have been used to estimate bacterial die-off in surface flow wetlands (Kadlec and Wallace, 2008).

In one study of a mixed-species surface flow wetland with a detention time of approximately 4 days several other types of microorganisms were examined. Results showed that *Cryptosporidium* was reduced by 53%, *Giardia* by 58%, and enteric viruses by 98% (Karpiscak et al., 1996).

22.7 SLUDGE PROCESSING

Primary, secondary, and even tertiary sludges generated during wastewater treatment are a major by-product of the treatment process. These sludges, in turn, are usually subjected to a variety of treatments. Raw sludge is sometimes subjected to *screening* to remove coarse materials including grit that cannot be broken down biologically. *Thickening* is usually done to increase the solids content of the sludge. This can be achieved via centrifugation which increases the solids content to approximately 12%. *Dewatering* can further concentrate the solids content to 20%–40%. This is normally achieved via filtration or by the

use of drying beds. *Conditioning* enhances the separation of solids from the liquid phase. This is usually accomplished by the addition of inorganic salts such as alum, lime, ferrous or ferric salts, or synthetic organic polymers known as polyelectrolytes. All of these processes reduce the water content of the sludge, which ultimately reduces transportation costs to the final disposal and/or utilization site.

Finally, *stabilization* technologies are available, reducing both the solids content of the sludge and inactivating pathogenic microbes present in the sludge.

22.7.1 Stabilization Technologies

22.7.1.1 Aerobic Digestion

This consists of adding air or oxygen to sludge in a 4 to 8-foot-deep open tank. The oxygen concentration within the tank must be maintained above 1 mg/L to avoid the production of foul odors. The mean residence time in the tank is 12–60 days depending on the tank temperature. During this process, microbes aerobically degrade organic substrate, reducing the volatilize solids content of the sludge by 40%–50% (EPA, 1992). Digestion temperatures are frequently moderate or mesophilic (30–40°C). By increasing the oxygen content, thermophilic digestion can be induced (>60°C). By increasing the temperature and the retention time, the degree of pathogen inactivation can be enhanced. Pathogen concentrations ultimately determine the treatment

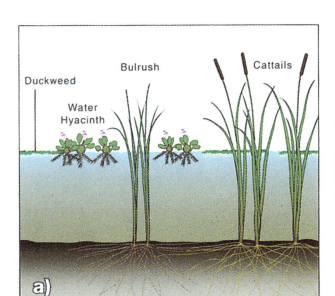

FIG. 22.25 (A) Common aquatic plants used in constructed wetlands. (B) An artificial wetland system in San Diego, California, utilizing water hyacinths. *(From Environmental Microbiology, 2nd Ed., Fig. 24.21, p. 519.)*

level of the product: Class B biosolids can contain many human pathogens (see Chapter 23). Class A biosolids, which result from more stringent and enhanced treatment, contain very low or nondetectable levels of pathogens. The degree of treatment, Class A versus Class B has important implications on the reuse potential of the material for land application (see Chapter 23). Aerobic digestion generally results in the production of Class B biosolids.

22.7.1.2 Anaerobic Digestion

This type of microbial digestion occurs under low redox conditions, with low oxygen concentrations. Carbon dioxide is a major terminal electron acceptor used and results in the conversion of organic substrate to methane and carbon dioxide. This process reduces the volatile solids

by 35%–60% (Bitton, 2011) and results in the production of Class B biosolids. The advantages and disadvantages of anaerobic digestion relative to aerobic digestion are shown in Information Box 22.3.

22.7.2 Sludge Processing to Produce Class A Biosolids

Class B biosolids that arise following digestion can be further treated to Class A levels prior to land application. The three most important technologies to achieve this goal are composting, lime treatment, and heat treatment.

22.7.2.1 Composting

Composting consists of mixing sludge with a bulking agent that normally has a high C:N ratio (Fig. 22.28). This is necessary because of the low C:N ratio of the sludge. The mixtures are normally kept moist but aerobic. These conditions result in very high microbial activity, and the generation of heat that increases the temperature of the composting material. Factors affecting the composting process are shown in Information Box 22.4. There are three main types of composting systems:

- The *Aerated Static Pile* process typically consists of mixing dewatered digested sludge with wood chips. Aeration of the pile is normally provided by blowers during a 21-day composting period. During this active composting period, temperatures increase to the mesophilic range (20–40°C) where microbial degradation occurs via bacteria and fungi. Temperatures subsequently increase to 40–80°C, with microbial populations dominated by thermophilic (heat tolerant) and spore-forming organisms. These high temperatures inactivate pathogenic microorganisms, and frequently result in a Class A biosolid product. Subsequently, the compost is cured for at least 30 days, during which time temperatures within the pile decrease to ambient levels.

- The *Windrow Process* is similar to the static pile process except that instead of a pile, the sludge and bulking agent are laid out in long rows of dimension: $2\,m \times 3\,m \times 80\,m$ (Fig. 22.29). Aeration for windrows is provided by turning the windrows several times a week. Once again, if the composting process is efficient, Class A biosolids are produced.

- In *Enclosed Systems* the composting is conducted in steel vessels of size 10–15 m high by 3–4 m diameter. For this type of composting, aeration via blowers and temperature of the composting are carefully controlled. This results in a high quality Class A compost, with little or no odor problems. However, costs of enclosed systems are higher.

FIG. 22.26 Aerial view of Sweetwater Recharge Facilities. Numbered blue areas are infiltration basins. *(Photo courtesy of the Water Reuse Association; From Environmental Microbiology, 2nd Ed., Case Study 24.1, p. 520.)*

Case Study 22.1 Sweetwater Wetlands Infiltration-Extraction Facility in Tucson, Arizona

Tucson, Arizona is located in the in the Sonoran Desert in the southwestern United States. Because of limited water supplies reclamation of wastewater is critical to meet water needs in the region. To meet these needs a system was built to provide tertiary effluents derived from and activated sludge/trickling filter system of sufficient quality to be used for landscape irrigation. The system is composed of several components that allows for various treatments and storage of tertiary effluent. A tertiary treatment plant filters the secondary effluent (to reduce turbidity and microorganisms) and then provides additional disinfection. The backwash from the filters is then discharged into an artificial wetland for treatment. When the water exists the wetland it is discharged into infiltration basins were it is further treated. In times of low reclaimed water demand (winter) the tertiary effluent may be discharged into the infiltration basins. The subsurface aquifer is then used as a storage facility, the water then being pumped to the surface (extraction) when needed during periods of peak demand.

The multiple barriers of conventional and natural technologies are design to enhance the removal of chemical and microbial contaminates. Filtration of the secondary wastewater during tertiary treatment allows for reduction of the larger protozoan parasites (which are more resistant to disinfection than enteric bacteria and viruses) and more effective disinfection. In the wetlands protozoan parasites settle out and bacteria and viruses are reduced by inactivation by sun light (UV light) and microbial antagonism. Infiltration of the water through the soil results in further removal of pathogens by filtration and adsorption to soil particle (especially viruses).

22.7.2.2 Lime and Heat Treatment

Lime stabilization involves the addition of lime as $Ca(OH)_2$ or CaO, such that the pH of digested sludge is equal to or greater than 12 for at least 2 h. Liming is very effective at inactivating bacterial and viral pathogens, but less so for parasites (Bitton, 2011). Lime stabilization also reduces odors and can result in a Class A biosolid product.

Heat treatment involves heating sludge under pressure to temperatures up to 260°C for thirty minutes. This process kills microbial pathogens and parasites, and also further dewaters the sludge.

22.7.2.3 The Cambi Thermal Hydrolysis Process

The Cambi process uses thermal hydrolysis as a pretreatment to anaerobic digestion. This increases the microbial degradation of organic volatile solids and increases the amount of biogas obtained. This process also

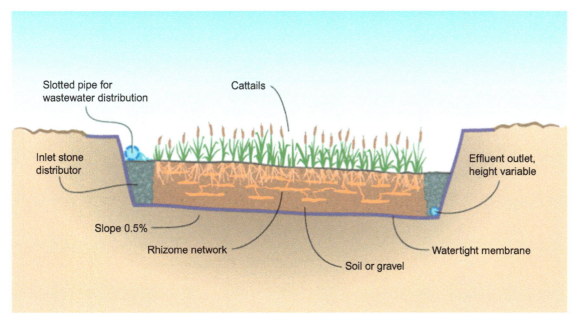

FIG. 22.27   Cross-section of a subsurface wetland. *(From Environmental Microbiology, 3rd Ed., Fig. 25.25, p. 603.)*

INFORMATION BOX 22.3 Advantages and Disadvantages of Anaerobic Digestion

Advantages:

* No oxygen requirement, which reduces cost
* Reduced mass of biosolids due to low energy yields of anaerobic metabolism (see also Chapter 3)
* Methane produced, which can be used to generate electricity
* Enhanced degradation of xenobiotic compounds
  Disadvantages:
* Slower than aerobic digestion
* More sensitive to toxicants

(Adapted from Bitton, G., 2011. Wastewater microbiology, fourth ed., New York, 2011, Wiley-Liss. Environmental Microbiology, 2nd Ed., Information Box 24.1, p. 521.)

INFORMATION BOX 22.4 Factors Affecting Efficient Composting

Temperature:	adequate aeration and moisture must be maintained to ensure temperatures reach 60 °C, to inactivate microbial pathogens
Aeration:	must be provided via blowers or by turning
Moisture:	must be neither too moist, which promotes anaerobic activity; nor too dry, which limits microbial activity
C:N ratio:	should be maintained around 25:1, to ensure adequate but not excessive amounts of nitrogen for the microbes

Surface area of shredded material should be used to increase substrate surface area for microbial
Bulking agent: metabolism

(From Environmental Microbiology, 2nd Ed., Information Box 24.2, p. 522)

FIG. 22.28   The wood bulking agent for composting. The wood is shredded to increase the surface area of bulking agent for composting. *(From Environmental Microbiology, 2nd Ed., Fig. 24.26, p. 522.)*

FIG. 22.29   Biosolid composting via the windrow process. Here three windrows are illustrated. *(From Environmental Microbiology, 2nd Ed., Fig. 24.27, p. 522.)*

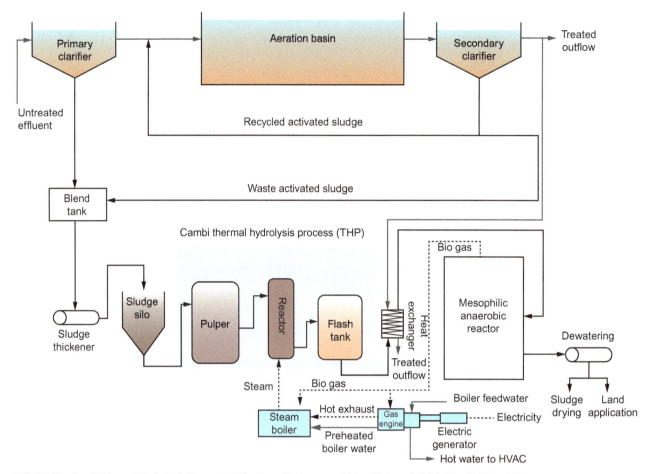

FIG. 22.30 Cambi Thermal Hydrolysis Process (THP). *(From Environmental Microbiology, 3rd Ed., Fig. 25.28, p. 604.)*

facilitates a higher degree of separation of solid and liquid phase after digestion. The process is depicted in Fig. 22.30. Sludge generated during primary and secondary treatment is dewatered to approximately 15%–20% dry solids content, preheated to 100°C in the pulper tank, then heated to 150–170°C under 8–9 bar pressure in the reactor. In the flash tank the sludge cools to about 100°C and the released steam is recirculated. Further cooling to 35°C occurs via the heat exchangers where more energy is recycled via the production of hot water. Following this, the sludge is subjected to mesophilic anaerobic digestion, leading to the production of biogas and Class A biosolids, since no pathogens can survive the steam treatment. A typical plant is shown in Fig. 22.31. Cambi is now well established in several countries within Europe, particularly the United Kingdom and Norway. The first thermal hydrolysis plant to be built in the United States is the Blue Plains treatment plant in Washington DC. This plant treats 450 dry tons of sludge per day, and the biogas produced will cover the entire steam needs of the plant saving $20 million a year from the energy saved.

FIG. 22.31 "Cambi" Thermal Hydrolysis—preferred technology for Cotton Valley and Whitlingham. *(From Environmental Microbiology, 3rd Ed., Fig. 25.29, p. 605.)*

QUESTIONS AND PROBLEMS

1. What are the three major steps in modern wastewater treatment?
2. Why is it important to reduce the amount of biodegradable organic matter and nutrients during sewage treatment?
3. When would tertiary treatment of wastewater be necessary?
4. What are some types of tertiary treatment?
5. What are the processes involved in the removal of heavy metals from wastewater during treatment by artificial wetlands.
6. What are the three types of land application of wastewater? Which one is most likely to contaminate the groundwater with enteric viruses? Why? What factors determine how far viruses will be transported in groundwater? How does nitrogen removal occur? Phosphorus removal?
7. What is the major contaminates in groundwater associated with the use of on-site treatment systems?
8. What factors may determine the concentration of enteric pathogens in domestic raw sewage?
9. Five milliliters of a wastewater sample is added to dilution water in a 300-mL BOD bottle. If the following results are obtained, what is the BOD after 3 days and 5 days?

Time (days)	Dissolved Oxygen (mg/L)
0	9.55
1	4.57
2	4.00
3	3.20
4	2.60
5	2.40
6	2.10

10. List some advantages and disadvantages of the wetland treatment of sewage.
11. What is the major mechanism of pathogen removal during activated sludge treatment?
12. What treatment processes would you need to obtain an $8 \log_{10}$ reduction of enteric viruses? *Giardia?* viruses?
13. What treatment process would you need to obtain a $8 \log_{10}$ reduction of enteric viruses from raw sewage. For *Giardia?*
14. How effective do you think sunlight is in killing *Cryptosporidium?* Enteric viruses?

REFERENCES

APHA, 1998. Standard Methods for Water and Wastewater. American Public Health Association, Washington, DC.

Bitton, G., 2011. Wastewater Microbiology, fourth Edition Wiley-Liss, New York.

Chang, I.S., LeClech, P., Jefferson, B., Judd, S., 2002. Membrane fouling in membrane bioreactors for wastewater treatment. J. Environ. Eng. 128, 1018–1029.

Edzwald, J.K., 2010. Dissolved air flotation and me. Water Res. 44, 2077–2106.

EPA (U.S. Environmental Protection Agency), 1992. Guidelines for water reuse. EPA/625/R-92/004, Washington, DC.

EPA (U.S. Environmental Protection Agency), 1977. Wastewater treatment facilities for sewered small communities. EPA–62511–77–009, Washington, DC.

Hammer, M.J., 1986. Water and Wastewater Technology. John Wiley and Sons, New York.

Kadlec, R.H., Wallace, S., 2008. Treatment Wetlands. CRC Press, Boca Raton, FL.

Karpiscak, M.M., Gerba, C.P., Watt, P.M., Foster, K.E., Falabi, J.A., 1996. Multi-species plant systems for wastewater quality improvements and habitat enhancement. Water Sci. Technol. 33, 231–236.

Leong, L.Y.C., 1983. Removal and inactivation of viruses by treatment processes for portable water and wastewater—a review. Water Sci. Technol. 15, 91–114.

Metcalf and Eddy, Inc., 2003. Wastewater Engineering. McGraw-Hill, New York.

Park, S., Seon, J., Byun, I., Cho, S., Park, T., Lee, T., 2010. Comparison of nitrogen removal and microbial distribution in wastewater treatment process under different electron donor conditions. BioResource Technol. 101, 2988–2995.

Rao, V.C., Metcalf, T.G., Melnick, J.L., 1986. Removal of pathogens during wastewater treatment. In: Rehm, H.J., Reed, G. (Eds.), Biotechnology. In: Vol. 8. VCH, Berlin, pp. 531–554.

Rose, J.B., Carnahan, R.P., 1992. Pathogen removal by full scale wastewater treatment. Report to Florida Department of Environmental Regulation, Tallahassee, FL.

Wang, Y-P., Liu, X-W., Li, W-W., Wang, Y-K., Shery, G.P., Zeng, R.J., and Yu, H-Q. 2012. A microbial fuel cell-membrane bioreactor integrated system for cost-effective wastewater treatment. Appl. Energy 98:230-235.

Yates, M.V., 1994. Monitoring concerns and procedures for human health effects. In: Wastewater Reuse for Golf Course Irrigation. CRC Press, Boca Raton, FL, pp. 143–171.

FURTHER READING

Ansari, S.A., Farrah, S.R., Chaudhry, G.R., 1992. Presence of human immunodeficiency virus nucleic acids in wastewater and their detection by polymerase chain reaction. Appl. Environ. Microbiol. 58, 3984–3990.

Brooks, J.P., Rusin, P.A., Maxwell, S.L., Rensing, C., Gerba, C.P., Pepper, I.L., 2007. Occurrence of antibiotic-resistant bacteria and endotoxin associated with the land application of biosolids. Can. J. Microbiol. 53, 1–7.

Enriquez, V., Rose, J.B., Enriquez, C.E., Gerba, C.P., 1995. Occurrence of Cryptosporidium and Giardia in secondary and tertiary wastewater effluents. In: Betts, W.B., Casemore, D., Fricker, C., Smith, H., Watkins, J. (Eds.), Protozoan Parasites and Water. Royal Society of Chemistry, Cambridge, UK, pp. 84–86.

EPA (U.S. Environmental Protection Agency), 1992. Technical support document for land application of sewage sludge. Vol. I EPA 822/R–93–001A.

EPA (U.S. Environmental Protection Agency), 1994. A plain English guide to the EPA Part 503 biosolids rule. EPA 832/R–93/003 U.S. Environmental Protection Agency.

EPA (U.S. Environmental Protection Agency), 1996. Technical support document for the round two sewage sludge pollutants. EPA–822–R–96–003.

EPA (U.S. Environmental Protection Agency), 1997. Exposure factors handbook, EPA/600/P–95/002. August 1997.

EPA (U.S. Environmental Protection Agency), 1998. Volume I—general factors, exposure factors handbook. Update to exposure factors handbook, EPA/600/8–89/043, May 1989.

EPA (U.S. Environmental Protection Agency), 1999a. Environmental regulations and technology: control of pathogens and vector attraction in sewage sludge. EPA/625/R–92/013 (Available from: http://www.epa.gov/ttbnrmrl/625/R-92/013.htm.

EPA (U.S. Environmental Protection Agency), 1999b. Environmental regulation and technology. Control of pathogens and vector attraction in sewage sludge EPA/652/R-92/013. Revised 1999. Office of Research and Development, U.S. Environmental Protection Agency, Washington, DC, 177 pp.

EPA (U.S. Environmental Protection Agency), 2000. A guide to field storage of biosolids and the organic by-products used in agriculture and for soil resource management. EPA/832–B–00–007, Washington, DC.

Feachem, R.G., Brandley, D.J., Garelick, H., 1983. Sanitation and Disease: Health Aspects of Excreta and Waste Management. Wiley, New York.

Gerba, C.P., Pepper, I.L., Whitehead, L.F., 2002. A risk assessment of emerging pathogens of concern in the land application of biosolids. in sludge management, regulation, utilization and disposal. Water Sci. Technol. 46, 225–230.

Koudela, B., Kucerova, S., Hudcovic, T., 1999. Effect of low and high temperatures on infectivity of Encephalitozoon cuniculi spores suspended in water. Folia Parasitol. 46, 171–174.

Madore, M.S., Rose, J.B., Gerba, C.P., Arrowood, M.J., Sterling, C.R., 1987. Occurrence of Cryptosporidium in sewage effluents and selected surface waters. J. Parasitol. 73, 702–705.

Neilson, J.W., Josephson, K.L., Pepper, I.L., Arnold, R.B., DiGiovanni, G.D., Sinclair, N.A., 1994. Frequency of horizontal gene transfer of a large catabolic plasmid (pJP4) in Soil. Appl. Environ. Microbiol. 60, 4053–4058.

Newby, D.T., Gentry, T.J., Pepper, I.L., 2000. Comparison of 2,4-dichlorophenoxyacetic acid degradation and plasmid transfer in soil resulting from bioaugmentation with two different pJP4 donors. Appl. Environ. Microbiol. 66, 3399–3407.

NRC (National Research Council), 1996. Use of Reclaimed Water and Sludge in Food Crop Production. National Academy Press, Washington, DC.

Pepper, I.L., Brooks, J.P., Gerba, C.P., 2006. Pathogens in biosolids. Adv. Agron. 90, 1–41.

Rose, J.B., Gerba, C.P., 1991. Use of risk assessment for development of microbial standards. Water Sci. Technol. 24, 29–34.

Rusin, P., Enriquez, C.E., Johnson, D., Gerba, C.P., 2000. Environmentally transmitted pathogens. In: Maier, R.M., Pepper, I.L., Gerba, C.P. (Eds.), Environmental Microbiology. Academic Press, San Diego, pp. 447–489.

Smid, T., Heederik, D., Houba, R., Quanjer, P.H., 1992. Dust and endotoxin related respiratory effects in the animal feed industry. Am. Rev. Respir. Dis. 146, 1474–1479.

Teunis, P.F.M., Nagelkerke, N.J.D., Haas, C.N., 1999. Dose-response models for infectious gastroenteritis. Risk Anal. 19, 1251–1260.

Whitmore, T.N., Robertson, L.J., 1995. The effect of sewage sludge treatment processes on oocysts of Cryptosporidium parvum. J. Appl. Bacteriol. 78, 34–38.

Chapter 23

Land Application of Organic Residuals: Municipal Biosolids and Animal Manures

I.L. Pepper, J.P. Brooks and C.P. Gerba

Chapter cover art: One method of land application of biosolids.

23.1 INTRODUCTION TO ORGANIC RESIDUALS

The term "organic residuals" includes several different waste categories. Among them are the organic fraction of municipal solid waste, animal wastes or manure, and municipal biosolids that comprise the organic solids remaining after sewage treatment. In the United States, approximately 450,000 animal feeding operations (AFOs), some of which are concentrated animal feeding operations (CAFOs), collectively produce over 100 million dry tons of manure per year (Burkholder et al., 2007). In contrast, approximately 16,500 municipal wastewater treatment plants operating in the United States produce a relatively small 7.2 million dry tons annually (NEBRA, 2007). Of these, the largest $\simeq 3300$ generate more than 92% of the total quantity of biosolids in the United States (NEBRA, 2007). Conversely, an average 68-kg human produces approximately 37 kg of waste per year and 6.5 million dry tons per year for all municipalities combined. Both types of residuals are used beneficially for crop production through land application. Overall, animal manures are applied to about 10% of available agricultural land with greater than 90% of the total available animal manures being land applied. In contrast, only 0.1% of available agricultural land is spread with biosolids, accounting for 55% of the available biosolids (NEBRA, 2007; Brooks et al., 2011). Though biosolids represent only a small fraction of total organic residuals produced, they are the most processed, most regulated, most studied, and most controversial with respect to disposal and beneficial reuse. In contrast, raw animal manures are not treated and are not regulated. In fact, certified organic farmers can use animal manures as a fertilizer and soil amendment, provided crops grown for human consumption are harvested at least 90 days after the last application (Organic Trade Association, 2012). The objective of this chapter is to compare and contrast the efficacy of land application of municipal biosolids and animal manures.

23.2 LAND APPLICATION OF BIOSOLIDS AND ANIMAL WASTES: A HISTORICAL PERSPECTIVE AND CURRENT OUTLOOK

In the United States, land application of municipal wastewater and biosolids has been practiced for its beneficial effects and for disposal purposes since the advent of modern wastewater treatment about 160 years ago. In England in the 1850s, "sewage farms" were established to dispose of untreated sewage. By 1875, about 50 farms were using land treatment in England, and many other farms were being used close to other major cities in Europe. In the United States, sewage farms were established by about 1900. At this same time, primary sedimentation and secondary biological treatment was introduced as a rudimentary form of wastewater treatment, and land application of "sludges" began. It is interesting to note that prior to wastewater treatment, "sludge" per se did not exist. Municipal sludge in Ohio was used as a fertilizer as early as 1907.

Environmental and Pollution Science. https://doi.org/10.1016/B978-0-12-814719-1.00023-9

Since the early 1970s, more emphasis has been placed on applying sludge to cropland at rates to supply adequate nutrients for crop growth (Hinesly et al., 1972). In the 1970s and 80s, many studies were undertaken to investigate the potential benefits and hazards of land application, in both the United States and Europe. Ultimately in 1993, U.S. federal regulations were established via the "Part 503 Sludge Rule." This document—"The Standards for the Use and Disposal of Sewage Sludge" (EPA, 1993) was designed to "adequately protect human health and the environment from any reasonably anticipated adverse effect of pollutants." As part of these regulations, two classes of treatment were defined as "Class A and Class B" biosolids, with different restrictions for land applications, based on the level of treatment. The distinction between sewage sludge and biosolids is described in Information Box 23.1, and it is important to note that the term biosolids implies treatment to defined levels. The requirements for Class A versus Class B biosolids are defined in Information Box 23.2.

INFORMATION BOX 23.1 Definitions

Sewage sludge: the solid, semisolid, or liquid residue generated during the treatment of domestic sewage in a treatment works.

Biosolids: two different definitions have been developed:

EPA: the primarily organic solid product yielded by municipal wastewater treatment processes that can be beneficially recycled (whether or not they are currently being recycled).

NRC (National Research Council), 2002: sewage sludge that has been treated to meet the land application standards in the Part 503 rule or any other equivalent land application standards or practices.

(From Environmental Microbiology, second ed., Info Box. 24.3, pg. 523)

INFORMATION BOX 23.2 Part 503 Pathogen Density Limits for Class A & B Biosolids

Standard Density Limits (Dry Weight)

Pathogen or Indicator

Class A

Salmonella <3 MPN/4 g total solids or

Fecal coliforms <1000 MPN/g and

Enteric viruses <1 PFU/4 g total solids

Class B

Fecal coliform density <2,000,000 MPN/g total solids

Adapted from EPA (U.S. Environmental Protection Agency): A guide to field storage of biosolids and the organic by-products used in agriculture and for soil resource management. EPA/832–B–00–007, Washington, DC, 2000

(From Environmental Microbiology, 2nd Ed., Info Box 24.4, pg. 524)

Land application increased when restrictions were placed on "ocean dumping." By the year 2000, 60% of all biosolids were land applied in the United States. Currently, most land application of biosolids in the United States utilizes Class B biosolids. However, due to public concerns over potential hazards, in some areas of the United States, land application of Class B biosolids has been banned. Thus by 2004, only 55% of all biosolids was applied to soil for agronomic, silvicultural, and/or land restoration processes. The remaining 45% was disposed of in municipal landfills or incinerated (NEBRA, 2007), with about two-thirds of the non-land-applied material being landfilled. Of the total applied to soils, 74% was on farmland for agricultural purposes (NEBRA, 2007). A recent report indicates that approximately 200 million farmers worldwide grow crops in fields fertilized with human waste (IWMI, 2010).

Contrary to municipal wastewater treatment sludge, CAFOs and their manures are a relatively new advancement in egg/dairy/meat production systems. Though manure has been around since the "dawn of time," the idea of a CAFO has not. AFOs are defined as a feedlot or a facility where animals are kept for greater than 45 days; cattle grazing on pasture are exempt (EPA, 2004). CAFOs are then designated based on numerical criteria such as greater than 300 cattle or 9000 broiler chickens (EPA, 2004). Thus all CAFOs are AFOs, but not all AFOs are CAFOs. Since the 1960s, the vast majority of animals raised for food and their products are produced in CAFOs. This movement has led to the concentration of most food animals into less than 20% of all AFOs. For all AFO and CAFO food animal production, it has always been the responsibility of the owner to dispose of the manure, with reliance on disposal to nearby fields, thereby keeping costs low. The vast majority of AFO owners apply manure to owned lands or rely on the sale or "giving away" of manure to other land owners. Manure land application has not been governed by any specific law or federal regulation; however, guidelines exist for suggested rates of manure for land application based on nutrient requirements, typically N or P, of the crop to be grown. Most states require nutrient management plans to be established prior to the establishment of a new CAFO. These plans essentially establish how the CAFO owner will dispose of the manure in both a quantitative (i.e., how much) and qualitative (i.e., which crop) manner.

Thus far, the level of scrutiny reserved for biosolids land application has not been similarly applied to manure. Anecdotally, the public has regarded manure as a "natural" material, and thus it has escaped intense criticism despite knowledge of pathogens, antibiotic-resistant bacteria, and nutrient runoff concerns that are all associated with land application of manures. In 1972, the Clean Water Act identified AFOs as potential pollutant sources, resulting in CAFO regulations being set in place in 1976. Increase in CAFO size necessitated revision of the regulations in

2008. Thus the current USEPA CAFO rule requires that CAFOs which discharge or propose to discharge waste, need to apply for a permit, along with the establishment of a nutrient management plan. The rule was challenged, and now an AFO can apply for an exception provided that the manure can be appropriately stored to prevent accidental release (i.e., runoff contaminated with manure) during a 24-h, 100-year storm event. These limits and rules are specifically designed to reduce discharge to surface water and do not govern the land application of manure to soil, unless there is a threat of effluent runoff to surface water. Guidelines for manure land application also suggest harvest delays when using manure on "organic" marketed food crops. These guidelines suggest a delay of between 90 and 120 days between land application and harvest, depending on the level of interaction between the manure/soil matrix and the food crop edible parts. Apart from these rules and guidelines, there are no other regulations for land application of manure.

Biosolids and manure are applied to agricultural and nonagricultural lands as a soil amendment because they improve the chemical and physical properties of soils, and because they contain nutrients for plant growth. Land application on agricultural land is used to grow food crops such as corn or wheat, and nonfood crops such as cotton. Nonagricultural land application includes forests, rangelands, public parks, golf courses, and cemeteries. Biosolids and manure are also used to revegetate severely disturbed lands such as mine tailings or strip mine areas.

23.2.1 Class A Versus Class B biosolids

A simplified schematic of how biosolids are produced is presented in Fig. 23.1. Biosolids are divided into two classes on the basis of pathogen content: Class A and Class B (Information Box 23.2). Class A biosolids are treated to reduce the presence of pathogens to below detectable levels and can be used without any pathogen-related restrictions at the application site. Class A biosolids can also be bagged and sold to the public. Class B biosolids are also treated to reduce pathogens, but still contain detectable levels of them. Class B biosolids have site restrictions to minimize the potential for human exposure, until environmental factors such as heat, sunlight, or desiccation have further reduced pathogen numbers. Class B biosolids cannot be sold or given away in bags or other containers or used at sites with public use.

23.2.2 Methods of Land Application of Organic Residuals

23.2.2.1 Land Application of Biosolids

The method of land application of biosolids essentially depends on the percent solids contained within them, which

INFORMATION BOX 23.3 Land Application Methods

% Solids	Nature of Biosolids	Method of Application
2	Liquid	Sprinkler system
8	Liquid	Spray or injection
>20	Cake	Spreaders or slingers

(From Environmental Microbiology, 2nd Ed., Info Box. 24.5, pg. 524)

determines whether the biosolids are liquid in nature, or a "cake" (Information Box 23.3). Methods of land application include:

- Injection. Liquid biosolids are injected to a soil depth of 30 cm. Injection vehicles simultaneously disc the field. Injection processes reduce odors, and bioaerosols, as well as the risk of runoff to surface waters.
- Surface application. Liquid or cake biosolids are surface applied and subsequently tilled into the soil.
- Slingers are also used to throw the material through the air as a means of land application.

23.2.2.2 Land Application of Animal Manures

Land application techniques for manure are far more varied than for land application of municipal biosolids. This is due, in part, to the variability associated with the various AFOs which can include manure from cattle, dairy, poultry, and swine industries. In addition, there are smaller operations specializing in niche foods such as ostrich, lamb, and bison, which contribute to AFO manure burden. Even within a large industry such as poultry, variability is considerable, given the specific subindustries such as egg production and turkey farms. Egg layer farms typically produce liquid manure, whereas turkey producers produce a litter/fecal matter solid mixture. Each AFO produces a different type and kind of manure, making the standardization afforded by the USEPA Part 503 biosolids Rule a milestone that is difficult to reach. Typically prior to land application, most AFOs store manure in a shed or lagoon for a period of time, depending on season and demand. Storage type depends on solid content, with shed storage reserved for solid manure, while lagoons are utilized for liquid or slurry manure. Composting or "unofficial composting" (typically consists of long-term manure storage, without temperature monitoring or any other handling) are common pretreatments prior to land application. Likewise, in lieu of land application, some poultry producers opt to combust or burn their litter for energy production.

Manure can vary in physical characteristics from a slurry (<2% solids) to a cake (>50% solids); thus land application can be quite varied. Pasture, cotton, food crops, and forage are the major crops using manure. As with biosolids, manure handling is based on solids content.

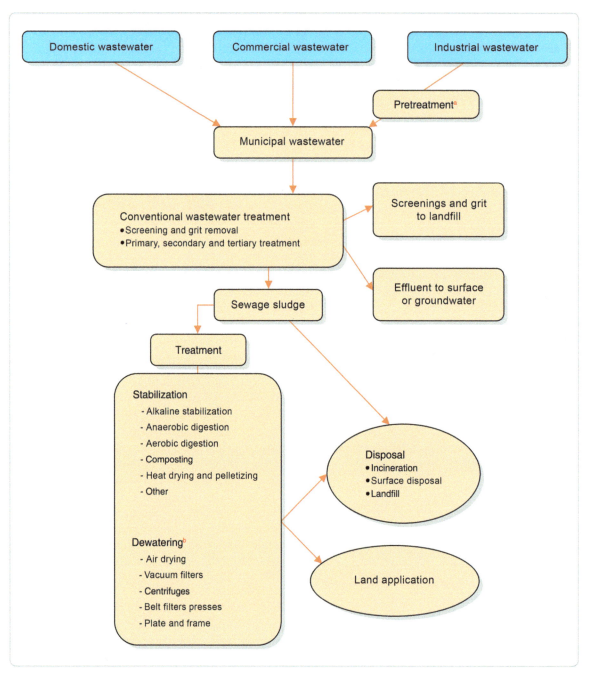

FIG. 23.1   Simplified schematic of biosolids production. *(From NRC (National Research Council): Biosolids applied to land: advancing standards and practices, Washington, DC, 2002, National Academy Press).*

Figs. 23.2–23.4, all illustrate methods of land application of animal manures, which can be summarized as:

- Injection. Liquid manure can be injected into soil, often for row crops.
- Subsurface banding. Dried or caked poultry litter can be banded into soil, again particularly useful for row crops.
- Slinging. Dry or solid manure can be surface spread onto pasture, hay, or row crops.
- Center pivot or reel-gun irrigation. Low solids liquid manure can be surface spread to pasture or hay lands.
- Surface deposition. Manure/feces are deposited on pasture lands during typical grazing periods.

FIG. 23.2 Liquid manure injection: coulters cut a path in the pasture with injectors applying swine liquid manure effluent below the surface in a furrow. *(From Environmental Microbiology, third ed.)*

FIG. 23.3 Sub-surface banding (dark section in soil probe at 2.5″ mark). *(From Environmental Microbiology, third ed.)*

FIG. 23.4 Liquid manure application using a center pivot system and a reel gun. *(From Environmental Microbiology, third ed.)*

23.3 BENEFITS OF LAND APPLICATION OF ORGANIC RESIDUALS

23.3.1 Organic Residuals as a Source of Plant Nutrients

Organic residuals including biosolids and manures contain all the elements essential for the growth of higher plants. Nitrogen and phosphorous, in particular, are abundant, making biosolids attractive as a fertilizer source. Nitrogen and phosphorus are typically present at concentrations of 1%–6% on a dry weight basis. Because plants need more nitrogen for growth than phosphorous, when biosolids are applied at a rate to supply sufficient nitrogen, this means that excess phosphorus is applied, which over time can accumulate in the soil.

In addition to nitrogen and phosphorous, residuals provide other nutrients such as Ca, Fe, Mg, K, Na, and Zn, in amounts adequate for crop needs. A typical analysis of a Class B biosolid is presented in Table 23.1.

TABLE 23.1 Analysis of Anaerobically Digested Sludge From Tucson, Arizona

Element	Concentration (Dry Weight Basis)
Metals	(mg/kg^{-1})
Copper	520
Nickel	13
Lead	59
Chromium	29
Cadmium	3.5
Zinc	1900
Silver	4.7
Arsenic	ND^a
Mercury	0.51
Molybdenum	12
Selenium	ND^a
Other Elements	(g100g^{-1})
Phosphorus	3.3
Calcium	3.6
Magnesium	0.45
Sodium	0.4
Organic carbon	16.6

Continued

TABLE 23.1 Analysis of Anaerobically Digested Sludge From Tucson, Arizona—cont'd

Element	Concentration (Dry Weight Basis)
Nitrogen	
Total Kjeldahl N	3.4
inorganic N	0.16
Total solids	2.8

[a]ND = Below detection limits.
Source: I.L. Pepper.

23.3.2 Organic Residuals Impact on Soil Physical and Chemical Properties

Soil organic matter enhances soil structure, through the formation of secondary aggregates (see Chapter 2). This results in increased soil porosity, which facilitate air and water movement through the soil. Continuous cropping of soils leads to degradation of both soil organic matter and hence soil structure. Applications of residuals to soils increase the soil organic content, improving soil structure. Land application improves soil physical properties such as water infiltration and retention, and reduces the soil's susceptibility to erosion. Chemically, land application of biosolids can be beneficial by increasing soil cation exchange capacity (CEC).

23.3.3 Reduced Pollution

In most instances, broad-scale land application of municipal wastes or manures at appropriate loading rates results in far less pollution than if the material had been concentrated by disposal at a single site. When loading rates are controlled, the soil has a chance to transform many waste components into plant-available nutrients. Thus plants are able to take up these nutrients and complete their natural cycle. These include carbon, nitrogen, sulfur, and phosphorus. Other important residuals components that are recycled include micronutrients such as zinc, iron, and copper. However, when found in high concentrations these metals can be toxic to plants. Additionally, the soil environment helps stabilize other potential pollutants found in biosolids such as lead, cadmium, zinc, arsenic, and so on, by sequestration to their solid phases. When this happens the pollutants are less likely to leach into groundwater and are also less likely to be taken up by plants. However, it is important to note that the metal contaminant itself is still retained in the soil and remains a potential source of pollution.

23.4 CHEMICAL HAZARDS OF LAND APPLICATION OF BIOSOLIDS AND MANURES

23.4.1 Nitrates and Phosphates

Biosolids and manures almost always contain large amounts of nitrogen (N) as nitrate (NO_3^-) and ammonium (NH_4^+). In addition, they contain organic forms of nitrogen that are readily transformed to NO_3^- via ammonification and subsequent nitrification. Nitrate ions are very soluble in water and ionic in nature (negatively charged). Therefore excess nitrates leach easily through soils and can reach aquifers (see Chapters 14 and 15). Nitrates are of public concern due to the potential for methemoglobinemia. This disease, also known as blue baby syndrome, can occur in young infants due to the drinking of water high in nitrates (>10ppm). Because of this potential hazard, biosolid land application rates should be matched with crop N requirements. In addition, for irrigated agriculture, irrigation rates should be managed to reduce excess surface runoff or excess subsurface leaching of NO_3^- (see also Chapter 14).

Biosolids also contain large amounts of phosphate (P) as HPO_4^{2-} or $H_2PO_4^-$. Since plants require more N than P, when biosolid loading rates are based on potential plant N uptake, excess P can accumulate in soil. During many years of continuous land application, soil P concentrations become excessive and can lead to eutrophication of estuaries via surface runoff. Interestingly, in the future, biosolid land application rates may have to be matched to crop P requirements, and thus additional N will need to be applied to crops as fertilizer.

23.4.2 Metals and Organics

All residuals have small amounts of essential trace metals needed for plant growth, but they also contain variable amounts of heavy metals. Moreover, even essential trace elements can be present in such high concentrations that they induce toxicities to plants or microorganisms. Biosolids can also contain organic compounds that can adversely affect public health. The U.S. EPA conducted an assessment to evaluate the potential risks associated with heavy metals and organics in biosolids for land application (EPA, 1992). The constituents considered in the assessment are presented in Table 23.2.

Metals of particular concern in biosolids include Zn, Cu, Cd, Ni, Pb, Hg, Mo, and As. The amount of metal contaminants in a particular sludge depends on the amount of industrial inputs into the municipal sewage system. It is therefore illegal to discharge excessive amounts of metal into municipal wastewater lines, and such wastes must often be treated on site, prior to disposal. The Part 503 rules contain provisions limiting the amount of select heavy metals that can be present in biosolids; the biosolid cannot be applied to the land if the concentration of any of the listed

TABLE 23.2 EPA Pollutants Selected for Risk Assessment

Inorganic Chemicals	Organic Chemicals
Arsenic	Aldrin and dieldrin
Cadmium	Benzol[a]pyrene
Chromium	Chlordane
Copper	DDT, DDD, DDE
Lead	Heptachlor
Mercury	Hexachlorobenzene
Molybdenum	Hexachlorobutadiene
Nickel	Lindane
Selenium	N-Nitrosodimethylamine
Zinc	Polychlorinated biphenyls
	Toxaphene
	Trichlorethylene

DDT, 1,1,1-trichloro-2,2-bis(*p*-chlorophenyl)ethane; *DDE*, 1,1-dichloro-2,2-bis(*p*-chlorophenyl)ethylene; *DDD*, 1,1-dichloro-2,2-bis(*p*-chlorphenyl)ethane.
Source: EPA (U.S. Environmental Protection Agency), Technical Support Document for Land Application of Sewage Sludge. Vol. I EPA 822/R–93–001A, 1992.

The soil environment also influences the availability and thus exposure risk of the metals associated with any particular sludge. Sludge-amended soils with high pH have lower plant-available metal concentrations than do sludge-amended low pH soils. This is because the water solubility of most metals increases as pH decreases. Thus one management strategy to reduce metal mobility and toxicity toward plants is to lime soils to a neutral or alkaline pH. The organic matter content of soils also affects metal availability. In general, soils with high organic matter content (>5%) exhibit relatively low metal uptake by plants, as metals are adsorbed and complexed by the polymer-like organic carbon structure of organic matter. However, when low-molecular-weight organic molecules (usually present in the early stages of plant tissue decay) form complexes with metals, their mobility and plant availability can be dramatically increased in the soil environment. Once metals are introduced into soil, their bioavailability and mobility can be manipulated by changing the valence state of the metal or by altering soil factors that influence their solubilities (see Chapter 8). The long-term fate of metals in the environment associated with land application of biosolids is difficult to predict. This is because metal pollutants do not biodegrade and therefore continue to accumulate in the soil environment.

Overall, concern over the potential public health hazard with regard to metals has decreased in the United States due to two major reasons. First, a large amount of research has been conducted on crop uptake of metals, including both food and nonfood crops, as well as the fate of metals in soil.

heavy metals in the sewage sludge exceeds the ceiling concentration for that metal (Table 23.3). There are also restrictions on the annual and cumulative amounts of metals that can accumulate due to land application, termed loading rates (Table 23.3).

TABLE 23.3 Pollutant Concentration Limits and Loading Rates for Land Application in the United States, Dry Weight Basis

Contaminant	Ceiling Concentration Limit (mg/kg)	Cumulative Pollutant Loading Rate Limit (kg/ha)	Pollutant Concentration Limit (mg/kg)	Annual Pollutant Loading Rate (kg/ha-yr)
Arsenic	75	41	41	2.0
Cadmium	85	39	39	1.9
Copper	4300	1500	1500	75
Lead	840	300	300	15
Mercury	57	17	17	0.85
Molybdenum[a]	75	—	—	—
Nickel	420	420	420	21
Selenium	100	100	100	5.0
Zinc	7500	2800	2800	140

[a]*Standards for molybdenum were dropped from the original regulation. Currently, only a ceiling concentration limit is available for molybdenum, and a decision about establishing new pollutant limits for this metal has not been made by the U.S. EPA.*
Source: EPA (U.S. Environmental Protection Agency): The standards for the use or disposal of sewage sludge. Final 40 CFR Part 503 Rules. EPA 822/Z–93/001 U.S. Environmental Protection Agency, 1993.

Contrary to the dire predictions of some scientists, the "time bomb theory" with respect to metals has not come to pass. Essentially, this theory held that since organic matter in soils is known to complex and accumulate metals, a "flood" of metals would be released when the organic material eventually degraded. In fact, an "aging" effect has been observed for metals wherein bioavailability decreases with time (Alexander, 2000). The second factor that has reduced fears with respect to metals is enhanced pretreatment technologies that reduce metal inputs into sewage and hence biosolids. Thus biosolid metal contents have decreased dramatically from the 1980s to the present. Despite this, scientists still have concerns about the long-term effects of metals on the ecosystem due to the fact that metals do not degrade.

As previously mentioned, biosolids can also contain organic compounds that can adversely affect public health. These include trace amounts of pesticides, polyaromatic hydrocarbons (PAHs), plasticizers, volatile organic compounds (hydrocarbons, chlorinated solvents), and various emerging contaminants. In contrast, animal manures typically do not contain industrial sources of organics. However, pesticides are used within CAFOs such as coumaphos, an organo phosphate used in animal dips, which can run off into surface waters. Pharmaceuticals are used in many CAFO operations, and thus these compounds are likely to be present in manure (see later).

Many organic compounds are degraded during wastewater treatment. However, the more refractory and insoluble compounds such as chlorinated hydrocarbons and pharmaceuticals do not degrade during treatment, particularly if they are sorbed to biosolids. As in the case of metals, it is the bioavailability and mobility of these compounds in the soil that determines their ultimate fate and also the potential public health hazard. Biodegradation and mobility are affected by soil type, in particular, soil texture, organic-matter content, pH, and soil moisture content, and critically the type and activity of microorganisms present in the soil. The complexity of each site and the interactions of the many factors determine whether or not these organic compounds degrade or accumulate in soil, and whether or not they have the potential to contaminate aquifers (see Chapters 8, 9, and 15).

The organic compounds considered in the EPA's risk assessment (Table 23.2) were not carried forward to the Part 503 rule regulations, based on the consideration that they were at that time banned or under various use restrictions. It was therefore hypothesized that levels of these compounds in biosolids would be low. Thus there are no specific regulations in the Part 503 rule governing hazardous organic compounds in biosolids for land application. However, the presence of PCBs in sludge is regulated through another federal law, such that no land application is allowed if the concentration is greater than 50 mg/kg. The presence of various emerging organic contaminants in biosolids is a growing concern, as discussed later.

23.5 POTENTIAL MICROBIAL HAZARDS ASSOCIATED WITH CLASS B BIOSOLIDS, ANIMAL MANURES, AND LAND APPLICATION

As described before, Class A biosolids are treated to reduce the presence of pathogens to below detectable levels. Conversely, both Class B biosolids and animal manures are known to contain pathogens including bacteria and protozoan parasites. Biosolids (but typically not animal manures) also contain human pathogenic viruses. Over the past 10–15 years a variety of potential microbial hazards associated with Class B biosolids and to a lesser extent animal manures have been identified. Many of these issues involved the potential for infection from pathogens associated with organic residues. These pathogens of concern are described in Section 23.7 and the risks associated with such pathogens in Chapters 13 and 29. However, in addition to pathogens, other potential hazards have centered on biological but nonpathogenic issues such as antibiotic-resistant bacteria, endotoxin, and prions.

Although manure is known to contain bacterial and parasitic pathogens, there have only been a few instances where human viral pathogens have been found in manure, including hepatitis E virus in pigs and norovirus in cattle. Normally, viral pathogens are exclusive to municipal wastes. Given that manure is generally not treated, bacterial counts tend to be greater in manures than in municipal treated biosolids. In addition to pathogens, manure is known to contain high levels of antibiotic-resistant pathogenic and commensal (i.e., normal, nonpathogenic) bacteria.

23.5.1 Antibiotic-Resistant Bacteria

A major area of concern with the general public has focused on the potential for antibiotic-resistant bacteria that reside in both animal manures and biosolids, due to the potential for subsequent transfer of the resistance to pathogens. The issue of antibiotic resistance in biosolids and manures is addressed in Chapter 26.

23.5.2 Endotoxin

Another issue associated with biosolids and manure is the presence of endotoxin. Endotoxin, or lipopolysaccharide (LPS) derived from the cell wall of gram negative bacteria is a highly immunogenic molecule present ubiquitously in the environment. Biosolids contain large populations of bacteria, and therefore are another potential source of endotoxin. Although most surfaces contain some traces of dust-associated endotoxin, it is primarily of concern as an aerosol, since most human endotoxin ailments are pulmonary associated (Sharif et al., 2004). Exposures to aerosolized endotoxin have been studied due to occupational exposures to cotton dust, composting plants, and feed houses (Castellan et al., 1987). Exposures to levels of endotoxin as little as 0.2 endotoxin unit (EU) per m^3 derived from poultry dust have been found to cause acute pulmonary ailments such as decreases in forced expiratory volume (Donham et al., 2000). Chronic effects such as asthma and chronic bronchitis have been found to be due to exposures of endotoxin from cotton dust as little as 10 EU per m^3 on a daily basis (Olenchock, 2001).

Endotoxin concentrations in a variety of environmental samples have been investigated, and data show that the endotoxin level in Class B biosolids is similar in magnitude to that of other wastes including animal manures and compost. Since the relevance of this to human health is via inhalation, the potential for aerosolization of endotoxin during land application of biosolids and manure has also caused concern. Brooks et al. (2006) showed that endotoxin values measured during biosolids application were comparable to those found in untreated agricultural soils. Therefore aerosolization of soil particles can result in endotoxin aerosolization, regardless of whether biosolids are involved. This is not surprising since bacterial concentrations in soil routinely exceed 10^8 per gram, with a majority of bacteria being gram negative. Soil particles containing sorbed microbes can be aerosolized and hence act as a source of endotoxin (see also Chapter 5).

A number of studies have investigated endotoxin in CAFOs, be it cattle, poultry, or swine (Brooks et al., 2010; Dungan and Leytem, 2009). The majority of studies report endotoxin levels greater than those recommended for farms (Dungan, 2010). However, the majority of endotoxin associated with CAFOs has been confined to open cattle/dairy farms (Dungan, 2010), and swine and poultry interior housing (Dungan, 2010; Brooks et al., 2010). For example, swine barns were found to have mean concentrations of endotoxin of 4385 EU per m^3 (Duchaine et al., 2001), while composting plants ranged from 10 to 400 EU per m^3 (Clark et al., 1983). Endotoxin release from open lot CAFOs (Dungan, 2010) and building exhaust fans (Brooks et al., 2010) has been shown to release endotoxin at levels of ~800 and 100 EU m^{-3}, respectively, with rapid decreases to near background levels just beyond the point source. It can be assumed that, as with municipal biosolids land application, the majority of aerosolized endotoxin will most likely arise from the dry soil surrounding the site; however, some manure, such as dry poultry litter, will be very prone to endotoxin release. Litter endotoxin levels are approximately 1 order of magnitude greater than typical Class B biosolids (Brooks et al., 2007). In all cases with endotoxin, the severity of the exposure is unknown since not all endotoxin is bioactive, and thus not all exposures are equal. Overall, land application of residuals and aerosolized endotoxin remains an area that is poorly understood by environmental microbiologists.

23.5.3 Prions

Prions are infectious proteins that can result in animal or human disease (see also Chapter 13). Transmissible spongiform encephalopathies (TSE) are a group of neurological prion diseases of mammals which in humans include Kuru, Creutzfeldt-Jakob Disease (CJD), sporadic Creutzfeldt-Jakob disease (sp CJD), and variant Creutzfeldt disease (VCJD) (Prusiner, 2004; Miles et al., 2013). Animal diseases such as bovine spongiform encephalopathy (BSE) are of particular concern. Prions have been detected in the environment at low concentrations (Nichols et al., 2009) and could originate from slaughterhouse wastes. Such wastes could reach wastewater treatment plants, and therefore interest has focused on whether or not prions survive wastewater treatment. If prions survived treatment, then they could end up within biosolids, with subsequent potential exposure to animals following land application. Adding to this concern is the fact that prions are reported to be very resistant to extreme physical conditions including irradiation and heat, and chemical treatment including acids, bases, and oxidizing agents (Taylor, 2000).

Originally it was reported that prions were capable of surviving a very common wastewater treatment, namely mesophilic anaerobic digestion (Kirchmayr et al., 2006; Hinckley et al., 2008). However, these studies used an immunoblot method of detection of the prions, which did not distinguish between infectious and noninfectious prions. Since then, Miles et al. (2013) developed an assay that only detects infectious prions. This assay used a standard scrapie cell assay linked to an Enzyme Linked Immuno-spot reaction (ELISPOT) for infectious prion detection. Using this assay and miniature anaerobic digestors (Fig. 23.5) the influence of various wastewater

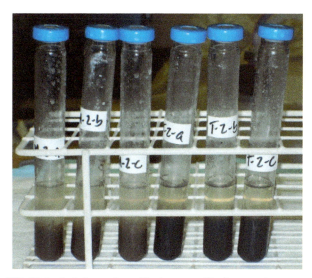

FIG. 23.5 Sealed test tubes utilized as anaerobic digestor microcosms. *(From Environmental Microbiology, third ed.)*

TABLE 23.4 Influence of Wastewater Treatment on Infectious Prion Inactivation

Treatment	Incubation Period	Decrease in Infectious Prions
Mesophilic anaerobic digestion	21 days	4.2 $\log_{10}$
Thermophilic anaerobic digestion	21 days	4.7 $\log_{10}$
Lime treatment of class B biosolids	2 hours	2.9 $\log_{10}$

Adapted from Miles, S.L., Sun, W., Field, J.A., Gerba, C.P., Pepper, I.L., 2013. Survival of infectious prions during wastewater treatment, J. Res. Sci. Technol. 10, 69–85.

treatments on infectious prion inactivation was evaluated (Table 23.4).

These data show a quantifiable reduction of infectious prions in wastewater during the normal period of anaerobic digestion (21 days), at both mesophilic and thermophilic temperatures. In addition, lime treatment of Class B biosolids was shown to be particularly effective in inactivating infectious prions. Overall, the data suggest that prions do not survive wastewater treatment, and that land application of biosolids is not likely to be a significant viable route of human animal exposure to prions.

23.6 PATHOGENS OF CONCERN IN ORGANIC RESIDUALS

Pathogenic bacteria and protozoa are known to reside within both Class B biosolids and animal manures. Pathogenic viruses can also be found in biosolids, but not animal manures. Note also that by definition, Class A biosolids do not contain detectable pathogens. Pathogens routinely associated with either organic residual are presented in Table 23.5.

Whereas human pathogenic viruses are found exclusively in Class B biosolids, concentrations of the bacterial pathogens are normally found in higher concentrations in animal manures than biosolids, most likely due to the fact that manures do not undergo treatment (Brooks et al., 2012a).

Manure is known to contain a wide and varied array of bacterial and parasitic pathogens, and depending on its origin, can be a source of *Campylobacter jejuni*, *Escherichia coli* O157:H7, *Salmonella* spp., *Listeria monocytogenes*, *Cryptosporidium parvum*, and *Giardia lamblia* (Guan and Holley, 2003; Hutchinson et al., 2004; McLaughlin et al., 2009). Two issues associated with pathogens found in residuals are *regrowth* and *reactivation*.

23.6.1 Regrowth and Reactivation of Pathogens Within Organic Residuals

23.6.1.1 Class B Biosolids

Regrowth and reactivation have both been documented as occurring in biosolids, but the two terms are not synonymous (Chen et al., 2011). Reactivation is defined as a large increase in fecal coliform or *E. coli* in biosolids collected immediately after centrifugation or other dewatering process, when compared with the feed into the dewatering equipment (Water Environment Federation (WERF), 2006; Higgins et al., 2007). Regrowth refers to an additional increase in the density of fecal indicators or *E. coli* upon storage of the biosolids of over a period of hours or days.

Reactivation is of concern since studies have documented a large increase in fecal coliforms of several orders of magnitude in a short period of time that would preclude increases due to normal growth that could occur due to binary fission. The phenomenon was first observed by Donald Hendrickson in 2001 (Hendrickson et al., 2004). Because it could not rationally be explained, reactivation was immediately controversial, and had implications for the designation of biosolids as Class A or B. This resulted in numerous studies on the process of reactivation, which indicated that reactivation did occur following dewatering by centrifugation, but not following dewatering with the use of a belt filter press (Erdal et al., 2003). Different hypotheses have been developed to explain the phenomenon of reactivation (Information Box 23.4).

Many studies have also evaluated the potential for growth and/or regrowth of indicators and pathogens in land amended biosolids with either Class A or B material (Jollis, 2006). These studies have resulted in a number of terms being coined to explain the increase in numbers (Information Box 23.5). Studies evaluating the growth and/or regrowth of *Salmonella* and fecal indicators have produced mixed results, some showing increased numbers following land application, and some showing no such increase in numbers. This is most likely due to the different ways in which studies have been conducted, including laboratory studies versus field studies. In addition, some studies have monitored numbers of organisms that survived wastewater treatment and are subsequently introduced into soil via biosolids, as compared to other studies where laboratory strains of organisms have been inoculated into biosolids and the numbers monitored (Zaleski et al., 2005a). Normally it is a thought that monitoring the organisms that survive treatment is the preferred option, with field studies being more "real-world" than laboratory studies. Whereas, the growth and regrowth of fecal indicators in biosolid-amended soil has frequently been noted, corresponding studies showing growth of *Salmonella* are far less frequent (Zaleski et al., 2005b). Also, regrowth of fecal indicators is

TABLE 23.5 Pathogens and Levels (Geometric Mean) Commonly Found in Waste Residuals

	Bovine	Poultry	Swine	Raw Sludge	Class B Biosolids
			CFU, PFU, MPNg^{-1}		
Campylobacter jejuni	~150	~340	~460	~3100	~2.0
E. coli	~170				
Listeria monocytogenes	~600	~180	~210	~2400	~25
Salmonella	~630	~60	~50	~2400	~25
Adenovirus				~130	~40
Enterovirus				~40	~4.0
Norovirus				~2.7×10^5	~1700
Cryptosporidium spp.	~7.0			~30	0.7

Modified from Brooks, J.P., McLaughlin, M.R., Adeli, A., Miles, D.M., 2012. Runoff release of fecal bacterial indicators as influenced by two poultry manure application rates and AlCl$_3$ treatment. J. Water Health 10, 619–628.

INFORMATION BOX 23.4 Main Hypotheses Developed to Explain Reactivation

- Clumping of bacteria when the biosolids were originally assayed, followed by desegregation of clumps into single cells following dewatering
- Formation of viable but nonculturable bacteria (VBNC) during wastewater treatment, and subsequent reactivation of the VBNC due to a signaling substance released into the centrate during centrifugation (Water Environment Federation (WERF), 2006)

To date, the VBNC hypothesis is the most likely explanation for reactivation. Use of quantitative polymerase chain reaction (qPCR) to enumerate *E. coli* showed that copy numbers were not significantly different before and after dewatering, which supports the VBNC concept (Higgins et al., 2007). Reactivation not only potentially affects the designation of biosolids as Class A or B, but also raises the possibility of reactivation of pathogens. Increased numbers of fecal coliforms in Class A biosolids have also been reported (Jollis, 2006).

INFORMATION BOX 23.5 Terms and Definitions Used for Increased Numbers of Pathogens and Indicators.

Growth:	Increase in detectable numbers of a known microbial population over time.
Regrowth:	Increase in numbers after a period of decline in numbers.
Recolonization:	Re-introduction of bacteria into biosolids followed by growth
Reactivation	Large rapid increase in numbers that cannot be ascribed to growth by binary fission.

frequently associated with increased moisture following rainfall events (Pepper et al., 1993).

Regrowth of *Salmonella* in Class A biosolids was observed after rainfall produced saturated conditions (Zaleski et al., 2005a). Subsequently, this was shown to be recolonization following contamination with bird feces, since the *Salmonella* serotypes identified prior to the increase in numbers were different than those identified after the rainfall event. This has implications for the storage of Class A biosolids which should be covered during storage for two reasons: (i) to prevent saturated conditions during rainfall events and (ii) to prevent recolonization by bird or animal feces.

23.6.1.2 Manure

Very few studies have demonstrated the regrowth of either indicator or pathogenic bacteria in manure. Composted manure has been demonstrated to support regrowth of *E. coli* O157:H7, particularly in compost with high moisture levels, above 30%, and low background bacterial counts. In other instances, regrowth of enterococci and *E. coli* has been demonstrated in poultry litter applied land following simulated rainfall (Brooks et al., 2012b) and cow pats in fields (Sinton et al. (2007). As in the situations with compost and biosolids, the driving factor behind regrowth was the presence of readily available organic nutrients and substrates and moisture. Finally, note that when pathogens are introduced into soil, some may adapt and be capable of survival within the soil, but only at the cost of the loss of pathogenicity (Ishii et al., 2006). Similarly, *E. coli* has been shown to lose virulence during manure storage (Duriez et al., 2008).

However, regardless of how low the incidence of infections from pathogens in soil is, people want to know how likely it is that *they* will become infected. To answer this question we can use the process of quantitative microbial risk assessment (see Chapter 29). A critical issue affecting risk is the assessment of exposure to a particular hazard.

23.7 ROUTES OF EXPOSURE

Sections 23.4–23.6 document that manures and biosolids, particularly Class B biosolids, do contain chemical and biological constituents that could result in adverse public health effects if significant exposure to a particular hazard occurs. The potential routes of exposure are identified in Information Box 23.6. Direct exposure could result from ingestion or inhalation of, or dermal contact with, harmful constituents associated with the biosolids. For land application of biosolids, this is limited by land application site restrictions. Indirect routes of exposure to pathogenic microbes or chemicals could be via inhalation of aerosols or ingestion of contaminated groundwater. Exposure can also occur from vectors such as flies and rodents. To prevent this, "vector attraction requirements" are enforced (NRC, 2002). These involve specific biosolid treatments and rapid incorporation of land-applied biosolids.

23.7.1 Direct Exposure Via Contact With Soil Following Land Application

The transport and fate of metals and organic compounds was discussed in several prior chapters. Pathogen survival in and transport through soil are considered together in this section. Human pathogens that are routinely found in domestic sewage sludge include viruses, bacteria, protozoan parasites, and helminths. Of those pathogens, viruses are the smallest and least complex, generally have a short survival in soil, and have the greatest potential for transport in soil. Survival of virus has been shown to be temperature dependent and decreases as temperature increases. Soil type affects virus survival, with longer

survival occurring on clay loam biosolid-amended soils compared with sandy loam biosolid-amended soils. Rapid loss of soil moisture also limits virus survival.

Like virus survival, bacteria survival in soil is affected by temperature, pH, and moisture. Soil nutrient availability also plays a role in bacteria survival. Lower temperatures usually increase survival, as do a neutral soil pH and soil at field capacity. Of the pathogenic bacteria, *Salmonella* and *E. coli* can survive for a long time in biosolid-amended soil—up to 16 months for *Salmonella*. In contrast, *Shigella* has a shorter survival time than either *Salmonella* or *E. coli*. Studies on indicator organisms have shown the total and fecal coliforms as well as fecal streptococci can all survive for weeks to several months depending on soil moisture and temperature conditions (Pepper et al., 1993).

Overall, pathogens from biosolids or manures can survive days, weeks, or even months depending on the specific organisms and environment. Therefore EPA in its 40 CFR 503 Rules (EPA, 1993) introduced site restrictions with durations based on subsequent land use to limit direct exposure to pathogens in soils amended with Class B biosolids. The duration of the site restrictions was intended to allow for die-off of pathogens in the amended soil. These site restrictions are presented in Table 23.6. If the restrictions are followed, EPA concluded that the level of protection from pathogens in Class B biosolids was equal to the level of protection provided by the unregulated use of Class A biosolids. No such site regulations apply to land-applied manures.

23.7.2 Indirect Exposure Via Aerosols

Human exposure to pathogens via air results from the formation of aerosolized biological particles that are referred to as bioaerosols. Until recently, little was known of the risk of infection from bioaerosols generated during land application of biosolids, and this topic was a concern for the efficacy of land application. However, recent national studies across the United States have demonstrated that the risk is far lower than previously thought (Brooks et al., 2005). Similar studies have not been conducted on land-applied manures, but bioaerosols resulting from manures should behave similarly to those resulting from biosolids.

The term biological aerosol is used to describe biological particles which have been aerosolized. These particles may contain microorganisms (bacteria, fungi, viruses) or biological remnants such as endotoxin and cell wall constituents such as peptidoglycan. Bioaerosol sizes range typically from 0.5 to 30 μm in diameter and are typically surrounded by a thin layer of water. In other instances, the biological particles can be associated with particulate matter such as soil or biosolids, depending on the place of origin. Bioaerosol particles in the lower

INFORMATION BOX 23.6 Potential Routes of Exposure to Hazardous Constituents during Land Application of Organic Residuals

Direct Exposure	Indirect Exposure
• Contact with stored biosolids or manures	• Aerosols
• Contact with land applied residuals at the application site	• Groundwater contamination

TABLE 23.6 Minimum Duration Between Application and Harvest/Grazing/Access for Class B Biosolids Applied to the Land

Criteria	Surface	Incorporation	Injection
Food crops whose harvested part may touch the soil/biosolids mixture (beans, melons, squash, etc.)	14 mo	14 mo	14 mo
Food crops whose harvested parts grow in the soil (potatoes, carrots, etc.)	20/38 mo[a]	38 mo	38 mo
Food, feed, and fiber crops (field corn, hay, sweet corn, etc.)	30 d	30 d	30 d
Grazing of animals	30 d	30 d	30 d
Public access restriction			
High potential[b]	1 yr	1 yr	1 yr
Low potential	30 d	30 d	30 d

[a]*The 20-month duration between application and harvesting applies when the biosolids that are surface applied stays on the surface for 4 months or longer prior to incorporation into the soil. The 38-month duration is in effect when the biosolids remain on the surface for less than 4 months prior to incorporation.*
[b]*This includes application to turf farms which place turf on land with a high potential for public exposure.*

Source: Adapted from EPA (U.S. Environmental Protection Agency): The standards for the use or disposal of sewage sludge. Final 40 CFR Part 503 Rules. EPA 822/Z–93/001 U.S. Environmental Protection Agency, 1993

spectrum of sizes (0.5–5 μm) are typically of most concern as these particles are more readily inhaled or swallowed.

Bioaerosols generated from the land application of biosolids or manures may be associated with soil or vegetation. In this situation the soil particle or vegetation is known as a "raft" for the biological particles contained with the aerosol. However, for soil particles to be aerosolized, the particles need to be fairly dry, and low soil moisture contents are known to promote microbial inactivation.

23.7.3 Indirect Exposure Via Groundwater Contamination

Pathogens originally present in biosolids or manures applied to land do have the potential to leach through soil and the vadose zone and contaminate groundwater. However, most soils limit the movement of microbes, and normally significant migration will only occur in coarse-textured soils. The potential to reach groundwater is also influenced by depth to groundwater and the amount of natural or artificial recharge occurring at the site. Viruses have the greatest potential to migrate through soil; however, they have been shown to tightly bind to and within biosolids resulting in little leaching (Chetochine et al., 2006). No direct cause and effect of pathogen groundwater contamination appears to have been identified in groundwaters near land where biosolids have been applied. Note that manures typically do not contain human pathogenic viruses.

23.8 EMERGING ISSUES FOR LAND APPLICATION

23.8.1 Fluorinated Organic Compounds

A specific class of fluorinated organic compounds are referred to as per- and polyfluoroalkyl substances (PFASs) (see Chapter 12). Of these chemicals, perfluorooctanoic acid (PFOA) and perfluorooctane sulfonic acid (PFOS) have been the most extensively produced and studied. PFOA and PFOS have been used extensively as key ingredients in stain-resistant and water-repellant materials such as carpets and clothing, stick-resistant cookware, and in firefighting foams. Because of the extensive use of these materials, PFOA and PFOS have been found in the blood of almost all people tested in the United States. This is a concern since peer-reviewed studies on laboratory animals indicate that high levels of exposure to PFOA and PFOS could result in adverse health effects.

EPA therefore established Drinking Water Health Advisory levels for both compounds. Biosolids are known to contain low levels of PFOA and PFOS, resulting in concern for potential groundwater contamination associated with land application. In particular, States in the northeastern USA such as Maine and New Hampshire view land application as a potential source of groundwater contamination due to shallow depths to groundwater of only a few meters.

In the early 2000s, both PFOS and PFOA were voluntarily phased out of production by major companies in the United States. However, products containing these compounds may still be imported, and there are large legacy sources of the compounds remaining. Thus it is unclear how concentrations of these compounds in biosolids will be affected by the phase out. In addition, there is concern over the fluorinated and partially fluorinated compounds being used as replacements for PFOA and PFOS. Additional research and assessment is needed to evaluate the risks of exposure to PFAS from biosolids, relative to other sources of exposure.

23.8.2   Microplastics

Microplastics are rapidly becoming a global environmental problem (see Chapter 25). These particles are frequently smaller than 5 μm and consist of polyethylene, polypropylene, and other polymers. To date, concern over microplastics has focused on levels in the oceans and their effects on marine life, and microplastics in soils have been largely overlooked. Microplastics are present in numerous personal care and cosmetic products such as lotions, soaps, facial and body scrubs, and toothpaste. Following rinsing down household drains, these microplastics end up in wastewater treatment plants. During wastewater treatment, microplastics accumulate in the solids portion which ultimately becomes biosolids. In contrast, effluents derived from wastewater treatment do not contain significant concentrations of microplastics (Carr et al., 2016). This is good news for the oceans, but perhaps not good news for agricultural soils. Nizzetto et al. (2016) estimated that between 110,000 and 730,000 tons of microplastics are added annually to agricultural soils in Europe and North America via land application. This is greater than the current burden of microplastics in oceans. Importantly, the effects of the microplastics accumulating in soil are unknown, including the effects on soil organisms, farm productivity, and food safety.

23.8.3   Emerging Viruses in Biosolids

In 2014 the World Health Organization reported cases of Ebola Virus Disease (EVD) in rural south eastern Guinea. This later resulted in the West Africa Ebola epidemic, which in turn resulted in eleven travel-associated cases of EVD in the United States. Ebola causes a fatal hemorrhagic fever in humans and is highly contagious even through casual person-to-person contact. Because of this, the news that Ebola-infected people were present in the United States resulted in something close to panic. One question of concern was: "Can the Ebola virus reach wastewater treatment plants via feces from infected individuals, and if so can the virus survive wastewater treatment and end up in biosolids?"

Fortunately studies on the survival of the Ebola virus during thermophilic and mesophilic anaerobic digestion using Ebola surrogates showed that inactivation of the viruses occurred during digestion (Sassi et al., 2018). However this "virus crisis" shows that public perception of biosolids and land application can quickly be swayed by fear of human pathogenic viruses.

23.8.4   Antibiotics and Biocides

In addition to the antibiotic-resistant bacteria and genes released through land application of biosolids and manure, there is also potential for release of antibiotics and other coselecting agents or biocides. Approximately 100,000–200,000 Mg of antibiotics per year are used worldwide in various industries, including healthcare (Franklin et al., 2016). In general, upwards of 70%–90% of antibiotics are passed, unchanged, through the body (both animal and human) and deposited in feces (Massé et al., 2014). It is unknown how much of the antibiotic dose is changed within the body, and how much of this is subsequently released, but what is known is that it varies based on antibiotic and animal combination. For instance, Massé et al. (2014) reviewed the literature and found that bulls release 17%–75% of administered chlortetracycline in feces in an unchanged status, while tylosin in pigs is released at a 40% rate with a portion of that as potent metabolites. Similar releases have been reported in hospital effluent as well (Kummerer, 2003). This indicates that land-applied manure and biosolids could be significant sources of environmental antibiotic pollution.

In addition to antibiotic release, there is also the possibility for the release of other agents selective for antibiotic resistance such as biocides (e.g., fungicides, antiseptics, and metals). The concern with biocides and antibiotic resistance lies in the ability for coselection, meaning that a biocide such as a Cu or triclosan may select for resistance to that biocide, but may also coselect for antibiotic resistance as a side effect, possibly because of mobile genetic elements or plasmids conferring resistance to both agents. The question is "are these changed or unchanged antibiotics and biocides active when in the environment?". This question remains largely unanswered, and it remains unclear to what extent these agents contribute to the global environmental antibiotic burden.

QUESTIONS AND PROBLEMS

1. What is the major component of biosolids?
2. Define the differences between sewage, sewage sludge, Class A and Class B biosolids.
3. Provide estimates of the amount of the major animal manures produced in (a) the United States and (b) China.
4. What are the major hazard differences between land application of Class B biosolids and animal manure?

5. Discuss the statement: "All land application of biosolids should be limited to Class A biosolids."
6. Discuss the statement: "Land application of animal manures should be subject to federal regulations."
7. What factors decrease the potential for contamination of groundwater by pathogens within biosolids?
8. Using the published literature, provide an update on the emerging issue of perfluorinated compounds within biosolids as a source of groundwater contamination.

REFERENCES

Alexander, M., 2000. Aging, bioavailability, and overestimation of risk from environmental pollutants. Environ. Sci. Technol. 34, 4259–4265.

Brooks, J.P., Tanner, B.D., Gerba, C.P., Haas, C.N., Pepper, I.L., 2005. Estimation of bioaerosol risk of infection to residents adjacent to a land applied biosolids site using an empirically derived transport model. J. Appl. Microbiol. 98, 397–405.

Brooks, J.P., Tanner, B.D., Gerba, C.P., Pepper, I.L., 2006. The measurement of aerosolized endotoxin from land application of Class biosolids in Southeast Arizona. Can. J. Microbiol. 52, 150–156.

Brooks, J.P., Rusin, P.A., Maxwell, S.L., Rensing, C., Gerba, C.P., Pepper, I.L., 2007. Occurrence of antibiotic-resistant bacteria and endotoxin associated with the land application of biosolids. Can. J. Microbiol. 53, 1–7.

Brooks, J.P., McLaughlin, M.R., Scheffler, B., Miles, D.M., 2010. Microbial and antibiotic resistant constituents associated with biological poultry litter within a commercial poultry house. Sci. Total Environ. 408, 4770–4777.

Brooks, J.P., Brown, S., Gerba, C.P., King, G.M., O'Connor, G.A., Pepper, I.L., 2011. Land Application of Organic Residuals: Public Health Threat or Environmental Benefit? American Society for Microbiology, Washington, DC.

Brooks, J.P., McLaughlin, M.R., Gerba, C.P., Pepper, I.L., 2012a. Land application of manure and Class B biosolids: an occupational and public quantitative microbial risk assessment. J. Environ. Qual. 41, 2009–2023.

Brooks, J.P., McLaughlin, M.R., Adeli, A., Miles, D.M., 2012b. Runoff release of fecal bacterial indicators as influenced by two poultry manure application rates and AlCl₃ treatment. J. Water Health 10, 619–628.

Burkholder, J., Libra, B., Weyer, P., Heathcote, S., Kolpin, D., Thorne, P.S., Wichman, M., 2007. Impacts of waste from concentrated animal feeding operations on water quality. Environ. Health Perspect. 115, 308–312.

Carr, S.A., Liu, J., Teroso, A.G., 2016. Transport and fate of microplastic particles in wastewater treatment plants. Water Res. 91, 174–182.

Castellan, R.M., Olenchock, S.A., Kinsley, K.B., Hankinson, J.L., 1987. Inhaled endotoxin and decreased spirometric values: an exposure-response relation for cotton dust. New Engl. J. *Med.* 317, 605–610.

Chen, Y.C., Murthy, S.N., Hendrickson, D., Araujo, G., Higgins, M.J., 2011. The effect of digestion and dewatering on sudden increases and regrowth of indicator bacteria after dewatering. Water Environ. Res. 83, 773–783.

Chetochine, A.S., Brusseau, M.L., Gerba, C.P., Pepper, I.L., 2006. Leaching of phage from Class B biosolids and potential transport through soil. Appl. Environ. Microbiol. 72, 665–671.

Clark, C.S., Rylander, R., Larsson, L., 1983. Levels of gram-negative bacterial, *Aspergillus fumigatus*, dust and endotoxin at compost plants. Appl. Environ. Microbiol. 45, 1501–1505.

Donham, K.J., Dumro, D., Reynolds, S.J., Merchant, J.A., 2000. Dose-response relationships between occupational aerosol exposures and cross-shift declines of lung function in poultry workers: recommendations for exposure limits. J. Occup. Environ. Med. 42, 260–269.

Duchaine, C., Thorne, P.S., Merizux, A., Grimard, Y., Whitten, P., Cormier, Y., 2001. Comparison of endotoxin exposure assessment by bioaerosol impinger and filter-sampling methods. Appl. Environ. Microbiol. 67, 2775–2780.

Dungan, R.S., Leytem, A.B., 2009. Airborne endotoxin concentrations at a large open-lot dairy in Southern Idaho. J. Environ. Qual. 38, 1919–1923.

Dungan, R.S., 2010. Fate and transport of bioaerosols associated with livestock operations and manures. J. Anim. Sci. 88, 3693–3706.

Duriez, P., Zhang, Y., Lu, Z., Scott, A., Top, E., 2008. Loss of virulence genes in *Escherichia coli* populations during manure storage on a commercial swine farm. Appl. Environ. Microbiol. 74, 3935–3942.

EPA (U.S. Environmental Protection Agency), 1992. Technical support document for land application of sewage sludge. Vol. I EPA 822/R–93–001A.

EPA (U.S. Environmental Protection Agency), 1993. The standards for the use or disposal of sewage sludge. Final 40 CFR Part 503 Rules. EPA 822/Z–93/001 U.S. Environmental Protection Agency.

EPA (U.S. Environmental Protection Agency), 2004. Risk assessment evaluation for concentrated animal feeding operations. EPA 600/R-04/042.

Erdal, Z.K., Mendenhall, T.C., Neely, S.K., Wagoner, D.L., Quigly, C., 2003. Implementing improvements in a North Caroliner residuals management program. In: Proc. WEF/AWWA/CWEA Joint Residuals and Biosolids Management Conference; Baltimore, MD.

Franklin, A.M., Aga, D.S., Cytryn, E., Durso, L.M., McLain, J.E., Pruden, A., Dungan, R.S., 2016. Antibiotics in agroecosystems: introduction to the special section. J. Environ. Qual. 45 (2), 377. https://doi.org/10.2134/jeq2016.01.0023.

Guan, T.Y., Holley, R.A., 2003. Pathogen survival in swine manure environments and transmission of human enteric illness: a review. J. Environ. Qual. 32, 383–392.

Hendrickson, D., Denard, D., Farrell, J., Higgins, M., Murthy, S., 2004. Reactivation of fecal coliforms after anaerobic digestion and dewatering. In: Proc. Water Environment Federation Annual Biosolids and Residuals Conference; Salt Lake City, Utah. Water Environment Federation: Alexandria, VA.

Higgins, M.J., Chen, Y.C., Murthy, S.N., Hendrickson, D., Farrell, J., Shafer, P., 2007. Reactivation and growth of non-culturable *E.coli* in anaerobically digested biosolids after dewatering. Water Res. 44, 665–673.

Hinckley, G.T., Johnson, C.J., Jacobson, K.H., Bartholomay, C., McMahon, K.D., McKenzie, D., Aiken, J.M., Pedersen, J.A., 2008. Persistence of pathogenic prion protein during simulated wastewater treatment processes. Environ. Sci. Technol. 42, 5254–5259.

Hinesly, T.D., Jones, R.L., Ziegler, E.L., 1972. Effects on corn by application of heated anaerobically digested sludge. Compost Sci. 13, 26–30.

Hutchinson, M.L., Walters, L.D., Avery, S.M., Synge, B.A., Moore, A., 2004. Levels of zoonotic agents in British livestock manures. Lett. Appl. Microbiol. 39, 207–214.

Ishii, S., Ksoll, W.B., Hicks, R.E., Sadowski, M.J., 2006. Presence and growth of naturalized *Escherichia coli* in temperate soils from Lake Superior watersheds. Appl. Environ. Microbiol. 72, 612–621.

IWMI (International Water Management Institute), 2010. Dreschel, P., Scott, C.A., Raschid-Sally, L., Redwood, M., Bahri, A. (Eds.), Wastewater Irrigation and Health: Assessing and Mitigating Risk in Low-Income Countries. Earthscan, London/Sterling, VA.

Jollis, D., 2006. Regrowth of fecal coliforms in Class A biosolids. Water Environ. Res. 78, 442–445.

Kirchmayr, R., Reichl, H.E., Schildorfer, H., Braun, R., Somerville, R.A., 2006. Prion protein: detection in 'spiked' anaerobic sludge and degradation experiments under anaerobic conditions. Water Sci. Technol. 53, 91–98.

Kummerer, K., 2003. Significance of antibiotics in the environment. J. Antimicrob. Chemother. 52 (1), 5–7. https://doi.org/10.1093/jac/dkg293.

Massé, D.I., Saady, N.M.C., Gilbert, Y., 2014. Potential of biological processes to eliminate antibiotics in livestock manure: an overview. Animals 4 (2), 146–163. https://doi.org/10.3390/ani4020146.

McLaughlin, M.R., Brooks, J.P., Adeli, A., 2009. Characterization of selected nutrients and bacteria from anaerobic swine manure lagoons on sow, nursery, and finisher farms in the Mid-South USA. J. Environ. Qual. 38, 2422–2430.

Miles, S.L., Sun, W., Field, J.A., Gerba, C.P., Pepper, I.L., 2013. Survival of infectious prions during wastewater treatment. J. Res. Sci. Technol. 10, 69–85.

NEBRA (North East Biosolids and Residuals Association). (2007). A national biosolids regulation, quality, end use, and disposal survey. Final Report, 20 July 2007.

Nichols, T.A., Pulford, B., Wyckoff, A.C., Meyerett, C., Michel, B., Gertig, K., Hoover, E.A., Jewell, J.E., Telling, G.C., Zabel, M.D., 2009. Detection of protease-resistant cervid prion protein in water from a CWD-endemic area. Prion 3, 171–183.

Nizzetto, L., Bussi, G., Futter, M.N., Butterfield, D., Whitehead, P.G., 2016. A theoretical assessment of microplastic transport in river catchmens and their retention by soils and river sediments. Environ. Sci.: Process Impacts 18 (8), 1050.

NRC (National Research Council), 2002. Biosolids Applied to Land: Advancing Standards and Practices. National Academy Press, Washington, DC.

Olenchock, S.A., 2001. Airborne endotoxin. In: Hurst, C.J., Crawford, R.L., Knudsen, G.R., McInerney, M.J., Stetzenbach, L.D. (Eds.), Manual of Environmental Microbiology, 2nd edition ASM Press, Washington, DC, pp. 814–826.

Organic Trade Association, 2012. Manure Facts. Washington, DC.

Pepper, I.L., Josephson, K.L., Bailey, R.L., Burr, M.D., Gerba, C.P., 1993. Survival of indicator organisms in Sonoran desert soil amended with sewage sludge. J. Environ. Sci. Health A A28 (6), 1287–1302.

Prusiner, S.B. (Ed.), 2004. Prion Biology and Diseases. Cold Spring Harbor Laboratory Press, Woodbury, NY.

Sassi, H.P., Ikner, L.A., Abd-Elmaksoud, S., Gerba, C.P., Pepper, I.L., 2018. Comparative survival of viruses during thermohilic and mesophilic anaerobic digestion. Sci. Total Environ. 615, 15–19.

Sharif El, N., Douwes, J., Hoe, P.H.M., Doekes, G., Nemery, B., 2004. Concentrations of domestic mite and pet allergens and endotoxin in Palestine. Allergy 59, 623–631.

Sinton, L.W.R., Braithwaite, R., Hall, C.H., Mackenzie, M.L., 2007. Survival of indicator and pathogenic bacteria in bovine feces on pasture. Appl. Environ. Microbiol. 73, 7917–17925.

Taylor, D.M., 2000. Inactivation of transmissible degenerative encephalopathy agents: a review. Vet. J. 159, 10–17.

Water Environment Federation (WERF), 2006. Reactivation and regrowth of fecal coliforms in anaerobically digested biosolids. Technical Practice Update Report, Alexandria, VA.

Zaleski, K.J., Josephson, K.L., Gerba, C.P., Pepper, I.L., 2005a. Potential regrowth and recolonization of *Salmonella* and indicators in biosolids and biosolids amended soil. Appl. Environ. Microbiol. 71, 3701–3708.

Zaleski, K.J., Josephson, K.L., Gerba, C.P., Pepper, I.L., 2005b. Survival, growth, and regrowth of enteric indicator and pathogenic bacteria in biosolids, compost, soil, and land applied biosolids. J. Residuals Sci. Technol. 2, 49–63.

FURTHER READING

Bengtsson-Palme, J., Larsson, D.G.J., 2016. Concentrations of antibiotics predicted to select for resistant bacteria: proposed limits for environmental regulation. Environ. Int. 86, 140–149. https://doi.org/10.1016/j.envint.2015.10.015.

EPA (U.S. Environmental Protection Agency), 2000. A guide to field storage of biosolids and the organic by-products used in agriculture and for soil resource management. EPA/832–B–00–007, Washington, DC.

Martinez, J.L., 2009. Environmental pollution by antibiotics and by antibiotic resistance determinants. Environ. Pollut. 157 (11), 2893–2902. https://doi.org/10.1016/j.envpol.2009.05.051.

Neilson, J.W., Josephson, K.L., Pepper, I.L., Arnold, R.B., DiGiovanni, G.D., Sinclair, N.A., 1994. Frequency of horizontal gene transfer of a large catabolic plasmid (pJP4) in Soil. Appl. Environ. Microbiol. 60, 4053–4058.

Unger, I.M., Goyne, K.W., Kennedy, A.C., Kremer, R.J., 2012. Antibiotic effects on microbial community characteristics in soils under conservation management practices. Soil Sci. Soc. Am. J. https://doi.org/10.2136/sssaj2012.0099.

Chapter 24

Drinking Water Treatment

C.P. Gerba and I.L. Pepper

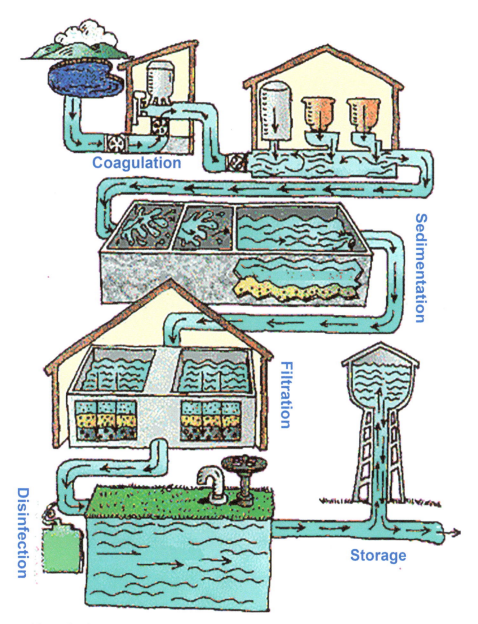

Coagulation

Sedimentation

Filtration

Disinfection

Storage

From U.S. Environmental Protection Agency.

Environmental and Pollution Science. https://doi.org/10.1016/B978-0-12-814719-1.00024-0

435

Rivers, streams, lakes, and aquifers are all potential sources of potable water. In the United States, all water obtained from surface sources must be filtered and disinfected to protect against the threat of microbiological contaminants. Such treatment of surface waters also improves values such as taste, color, and odor. In addition, groundwater under the direct influence of surface waters such as nearby rivers must be treated as if it were a surface water supply. In many cases however, groundwater needs either no treatment or only disinfection before use as drinking water. This is because soil itself acts as a filter to remove pathogenic microorganisms, decreasing their chances of contaminating drinking water supplies.

At first, slow sand filtration was the only means employed for purifying public water supplies. Then, when Louis Pasteur and Robert Koch developed the Germ Theory of Disease in the 1870s, things began to change quickly. In 1881, Koch demonstrated in the laboratory that chlorine could kill bacteria. Following an outbreak of typhoid fever in London, continuous chlorination of a public water supply was used for the first time in 1905 (Montgomery and Consulting Engineers Inc., 1985). The regular use of disinfection in the United States began in Chicago in 1908. The application of modern water treatment processes had a major impact on water-transmitted diseases such as typhoid in the United States. The following sections describe conventional water treatment that is practiced in the public sector (e.g., municipal water supplies).

24.1 WATER TREATMENT PROCESSES

Modern water treatment processes provide barriers, or lines of defense, between the consumer and waterborne disease. These barriers, when implemented as a succession of treatment processes, are known collectively as a *treatment process train* (Fig. 24.1). The simplest treatment process train, known as *chlorination*, consists of a single treatment process, disinfection by chlorination (Fig. 24.1A). The treatment process train known as *filtration*, entails chlorination followed by filtration through sand or coal, which removes particulate matter from the water and reduces turbidity (Fig. 24.1B). At the next level of treatment, *in-line*

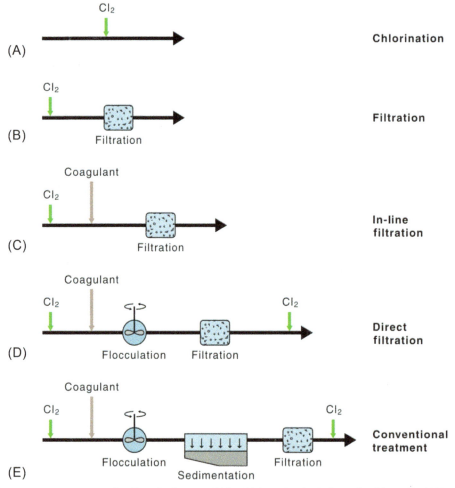

FIG. 24.1 Typical water treatment process trains. From Environmental Microbiology, Academic Press, San Diego, CA, 2000.

filtration, a coagulant is added prior to filtration (Fig. 24.1C). Coagulation alters the physical and chemical state of dissolved and suspended solids and facilitates their removal by filtration. More conservative water treatment plants add a flocculation (stirring) step before filtration, which enhances the agglomeration of particles and further improves the removal efficiency in a treatment process train called *direct filtration* (Fig. 24.1D). In direct filtration, disinfection is enhanced by adding chlorine (or an alternative disinfectant, such as chlorine dioxide or ozone) at both the beginning and end of the process train. The most common treatment process train for surface water supplies, known as conventional treatment, consists of disinfection, coagulation, flocculation, sedimentation, filtration, and disinfection (Fig. 24.1E).

As already mentioned, *coagulation* involves the addition of chemicals to facilitate the removal of dissolved and suspended solids by sedimentation and filtration. The most common primary coagulants are hydrolyzing metal salts, most notably alum [$Al_2(SO_4)_3\ 14H_2O$], ferric sulfate [$Fe_2(SO_4)_3$], and ferric chloride ($FeCl_3$). Additional chemicals that may be added to enhance coagulation are charged organic molecules called polyelectrolytes; these include high-molecular-weight polyacrylamides, dimethyldiallyl ammonium chloride, polyamines, and starch. These chemicals ensure the aggregation of the suspended solids during the next treatment step, flocculation. Sometimes polyelectrolytes (usually polyacrylamides) are added after flocculation and sedimentation as an aid in the filtration step. Coagulation can also remove dissolved organic and inorganic compounds. Hydrolyzing metal salts added to the water may react with the organic matter to form a precipitate, or they may form aluminum hydroxide or ferric hydroxide floc particles on which the organic molecules adsorb. The organic substances are then removed by sedimentation and filtration, or filtration alone if direct filtration or in-line filtration is used. Adsorption and precipitation also remove inorganic substances.

Flocculation is a purely physical process in which the treated water is gently stirred to increase interparticle collisions, thus promoting the formation of large particles. After adequate flocculation, most of the aggregates settle out during the 1–2h of sedimentation. Microorganisms are entrapped or adsorbed to the suspended particles and removed during sedimentation.

Sedimentation is another purely physical process, involving the gravitational settling of suspended particles that are denser than water. The resulting effluent is then subjected to rapid filtration to separate out solids that are still suspended in the water. Rapid filters typically consist of 50–75cm of sand and/or anthracite having a diameter between 0.5 and 1.0mm (Fig. 24.2). Particles are removed as water is filtered through the medium at rates of 4–24L/min/10dm^2. Filters need to be backwashed on a regular basis to remove the buildup of suspended matter. This

FIG. 24.2 Drinking water treatment plant showing sand filter beds in the foreground and tanks containing alum flocculant in the background. Photo courtesy C.P. Gerba.

backwash water may also contain significant concentrations of pathogens removed by the filtration process. Rapid filtration is commonly used in the United States. Another method, slow sand filtration, is also used. Employed primarily in the United Kingdom and Europe, this method operates at low filtration rates without the use of coagulation. Slow sand filters contain a layer of sand (60–120cm deep) supported by a gravel layer (30–50cm deep). The hydraulic loading rate is between 0.04 and 0.4m/h. The buildup of a biologically active layer, called a schmutzdecke, occurs during the operation of a slow sand filter. This eventually leads to head loss across the filter, requiring removing or scraping the top layer of sand. Factors that influence pathogen removal by filtration are presented in Table 24.1.

Taken together, coagulation, flocculation, sedimentation, and filtration effectively remove many contaminants as presented in Table 24.2. Equally important, they reduce turbidity, yielding water of good clarity and hence enhanced disinfection efficiency. If not removed by such methods, particles may harbor microorganisms and make final disinfection more difficult. Filtration is an especially

TABLE 24.1 Factors Effecting the Removal of Pathogens by Slow Sand Filters

Factor		Removal
↑	Temperature	↑
↓	Sand grain size	↑
↑	Filter depth	↑
↓	Flow rate	↑
↑	Well-developed biofilm layer	↑

From Environmental Microbiology, Academic Press, San Diego, CA, 2000.

TABLE 24.2 Coagulation, Sedimentation, Filtration: Typical Removal Efficiencies and Effluent Quality

Organisms (% Removal)	Coagulation and Sedimentation (% Removal)	Rapid Filtration (% Removal)	Slow Sand Filtration (% Removal)
Total coliforms	74–97	50–98	>99.999
Fecal coliforms	76–83	50–98	>99.999
Enteric viruses	88–95	10–99	>99.999
Giardia	58–99	97–99.9	>99
Cryptosporidium	90	99–99	99

Adapted from Pollution Science, Academic Press, San Diego, CA, 1996.

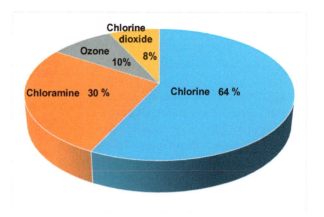

FIG. 24.3   Chemical methods of disinfection in the United States, 2007. *(Raw data from: Committee Report: Disinfection Survey, Part 1—Recent changes, current practices, and water quality. Journal of the American Water Works Association, October 2008.)*

important barrier in the removal of the protozoan parasites *Giardia lamblia* and *Cryptosporidium*. The cysts and oocysts of these organisms are very resistant to inactivation by disinfectants, so disinfection alone cannot be relied on to prevent waterborne illness. Because of their larger size, *Giardia* and *Cryptosporidium* are removed effectively by filtration. Conversely, because of their smaller size, viruses and bacteria can pass through the filtration process. Removal of viruses by filtration and coagulation depends on their attachment to particles (adsorption), which is dependent on the surface charge of the virus. This is related to the isoelectric point (the pH at which the virus has no net surface charge) and is both strain and type dependent. The variations in surface properties explain why different types of viruses are removed with different efficiencies by coagulation and filtration. Thus disinfection remains the ultimate barrier to these microorganisms.

24.2   DISINFECTION

Disinfection plays a critical role in the removal of pathogenic microorganisms from drinking water. The proper application of disinfectants is critical to kill pathogenic organisms.

Generally, disinfection is accomplished through the addition of an oxidant. Chlorine is by far the most common disinfectant used to treat drinking water, but other oxidants, such as chloramines, chlorine dioxide, and even ozone, are also used (Fig. 24.3).

Inactivation of microorganisms is a gradual process that involves a series of physicochemical and biochemical steps. In an effort to predict the outcome of disinfection, various models have been developed on the basis of experimental data. The principal disinfection theory used today is still the *Chick–Watson Model*, which expresses the rate of inactivation of microorganisms as a first-order chemical reaction.

$$N_t/N_o = e^{-kt} \qquad (24.1)$$

or

$$\ln N_t/N_o = -kt \qquad (24.2)$$

where N_o = number of microorganisms at time 0, N_t = number of microorganisms at time t, k = decay constant (1/time) and t = time.

The logarithm of the survival rate (N_t/N_o) plots as a straight line versus time (Fig. 24.4). Unfortunately,

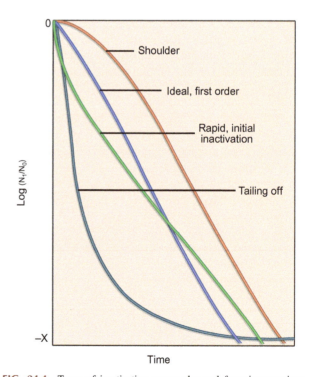

FIG. 24.4   Types of inactivation curves observed for microorganisms. *(From Environmental Microbiology, Academic Press, San Diego, CA, 2000.)*

laboratory and field data often deviate from first-order kinetics. Shoulder curves may result from clumps of organisms or multiple hits of critical sites before inactivation. Curves of this type are common in disinfection of coliform bacteria by chloramines (Montgomery and Consulting Engineers Inc., 1985). The tailing-off curve, often seen with many disinfectants, may be explained by the survival of a resistant subpopulation as a result of protection by interfering substances (suspended matter in water), clumping, or genetically conferred resistance.

In water applications, disinfectant effectiveness can be expressed as $C \cdot t$, where C is the disinfectant concentration and t is the time required to inactivate a certain percentage of the population under specific conditions (pH and temperature)

Typically, a level of 99% inactivation is used when comparing $C \cdot t$ values. In general, the lower the $C \cdot t$ value, the more effective the disinfectant. The $C \cdot t$ method allows a general comparison of the effectiveness of various disinfectants on different microbial agents (Tables 24.3–24.6). It is used by the drinking water industry to determine how much disinfectant must be applied during treatment to achieve a given reduction in pathogenic microorganisms.

TABLE 24.3 $C \cdot t$ Values for Chlorine Inactivation of Microorganisms in Water (99% Inactivation)[a]

Organism	°C	pH	$C \cdot t$
Bacteria			
E. coli	5	6.0	0.04
E. coli	23	10.0	0.6
L. pneumophila	20	7.7	1.1
Mycobacterium avium	23	7.0	51–204
Viruses			
Polio 1	5	6.0	1.7
Coxsackie B5	5	8.0	9.5
Protozoa			
G. lamblia cysts	5	6.0	54–87
G. lamblia cysts	5	7.0	83–133
G. lamblia cysts	5	8.0	119–192
Cryptosporidium oocysts	25	7.0	9740–11,300

[a]In buffered distilled water.

From Sobsey, M.D., 1989. Inactivation of health-related microorganisms in water by disinfection processes, Water Sci. Technol. 21, 179–195; Rose, J.B., Lisle, J.T., Lechevallier, M., 1997. Cryptosporidium and Cryptosporidiosis. CRC Press, Boca Raton, FL, pp. 93–109; Gerba, C.P., Nwachuko, N., Riley, K.R., 2004. Disinfection resistance of waterborne pathogens on the United States environmental protection agency's contaminant candidate list (CCL). J. Water Supply Res. Technol. AQUA 52, 81–94.

$C \cdot t$ values for chlorine for a variety of pathogenic microorganisms are presented in Table 24.3. The order of resistance to chlorine and most other disinfectants used to treat water is protozoan cysts > viruses > vegetative bacteria. To obtain the proper $C \cdot t$, contact chambers (Fig. 24.5) are used to retain the water in channels before entering the drinking water distribution system or sewage discharge.

24.3 FACTORS AFFECTING DISINFECTANTS

Numerous factors determine the effectiveness and/or rate of kill of a given microorganism. Temperature has a major effect, because it controls the rate of chemical reactions. Thus as temperature increases, the rate of kill with a chemical disinfectant increases. The pH can affect the ionization of the disinfectant and the viability of the organism. Most waterborne organisms are adversely affected by pH levels below 3 and above 10. In the case of halogens such as chlorine, pH controls the amount of HOCL (hypochlorous acid) and −OCl (hypochlorite) in solution. HOCl is more effective than −OCl in the disinfection of microorganisms. With chlorine, the $C \cdot t$ increases with pH. Attachment of organisms to surfaces or particulate matter in water such as clays and organic detritus aids in the resistance of microorganisms to disinfection. Particulate matter may interfere by either acting chemically to react with the disinfectant, thus neutralizing the action of the disinfectant, or physically shielding the organism from the disinfectant (Stewart and Olson, 1996).

Repeated exposure of bacteria and viruses to chlorine appears to result in selection for greater resistance (Bates et al., 1977; Haas and Morrison, 1981). However, the enhanced resistance has not been great enough to overcome concentrations of chlorine applied in practice.

24.4 HALOGENS
24.4.1 Chlorine

Chlorine and its compounds are the most commonly used disinfectants for treating drinking and wastewater (Fig. 24.6). Chlorine is a strong oxidizing agent that, when added as a gas to water, forms a mixture of hypochlorous acid (HOCl) and hydrochloric acids.

$$Cl_2 + H_2O \leftrightarrow HOCl + HCl \tag{24.3}$$

In dilute solutions, little Cl_2 exists in solution. The disinfectant's action is associated with the HOCl formed. Hypochlorous acid dissociates as follows:

TABLE 24.4 $C \cdot t$ Values for Chlorine Dioxide in Water

Microbe	ClO$_2$ Residual (mg/L)	Temperature (°C)	pH	% Reduction	$C \cdot t$
Bacteria					
E. coli	0.3–0.8	5	7.0	99	0.48
Mycobacterium	0.1–0.2	23	7.0	99.9	0.1–11
Viruses					
Polio 1	0.4–14.3	5	7.0	99	0.2–6.7
Rotavirus SA11 Dispersed	0.5–1.0	5	6.0	99	0.2–0.3
Cell-associated	0.45–1.0	5	6.0	99	1.0–2.1
Hepatitis A	0.14–0.23	5	6.0	99	1.7
Protozoa					
G. muris	0.1–5.55	5	7.0	99	10.7
G. muris	0.26–1.2	25	5.0	99	5.8
G. muris	0.15–0.81	25	9.0	99	2.7
Cryptosporidium	4.03	10	7.0	95.8	7.8

Adapted from Sobsey, M.D., 1989. Inactivation of health-related microorganisms in water by disinfection processes, Water Sci Technol 21, 179–195; Rose, J.B., Lisle, J.T., Lechevallier, M., 1997. Cryptosporidium and Cryptosporidiosis, CRC Press, Boca Raton, FL, pp. 93–109. From Environmental Microbiology, Academic Press, San Diego, CA, 2000.

TABLE 24.5 $C \cdot t$ Values for Chloramines in Water (99% Inactivation)[a]

Microbe	°C	pH	$C \cdot t$
Bacteria			
E. coli	5	9.0	113
Viruses			
Polio 1	5	9.0	1420
Hepatitis A	5	8.0	592
Coliphage MS2	5	8.0	2100
Rotavirus SA-11, dispersed	5	8.0	4034
Rotavirus SA-11, cell-associated	5	8.0	6124
Protozoa			
G. muris	3	6.5–7.5	430–580
G. muris	5	7.0	1400
Cryptosporidium	25	7.0	>7200

[a]In buffered distilled water.

Adapted from Sobsey, M.D., 1989. Inactivation of health-related microorganisms in water by disinfection processes. Water Sci. Technol. 21, 179–195; Rose, J.B., Lisle, J.T., Lechevallier, M., 1997. Cryptosporidium and Cryptosporidiosis, CRC Press, Boca Raton, FL, pp. 93–109; Gerba, C. P., Nwachuko, N., Riley, K.R., 2004. Disinfection resistance of waterborne pathogens on the United States environmental protection agency's contaminant candidate list (CCL). J, Water Supply Res. Technol. AQUA 52, 81–94. From Environmental Microbiology, Academic Press, San Diego, CA, 2000.

TABLE 24.6 $C \cdot t$ Values for Ozone Inactivation of Microorganisms in Water (99% Inactivation)

Organism	°C	pH	$C \cdot t$
Bacteria			
E. coli	23	7.2	0.006–0.02
Mycobacterium	1	7.0	0.10–0.12
Viruses			
Polio 1	5	7.2	0.2
Polio 2	25	7.2	0.72
Rota SA11	4	6.0–8.0	0.019–0.064
Protozoa			
G. lamblia	5	7.0	0.53
Cryptosporidium	7	–	7.0
Cryptosporidium	22	–	3.5

From Sobsey, M.D., 1989. Inactivation of health-related microorganisms in water by disinfection processes. Water Sci. Technol. 21, 179–195; Rose, J. B., Lisle, J.T., Lechevallier, M., 1997. Cryptosporidium and Cryptosporidiosis, Boca Raton, FL. From Environmental Microbiology. Academic Press, San Diego, CA, 2000.

FIG. 24.5 Chlorine contact chambers at a sewage treatment plant.

FIG. 24.6 Chlorine storage tanks at a wastewater treatment plant.

$$HOCl \leftrightarrow H^+ + OCl^- \qquad (24.4)$$

The preparation of hypochlorous acid and OCl^- (hypochlorite ion) depends on the pH of the water (Fig. 24.7). The amount of HOCl is greater at neutral and lower pH levels, resulting in greater disinfection ability of chlorine at these pH levels. Chlorine as HOCl or OCl^- is defined as *free available chlorine*. HOCl combines with ammonia and organic compounds to form what is referred to as *combined chlorine*. The reactions of chlorine with ammonia and nitrogen-containing organic substances are of great importance in water disinfection. These reactions result in the formation of monochloramine, dichloramine, trichloramine, and so on.

Monochloramine:

$$NH_3 + HOCl \leftrightarrow NH_2Cl + H_2O \qquad (24.5)$$

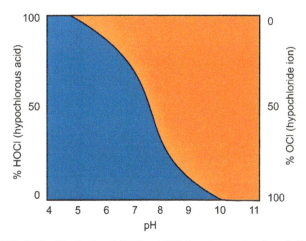

FIG. 24.7 Distribution of HOCl and OCl^- in water as a function of pH. *(From Bitton, G., 2000. Wastewater Microbiology, Wiley-Liss, New York; Environmental Microbiology, Academic Press, San Diego, CA, 2000.)*

Dichloramine:

$$NH_2Cl + HOCl \leftrightarrow NHCl_2 + H_2O \qquad (24.6)$$

Trichloramine:

$$NHCl_2 + HOCl \leftrightarrow NCl_3 + H_2O \qquad (24.7)$$

Such products retain some of the disinfecting power of hypochlorous acid but are much less effective at a given concentration than chlorine.

Free chlorine is quite efficient in inactivating pathogenic microorganisms. In drinking water treatment, 1 mg/L or less for about 30 min is generally sufficient to significantly reduce bacterial numbers. The presence of interfering substances in wastewater reduces the disinfection efficacy of chlorine, and relatively high concentrations of chlorine (20–40 mg/L) are required (Bitton, 2011). Enteric viruses and protozoan parasites are more resistant to chlorine than bacteria and can be found in secondary wastewater effluents after normal disinfection practices. *Cryptosporidium* is extremely resistant to chlorine. A chlorine concentration of 80 mg/L is necessary to cause 90% inactivation following a 90-min contact time (Korich et al., 1990). Chloramines are much less efficient than free chlorine (about 50 times less efficient) in inactivation of viruses.

Bacterial inactivation by chlorine is primarily caused by impairment of physiological functions associated with the bacterial cell membrane. Chlorine may inactivate viruses by interaction with either the viral capsid proteins or the nucleic acid (Thurman and Gerba, 1988).

24.4.2 Chloramines

Inorganic chloramines are produced by combining chlorine and ammonia (NH_4) for drinking water disinfection.

The species of chloramines formed (see Eqs. 24.5–24.7) depend on a number of factors, including the ratio of chlorine to ammonia-nitrogen, chlorine dose, temperature, and pH. Up to a chlorine-to-ammonia mass ratio of 5, the predominant product formed is monochloramine, which demonstrates greater disinfection capability than other forms, that is, dichloramine and trichloramine. Chloramines are used to disinfect drinking water by some utilities in the United States, but because they are slow acting, they have mainly been used as secondary disinfectants when a residual in the distribution system is desired. For example, when ozone is used to treat drinking water, no residual disinfectant remains. Because bacterial growth may occur after ozonation of tap water, chloramines are added to prevent regrowth in the distribution system. In addition, chloramines have been found to be more effective in controlling biofilm microorganisms on the surfaces of pipes in drinking water distribution systems because they interact with capsular bacterial polysaccharides (LeChevallier et al., 1990).

Because of the occurrence of ammonia in sewage effluents, most of the chlorine added is converted to chloramines. This demand on the chlorine must be met before free chlorine is available for disinfection. As chlorine is added, the residual reaches a peak (formation of mostly monochloramine) and then decreases to a minimum called the *breakpoint* (Fig. 24.8). At the breakpoint, the chloramine is oxidized to nitrogen gas in a complex series of reactions summarized in Eq. (24.8).

$$2NH_3 + 3HOCl \leftrightarrow N_2 + 3H_2O + 3HCL \qquad (24.8)$$

Addition of chlorine beyond the breakpoint ensures the existence of a free available chlorine residual.

24.4.3 Chlorine Dioxide

Chlorine dioxide is an oxidizing agent that is extremely soluble in water (five times more than chlorine) and, unlike chlorine, does not react with ammonia or organic compounds to form trihalomethane, which is potentially

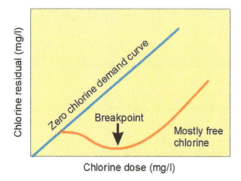

FIG. 24.8 Dose-demand curve for chlorine.

carcinogenic. Therefore it has received attention for use as a drinking water disinfectant. Chlorine dioxide must be generated on site because it cannot be stored. It is generated from the reaction of chlorine gas with sodium chlorite:

$$2NaClO_2 + Cl_2 \leftrightarrow 2ClO_2 + 2NaCl \qquad (24.9)$$

Chlorine dioxide does not hydrolyze in water but exists as a dissolved gas.

Studies have demonstrated that chlorine dioxide is as effective as or more effective in inactivating bacteria and viruses in water than chlorine (Table 24.4). As is the case with chlorine, chlorine dioxide inactivates microorganisms by denaturation of the sulfhydryl groups contained in proteins, inhibition of protein synthesis, denaturation of nucleic acid, and impairment of permeability control (Stewart and Olson, 1996).

24.4.4 Ozone

Ozone (O_3), a powerful oxidizing agent, can be produced by passing an electric discharge through a stream of air or oxygen. Ozone is more expensive than chlorination to apply to drinking water, but it has increased in popularity as a disinfectant because it does not produce trihalomethanes or other chlorinated by-products, which are suspected carcinogens. However, aldehydes and bromates may be produced by ozonation and may have adverse health effects. Because ozone does not leave any residual in water, ozone treatment is usually followed by chlorination or addition of chloramines. This is necessary to prevent regrowth of bacteria because ozone breaks down complex organic compounds present in water into simpler ones that serve as substrates for growth in the water distribution system. The effectiveness of ozone as a disinfectant is not influenced by pH and ammonia.

Ozone is a much more powerful oxidant than chlorine (Tables 24.3 and 24.6). Ozone appears to inactivate bacteria by the same mechanisms as chlorine-based disinfection: by disruption of membrane permeability (Stewart and Olson, 1996), impairment of enzyme function and/or protein integrity by oxidation of sulfhydryl groups, and nucleic acid denaturation. *Cryptosporidium* oocysts can be inactivated by ozone, but a $C \cdot t$ of 1–3 is required. Viral inactivation may proceed by breakup of the capsid proteins into subunits, resulting in release of the RNA, which can subsequently be damaged.

24.4.5 Ultraviolet Light

The use of ultraviolet disinfection of water and wastewater has seen increased popularity because it is not known to produce carcinogenic or toxic by-products, or taste and odor problems. Also, there is no need to handle or store

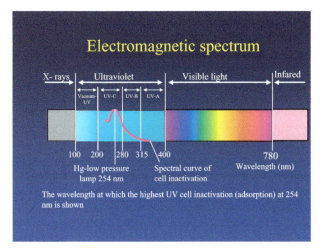

FIG. 24.9 The wavelength at which the highest UV cell inactivation (adsorption) at 254 nm is shown.

toxic chemicals. A wavelength of 254 nm is most effective against microorganisms because this is the wavelength absorbed by nucleic acids (Fig. 24.9). Unfortunately, it has several disadvantages, including higher costs than halogens, no disinfectant residual, difficulty in determining the UV dose, maintenance and cleaning of UV lamps, and potential photoreactivation of some enteric bacteria (Bitton, 2011) (Fig. 24.10). However, advances in UV

technology are providing lower cost, more efficient lamps, and more reliable equipment. These advances have aided in the commercial application of UV for water treatment in the pharmaceutical, cosmetic, beverage, and electronic industries in addition to municipal water and wastewater application.

Microbial inactivation is proportional to the UV dose or intensity, I, which is expressed in microwatt-seconds per square centimeter (μW s/cm^2) or

$$\text{UV dose} = I \cdot t \qquad (24.10)$$

where $I = \mu$W/cm^2, $t =$ exposure time.

In most disinfection studies, it has been observed that the logarithm of the surviving fraction of organisms is nearly linear when it is plotted against the dose, where dose is the product of concentration and time ($C \cdot t$) for chemical disinfectants, or intensity and time ($I \cdot t$) for UV. A further observation is that constant dose yields constant inactivation. This is expressed mathematically in Eq. (24.11).

$$\log N_s/N_i = \text{functional} \,(I\,t) \qquad (24.11)$$

where N_s is the density of surviving organisms (number/cm^3), N_i is the initial density of organisms before exposure (number/cm^3).

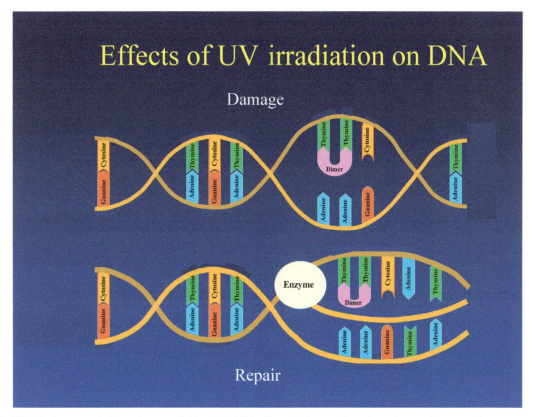

FIG. 24.10 UV light damages cells by causing cross-linking of the DNA of bacteria, but some bacteria produce repair enzymes that can remove the cross-linking of the nucleotides in the DNA.

Because of the logarithmic relationship of microbial inactivation versus UV dose, it is common to describe inactivation in terms of log survival, as expressed in Eq. (24.12). For example, if one organism in 1000 survived exposure to UV, the result would be a –3 log survival, or a 3 log reduction.

$$\text{Log}_{10}\text{ survival} = \log_{10} N_s/N_i \qquad (24.12)$$

Determining the UV susceptibility of various indicator and pathogenic waterborne microorganisms is fundamental in quantifying the UV dose required for adequate water disinfection. Factors that may affect UV dose include cell clumping and shadowing, suspended solids, turbidity, and UV absorption. UV susceptibility experiments described in the literature are often based on the exposure of microorganisms under conditions optimized for UV disinfection. Such conditions include filtration of the microorganisms to yield monodispersed, uniform cell suspensions and the use of buffered water with low turbidity and high transmission at a wavelength of 254 nm. Thus, in reality, higher doses are required to achieve the same amount of microbial inactivation in full-scale flow through operating systems.

The effectiveness of UV light is decreased in wastewater effluents by substances that affect UV transmission in water. These include humic substances, phenolic compounds, lignin sulfonates, and ferric iron. Suspended matter may protect microorganisms from the action of UV light; thus, filtration of wastewater is usually necessary for effective UV light disinfection.

Ultraviolet radiation damages microbial DNA or RNA at a wavelength of approximately 260 nm. It causes thymine dimerization (Fig. 24.11), which blocks nucleic acid replication and effectively inactivates microorganisms. The initial site of UV damage in viruses is the genome, followed by structural damage to the virus protein coat. Viruses with high-molecular-weight double-stranded DNA or RNA are easier to inactivate than those with low-molecular-weight double-stranded genomes. Likewise, viruses with single-stranded nucleic acids of high molecular weight are easier to inactivate than those with single-stranded nucleic acids of low molecular weight. This is presumably because the target density is higher in larger genomes. However, viruses with double-stranded genomes are less susceptible than those with single-stranded genomes because of the ability of the naturally occurring enzymes within the host cell to repair damaged sections of the double-stranded genome, using the nondamaged strand as a template (Roessler and Severein, 1996) (Fig. 24.12).

A minimum dose of 16,000 μW s/cm^2 has been recommended for treating drinking water, as this results in a 99.9% reduction in coliforms and is very effective against the protozoan parasite *Cryptosporidium*. However, this level is not sufficient to inactivate enteric viruses (Table 24.7). Filtration can be applied before UV light disinfection to improve performance (Fig. 24.13).

24.5 DISINFECTION BY-PRODUCTS

All chemical disinfectants produce organic and/or inorganic disinfection by-products (DBPs), which may be carcinogenic or otherwise deleterious. A summary of the types of disinfectant by-products is presented in Table 24.8. The most widely recognized chlorination by-products include chloroform, bromodichloromethane, dibromochloromethane, and bromoform. These compounds are collectively known as the trihalomethanes (THM) (Fig. 24.14), and the term total trihalomethane (TTHM) refers to their combined concentrations. These compounds are formed by the reaction

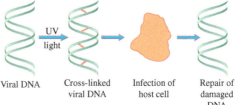

Host cell repair of double stranded
DNA viruses

Viral DNA   Cross-linked   Infection of   Repair of
             viral DNA   host cell   damaged
                                             DNA

FIG. 24.12 Viral repair in double-stranded DNA viruses.

Thymine   Thymine   Thymine dimer

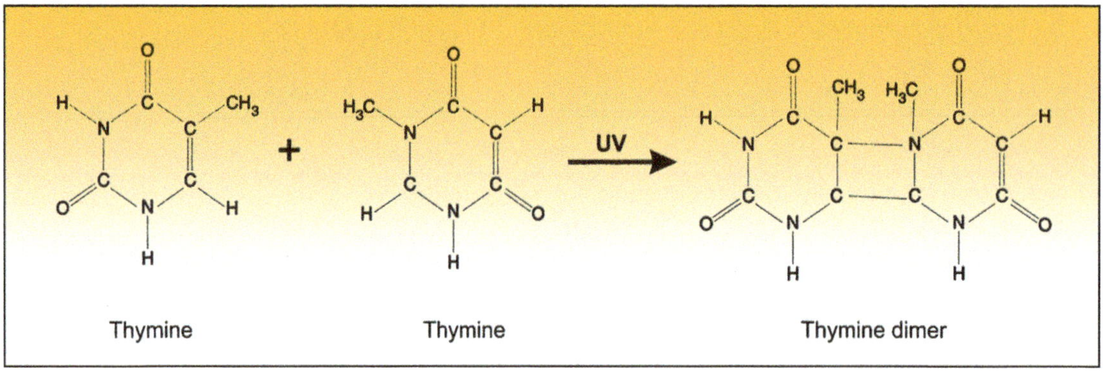

FIG. 24.11 Formation of thymine dimers in the DNA. From Environmental Microbiology, Academic Press, San Diego, CA, 2000.

TABLE 24.7 UV Dose Requirements (mJ/cm^2) for Inactivation of Microorganisms

Target	Log Inactivation							
0.5	0.5	1.0	1.5	2.0	2.5	3.0	3.5	4.0
Protozoa								
Giardia cysts[1]	1.5	2.1	3.0	5.2	7.7	11	15	22
Cryptosporidium oocysts[1]	1.6	2.5	3.9	5.8	8.5	12	15	22
Viruses								
"Viruses"[1]	39	58	79	100	121	143	163	186
Adenovirus type 40[2]		56		111		167		
Poliovirus[2]		7		15		22		30
Adenovirus type 41[3]								112
Hepatitis A[3]								21
Coxsackievirus B5[3]								36
Poliovirus type 1[3]								27
Rotavirus SA11[3]								36
Bacteria								
B subtilus spores[1]		28		39		50		62
E coli[1]		3		4.8		6.7		8.4
Streptococcus faecalis[2]		9		16		23		30
Vibrio cholerae[2]		2		4		7		9
Enterobacter cloacae[3]								10 (33)
Enterocolitica faecium[3]								17 (20)
Campylobacter jejuni[3]								4.6
Clostridium perfringens[3]								23.5
E. coli 0157:H7[3]								6 (25)
E. coli wild type[3]								8.1
Klebsiella pneumoniae[3]								20 (31)
Legionella pneumophila[3]								9.4
Mycobacterium smegmatis[3]								20 (27)
Pseudomonas aeruginosa[3]								11 (19)
Salmonella typhi[3]								8.2
Shigella dysenteriae ATTC29027[3]								3
Streptococcus faecalis								11.2
Vibrio cholerae								2.9 (21)

[1]*USEPA UV Manual 2006.*
[2]*Hijnen WAM, Beerendonk EF and Medema GJ, 2006.*
[3]*Bolton JR and Cotton CA, 2008 - values in brackets include photoreactivation data.*

From Water Treatment Manual: Disinfection. U.S. Environmental Protection Agency, 2011. https://www.epa.ie/pubs/advice/drinkingwater/Disinfection2_web. pdf.

FIG. 24.13 UV light disinfection of drinking water. Available from: www.mindfully.org

of chlorine with organic matter—largely humic acids—naturally present in the water.

Haloacetic acids are another group of by-products produced when chlorine and other disinfectants are used (Fig. 24.14). Monochloramine produces lower THM concentrations than chlorine but produces other DBPs, including cyanogen chloride. The five haloacetic acid constituents are monochloroacetic acid, dichloroacetic acid, trichloroacetic acid, monobromoacetic acid, and dibromoacetic acid, and they are referred to as HAA5.

Ozone oxidizes bromide to produce hypohalous acids, which react with precursors to form brominated THMs.

A range of other DBPs, including aldehydes and carboxylic acids, may be formed. Of particular concern is bromate, formed by oxidation of bromide. Bromate is mutagenic and is carcinogenic in animals.

Many of these by-products are classified as possible carcinogens. However, the results of numerous epidemiological studies of populations consuming chlorinated drinking water in the United States show that the risks of cancer appear low (Craun, 1993). A *maximum contaminant level* (MCL) has been established for TTHM of 80 µg/L and an MCL of 60 µg/L has been established for HAA5. Meanwhile, it is fair to say that the risks of illness and death posed by waterborne microorganisms outweigh the risk from low levels of potentially toxic chemicals produced during water treatment (Craun, 1993).

The formation of THMs during chlorination can be reduced by removing precursors prior to contact with chlorine—for example, by installing or enhancing coagulation (using high coagulant doses), or reducing the amount of organic matter (pretreatment with activated charcoal). UV irradiation is also an alternative to chemical disinfection, but does not provide any residual disinfection. However, lower doses of chemical disinfectants may be added.

24.6 RESIDENTIAL WATER TREATMENT

Conventional water treatment processes are highly effective at removing contaminants from drinking water sources, and waterborne outbreaks are rare in the United States. However, they do still occur (Fig. 24.15 and Table 24.9). Even well-operated, state-of-the-art treatment plants cannot ensure that drinking water delivered at the consumer's tap is entirely free of harmful microbes and chemicals. Numerous studies reporting the presence of disease-causing microbes and toxins in finished water, designated for human consumption, have led to decreased consumer confidence. This has resulted in recommendations from health agencies that certain populations use additional treatment measures for water purification at the point of consumption. *Point-of-use* (POU) residential water purification systems can limit the effects of incidental contamination of drinking

TABLE 24.8 Primary Disinfectant Byproducts

Disinfectant	Byproduct
Chlorine	Trihalomethanes, haloacetic acids
Chloramine	Nitrite
Chlorine dioxide	Chlorite, chlorate
Ozone	Bromate

From Water Treatment Manual: Disinfection. U.S. Environmental Protection Agency, 2011. https://www.epa.ie/pubs/advice/drinkingwater/Disinfection2_web.pdf

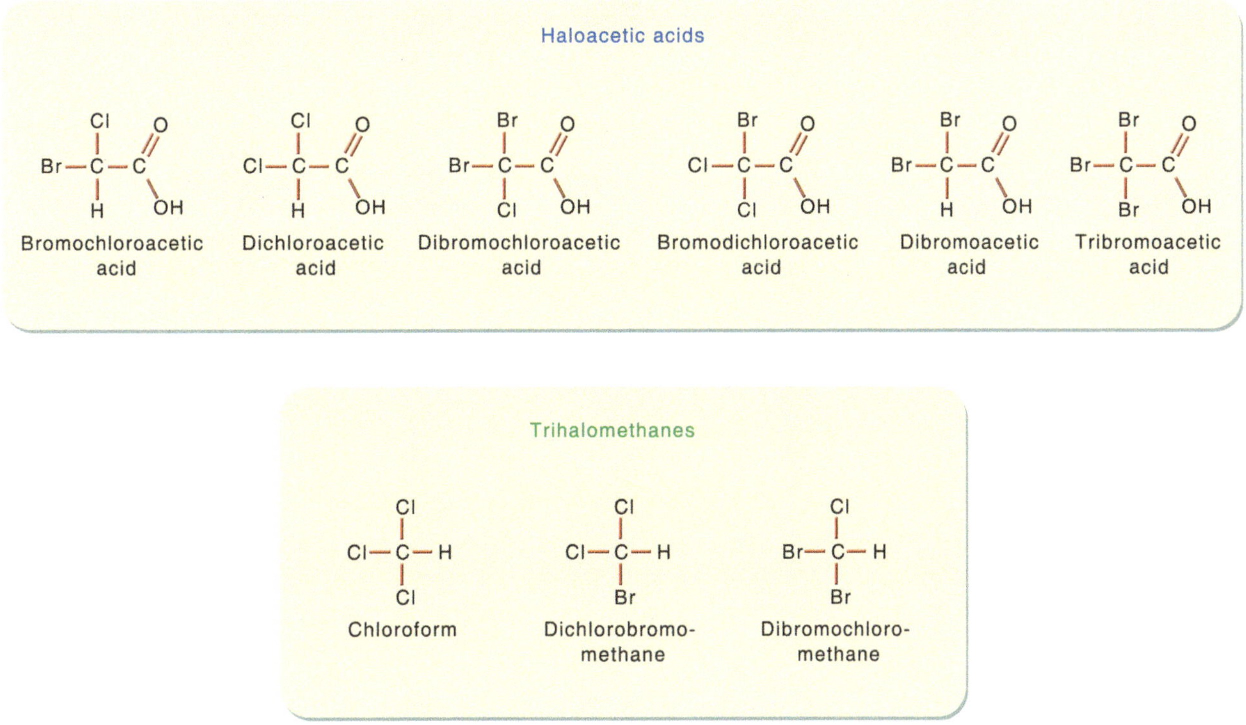

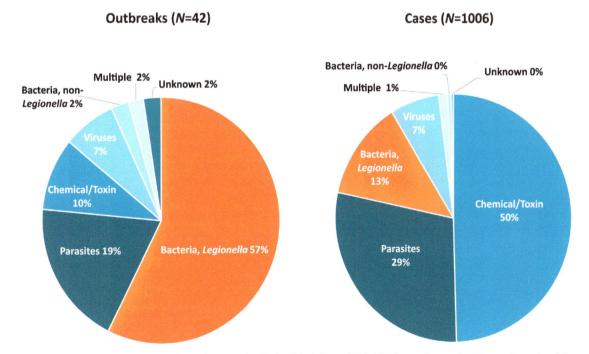

FIG. 24.14 Haloacetic acids and trihalomethanes produced during disinfection with chlorine.

Etiology of drinking water outbreaks and outbreak-related cases — waterborne disease and outbreak surveillance system, 2013–14

FIG. 24.15 Cause of drinking-water related waterborne outbreaks in United States 2013–14. *(From: Centers for Disease Control and Prevention; Benedict KM, Reses H, Vigar M, et al. Surveillance for Waterborne Disease Outbreaks Associated with Drinking Water — United States, 2013–2014. MMWR Morb Mortal Wkly Rep 2017;66:1216–1221. https://doi.org/10.15585/mmwr.mm6644a3.)*

TABLE 24.9 Drinking Water Outbreaks and Water Supply Deficiencies, 1971–2002

Deficiency	Outbreaks			
	Community	Noncommunity	Individual	Total Outbreaks
Deficiency in water treatment	152	133	8	293
Distribution system deficiency	100	25	8	133
Miscellaneous/unknown deficiency	22	33	18	73
Untreated ground water	35	144	60	239
Untreated surface water	6	13	19	38
Total Outbreaks	315	348	113	776

From Calderon, R.L., 2004. Measuring benefits of drinking water technology: "Ten" years of drinking water epidemiology. NEWWA Water Quality Symposium, 20 May 2004. Boxborough, MA.

water due to source water contamination, treatment plant inadequacies, minor intrusions in the distribution system, or deliberate contamination posttreatment (i.e., a bioterrorism event). Multistage POU water treatment systems are available for the removal of a wide variety of contaminants, such as arsenic, chlorine, microbes, and nitrates and have the benefit of providing a final barrier for water treatment closest to the point of consumption (Fig. 24.16).

Current treatment technologies are capable of addressing most contaminants of concern in drinking water; however, the site of application is critical. An appropriate barrier at the point of use can minimize health risks resultant from treatment failures, untreated source waters, and distribution system contamination, and could be a life-saving choice for susceptible populations.

Choosing the appropriate POU device for individual water quality needs is a difficult task, as water is an ever-changing entity, and delivery of contaminated water can occur randomly and without warning. Many POU treatment devices are designed to improve water esthetics, such as taste, color, and odor, but do not remove other harmful contaminants, such as *Cryptosporidium* or viruses (Table 24.10).

Everyone is at risk of waterborne disease, but the immunocompromised are generally at increased risk. It is estimated that up to 25% of the U.S. population is immunocompromised, including the very young (<5 years), the elderly (>55 years), pregnant women, and persons subject to certain medical interventions (radiation treatment, chemotherapy, transplant therapy). In addition, people with previous illnesses (diabetes, cancer), or prior infections (AIDS patients) are also at higher risk. The U.S. EPA and Centers for Disease Control and Prevention (CDC) advise severely immunocompromised individuals to purify their water by boiling for one minute, as a safeguard against waterborne exposures to *Cryptosporidium*. As an alternative, the U. S. EPA and CDC recommend POU devices with reverse-osmosis treatment, labeled as absolute one-micrometer filters, or that have been certified by NSF International under standard 53 for "Cyst Removal." The CDC further recommends that homeowners with individual groundwater wells purchase appropriately designed POU devices (Blackburn et al., 2004).

24.7 WATER SAFETY

The increase in terrorism events over the past few decades has resulted in speculation about the potential for terrorist attacks on public water supplies and water safety in general. Specific concern centered on the potential for deliberate water intrusion events by terrorists. However, contaminants can also be found in water due to accidental intrusion events resulting from pipe breakages and leaks,

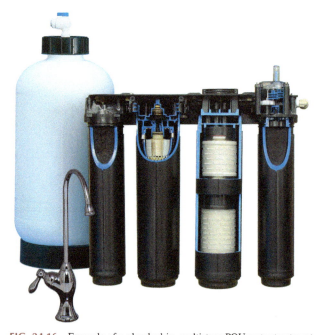

FIG. 24.16 Example of a plumbed-in, multistage POU water treatment device. Photo courtesy of Kinetico, Inc. and Pall Corp, 2004.

TABLE 24.10 Common POU Treatment Options

POU System	Primarily Removes	Advantages	Limitations
Activated carbon	Chlorine, VOCs, pesticides, radon, some metals	Taste and odor improvement, inexpensive	Bacterial regrowth potential
Reverse osmosis	Inorganic chemicals; microbes, nitrates, some organics	Removes wide range of contaminants	Expensive; requires professional maintenance; wastes 2–3x water produced
Distillation	Inorganic chemicals; pathogens, dissolved minerals, trace metals, many organics	Removes wide range of contaminants	Concentrates some organic chemicals, small capacity
Particle filters	Sand, rust	Inexpensive	Not designed for removal of most health-related contaminants
Ion exchange	Softening, iron removal, scale reduction	Improves soap/detergent use	Not designed for removal of most health-related contaminants
Disinfectants (ozone, UV, chlorine, iodine)	Microbes	Simple, inexpensive	Efficacy varies with pathogen and source water quality

or from incremental treatment failure at water utility plants. This in turn resulted in the development of new strategies for monitoring the quality of water delivered to the community. One approach involves implementing hydraulic sensors at discrete locations to monitor the water pressure within distribution lines and the amount of water passing through a particular point in the distribution system. These hydraulic sensors coupled to acoustic sensors are capable of detecting leaks in the distribution system that could impact water quality through intrusion events. A different approach is the use of in-line sensors capable of detecting chemical and microbial contaminants in real time.

24.7.1 Monitoring Community Water Quality

Utilities in the United States are trending toward monitoring community water quality through the development of *Water Quality Information Networks*. These networks are based on the concept of developing water quality zones within the distribution system. Each water quality zone has a set number of dedicated sampling stations and *points of entry* (POEs). POEs are normally the routes of individual sources of water into a given community. The individual sources of water for a community can be as simple as a major river serving a community or a complicated well system such as the one serving Tucson, Arizona. In Tucson, there are many wells supplying groundwater to the community. An additional source of water is supplied by the Central Arizona Project (CAP), which brings Colorado River water to Tucson via a surface water canal. The

Colorado River water is recharged into an aquifer, where it blends into groundwater prior to being pumped to the surface for use. This example illustrates that the Water Quality Information Network for a given community will govern the number of POEs and the necessary number of dedicated sampling stations that provide the appropriate monitoring capabilities. Overall the drinking water is monitored according to State and Federal Regulations and guidelines. In addition, nonregulated contaminants may also be monitored as in the case of emerging contaminants of concern.

24.7.2 Real-Time Sensors

Currently there are many in-line sensors (within distribution lines) for real-time monitoring of chemicals that are likely to be in water that allow for a chemical fingerprint of the water to be established. These include measurements such as pH, free chlorine, total organic carbon, salinity, turbidity, and total oxygen. In theory, if the chemical fingerprint changes due to a water intrusion event or treatment failure, this change in water quality is detected in real time. This real-time trigger event can be followed by near real-time technologies to determine the actual identity of the contaminant. Real-time detection can be defined as detection within 5 min, while near real-time detection requires 1–2 h (Information Box 24.1).

In-line sensors are available for general, organic, and inorganic parameters (Table 24.11). There is also a need for the real-time detection of microbes, but real-time sensors for microorganisms are less well developed. This is important since even a one-time exposure to an

INFORMATION BOX 24.1 New Real-Time Monitoring Approaches

1. Potentially, water monitoring detection systems need to be placed at key multiple stations throughout the water distribution network.
2. Efficient placement of such stations is necessary to allow discrete sections of the network to be monitored individually, and if need be, shut down.
3. A two-stage detection system is envisioned.
   (a) A primary "trigger" mechanism that identifies any addition (chemical or biological) to water within the distribution system. This must be instantaneous or "real-time."
   (b) Secondary detection technology that allows for the contaminant to be identified during a period of a few hours, while the discrete segment of the distribution system is shut down (near real-time).
   (c) Depending on the identification of the contaminant, specific water treatment (for chemicals) or disinfection (for biologicals) options are implemented prior to re-opening the distribution network.

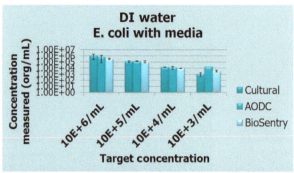

BioSentry results for *E.coli*

BioSentry output coincides well with all concentrations

FIG. 24.17 *E. coli* detected by (A) cultural dilution and plating (24h), (B) acridine orange direct (AODC) staining (6h) and MALS (instantaneous). MALS output coincides well with the other two assays, demonstrating real-time detection.

organism was detected in the water using three different assays. Disadvantages of MALS include the fact that it cannot distinguish between (a) colloidal and biological particles, and (b) viable and nonviable organisms.

24.7.2.2 Real-Time Microbial Detection Via ATP Production

ATP-based technologies rely on measurement of ATP produced by microorganisms that is converted into light, and subsequently into microbial equivalents. Hence detection via measurement of ATP provides a surrogate for the total microbial load in the water. Commercial kits are available to generate the light. Specifically, luciferin is added to the water sample which in the presence of the enzyme luciferase, generates light which can be detected.

$$ATP + O_2 + luciferin = AMP + PPi + oxyluciferin + light \quad (24.13)$$

The light is detected via a luminometer and microbial equivalents are based on one bacterium containing 1 femtogram of ATP.

Cellular ATP (cATP) represents the amount of ATP contained within living cells and is a direct indication of total living biomass quantity.

$$cATP\left(pg\frac{ATP}{mL}\right) = \frac{RLU_{cATP}}{RLU_{ATP1}} \times \frac{10,000\,(pg\,ATP)}{V_{sample}\,(mL)} \quad (24.14)$$

Note: When applicable, subtract RLU_{bg} from RLU_{cATP} prior to executing the above calculation.

To communicate results on the same basis as traditional culture tests, cATP results are converted into Microbial Equivalents (ME's). This is based on the established conversion that 1 *E. coli*-sized bacteria contains 0.001 pg (1 fg) of ATP.

TABLE 24.11 Examples of Available In-Line Sensors

General	Organic	Inorganic
Temperature	Total organic carbon	Chlorine
Conductivity	Dissolved organic carbon	Nitrate
Turbidity	Fluorescence[a]	As, Pb
pH	Ultraviolet[a]	

[a]Surrogate for trace organic compounds.

introduced microbial pathogen can lead to infection. Potential sensors for microbes include Multi Angled Light Scattering (MALS) and ATP-based technologies.

24.7.2.1 Real-Time Microbial Detection Via MALS

MALS relies on detection of biological particulates by impact with laser light illuminating the water. The impact results in light scattering which is dependent on the size and shape of the particle. If the light scattering matches images in a computerized data base, then the particle is recognized as a bacterium or a spore in real time. An example of real-time detection is shown in Fig. 24.17. Here, *E. coli* was spiked into water at different concentrations and the

$$cATP\,(ME/mL) = cATP\left(pg\,\frac{ATP}{mL}\right) \times \frac{1\,ME}{0.001\,pg\,ATP}$$

$$(24.15)$$

Commercial kits are available that are portable allowing for field evaluations of total microbial loads. A further advantage is that the assay takes approximately 2 min to conduct and is therefore essentially real time. Finally note that all viable microbes in the water are accounted for including bacteria, actinomycetes and fungi, and including viable but nonculturable microbes.

24.8 RECLAIMED WATER FOR POTABLE REUSE

Sewage or wastewater treated at wastewater treatment plants result in biosolids and effluent. The effluent can receive additional disinfection and is termed "reclaimed water" which can be beneficially reused.

Recycled or reclaimed water can be used for many different purposes including irrigation, industrial processes, and toilet flushing, but in addition it can be used for potable reuse.

24.8.1 Potable Reuse

Potable reuse refers to the process of augmenting surface or groundwaters with recycled water to aid in water supply sustainability. This is practiced in many parts of the world including the United States, Singapore, Australia, Saudi Arabia, and the United Kingdom (Rock et al., 2015). Unplanned or incidental potable reuse occurs when wastewater is discharged from a wastewater treatment plant into a river and is subsequently used as a drinking water source for a downstream community. For example, this occurs for downstream communities on the Mississippi and Ohio Rivers in the United States. In contrast to this, "planned" potable reuse can be direct or indirect.

24.8.1.1 *Indirect Potable Reuse (IPR)*

Planned Indirect Potable Reuse (IPR) involves the intentional discharge of treated reclaimed water into bodies of water used as potable sources. Prior to discharge, the reclaimed water is subjected to additional advanced treatment (see Section 24.8.2). Normally such discharge occurs upstream of the drinking water treatment plant. Planned reuse indicates that there is an intent to reuse the water for potable use. The point of return can either be into a major water supply reservoir, or a stream feeding a reservoir, or into a supply aquifer (Managed Aquifer Recharge or MAR). In the case of MAR, passage of the water through soil and the vadose zone provides an

Case Study 24.1 The Orange County Sanitation District IPR System

- Known as the Groundwater Replenish System
- Treats 70 mgd
- 35 mgd pumped into injection wells to create a seawater intrusion barrier
- The other 35 mgd pumped into percolation basins close to the city of Anaheim
- Wastewater effluent subjected to additional advanced treatment: microfiltration, reverse osmosis, advanced oxidation via UV and H_2O_2 and finally MAR
- Upon withdrawal from the aquifer the water is subjected to chlorine disinfection prior to distribution to consumers
- Meets potable water needs of 600,000 residents

additional level of treatment that removes contaminants via filtration processes associated with the environmental buffer (see Section 24.8.2). Locations of large planned IPR facilities in the United States include many cities such as San Diego, Tampa, and Denver. The largest implemented IPR system in the United States is in the Orange County Water District in California. Other countries where IPR is practiced include Australia, Singapore, and Namibia in South Africa. Reuse in Orange County is known as the Groundwater Replenishment System (Case Study 24.1).

24.8.1.2 *Direct Potable Reuse (DPR)*

This is the so-called toilet-to-tap concept in which wastewater is subjected to both conventional and advanced treatment, and supplied to consumers through pipe-to-pipe connections without recharge via an environmental buffer. In some counties such as Australia, the definition of DPR has been expanded to include: injection of recycled water directly into the potable water supply distribution system downstream of the water treatment plant, or into the raw water supply immediately upstream of the water treatment plant. Thus injection could be either into a service reservoir or directly into a water pipeline. The key distinction with indirect potable reuse is that there is no separation temporally or spatially between the recycled water introduction and its distribution to consumers.

Abroad the best examples of DPR include Windhoek, Namibia in South Africa and the use of NEWater in Singapore, where 30% of the water supply consists of treated wastewater (Rock et al., 2015). In the United States, Texas led the way in terms of the first DPR facilities in Big Springs (2013) and Wichita Falls (2014). In both cases, wastewater is subjected to advanced treatment and then mixed with additional nonwastewater sources of water. Following conventional drinking water treatment the water is subsequently distributed to consumers. In Big Springs, the mix is 20% recycled water and 80% raw water from a neighboring lake that ultimately supplies 250,000 people. In Wichita Falls, the mix is 50% recycled water and 50% water from a brackish lake, which is distributed to 150,000 consumers.

24.8.2 Advanced Treatment of Recycled Water

Advanced treatment of water normally consists of sequential passage through a multiple barrier treatment train that consists of several technologies. Because these variable technologies use different mechanisms of contaminant removal, the resultant product emanating from the end of the train is highly purified. Reclaimed water routinely contains conventional and emerging chemical and microbial contaminants that must be removed prior to reuse, in order to ensure safe, potable water. The multibarrier treatment trains typically consist of technologies such as microfiltration, reverse osmosis (RO), advanced oxidation, or granular activated carbon (GAC) (Information Box 24.2). For example in the Orange County System (GWRS), the treatment train consists of microfiltration, reverse osmosis, and advanced oxidation prior to discharge into injection wells or percolation basins.

INFORMATION BOX 24.2

Multi-Barrier Treatment Train Technologies

Microfiltration/Ultrafiltration
MF and UF units are commonly referred to as low-pressure membranes and have been shown to be very effective for the filtration of several waterborne microorganisms such as bacteria, viruses, *Cryptosporidium*, *Giardia*, total coliforms, and phages such as MS2 (Jacangelo et al., 1995)

Reverse Osmosis
Reverse osmosis is a water purification technology that uses a semipermeable membrane to remove ions, molecules, and larger particles from water. In reverse osmosis, an applied pressure is used to overcome osmotic pressure. RO membranes are also known as tight membrane or high-pressure membrane processes that are capable of removing trace organics (Kim et al., 2005)

UV–Advanced Oxidation Processes
UV light has been known to be a water disinfectant for several decades. The application of higher doses of UV can also lead to the destruction of chemical contaminants through UV–photolysis (Rosario-Ortiz et al., 2010). Ozone is also a powerful disinfectant that can remove color, turbidity, and inactivate microbes at very low concentrations in water (Zuma et al., 2009). Ozonation has been shown to attenuate most trace organics in wastewater at doses between 1 and 6 mg/

INFORMATION BOX 24.2 —cont'd

L. Addition of hydrogen peroxide to ozone increases the oxidation rates considerably and can be considered when additional attenuation is required (Ternes et al., 2003). Although ozone appears to give excellent removal of a wide variety of trace organics, it can also result in several potentially toxic by-products, like NDMA, bromate, and transformation products (Andrzejewski et al., 2008)

Granular Activated Carbon (GAC)
GAC has been shown to be an effective filtration process for the removal of a variety of compounds of emerging concern (Corwin and Summers, 2012).

INFORMATION BOX 24.3 Recycled Water and the "Yuck" Factor

The concept of reusing water for potable purposes has for many people induced a fear and repugnance now colloquially known as the "yuck factor." The yuck factor was coined by University of Pennsylvania bioethicist Arthur Caplan to describe the instinctive adverse response to the concept of converting wastewater into drinking water. The yuck factor creates such strong feelings that it is difficult to overcome. In fact, the yuck factor creates feelings similar to the fear of eating genetically modified food crops, where opponents of such modified food exploited the gut reaction by calling it "Frankenfood" (Schmidt, 2008). Hence even when presented with scientific facts that document the safety of recycled water, changing the opinions of the public is hard to do. Overall, many studies and projects have evaluated the safety of using recycled water to augment potable sources, and there is no scientific documented adverse effect of such practices on human health.

24.8.3 Public Perception of Portable Reuse

Despite the fact that product water following advanced treatment results in water with a higher water quality than any surface or groundwater, the so-called toilet-to-tap perception associated with DPR has generally caused an adverse public reaction known as the "yuck factor" (Information Box 24.3).

However, public perception generally changes when drought causes extreme water shortages such that DPR is the only resort, as in the case of Big Springs, Texas. Despite that, education and communication efforts are clearly needed to increase transparency of and confidence in potable reuse. Innovative demonstration projects have been developed to enhance public perception, including using advanced-treated reclaimed water for the purpose of brewing beer Case Study 24.2.

Case Study 24.2 Converting Reclaimed Water to Beer

In 2016, the University of Arizona Water and Energy Sustainable Technology Center (WEST) joined a public/private partnership led by Pima County Wastewater to compete for the Arizona Water Innovation Challenge. The team ended up winning the $300,000 prize through the concept of developing a mobile advanced treatment train on the bed of an 18-wheeler (Fig. 24.18). Once built, the rig was driven to various water reclamation facilities in different cities in Arizona, and subsequently utilized to advance treat reclaimed water from the treatment plants to potable water standards. The treated water was then used to brew beer at microbreweries. This concept attracted much local and national media attention, which was used to fulfill the main goal of the project which was to enhance public perception of the use of reclaimed water for potable reuse. Clearly, of paramount importance was the need to ensure that the treated water was safe to drink, which necessitated comprehensive testing for both chemical and microbial contaminants, and close coordination with the Arizona Department of Environmental Quality. Ultimately the semi-truck was driven to various events and festivals and tours of the advanced treatment train were given to provide educational opportunities for the public.

Overall the project was a great success with more than 5000 people touring the rig and multiple national TV media events. Of course, a major reason for the success of the project was FREE BEER! Use of reclaimed water for potable reuse is feasible if managed appropriately and in some cases represents the only possible new source of water in water-scarce regions.

QUESTIONS AND PROBLEMS

1. Which pathogenic microorganisms are the most difficult to remove by conventional water treatment and why?
2. Describe the major steps in the conventional treatment of drinking water.
3. Why are all microorganisms not inactivated according to first-order kinetics?
4. How long would you have to maintain a residual of 1.0 mg/L of free chlorine to obtain a $C \cdot t$ of 18? A $C \cdot t$ of 0.2?
5. Why is chlorine more effective against microorganisms at pH 5.0 than at pH 9.0?
6. What factors interfere with chlorine disinfection? Ultraviolet disinfectant?
7. What is the main site of UV light inactivation in microorganisms? What group of microorganisms are the most resistant to UV light? Why?
8. Why does suspended matter interfere with the disinfection of microorganisms?
9. What are some options for reducing the formation of disinfection by-products formed during drinking water treatment?
10. How much of a $\log_{10}$ reduction and % reduction of *Escherichia coli* will occur with a UV dose of 16,000 μW s/cm^2?
11. Outline how we ensure safe potable water.
12. In what ways are indirect potable reuse and direct potable reuse similar? In what ways to they differ?

REFERENCES

Andrzejewski, P., Kasprzyk-Hordern, B., Nawrocki, J., 2008. N-nitrosodimethylamine (NDMA) formation during ozonation of dimethylamine-containing waters. Water Res. 42, 863–870.

Bates, R.C., Shaffer, P.T.B., Sutherland, S.M., 1977. Development of poliovirus having increased resistance to chlorine inactivation. Appl. Environ. Microbiol. 33, 849–853.

Bitton, G., 2011. Wastewater Microbiology, 4th ed. Wiley-Liss, New York.

Blackburn, B.G., Craun, G.F., Yoder, J.S., Hill, V., Calderon, R.L., Chen, N., Lee, S.H., Levy, D.A., Beach, M.J., 2004. Surveillance for waterborne-disease outbreaks associated with drinking water— United States, 2001–2002. MMWR Surveill. Summ. 53, 23–45.

Corwin, C.J., Summers, R.S., 2012. Controlling Trace Organic Contaminants with GAC Adsorption. JAWWA 104, 43–44.

(A)

(B)

FIG. 24.18 (A) The 18 wheeler rig housing the advanced water-treatment train designed to promote direct potable reuse of wastewater. (B) The advanced water-treatment train consisting of (left to right): microfiltration; reverse osmosis; advanced oxidation; biological activated carbon and chlorine disinfection.

Craun, G.F., 1993. Safety of Water Disinfection: Balancing Chemical and Microbial Risks. ILSI Press, Washington, DC.

Haas, C.N., Morrison, E.C., 1981. Repeated exposure of *Escherichia coli* to free chlorine: production of strains possessing altered sensitivity. Water Air Soil Pollut. 16, 233–242.

Jacangelo, J.G., Adham, S.S., Laine, J.M., 1995. Mechanism of Cryptosporidium, Giardia, and MS2 virus removal by MF and UF. J. Am. Water Works Assoc. 87, 107–121.

Kim, T.U., Amy, G., Drewes, J.E., 2005. Rejection of trace organic compounds by high pressure membranes. Water Sci. Technol. 51, 335–344.

Korich, D.G., Mead, J.R., Madore, M.S., Sinclair, N.A., Sterling, C.R., 1990. Effect of zone, chlorine dioxide, chlorine, and monochloramine on *Cryptosporidium parvum* oocyst viability. Appl. Environ. Microbiol. 56, 1423–1428.

LeChevallier, M.W., Lowry, C.H., Lee, R.G., 1990. Disinfecting biofilm in a model distribution system. J. Am. Water Works Assoc. 82, 85–99.

Montgomery, J.M., Consulting Engineers Inc., 1985. Water Treatment Principles and Design. John Wiley & Sons, New York.

Rock, C., Gerba, C.P., Pepper, I.L., 2015. Recycled water treatment and reuse. In Environmental Microbiology, 3rd Edition. Elsevier Academic Press.

Roessler, P.F., Severein, B.F., 1996. Ultraviolet light disinfection of water and wastewater. In: Hurst, C.J. (Ed.), Modeling Disease Transmission and Its Prevention by Disinfection. Cambridge University Press, Cambridge, England, pp. 313–368.

Rosario-Ortiz, F.L., Wert, E.C., Snyder, S.A., 2010. Evaluation of UV/H_2O_2 Treatment for the oxidation of pharmaceuticals in wastewater. Water Res. 44, 1440–1448.

Stewart, M.H., Olson, B.H., 1996. Bacterial resistance to potable water disinfectants. In: Hurst, C.J. (Ed.), Modeling Disease Transmission and Its Prevention by Disinfection. Cambridge University Press, Cambridge, England, pp. 140–192.

Ternes, T.A., Stuber, J., Herrmann, N., Mcdowell, D., Ried, A., Kampmann, M., Teiser, B., 2003. Ozonation: a tool for removal of pharmaceuticals, contrast media and musk fragrances from wastewater? Water Res. 37, 1976–1982.

Thurman, R.B., Gerba, C.P., 1988. Molecular mechanisms of viral inactivation by water disinfectants. Adv. Appl. Environ Microbiol. 33, 75–105.

Zuma, F., Lin, J., Jonnalagadda, S.B., 2009. Ozone-initiated disinfection kinetics of *Escherichia coli* in water. J. Environ. Sci. Health A 44, 48–56.

FURTHER READING

Calderon, R.L., 2004. Measuring benefits of drinking water technology: "Ten" years of drinking water epidemiology. NEWWA Water Quality Symposium, Boxborough, MA 20 May 2004.

Gerba, C.P., Nwachuko, N., Riley, K.R., 2004. Disinfection resistance of waterborne pathogens on the United States environmental protection agency's contaminant candidate list (CCL). J. Water Supply Res. Technol. AQUA 52, 81–94.

Rose, J.B., Lisle, J.T., Lechevallier, M., 1997. Waterborne *Cryptosporidiosis*: Incidence, outbreaks, and treatment strategies. In: Fayer, R. (Ed.), *Cryptosporidium* and *Cryptosporidiosis*. CRC Press, Boca Raton, FL, pp. 93–109.

Sobsey, M.D., 1989. Inactivation of health-related microorganisms in water by disinfection processes. Water Sci. Technol. 21, 179–195.

Part IV

Global Systems and the Human Dimensions to Environmental Pollution

Chapter 25

Pollution and Environmental Perturbations in the Global System

J. Maximillian, M.L. Brusseau, E.P. Glenn and A.D. Matthias

CREDIT: NASA/NOAA/GSFC/SUOMI NPP/VIIRS/NORMAN KURING

The Earth. *From https://climate.nasa.gov/blog/674/*

25.1 INTRODUCTION

The environmental impacts and pollution processes discussed in the preceding chapters focused primarily on smaller (local) spatial scales, such as point-source pollution originating from a hazardous waste site or non-point-source pollution associated with stormwater runoff from agricultural fields. There are other types of environmental disturbances that occur at much larger scales, encompassing areas from the size of watersheds, to entire regions of a continent, to the entire planet. These disturbances are referred to as *Global Change*. Examples of major global change phenomena of current concern are presented in Information Box 25.1.

It is important to understand that Global Change is not synonymous with global climate change. Many media reports and other public discussions often equate or confuse the two. Global climate change is one example of a global

change phenomenon. A particular significant aspect of global climate change is that it is likely increasing the intensity or rate of other global change phenomena. This will be further discussed in a later section.

The large-scale disturbances presented in Information Box 25.1 are caused by the aggregate and cumulative impacts of large-scale human activities. These include population growth, industrialization, economic development, and urbanization. While the focus of this chapter is on human-induced change, it is recognized that there are also natural sources of global change. The main difference between natural and human-induced global change is in the time scale of events. Natural changes typically take place over geologic time scales—thousands or millions of years. Conversely, human activities are causing disturbances over time scales of decades to a few hundreds of years. This chapter highlights major examples of global change that will challenge the integrity of natural and human systems over the next few decades. Approaches for mitigating, managing, and preventing these changes and their accompanying impacts are discussed in Chapter 32.

25.2 MAJOR GLOBAL CHANGE ISSUES

25.2.1 Extreme Events and Natural Disasters

Extreme events are natural phenomena that generally occur over very short time scales of hours to days to weeks. Extreme events include tropical cyclones (hurricanes, typhoons), tornadoes, floods, heat and cold waves, wildfires, earthquakes, tsunamis, and volcanic eruptions. These events can have significant, sometimes catastrophic direct impacts on the environment and humans. Examples include destruction of buildings and other infrastructure, destruction of habitat, and loss of life. Major extreme events are referred to as natural disasters.

The direct impacts on infrastructure may have secondary impacts that cause environmental pollution. An example of this indirect effect is if the high winds of a hurricane caused oil storage tanks to fail, which then resulted

Environmental and Pollution Science. https://doi.org/10.1016/B978-0-12-814719-1.00025-2

Global Change Issue	Impact
Climate Change	Heat-trapping gases emitted by fossil fuel burning are increasing the earth's surface temperature, with numerous potentially negative impacts on ecosystems and human welfare.
Coral Reef Destruction	Coral reefs are thought to be home to 24% of all marine life, yet 70% of the earth's coral reefs could be lost over the next few decades.
Deforestation	Tropical rainforests are being cleared for agriculture and settlement around the globe, with negative effects on biodiversity. Forest clearing also increases greenhouse warming and increased erosion.
Desertification	Over 70% of the world's semiarid zone has been moderately to severely damaged by overgrazing and unsustainable agricultural practices. Desertification interacts with natural drought cycles to produce starvation and refugee crises in Africa.
Drought	Severe, long-term precipitation deficits combined with population growth and water pollution are limiting the availability of drinking water and water for agriculture for many communities and populations.
Species Extinctions	Poaching, habitat destruction, spread of exotic species, pollution, and global warming are driving many species to extinction. Over 19,000 plant species and 5000 animal species are classified as endangered.
Upper Atmosphere Ozone Depletion	Chlorofluorocarbons (CFCs) injected into the atmosphere from aerosol spray cans and air conditioners lowered ozone levels, allowing damaging ultraviolet light to penetrate to the earth's surface.

in the release of oil into the environment. Another example is illustrated by Hurricane Sandy that hit the eastern seaboard of the United States in 2012. After it hit, many waterways were contaminated by raw sewage from flooded wastewater treatment plants. These pollution-causing impacts may pose a risk to the health of humans and other lifeforms. In this manner, these extreme events can have lasting economic and human health effects.

The global economic impact (dollar loss) of extreme events for the year 2015 is presented in Fig. 25.1 by category. It is observed that total losses summed to more than $100 billion. Also presented are the annual averages over a 10-year period. Severe weather events (cyclones, flooding, other weather-related events) are the predominant types of events contributing to economic loss. Annual human fatalities due to extreme events range from several thousand to a few hundred thousand (Fig. 25.2). From 2000 to 2015, the annual number of major extreme events has averaged almost 300.

25.2.2 Drought

Drought is a deficiency in precipitation over an extended period. The United States loses several $billion annually because of drought. Drought can threaten the availability of drinking water supplies, the health of ecosystems, and the availability and cost of food. The analysis of a drought involves determining the onset and end time of the drought, its duration, areal extent, severity, and frequency in reference to historical records. Studies show that the drought recovery time, the length of the period it takes for the ecosystem to recover from drought events, is getting longer. Furthermore, the interval between drought events is getting shorter. The shorter drought intervals and longer recovery times will likely lead to more severe consequences of future drought events.

One major impact of extended droughts is the reduction in size and volume of lakes. Lakes supply food through direct fishing and irrigation of farmland, supply water for domestic use and energy production, provide habitat for biodiversity, control floods, and provide recreation. Severe droughts, in combination with overuse and pollution, have caused significant depletion of many lakes. About nine lakes no longer exist or have lost over 90% of their water—Lake Urmia, Iran; Lake Waiau, Hawaii; Dead Sea, Israel, the West Bank, and Jordan; Scott Lake, Florida; Aral Sea, Kazakhstan and Uzbekistan; Lake Peigneur, Louisiana; Lake Cachet 2, Chile; Cachuma Lake, California; and Lake Chad, Chad, Cameroon, Niger, and Nigeria (Zielinski, 2014). Lake Chad, shown in Fig. 25.3, is an example of a shrinking lake. This has had major impacts on the ecosystem and human well-being in the region (Information Box 25.2).

25.2.3 Deforestation

Deforestation, the cutting down of forests at a large scale, has occurred since prehistoric times. All of Europe was once densely forested, but the trees were cut down to provide wood and to clear the land for agriculture. The United States experienced a wave of deforestation as European settlers began spreading from east to west across the continent in the 1700s.

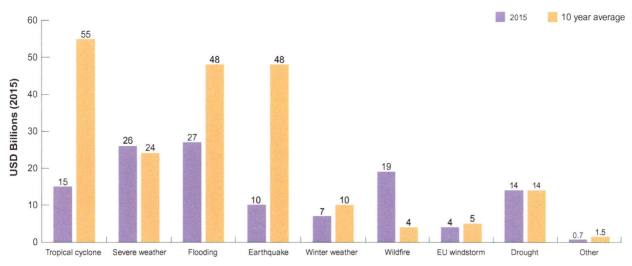

FIG. 25.1 Economic loss due to extreme events in 2015. *(Reproduced with permission from The Aon Center, http://thoughtleadership.aonbenfield.com/Documents/20160113-ab-if-annual-climate-catastrophe-report.pdf.)*

Today, rapid deforestation is taking place in the tropics, where rainforests in South America, Africa, and Asia are being cleared for agriculture and other purposes (Fig. 25.4). At present there are about 2000 million ha of tropical forests in the world, but they are being cleared at a rate of about 14–16 million ha per year, mainly for agriculture. Often, the trees are not used, but are burned or left in piles to decompose. Rapid deforestation is also taking place in the subpolar region of the Northern Hemisphere, which covers about one-tenth of the hemisphere's land surface.

These boreal forests include spruce, cedar, larch, oak, hemlock, fir, pine, and aspen trees, and they provide rich ecosystems for plants and wildlife. Boreal and other northern forests cover 30% of Canada and 45% of Russia. The boreal forests are being clear-cut for lumber and are also susceptible to fire. NASA scientists estimate that Russia's boreal forests are being destroyed at the same rate as tropical rainforests.

The local-scale effects of deforestation were discussed in Chapter 14. Although they only occupy about 10% of the earth's surface area, forests are estimated to support over half of the estimated species of life on earth (up to 80 million). Many of these species can only live in forest environments, so much of this species biodiversity is lost when forests are cleared. Forests also hold an immense amount of carbon. Typically, each ha of forest stores about 180 tons of carbon, most of which is released into the atmosphere as CO_2 when a forest is logged or burned. Forest clearing is estimated to release $1.6\,gt\,yr^{-1}$ of carbon to the atmosphere (compared to $6.2\,gt$ from fossil fuel burning). Forest clearing also has an impact on regional climate, leading to drier and less productive ecosystems after the forests are gone.

Deforestation of tropical and boreal forests has become a current major source of carbon dioxide in the atmosphere, as described in the preceding paragraph. Conversely, temperate zone forests over much of the globe are regenerating and thus have become a net sink (depository) for carbon. Photosynthesis in trees converts atmospheric carbon dioxide into plant cellulose, drawing carbon out of the atmosphere and storing it as biomass. Forests in the United States and Europe are experiencing new growth, mainly because they are no longer harvested as a primary fuel source as they were 100 years ago. However, forest fires, disease, and other threats to these forests make them an uncertain form of carbon storage for the future.

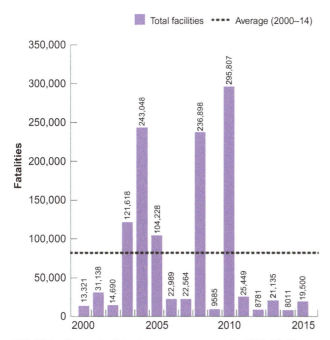

FIG. 25.2 Human fatalities due to extreme events for 2000–15. *(Reproduced with permission from The Aon Center, http://thoughtleadership. aonbenfield.com/Documents/20160113-ab-if-annual-climate-catastrophe-report.pdf.)*

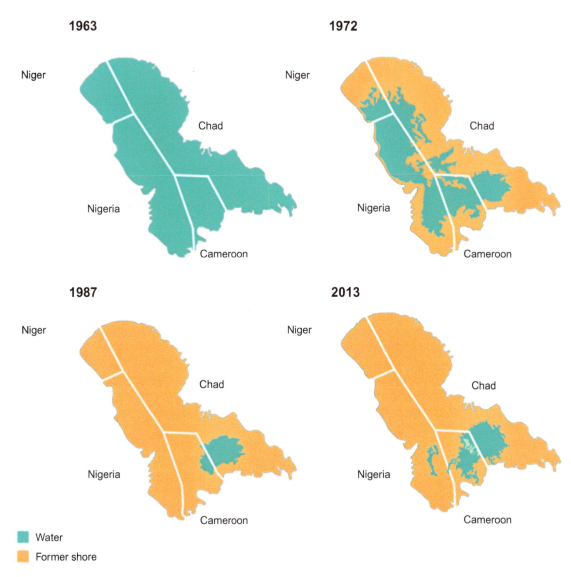

1963

1972

1987

2013

■ Water

■ Former shore

Source: United Nations Environment Program and DIVA-GIS in Kingsley, P. "The small African region with more refugees than all of Europe." *The Guardian*. 26 November 2016. Available at: http://www.theguardian.com/world/2016/nov/26/boko-haram-nigeria-famine-displacement-refugees-climate-change-lake-chad.

FIG. 25.3   Shrinking Lake Chad.

25.2.4   Desertification

The world's *drylands* (arid, semiarid, and subhumid zones where potential evaporation exceeds precipitation) cover about 40% of the earth's land area. In their natural state, these lands vary from bare soil, to grasslands, to mixed grass and shrublands, and to thorn forests, depending on how much rain they receive. However, they have been extensively degraded by human activity, mainly by overgrazing and unsustainable agricultural practices. Land degradation in the drylands is called *desertification*, although this term can be confusing. It conjures up images of desert dunes spreading into fertile areas, which is only part of the problem. More commonly, desertification takes place within a dryland region in a mosaic pattern, rather than at the margins. In fact, the ebb and flow of deserts at their margins can be a natural response to climate variations, rather than an effect of humans on the land. An example of desertification is the conversion of grasslands to shrublands through overgrazing. Too many grazing animals stocked on a range eat the more palatable grasses, which are replaced in time by grazing-resistant shrubs. Grasses tend to hold the soil in place, whereas shrubs tend to be separated by areas of bare soil, which is subject to wind and water erosion. Hence, the landscape is converted from grassland to shrub-desert, even though the climate might remain the same.

The Lake Chad Basin covers parts of Algeria, Cameroon, Central African Republic, Chad, Libya, Niger, Nigeria, and Sudan. Thus Lake Chad is a transboundary lake. Lake Chad once was the world's sixth largest lake, and Africa's largest water reservoir in the Sahel region, comprising about 25,000 km^2—the size of the US state of Maryland. More than 50 million people from over 70 ethnic groups depends on the lake for their livelihood.

High climatic irregularity and occasional extreme droughts, unsustainable management of natural resources, population increases, diversion of water from The River Chari (which provides 90% of the inflow to the lake), farming of high-water-intensity food crops, and incursion of invasive grass species have combined to threaten the lake. As a result, the lake has shrunk to ~2500 km^2 (see Fig. 28.3).

Impacts of reduced water levels

1. Increased alkalinity, increased anoxic conditions, and worsened effects of eutrophication
2. Increased distance to the lake up to 20 km from the edge
3. Decline in fishing economies
4. Migration of farmers and cattle herders in search of land resources and jobs
5. Increase in unemployment and poverty which led to the emergence of extremists
6. Conflicts between farmers and fishermen over water use among countries
7. Interstate conflicts

The Lake Chad Basin Commission, an intergovernmental organization, was formed to oversee water and other natural resource usage in the basin. There are eight member governments—Cameroon, Chad, Niger, Nigeria, Algeria, the Central African Republic, Libya, and Sudan.

Some recommended solutions provided by various entities

1. Build a dam and 60 miles of canals to pump water uphill from the Congo River to the Chari River and then on to Lake Chad.
2. Environmental education on efficient use of water resources and appropriate fishing practices
3. Transboundary protected area designation

Another type of desertification is caused by converting semiarid land to dryland crop production. The Dust Bowl that arose in the Great Plains region of the United States and in western Canada in the 1930s is an example of this type of land degradation. This region has a variable climate and is only marginally suitable for agriculture. However, in the first part of the 19th century, much of the land was converted into dryland grain farms (mostly wheat). A wheat crop requires more water than the annual precipitation in this region. Therefore the farmers practiced a type of cultivation called dust mulching. The soil was plowed and pulverized to a fine powder and left fallow for a year. During the fallow period, rain fell on the fine dust surface and

soaked into the soil. Then in the second year, a wheat crop was planted and it produced grain, using water stored in the soil from the fallow period plus what fell on the field during the second year.

Unfortunately, when a period of drought came in the early 1930s, the huge areas of cleared fields began to blow and created region-wide dust storms that covered over farms and even farmhouses. One of the worst dust storms occurred on April 14, 1935. Dust rose several thousand feet into the air across the Great Plains, in one large storm involving New Mexico, Texas, Colorado, Kansas, and Oklahoma. Ships entering New York harbor from Europe arrived with a fine coat of soil that was blown out to sea from the Midwest.

Other types of desertification are caused by salinization of land through irrigation with saline water, and destruction of riparian ecosystems through diversion of river water for human use and construction of dams that interfere with the normal hydrological cycle of rivers.

The United Nations Environment Program estimates that 80% of the drylands have been affected by desertification and that 40% have been moderately to severely impacted. A map showing the vulnerability of different regions to desertification is presented in Fig. 25.5. Some types of desertification are difficult or even impossible to reverse. For example, replacement of grasslands with shrublands may encourage gully erosion to take place, leading to the formation of deep cuts over the land surface. Like deforestation, desertification is not new. The area of the Middle East known as the Fertile Crescent, now consisting of southern Iraq and part of Iran, where crops were first domesticated and human civilization first arose, has been largely converted to desert, covered by blowing sands.

Since drylands do not store as much biomass on their surface as forests, it has been thought that desertification may not contribute much to the greenhouse effect. However, this might be an erroneous conclusion. Semiarid and subhumid regions store a large amount of carbon in their soils, and this carbon can be released during desertification, adding carbon dioxide to the atmosphere. Conversely, some forms of desertification, such as replacement of grasses with trees and shrubs, may actually lead to greater storage of carbon on the land and in the soil. A better estimate of the role of drylands in the global carbon budget is needed.

25.2.5 Soil Degradation

Soil degradation is the loss of land's production capacity in terms of loss of soil fertility, soil biodiversity, and degradation. Soil degradation causes include agricultural, industrial, and commercial pollution; loss of arable land due to urban expansion, overgrazing, and unsustainable agricultural practices; and long-term climatic changes. According to a recent report to the United Nations, almost one-third of

FIG. 25.4 Example of forest clearing in the Amazon rainforest in northern Brazil. *(From Daniel Beltrá/Greenpeace. https://e360.yale.edu/features/business-as-usual-a-resurgence-of-deforestation-in-the-brazilian-amazon.)*

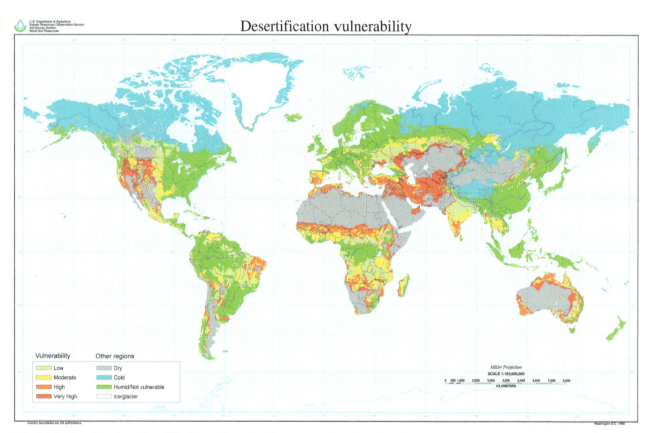

FIG. 25.5 Desertification vulnerability Map. *(From https://www.nrcs.usda.gov/wps/portal/nrcs/detail/national/nedc/training/soil/?cid=nrcs142p2_054003.)*

the world's farmable land has disappeared in the last four decades. It was also reported that all of the World's topsoil could become unproductive within 60 years if current rates of loss continue. The issues of soil health and impacts on human well-being are discussed in detail in Chapter 27.

25.2.6 Depletion of Ocean Fish Stocks

One of the first global environmental disasters was the collapse of many of the major open-ocean fisheries, starting in the 1950s with the advent of modern fishing technologies.

Until then, fishing was mainly conducted by local fleets of boats using traditional fishing techniques. Starting in the 1950s, large factory ships that could stay at sea for weeks and process and freeze the catch on board were deployed by the top fishing nations (China, Russia, Peru, Chile, the United States, Norway, and Japan) to roam widely over the oceans. These ships were equipped with increasingly sophisticated fish-finding and capture technologies. Today, open-ocean fishers use GPS-assisted sonar on scout boats or helicopters to locate schools of fish. These are harvested using long-lines (lines of baited hooks several kilometers long), drift nets (large nets that float near the surface), or large-capacity trawl nets that scoop up shrimp and bottom-feeding species. Because modern fishing gear is expensive, fishers are motivated to increase their catch to pay for the added expense of catching fish.

At the same time as the fishing became mechanized, new international markets developed for marine species that were formerly of only local interest. These included many coastal species that reproduced slowly, so the stocks were quickly depleted. Abalone, lobsters, crabs, a wide variety of reef fish, and even sea urchins (harvested for their eggs) became international favorites. Even the lowly seahorse has been harvested to near *extirpation* (local extinction) on many Asian reefs, because it is used as a cure for impotency and asthma in traditional Chinese medicine.

The first large fishery collapse occurred in the 1990s with the loss of the north Atlantic cod fishery (Fig. 25.6), which had been in existence since the 1600s. The collapse of this fishery was preceded by intense negotiations between nations on methods to preserve and regulate the fishery, but each nation saw it in their self-interest to demand their fair share of a declining resource. By the middle of the 1990s, there were insufficient fish left to support a commercial fishing fleet. Fishers around the north Atlantic lost their livelihood. Ten years after this collapse, with a near-total ban on cod fishing in place, stocks have still not recovered sufficiently to support commercial fishing. The cod wars (disputes between nations over the size of the catch) and the subsequent collapse of the fishery showed that modern fishing methods are unsustainable and that the international community lacks a mechanism by which common fishing grounds can be regulated.

The United Nations' Food and Agriculture Office (FAO) estimated that, in 2000, 9%–10% of the world's marine fish stocks had been depleted or were recovering from depletion; 15%–18% were overexploited and headed for depletion; and 47%–50% were fully exploited and were at or near their peak of production; while only 25%–27% were moderately exploited and had some potential for the catch to increase. With 71%–78% of the world's commercial fish species fully exploited, overexploited, or depleted, FAO concluded that commercial fishing is not a sustainable food resource and is urgently in need of

international and national-level management. This has serious implications for human well-being, given that by 2013, fish accounted for about 17% of the global population's intake of animal protein to 3.1 billion people according to the FAO.

Overfishing has environmental consequences beyond the effect on the target species. One problem is the *by-catch*—the marine species that are taken inadvertently with the target species. For example, by-catch in tuna nets is by far the most serious threat to dolphins and other cetaceans. FAO estimates the by-catch of marine species at 20 million tons per year. In some fisheries, such as shrimp trawling, the by-catch is many times the size of the main catch. The by-catch is generally thrown back into the sea dead. In many fisheries, the by-catch includes juveniles of commercial species; for example, in the Gulf of Mexico, by-catch with shrimp has undermined the populations of red snappers. Most of the modern fishing methods are nonselective and produce a by-catch. Longlines catch seabirds, sea turtles, and nontargeted fish, while gillnets catch seabirds and turtles, and continue to kill marine life when the lines are lost or abandoned at sea, a process called "ghost fishing." Bottom trawling not only produces a by-catch, but it also disrupts the bottom habitat, sweeping away bottom structure and *benthic* (bottom-dwelling) organisms of all types.

Overfishing can have broader effects on the ecosystem. As stocks of the most desirable species become depleted, fishers have turned to a practice called "fishing down the food chain." Top predators, such as billfish and tuna, tend to be removed first. Then the less desirable fish lower on the food chain are exploited. This practice simplifies the marine *food chain* (the sequence of predator–prey relationships in the ecosystem), affecting future fish and marine mammal populations. The overharvesting of haddock, cod, and mackerel in Alaska, for example, has reduced populations of Steller sea lions, which feed on these fish. Fishing for krill (small, red marine shrimp about 6 cm long) in the Antarctic has placed stress on the species that depend on them for food, including penguins, whales, squid, and fish. When plant-eating fish are removed from tropical reefs, seaweeds often proliferate, covering over corals.

The size of annual harvest by marine capture fisheries peaked in about 1995, despite increased investment in fishing gear and greater effort per unit catch. Increasingly, *aquaculture* (raising marine species artificially in ponds, pens, or cages) is augmenting the wild catch. When farms are sited in sensitive ecosystems such as mangrove swamps and salt marshes, aquaculture has produced great environmental damage. The *effluent* (wastewater containing feces and uneaten feed) from fish farms can produce *eutrophication* (overfertilization) of coastal waters, leading to harmful algae blooms and loss of oxygen in the water, producing fish kills. Carnivorous aquaculture species such as marine shrimp require large amounts of fish-derived protein

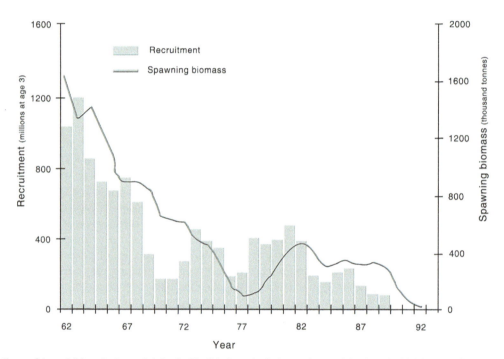

FIG. 25.6 Collapse of the cod fishery in the north Atlantic. The fish shown is *Gadus morhua*, or Atlantic cod, which has been fished commercially since the 16th century. Overfishing led to a collapse of the fishery in 1992, and it has not been fished commercially since then. The graph shows the decrease in Canadian cod stocks, but similar decreases occurred throughout the North Atlantic. *(Image from http://www.erin.utoronto.ca/~w3env100y/env/ENV100/ hum/cod.htm.)*

in their diets; many fish species that were previously not exploited are now harvested by capture fisheries to provide feed for aquaculture species.

25.2.7   Coastal Degradation and Coral Reefs

Interacting with overfishing to place stress on marine ecosystems are the problems of pollution and overdevelopment taking place along the world's coastlines. In 1995, 2.2 billion people (39% of the world population) lived within 100 km of a coastline. The trend is toward an even greater concentration of population along coastlines. In the United States, 54% of

Americans live in counties along the Atlantic or Pacific Oceans, the Gulf of Mexico, or the Great Lakes. Populations are expanding rapidly in the coastal states (e.g., Florida and California), so that by 2025, 75% of Americans are expected to live in coastal counties. In China, nearly 700 million people (56% of the population) live in coastal regions, and the trend is expected to continue because the 14 "economic free zones" and 5 "special economic zones," areas targeted by the Chinese government for economic expansion, are in coastal provinces. Similar trends exist around the world.

This high concentration of people along shorelines produces a number of environmental problems. Municipal and industrial effluents add nitrogen, phosphorous, and industrial chemicals to the coastal environment, leading to eutrophication that produces a degradation of the marine food web. Natural ecosystems such as bays, salt marshes, and estuaries are transformed by construction of ports, breakwaters, marinas, and aquaculture facilities. In their natural state, these coastal ecosystems provide protected breeding and nursery grounds for fish and crustaceans, and feeding stations for a wide variety of birds. These functions are lost when they are developed for human use.

Overfishing, pollution, and coastal development are affecting several key coastal ecosystems in the tropics and subtropics. *Seagrass meadows* (beds of marine grasses growing in shallow, sandy areas near coastlines) cover less than 0.2% of the global ocean, but fulfill a critical role in the coastal ecosystem. For example, they are feeding grounds for fish, crustaceans, sea turtles, and marine mammals such as the manatee. Widespread seagrass losses have been caused by human activities including dredging, fishing, anchoring, oil spills, siltation and eutrophication from land runoff, construction projects, and food web alterations. These stresses are expected to be worsened by negative effects of climate change discussed earlier, such as sea level increase, increased storms, and higher levels of ultraviolet radiation.

Coral reef ecosystems are also in decline throughout the tropics. Principal threats are similar to those facing seagrass meadows—overharvesting, pollution, disease, and climate change. The Great Barrier Reef on the north coast of Australia, widely regarded as the most pristine reef in the world, already shows system-wide symptoms of decline in productivity and species richness. Furthermore, coral reefs are susceptible to damage by *coral bleaching*. The hard skeleton of a coral reef is made up of calcium carbonate secreted by millions of small animals called *coral polyps* that filter food particles from the water. Living in symbiosis with the polyps are microscopic, single-celled algae called *zooxanthellae*. These are essential to the health of the polyp and the coral reef, since they provide food to the polyps through photosynthesis and help in the deposition of carbonate. When the coral is put under heat stress, the zooxanthellae (which are equipped with flagella) leave the coral, which become white (bleached) in appearance. While not necessarily irreversible, coral bleaching can lead to large-scale die-offs of coral, and it is expected to increase in severity with global warming. Another problem for coral reefs is silt runoff from land, especially when formerly forested land is converted to grazing or is developed for human settlement. The silt disperses in the water, forming a fine sediment called "marine snow" that settles on corals and eventually kills them.

Mangrove swamps are also threatened by human activities. They have traditionally been harvested for wood. In the past 20 years, shrimp farms have been built along many tropical coastlines, and mangrove swamps have been

favored sites for development. A rough estimate is that about half of the world's seagrass meadows, coral reefs, and mangrove swamps have already been lost to development. These losses exacerbate the problem of overfishing and loss of marine biodiversity.

25.2.8 Persistent Contaminants

Certain types of pollutants are extremely persistent in the environment. These persistent pollutants have spread to remote areas such as polar regions and the ocean. Their presence in these environments is causing impairment to ecosystem functions and affecting the health of wildlife.

Certain organic chemicals that are very persistent are labeled *persistent organic pollutants* or POPs and are managed under the 2001 Stockholm Convention (Chapter 12). Example POPs include polychlorinated biphenyls, DDT, and hexachlorobenzene. These compounds travel via long-range transport in the atmosphere from their sources in industrialized parts of the world to the polar regions, in the process known as the grasshopper effect or global distillation effect. For example, during the summer, chemicals volatilize from the soil and are transported as gases and aerosols by the southerly atmospheric currents. When the vapor comes into contact with the cold climate of the polar regions, it condenses and the pollutants are deposited as contaminated rain or snow. Once deposited, these compounds become incorporated into the food chain through bioaccumulation and biomagnification processes.

Plastic wastes are another example of persistent pollutants, and they have become a major concern in particular for ocean environments. It is estimated that more than 5.25 trillion items of plastic weighing more than 268,000 tons are floating in the oceans (Eriksen et al., 2014). The main sources of waste are ocean vessels and rivers. International shipping regulation prohibits the dumping of waste into the ocean since 1990. Although illegal dumping still occurs to some extent. Rivers transport the largest percentage of the waste into the ocean. The amount of waste deposited into the ocean by a single river varies depending on the population density, levels of urbanization and industrialization within catchment areas, rainfall rates, and the presence of artificial barriers such as weirs and dams. Fig. 25.7 shows plastic waste traveling by river from inland into the ocean. Asia accounts for more than two-thirds (67%) of the global annual input estimated at 1.21 million tons. Conversely, Central and North America contribute only 1% (Barnes et al., 2009).

Once in the ocean, heavier polyvinyl chloride plastics sink or wash up on shore. The buoyant polyethylene and polypropylene plastics are carried by oceanic currents and atmospheric winds out to the open ocean where they are accumulated by ocean gyres. NOAA describes a gyre

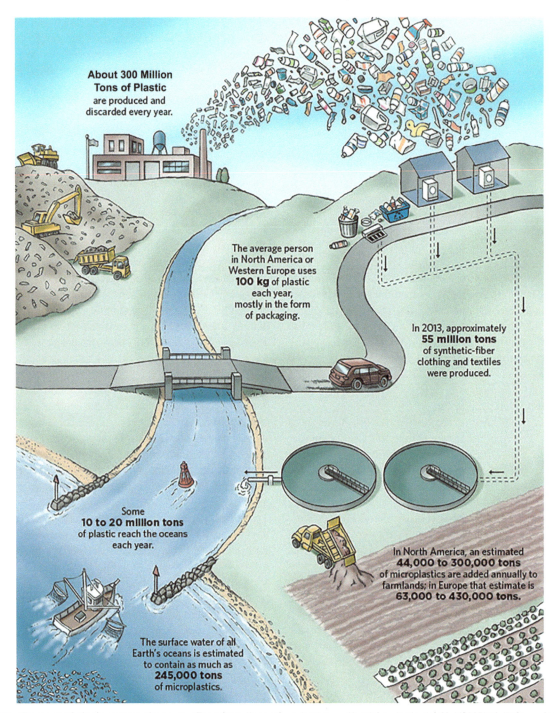

FIG. 25.7 Transport of plastic waste to the ocean. *(From https://www.the-scientist.com/?articles.view/articleNo/49507/title/Infographic-Plastic-Pollution/, © Al Granberg.)*

as "a large system of rotating ocean currents that spiral around a central point, clockwise in the Northern Hemisphere and counterclockwise in the Southern Hemisphere. Worldwide, there are five major subtropical oceanic gyres: the North and South Pacific Subtropical Gyres, the North and South Atlantic Subtropical Gyres, and the Indian Ocean Subtropical Gyre." The waste debris accumulated in these gyres is popularly known as a "garbage patch."

The prolonged exposure to UV light and wave actions degrades and fragments the plastics into small particles. Those that are less than five millimeters known as microplastics and the ones that are larger than five millimeters, macroplastics. When ingested, microplastics may clog the feeding appendages or the digestive system of ocean wildlife. Wildlife may also become entangled within clumps of plastic.

25.2.9   Ozone in the Upper Atmosphere

Ozone is a molecule composed of three oxygen atoms, as opposed to diatomic oxygen, which is composed of two oxygen molecules and makes up most of the oxygen in the atmosphere. Human activities are creating two types of ozone problems: they are depleting the amount of ozone in the upper atmosphere, which screens out harmful ultraviolet radiation (*good ozone*), and they are creating ozone at ground level, which is a major component of urban smog (*bad ozone*) (Chapter 17).

Good ozone is formed in the stratosphere when incoming ultraviolet radiation is absorbed by ordinary diatomic oxygen. An ozone molecule can absorb further ultraviolet radiation that regenerates diatomic oxygen. Hence, diatomic oxygen and ozone are interconverted by photochemical processes, but there is no overall loss of ozone when it absorbs ultraviolet light. The *ozone layer* in the stratosphere is vital in protecting life on earth from the harmful effects of ultraviolet radiation (see Chapters 4 and 17).

In the 1970s, scientists from the British Antarctic Service first noticed an apparent decline in the amount of ozone in the stratosphere above the Antarctic in winter. They confirmed their findings with careful, annual measurements from the 1980s to the present (Fig. 25.8). The cause of the ozone loss was found to be caused by reactions with chlorine in the upper atmosphere, and the chlorine was found to have originated from human release of *chlorofluorocarbons* (CFCs), extremely long-lived gases used in air conditioning systems and as propellants for aerosol spray cans. CFCs are also potent greenhouse gases, but in the amounts that have been emitted to the atmosphere so far, their main damage has been to the ozone layer. Today, an *ozone hole* forms over the South Pole (and more recently the North Pole) each year, due to the *Coriolis effect*, the pattern of global wind circulation caused by the spinning of the earth. In the Southern Hemisphere, these circumpolar winds produce a polar vortex that isolates the air within it during the south pole winter. Then the intense cold leads to the formation of polar stratospheric clouds, and it is within the ice crystals in these clouds that the chemical conditions necessary for ozone destruction by CFCs can take place. The precursors needed to form reactive compounds from CFCs are formed in the ice particles during the dark period, and when sunlight returns, ultraviolet light energizes the destruction of ozone from the by-products of CFCs. Although the destruction of ozone is local, the depletion affects the total amount of ozone in the atmosphere. Hence, ozone destruction is a global concern.

25.2.10   Species Extinction

Species diversity, also referred to as biodiversity, is an essential part of proper ecosystem structure and functioning, and to the services ecosystems provide (Chapter 6). Speciation, the formation of new species, and extinction, the permanent loss of species, are the major components of changes in the earth's biological diversity. Over the entire period of life on earth, a few billion years, plant and animal species have disappeared and new species have appeared as a result of natural evolutionary processes. The background rate of extinction (or normal extinction rate) is a measure of the rate of species disappearance due to natural processes operating under normal conditions. The rate has been determined to approximately five extinctions per year averaged over geologic time.

Much greater rates of extinction have occurred during specific periods in earth's history. These events, called mass extinctions, took place approximately 440, 360, 250, 200, and 65 million years ago. The causes of these extinction events were natural factors such as volcanic eruptions and asteroid impacts. These natural processes produced major changes in climate and habitats to which many species could not adequately adapt. These events are estimated to have caused the extinction of 50%–96% of species present at each time.

The current extinction rate is estimated to be anywhere from 10 to 1000 times the background rate of extinction, with the rate varying across taxonomic groups. This is the result of a myriad of human activities and their impacts, discussed in other sections of this chapter. The rate at which

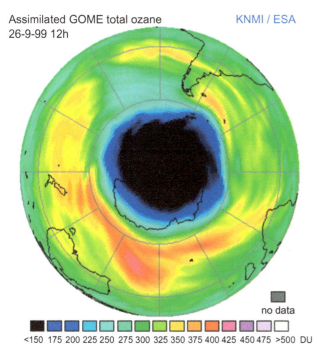

FIG. 25.8   Satellite image of the ozone hole over Antarctica in September 1999. Darker colors denote lower ozone values. Chlorofluorocarbons in the upper atmosphere react with ozone in stratospheric ice clouds over the South Pole to temporarily deplete the atmosphere of ozone each summer. (*Source: © KNMI/ESA; see the GOME Fast Delivery Service page for details. Available from: http://www.xs4all.nl/~josvg/KNMI/hole1999.*)

the climate is changing is accelerating the rate of species extinction by reducing the time plants and animals have to adapt to changes in ecosystem conditions. Further complicating this issue is the fact that biodiversity is unevenly distributed across the planet, where more than 1/3 of the known land plants and vertebrates are confined to 2% of the planet surface, the so-called biodiversity hotspots. These areas are usually located where population growth rates are high, climate impacts are projected to be particularly significant, the rate of urbanization is fast, and economic and technical capabilities to protect species from these changes are limited. For example, Madagascar and Indonesia have very high species diversity but also have the largest number of endangered and threatened species due to anthorpogenic activities.

25.3 GLOBAL CLIMATE CHANGE

25.3.1 The Greenhouse Effect

Air temperatures near the earth's surface are determined by the balance between incoming and outgoing energy flows through the atmosphere (Fig. 25.9). A small amount of the surface energy balance is due to heat flow from earth's molten interior to the surface, but most of the energy affecting the surface comes from incoming solar radiation from the sun. The sun is extremely hot, with a temperature of approximately 6000°C. It emits relatively high-energy (short wavelength) radiation, ranging from ultraviolet (100–400 nm) and visible (400–700 nm), to near infrared (700–4000 nm) wavelengths. Much of the ultraviolet radiation is absorbed by ozone (O_3) and molecular oxygen

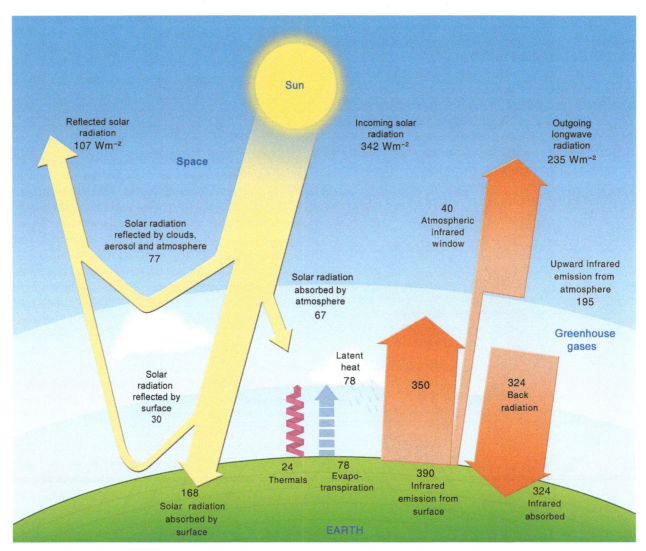

FIG. 25.9 The earth's surface energy balance. Solar radiation enters the atmosphere relatively easily, but part of the return infrared radiation emitted by the earth's surface is reabsorbed. Emissions of carbon dioxide, methane, and nitrous oxide from human activities are increasing the amount of heat trapped in the atmosphere, leading to global warming via the so-called greenhouse effect. *(From Trenberth, K.E., Houghton, J.T., Meira Filho, L.G., 1996. The climate system: an overview. In: Houghton, J.T., Meira Filho, L.G., Callander, B., Harris, N., Kattenberg, A., Maskell, K. (Eds.), Climate Change 1995. The Science of Climate Change. Contribution of WG 1 to the Second Assessment Report of the Intergovernmental Panel on Climate Change. Cambridge University Press, pp. 51–64 (Chapter 1). http://www.atmos.washington.edu/~dennis/Energy_Flow_small.gif.)*

(O_2) in the stratosphere layer at the top of the atmosphere (see Chapter 4). A small percentage of the visible and infrared radiation is reflected back to space or absorbed by small particulates or aerosols in the atmosphere. These particles can originate as dust, soot, sea salts, and a variety of chemicals emitted into the air by human activities, plants, soils, volcanoes, and oceanic processes. Clouds typically reflect 40%–80% of incoming radiation back into space, and they absorb 5%–15%. Therefore only about a half of the incoming solar radiation reaches the earth's surface.

Some of the radiation striking the earth's surface is reflected back into the atmosphere, but most is absorbed. Some of this energy is reemitted to the atmosphere as long wavelength radiation (4000–50,000 nm), but most is consumed in the evaporation of water from plants, lakes, and oceans. This evaporation releases *latent heat* energy, because the water does not change temperature in passing from the liquid to the vapor stage, but represents potential energy. As this vapor rises in the atmosphere, it eventually recondenses as water or ice crystals, releasing the latent energy of evaporation back into the atmosphere as heat. Radiation absorbed by clouds and water in the atmosphere is also reemitted as long wavelength radiation. Normally, about two-thirds of the solar energy that enters the atmosphere as ultraviolet, visible, and near infrared energy is reemitted back to space as reflected radiation or heat energy, and the remainder stays in the atmosphere, warming the earth's surface. Anything that perturbs this delicate balance can affect the surface temperature on earth.

The atmosphere contains trace gases that efficiently absorb outgoing long wavelength, infrared radiation, and that therefore contribute greatly to warming the atmosphere. The most important of these is water vapor, followed by carbon dioxide, methane, and nitrous oxide. These heat-trapping molecules are called *greenhouse gases* and occur naturally. Opposing the action of these gases are trace gases derived from sulfur compounds emitted by soil, plants, volcanoes, and the oceans. These gases form sulfate aerosols in the atmosphere that reflect incoming solar radiation, leading to a potential lowering of atmospheric temperature. Except for water, the concentrations of all these trace gases in the atmosphere have been greatly perturbed by human activities over the past century, potentially affecting the atmospheric energy balance and global temperatures.

Svante Arrhenius, a professor of chemistry at Stockholm, Sweden's Hogskola, was better known for developing the electrolytic dissociation theory, for which he received the Nobel Prize in 1903. However, he had a lively interest in all aspects of physics, and he began to wonder how fossil fuel burning, which was increasing exponentially in Europe at the time, might affect global climate. After months of calculations, done by hand, he concluded that a doubling of atmospheric carbon dioxide levels would increase global temperatures by 5°C, an estimate that is still valid today. This theoretical possibility came to be known as the greenhouse effect, because it superficially resembles the way heat is trapped in a greenhouse during the day.

Little practical attention was paid to the possibility of atmospheric warming until the 1980s, when Charles Keeling began to publish evidence on the rise of atmospheric carbon dioxide based on his measurements made on the top of Mauna Loa volcano in Hawaii. This station in the middle of the Pacific Ocean provides a well-mixed sample of air from the entire Northern Hemisphere. Charles Keeling was a professor of oceanography at the Scripps Institution of Oceanography in California at a time when scientists were unsure whether carbon dioxide emissions from fossil fuel burning would actually accumulate in the atmosphere or would be quickly absorbed into vegetation and by the oceans. He worked out that carbon dioxide measurements made on the tops of mountains, far from industrial sources, could be used to track changes in atmospheric levels over time. In 1958, he persuaded the National Science Foundation to set up a monitoring station on the top of Hawaii's Mauna Loa mountain. However, they stopped funding the work in the early 1960s, calling his results "routine." He found other funds to continue, and by the 1980s, it became obvious that there was a steep upward trend in atmospheric carbon dioxide levels, alerting scientists and policy makers to the imminent possibility of an increase in global temperature. President Bush awarded him the National Medal of Science in 2002. The National Science Foundation, which administers the award, declared that the Mauna Loa measurements were "... some of the most important data in the study of global climate change." The type of plot he provided, now called a Keeling curve, showed a typical seasonal pattern of rise and fall of carbon dioxide over each annual cycle, superimposed on a steady interannual increase in carbon dioxide (Fig. 25.10). When the Keeling curve for Mauna Loa is extended back in time from measurements of carbon dioxide in gas bubbles trapped in Antarctic and Greenland ice cores, it is estimated that atmospheric carbon dioxide has increased from a concentration of 270 ppm (parts per million) in the year 1600 to almost 400 ppm in 2015.

Currently, human activities add about 7–9 gt (gigatons) of carbon to the atmosphere each year in the form of carbon dioxide. About 80% of this is from the burning of fossil fuels, and the remainder is from the clearing of forests for agriculture, especially in the tropics. Of the amount derived from fossil fuel burning, about 60% is from the burning of coal and natural gas to generate electricity, while 40% is from the burning of liquid fuels, mainly to propel cars and trucks.

Fig. 25.11 shows the increase in carbon dioxide as well as the two other main greenhouse gases, methane and nitrous oxide, in earth's atmosphere over the last 1000 years. There has been a clear and dramatic increase in these gases

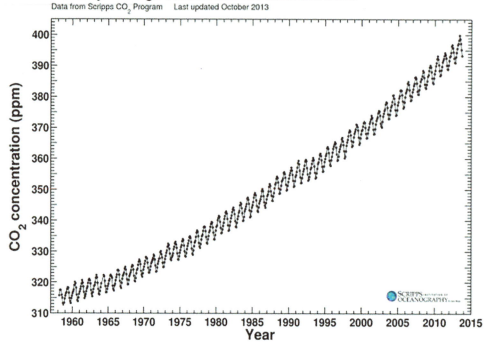

FIG. 25.10 Keeling's curve of increasing CO_2 concentration. (*Image from http://scrippsco2.ucsd.edu/history_legacy/keeling_curve_lessons, Scripps CO2 program.*)

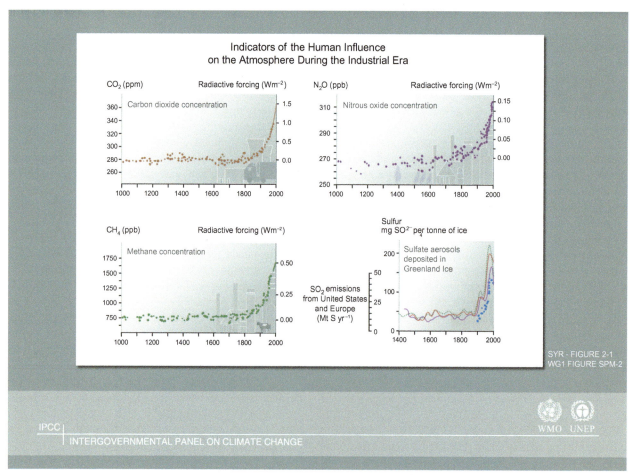

FIG. 25.11 Increase in levels of the major greenhouse gases in earth's atmosphere due to human activities. (*From IPCC, 2001. Climate change 2001: The scientific basis. Intergovernmental Panel on Climate Change. Available from: http://www.grida.no/climate/ipcc_tar/wg1/index.htm; http://www.ipcc. ch/present/graphics.htm.*)

since 1800 due to human activities. *Anthropogenic* (human-caused) methane emissions originate from the burning of fossil fuel and are also produced by livestock, landfills, and rice paddies. Nitrous oxide originates from fossil fuel burning as well, but most of it comes from the wide-scale use of chemical fertilizers to boost crop yields around the world. Excess nitrogen, not taken up by crops, is converted to nitrous oxide and dinitrogen gas by soil microbes via denitrification. The amounts of methane and nitrous oxide emitted from human activities are much lower than carbon dioxide, but they respectively have 26 and 206 times greater heat-trapping ability than carbon dioxide. Currently, 70% of the global warming potential that can be attributed to human activities is due to carbon dioxide emissions, while 23% is due to methane and 7% to nitrous oxide emissions.

Scientists are very certain that the recent rise in carbon dioxide levels is due to human activities. It is relatively easy to know the amount of fossil fuel that is burned each year, because records are kept on the amount of coal and gas that are burned for electricity and on the production of liquid fuels for the transportation sector. It is also relatively easy to measure the increase in carbon dioxide and other greenhouse gases in the atmosphere. In addition to the Mauna Loa observatory, gas measurements are now made around the world, including the Antarctic, and they all show similar increases in greenhouse gases that over time roughly track the amount calculated to have been released by fossil fuel burning. However, only about three-fourths of the amount of carbon dioxide that is expected to enter the atmosphere and oceans is actually detected. The other quarter is referred to as the *missing carbon sink*. Scientists are greatly interested in solving the missing carbon problem. Do their models on oceanic uptake of carbon dioxide underestimate actual uptake? If so, global climate change may not be as severe as expected. Or does the missing carbon represent extra biomass production in forests due to the photosynthesis effect? If so, it is important to protect these forests from fire and clear-cutting or the carbon will be quickly released back to the atmosphere, making the global climate change more severe than expected.

The carbon dioxide emissions entering the atmosphere today will contribute to global climate change for the next several centuries. Experts project that at present rates of emissions, atmospheric carbon dioxide levels will double over the next century and will peak at 1700 ppm in the year 2400 (over 5 times preindustrial levels). This amount of carbon dioxide is projected to produce a temperature rise of 2–4°C at the earth's surface over the next 100 years and a rise of 4–8°C at the peak of carbon dioxide levels. These elevated temperatures, and their effects at sea level, are expected to persist for many centuries after carbon dioxide emissions have stabilized (Fig. 25.12). Currently,

we are just at the beginning of the rise in greenhouse gas levels in the atmosphere.

25.3.2 Effects of Greenhouse Gas Emissions on the Global Climate

There is great uncertainty about the amount of atmospheric warming, if any, that has already taken place due to greenhouse gas emissions. There is even more uncertainty about the impacts of future warming on global climate systems. The rise in atmospheric carbon dioxide levels has been documented by careful measurements over time, whereas climate projections are based on models and other indirect methods that are subject to error. Nevertheless, there are some logical consequences that follow from a warming of global temperature. These include shifts in regional weather patterns due to unequal heating at the equator and the poles; partial melting of the Antarctic ice shield, resulting in a rise in sea level around the world; and shifts in the distribution of vegetation zones, with impacts on agriculture and natural ecosystems. The evidence for an actual increase in global temperature, and possible consequences over the next 100 years, are discussed briefly as follows.

The problem of documenting a change in surface temperature is much more difficult than the problem of documenting a rise in atmospheric carbon dioxide levels. The atmosphere at the top of Mauna Loa and other reporting stations is well mixed, and carbon dioxide levels are fairly consistent from station to station. Conversely, surface temperatures vary widely from one location on earth to another, depending on the type of landscape, the time of day and season of year, and method of measurement. For example, the air over a plowed but unplanted field on a summer day might be 10°C warmer than over a fully vegetated adjacent field. The air over cities is warmer than over rural areas, due to the *heat island effect* (absorption of heat by streets and concrete surfaces), and as most historical temperature records were taken near cities, there could be an apparent increase in temperature from the growth of cities quite apart from any effect of greenhouse gases. Ocean temperatures might be a more reliable means to look for global warming; however, over the years, methods for measuring sea temperatures changed from bringing wooden or canvas buckets of water on deck for measurement (which allows the water to cool by evaporation) to making measurements directly in the water. These changing techniques add bias to the temperature record. Indirect methods of measuring surface temperature, such as using the thermal bands on satellites, present technical problems and often do not agree with the results of surface measurements.

Despite the uncertainties, combined annual land-surface and sea-surface temperature databases show a temperature

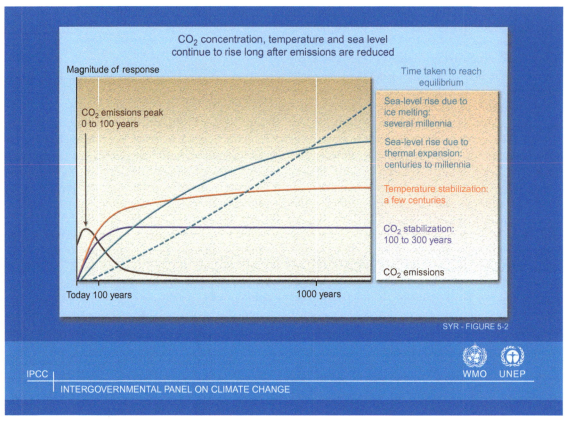

FIG. 25.12   Long-term projections of the effect of fossil fuel burning on atmospheric carbon dioxide levels, surface temperature, and sea level. Notice that carbon dioxide emissions will decrease after 100 years, but the effects will last for hundreds or thousands of years. *(From IPCC, 2001. Climate change 2001: The scientific basis. Intergovernmental Panel on Climate Change. http://www.grida.no/climate/ipcc_tar/wg1/index.htm; http://www.ipcc.ch/present/graphics.htm.)*

rise of 0.65°C plus or minus 0.15°C over the past 150 years (Fig. 25.13). Many scientists believe these data sets confirm the existence of a global warming trend with a certainty of 95% or greater. Other scientists are trying to confirm global warming by using a wide variety of *surrogate measurements* (indirect tests of a warming signal). For example, mountain glaciers around the world retreated in the 20th century at a rate that is consistent with a warming of about 0.6–1.0°C. The growing season for wild and cultivated plants in the Northern Hemisphere increased by 12 days during the period 1981–1991, also indicating a warming trend. Furthermore, a study of plant distribution on 30 alpine peaks showed that the distribution of many species has shifted upward in elevation, while the distribution of a number of butterfly species has shifted toward the pole, and mosquito-borne diseases such as dengue fever and malaria are reported at ever-higher elevations. Scientists using different techniques have estimated a rate of sea level rise of about $1.8\,cm\,yr^{-1}$, expected from the melting of ice caps and the thermal expansion of seawater due to global warming (Fig. 25.14). In total, while we are still at the beginning of the expected global temperature rise due to

emission of greenhouse gases, scientists are reasonably certain they have already detected a rise in temperature over the past 150 years.

Overall, the range of the mean global temperature increase over the next 100 years is expected to be in the range of 2–4°C. However, a greater mean annual warming at higher latitudes than near the equator is expected, called a *polar amplification* of the warming. This is in part because, as snow melts, more radiation will be absorbed rather than reflected back to space in the polar regions, leading to greater heating of the land surface. Whereas a rise of less than 1°C is expected at the equator, at 80°N or S latitude a rise of 3°C or more is anticipated. For similar reasons, greater warming in winter than in summer is expected at high latitudes, whereas greater warming in summer is expected in arid and semiarid regions, where soils become drier in summer. A general tendency for drying of mid-latitude soils is expected and could have implications for agriculture.

The changes in global temperature distribution could have profound effects on the global climate cycles, because climate patterns are driven in part by differences in surface

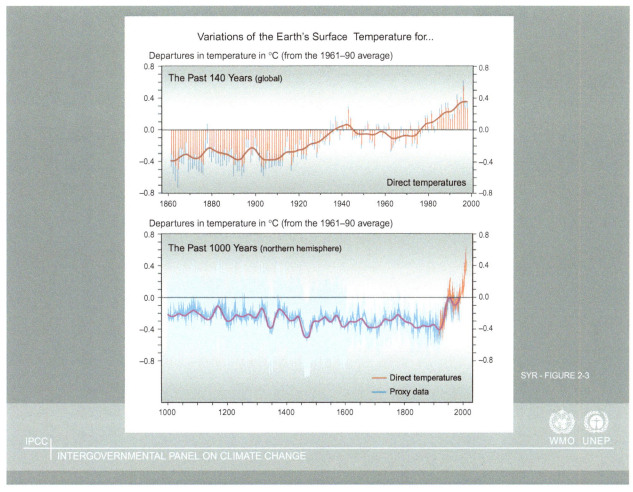

FIG. 25.13 Variations in the earth's surface temperature over the past 1000 years, inferred from a number of data sources. *(From IPCC, 2001. Climate change 2001: The scientific basis. Intergovernmental Panel on Climate Change. Available from: http://www.grida.no/climate/ipcc_tar/wg1/index.htm; http://www.ipcc.ch/present/grahics.htm.)*

temperature at different latitudes. Although the direction of change is difficult to predict, it can be expected that monsoon rains, tropical cyclones and hurricanes, precipitation patterns over the continents, and the frequency of extreme weather anomalies such as droughts and floods will be affected by global warming. Also, it is expected that sea level will rise around the globe due to melting of the polar ice caps. A mean global temperature rise of 2–4°C is expected to raise the sea level by 25–75 cm by the year 2100. This would impact large areas of coastal land around the world, including cities, agricultural areas, and natural coastal ecosystems such as coral reefs, salt marshes, and coastal forests.

The projected effects and alterations associated with global climate change discussed before have the potential to significantly impact human health and well-being. The potential effects of global climate change on human health are discussed in Chapter 26.

25.4 DRIVERS OF HUMAN-INDUCED GLOBAL CHANGE

25.4.1 Population Growth

Population growth is the increase in the number of people per area due to natural increase and net migration. Natural increase is the difference between birth and death rates, and net migration is the difference between immigration and emigration. According to the United Nations, the combination of economic development, green revolution, and advances in public health and medical services has reduced child mortality rates by 75%, increased life expectancy by 67%, and decreased death rates by 55% since the 1950s. As a result, global population increased from 2.5 billion in the 1950s to 7.5 billion in 2017. The population increased by 200% in 67 years and the population is expected to reach 9.7 billion in 2050 despite a reduction in fertility and birth rates.

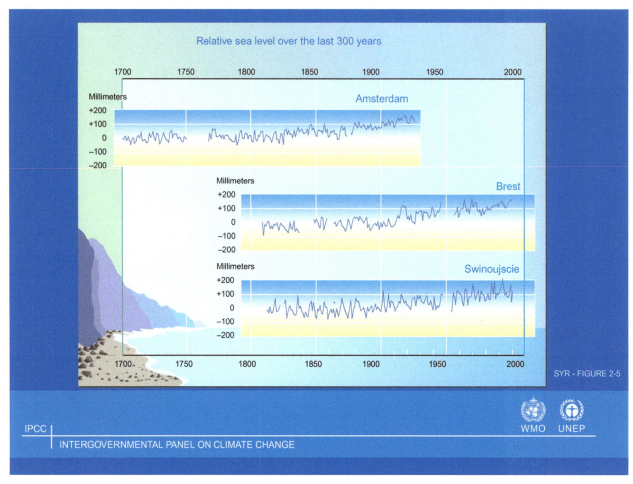

FIG. 25.14   Rise in sea level since 1700 at three locations in Europe. Global warming increases sea level through the thermal expansion of water and the melting of polar ice caps. *(IPCC, 2001. Climate change 2001: The scientific basis. Intergovernmental Panel on Climate Change. Available from: http://www.grida.no/climate/ipcc_tar/wg1/index.htm.)*

This increase in population will disproportionately come from developing countries, and within Asia and Africa in particular. The challenge is not only the increase in the number of people but also the fact that countries are at different demographic transitions. The demographic transition is a model that describes a countries' population changes—age structure, birth and death rates—as they go through the industrialization process or economic evolution over time. While some countries such as the United Kingdom, Germany, and Japan are late in the demographic transitions, the majority of the African countries and other developing countries are still in the early stages. The demographic transition stage influences how a society prioritizes social, economic, and environmental issues. A country with a large number of the population below 25 years of age is most likely to prioritize socioeconomic issues such as jobs, education, and health services over environmental issues, and a country with the high proportion of people above 64 years of age will prioritize retirement services. Similarly, a country with a large population of 25–64 years of age will prioritize job creation, construction, and production of good and services.

Another dimension of population growth is the increase in per capita consumption of natural resources. For example, per capita, food consumption has increased from an average of 2370 kilocalories (kcal) per person per day in the late 1970s to 2770 kcal per person per day in 2012. While global population has grown by 37% since 1990, food production has increased by 40% according to the FAO. This increase in food production has resulted in greater deforestation and other issues as agricultural land use has expanded.

Increases in population size and per capital consumption mean an increase in the extraction of natural resources to meet energy, food, and domestic commodities demands, and an increase in the generation of wastes and pollutants. The increase in resource extraction and waste generation threatens the regenerative and assimilative capacities of ecosystems. These issues combine to place pressure on governments and societies on deciding between taking care of the environment and meeting the demands of the growing population and changing consumption patterns. This challenges the abilities of countries to compromise on the prioritization of environmental issues such as climate change. Research has shown that rapid population growth under changing environmental conditions makes resource users insecure and vulnerable to violence.

25.4.2 Industrialization and Economic Evolution

Economic evolution examines an economy as an evolutionary system. It looks at the changes in the institutions, habit of thought, mechanisms of specialization and technology, as well as the cause and effects of economic development. Economic evolution is evidenced by the stark differences between preindustrial, industrial, and modern societies' lifestyles, aspirations, and sociopolitical conditions (Alam, 2003). The muscle-driven, organic and land-based economy characterized the preindustrial era—before 1750. Constrained by transportation and communication, there was limited travel and exchange of ideas and information. For example, in the United States it took one month for a letter to travel from Boston to Williamsburg, Virginia and in the winter 2 months. The family was the primary unit of production and it took a long time to produce a product. For example, the spinning wheel circa 1000 AD spun one thread at a time whereas the spinning jenny invented in the late 1700's spun eight threads. The dawn of industrialization transformed the preindustrial era to the power-driven machinery and mineral-based economy.

The new economy's main resources were steel and iron for raw materials and coal and petroleum for energy. Product standardization, labor specialization, technological advancement, and the factory system led to mass production in terms of both numbers and diversity. Improved transportation and communication accelerated the distribution of products and exploration of new areas for expansion, settlement, and extraction. This also facilitated the exchange of information and ideas. Production changed from family to manufacturing companies and machines started to replace manual labor. This means more natural resources and financial capitals were needed to feed the newly discovered engine of growth.

The advancement of science and social concerns led to the discovery of the negative impacts—externalities—of the industrial economy to the health of the people and environment (Chapter 26). The major impacts were natural resources depletion and pollution, as has been covered in other chapters of this text. Even today, the integration of social and ecological concerns in the production of goods and service continually challenges economic undertakings.

The initial stage of the industrial revolution from 1800 to 1940s took place in Britain, France, United States, Germany, and Russia. The industrialization footprint expanded to other parts of the world through the export of manufacturing products to nonindustrialized and import of raw materials and energy resources from nonindustrialized areas. For example, by 1913 steel, machinery, and automobiles accounted for 28% of American exports and it was 51% by 1929 (Jones, 2016). In addition to products and raw materials, other things that took place include the spread of investment, technology, and communications, the growth of international specialization, the explosion of migration, and international cooperation as well as the export of environmental impacts.

The result was the decentralization of economic activities depending on the comparative advantages (cheap labor, raw materials, regulatory and institutional mechanisms, etc.), and hence the birth of the global assembly line or global supply chain and Multinational Corporations. Hence, the Ecological and Economic Footprint of an entity is a composite of multiple and varied institutions, hierarchies of shareholders and stakeholders, multiple countries, and diversity of ideas and skills. The main outcomes of industrialization and globalization are the improvement in living standards and health services, reduction in child mortality, and increase in income due to economic growth, which in turn led to population growth and rapid urbanization.

25.4.3 Urbanization

The word urban means an area that is densely populated and characterized by industrial and service industries. Urban areas include cities, towns, suburbs, and metropolitan areas. Urbanization is the increase in the proportion of people living in urban centers and the associated conversion from rural to urban lifestyles. The main causes of urbanization are rural–urban migration and a natural increase due to the decline in death rates. Also, climatic changes, extreme events, and conflicts are pushing people to urban areas.

Since 1950s urbanization has grown nearly tenfold according to the United Nations, and now about 55% of the global population lives in urban centers compared to 30% in 1950. The percentage is expected to increase to 60%

in 2030 (UN 2016). About 90% of this increase will be concentrated in Africa and Asia. Urbanization typically improves the socioeconomic conditions of people, increasing literacy, education, health, access to social services, and political participation. Additionally, urbanization stimulates economic growth.

However, rapid and unmanaged urbanization threatens economic growth and sustainability, and can lead to environmental pollution issues (Chapter 18) and social–political conflict. This can happen when urban population growth outpaces the ability of governments to provide necessary infrastructure and employment. This, in turn can lead to slum formation (UN, 2016), with residents having poor access to solid waste and wastewater management services, clean drinking water, quality education, health services, and transportation facilities. The United Nations estimates that today, one in every eight people live in slums. Urbanization of coastal areas is of particular concern given projected raises in sea level and ongoing coastal degradation discussed previously in this chapter.

QUESTIONS AND PROBLEMS

1. How can we balance between growing population and consumption?
2. Give some additional examples of positive and negative feedback effects related to global change issues.
3. How can we ensure sustainable urbanization?
4. What actions were taken to deal with the issue of the ozone hole?
5. Which different methods are used to track changes in global temperature? What limitations are associated with each method?
6. How might a rising sea level affect society in the future? Give some specific examples.

REFERENCES

Alam, M.S., 2003. A short history of the global economy since 1800. MPRA Paper No. 1263. Available from: https://mpra.ub.uni-muenchen.de/1263/.

Barnes, D.K.A., Galgani, F., Thompson, R.C., Barlaz, M., 2009. Accumulation and fragmentation of plastic debris in global environments. Philos. Trans. R. Soc. Lond. Ser. B Biol. Sci. 364 (1526), 1985–1998. https://doi.org/10.1098/rstb.2008.0205. https://www.ncbi.nlm.nih.gov/pmc/articles/PMC2873009/.

Eriksen, M., Lebreton, L.C.M., Carson, H.S., Thiel, M., Moore, C.J., Borerro, J.C., …Reisser, J., 2014. Plastic pollution in the world's oceans: more than 5 trillion plastic pieces weighing over 250,000 tons afloat at sea. PLoS One 9 (12), e111913. https://doi.org/10.1371/journal.pone.0111913.

Jones, C.I., 2016. The Facts of Economic Growth. In: Handbook of Macroeconomics (Volume 2A). Elservier, B.V. https://doi.org/10.1016/bs.hesmac.2016.03.002.

UN, 2016. Urbanization and development: emerging futures. In: World Cities Report 2016. http://wcr.unhabitat.org/wp-content/uploads/2017/02/WCR-2016-Full-Report.pdf.

Zielinski, S., 2014. A world of vanishing lakes. http://smithsonian.com.

FURTHER READING

FAO, 2016a. State of the World's Forests 2016. Forests and agriculture: land-use challenges and opportunities. Rome. 107 pp.

FAO, 2016b. The State of World Fisheries and Aquaculture 2016. Contributing to Food Security and Nutrition for All. Rome. 200 pp.

IPCC, 2014. In: Core Writing Team, Pachauri, R.K., Meyer, L.A. (Eds.), Climate Change 2014: Synthesis Report. In: Contribution of Working Groups I, II and III to the Fifth Assessment Report of the Intergovernmental Panel on Climate Change, IPCC, Geneva, Switzerland, p. 151.

United Nations, Department of Economic and Social Affairs, Population Division, 2015. World Urbanization Prospects: The 2014 Revision, ST/ESA/SER.A/366.

United Nations, Department of Economic and Social Affairs, Population Division, 2016. The world's cities in 2016—data booklet. ST/ESA/SER.A/392.

Chapter 26

Environmental Impacts on Human Health and Well-Being

M.L. Brusseau, M. Ramirez-Andreotta, I.L. Pepper and J. Maximillian

AIR POLLUTION
including indoors and outdoors

INADEQUATE
WATER, SANITATION
and hygiene

CHEMICALS
and biological agents

RADIATION
ultroviolet and ionizing

COMMUNITY
NOISE

OCCUPATIONAL
RISKS

AGRICULTURAL
PRACTICES
including pesticide-use,
wate-water rouse

BUILT
ENVIRONMENTS
including housing
and roads

CLIMATE
CHANGE

World Health
Organization

#EnvironmentalHealth

Environment-related sources of risks to human health and well-being. *World Health Organization.*

Environmental Health is the field of science that studies how the environment influences human health and disease. Therefore contrary to what the name implies, "Environmental Health" is not focused on the health and well-being of the environment. Rather, it refers to human health and well-being as impacted by the environment (see Information Box 26.1). Earlier chapters in this book have focused on the properties and conditions of impaired or contaminated environments, the hydrobiogeochemical processes governing pollutant transport and fate, and methods for mitigation, management, and remediation of pollution. However, it is also very important to consider the human dimensions of environmental pollution.

Human health impacts can originate from pollution events and from environmental disturbances such as those discussed in Chapter 25. Pollution and other environmental disturbances may have direct impacts on human health through causing illness, disease, or death as covered in Chapter 28. They may also have indirect impacts. As presented in Chapter 6, the environment provides numerous services that are critical to human life (which we term ecosystem services). When an environment becomes contaminated or is otherwise disrupted or perturbed, those services may be impaired, which may have adverse outcomes for human well-being. An example would be the development of water scarcity conditions due to persistent extreme drought.

Environmental and Pollution Science. https://doi.org/10.1016/B978-0-12-814719-1.00026-4

INFORMATION BOX 26.1 Definitions of Environmental Health

The World Health Organization: Environmental health addresses all the physical, chemical, and biological factors external to a person, and all the related factors impacting behaviors. It encompasses the assessment and control of those environmental factors that can potentially affect health. It is targeted towards preventing disease and creating health-supportive environments. From: http://www.searo.who.int/topics/environmental_health/en/.

The National Institute of Environmental Health Science:

Environmental Health is the field of science that studies how the environment influences human health and disease. "Environment," in this context, means things in the natural environment like air, water, and soil, and also all the physical, chemical, biological, and social features of our surroundings. The man-made, or "built," environment includes physical structures where people live and work such as homes, offices, schools, farms, and factories, as well as community systems such as roads and transportation systems, land use practices, and waste management. Consequences of human alteration to the natural environment, such as air pollution, are also parts of the man-made environment. The social environment encompasses lifestyle factors like diet and exercise, socioeconomic status, and other societal influences that may affect health.

The National Environmental Health Association: Environmental health and protection refers to protection against environmental factors that may adversely impact human health or the ecological balances essential to long-term human health and environmental quality, whether in the natural or man-made environment.

It is critical to recognize the interconnections between the environment, the ecosystem services it provides, and human health. A primary specific example of the critical interconnectedness of these systems is the soil health:human health nexus, discussed in Chapter 27.

26.1 CAUSES OF HUMAN MORTALITY

The World Health Organization tracks global human mortality and estimates the deaths caused by different factors. The most recent data available are presented in Table 26.1. The top four causes of death are cardiovascular disease, cancer, respiratory infections and disease, and infectious diseases.

Many individual diseases have multiple causes. For example, major possible causes of stroke include high blood pressure, heart disease, diabetes, tobacco use, and obesity. These causes in turn can be influenced by other factors. Thus it is important to determine the major risk factors that influence human mortality. An analysis of risk factors helps

to delineate priorities and define opportunities for research, prevention, and policy.

The major risk factors for causes of death globally are presented in Table 26.2 for 2004 and Table 26.3 for 2015. Environmental factors in aggregate comprise the number one risk factor for both periods, with approximately 13 million attributable deaths. This represents 23% of all deaths. It is clearly established that environmental pollution is an important causative agent of many human illnesses, diseases, and deaths (GBD, 2016; Prüss-Üstün et al., 2016). Pollution impacts contribute to noncommunicable diseases including asthma, cancer, neurodevelopmental disorders, birth defects, heart disease, stroke, and chronic obstructive pulmonary disease. For example, environmental-related factors are estimated to cause 27% of deaths associated with cardiovascular diseases, 18% of cancer deaths, and 28% of deaths caused by respiratory infections and disease (Table 26.1). Environmental factors also contribute to infectious diseases such as diarrhea and vector-driven diseases.

The contributions of specific environmental factors to human mortality vary greatly (Table 26.4). Air pollution, including both ambient (outdoor) and indoor sources, contributes by far the most, with more than 7 million estimated deaths annually. This represents almost 60% of the 12.6 million total deaths in 2015 attributed to environmental factors. It is estimated that approximately 9 out of every 10 people breathe outdoor air polluted above World Health Organization guideline levels (WHO, 2016). Note that only particulate matter and ozone are included in the analysis for ambient air pollution, and that the effects of nitrogen and sulfur oxides and volatile organic compounds are not included. Similarly, only particulate matter from cooking and heating are included for the indoor air pollution analysis. Water pollution contributes the second most, with 1.8 million deaths (~15% of the total). The analysis includes only the impacts of microbial contamination and poor sanitation, and does not account for the impacts of chemical pollutants.

Combined, approximately 75% of the total annual deaths associated with environmental factors are attributed to air and water pollution specifically. The number of deaths attributed to air and water pollution annually (~9 million) is greater than those caused by tobacco use, high-sodium diet, being overweight, alcohol use, and unsafe sex, among others. In fact, more deaths are attributed to air and water pollution than any other single factor except for high blood pressure (Tables 26.3 and 26.4).

The human health impacts caused by pollution can have severe economic and social effects. Illnesses result in loss of productivity and possibly unemployment, which results in economic loss and negative social impacts. In addition, there are significant costs associated with treatment of the illnesses, which results in increased healthcare costs. The human welfare costs of pollution include: (1) direct medical

TABLE 26.1 Global Estimates of the Causes of Human Mortality for 2016

Cause[a]	Deaths	% of Total[b]	% Environmental Contribution[c]
Total	56,427,684	100	23
Cardiovascular diseases [heart disease, stroke]	17,858,012	31.7	27
Cancer	9,181,488	16	18
Respiratory infections & disease [bronchitis, asthma, emphysema, influenza, pneumonia]	6,772,422	12	28
Infectious diseases [diarrhea, tuberculosis, STDs, dengue, rabies]	5,045,215	8.9	33
Injuries [road accidents, falls, drowning, self-harm, violence]	4,883,194	8.7	40
Neonatal & congenital conditions [preterm birth, heart anomalies]	2,829,658	5	11
Neurological conditions [Alzheimers, Parkinsons]	2,538,460	4.5	5
Digestive diseases [cirrhosis, ulcers, appendicitis]	2,529,980	4.5	0
Genitourinary diseases [kidney disease]	1,447,991	2.6	2
All other	3,341,264	6	-

[a]Representative diseases for each category are listed in the brackets.
[b]Rounded.
[c]Approximate percentage of the total global deaths in each category attributable to environmental factors for year 2012; compiled from Prüss-Üstün et al. (2016).

Source: Compiled from WHO: Global health estimates 2016: deaths by cause age, sex, by country and by region, 2000–2016, 2018a. Available from: http://www.who.int/healthinfo/global_burden_disease/estimates/en/.

TABLE 26.2 Top 10 Risk Factors for Causes of Human Mortality Globally for Year 2004

Risk Factor	Deaths (millions)	% of Total
Environmental Factors[a]	13.3	23.3
High blood pressure	7.5	12.8
Tobacco use	5.1	8.7
High blood glucose	3.4	5.8
Physical inactivity	3.2	5.5
Overweight and obesity	2.8	4.8
High cholesterol	2.6	4.5
Unsafe sex	2.4	4.0
Alcohol use	2.3	3.8
Childhood underweight	2.2	3.8
All other	13.4	23

[a]Data for year 2002, from Prüss-Üstün et al. (2016).

From World Health Organization, WHO, 2009. Global health risks: mortality and burden of disease attributable to selected major risks.

TABLE 26.3 Top 10 Risk Factors for Causes of Human Mortality Globally for Year 2015

Risk Factor	Deaths (millions)	% of Total
Environmental Factors[a]	12.6	22.7
High blood pressure	10.7	19.2
Tobacco use	6.4	11.5
High blood glucose	5.2	9.3
High sodium diet	4.3	7.7
Overweight and obesity	4.3	7.7
High cholesterol	4.3	7.7
Alcohol use	2.3	4.1
Physical inactivity	1.6	2.9
Unsafe sex	1.5	2.7
All other	4.2	7.5

[a]Data for year 2012, from Prüss-Üstün et al. (2016).

Source: From GBD 2015 Risk Factors Collaborators: Global, regional, and national comparative risk assessment of 79 behavioural, environmental and occupational, and metabolic risks or clusters of risks, 1990–2015: a systematic analysis for the Global Burden of Disease, Lancet 388:1659–724, 2016.

TABLE 26.4 Estimated Annual Mortality Associated With Select Environmental Factors

Factor	Deaths (millions)	Data Source
Air Pollution—outdoor [particulate matter, ozone]	4.5	GBD, 2016
Air Pollution—indoor	2.8	GBD, 2016
Water pollution [sanitation effects]	1.8	GBD, 2016
Lead exposure [home, occupational]	0.5	GBD, 2016
Global climate change	0.2	World Health Organization, WHO, 2002
Hazardous waste sites	0.2	Extrapolation of data reported in Chatham-Stephens et al. (2013) and Pure Earth (2016)
Natural disasters	0.08	Aon, 2015
Residential radon	0.06	GBD, 2016

expenditures, including hospital, physician, and medication costs, long-term rehabilitation or home care, and nonclinical services such as management, support services, and health insurance costs; (2) indirect health-related expenditures, such as time lost from school or work, costs of special education, and the cost of investments in the health system (including health infrastructure, research and development, and medical training); (3) diminished economic productivity in persons permanently impaired by pollution effects; and (4) losses in productivity resulting from premature death.

Human welfare costs associated with pollution have been estimated to total several trillion dollars annually. For example, it is estimated that the human welfare costs associated with air pollution alone exceed $5 trillion annually (UNEP, 2017). These costs are typically not considered when assessing the impacts of environmental pollution to society. Note that these costs do not include the economic impacts of air pollution to ecosystems, wildlife, or crop production, or the impacts associated with interactions between air pollution and global climate change.

The stark statistics presented in this section illustrate the tremendous impacts of environmental pollution on human health and well-being. They also demonstrate the critical importance of addressing pollution issues. Characterizing and resolving the impacts of environmental pollution on human health and well-being is accomplished through the science of environmental health.

26.2 ENVIRONMENTAL HEALTH SCIENCE

This chapter is focused on environmental public and human health and how the environment can impact health. Public health is concerned with the health of all of the members of a community. This is in contrast to the field of medicine, which focuses on the health of individuals. The field of public health focuses on preventing illness and is a combination of science and policy. Numerous individual fields of study contribute to the broad field of public health, including epidemiology, statistics, biomedical and toxicological sciences, environmental health science, environmental science, and the social and behavioral sciences. For this chapter, we will focus on environmental health science.

We live in a world of exposures where chemicals and potential toxins are all around us. The World Health Organization's categorization of the different types of environmental health risks is presented in the chapter cover art plate. The field of environmental health works to understand how the environment affects people, with the ultimate goal of protecting environmental public health. An environmental health scientist asks:

- What is the fate and transport of a chemical in the environment?
- How are we exposed to chemicals and/or potential toxins (exposure route)?
- How much are we exposed to (dose)?
- How and why do some chemicals and/or toxins harm us?
- How and why does susceptibility vary?

An environmental health scientist may focus their efforts on (Figs. 26.1 and 26.2):

- Fate and transport of the chemical in the environment (source and environmental concentration/condition; activities of traditional environmental science): Relies mainly on environmental science to determine the fate of contaminants (see Chapters 7–9)
- Exposure science: Addresses the contact of humans and other organisms with chemical, physical, or biologic stressors over space and time and the fate of these stressors within the ecosystem and organisms—including humans (NRC, 2012). The exposure assessment is a component of the risk assessment process (see Chapter 29).
- Toxicology (Dose): Deposition of the toxicant in the body, which depends on the dose, absorption rate,

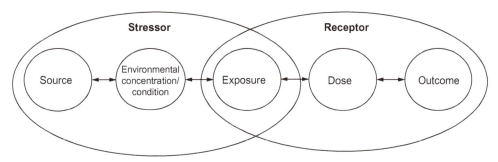

FIG. 26.1 The classic environmental health continuum. *(Adopted from U.S. Environmental Protection Agency, 2009. A Conceptual Framework for U.S. EPA's National Exposure Research Laboratory. EPA/600/R-09/003. National Exposure Research Laboratory, Office of Research and Development, U.S. Environmental Protection Agency [online]. Available: http://www.epa.gov/nerl/documents/nerl_exposure_framework.pdf.)*

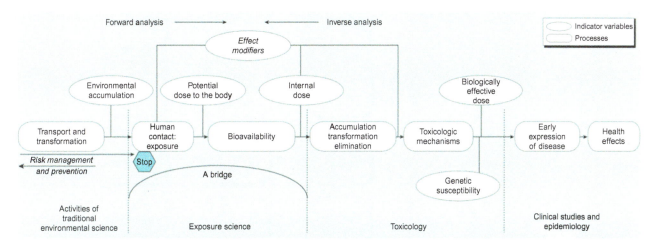

FIG. 26.2 Detailed composition of the field of environmental health. Process continuum from contaminant emissions to a health effect and application to risk reduction and prevention strategies. *(Adopted from Lioy, P.J., 1990. The analysis of total human exposure for exposure assessment: a multi-discipline science for examining human contact with containments. Environ. Sci. Technol. 24, 938–945.)*

metabolism, distribution, storage, and rate of excretion. These processes will influence the receptor's response (see Chapter 28).

- Epidemiology and Clinical Studies (Outcomes): Study of patterns of disease occurrence in human populations and the factors that influence these patterns. Environmental Epidemiology is concerned with characterizing how large numbers of people are exposed involuntarily to environmental agents and the resultant impacts.
- Statistics (Outcome): The collection, analysis, interpretation, and presentation of data.

The term "environment" can refer to either the physical (e.g., outdoor), built (e.g., indoor), or social environment. It also includes occupational health, workplace safety, and indoor living environments. Healthy (or unhealthy) housing is important to the field of environmental health as families may be exposed to mold, rodents, insects, lead (a heavy metal known to cause irreversible neurological damage), and indoor air pollution. The sources and impacts of indoor pollution are covered in Chapter 18. The types of environmental health hazards we are concerned with can be

biological (e.g., microbial, vector-borne diseases), chemical (e.g., solvents, pesticides), physical (e.g., particulates), and radiation. When we speak of exposures, we are referring to contaminants that may exist in an environmental medium, such as air (indoor and outdoor), water, food, and/or soils.

The presence of contamination in an environmental medium can serve as an exposure point, which becomes an exposure route if it connects to a human being. Fig. 26.3 demonstrates the numerous potential exposure pathways that exist, using the example of a leaking drum containing hazardous waste (ATSDR, 2005).

More recently, the term "exposure" has expanded in definition and the term *exposome* has been introduced (see Information Box 26.2). The exposome is composed of every exposure to which an individual is subjected from conception to death. Therefore it requires consideration of both the nature of those exposures and their changes over time. For ease of description, three broad categories of nongenetic exposures may be considered: internal, specific external, and general external (Fig. 26.4).

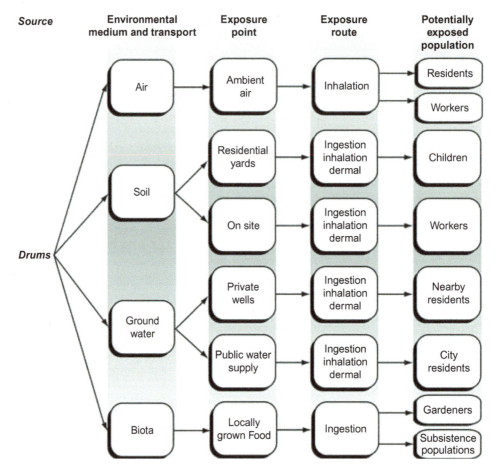

FIG. 26.3 Conceptual model for exposure pathway evaluation. *(Adopted from Agency for Toxic Substances and Disease Registry (ATSDR): Public Health Assessment Guidance Manual. Chapter 6: Exposure Evaluation: Evaluating Exposure Pathways. 2005. Available from: https://www.atsdr.cdc.gov/hac/phamanual/ch6.html.)*

INFORMATION BOX 26.2

Exposome is defined as the totality of exposures and associated biological responses one experiences across their life. It includes any factor of interest that may impact health (e.g., biological materials, noise, neighborhood factors, access to healthy foods).

26.3   HISTORY OF ENVIRONMENTAL HEALTH

In the late 19th century, individuals and families migrated to city centers for work (e.g., factory jobs) and other opportunities after the industrial revolution. These urban centers were growing fast and people were living closer together. A consequence of having densely populated areas was waste generation (human and industrial) and environmental pollution, primarily air and water. These densely packed areas lacked basic garbage collection, sewage systems, wastewater treatment facilities, and potable water. As populations grew,

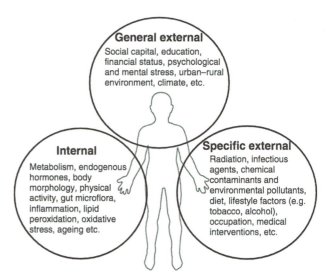

FIG. 26.4 Three different domains of the exposome are presented with nonexhaustive example for each of these domains. *(Adopted from Wild, C. W., 2012. The exposome: from concept to utility. Int. J. Epidemiol. 41(1), 24–32. Available from: https://doi.org/10.1093/ije/dyr236.)*

the amount of pollution and waste grew as well. As a result, more and more people were afflicted with environment-related illness and disease. These conditions fostered the development of environmental public health.

As factory work became more and more prominent, the impacts of unsafe workplace conditions on human health and well-being started to become more apparent. This initiated the field of occupational health. The fields of environmental health and occupational health grew together.

It is often difficult to recognize the health effects stemming from exposure of the general population to environmental substances. This is a result of several factors. First, exposure levels may be relatively low, and the impacts of exposure may not be apparent until many years later. Second, people respond differently to the same exposure due to inherent biological differences and to differences in lifestyle. Third, it is often difficult to trace a given health impairment to a specific environmental source. In contrast, occupational exposures are often much easier to characterize because exposures are more definitive and levels are typically higher. Because illnesses caused by occupational exposures are easier to recognize, workers are often in effect the guinea pigs that call attention to environmental hazards.

In general, the health of those in developed nations has improved a tremendous amount over the past century. For example, life expectancy among the U.S. population increased by 62%, from 47.3 years in 1900 to 76.8 in 2000. These unprecedented improvements in population health status are observed at every stage of life (National Center for Health Statistics, 2011). Much of the improvement is due to improved living conditions brought about by public health. Information Box 26.3 lists the Center for Disease Control's top 10 list of great public health achievements of the 20th century. Four out of the 10 (see bolded items in Information Box) can be attributed to advances in environmental and occupational health.

INFORMATION BOX 26.3

The Centers for Disease Control and Prevention Top 10 List of Great Public Health Achievements of the 20th Century:
1. Vaccine-Preventable Diseases
2. Prevention and Control of Infectious Diseases
3. Tobacco Control
4. Maternal and Infant Health
5. Motor Vehicle Safety
6. Cardiovascular Disease Prevention
7. Occupational Safety
8. Cancer Prevention
9. Childhood Lead Poisoning Prevention
10. Public Health Preparedness and Response

Source: U.S. Centers for Disease Control and Prevention, https://www.cdc.gov/mmwr/preview/mmwrhtml/mm6019a5.htm.

Classic environmental health success stories include John Snow's discovery of the cause of cholera outbreaks in London, Jane Addams work to prove the dangers to human health from poor sanitation and occupational exposures (e.g., Hull House in Chicago, 1889), and Alice Hamilton who listened to workers' accounts of the workplace experience to help develop her hypothesis on why certain occupations and industrial processes were hazardous. Two examples of hallmark environmental health case studies are presented for mercury and lead.

26.3.1 Mercury

Most human exposure to elemental mercury occurs through the inhalation of mercury vapors. Some absorption through the skin may occur from contact with contaminated air. Workers in industries where elemental mercury is used (e.g., light bulb manufacturing) are exposed to levels much higher than the general public. Inorganic mercury (e.g., mercuric chloride) can be present in water or soil and can be taken up by ingestion. Certain microbes in aquatic systems can transform inorganic mercury into the more toxic organic methylmercury (MeHg).

Around 1865 it was observed that mercury exposure in a factory setting led to neurological symptoms. As stated before, Alice Hamilton conducted research alongside workers to understand their occupational exposures. Her efforts uncovered that workers making felt hats were being exposed to mercury, leading to mad hatter disease (Information Box 26.4). This occupational exposure case led to an understanding of the health effects of mercury.

Mercury continues to be a primary environmental health concern (Information Box 26.5). Emissions from coal-burning power plants is a major source of mercury, leading to air pollution, and subsequently to water pollution, and ultimately to contamination of fish. Fish consumption is the main source of human exposure to methylmercury.

In May 1956 a chemical plant discharged waste water that contained high levels of methylmercury into Minamata

INFORMATION BOX 26.4 Mercury: Alice's Mad Hatter and Work-Related Illness

- The Hatter's erratic, agitated behavior in the classic story in Lewis Carroll's *Alice in Wonderland* refers to a real industrial hazard during 1865 in Britain.
- Hat-makers exhibited slurred speech, tremors, irritability, shyness, depression, and other neurological symptoms leading to the term "mad as a hatter."
- The symptoms were associated with chronic occupational exposure to mercury: poorly ventilated rooms using hot solutions of mercuric nitrate to shape wool felt hats.
- In 1941, the U.S. Public Health Service ended mercury's use by hat manufacturers in 26 states.

INFORMATION BOX 26.5

Methylmercury—easily taken up by animals (e.g., fish, shellfish) and stored. Predatory fish that feed on smaller fish and shellfish further concentrate the methylmercury (*bio-magnification*). Humans can be exposed by eating these mercury-contaminated animals.

Short-Term, High-Level Exposure (Generally only in occupational settings) can lead to:

- Organ damage (primarily kidney), nausea, vomiting, diarrhea, tremors, memory loss, and vision/hearing changes
- Increase in blood pressure and heart rate
- Irritation of the skin, lungs, and eyes
  Long-Term, Low-Level Exposure can lead to:
- Negative effects on the central nervous system, motor ability, mood, concentration, short-term memory, speech and vision
- Cardiovascular and immunological effects
  Available from: https://superfund.arizona.edu/sites/superfund.arizona.edu/files/mercury_june_2013_final.pdf

Bay, Japan, where residents frequently fished. Residents who consumed fish and shellfish were exposed to elevated levels of methylmercury between 5 and 36 parts per million (ppm). For comparison, the US EPA/FDA regulatory limit for methylmercury in fish fillets is 0.3 ppm. This acute exposure led to what is now called Minamata disease, sometimes referred to as Chisso-Minamata disease. This is a neurological syndrome caused by severe mercury poisoning. In some cases, the Hg content in the hair of inhabitants of the Shiranui Sea coastline reached as high as 705 ppm. For comparison, the US EPA has set the Reference Dose

for mercury content in hair at 1.0 ppm. As of March 2001, 2265 victims had been officially recognized (1784 of whom had died) and over 10,000 had received financial compensation from Chisso (Boston University, n.d.). By 2004, Chisso Corporation was ordered to clean up its contamination and had paid $86 million in compensation; lawsuits and claims for compensation continue to this day (Boston University, n.d.).

26.3.2 Lead

Lead is a metal in the earth's crust that is found with other metals such as zinc, silver, and copper. It has been used by humans for many purposes throughout the years, leading to many potential routes of exposure. Lead-based solder that had been used to connect copper water pipes was banned in the 1980s, but may still be a source of lead in drinking water in older homes. In the United States, lead was used as a gasoline additive, but was banned beginning in 1973 and eliminated by 1996. Lead was widely used in paints. Its use for consumer products was banned in many countries starting in the mid-1970s; however it is still used in some countries.

Humans can be exposed to lead by ingesting contaminated food, water, and soil, swallowing or breathing chips or dust from lead-based paints, working in a job where lead is used, using healthcare products or folk remedies that contain lead, and engaging in certain hobbies in which lead is used. The various sources of lead are depicted in Fig. 26.5. In general, exposure to lead can lead to:

- Neurological effects
- Gastrointestinal Effects

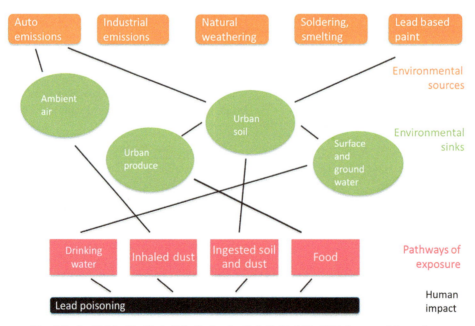

FIG. 26.5 Environmental Lead Cycle. *(Original in Clark, H.F., Brabander, D.J., Erdil, R.M., 2006. Sources, sinks, and exposure pathways of lead in urban garden soil. J. Environ. Qual. 35 (6), 2066–2074. https://doi.org/10.2134/jeq2005.0464, redrawn by Ramirez-Andreotta.)*

- Reproductive Effects
- Blood cell Synthesis
- Renal Effects

As presented in Table 26.4, approximately 500,000 deaths per year are attributed to lead exposure.

Due to the exposure pathways, health effects, and level of sensitivity, we are mainly concerned with children's exposure to lead. According to the Center for Disease Control (2014):

Lead-based paint and lead contaminated dust are the most hazardous sources of lead for U.S. children. Lead-based paints were banned for use in housing in 1978. All houses built before 1978 are likely to contain some lead-based paint. However, it is the deterioration of this paint that causes a problem. Approximately 24 million housing units have deteriorated leaded paint and elevated levels of lead-contaminated house dust. More than 4 million of these dwellings are homes to one or more young children. No safe blood lead level in children has been identified. Even low levels of lead in blood have been shown to affect IQ, ability to pay attention, and academic achievement. And, the effects of lead exposure cannot be corrected.

An infographic developed by the CDC for the issue of children and lead is presented in Fig. 26.6.

26.4 COMMUNITY ENVIRONMENTAL HEALTH

Environmental human health issues can be discussed at two scales, the community level and the global level. Community environmental health has evolved to focus on local-scale environmental health issues. In other words, the health of all of the residents of a specific community. Global-scale human health issues are typically associated with large-scale environmental disturbances and are discussed in the following section.

Community-scale health issues often revolve around hazardous waste sites (see Chapter 12 and 15). There are hundreds of thousands of these sites across the world, the vast majority of which are located within population centers. For example, twenty-five percent of Americans live within 3 miles of a hazardous waste site [U.S. Government Accountability Office, 2013]. As a result, we could say that one's zip code can be more important than their genetic code in terms of environmental health status (Graham, 2016). Thus these sites are a critical concern for community environmental health.

The primary human health concerns associated with hazardous waste sites are chronic (long-term) exposure of people living near the site to various toxic contaminants. Exposure typically occurs through consumption of contaminated water or inhalation of contaminated air or dust. It is difficult to determine direct cause-effect relationships between exposure and adverse human health outcomes at

FIG. 26.6 A CDC infographic describing how lead can affect a child's health and preventing childhood lead poisoning. (*Available from: https://www.cdc.gov/nceh/lead/publications/lead-infographic-final-full.pdf.*)

these sites for several reasons, including lack of historical exposure data, incomplete health assessments, and the potential contributions of other environmental and behavioral factors. A review of the scientific literature reported for human health outcomes associated with exposure to hazardous waste sites indicated potential causal relationships for liver, bladder, breast, and testis cancers, non-Hodgkin lymphoma, asthma, various congenital anomalies, low birth weight, and preterm birth (Fazzo et al., 2017). A study was conducted to estimate the disease burden risk for people living near hazardous waste sites in India, Indonesia, and the Philippines (Chatham-Stephens et al., 2013). The burden of disease and death was evaluated for 373 sites. Extrapolation of the results to the total estimated number of sites in the three countries (5000 unscreened and 373 screened sites) produced an estimated 66,747 deaths from cancer, specifically liver and lung cancer, projected to occur from exposures at all of the sites combined.

The United States has a history of environmental pollution disasters associated with hazardous waste sites that have significantly affected communities and their health. To name a few: Hinkley, California hexavalent chromium groundwater contamination case (as depicted in the movie "Erin Brockovich"); Woburn, Massachusetts trichloroethene groundwater contamination case (as depicted in the movie "A Civil Action"); and Love Canal, Niagara Falls, New York. See Information Box 26.6 for an overview of the Love Canal case.

The passage of relevant environmental laws and their associated regulations (Chapters 30 and 31) has greatly reduced the exposure of communities to hazardous wastes in most economically developed countries. However, exposures still occur, often through the advent of emerging contaminants (Chapter 12). A prime example is that of the PFAS issue discussed in Chapter 12, which has become a concern in several communities in the United States. Conversely, the issue of hazardous wastes sites and human health impacts is a growing concern for developing countries (Landrigan et al., 2017).

While chronic exposure associated with living near hazardous waste sites is a predominant source of community-scale health impacts, other sources exist. One source is the release of toxic elements as a result of accidents at industrial facilities that use hazardous materials. A prime example is the Bhopal disaster that occurred on December 3, 1984, where more than 40 tons of methyl isocyanate gas leaked from a pesticide plant in Bhopal, India. It immediately killed a minimum of 3800 people and caused significant morbidity and premature death for many thousands more. Another source is the result of poor management of public utilities and services, a prime example being the operation of public water supply systems. The recent case of drinking water quality issues in Flint, Michigan, provides an illustration of this issue. There also are natural sources of

INFORMATION BOX 26.6 The Phases of Love Canal

Phase 1

- 1890: W.T. Love began the development of tract for a "model city."
- Included construction of a canal between the upper and lower Niagara Rivers to allow hydroelectric power generation. Project was not completed; leaving partial canal (~1.6 km long).

Phase 2

- 1920s: The City of Niagara Falls began using canal as municipal waste landfill.
- 1942: Hooker Chemical Company given permission to use canal as disposal site for chemical wastes stored in drums.
- 1952: They later purchased the site and stopped use.

Phase 3

- 1953: Hooker deeded site to local school board after pressure to do so. Notified them of disposal activities.
- 1954–1957: A school, 800 private houses, and 240 low-income apartments were constructed at the site.

Phase 4

- 1976–1978: Investigations by reporters from local newspaper publicized the contamination and associated apparent health issues.
- 1978: Lois Gibbs canvasses neighborhood to document health effects in newborns and toddlers living in area.
- 1978: President Jimmy Carter announced a federal health emergency, called for the allocation of federal funds, and ordered the Federal Disaster Assistance Agency to assist the City of Niagara Falls to remedy the Love Canal site. This was the first time that emergency funds were used for a situation other than a natural disaster. ~1000 families were evacuated.
- 1980: Comprehensive Environmental Response, Compensation and Liability Act (Superfund) legislation enacted (see Chapter 30)
- 1983: Love Canal is the first site listed on the Superfund National Priorities List.

Phase 5

- 2004: Love Canal removed from Superfund list.
- Total site remediation costs = ~$400 million

contamination, as noted in Chapter 12, that can cause community-scale health issues. The specific case of how natural sources of contamination can affect human health is discussed in Chapter 27.

26.5   GLOBAL ENVIRONMENTAL HEALTH

In contrast to community environmental health, global environmental health deals with topics that affect populations across larger scales. These typically involve the human health impacts of global-scale phenomena—specifically those phenomena that were discussed in Chapter 25, in addition to pervasive pollution issues.

In most economically developed countries, the impacts of environmental pollution have been reduced significantly through regulatory action and technological advances. This has resulted in remaining pollution issues occurring typically at the community level, often associated with legacy contamination. Conversely, pollution remains pervasive for many developing countries, affecting large portions of the populace. According to a recent study, nearly 92% of pollution-related deaths occur in low- and middle-income countries (Landrigan et al., 2017). These countries are distributed primarily in Africa, Asia, and Eastern Europe. Air pollution and water pollution issues are often of primary critical concern for these areas.

The combination of pervasive environmental pollution with large-scale environmental disturbances such as drought can lead to widespread scarcities of resources critical to human health and well-being. This issue of resource scarcity is of critical concern as the global human population continues to grow. Two primary examples of critical scarcity issues are food and water.

26.5.1 Food Security

Food scarcity is a subset of *food security*, a ubiquitous term that has been defined in multiple ways, with over 200 definitions in published writings (Information Box 26.7). Regardless of how it is defined, food scarcity results in undernourished people in all regions of the world. Trends in the numbers of undernourished people in the world are shown in Fig. 26.7. In 2000, there were approximately 900 million people in the world that were undernourished. This number reached a high or almost 930 million in the early 2000s, but then declined from 2004 to 2015. However, since then, numbers have increased to 815 million, representing 11% of the world population. In particular, food security has become worse in parts of sub-Saharan Africa

and parts of Asia. "Zero hunger" is Goal 2 of the 17 Sustainable Development Goals proposed by the United Nations in 2015 (see Chapter 32).

Reasons for food scarcity are often linked to extreme weather events such as droughts or floods. Another major factor is internal and external conflict. A third major factor is loss of arable land and poor soil quality due to erosion and pollution. Fundamentally, food security and undernutrition is linked to the supply of and demand for food, much of which is dependent on crop production.

Climate, soil quality, and terrain significantly limit where crops can be grown productively. Some of these constraints can be mitigated with inputs like irrigation and fertilizer and the use of advanced technologies. One of the major triumphs of the 20th century was the "Green Revolution," which resulted in dramatic increases in crop yields and food production through the use of fertilizers and modern agricultural technology. This led to a decrease in the number of people in the world that were hungry despite a doubling of the world population (Godfray et al., 2010). However, the Green Revolution led to monoculture farming—large expanses of a single crop, and the use of large quantities of fertilizers and pesticides. These practices have resulted in environmental pollution and negative effects on ecosystem biodiversity. Efforts are underway to improve upon large-scale farming methods, while in addition local sourcing of food is becoming more widespread through, for example, development of small local farms, urban agriculture, and community gardens.

As populations continue to increase, the challenge of producing enough food while maintaining healthy environments will be enormous. We also need to focus on the threats of global climate change to food security. Here we outline some of the major issues that are likely to impact food security.

Global Climate Change and Food Security: All quantitative assessments indicate that global climate change will

INFORMATION BOX 26.7 Examples of Food Security Terminology

Food Security refers to conditions when all people, at all times have physical, social, and economic access to sufficient, safe, and nutritious food that meets their dietary needs and food preferences for an active and healthy life.

Household Food Security is the application of this concept to the family level.

Food Insecurity exists when people do not have adequate physical, social, or economic access to food as defined above.

Food Scarcity occurs when not enough food is produced. Prolonged food scarcity results in hunger and extreme food scarcity can result in famine.

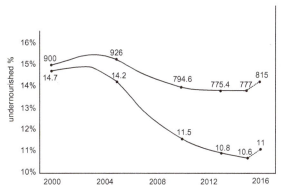

FIG. 26.7 Number of undernourished people in the world (millions). *(Adapted from: The State of Food Security and Nutrition in the World 2017. Rome, 2017, Food and Agriculture Organization of the United Nations.)*

adversely affect food security (Schmidhuber and Tubiello, 2007). If greenhouse gas emissions continue unabated, changes in temperature and precipitation will bring changes in land suitability and crop yield. Increased temperatures may also expand the range of agricultural pests and the ability of pests to survive winter and attack spring crops. The predicted regional shifts in land suitability and crop yields do not affect all regions of the world equally. Although total land for cropping and total prime land may stay the same globally, increases in suitable cropland at higher latitudes (developed countries) may be offset by decreases in suitable cropland at lower latitudes (developing countries). Overall, Africa is likely to see the largest losses in suitable cropland (Schmidhuber and Tubiello, 2007).

Increases in atmospheric carbon dioxide will also influence global food security. Higher CO_2 concentrations can potentially increase photosynthetic rates, enhancing plant growth, biomass accumulation, and final crop yield. However, the quality and nutritional value of crops may not increase correspondingly with higher yields.

Extreme weather events such as hurricanes, cyclones, floods, and droughts can also influence the stability of food production, adversely affecting food security. As global climate change continues, these dramatic weather events and climate fluctuations are likely to become more numerous and severe.

Carbon Sequestration and Global Food Security: Crop production can be severely reduced due to loss of soil organic carbon as soil is degraded (Lal, 2004). Conversely, soil carbon sequestration can result in significantly enhanced crop yields (Information Box 26.8). Strategies to increase the soil carbon pool include soil restoration, no-till farming, cover crops, nutrient management, land application of animal manures and biosolids, and water conservation and harvesting (see Chapter 27). An additional benefit of carbon sequestration is the potential to offset fossil fuel emissions of global warming gases (see Chapter 32).

Soil Phosphorus and Global Food Security: Crop production is heavily dependent on fertilizer applications, particularly of the macronutrients N, P, and K. However,

INFORMATION BOX 26.8 Influence of Soil Carbon Sequestration on Yields Obtained From Degraded Cropland Soils

Crop	Potential Yield Increase From 1 Ton per Acre of Additional Soil Carbon
Wheat	20–40 kg/ha
Maize	10–20 kg/ha
Cowpeas	0.5–1 kg/ha

Source: Adapted from Lal, R., 2004. Soil carbon sequestration impacts on global climate change and food security. Science 304, 1623–1627.

INFORMATION BOX 26.9 Multiple Causes of Honey Bee Losses

- Infection by the ectoparasitic mit *Varrou destructor*
- Infection by honey bee pathogens including several viruses
- Land-use change resulting in loss of habitat
- Increased pesticide usage
- Environmental pollution
- Climate change

phosphorus is derived from phosphate rock, which is a non-renewable resource that may be depleted in 50–100 years. With the demand for phosphate increasing, and the quality and quantity of remaining phosphate rock decreasing, scarcity of phosphate could adversely affect food security in the future.

Global Pollinators and Global Food Security: Globally the most important pollinator worldwide is the honey bee *Apis mellifera*. This species is a primary vector for crop pollination (Potts et al., 2010). However, there have been severe regional declines in domestic honey bee stocks in the United States and Europe. In the United States, there has been a loss of 59% of colonies between 1947 and 2005, and in Europe, a loss of 25% between 1985 and 2005 (Potts et al., 2010). This has raised concerns about the impact of pollinator declines on crop production. There are many causes of the decline of the honey bee, and it is believed that these drivers of loss are interactive (Information Box 26.9).

Water Scarcity and Global Food Security: It is well documented that access to clean water to support human activities is becoming scarcer. Safe potable water is severely limited in many parts of the world, as will be discussed in the following paragraphs. The limited availability of safe potable water in itself poses direct threats to human health and well-being. Water scarcity also adversely affects global food security. Crops require water, and thus water scarcity will limit crop production. This issue is a critical concern for arid and semiarid regions of the world. Crop production in many arid areas of the world is made possible through irrigation, which uses surface waters such as the River Yangtze in China or groundwater in the Southwest USA. Sustaining the water resources required for global food security will only be successful if sustainable water management policies are followed coupled to technologies that enhance irrigation efficiency.

26.5.2 Water Security

Water scarcity and security has been and will continue to be a key global-scale environmental issue. Water security is defined by the United Nations Water program as "the capacity of a population to safeguard sustainable access to adequate quantities of acceptable quality water for sustaining livelihoods, human well-being, and socio-economic development, for ensuring protection against water-borne pollution and water-related disasters, and for preserving ecosystems in a climate of peace and political stability." Water scarcity is the lack of a sufficient supply of usable water. Clean water and sanitation comprises Goal 6 of the 17 Sustainable Development Goals proposed by the United Nations in 2015 (see Chapter 32).

Water is essential for human life in a number of ways. In addition to its importance for bodily functions, it supports food production as well as energy production and distribution. The interconnectedness between water, food, and energy, all required to support human life and well-being, is recognized with the concept of the Water-Energy-Food Nexus, discussed in Chapter 32.

A recent study employing NASA satellite data examined the distribution of freshwater supplies across the globe, including surface waters, soil moisture, groundwater, and snow and ice (Rodell et al., 2018). It was determined that humid regions are becoming wetter while arid/semiarid regions are becoming drier. High levels of water scarcity were identified in several regions around the world. Areas in northern and eastern India, the Middle East, northern Africa, southwest United States, and Australia were among locations where significant declines in the availability of freshwater are causing current problems.

According to a recent report by the United Nations Water program (United Nations Water, 2018), an estimated 3.6 billion people (nearly half the global population) currently live in areas that are potentially water scarce at least one month per year. It is estimated that this population could increase to 4.8–5.7 billion by 2050. Access to safe water is clearly essential for human health and well-being. Limited availability hence poses severe threats to such health and well-being. These issues were illustrated with the example of Lake Chad in Chapter 25 (Information Box 25.2).

Water scarcity is a function of several interrelated factors. Global water use has increased by a factor of six over the past 100 years and continues to grow steadily at a rate of about 1% per year (United Nations Water, 2018). This is a result of population growth, economic development, and changing consumption patterns. We can anticipate that global water demand will continue to increase in conjunction with continued population growth and economic development. While demand has increased, the quantity of available fresh water has decreased through global climate change, as illustrated by the recent study cited before. This is exemplified by the severe, long-term droughts affecting several regions of the world. In addition, water pollution has led to reduced water quality for many regions, further reducing the amount of clean freshwater available. For example, water quality has declined in most

rivers in Africa, Asia, and Latin America (United Nations Water, 2018). This combination of factors poses significant threats to water security for many regions of the world.

26.5.3   Global Climate Change

Global climate change is anticipated to exacerbate the impacts of pollution and other environmental disturbances on human health and well-being. Prime examples were discussed before with respect to food and water security. In addition, the many impacts of global climate change are likely to pose other, additional risks to human health and well-being. An illustration of the numerous potential impacts of global climate change on human health is presented in Fig. 26.8. The case of the influence of climate change on the incidence of vector-borne diseases is highlighted in Information Box 26.10.

It is estimated that global climate change caused approximately 150,000 human deaths in 2000 (WHO, 2002). Climate change was estimated to be responsible for approximately 2.4% of worldwide diarrhea, 6% of malaria in some middle income countries, and 7% of dengue fever in some industrialized countries. A recent study estimated that global climate change will cause approximately 250,000 additional human deaths annually between 2030 and 2050 (WHO, 2014). The total includes 38,000 due to heat exposure, 48,000 due to diarrhea, and 60,000 due to malaria. The study did not evaluate all environmental causes of death that could potentially be influenced by climate change, such as river flooding and water scarcity. It is estimated that the human welfare costs of global climate change will range between $2 and $4 billion per year by 2030 (WHO, 2018b).

26.6   ANTIMICROBIAL RESISTANCE

Antimicrobial resistance (AMR) is the development of the ability by bacteria to resist treatment. As a result, options for treating various bacterial infections in humans and other organisms become more limited, which can lead to increased rates of morbidity and mortality. AMR is considered to be a major current human health issue globally. The U.S. Centers for Disease Control estimated that a minimum of 2 million people in the United States had antibiotic-resistant infections, resulting in 23,000 deaths, in 2017 (CDC, 2013). The human welfare costs for AMR in the United States are estimated to approach $55 billion a year. An estimated 25,000 annual deaths were attributed to antibiotic-resistant bacterial infections in the European Union, Iceland, and Norway (ECDC, 2009). The World Health Organization reported that multidrug-resistant tuberculosis was responsible for 240,000 deaths globally in 2016 (WHO, 2017). The issue of AMR is an example of a human health issue that is not a direct result of environmental pollution, but rather

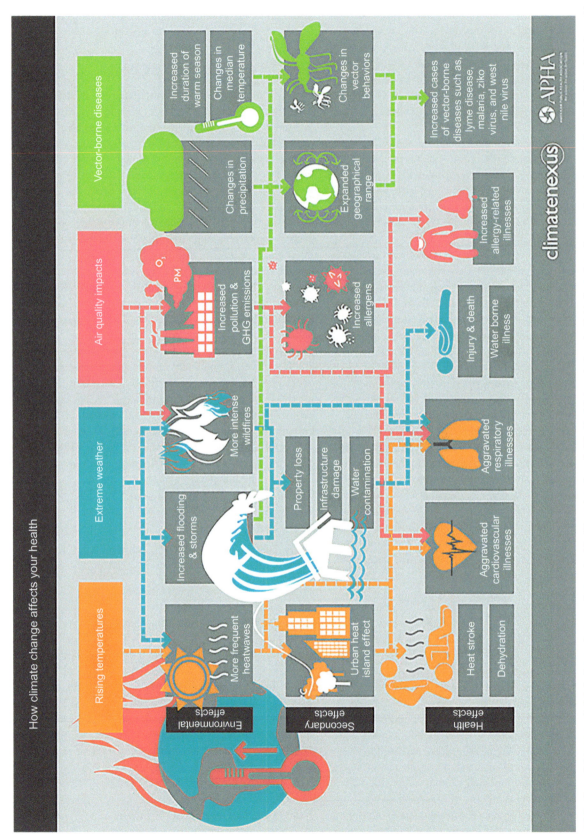

FIG. 26.8   Potential impacts of global climate change on human health. *(From American Public Health Association. Available from: https://www.apha.org/news-and-media/multimedia/infographics/how-climate-change-affects-your-health.)*

INFORMATION BOX 26.10 Global Climate Change and Vector-Borne Diseases

Vector-borne diseases are illnesses that are transmitted by vectors, which include mosquitoes, ticks, and fleas. These vectors can carry infective pathogens such as viruses, bacteria, and protozoa, which can be transferred from one host (carrier) to another. The seasonality, distribution, and prevalence of vector-borne diseases are influenced significantly by climate factors, primarily high and low temperature extremes and precipitation patterns.

Climate change is likely to have both short- and long-term effects on vector-borne disease transmission and infection patterns, affecting both seasonal risk and broad geographic changes in disease occurrence over decades. While climate variability and climate change both alter the transmission of vector-borne diseases, they will likely interact with many other factors, including how pathogens adapt and change, the availability of hosts, changing ecosystems and land use, demographics, human behavior, and adaptive capacity. These complex interactions make it difficult to predict the effects of climate change on vector-borne diseases.

Source: *Reproduced from: The Impacts of Climate Change on Human Health in the United States, U.S. Global Change Research Program, 2016.*

List of Diseases Caused by Mosquitoes, Ticks, and Fleas

Information from the Center for Disease Control

Mosquito-Borne

Chikungunya	Malaria
Dengue	St. Louis Encephalitis
Eastern Equine Encephalitis	West Nile
Japanese Encephalitis	Yellow Fever
La Crosse Encephalitis	Zika
Tick-Borne	
Anaplasmosis/Ehrlichiosis	Powassan
Babesiosis	Spotted Fever Rickettsia
Lyme Disease	Tularemia
Flea-Borne	
Plague	Typhus

It is estimated that these and other parasitic and vector diseases caused approximately 130,000 deaths in 2016 (WHO, 2018a).

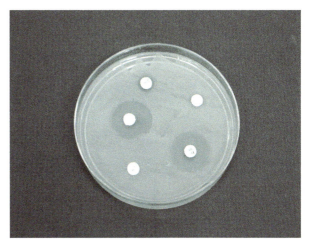

FIG. 26.9 Zone of inhibition of bacterial growth on a spread plate. The inhibition is due to diffusion of antibiotics from antibiotic filter disks. *(Photo courtesy: Josephson, K.L., 2004. From Environmental Microbiology: A Laboratory Manual, second ed. Elsevier Academic Press, San Diego, CA.)*

results from the intersection of human activity and natural environmental systems.

Antibiotics are natural compounds produced by microorganisms that kill or inhibit other microorganisms. The influence of antibiotics on bacterial growth is shown in Fig. 26.9. Perhaps the most effective antibiotic discovered was penicillin, first isolated by Sir Alexander Fleming in 1929, from the fungus *Penicillium*. *Penicillium* was highly effective in combating staphylococcal and pneumococcal infections. Later, another potent antibiotic was discovered by Selman Waksman in 1943. This antibiotic streptomycin was isolated from the actinomycete *Streptomyces griseus*, a feat for which Waksman was given the Nobel Prize in 1952.

Bacteria are prokaryotic organisms with the ability to metabolize and replicate very quickly. They are also very adaptable genetically. When confronted with an antibiotic, there need only be one bacterial cell with a genetic or mutational change that confers resistance to that antibiotic that subsequently allows for the proliferation of antibiotic-resistant bacteria (ARB) that contain antibiotic-resistant genes (ARGs). Thus the more antibiotics are used, the greater the likelihood of antibiotic-resistant strains developing. The conundrum then is that the more an antibiotic is used to prevent infectious disease, the less effective it will become over time. In addition, public health risks increase significantly when bacteria accumulate resistance to multiple antibiotics, making them particularly difficult to control as in the case of methicillin-resistant *Staphylococcus aureus* (MRSA).

AMR or *antibiotic drug resistance* is the acquired ability of an organism to resist the effect of a chemotherapeutic agent to which it is normally susceptible. In general, there are three ways in which a microbe can become resistant:

(a) Selective pressure: In some cases a fraction of a bacterial population may have a native, inherent ability to resist a given antimicrobial agent. These bacteria will then survive treatment and propagate, and their progeny will eventually become the dominant members of the population.

(b) Mutational change (spontaneous): One way microbes can become more tolerant of an antibiotic is to alter the target of an agent within the cell. For example, spontaneous mutations in the genes encoding ribosomal RNAs can prevent antibiotics such as tetracycline from binding and blocking gene translation.

(c) Gene transfer of foreign DNA: Host bacteria can obtain resistance to specific antibiotics via horizontal gene transfer from other antibiotic-resistant bacteria. Genes encoding resistance can be located on a plasmid or on a chromosome. ARGs can potentially make human pathogenic bacteria resistant to clinically used antibiotics. In addition, mobile genetic elements (MGEs) such as plasmids or transposons can mobilize ARGs, resulting in horizontal gene transfer (Brooks *et al.*, 2007). Such transfer can be via conjugation, transformation, or transduction.

The original source of antibiotics and ARBs was soil microorganisms, which resulted in ARBs being detected in both soil and waters. However, another anthropogenic source of ARBs is human and animal wastes. This has resulted in ARBs and ARGs being detected in domestic wastewater, agricultural waste releases from concentrated animal feedlot operations (CAFOs), biosolids and animal manures, and hospital waste discharged into sewers. The role of anthropogenic activity in the incidence of ARB and ARG has led to the introduction of the term "environmental antibiotic resistance." Select sources of antibiotic-resistant bacteria are as follows.

Soils: The vast majority of antibiotics are natural products synthesized by soil microorganisms (Pepper et al., 2018). These antibiotics are produced by diverse populations of soil microorganisms including bacteria, fungi, and actinomycetes (Pepper, 2013). These natural products can be used by indigenous soil microbes as a form of self-defense against neighboring soil microorganisms. Interestingly, the gene clusters that result in antibiotic production are always present in the soil environment, but are only expressed under very specific conditions, such as interactions with other microbes (Zhu et al., 2014). Studies have shown that even pristine, undisturbed soils contain ARB, and that "environmental antibiotic resistance" has been a naturally occurring factor in nature for more than 3 billion years (Gaze et al., 2013). Culture-based estimates of ARB extracted from soils that were resistant to tetracycline, ciprofloxacin, cephalothin, and ampicillin range from 10^6 to 10^7 colony forming units (CFU) per gram of soil (Brooks et al., 2007).

Water and wastewater: Similar to soils, a variety of environmental waters contain ARB and ARG including rivers, marine environments, and potable sources of water. Total bacterial numbers and diversity tend to be less in water sources than in soils, and organic carbon availability directly influences the levels of heterotrophic plate count (HPC) bacteria cultured from environmental waters.

Foodstuffs: Produce items traditionally consumed raw are known to contain significant numbers of ARB (Holvoet et al., 2013). Meat, poultry, and leafy greens can all be sources of ARBs (Oliveira *et al.*, 2012).

The sources of these ARBs can be from the use of antibiotics in food production, resulting in antibiotic resistance in foodborne pathogens (Friedman, 2015). Other sources of ARBs in vegetables are contaminated soil and water (Oliveira et al., 2012). For example, Holvoet et al. (2013) detected *E. coli* ARB in irrigation water and lettuce in the Netherlands, and suggested that cattle may be a source of the contamination events. They emphasized the use of good agricultural practices in preventing *E. coli* exposure to produce crops. Chajecka-Wierzchowska et al. (2016) detected enterococcus strains in ready-to-eat meats that were resistant to streptomycin, erythromycin, fosfomycin, rifampicin, tetracycline, and tigecycline. Overall, consumption of foodstuffs, particularly those eaten raw can lead to human exposure to both ARBs and ARGs.

Municipal and animal wastes: Biosolids, treated effluents, and animal manures all contain ARBs and ARGs. This in turn has led to speculation about the impact of land application of these materials and discharge of effluent into rivers, on the levels of antibiotic resistance in the environment. The concern has centered on whether such resistance could increase the number of antibiotic-resistant pathogens capable of infecting humans (CDC, 2013). However, calculation of the number of added ARBs and ARGs from land application at typical rates shows that the numbers added relative to indigenous resistant bacteria already present in soil appear to be very low. The impact of land application of Class B biosolids was evaluated by Brooks et al. (2007). They found no significant increase in soil ARBs following biosolid land application. In contrast, ARG levels have been documented to increase following swine manure land application in several studies. Sui et al. (2016) reported levels of up to 10^8 ARGs per g soil, but these enhanced levels quickly dissipated. Both Heuer and Smalla (2007) and Fahrenfeld et al. (2014) reported that ARG levels increased after swine manure applications, but decreased to background levels within 2 months.

Hospitals: Antibiotics are frequently given to patients during hospital stays in order to prevent infections, particularly following surgical procedures. In developed countries, most of the problems related to antibiotic resistance are generated in hospitals due to the intensive use of antibiotics (Levy and Marshall, 2004). Methicillin-resistant *Staphylococcus aureus* is probably the most important issue related to *hospital-acquired infections* (HAI) and results in thousands of deaths annually. Measures to manage and prevent the spread of drug resistance in hospitals and HAI are shown in Information Box 26.11. Due to the large use of antibiotics within hospitals, effluents generated from hospitals can pose significant risks with respect to the spread of antibiotic resistance. Antibiotic residues as well as ARB and ARG are found in hospital effluents and make their way to wastewater treatment plants (Harris et al., 2012).

Resistant enterococci have been identified in hospital effluents at concentrations of 10^3 CFU per mL (Verlicchi et al., 2013). Harris et al. (2014) documented antimicrobial-resistant *E. coli* in a hospital effluent wastewater at levels of $\simeq 10^5$ CFU per 100 mL. However, the same study suggested that the hospital effluent was not the causative factor in the development and persistence of antibiotic-resistant *E. coli* within the receiving wastewater treatment plant. Yilmaz et al. (2017) characterized hospital wastewaters in Turkey and found bacterial isolates with multiple antibiotic resistances including ciprofloxacin, trimethoprim, and ceftazidime. Overall, the contribution of hospital effluents to the prevalence of resistance in WWTP effluents is not well understood.

Antibiotic-resistant bacteria are widely dispersed throughout the environment and have been present on Earth for billions of years. Generally, it is only in specialized environments that their presence adversely affects human health and well-being. In hospitals, the high usage of antibiotics and the fact that sterilized surfaces reduce biotic competition helps to generate ARB. This enhances the potential for both antibiotic-resistant pathogens and HAI.

While the development of AMR is a natural process, human activity has exacerbated its rate of occurrence and magnitude of effects. Several factors are responsible for this. One major factor is the inappropriate and misuse of antibiotics in treatment of humans. For example, antibiotics are often used for illnesses for which they are not effective, such as common viral infections (colds and influenza). Another cause of AMR is the use of antimicrobial agents in products for which they are not necessary or effective. One prime example of this is the use of triclosan and other antibiotics in soaps and washes. Antimicrobial washes became widespread consumer products, despite the fact that there is no scientific evidence that they are any better than plain soap and water. Because of concern over their potential to contribute to AMR, and the fact that there is no evidence that they are more effective than soap and water, the U.S. Food and Drug administration in 2016 banned the use of triclosan and 18 other active ingredients in consumer wash products. Another example is the use of antibiotics in the feed of healthy animals as a purported means to enhance growth. However, there is no clear understanding of how this might work, and scientific studies indicate minimal growth gains at best (e.g., Angulo et al., 2005; Graham et al., 2007). Efforts are now underway to reduce and stop such use.

Several organizations have developed guidelines to mitigate the advent and severity of AMR. An example is presented in Information Box 26.12. We see that responsibility lies across the full spectrum of stakeholders, from the public, to industry, to healthcare professionals, to policy makers.

26.7 SOCIAL VULNERABILITY AND RESILIENCE

When characterizing environmental health or developing interventions, it is important to consider how humans—individuals, communities, and societies—respond psychologically and socially to environmental disturbances. This idea is represented by the interrelated concepts of social vulnerability and resilience. *Social vulnerability* describes the susceptibility of individuals and communities to experiencing adverse effects from an environmental disturbance. *Resiliency* refers to the ability of individuals and communities to cope with, adapt to, and recover from an environmental disturbance. In other words, how well is an individual or community able to "bounce back" by recovering from adversity using its own resources.

Ecological resilience was discussed in Chapter 6. Social resiliency can be thought of in a similar manner. An ecological or human system has certain inherent vulnerabilities to change, depending upon its properties and conditions. Exceeding the limits of resiliency will cause irreparable changes to the system. There are many factors that influence social vulnerability and resilience, including demographics, social connectedness aspects (sense of community, communication capacity), economic equality, and infrastructure support (e.g., equitable access to schools, healthcare, etc.).

The Social Vulnerability Index (SoVI[®]) has been developed to measure the social vulnerability of U.S. counties to environmental hazards. The index is a comparative metric that facilitates the examination of the differences in social vulnerability among counties. The index synthesizes 29 socioeconomic variables that the research literature suggests contribute to a reduction in a community's ability to prepare for, respond to, and recover from hazards. A map of county-by-county SoVI for the United States for the period 2010–2014 is presented in Fig. 26.10. We see a great geographic disparity in vulnerability distribution within and among states. As we might anticipate, a recent study by Bergstrand et al. (2015) found a significant correlation between vulnerability and resilience, indicating that counties that are more susceptible to harm (more vulnerable) also lack the means to rebound effectively (less resilient).

Community resilience is defined as the sustained ability of a community to withstand and recover from adversity (e.g., economic stress, pandemic influenza, anthropogenic or natural disasters). A community's capacity is dependent upon its access to human, economic, political, and social capital. The community resilience literature strongly embraces the importance of a sustained commitment to improving connectedness (both social networks and information linkages) between individuals, organizations, and formal governmental agencies as a primary objective of building community resilience.

Chandra et al. 2011 describe the five core components of community resilience:

- Physical and psychological health
- Social and economic equity and well-being,
- Effective risk communication
- Integration of organizations (governmental and nongovernmental)
- Social connectedness

Abramson et al. (2015) proposed the Resilience Activation Framework in an attempt to describe the mechanisms by which access to social resources activate and sustain resilience capacities for optimal mental health outcomes postdisaster can lead to the development of effective preventive and early intervention programs. In general, community resilience attributes interact with individual resilience attributes (Fig. 26.11). They include:

- Human capital = access to a healthy and capable population
- Economic capital = access to money and other financial instruments and assets
- Political capital = access to both capable governance and to those institutions that influence the distribution of resources
- Social capital = community's access to local institutions and networks that promote collective cohesion and self-efficacy

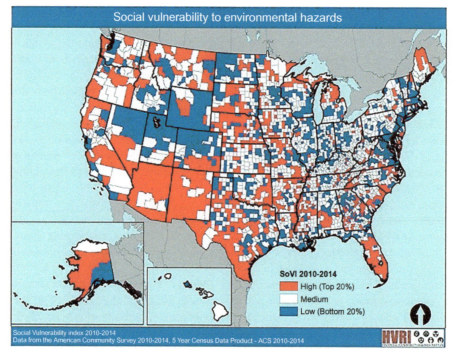

FIG. 26.10 Social Vulnerability Index for the United States—2010–2014. *(Available from: http://artsandsciences.sc.edu/geog/hvri/sovi%C2%AE-0.)*

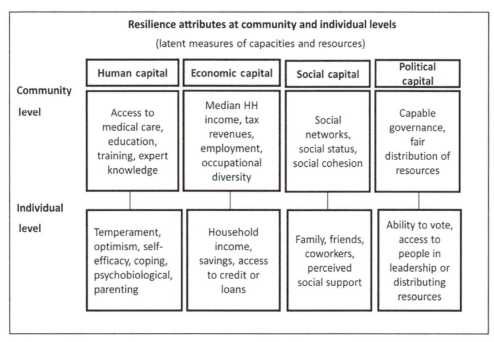

Resilience attributes at community and individual levels
(latent measures of capacities and resources)

	Human capital	Economic capital	Social capital	Political capital
Community level	Access to medical care, education, training, expert knowledge	Median HH income, tax revenues, employment, occupational diversity	Social networks, social status, social cohesion	Capable governance, fair distribution of resources
Individual level	Temperament, optimism, self-efficacy, coping, psychobiological, parenting	Household income, savings, access to credit or loans	Family, friends, coworkers, perceived social support	Ability to vote, access to people in leadership or distributing resources

FIG. 26.11 Resilience Attributes at Community and Individual Levels. *(Adopted from Abramson, D.M., Grattan, L.M., Mayer, B., et al., 2015. The resilience activation framework: a conceptual model of how access to social resources promotes adaptation and rapid recovery in post-disaster settings. J Behav Health Ser Res 42(1), 42–57.)*

Work is ongoing to better understand the issues of social vulnerability and resilience. Concomitantly, efforts are underway to improve resiliency to better prepare communities for future environmental disturbances.

26.8 THE INTERCONNECTION BETWEEN ENVIRONMENTAL ISSUES AND SOCIOPOLITICAL CONFLICT

Conflict occurs when two groups with subjective rule sets interact wishing to impose their pattern of organization on the other. Conflict may arise because of competition, incompatibility of interests, and irreconcilable differences. Conflicts can be nonviolent or violent. Violent conflict involves the use of weapons to gain desired outcomes. Nonviolent conflict is a state of disagreement that results into no physical harm but it can engender ill-feelings, distrust, and disaffection toward the perceived opponent, and it has the potential of degenerating into violent conflict if not properly managed (Laleye, 2015, p. 100).

Environmental issues result from environmental changes that are caused by natural variability and/or anthropogenic activities. In some instances, environmental issues are categorized as a threat to security because human health and well-being is influenced by accessibility to and quality of environmental resources and the functioning of environmental systems. Examples of this were discussed before with respect to food and water security. Threats to the availability of natural resources and services can have profound

impacts on human health and well-being. The Lake Chad case study discussed in Chapter 25 illustrates this situation. Water from Lake Chad is crucial to the 30 million fishermen and farmers for their livelihood. Water shortages and resultant loss of jobs deprived vulnerable people of their sources of livelihoods.

Once an environmental issue becomes a threat to security, it can lead to sociopolitical conflict. This is especially true for scarce and less substitutable natural resources such as water. For example, water scarcity was identified as one of the significant causes of armed conflicts between 1989 and 2003 (Eriksson et al., 2003). The causal nature of environmental issues and sociopolitical conflicts stems from three conditions: (1) allocation of resources to different users and uses, (2) importance of resources in livelihood sustenance (discussed in the preceding paragraph), and (3) inadequate resource governance.

Recourse allocation refers to the distribution of natural resources to different users and for different uses, and to the distribution of financial and human capital to address environmental issues. Resource utilization involves assigning shares of natural resources to different users and uses such as water for irrigation between upstream and downstream users. For example, for the case of Lake Chad, in 1992 there were clashes between upstream (Nigeria) and downstream (Niger) communities over access to the waters from the Tiga and Challawa Gorge dams at the south-west end of Lake Chad (Odada et al., 2006). The nature and magnitude of conflict or tension will differ between private and common property resources. Another form of resource conflict may emerge from disagreements on the alternative (end) uses of a natural resource, particularly land. For example, one group of people may want to use the land for agriculture and another group for eco-tourism.

Inadequate resource governance refers to the institutional, technical, and administrative capacities necessary for resource management and conservation. Lack of clear and effective institutional arrangements can lead to poor management and foment conflict. In Lake Chad, fishing, farming, and upstream water diversion, as well as dumping of household waste take place with minimal governmental intervention and regulation. The Lake Chad is a classic case of the tragedy of the commons (Chapter 6).

Environmental issues can also exacerbate or catalyze sociopolitical conditions due to resource management, ownership, and benefit sharing. For example, in Sudan, the conflict between north and south started in the 1940s because of religious, ethnolinguistic, and class reasons. However, when oil was discovered in central Sudan in 1978, the war intensified due to competition for ownership and shares in the benefits of the country's oil and gas reserves.

Conflicts can damage resources in terms of quantity and quality. In addition, the reduction and/or degradation of resources due to conflict can trigger migration of people to another area, imposing greater resource demand on that region. This kind of migration can create social and political instability and raise tensions between newcomers and hosts (Carius et al., 2004).

26.9 SOLUTIONS TO ENVIRONMENTAL HEALTH ISSUES

As described before, many human health issues originate from or are exacerbated by a pollution event. It then stands to reason that improving the health of the environment can improve human health and well-being. The World Health Organization identified nine ways in which we can improve the environment to improve human health and well-being (Fig. 26.12).

FIG. 26.12 Nine ways in which we can improve the environment to improve human health and well-being. *(World Health Organization)*

Human health and well-being are also significantly influenced by the availability or scarcity of natural resources and ecosystem services. As human population continues to increase, the demand on these services and the impacts to the environment are going to continue to increase. Maintaining and improving human health and well-being under these conditions requires sustainable management of resources and human-built systems. This topic is covered in Chapter 32.

26.10 QUESTIONS AND PROBLEMS

1. What is the meaning of "environmental health"?
2. Why are workers considered Guinea pigs in terms of health effects from exposure to contaminants. Discuss a case study example not presented in the text.
3. What causes conflict over environmental issues?
4. Discuss an example of how global climate change may impact human health and well-being.
5. Use the SoVI to evaluate the vulnerability of your home county and state.
6. Describe other examples of antibiotic resistance besides MSRA. How can we reduce the incidence of antibiotic-resistant bacteria in the environment?
7. What is food security, and what factors influence it?
8. What can we do to improve water security?
9. Describe how air pollution may contribute to cardiovascular disease or stroke.

REFERENCES

Abramson, D.M., Grattan, L.M., Mayer, B., et al., 2015. The resilience activation framework: a conceptual model of how access to social resources promotes adaptation and rapid recovery in post-disaster settings. J. Behav. Health Ser. Res. 42 (1), 42–57. https://doi.org/10.1007/s11414-014-9410-2.

Agency for Toxic Substances and Disease Registry (ATSDR), 2005. Public Health Assessment Guidance Manual. In: Chapter 6: Exposure Evaluation: Evaluating Exposure Pathways. Available from: https://www.atsdr.cdc.gov/hac/phamanual/ch6.html.

Angulo, F.J., Collignon, P., Wegener, H.C., Braam, P., Butler, C.D., 2005. The routine use of antibiotics to promote animal growth does little to benefit protein undernutrition in the developing world. Clin. Infect. Dis. 41, 1007–1013.

Aon (2015). 2015 annual global climate and catastrophe report. Available from: http://thoughtleadership.aonbenfield.com/Documents/20160113-ab-if-annual-climate-catastrophe-report.pdf.

Bergstrand, K., Mayer, B., Brumback, B., Zhang, Y., 2015. Assessing the relationship between social vulnerability and community resilience to hazards. Soc. Indic. Res. 122, 391–409.

Boston University, n.d. Minamata disease. Available from: http://www.bu.edu/sustainability/minamata-disease/. Accessed 25 May 2018.

Brooks, J.P., Maxwell, S.L., Rensing, C., Gerba, C.P., Pepper, I.L., 2007. Occurrence of antibiotic-resistant bacteria and endotoxin associated with the land application of biosolids. Can. J. Microbiol. 53, 616–622.

Carius, A., Dabelko, G.D., Wolf, A.T., 2004. Water, conflict, and cooperation. ECSP Report, Issue 10 (Policy brief—the United Nations and Environmental Security). 2004. URL: http://www.wilsoncenter.org/sites/default/files/ecspr10_unf-caribelko.pdf.

CDC, 2013. Antibiotic Resistance Threats in the United States. Center for Disease Control and Prevention. U.S. Department of Health and Human Services, Washington, DC, USA, pp. 1–114.

Center for Disease Control, 2014. Prevention tips. Available from: https://www.cdc.gov/nceh/lead/tips.htm. (Accessed 25 May 2018).

Chajecka-Wierzchowska, W., Zadernowska, A., Laniewoka-Trokenheim, K., 2016. Diversity of antibiotic resistance genes in *Enterococcus* strains isolated from ready-to-eat-meat products. J. Food Sci. 81, M2799–M2807.

Chandra, A., Acosta, J., Howard, S., et al., 2011. Building community resilience to disasters: a way forward to enhance national health security. Rand Health Quat. 1 (1), 6.

Chatham-Stephens, K., Caravanos, J., Ericson, B., Sunga-Amparo, J., Susilorini, B., Sharma, P., Landrigan, P.J., Fuller, R., 2013. Burden of disease from toxic waste sites in India, Indonesia, and the Philippines in 2010. Environ. Health Perspect. 791–796. https://doi.org/10.1289/ehp.1206127.

Eriksson, M., Wallensteen, P., Sollenberg, M., 2003. Armed conflict: 1989–2002. J. Peace Res. 40, 593–607.

European Centre for Disease Prevention and Control, ECDC, 2009. ECDC/EMEA Joint Technical Report: The bacterial challenge: Time to react. ECDC, Stockholm, .p. 2009. Available from: http://ecdc.europa.eu/en/publications/Publications/0909_TER_The_Bacterial_Challenge_Time_to_React.pdf.

Fazzo, L., Minichilli, F., Santoro, M., Ceccarini, A., Della Seta, M., Bianchi, F., Comba, P., Martuzzi, M., 2017. Hazardous waste and health impact: a systematic review of the scientific literature. Environ. Health 16, 107.

Friedman, M., 2015. Antibiotic resistant bacteria: prevalence in food and inactivation by food compatible compounds and plant extracts. J. Agric. Food Chem. 63, 3805–3822.

Gaze, W.H., S.M. Krone, D.G. Joakim Larrson, X.Z. Li, J.A. Robinson, P. Simonet, K. Smalla, M. Timinouni, E. Topp, Wellington, G.D., Wright, Y-G. Zhu. (July 2013) Influence of humans on evaluation and mobilization of environmental antibiotic resistance. CDC 19, (7), https://doi.org/10.3201/eid1907.120871.

GBD 2015 Risk Factors Collaborators, 2016. Global, regional, and national comparative risk assessment of 79 behavioural, environmental and occupational, and metabolic risks or clusters of risks, 1990–2015: a systematic analysis for the Global Burden of Disease. Lancet 388, 1659–1724.

Godfray, H.C.J., Beddington, J.R., Crute, I.R., Haddad, L., Lawrence, D., Muir, J.F., Pretty, J., Robinson, S., Thomas, S.M., Toulimin, C., 2010. Food security: the challenge of feeding 9 billion people. Science 327, 812–818.

Graham, G.N., 2016. Why your ZIP code matters more than your genetic code: promoting healthy outcomes from mother to child. Breastfeed Med. 11, 396–397. https://doi.org/10.1089/bfm.2016.0113.

Graham, J.P., Boland, J.J., Silbergeld, E., 2007. Growth-promoting antibiotics in food animal production: an economic analysis. Public Health Rep. 122, 79–87.

Harris, S., Morris, C., Morris, D., Cormican, M., Cummins, E., 2014. Antimicrobial resistant *Escherichia coli* in the municipal wastewater system: Effect of hospital effluent and environmental fate. Sci. Total Environ. 468-469, 1078–1085.

Harris, S.J., Cormican, M., Cummins, C., 2012. Antimicrobial residues and antimicrobial-resistant bacteria: impact on the microbial environment and risk to human health—a review. Hum. Ecol. Risk Assess. Int. J. 18, 767–809.

Heuer, H., Smalla, K., 2007. Manure and sulfadiazine synergistically increased bacterial antibiotic resistance in soil over at least 2 months. Environ. Microbiol. 9, 657–666.

Holvoet, K., Sampers, I., Callens, B., Dewulf, J., Uyttendaelw, M., 2013. Moderate prevalence of antimicrobial resistance in *Escherichia coli* isolates from lettuce, irrigation water, and soil. Appl. Environ. Microbiol. 79, 6677–6683.

Lal, R., 2004. Soil carbon sequestration impacts on global climate change and food security. Science 304, 1623–1627.

Laleye, S.A., 2015. Resolving socio-political conflict in Africa. In: Fiala, A. (Ed.), The Peace of Nature and the Nature of Peace: Essays on Ecology, Nature, Nonviolence, and Peace. Brill Online. pp. 109–127. https://doi.org/10.1163/9789004299597.

Landrigan, P.J., et al., 2017. The Lancet Commission on pollution and health. Lancet. https://doi.org/10.1016/S0140-6736(17)32345-0.

Levy, S.B., Marshall, B., 2004. Antibacterial resistance worldwide: causes, challenges and responses. Nat. Med. 10, S122–S129.

National Center for Health Statistics, Health, United States, 2010: With Special Feature on Death and Dying. Hyattsville, MD: CDC, National Center for Health Statistics, 2011. Available from: http://www.cdc.gov/nchs/hus.htm. Accessed 16 May 2011.

National Research Council, 2012. Exposure Science in the 21st Century: A Vision and a Strategy. The National Academies Press, Washington, DC. Available from: https://doi.org/10.17226/13507.

Odada, E., Oyebande, L., Oguntola, J., 2006. Lake Chad: experience and lessons learned. Retrieved from http://www.worldlakes.org/uploads/06_lake_chad_27february2006.pdf.

Oliveira, M., VInas, L., Usall, J., Anguera, M., Abadias, M., 2012. Presence and survival of *Escvherichia coli* 0157:H7 on lettuce leaves and in soil treated with contaminated compost and irrigation water. Int. J. Food Microbiol. 156, 133–140.

Pepper, I.L., 2013. The soil health: human health nexus. Crit. Rev. Environ. Sci. Technol. 43, 2617–2652.

Pepper, I.L., Brooks, J.P., Gerba, C.P., 2018. Antibiotic resistant bacteria in wastes: is there reason for concern? Environ. Sci. Technol. 52, 3949–3959.

Potts, S.G., Biesmeijer, J.C., Kremen, C., Newmann, P., Schweiger, O., Kunin, W.E., 2010. Global pollinator declines: trends, impacts and drivers. Trends Ecol. Evol. 25, 345–353.

Prüss-Üstün, A., Wolf, J., Corvalan, C., Bos, R., Neira, M., 2016. Preventing Disease Through Healthy Environments. World Health Organization, Geneva.

Pure Earth, World Worst Pollution Problems, 2016, Pure Earth, New York, NY.

Rodell, M., Famiglietti, J.S., Wiese, D.N., Reager, J.T., Beaudoing, H.K., Landerer, F.W., Lo, M.-H., 2018. Emerging trends in global freshwater availability. Nature 557, 651–659.

Schmidhuber, J., Tubiello, F.N., 2007. Global food security under climate change. Proc. Natl. Acad. Sci. U.S.A. 104, 19703–19708.

Sui, Q., Zhang, J., Chen, M., Tong, J., Wang, R., Wei, Y., 2016. Distribution of antibiotic resistance genes (ARGs) in anaerobic digestion for land application of swine wastewater. Environ. Pollut. 213, 751–759.

U.S. Government Accountability Office, 2013. Key issues: hazardous waste. Available from: www.gao.gov/key_issues/hazardous_waste/issue_summary#t=1. (Accessed 11 April 2017).

United Nations Water, 2018. Nature-Based Solutions for Water. United Nations Educational, Scientific and Cultural Organization, Paris, France.

Verlicchi, P., Aukidy, M., Galletti, A., Petrovic, M., Barcelo, D., 2013. Hospital effluent: investigation of the concentrations and distribution of pharmaceuticals and environmental risk assessment. Sci. Total Environ. 430, 109–111.

World Health Organization, WHO, 2002. The world health report 2012: reducing risks, promoting healthy life. .

World Health Organization, WHO (2014). Quantitative risk assessment of the effects of climate change on selected causes of death, 2030s and 2050s.

World Health Organization, WHO (2016). Ambient air pollution: a global assessment of exposure and burden of disease. Available from: http://apps.who.int/iris/bitstream/10665/250141/1/9789241511353-eng.pdf.

World Health Organization, WHO, 2017. Multidrug-resistant tuberculosis fact sheet. Available from: http://www.who.int/tb/challenges/mdr/MDR-RR_TB_factsheet_2017.pdf.

World Health Organization, WHO, 2018a. Global health estimates 2016: deaths by cause age, sex, by country and by region, 2000–2016. Available from: http://www.who.int/healthinfo/global_burden_disease/estimates/en/.

World Health Organization, WHO, (2018b). Climate change and health. Available from: http://www.who.int/en/news-room/fact-sheets/detail/climate-change-and-health. Accessed July 2018.

Yilmaz, G., Kaya, Y., Vergili, I., Beril Gonder, Z., Ozhan, G., Celik, B.O., Altinkum, S.M., Bagdatli, Y., Boergers, A., Tuerk, J., 2017. Characterization and toxicity of hospital wastewaters in Turkey. Environ. Monit. Assess. 189, 55. https://doi.org/10.1007/s10661-016-5732.2.

Zhu, H., Sandiford, S.K., van Wezel, G.P., 2014. Triggers and cues that activate antibiotic production by actinomycetes. Biotechnology 41, 371–386.

FURTHER READING

Center for Disease Control, n.d. Preventing childhood lead. Available from: https://www.cdc.gov/nceh/lead/publications/lead-infographic-final-full.pdf. Accessed 25 May 2018.

Cordell, D., Drangert, J.-O., White, S., 2009. The story of phosphorous: global food security and food for thought. Global Environment Change-Human and Policy Dimensions 19, 292–305.

Fahrenfeld, K., Knowlton, K., Krometis, L.A., Hession, W.C., Xia, K., Lipscomb, E., Libuit, K., Green, B.L., Prudent, A., 2014. Effect of manure application on abundance of antibiotic resistance genes and their attenuation rates in soil: field-scale mass balance approach. Environ. Sci. Technol. 48, 2643–2650.

Finley, R.L., Collignon, P., Larrson, D.G., McEwen, S., Li, X.Z., Gaze, W.H., Reid-Smith, R., Timinouni, M., Graham, D.W., Topp, E., 2013. The scourge of antibiotic resistance: the important role of the environment. Clin. Infect. Dis. 57, 704–710.

Lioy, P.J., 1990. The analysis of total human exposure for exposure assessment: a multi-discipline science for examining human contact with containments. Environ. Sci. Technol. 24, 938–945.

Rizzo, L., Manaia, C., Merlin, C., Schwartz, T., Dagot, C., Ploy, M.C., Michael, I., Ratta-Kassinos, D., 2013. Urban wastewater treatment plants as hotspots for antibiotic resistant bacteria and genes spread into the environment: a review. Sci. Total Environ. 447, 345–360.

U.S. Centers for Disease Control and Prevention, 1999. Ten great public health achievements-United States, 1900–1999. Morb. Mortal. Wkly Rep. 48, 241–243.

U.S. Environmental Protection Agency (EPA), 2009. A Conceptual Framework for U.S. EPA's National Exposure Research Laboratory. EPA/600/R-09/003. National Exposure Research Laboratory, Office of Research and Development, U.S. Environmental Protection Agency. Available from: http://www.epa.gov/nerl/documents/nerl_exposure_framework.pdf. Accessed 25 May 2018.

United States Environmental Protection Agency (EPA), Overview of the Brownfields program. (April 28, 2017, Available from: https://www.epa.gov/brownfields/overview-brownfields-program. Accessed 1 May 2017.

Wild CW. 2012. The exposome: from concept to utility, Int. J. Epidemiol., 41(1) 2012, 24–32. Available from: https://doi.org/10.1093/ije/dyr236

World Health Organization, WHO (2009). Global health risks: mortality and burden of disease attributable to selected major risks.

Chapter 27

Medical Geology and the Soil Health-Human Health Nexus

M.L. Brusseau and I.L. Pepper

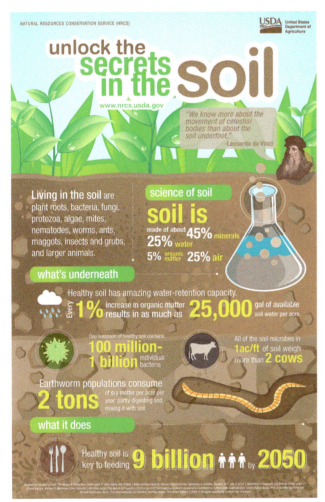

Infographic from the U.S. Department of Agriculture illustrating the properties of a healthy soil and its benefits to humans. *(Infographic. Available from: https://www.nrcs.usda.gov/wps/portal/nrcs/detail/national/soils/health/?cid=stelprdb1143889)*

As discussed in Chapter 6, the environment provides numerous benefits and services that are essential for human health and well-being. However, the environment can also negatively affect humans. In some instances, certain geologic materials and processes can have adverse impacts on human health and ecosystem health. For example, the presence of elevated levels of elements such as arsenic in sediment may result in groundwater contamination, which can cause human disease when that groundwater is ingested. For another example, volcanic eruptions can release vast quantities of elements to the environment, some of which are toxic to humans and other lifeforms. In addition, the very soil that we use to grow crops and build houses upon can have adverse impacts to humans. These topics are discussed in this chapter.

27.1 MEDICAL GEOLOGY

Medical Geology is a relatively new scientific discipline that examines the impacts that geologic materials and processes have on human and ecosystem health. The connection between the environment and human health has long been known. For example, the following two quotes have been attributed to Hippocrates (460–370 BC), a Greek physician who is called the Father of Medicine:

> *If you want to learn about the health of a population, look at the air they breath, the water they drink, and the places where they live. (https://kids.niehs.nih.gov/topics/environment-health/index.htm)*

> *Whoever wishes to investigate medicine properly, should proceed thus... We must also consider the qualities of the waters, for as they differ from one another in taste and weight, so also do they differ much in their qualities. (http://classics.mit.edu/Hippocrates/airwatpl.1.1.html)*

These quotes illustrate an essential concept of medical geology—that the air that we breathe, the water that we drink, and the food that we eat contain elements that originate from the geological environment, and that these elements can be both beneficial and detrimental to the health of humans and other life.

Environmental and Pollution Science. https://doi.org/10.1016/B978-0-12-814719-1.00027-6

27.1.1 Geologic Elements and Processes

The Earth's crust, and the rocks and sediments contained therein, contain many elements. Natural and human-induced weathering of these materials can lead to the release of these elements into water, soil, and air. Once this occurs, humans and other organisms can be exposed directly or indirectly. Examples of the levels of selected elements present in geologic materials are presented in Table 27.1.

Copper, listed in the table, is an essential micronutrient vital to human health. For example, copper is essential to the proper functioning of organs and metabolic processes. Conditions linked to copper deficiency include osteoporosis, osteoarthritis, rheumatoid arthritis, cardiovascular disease, colon cancer, and chronic conditions involving bone, connective tissue, heart, and blood vessels. Copper cannot be formed by the human body, and therefore it must be ingested from dietary sources (food and drinking water). The copper present in water and in foodstuffs originated primarily from geological materials. While copper is an essential nutrient, exposure to elevated levels beyond those required for maintaining health can lead to adverse health effects. Excess copper intake can cause nausea and diarrhea and can lead to tissue injury and disease.

The copper example is representative of the case for many geologic elements (e.g., calcium, iron, magnesium, potassium, selenium, iodine). As noted by Swiss alchemist and physician Paracelsus (1493–1541) almost 500 years ago:

All substances are poisons; there is none that is not a poison. The right dose differentiates a poison from a remedy. (https:// www.sciencelearn.org.nz/resources/365-all-in-the-dose)

Hence, many elements supplied via geological materials that are essential for maintaining human health can also be toxic under certain conditions. Other elements present in geological materials serve no known health benefits and are toxic at even relatively low levels.

It is important to recognize that the interactions between geological materials and lifeforms can happen in numerous

TABLE 27.1 Compilation of Average Geochemical Background Data for the Earth's Crust and Selected Rock Types

Compilation of Average Geochemical Background Data for the Earth's Crust and Selected Rock Types

	Hg (μg kg^{-1})	Pb (mg kg^{-1})	Cd (mg kg^{-1})	Cr (mg kg^{-1})	Ni (mg kg^{-1})	As (mg kg^{-1})	Cu (mg kg^{-1})	Zn (mg kg^{-1})
Earth's crust								
	80	13	0.2	100	75	2	55	70
	90	12	0.2	110	89	2	63	94
Upper continental crust								
		20	0.1	35	20	1.5	25	71
	80	13	0.2	77	61	1.7	50	81
Igneous rocks								
Ultramafic	4	1	0.1	1600	2000	1	10	50
Mafic	13	6	0.2	170	130	2	87	105
Intermediate	21	15	0.1	22	15	2	30	60
		10		55	30		60	
Felsic	39	19	0.1	4	5	1	10	39
Sedimentary rocks								
Sandstone	57	14	0.02	120	3	1	15	16
Limestone	46	16	0.05	7	13	2	4	16
Shale	270	80	0.2	423	29	9	45	130
Black shale		15	4.0	18	68	22	50	189
		100		700	300		200	1500

Source: From Chapter 2 of Essentials of Medical Geology, Selinus, 2005.

ways. The following sections will highlight some examples of major interactions and consequences, focusing on human health.

27.1.2 Geologic Sediments and Drinking Water

As presented in Table 27.1, geologic materials contain many elements. These elements can dissolve into water as surface water or groundwater contact these materials over long periods. In fact, the composition and quality of natural waters is controlled by the geologic materials they contact. These dissolved constituents can positively or adversely impact human health when the water is consumed.

For example, fluoride-containing minerals are a natural source of fluoride in groundwater. At moderate concentrations in water (0.7–1.2 mg/L) fluoride helps prevent dental caries and can improve bone matrix integrity (Burt and Tomar, 2007). Excess concentrations (>4 mg/L) can be detrimental resulting in fluorosis. Similarly, calcium and magnesium-containing minerals provide sources of essential human nutrients.

Perhaps the most infamous example of a geologic constituent creating adverse human health effects is arsenic. Long-term chronic arsenic exposure can result in skin disease (hyperkeratosis), multiple cancers (skin, lung, bladder, and kidney), atherosclerosis, and peripheral vascular disease. Worldwide, use of contaminated water, particularly groundwater, is the leading cause of exposure to arsenic (National Research Council, 2007). Ingestion of contaminated water via its use as drinking water is a primary source of exposure. However, other exposure routes include using contaminated water for bathing, food preparation, and irrigation of food crops.

Current Environmental Protection Agency (EPA) and World Health Organization (WHO) maximum contaminant levels for arsenic in drinking water are 10 μg/L. Maps showing measured and estimated levels of arsenic in groundwater for the United States and the World are presented in Figs. 27.1 and 27.2, respectively. Inspection of these maps indicates that many millions of people are affected by arsenic-contaminated groundwater. The World Health Organization estimates that at least 140 million people in 50 countries have been drinking water containing arsenic at levels above the 10 μg/L level. They consider arsenic to be the most significant chemical contaminant in drinking water globally. Arsenic enters groundwater through contact with arsenic-bearing minerals (see Table 27.2).

One of the largest impacts of arsenic exposure has occurred in Bangladesh. The tragedy of arsenic poisoning in Bangladesh occurred when tube wells were drilled in the 1970s to allow access to groundwater as a source of drinking water (see Case Study 12.1). This was done to prevent the drinking of contaminated surface waters that had resulted in dysentery, typhoid, and cholera diseases. Sediments in the subsurface contained arsenic minerals such as arseno-pyrite, which is stable under anaerobic conditions. At low redox potentials, sulfate is reduced to sulfide and arsenic is precipitated onto iron sulfide as arseno-pyrite. There are two theories for the mechanism of arsenic contamination of groundwater (National Research Council, 2007). The first states that excessive groundwater extraction allowed greater diffusion of oxygen into arsenic-containing sediments and oxidation of insoluble arseno-pyrite to the soluble hydrated iron arsenate mineral known as pitticite and is known as the oxyhydroxide reduction theory, where arsenic is desorbed from ferric hydroxide under reducing conditions. Soil microorganisms also transform arsenic species found in soil. Specifically, some soil microbes use arsenate as a terminal electron acceptor under anaerobic conditions, converting arsenate to arsenite, which is the more toxic and mobile species most likely to contaminate groundwater (National Research Council, 2007). Regardless of the mechanism, arsenic contamination of groundwater due to arsenic remains the largest mass poisoning in recorded history (Chowdhury et al., 2000).

27.1.3 Geogenic Dust

Aerosols and particulate matter are produced by several geologic processes. Examples include wind-generated dust, volcanic ash (see next section), marine aerosols, and forest fires. The airborne materials generated from these activities can travel tens to hundreds to thousands of kilometers from the source.

Aerosols and particulate matter can have both direct and indirect impacts on human health. A primary direct impact is respiratory effects caused by inhalation. The particles themselves, and/or contaminants associated with the particles, can be hazardous. Indirect effects include deposition onto land or vegetation surfaces, which can have multiple consequences for human exposure.

More detailed discussion of airborne matter is presented in Chapters 11 and 17.

27.1.4 Volcanism

Volcanism and related processes are the primary means by which metals are brought to the Earth's surface. Volcanic activity can be a major source of geologic elements. For example, it is estimated that about 50% of SO_2 introduced into the environment is from natural sources, primarily volcanic activity. The eruption of Pinatubo in 1991 ejected approximately 10 billion tons of magma. Along with the

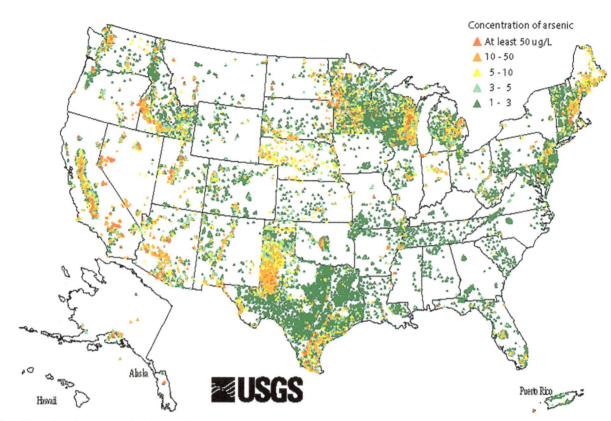

FIG. 27.1   Arsenic concentrations in groundwater based on 31,350 samples collected in 1973–2001 by the USGS. *(From: Ryker, S.J., 2001. Mapping arsenic in groundwater. Geotimes 46(11), 34–36.)*

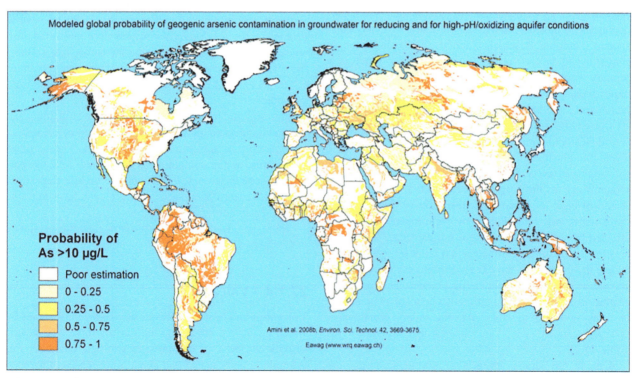

FIG. 27.2   Risk for arsenic concentrations to exceed 10ug/L in groundwater worldwide. *(From: Amini, M., Abbaspour, K.C., Berg, M., Winkel, L., Hug, S.J., Hoehn, E., Yang, H., Johnson, C.A., 2008. Statistical modeling of global geogenic arsenic contamination in groundwater. Environ. Sci. Technol. 42 (10), 3669–3675.)*

TABLE 27.2 Typical Ranges of Arsenic Concentrations in Common Rock-Forming Minerals

Mineral	Arsenic Concentration Range $(mg\,kg^{-1})$
Sulfide minerals	
Pyrite	100–77,000
Pyrrhotite	5–100
Marcasite	20–126,000
Galena	5–10,000
Sphalerite	5–17,000
Chalcopyrite	10–5000
Oxide minerals	
Hematite	up to 160
Fe(III) oxyhydroxide	up to 76,000
Magnetite	2.7–41
Ilmenite	<1
Silicate minerals	
Quartz	0.4–1.3
Feldspar	<0.1–2.1
Biotite	1.4
Amphibole	1.1–2.3
Olivine	0.08–0.17
Pyroxene	0.05–0.8
Carbonate minerals	
Calcite	1–8
Dolomite	<3
Siderite	<3
Sulfate minerals	
Gypsum/anhydrite	<1–6
Barite	<1–12
Jarosite	34–1000
Other minerals	
Apatite	<1–1000
Halite	<3–30
Fluorite	<2

From Chapter 11 of Essentials of Medical Geology, Selinus, 2005.

magma, many elements were introduced into the atmosphere, including approximately (Selinus, 2004):

- 20 million tons of SO_2
- 2 million tons of Zn
- 1 million tons of Cu
- 550,000 tons of Cr
- 100,000 tons of Pb
- 30,000 tons of Ni
- 5,500 tons of Cd
- 800 tons of Hg

In addition to introducing elements into the environment, volcanic activity is also a major source of particulate matter (volcanic ash). Plumes of ash in some cases can affect areas hundreds of kilometers away from the volcano. Inhalation of ash particles can lead to adverse respiratory effects. In addition, the presence of ash can disrupt human activities such as air travel.

27.2 SOIL HEALTH-HUMAN HEALTH NEXUS

The single most prevalent source of geologic-human interactions is the soil, the thin veneer of material that covers much of the Earth's surface. This fragile skin is frequently less than a meter thick but is absolutely vital for human (and most other) life as we know it. Soil is the most complicated biomaterial on the planet other than perhaps humans. Interestingly, the zone of maximum life, activity, and diversity on Earth is terrestrial environments where soils and humans meet at the surface of the Earth. Upward or downward movement away from this interface decreases all three parameters. Based on these facts, perhaps it is not surprising that soils affect human health, and that humans affect soil health. This concept is termed the "Soil Health-Human Health Nexus."

Soils impact human health in a multitude of ways including what we eat, drink, and breathe. In addition, soils can infect us with disease-causing microorganisms, and yet can also be a source of natural products that fight diseases and otherwise improve human health. As well, soil serves as a living filter that can remove harmful contaminants from the environment, reducing the levels to which humans and other life are exposed. These are examples of the "ecosystem services" provided by soil.

27.2.1 What Infects Us

Soils can be a source of bacterial, fungal, protozoan, and viral human pathogens. These microbial pathogens can be normal soil residents (*geo-indigenous*) which survive within soils as residents, or can be introduced into soils where they survive for short periods of time, prior to inactivation or death due to biotic and abiotic stress (Pepper, 2013). These introduced microbes can be termed *Geo-Treatable*.

Geo-indigenous pathogens are those found in soils that are capable of metabolism, growth, and reproduction

(Pepper et al., 2009). They are found in all soils and include prokaryotic and eukaryotic organisms (Table 27.3). Note that there are no indigenous human pathogenic viruses found in soil, because there is no natural host for such viruses in the soil. Overall, the microbes listed here can cause a variety of diseases, some of which can be fatal.

Geo-treatable pathogens that are not indigenous to soil can be introduced into soil either deliberately via anthropogenic activities or accidently via animal or bird feces. Deliberate introduction can be through land application of animal manures or biosolids (see Chapter 23). Examples of geo-treatable pathogens are given in Table 27.4.

Human exposure to geo-treatable pathogens can be through direct contact with soil containing the pathogens or indirectly via contamination of leafy food crops via irrigation of manure amended soils. Typically, such outbreaks are reported on the national news several times a year. Leafy food crops such as lettuce or spinach are particularly prone to such contamination.

In contrast to geo-indigenous pathogens, which can live in a soil environment indefinitely, geo-treatable pathogens are normally inactivated due to biotic and abiotic stresses. Biotic stress is caused by competition with the billions of microbes indigenous to the soil for food substrates and growth factors. Other negative biotic impacts on introduced microbes include secretion of antimicrobial substances such as antibiotics produced by microbes (see also Chapter 26). Finally, many introduced microbes are subject to infection or predation by phage or protozoa. Abiotic stress is due to the fluctuation of environmental parameters such as soil moisture, temperature, and microsite pH. Such

environmental fluctuations can be particularly severe on enteric microorganisms such as *E. coli*, which are attuned to environments such as the human gut, where the environment tends to be stable rather than fluctuating. The combination of biotic and abiotic stresses normally leads to the death of enteric bacteria within a few weeks following their introduction into soil. Enteric viruses usually survive slightly longer up to a few months. Based upon survival data of introduced microbes, federal regulations for land application of Class B biosolids (which contain detectable pathogens) include site restrictions of several months to allow for the inactivation or death of the pathogens prior to potential human exposure (see also Chapter 23).

27.2.2 What Heals Us

As well as containing pathogenic microbes that can infect us, soils also contain organisms that provide a treasure chest of natural products critical to maintaining or even improving human health. The earliest of these classes of compounds to be discovered were the antibiotics. Antibiotics are compounds produced by soil microorganisms that kill or inhibit other microorganisms, and soils were the source of the first known antibiotics. Penicillin was isolated from the soilborne fungus *Penicillium* by Sir Alexander Fleming in 1929 (Fleming and Lond, 1942). In 1943, Selman Waksman discovered streptomycin, a feat for which he received the Nobel Prize. This antibiotic was isolated from *Streptomyces griseus*, and, since then, soil actinomycetes have been shown to be a prime source of antibiotics (Demain and Fang, 2000). Because antibiotics

TABLE 27.3 Human Geo-Indigenous Soil Pathogens

Type of Organism	Affliction	Incidence in Soil
Human virus	NA	Never indigenous: no host
Bacteria		
Bacillus anthracis	Anthrax	Routinely found in most soils
Legionella spp.	Legionnaire's disease	Found in soil composts and potting soil
Clostridium perfringens	Minor infections and gas gangrene	Common soil organism
Burkholderia pseudomallei	Melioidosis	Endemic in southeast Asia
Fungi		
Coccidioides immitis	Valley fever	Highly prevalent in southwestern United States
Histoplasma capsulatum	Respiratory infections	Prevalent in the midwestern and southern United States
Protozoa		
Naegleria fowleri	Brain encephalitis	Found in soil and water
Balamuthia mandrillaris	Brain encephalitis	Found in soil and water

Modified from Pepper, I.L., Gerba, C.P., Newby, D.T., Rice, C.W., 2009. Soil: a public health threat or savior? Crit. Rev. Environ. Sci. Technol. 39, 416–432.

TABLE 27.4 Examples of Geo-Treatable Pathogens Introduced into Soil

Bacteria	Protozoa
Salmonella	*Cryptosporidium parvum*
Escherichia coli	*Giardia, lamblia*
Shigella	*Entamoeba histolytica*
Viruses	**Helminths**
Adenovirus	*Ascaris lumbricoides*
Norovirus	*Ascaris suum*
Reovirus	*Taenia solium*
Enterovirus	

Source: New original table.

are produced by soil microorganisms, soils are also a source of antibiotic-resistant bacteria. This has led to the term "environmental antibiotic resistance" (see also Chapter 26).

In addition to antibiotics, soils are also a source of other natural products beneficial to human health. Fungi are also present in soil with immense diversity and both bacterial and fungal endophytes are a rich source of natural products. Endophytes are bacterial or fungal microbes that colonize plant roots without pathogenic effects. Endophytes produce metabolites that not only protect plant roots, thereby improving plant health, but can also improve human health. Endophytes have been shown to produce novel antibiotics, antimycotics, immunosuppressants, and anticancer agents (Strobel and Daisey, 2003). Very recently, a new technology known as genomic mining has resulted in new discoveries of useful natural products. Genomic mining is the identification of protein-encoding regions of a genome and the assignment of functions to these genes on the basis of similarity to sequences of other genes of known structure and function. The technology allows for the identification of new potential drug products, resulting from gene clusters that are not normally expressed under laboratory conditions. These new approaches bode well for future sources of new natural products that maintain and even improve human health. In addition to microorganisms, plants grown in soil are also being mined for potential drug products.

27.2.3 What We Drink

Soils can dramatically affect human health by influencing what we drink. Soils may have certain minerals that contain toxic elements that can transfer to water, such as arsenic as discussed previously. In addition, human activity may add toxicants to soil, which are then transferred to water or air.

For example, selenium gained infamy in the United States in the 1980s due to selenium toxicity in the Kesterson Reservoir in the San Joaquin Valley in California (National Research Council, 2007). Essentially, agricultural irrigation of soils that contained high levels of selenium resulted in solubilization of the selenium and accumulation in drainage water collected in the reservoir. Birth defects in birds and die-off of waterfowl and fish within the reservoir resulted from the elevated selenium levels (Ohlendorf et al., 1990). In June 1986 the reservoir was closed to drainage water to prevent further selenium inputs (Wu, 2004).

Anthropogenic activities result in the production of inorganic and organic chemicals that end up in soil (see Chapters 14 and 15). This can lead to subsequent adverse human health effects that result, for example, from consumption of contaminated groundwater used as potable water sources or via the consumption of fruits and vegetables grown in the contaminated soil (see next section). An example of a widespread inorganic contaminant is nitrate. Nitrates are introduced into soils deliberately as inorganic fertilizers or via land application of animal manures and biosolids. Regardless of whether inorganic or organic forms of nitrate are added, soil microbial processes such as ammonification and nitrification result in nitrate as the major soil constituent. Nitrates are highly soluble and can easily migrate to groundwater. In humans, nitrate is reduced to nitrite, which in infants combines with hemoglobin inhibiting oxygen transport. The result is blue baby syndrome or methemoglobinemia, which can be fatal. Therefore an international standard of 10 ppm of NO_3-N for groundwater has been set.

Organic contamination of soil and water has also been well documented as in the case of chlorinated solvents like trichloroethene (TCE) and organic pesticides. Information on organic chemicals is discussed in Chapter 12, and their impacts on groundwater pollution are discussed in Chapter 15. More recently, emerging contaminants of concern have become an issue (see Chapter 12).

27.2.4 What We Eat

Soils can impact human health in terms of what we eat in two ways: either by direct ingestion of soil itself or through ingestion of food that was originally grown in soil. The practice of deliberate consumption of soil is known as geophagia and has been documented since historical times. Incidental ingestion of soil has been estimated as ≃50 mg/day for children (Stanek and Calabrese, 1995). Beneficial health effects of geophagia can result from intake of essential micronutrients. Adverse health effects can result from exposure to either pathogens or toxic chemical contaminants found in soil.

Soil is also fundamentally important for food production and quality since the vast majority of food for human or animal consumption is grown in soil. The interface between the soil and plant roots is mediated by the rhizosphere which contains both rhizosphere microorganisms and endophytes. The rhizosphere encompasses the millimeters of soil surrounding plant roots and is characterized by extremely high soil microbial populations, typically orders of magnitude higher than those found in bulk nonrhizosphere soil. Overall the relationship between plant growth and soil microorganisms is complex and vitally affects vegetative growth and hence food production. Soil quality determines the nutritional value and safety of the foods grown. For example, plants can take up heavy metals from contaminated soils, after which the metals enter the food chain. Soils contaminated with geo-treatable pathogens can also adversely impact human health through adhesion to food produce grown in such soils.

Efforts to avoid potential risks from anthropogenic constituents have prompted renewed interest in organic gardening and organic foods. There are many definitions of organic gardening, but essentially it consists of growing crops without inorganic fertilizers or pesticides. Instead, organic sources of nutrients are supplied and pests are controlled using natural organic products or management practices such as planting strategies and use of natural biocontrol agents. Crop production using organic fertilizers is normally very successful because nutrients are released slowly over time, acting as a slow release fertilizer. Crop quality of organic foods is usually excellent, but claims of nutritional and improved health effects over conventionally fertilized crops are more controversial. Dangour et al. (2010) recently completed a systematic review of nearly 100,000 studies evaluating nutritionally related health effects of organic foods. The authors concluded that evidence is lacking for positive nutrition-related health effects resulting from the consumption of organically produced foodstuffs. That being said, it is clear that plants grown in soils for human consumption are critical in maintaining human health, regardless of whether they are grown organically or conventionally.

27.2.5   What We Breathe

All humans are aerobic and therefore require oxygen as a terminal electron acceptor. This is acquired from the air that we breathe that contains approximately 21% oxygen. Soils can adversely impact the quality of the air we breathe either directly or indirectly. Direct effects include suspended particulates and gases originating from Earth materials. Indirect effects include the inhalation of elements or microorganisms attached to soil particles—the latter of which are known as bioaerosols. The generation of particulates is covered in Chapter 11. Aspects of bioaerosols are covered in Chapter 23.

27.2.6   The Impact of Humans on Soil Health

Soils develop over hundreds or thousands of years via the five soil forming factors: parent materials, climate, vegetation and organisms (including humans), topography, and time. Undisturbed, soils develop an exquisitely complex architecture based on texture, structure, horizonation, organic matter, and microbial communities. Undisturbed pristine soils also develop pseudo-equilibria that adjust to seasonal changes, and control both microbial populations and activities, and the vegetation that becomes established on the surface soil. Ultimately, ecosystems become established, and the terrestrial environment can be thought of as a large organelle with an "amoeba defense" that degrades natural organic residues introduced into it. This concept is a modification of the original Gaia hypothesis in which James Lovelock proposed that the Earth behaves as a super organism (Lovelock, 1995).

Human activities have direct and indirect impacts on soils. Such activities include farming, deforestation, urbanization, and direct introduction of contaminants. For example, a soil that developed over hundreds or even thousands of years can be obliterated in minutes through the use of a bulldozer, or strip-mining activities. Humans can also "infect" soils through the addition of contaminants including metals or pesticides or other toxic organics. Inputs of water into soils can also adversely affect soils due to anthropogenic activities. For example, acid rain can result in acidic soils, whereas irrigation with high salinity water can result in alkaline soils.

The concept of soil health, sometimes referred to as soil quality, has been widely discussed and even debated. Doran (2002) defined soil health as: "the capacity of a living soil to function, within natural or managed ecosystems boundaries, to sustain plant and animal productivity, maintain or enhance water and air quality, and promote plant and animal health." A different concept is to think of soil as a living entity and to think of soil health as analogous to human health. Both must be in a state of well-being with respect to their physical, chemical, and biological characteristics. Likewise neither should be diseased nor compromised, and ideally, both should function sustainably at optimal potential (Pepper, 2013). Human activities can maintain or even improve soil health, but often such activities are detrimental.

Soil health is an assessment of how well soil performs all of its beneficial functions now and potentially in the future (i.e., the services it provides)—if the soil is to be sustainable. To this point in time, soil health cannot be determined by measuring a single parameter such as crop yield. Since soil health cannot be easily measured directly, the current approach is to use indicators, which are measurable and quantifiable properties. However, currently there is no consensus on the ideal suite of indicators that should be used. Examples of potential soil health indicators are shown in Information Box 27.1. These indicators can be a combination of physical, chemical, and biological parameters, which need to be validated by the scientific community (Soil Health Institute, 2017).

Humans and anthropogenic activities can also be used to "heal" or improve soils. Typically, such activities involve soil amendments including organic composts or inorganic amendments such as sulfur or lime to adjust soil pH. Such "healing" or restoration activities are discussed in Chapter 20.

27.2.7 Improved Human Health Through Soil Health Maintenance

The obvious and most direct manner in which we can improve human health through soil health maintenance is via the production of large amounts of healthy foods including cereals and grains, pulse crops, vegetables, and fruits. To accomplish this, soil health maintenance should focus on humans creating an optimum environment for plant growth. It would also require humans to be mindful with respect to pollution prevention. As part of this strategy, organic wastes should only be applied to soils at appropriate agronomic rates that allow for the material to be incorporated into the soil following microbial degradation. With respect to inorganic compounds, wastes with excessive amounts of heavy metals such as Zn, Cd, or Hg should not be applied to soils. The green revolution of the 20th century resulted in vast increases in crop yields for many decades. However more recently, attention has turned to sustainable practices that protect soil health including reduced tillage practices and best-management practices for fertilizer application. The importance of soil health maintenance on improved human health through safe food production cannot be overemphasized.

A major indirect effect of soil health on human health is the influence of soils on global climate change. Soils can be a source of CO_2 due to microbial respiration, or a sink for CO_2 due to enhanced photosynthetic activity and carbon sequestration. Relatively small changes in soil carbon storage could significantly affect the global carbon balance. Human activities including intensive farm tillage practices enhance soil organic matter degradation, releasing CO_2 into the atmosphere. In contrast, setting aside land for conservation, reduced tillage, and enhanced crop productivity (CO_2 uptake) may significantly enhance soil carbon sequestration. Thus human activities could enhance soil health and moderate global climate change.

The mutualistic relationship between soil health and human health is now well documented by peer reviewed literature. Clearly, soils influence all aspects of our daily lives and yet the importance of soil with respect to human health is not well recognized by the general public. The unawareness is, perhaps, surprising given numerous testimonials, including President Franklin D. Roosevelt ("The Nation that destroys its soil destroys itself") and Leonardo da Vinci ("Even the richest soil, if left uncultivated will produce the rankest weeds"). History has taught us that soil health must be maintained because human life on Earth without soil would be impossible.

QUESTIONS AND PROBLEMS

1. Identify some of the most recently discovered novel antibiotics that may benefit human health?
2. State reasons why soil is important to humans.
3. Give an example of medical geology NOT covered in the text.
4. Discuss what can be done to improve soil health.

REFERENCES

Burt, B.A., Tomar, S.L., 2007. Changing the face of America: water fluoridation and oral health. In: Ward, J.W., Warren, C. (Eds.), The History and Practice of Public Health in Twentieth Century America. Oxford University Press, Oxford, England.

Chowdhury, U.K., Biswas, B.K., Chowdhury, T.R., Samanta, G., Mandal, B.K., Basu, G.C., Chanda, C.R., Lodh, D., Saha, K.C., Mukherjee, S.K., Roy, S., Kabir, S., Quamruzzaman, Q., Chakraborti, D., 2000. Groundwater arsenic contamination in Bangladesh and West Bengal. India. Environ. Health Perspect. 108, 393–397.

Dangour, A.D., Dodhia, S.K., Hayter, A., Aikanhead, A., Allen, E., Uauy, R., 2010. Nutrition-related health effects of organic foods: a systematic review. Am. J. Clin. Nutr. 92, 203–210.

Demain, A.L., Fang, A., 2000. The natural functions of secondary metabolites. Adv. Biochem. Eng. Biotechnol. 69, 1–39.

Doran, J.W., 2002. Soil health and global sustainability: translating science into practice. Agric. Ecosyst. Environ. 88, 119–127.

Fleming, A., Lond, M.B., 1942. In vitro tests of penicillin potency. Lancet 1, 732–733.

Lovelock, J., 1995. The Ages of Gaia. W.W. Norton, New York.

National Research Council, 2007. Earth Materials and Health. The National Academies Press, Washington, DC.

INFORMATION BOX 27.1 Potential Indicators of Soil Health

Adapted from Soil Health Institute: Enriching soil, enhancing life: an action plan for soil health. Soil Health Institute, May 2017.

Indicator Type	Parameter
Physical	Texture; water stable aggregates; bulk density
Chemical	N P K nutrient analysis; micronutrients; pH; Cation exchange capacity (CEC); soil organic carbon
Biological	Short term Carbon mineralization; Nitrogen mineralization

Ohlendorf, H.M., Hothem, R.L., Bunck, C.M., Marois, K.C., 1990. Bioaccumulation of selenium in birds at Kesterson Reservoir, California. Archives Environ. Contam. Toxicol. 19, 495–507.

Pepper, I.L., 2013. The soil health-human health nexus. Crit. Rev. Environ. Sci. Technol. 43, 2617–2652.

Pepper, I.L., Gerba, C.P., Newby, D.T., Rice, C.W., 2009. Soil: a public health threat or savior? Crit. Rev. Environ. Sci. Technol. 39, 416–432.

Selinus, O., 2004. Medical geology: an emerging specialty. Terrae 1 (1), 8–15.

Selinus, O., 2005. Essentials of Medical Geology. Elsevier Academic Press, San Diego, CA.

Soil Health Institute, May 2017. Enriching Soil, Enhancing Life: An Action Plan for Soil Health. Soil Health Institute, Washington, DC.

Stanek 3rd, E.J., Calabrese, E.J., 1995. Daily estimates of soil ingestion in children. Environ. Health Perspect. 103, 276–285.

Strobel, G., Daisey, B., 2003. Bioprospecting for microbial endophytes and their natural products. Microbiol. Mol. Biol. Rev. 67, 491–502.

Wu, L., 2004. Review of 15 years of research on ecotoxicity and remediation of land contaminated by agricultural drainage sediment rich in selenium. Ecotoxicol. Environ. Saf. 57, 257.

FURTHER READING

Amini, M., Abbaspour, K.C., Berg, M., Winkel, L., Hug, S.J., Hoehn, E., Yang, H., Johnson, C.A., 2008. Statistical modeling of global geogenic arsenic contamination in groundwater. Environ. Sci. Technol. 42 (10), 3669–3675.

Ryker, S.J., Nov. 2001. Mapping arsenic in groundwater. Geotimes 46 (11), 34–36.

Chapter 28

Environmental Toxicology

C.P. Gerba

Flow through system for fish toxicity testing. *(Source: Broxham Environmental Laboratory. Available from: www.brixham-lab.com)*

28.1 HISTORY OF MODERN TOXICITY IN THE UNITED STATES

In *toxicology*, we study both the adverse effects of chemicals on health and the conditions under which those effects occur. A natural outgrowth of biology and chemistry, toxicology began to assume a well-defined shape in just the past five to six decades. Newer still, environmental toxicology is concerned with the effects of chemical contaminants on various ecological systems, both large and small. Here, we focus on some of the basic principles of toxicology as related to environmental contaminants.

The first federal law for regulating potentially toxic substances was the Pure Food and Drug Act, passed by Congress in 1906. Much of the impetus for this law came from the work of Harvey Wiley, who was the chief chemist of the Department of Agriculture under Theodore Roosevelt. Wiley and his "Poison Squad" had a very personal interest in their work—he and his team of chemists not

infrequently dosed themselves with suspect chemicals to test for their deleterious effects (Rodricks, 2007).

The systematic study of toxic effects in laboratory animals (other than chemists) began in the 1920s, spurred by concerns about the unwanted side effects of food additives, drugs, and pesticides. (In this era DDT and related pesticides first became available.) The 1930s saw issues raised about occupational cancers and other chronic diseases resulting from chemical exposures. These concerns and issues prompted increased legislative activity culminating in the modern version of the Food, Drug, and Cosmetic Act. This law, enacted by Congress in 1938, was passed in response to a tragic episode in which more than 100 people died from acute kidney failure after ingesting contaminated sulfanilamide—the antibiotic had been improperly prepared in a diethylene glycol solution (Rodricks, 2007).

The real growth of toxicology largely paralleled that of the chemical industry, especially after World War II. The thousands of new compounds produced by chemical manufacturers created a need for information about their possible harmful effects. This growth received a significant stimulus from public opinion. Sporadically during the 1940s and 1950s, the public was presented with a series of seemingly unconnected announcements about poisonous pesticides in their foods, food additives of dubious safety, chemical disasters in the workplace, and air pollution episodes that claimed thousands of victims in urban centers throughout the world. Then, in 1962, marine biologist Rachel Carson (1907–1964) drew together these various environmental horror stories in her book, *Silent Spring*. Concerned by the presence of synthetic chemical toxins in the environment, the public responded with predictable outrage, which (among other things) fostered renewed interest in the science of toxicology. It also helped pave the way for the introduction of several major federal environmental laws in the late 1960s and early 1970s, and for the creation of the EPA in 1970 (Chapter 30). See Chapter 31 for more details on the history of the environmental movement.

The relative newness of the science of toxicology is reflected in the fact that, even today, we have little solid

Environmental and Pollution Science. https://doi.org/10.1016/B978-0-12-814719-1.00028-8

information about the toxicity of a large number of chemicals. Of the 6,000,000 known chemicals, about 50,000 are in common use, and detailed chronic toxicity tests have been performed on only a few hundred of these. Even for those that have been tested, many questions remain about the interpretation of the results obtained, including serious reservations about the applicability of laboratory test results to human populations in everyday situations. In many cases, then, we lack a basic understanding of how toxicants act.

28.2 TOXIC VERSUS NONTOXIC

The term *safe* commonly means "without risk." But this common definition has no meaning in scientific study. Scientists cannot ascertain conditions under which a given chemical exposure is absolutely without risk of any type. Conversely, they can describe conditions under which risks are so low that they have no practical consequence to a specific population. In technical terms, the safety of chemical substances—whether in food, drinking water, air, or the workplace—has typically been defined as a condition of exposure under which there is a "practical certainty" that no harm will result to exposed individuals. In terms of mortality, this is usually accepted as a risk of 1:1,000,000 chance of dying during a lifetime (see Chapter 29).

Another fundamental concept is the classification of chemical substances as either *safe* or *unsafe* (or as *toxic* and *nontoxic*). This type of classification can be highly problematic. All substances, even those that we consume in high amounts every day, can be made to produce a toxic response under some conditions of exposure. In this sense, all substances can be "toxic." Thus safety involves not simply the degree of toxicity of a substance, but rather the degree of risk under given conditions. In other words, we ask, "What is the probability that the toxic properties of a chemical will be expressed under actual or anticipated conditions of human or animal exposure?" The science of risk assessment attempts to link toxicological information on adverse effects to the probability of toxic effects during likely exposure scenarios (see Chapter 29).

28.3 EXPOSURE AND DOSE

Humans and other organisms can be exposed to substances in different environmental media—air, water, soil, or food—or they may have direct contact with a sample of the substance. The *exposure concentration* is the amount of a substance present in the medium with which an organism has contact. The *dose* is the amount of the chemical that is received by the target (organ). The exposure concentration may differ from the dose owing to biochemical transformations in living organisms.

Suppose, for example, a substance is present in drinking water. The amount of this substance in the water is the exposure concentration. For many environmental substances, this amount ranges from less than 1 microgram (μg) to greater than 1 milligram (mg) and is usually reported as milligrams or micrograms of the substance present in 1 L of water (i.e., in $mg\,L^{-1}$ or $\mu g\,L^{-1}$).[1]

An individual's intake—or *dose*—of this substance depends on the amount present in a given volume of water and on the amount of water consumed in a given period of time. Given the concentration of the substance in water (say, in ppm) and the human consumption of water per unit of time, it is possible to estimate the total amount of the substance an individual will consume through use of contaminated water. For instance, adults are assumed to consume 2 L of water each day through all uses (see Table 29.4). Thus if a substance is present at $10\,mg\,L^{-1}$ ($=10\,ppm$) in water, the average daily individual intake of the substance is

$$10\,mg\,L^{-1} \times 2\,L\,day^{-1} = 20\,mg\,day^{-1}$$

Toxicity measures must also take body size differences into account, usually by dividing daily intake by the weight of the individual. That is, the toxicity of a substance is usually dependent upon concentration per unit of body weight. Thus for a man of average weight (usually assumed to be 70 kg), the daily dose of this substance is

$$20\,mg\,day^{-1}/70\,kg = 0.29\,mg\,kg^{-1}\,day^{-1}$$

For a person of lower weight, such as a female or child, the *daily dose* at the same intake rate would be larger. For example, a 50-kg woman ingesting this substance would receive a dose of

$$20\,mg\,day^{-1}/50\,kg = 0.40\,mg\,kg^{-1}\,day^{-1}$$

Using the same equation, a child of 10 kg would receive a dose of $2.0\,mg\,kg^{-1}\,day^{-1}$. However, children drink less water each day than do adults (say, 1 L), so a child's dose would be

$$10\,mg\,L^{-1} \times 1\,L\,day^{-1}/10\,kg = 1.0\,mg\,kg^{-1}\,day^{-1}$$

In general, the smaller the body size, the greater the dose (in $mg\,kg^{-1}\,day^{-1}$) received from drinking water. This is also true of animals used for toxicity studies. Usually rats or mice will receive a much higher dose of drinking water contaminants than humans because of their much smaller body size.

Because each medium (air, soil, water) of exposure must be treated separately, some calculations are more complex than those of dose per liter of water. Many calculations may simply be additive. For example, a human may be simultaneously exposed to the same substance through several media (e.g., through inhalation, ingestion, and dermal

---

1. These two units are sometimes expressed as the more ambiguous units of parts per million (ppm) or parts per billion (ppb), respectively.

contact). Thus if an individual can both ingest and inhale (say, in the shower) some volatile compound in tap water, the *total dose* received by that individual is the sum of doses received through each individual route. In some cases, however, it is inappropriate to add doses in this fashion because the toxic effects of a substance may depend on the route of exposure. For example, inhaled chromium is carcinogenic to the lung, but it appears that ingested chromium is not. In general, though, as long as a substance acts at an internal body site (i.e., acts systematically rather than at a particular point of initial contact), it is usually acceptable procedure to add doses received from several routes.

Absorption, or *absorbed dose*, is another factor that requires special attention when considering dose and exposure. When a substance is ingested in food or drinking water, it enters the gastrointestinal tract. When it is present in air (e.g., as a gas, aerosol, particle, dust, or fume), it enters the upper airways and lungs. A substance may also come into contact with the skin and other body surfaces such as a gas, liquid, or solid. Some substances may cause toxic injury at the point of initial contact (the skin, gastrointestinal tract, upper airways, lungs, or eyes). Indeed, at high concentrations, most substances do cause at least irritation at these points of contact. However, for many substances, toxicity occurs after they have been absorbed, that is, after they pass through certain barriers (e.g., the wall of the gastrointestinal tract or the skin itself), enter blood or lymph, and gain access to the various organs or systems of the body. Some chemicals may be distributed in the body in various ways and then excreted. However, some chemical types—usually lipid-soluble substances such as the pesticide dichlorodiphenyltrichloroethane (DDT)—can be stored for long periods of time, usually in body fat.

Substances vary widely in extent of absorption. The fraction of a dose that passes through the wall of the gastrointestinal tract may be very small (1%–10% for some metals) or it may be substantial (close to 100% for certain types of organic molecules). Absorption rates also depend on the medium in which a chemical is present; a substance present in water might be absorbed differently than the same substance present in, say, a fatty diet. Absorption rates also vary among animal species and among individuals within a species. Ideally, an estimation of a *systemic dose* should consider absorption rates. Unfortunately, data on absorption are limited for most substances, especially in humans, so absorption is not always included in dose estimation. In some cases, dose estimates may be crudely adjusted on the basis of the molecular characteristics of a particular substance and/or general principles of absorption. In many cases, however, absorption is simply considered to be complete by default.

The technique of *extrapolation*, or drawing inferences, from experimentally observed results can also be a major factor in predicting the likelihood of toxicity, say, from one route of exposure to other routes, or from one organism to another. Experiments for studying toxicity typically involve intentional administration of substances to subjects (usually mice or rats) through ingested food or inhaled air, or through direct application to skin. In other cases, they may include other routes of administration, such as injection under the skin (subcutaneous), into the blood (usually intravenous), or into body cavities (intraperitoneal). Such toxicity studies in experimental animals are of greatest value when experimental exposures mimic the mode of human exposure. Thus if both animals and humans are exposed to a contaminant via drinking water, it is generally assumed that the data in animals can be applied directly to humans. But when experimental routes differ from human routes (e.g., animal exposure via injection; human exposure via drinking water), a correction or safety factor must be used to apply such data to human exposures (see Section 29.2.3).

28.4 EVALUATION OF TOXICITY

Information on the toxic properties of chemical substances is obtained through plant, bacterial, and animal studies; controlled epidemiological investigations of exposed human populations or microcosms; and clinical studies or case reports of exposed humans or ecosystem studies (e.g., oil spills). Other information bearing on toxicity derives from experimental studies in systems other than whole animals (e.g., isolated organics in cells or subcellular components) and from analysis of the molecular structures of the substances of interest. These last two sources of information are generally less certain as indicators of toxic potential.

Many types of toxicity studies can be conducted to identify the nature of health damage produced by a substance and the range of doses over which such damage is produced. Each of the many different types of toxicological studies has a different purpose. The usual starting point for such investigations is a study of the *acute (single-dose) toxicity* of a chemical in experimental animals, plants, or bacteria. Acute toxicity studies are used to calculate doses that will not be lethal to any organism and can be used in toxicity studies of longer duration. Moreover, such studies provide an estimate of the compound's comparative toxicity and may indicate the target organ system (e.g., kidney, lung, or heart) affected in an animal. Once the acute toxicity is known, organisms may be exposed repeatedly or continuously for several weeks or months in *subchronic toxicity* (Table 28.1) studies, or for close to their full lifetimes in *chronic toxicity* (Table 28.2) studies.

When toxicologists examine the lethal properties of a substance, they estimate its LD_{50}, which is the lethal dose for 50% of an exposed population (Fig. 28.1). A group of well-known substances and their LD_{50} values are listed in

TABLE 28.1 Subchronic Toxicity Tests Are Employed to Determine Toxicity Likely to Arise From Repeated Exposures of Several Weeks to Months

Species

Rodents (*usually rats*) preferred for oral and inhalation studies; rabbits for dermal studies; nonrodents (*usually dogs*) recommended as a second species for oral tests

Age

Young adults

Number of animals

10 of each sex for rodents, 4 of each sex for nonrodents per dose level

Dosage

Three dose levels plus a control group; include a toxic dose level plus NOAEL[a]; exposures are 90 days

Observation period

90 days (*same as treatment period*)

[a]*See Section 28.4.6.*

TABLE 28.2 Chronic Toxicity Tests Determine Toxicity From Exposure for a Substantial Portion of a Subject's Life

Species

Two species recommended; rodent and nonrodent (*rat and dog*)

Age

Young adults

Number of animals

20 of each sex for rodents, 4 of each sex for nonrodents per dose level

Dosage

Three dose levels recommended; includes a toxic dose level and NOAEL[a]; exposures generally for 12 months; FDA requests 24 months for food chemicals

Observation period

12–24 months

[a]*See Section 28.4.6.*

Table 28.3. LD$_{50}$ studies reveal one of the basic principles of toxicology:

Not all individuals exposed to the same dose of a substance will respond in the same way.

Thus the same dose of a substance that leads to the death of some individuals will only impair other organisms and not affect other organisms at all.

The premise underlying animal toxicity studies is the long-standing assumption that effects in humans can be inferred from effects observed in animals. This principle of extrapolating animal data to humans has been widely accepted in scientific and regulatory communities. For instance, all of the chemicals that have been demonstrated to be carcinogenic in humans are carcinogenic in some, although not all, animal species typically used for toxicological studies. In addition, the acutely toxic doses of many chemicals are assumed to be similar in humans and a variety of experimental animals. This inference is based on the evolutionary relationships between animal species. That is, at least among mammals, the basic anatomical, physiological, and biochemical parameters are expected to be much the same across species.

On the whole, the general principle of making such *interspecies inferences* is well founded. But some exceptions have been noted; for example, guinea pigs are much more sensitive to dioxin (2,3,7,8-tetrachlorodibenzo-*p*-dioxin) than are other laboratory animals. Many of these exceptions arise from differences in the ways various species handle exposure to a chemical and to differences in *pharmacokinetics*, which includes the rates at which a specific chemical is distributed among tissues, the manner in which it is excreted, and the types of metabolic changes it causes. Because of these potential differences, it is essential to evaluate all interspecies differences carefully when inferring human toxicity from animal toxicological studies.

In the particular case of long-term animal studies conducted to assess the carcinogenic potential of a compound, certain general observations increase the overall strength of the evidence that the compound is carcinogenic. Thus, for example, an increase in the number of tissue sites affected by the agent is a strong indicator of carcinogenicity, as is an increase in the number of animal species, strains, and sexes showing a carcinogenic response. Other observations that affect the strength of the evidence may involve a high level of statistical significance of the increase of tumor incidence in treated versus control animals, as well as clear-cut dose–response relationships in the data evaluated, such as dose-related shortening of the time-to-tumor occurrence or time-to-death with tumor and a dose-related increase in the proportion of tumors that are malignant.

28.4.1 Manifestations of Toxicity

Toxic effects can take various forms. A toxic effect can be immediate, as in strychnine poisoning, or delayed, as in lung cancer. Indeed, cancer typically affects an individual many years after continuous or intermittent exposure to a carcinogen. An effect can be local (i.e., at the site of application) or systemic (i.e., carried by the blood or lymph to different parts of the body). When examining toxic effects,

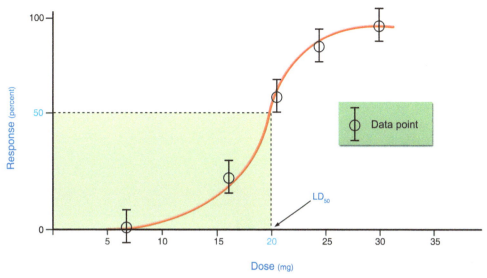

FIG. 28.1 LD$_{50}$ is the dose (in mg) lethal to 50% of the animals administered the dose.

TABLE 28.3 Approximate Oral LD$_{50}$ in a Species of Rat for a Group of Well-Known Chemicals

Chemical	LD$_{50}$ (mg kg^{-1})
Sucrose (table sugar)	29,700
Ethyl alcohol	14,000
Sodium chloride (common salt)	3000
Vitamin A	2000
Vanillin	1580
Aspirin	1000
Chloroform	800
Copper sulfate	300
Caffeine	192
Phenobarbital, sodium salt	162
DDT	113
Sodium nitrite	85
Nicotine	53
Aflatoxin B1	7
Sodium cyanide	6.4
Strychnine	2.5

there are several important factors to consider, some of which are dosage related.

1. The *severity* of injury can increase as the dose increases and vice versa. Some organic chemicals, for example, are known to affect the liver. High doses of such a chemical (e.g., carbon tetrachloride) will kill liver cells—perhaps killing so many cells that the whole liver is destroyed, so that most or all of the experimental animals die. As the dose is lowered, fewer cells are killed, but the liver exhibits other forms of injury that indicate impairment in cell function and/or structure. At still lower doses, no cell deaths may occur, and only slight changes are observed in cell function or structure. Finally, a dose may be so low that no effect is observed or the biochemical alterations that are present have no known adverse effects on the health of the animal. One of the goals of toxicity studies is to determine this dose level, known as the no-observed-adverse-effect level (NOAEL) (see Section 29.2.3).

2. The *incidence* of an effect, but not its severity, may increase with increasing dosage. In such cases, as the dose increases, the fraction of experimental organisms experiencing diverse effects (i.e., disease or injury) increases. At sufficiently high doses, all experimental subjects will experience the effect. Thus increasing the dose increases the probability (i.e., the risk) that an abnormality will develop in an exposed population.

3. Both the severity and the incidence of a toxic effect may increase as the level of exposure increases. The increase in severity is a result of increased damage at higher doses, while the increase in incidence is a result of differences in individual sensitivity. In addition, the site at which a substance acts (e.g., liver, kidney) may change as the dosage changes. Many toxic effects, including cancer, fall in this category. Generally, as the duration of exposure increases, the critical NOAEL dose decreases; in some cases, new effects not seen with exposures of short duration appear after long-term exposure.

4. The *seriousness* of a toxic effect must also be considered. Certain types of toxic damage, such as asbestosis caused by inhalation of asbestos fibers, are clearly adverse and are a definite threat to health. However, the health significance of other types of effects observed during toxicity studies may be ambiguous. For example, at a given dose, a chemical may produce a slight increase in body temperature. If no other effects are observed at this dose, researchers cannot be sure that a true adverse response has occurred. Determining whether such slight changes are significant to health is one of the critical issues in assessing safety.

5. Toxic effects also vary in degree of reversibility. In some cases, an adverse health effect will disappear almost immediately following cessation of exposure. At the other extreme, some exposures will result in a permanent injury—for example, a severe birth defect arising from exposure to a substance that irreversibly damaged the fetus at a critical moment of its development. Furthermore, some tissues, such as the liver, can repair themselves relatively quickly, while others, such as nerve cells, have no ability to repair themselves. Most toxic responses fall somewhere between these extremes.

28.4.2 Toxicity Testing

Any organism can be used to assess the toxicity of a substance. The choice of test organism depends on several factors, including budget, time, and the organism's occurrence in a given environment. The simplest and least costly tests are performed with unicellular animals or plants and may last only a few hours or days. In a water environment, for example, *Daphnia* or algae may be used in testing for potential aquatic pollutants. Short-term tests may look at the death or immobilization of swimming *Daphnia* (Fig. 28.2), while longer term tests may look at the growth of the organisms (increase in biomass) or numbers of offspring. Animals higher in the food chain are also important in aquatic toxicity tests, and experiments involving fish, amphibians, and other macroinvertebrates are familiar standbys (Table 28.4). For terrestrial toxicity tests, higher plant, rodent, or bird toxicity tests can be used. Strains of genetically characterized rats and mice, for example, are often used in such studies. Avian toxicity tests have also been developed; for instance, birds are frequently used in evaluating the effects of pesticides on nontarget species.

In environmental toxicology, researchers often perform toxicity tests in artificially contained communities to assess the environmental impacts of toxic substances after release into the environment. These artificial communities, which serve as laboratory models of natural ecosystems, are referred to as *microcosms*. While many microcosms are elaborate systems that effectively mimic whole ecosystems,

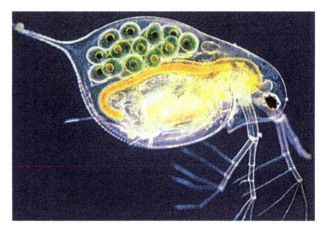

FIG. 28.2 *Ceriodaphnia dubia*, an invertebrate used for toxicity testing. *(Source: South Dakota Department of Natural Resources. Available from: www.state.5d.us.)*

TABLE 28.4 Organisms Commonly Used in Toxicity Testing

Types of Organisms	Organism
Invertebrates	*Daphnia magna*
	Crayfish
	Mayflies
	Midges
	Planaria
Aquatic vertebrates	Rainbow Trout
	Goldfish
	Fathead minnow
	Catfish
Algae	*Chlamydomonas reinhardtii* (green algae)
	Microcystis aeruginosa (blue-green algae)
Mammals	Rats
	Mice
Avian species	Bobwhite
	Ring-necked pheasant

From Pollution Science, Academic Press, San Diego, CA, 1996.

some microcosms may be nothing more than a set of glass jars containing soil or water with sediment at the bottom (Fig. 28.3). But even simple glass jars allow researchers to examine the effect of substances on multispecies, such as algae, bacteria, and microinvertebrates.

FIG. 28.3 Laboratory microcosms. *(Source: M.L. Brusseau.)*

Toxicity experiments vary widely in design and protocols. Some tests and research-oriented investigations are conducted using prescribed study designs, as is the case with carcinogenicity assays in fish. In connection with pre-market testing requirements for certain classes of chemicals, however, regulatory and public agencies have developed relatively few standardized tests for various types of toxicity.

Rats and mice are the most commonly used laboratory animals for toxicity testing. These rodents are inexpensive and can be handled relatively easily; moreover, the genetic background and disease susceptibility of these species are well established. In addition, the full life span of these small rodents is complete in 2–3 years; thus the effects of lifetime exposure to a substance can be measured relatively quickly. Other rodents such as hamsters and guinea pigs are also common laboratory subjects, as are rabbits, dogs, and primates. Usually, the choice of experimental animal depends on the system being studied. Reproductive studies, for example, often use primates such as monkeys or baboons because their reproductive systems are similar to that of humans. Similarly, rabbits are often used for testing dermal toxicity because their shaved skin is more sensitive than that of other animals.

Animals are usually exposed by a route that is as close as possible to the route by which humans will be exposed. In some cases, however, it may be necessary to use other routes or conditions of dosing to achieve the desired experimental dose. For example, some substances are administered by stomach tube (gavage) because they are too volatile or unpalatable to be placed in the animals' feed at the high levels needed for toxicity studies.

A toxicity experiment is of limited value unless researchers find a dose of sufficient magnitude to cause some type of adverse effect within the duration of the experiment. If no effects are seen at any dose administered, the toxic properties of the substance cannot be characterized; thus experiments may be repeated at higher doses or for longer times until distinct adverse effects are observed. The most distinctive adverse effect is, of course, death. Therefore researchers frequently begin their experiments by determining the LD_{50}, since the endpoint of this experiment (death) is easily measured. Next, researchers usually look at the effects of lower doses administered over longer periods to find the range of doses over which adverse effects occur and to identify the NOAEL for these effects.

Studies may be characterized according to the *duration of exposure*. Acute toxicity studies involve a single dose or exposures of very short duration (e.g., 8 h of inhalation). Chronic studies involve exposures for nearly the full lifetime of the experimental animals, while subchronic studies vary in duration between these two extremes. Although many different dose levels are needed to develop a well-characterized dose–response relationship, practical considerations usually limit the number to two or three, especially in chronic studies. Experiments involving a single dose are frequently reported, but these leave great uncertainty about the full range of doses over which effects are expected.

28.4.3 Toxicity Tests for Carcinogenicity

One of the most complex and important of the specialized tests is the *carcinogenesis bioassay*. This type of experiment is used to test the hypothesis of carcinogenicity, that is, the capacity of a substance to produce malignant tumors.

Usually, a test substance is administered over most of the adult life of a laboratory animal, then the animal is observed for formation of tumors. In this kind of testing, researchers generally administer high doses of the chemical to be tested—specifically, the *maximum tolerated dose* (MTD), which is the maximum dose that an animal can tolerate for a major portion of its lifetime without significant impairment of growth or observable toxic effect other than carcinogenicity. The MTD and one-half of that, or MTD_{50},

TABLE 28.5 Carcinogenicity Tests Are Similar to Chronic Toxicity Tests. However, They Extend Over a Longer Period of Time and Require Larger Groups of Animals

Species

Testing in two rodent species, the rat and mouse, preferred due to relatively short life spans

Age

Young adults

Number of animals

50 of each sex per dose level

Dosage

Three dose levels recommended; highest should produce minimal toxicity; exposure periods are at least 18 months for mice and 24 months for rats

Observation period

12–24 months for mice and 24–30 months for rats

are the usual doses used in a National Cancer Institute (NCI) carcinogenicity bioassay, so that the animals survive in relatively good health over their normal lifetime. The main reason for using the MTD as the highest dose in a bioassay is that these very high doses help to overcome the statistical insensitivity inherent in small-scale experimental studies.

Owing largely to cost considerations, experiments are carried out with relatively small groups of animals—typically, 50 or 60 animals of each species and sex at each dose level (Table 28.5), including the control group. At the end of such an experiment, the incidence of cancer (including tumor incidence in control animals) is tabulated and plotted as a function of dose. Then the data are analyzed to determine whether any observed differences in tumor incidence (the fraction of animals having a tumor of a certain type) are due to exposure to the substance under study or to random variations. In an experiment of this size, assuming none of the control animals develop tumors, the lowest incidence of cancer that is detectable with statistical reliability is in the range of 5%, which is equivalent to 3 out of 60 animals developing tumors. If control animals develop tumors (as they frequently do), the lowest range of statistical sensitivity is even higher. A cancer incidence of 5% is very high; but ordinary experimental studies are not capable of detecting lower rates, and most are even less sensitive.

28.4.4   Epidemiological Studies

Information on adverse health effects in human populations is obtained from four major sources: (1) summaries of self-reported symptoms in exposed persons; (2) case reports prepared by medical personnel; (3) correlation studies, in which differences in disease rates in human populations are associated with differences in environmental conditions; and (4) epidemiological studies. The first three of these sources are characterized as descriptive epidemiology, while the fourth category—*epidemiological studies*—is generally reserved for studies that compare the health status of a group of persons who have been exposed to a suspected agent with that of a nonexposed control group. (*Note:* Such studies cannot identify cause-and-effect relationships between exposure to a substance and particular diseases or conditions; however, they can draw attention to previously unsuspected problems and generate hypotheses that can be tested further.)

Most epidemiological studies are either case–control studies or cohort studies. *Case–control studies* (Fig. 28.4) first identify a group of individuals who have a specific disease, then attempt to ascertain commonalities in exposures that the group may have experienced. For example, the carcinogenic properties of diethylstilbestrol (DES), a drug once used to prevent miscarriages, were brought to light through studies of women afflicted with certain types of vaginal and cervical cancer. *Cohort studies* (Fig. 28.5), on the other hand, begin by examining the health status of individuals known to have had a common exposure. These studies then attempt to determine whether any specific condition is associated with that exposure by comparing the exposed group's health with that of an appropriately matched control population. For example, a cohort study of lab workers exposed to benzene revealed an excessively high incidence of leukemia, thereby providing strong evidence in support of a benzene leukemogenesis hypothesis. Generally, epidemiologists have used individuals who belong to an identifiable group, such as those in certain occupational settings or patients treated with certain drugs, to conduct such studies—hence the name "cohort."

Convincing results from epidemiological investigations can be enormously beneficial, because the data provide information about humans under actual conditions of exposure to a specific agent. Therefore results from well-designed, properly controlled studies are usually given more weight than results from animal studies. Although no study can provide complete assurance that a chemical is harmless, negative data from epidemiological studies of sufficient size can assist in establishing the maximum level of risk due to exposure to the agent.

Obtaining and interpreting epidemiological results, however, can be quite difficult. Appropriately matched control groups are difficult to identify, because the factors that lead to the exposure of the study group (e.g., occupation or residence) are often inextricably linked to the factors that affect health status (e.g., lifestyle and socioeconomic

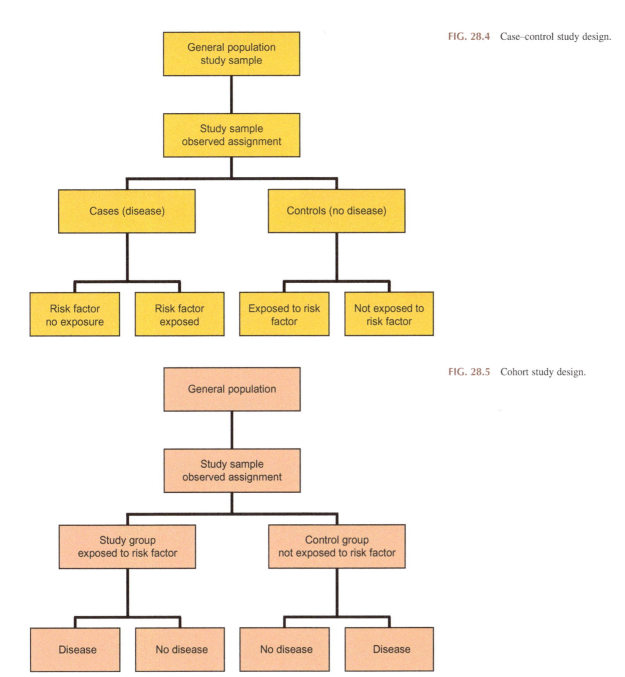

FIG. 28.4 Case–control study design.

FIG. 28.5 Cohort study design.

status). Thus controlling for related risk factors (i.e., cigarette smoking) that have strong effects on health is difficult. Moreover, the statistical detection power of epidemiological studies depends on the use of very large populations; and such data may be hard to come by or incomplete. Few types of health effects—other than death—are recorded systematically in human populations, and even the information on cause of death is limited in reliability. For example, infertility, miscarriages, and mental illness are not, as a rule, systematically recorded by public health agencies, while death is often attributed to "heart failure," whatever its proximate cause.

In addition, accurate data on the degree of exposure to potentially hazardous substances are only rarely available, especially when exposures have taken place years or decades earlier. Establishing dose–response relationships is therefore frequently impossible. Nor can current data, however carefully obtained, immediately help researchers who are investigating slowly developing diseases such as cancer; rather, epidemiologists must wait many years to determine the presence or absence of an effect. Meanwhile, exposure to suspect agents could continue during these extended periods of time, thereby increasing risk further.

For these and other reasons, interpretations of epidemiological studies are sometimes subject to extreme uncertainties. Independent confirmatory evidence is usually necessary, including supporting results from a second epidemiological study or supporting data from experimental studies in animals. Such confirmatory evidence is particularly necessary in the case of negative findings, which must be interpreted with great caution (EPA, 1989).

For example, suppose we have a drinking water contaminant that is known to cause cancer in 1 out of every 100 people exposed to $10\,mg\,L^{-1}$. Further suppose that the average time required for cancer to develop from $10\,mg\,L^{-1}$ of exposure is 30 years (which is not uncommon for a carcinogen). After our townspeople have been exposed to the drinking water contaminant for 15 years, we conduct a study. For this study, we collect death certificates of 20 people exposed to the contaminant, but we have little information on actual exposure. We know that some of the deceased were exposed when the contaminant was first introduced into the water supply and that others were exposed several years later. When we turn to the health records, we find that they are incomplete. Finally, the results of our study reveal that 20 cancer deaths is not an excessive number when compared to an appropriate control group. Is it then correct for us to conclude that our known carcinogen is not carcinogenic?

28.4.5 Short-Term Tests for Toxicity

The lifetime animal study is the primary method used for detecting the carcinogenic properties and general toxicity of a substance. Short-term tests for toxicity, however, are used to measure effects that appear to be correlated with specific toxic effects. For example, those for carcinogenicity include assays for gene mutations in bacteria, yeast, fungi, insects, and mammalian cells; mammalian cell transformation assays; assays for DNA damage and repair, and in vitro (outside the animal) or in vivo assays (within the animal) for chromosomal mutations in animal cells. There are also a number of short-term toxicity assays that are based on inhibiting the functions of necessary enzymes in organisms, such as ATPases, phosphatases, and dehydrogenase. Phosphatase measurements, for example, can be used to assess the activity of specific substances, such as the toxicity of heavy metals in soils. In addition, short-term bioassays can use enzymes or microorganisms to assess general toxicity of environmental samples (Bitton, 1999).

Several tests involving whole animals are also available. These tests, which are usually of intermediate duration, include the induction of skin and lung tumors in female mice, breast cancer in certain species of female rats, and anatomical changes in the livers of rodents.

Many carcinogenic (cancer-causing), mutagenic (mutation-causing), and teratogenic (defect-causing) agents act in the same way: they cause changes in DNA that eventually affect cell development. Because of this relationship, initial screening for such substances can often be accomplished quickly by testing their capacity to cause mutations in a particular strain of bacteria—*Salmonella typhimurium*. This unique strain of bacteria requires the essential amino acid histidine to grow, so it can only grow on histidine-free media if it has first mutated. Thus if we observe these bacteria growing on histidine-free media after exposure to a test chemical, we can safely assume that the chemical caused the bacteria to mutate. The test chemical is therefore likely to be a carcinogen, mutagen, or teratogen. This short-term test is called the *Ames test* after its developer, Dr. Bruce Ames of the University of California at Berkeley (Fig. 28.6).

Another short-term test is a bioassay based on the light output of the bioluminescent bacterium, *Photobacterium phosphoreum*. This bioassay has been used to assess the general toxicity of wastewater effluents, industrial wastes, sediment extracts, and hazardous waste leachates. As toxic substances diminish the viability of the bacterium, bioluminescent activity decreases, and the light output can be quantitatively measured by an instrument (Fig. 28.7).

FIG. 28.6 The Ames Test. *(From https://en.wikipedia.org/wiki/Ames_test#/media/File:Ames_test.svg)*

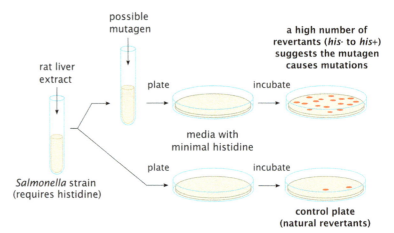

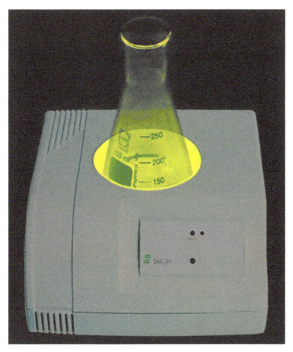

FIG. 28.7 Tox Tracer System is a bioassay based on the principle of inhibition of the natural bioluminescence of the marine bacterium *Vibrio fischeri*. *(Available from: www.skalar.com.)*

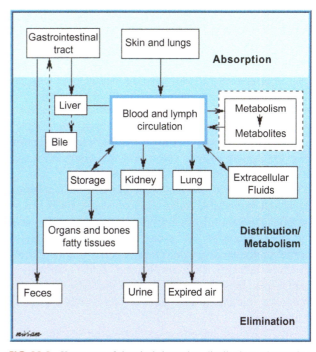

FIG. 28.8 Key routes of chemical absorption, distribution and excretion.

28.4.6 Threshold Effects

Commonly accepted theory suggests that most biological effects of noncarcinogenic chemical substances occur only after a certain concentration or level is achieved. This level, known as the threshold dose, is approximated by the NOAEL (see Section 29.2.3). Another widely accepted premise, at least in the setting of public health standards, is that the human population is likely to have a much more variable response to toxic agents than do the small groups of well-controlled, genetically homogeneous animals that are routinely used in experiments. The NOAEL is itself subject to some uncertainty owing to variabilities in the data from which it was obtained. For these reasons, public health agencies divide experimental NOAELs by large uncertainty factors, known as safety factors, when examining substances that display threshold effects (see Section 29.2.3). The magnitude of these safety factors varies according to the following: the nature and quality of the data from which the NOAEL is derived; the seriousness of the toxic effects; the type of protection being sought (e.g., protection against acute, subchronic, or chronic exposures); and the nature of the population to be protected (i.e., the general population versus identifiable subpopulations expected to exhibit a narrower range of susceptibilities). Safety factors of 10, 100, 1000, and 10,000 have been used in various circumstances.

At present, only agents displaying carcinogenic properties are treated as if they display no thresholds (see Section 29.2.3). Thus the dose–response curve for carcinogens in the human populations achieves zero risk only at zero dose; as the dose increases above zero, the risk immediately becomes finite and thereafter increases as a function of dose. Risk in this case is the probability of producing cancer, and at very low doses the risk can be extremely small.

28.5 RESPONSES TO TOXIC SUBSTANCES

In general, an organism's response to a toxic chemical depends on the dose administered. However, once a toxicant enters the body, the interplay of four processes—absorption, distribution, excretion, and metabolism—determines the actual effect of a toxic chemical on the target organ, which is the part of the body that can be damaged by that particular chemical. Carbon tetrachloride, for example, affects the liver and kidneys, while benzene affects the blood cell forming system of the body. Fig. 28.8 summarizes routes of absorption, distribution, and excretion.

28.5.1 Absorption

Absorption of toxicants across body membranes and into the bloodstream can occur in the gastrointestinal (GI) tract, in the lungs, and through the skin. Contaminants present in drinking water, for example, enter the body primarily through the GI tract. Once they enter the GI tract, most chemicals must be absorbed to exert their toxic effect. Owing largely to differences in solubility, some compounds are absorbed more readily than others. Lipid-soluble, nonionized organic compounds, such as DDT and polychlorinated biphenyls (PCBs), are more readily absorbed by diffusion in the GI tract than are lipid-insoluble, ionized compounds, such as lead and cadmium salts. The GI tract also employs specialized active transport systems for compounds, such as sugars, amino acids, pyrimidines, calcium, and sodium. Although these active transport systems do not generally play a major role in absorption of toxicants, they can contribute to their absorption in some cases; lead, for example, can be absorbed via the calcium transport system.

The behavior of toxicants in the GI tract also depends on the action of digestive fluids. These digestive fluids can be beneficial or harmful. For example, snake venom, a protein that is quite toxic when injected, is nontoxic when administered orally because stomach enzymes attack the protein structure, breaking it down into amino acids. However, in the GI tract, these enzymes can also contribute to the conversion of nitrates to carcinogenic compounds known as nitrosamines.

Age is also an important factor affecting the intestine's ability to act as a barrier to certain toxicants. The GI tract of newborns, for instance, has a higher pH and a higher number of *E. coli* bacteria than that in adults. These conditions promote the conversion of nitrate, a common drinking water pollutant from agricultural runoff, into the more toxic chemical nitrite. The resulting nitrite then interferes with the blood's ability to carry oxygen, causing methemoglobinemia or "blue-baby syndrome." Lead is also absorbed more readily in newborns than in adults. (*Note*: Even though a chemical has been absorbed through the GI tract, it can still be excreted or metabolized by the intestine or liver before it reaches systemic circulation.)

The lungs are anatomically designed to absorb and excrete chemicals, as is shown by their continuous absorption of oxygen and excretion of carbon dioxide. The alveoli have a large surface area (50–100 m^2) and are well supplied with blood, and the blood is very close (10 μm) to the air space within the alveoli. These characteristics make the lungs particularly good vehicles for the absorption of toxicants. Toxicants may have to pass through as few as two cells to travel from the air into the bloodstream.

In contrast, the skin is relatively impermeable to toxicants. However, some toxicants, such as carbon tetrachloride, can be absorbed through the skin in sufficient quantities to cause live injury. In addition, a few chemicals, such as dimethyl sulfoxide (DMSO), have been shown to penetrate the skin fairly readily. Absorption through the skin is possible through the hair follicles, through the cells of the sweat glands and sebaceous glands, and through cuts or abrasions, which increase the rate and degree of absorption. The sole means of absorption through the skin appears to be passive diffusion.

28.5.2 Distribution

Distribution of a toxicant to various organs depends on the ease with which it crosses cell membranes, its affinity for various tissues, and the blood flow through the particular organ. A toxicant's site of concentration is not necessarily the target organ of toxicity. For example, lead can be stored harmlessly in bone, and many lipid-soluble toxicants (such as the chlorinated hydrocarbon insecticides) are stored in fat, where they cause relatively little harm. However, a stored contaminant can be released back into the bloodstream under various conditions. Thus fat-stored chlorinated pesticides can be released during starvation, dieting, or illness, when fat is consumed.

A number of anatomical barriers in the body are thought to prevent or hinder the entrance of certain toxicants into organs. However, these barriers are not impenetrable walls. The so-called blood–brain barrier, for example, does not prevent toxicants from entering the *central nervous system* (CNS); rather, the physiological conditions at the blood–brain interface make it more difficult for some toxicants to leave the blood and enter the CNS. In general, lipid-soluble toxicants can cross the blood–brain barrier, but some water-soluble toxicants cannot. There is also the "placental barrier," which is even less of a barrier; the fact is that any chemical absorbed into the mother's bloodstream can and will cross her placenta and enter the bloodstream of the fetus to some degree.

28.5.3 Excretion

Chemicals can be excreted from the body in several ways. The kidney removes toxicants from the blood in the same way that the end products of metabolism are eliminated, that is, by glomerular filtration, passive tubular diffusion, and active secretion. Glomerular filtration is simply a filtration process in which compounds below a certain molecular weight (and hence bulk) pass through pores in a part of the kidney known as the glomeruli. All compounds whose molecular weight is less than 60,000 can filter through the glomeruli unless they are bound to plasma proteins. The molecular weight of most toxicants in drinking water is between 100 and 500; thus these compounds easily pass through the glomeruli. The toxicants then pass through collecting ducts and tubules, through which water-soluble toxicants may be excreted with urine. However, lipid-soluble toxicants can defeat the excretion process at this point by moving (via passive diffusion) through the tubule wall and back into the bloodstream.

The liver eliminates toxicants through the bile, which passes into the intestine through the gall bladder and bile duct, and finally exits from the body via the feces. As in the kidney, the transport mechanisms used are passive diffusion and carrier-mediated transport. Toxicants that have been excreted into the intestine through the bile can be reabsorbed (especially if they are lipid soluble) into the bloodstream while in the intestine.

Toxicants are also excreted through several other routes, including the lungs, GI tract, cerebrospinal fluid, milk, sweat, and saliva. Milk, for example, has a relatively high concentration of fat (3.5%); thus lipid-soluble compounds such as DDT and PCBs can concentrate in milk. In addition, because milk is slightly acidic, with a pH of 6.5, basic compounds can also concentrate in it. In this way, toxicants may be passed from mother to child, or from cows to humans.

An important concept in excretion is a toxicant's half-life $T_{1/2}$, which is the time it takes for one-half of the chemical to be eliminated from the body. Thus if a chemical has a half-life of 1 day, 50% of it will remain with the body one day after absorption, 25% will remain after two days, 12.5% after 3 days, and so on. The concept of the half-life is important, because it indicates how long a compound will remain within the body. Generally a compound is considered eliminated after a period of seven half-lifes.

28.5.4 Metabolism

Because lipid-soluble compounds can cross cell membranes to be reabsorbed in the kidney and intestine, they are subject to metabolic processes, which are the biochemical reactions by which cells transform food into energy and living tissue. Metabolism is the sum of biochemical changes occurring to a molecule within the body. In many cases, the body metabolizes these toxicants into water-soluble compounds, which can be excreted easily. However, in some instances, metabolism of a chemical creates a more toxic chemical or does not change the chemical's toxicity.

Two types of reactions occur in metabolism: (1) relatively simple reactions involving oxidation, reduction, and hydrolysis; and (2) more complex reactions involving conjugation and synthesis. All of these reactions occur primarily in the liver. Oxidation is the mechanism of metabolism for many compounds. When considering metabolism, researchers also look at species, strain, and gender differences. Age is also an important factor in both humans and laboratory animals; both the very young and very old are more susceptible to certain chemicals.

28.5.5 Biotransformation of Toxicants

Biotransformation is the process by which substances that enter the body are changed from hydrophobic to hydrophilic molecules to facilitate elimination from the body. This process usually generates products with few or no toxicological effects. Biotransformations sometimes yield toxic metabolites, through a process known as *bioactivation*. The chemical reactions responsible for changing a lipophilic toxicant into a chemical form are known as *Phase*

I and *Phase II biotransformations*. The two groups are defined based on the reactions that are catalyzed. Phase I reactions transform hydrophobic chemicals to more polar products via oxidation, hydrolysis, or similar reactions (Fig. 28.9). Phase II processes involve conjugation reactions that add polar functional groups, such as glucose or sulfate, to the Phase I products, to produce what are often even more polar metabolites. Thus these become even more water soluble and can be readily excreted. Many biotransformation enzymes exhibit broad substrate specificity, providing a mechanism for enhancing the excretion of a wide range of hazardous compounds (Watts, 1998).

At physiologic pH, a toxicant or its metabolites that are water soluble will undergo dissociation into ions or become ionized. Ionized molecules are the molecules that *react* in living systems. These ionized molecules (e.g., toxic metabolites), with their positively or negatively charged regions, are the molecules that are more readily transported across cell membranes. On occasion, biotransformation produces intermediate or final metabolites possessing toxic properties not found in the original parent chemical. The liver is the most important organ of bioactivation because of the high concentration of enzymes that catalyze biotransformation reactions.

28.5.6 Phase I Transformations

During Phase I biotransformation reactions, a small polar group is either exposed ("unmasked") in the toxicant or added to the toxicant (Table 28.6). The polar group enhances the solubility of the toxicant in water, which favors elimination. The reactions are catalyzed by nonspecific enzyme systems.

Toxicants undergoing Phase I biotransformation will result in metabolites that are sufficiently ionized, or hydrophilic, to be either readily eliminated from the body without further biotransformation reactions or rendered as an intermediate metabolite ready for Phase II biotransformation. Some intermediate or final metabolites may be more toxic than the parent chemical.

28.5.7 Phase II Transformations

On completion of a Phase I reaction, the new intermediate metabolite produced contains a reactive chemical group (e.g., hydroxyl, —OH; amino, —NH$_2$; or carboxyl, —COOH). For many intermediate metabolites, the reactive sites, which were either exposed or added during Phase I biotransformation, do not confer sufficient hydrophilic properties to permit elimination from the body. These metabolites must undergo additional biotransformation, called a Phase II reaction (Hughes, 1996).

During Phase II reactions, a molecule provided by the body must be added to the reactive site produced during

FIG. 28.9 The various steps of biotransformation. This figure gives a general overview of biotransformation, showing Phase I and Phase II reactions. The final block represents the various excretory processes. Some lipophilic substances cannot be processed by the biotransformation system, for example, because they are not suitable substrates for the enzymes of the system. Examples of such substances are polychlorinated biphenyls (PCBs) and DDT. Such substances will therefore accumulate in the body, in particular in body fat.

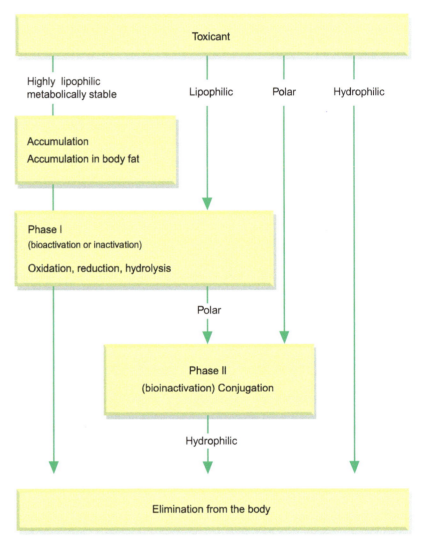

Phase I. Phase II reactions are referred to as conjugation reactions. These reactions produce a conjugate metabolite that is more water soluble than the original toxicant or Phase I metabolite. In most instances, the hydrophilic Phase II metabolite can be readily eliminated from the body (Hughes, 1996).

One of the most common molecules added directly to the toxicant or its Phase I metabolite is glucuronic acid, a molecule derived from glucose, a common carbohydrate that is the primary source of energy for cells.

28.6   CARCINOGENS

Carcinogens are agents that cause cancer, which is the uncontrolled growth of cells. Every human being is made up of approximately 100 trillion cells, and any one of these cells can be transformed to a malignant or cancerous cell by a variety of agents, which may be chemical (e.g., disinfection by-products), biological (e.g., cancer-causing viruses), or physical (e.g., ultraviolet light,

gamma irradiation) in origin. Approximately 100 different types of cancer, which can be found in every organ and system of the body, have been identified.

As of the mid-1990s in the United States, cancer is second only to heart disease as a cause of death. Figs. 28.10 and 28.11 show an estimate of yearly cancer deaths in the United States, broken down by site and sex. More than 500,000 people die from cancer each year, with lung cancer by far the leading killer. In fact, lung cancer has increased more than 200% during the last 35 years. Cancer ultimately kills one out of about every four Americans. However, aside from the increased incidence of lung cancer, the incidence of all other forms of cancer has collectively declined by about 13% over the past 30 years.

Carcinogens trigger uncontrolled cell growth in many different ways. In general, we can think of carcinogens as initiators or promoters, depending upon the stage of carcinogenesis in which they are active.

Initiation—the first stage of carcinogenesis, or conversion of a normal cell to a cancer cell—is a rapid,

TABLE 28.6 Representative Phase I Biotransformation Reactions

Reaction	Example
Nitrogen oxidation	$RNH_2 \rightarrow RNHOH$
Sulfur oxidation	$\begin{array}{ccc} R_1 & & R_1 \\ \backslash & & \backslash \\ S & \rightarrow & S{=}O \\ / & & / \\ R_2 & & R_2 \end{array}$
Carbonyl reduction	$\begin{array}{ccc} RCR' & \rightarrow & RCHR' \\ \parallel & & \mid \\ O & & OH \end{array}$
Hydrolysis (Esters)	$R_1COOR_2 \rightarrow R_1COOH + R_2OH$
Desulfuration	$\begin{array}{ccc} R_1 & & R_1 \\ \backslash & & \backslash \\ C{=}S & \rightarrow & C{=}O \\ / & & / \\ R_2 & & R_2 \end{array}$
Dehydrogenation	$RCH_2OH \rightarrow RCHO$

From Hughes, W.W., 1996. Essentials of Environmental Toxicology. Taylor and Francis, Washington, DC. Reproduced by permission of Taylor and Francis Group, LLC, a division of Informa plc.

essentially irreversible change caused by the interaction of a carcinogen with cell DNA (Fig. 28.12). This step in the development of cancer involves an *initiator*, a type of carcinogen that structurally modifies a gene that normally controls cell growth. A gene is a specific segment of a DNA molecule, so these altered growth-regulating genes are called *oncogenes* (from *onco*, meaning tumor, and gene). For a cell to begin to grow uncontrollably, at least two different growth-regulating genes must be altered. Such changes prime the cell for subsequent neoplastic development (from neo, meaning new, and *plastic,* referring to something formed, literally, new growth).

Some chemicals are initiators in their own right. For instance, formaldehyde, a widely used chemical in industrial glues, is thought to be an initiator. We know that vapors of formaldehyde trigger the development of malignant tumors in the respiratory tracts of rats, even though this compound has not been shown to cause cancer in humans. In other cases, however, an initiator can be a metabolic by-product. Thus benzo(a)pyrene, a natural product of the incomplete combustion of organic materials (including tobacco), is not itself an initiator, but the body metabolizes it to a related chemical, benzopyrene-7,8-diol 9,10-epoxide, which is an initiator *bioactivation* (Fig. 28.13).

The *promotion* process triggers the progressive multiplication of abnormal cells known as neoplastic development. *Promoters*, then, are carcinogens that activate the oncogenes, which would otherwise remain dormant. These carcinogens may act in several ways. For example, normal cells appear to prevent the activation of an oncogene in an adjacent initiated cell. A substance may

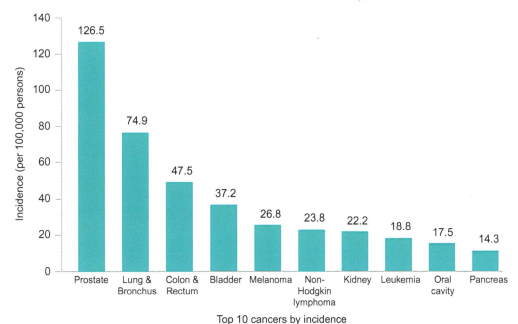

National incidence rates for men, 2009–13

FIG. 28.10 Cancer incidence rates among men for the period 2009–2013. *(Source: Annual Report to the Nation on the Status of Cancer 1975–2014. Available from https://seer.cancer.gov.)*

FIG. 28.11 Cancer incidence rates among women for the period 2009–2013. *(Source: Annual Report to the Nation on the Status of Cancer 1975–2014. Available from https://seer.cancer.gov.)*

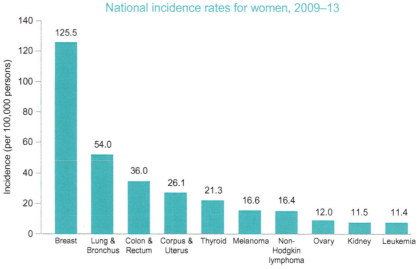

National incidence rates for women, 2009–13

Top 10 cancers by incidence

FIG. 28.12 The three steps in cancer development. *(From Sullivan, J.B., Krieger, G.R., 1992. Hazardous materials toxicology: clinical principles of environmental health. Williams and Wilkins, Baltimore, MD.)*

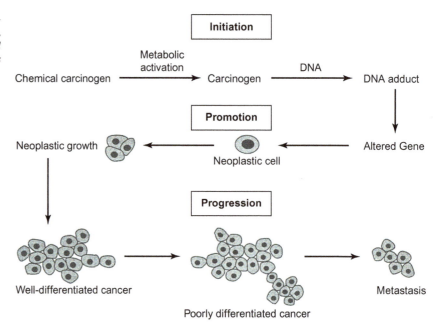

act as a promoter by killing the normal cells that surround an initiated cell. Alternatively, promoters may activate oncogenes by inhibiting the action of *suppressor genes*, which prevent oncogenes from initiating uncontrolled cell growth. If suppressor genes are inactivated, oncogenes can then spur tumor formation. The final stage in carcinogenicity is the growth of neoplastic cells followed by *progression,* or spreading throughout the body. This final step may be subdivided into two stages: invasion and metastasis. *Invasion* is the localized movement of neoplastic cells into adjoining tissues, and *metastasis* is the more distant movement throughout the body. Invasion and metastasis are consistent with a physiological system that is out of control.

Many compounds, some of them seemingly innocuous, can act as promoters. For instance, dietary factors such as salts and fats apparently act as promoters by killing normal cells. Ingestion of such promoters is not immediately harmful, but lifelong exposure significantly increases the risk of cancer. Thus people whose diets are high in salts or fats are more likely to develop stomach cancer and colon cancer, respectively. Fortunately, removing promoters from the area of an initiated cell that has not yet completed the promotion stage prevents the formation of a cancer cell.

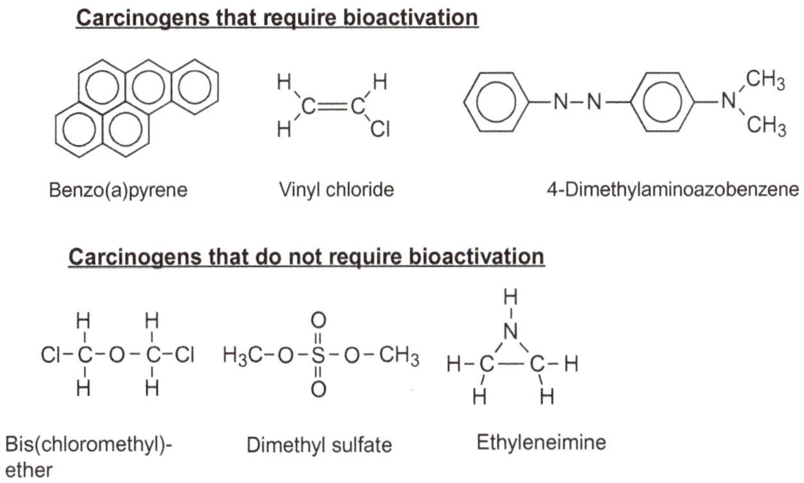

Carcinogens that require bioactivation

Benzo(a)pyrene   Vinyl chloride   4-Dimethylaminoazobenzene

Carcinogens that do not require bioactivation

Bis(chloromethyl)-   Dimethyl sulfate   Ethyleneimine
ether

FIG. 28.13 Examples of the major classes of naturally occurring and synthetic carcinogens, some of which require bioactivation and others of which act directly.

Therefore if we reduce the salt and fat we consume, we can greatly reduce the risk of developing cancer.

Carcinogens and toxins differ in one very important respect: the incidence of cancer (number of cases per million population) is dose dependent, but the severity of the response (cancer) is independent of dose. This means that we would expect more cases of cancer to develop as a population is exposed to higher levels of a carcinogen, just as we would expect more cases of poisoning in a population exposed to higher levels of a toxin. But while the severity of toxic response is also dose dependent, the dose of the carcinogenic agent has little to do with the severity of the disease once an individual has contracted cancer. This distinction between toxins and carcinogens explains why exposure regulations for substances classified as carcinogens are much more stringent than for toxins.

28.7 MUTAGENS

Like carcinogens, mutagens and teratogens affect DNA. *Mutagens* cause *mutations*, which are inheritable changes in the DNA, sequences of chromosomes. Mutations involve a random change in the natural functioning of chromosomes or their component genes; such changes rarely benefit the organism's offspring.

A mutation is a change in the genetic code that may or may not have an effect on the organism. Harmful effects from mutations depend on the type of cell that is affected and whether the mutation leads to metabolic malfunctions. Mutations that occur in somatic cells (nonreproductive cells of the body) may or may not prove to be a threat to the organism. Mutations occur naturally, most commonly from ionizing radiation. Humans and other organisms have enzymes that can repair damaged DNA. However, not all mutations are repaired, and some mutations will cause the cell's metabolism to become out of control, which may result in cancer.

Numerous types of mutations have been documented. The simplest type of genetic damage results when there is a mutation of the DNA sequence. Referred to as a *point mutation*, this mutation represents a change in the chromosome involving a single nucleotide (base) within the gene. These changes may result in the substitution, deletion, or insertion of a base (Fig. 28.14).

A *base substitution* occurs when a nucleotide is substituted for a normally occurring base. If the substituted base does not alter the amino acid coded for in that position, then it will have no effect on the protein being made from the DNA. This outcome is possible, since each amino acid is coded for by more than one codon (see Fig. 28.14). Two additional outcomes, *missense* and *nonsense*, result when the mutated triplet codon codes for a different amino acid or stop signal, respectively. Base substitutions usually do not result in significant mutagenic activity. Because of the redundancy of the genetic code, one base substitution will not likely result in a major change in the translation of the genetic information. Moreover, if a mutation results in an inappropriate amino acid in a protein that functions as an enzyme, it may not even change the enzymatic activity if it is not at or near the active site on the enzyme.

Frameshift mutations are the result of the addition between base pairs, which shifts the triplet code down the DNA strand. Such shifts essentially change the entire coding of proteins, with consequent high potential for malfunction of the protein. The effects of some physical and chemical mutagens are shown in Table 28.7.

28.8 TERATOGENS

Teratogens affect the DNA in a developing fetus, often causing gross abnormalities or severe deformities such as the shortening or absence of arms or legs.

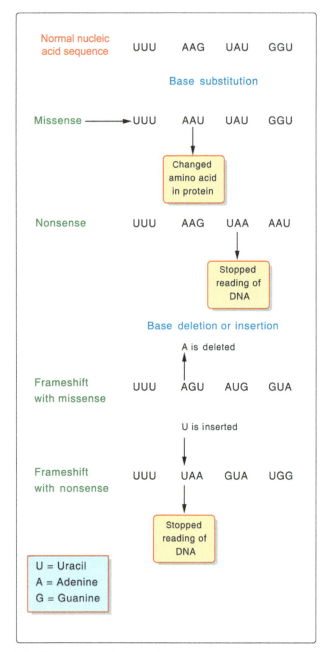

FIG. 28.14   Types of base mutations.

TABLE 28.7 The Effects of Selected Physical and Chemical Mutagens

Mutagen	Effect on DNA/RNA	Type of Mutation
Ultraviolet radiation	C, T, and U dimers that cause base substitutions, deletions, and insertions	No effect, missense, or nonsense
X rays	Breaks in DNA	Chromosomal rearrangements and deletions
Acridines (tricyclic ring present in dyes)	Adds or deletes a nucleotide	Missense or nonsense
Alkylating agents	Interferes with specificity of base pairing (e.g., C with T or A, instead of G)	No effect, missense, or nonsense
5-Bromouracil	Paris with A and G, replacing AT with GC, or GC with AT	No effect, missense, or nonsense

C = cytosine.
T = thymine.
U = uracil.

For example, the rubella virus, which causes a mild viral infection (German measles), is a teratogen during the first trimester of pregnancy. This virus can cross the transplacental barrier to produce cardiac defects and deafness in the offspring.

28.9   CHEMICAL TOXICITY: GENERAL CONSIDERATIONS

The toxic effects of a substance on a particular individual depend on both the chemical and the individual. However, the variability in the toxic potential of different compounds greatly exceeds the variability in toxic response from individual to individual. That is, if we expose a particular individual to several chemicals, we would see that some substances cause toxic effects in minute amounts, whereas others must be present in huge quantities. The range for toxic effects is enormous: the toxicity of one chemical can be millions or billions of times greater than that of another chemical. Thus it can take millions or billions of times as much of one chemical to cause the same effect as another. The range of human variability is not nearly so great. If a particular chemical causes an effect in one individual when a particular amount is administered, it is not likely that an amount a billion times less will cause a toxic effect in another individual. The

Perhaps the most famous (or infamous) teratogen is thalidomide, a sedative that was taken by thousands of pregnant women during the early 1960s. Sometimes, however, the deleterious effect of a teratogen does not appear until many years after the mother has been exposed. This was the case for diethylstilbestrol (DES), a drug that was prescribed for pregnant women in the United States for more than 30 years. Developed to prevent miscarriages, DES has been implicated in cervical and vaginal abnormalities in the daughters of women who had used DES during pregnancy. Nor are drugs the only teratogens.

exact range of human variability is not well established, but it is probable that it is closer to a tenfold than a billionfold.

In considering toxicity, we cannot make a distinction between human-made (synthetic) and naturally occurring chemicals, that is, everything is chemical in composition. It is not the source of the chemical that is important, but its characteristics. In Fig. 28.15, for example, we see the structures of some natural and synthetic compounds, each of which is considered a toxin at certain dosages. However, most chemicals, synthetic or natural, are not very toxic.

The molecular shape or structure of a chemical is one of the most important characteristics to consider in determining its toxicity. Current theory suggests that a living organism "recognizes," and hence reacts to, most chemicals that enter the body by their shape. These body-recognition responses can be very sensitive to subtle differences in shape or conformation. Two molecules, for example, might be very similar in structure, but exhibit slightly different configurations or three-dimensional isomers, one of which one may induce a toxic response in the body, while the other will not (Fig. 28.16). In fact, toxicity is frequently the result of recognition gone wrong, in which, at a cellular or chemical level, the body "prefers" the toxin over its structurally similar counterpart.

A second important characteristic in determining the degree of toxicity of a chemical is its solubility in different solvents. In particular, compounds are divided into those that are polar and hence soluble in water or water-like (polar) solvents, and those that are nonpolar and are soluble in fat (oil) or fat-like

High toxicity

Low toxicity

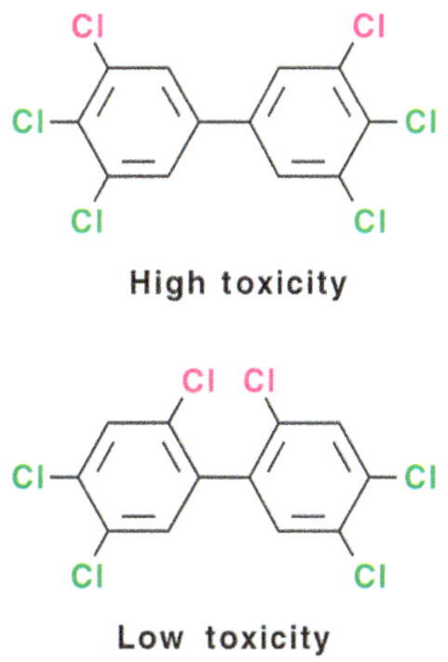

FIG. 28.16 Example of structurally similar chemicals with different toxic potency. *(From Pollution Science, Academic Press, San Diego, CA, 1996.)*

(nonpolar) solvents. This difference is very important in determining how easily a chemical can enter the body, how it is distributed inside the body, and how easily it can be excreted. Animals are most efficient at excreting polar compounds, so an ionic compound like sodium chloride, which is very soluble in water, is easily excreted. But a nonpolar compound like the pesticide DDT, which dissolves in fat, is not so easily eliminated. In some cases, the body is capable of converting nonpolar compounds into polar variants, thus facilitating their removal from the body. Unfortunately, this is not true of DDT, which can remain in the body for longer periods of time.

Target organ toxicity (Fig. 28.17) is defined as adverse effects manifested in specific organs in the body. Toxicity is unique for each organ, since each organ is a unique assemblage of tissues, and each tissue is a unique assemblage of cells. Toxicity may be enhanced by distribution features that deliver a high concentration of the toxicant to a specific organ or by inherent features of the cells and tissues of the organ that render it highly susceptible to the toxicant (Table 28.8).

28.10 CHEMICAL TOXICITY: SELECTED SUBSTANCES

28.10.1 Heavy Metals

Heavy metals, including lead, mercury, cadmium, chromium, arsenic, and selenium, constitute a major category of inorganic pollutants. Heavy-metal contamination often originates from manufacturing and the use of various synthetic products (e.g., pesticides, paints, and batteries), but these chemicals may also occur naturally (see Chapter 12). Heavy metals, many of which are toxic to plants as well as animals, tend to be mobile in the food chain, which means they can be bio-concentrated in animals, including humans, who are at the top of the food chain. The toxicity of some heavy metals such as lead and mercury has been known for centuries, but was not fully

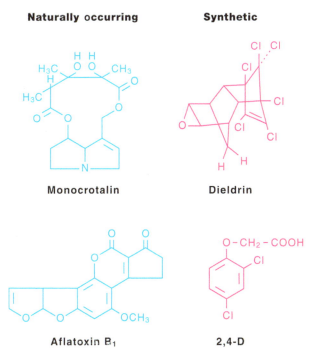

Naturally occurring **Synthetic**

Monocrotalin **Dieldrin**

Aflatoxin B₁ **2,4-D**

FIG. 28.15 Structures of natural and synthetic toxic chemicals found in the environment. *(From Pollution Science, Academic Press, San Diego, CA, 1996.)*

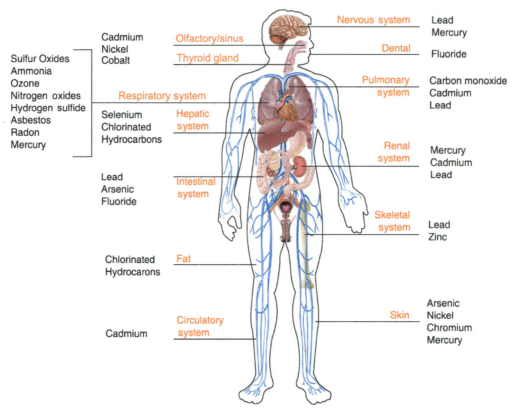

FIG. 28.17 Target organ toxicity.

TABLE 28.8 Additional Examples of Target Organ Toxicity

Organ	Toxicant	Mechanism	Toxicity
Heart	Fluorocarbons (Freon)	Sensitizes heart to epinephrine	Decreased heart rate, contractility, and conduction
	Carbon monoxide (CO)	Interferes with energy metabolism	Myocardial infarction, increase or decrease in heart rate
	Cobalt (Co)	Competes with Ca^{2+}	Heart failure
Testis	Lead (Pb)	Mutations in sperm	Decreased male fertility, increased spontaneous abortions in females
	Carbon disulfide (CS_2)	CNS effect on ejaculation	Reduced sperm counts
Ovary/uterus	Solder fumes	Unknown	Increased spontaneous abortions
	Polycyclic aromatic hydrocarbons (PAH)	Unknown	Damaged oocytes
Eye	Busulfan (chemotherapeutic agent)	Alters mitosis in lens cells	Formation of cataracts
	Methanol (CH_3OH)	Produces optic atrophy	Permanent visual impairment, blindness

From Hughes, W.W., 1996. Essentials of Environmental Toxicology. Taylor and Francis, Washington, DC. Reproduced by permission of Taylor and Francis Group, LLC, a division of Informa plc.

understood until recent times. Because of their widespread use and/or occurrence in nature, some heavy metals are of particular concern.

Lead borders on ubiquitous (Fig. 28.18). It can be found in drinking water, where it comes from several sources. The most significant of these sources is lead solder and piping in water distribution systems, particularly when in contact with corrosive water. While the use of lead solder and pipes in repairs and construction of water systems has been banned, many existing water distribution systems still contain lead materials. Lead is also found in food, tetraethyl lead in gasoline (which ends up in the air, soil, and water), lead-based paint, and improperly glazed earthenware. In addition, it can come from industrial sources, such as smelters and lead-acid battery manufacturing. Since the toxic effects of lead depend on total exposure, environmental assessment must take into account all of these sources.

Compared with fat-soluble substances, lead is relatively poorly absorbed, as lead compounds are water soluble. In adults, only about 10% of the lead ingested through the GI tract is absorbed into the bloodstream. Lead is initially distributed to the kidney and liver and then redistributed, mostly to bone (about 95%). Moreover, lead does not readily enter the central nervous system in adults because the blood–brain barrier can keep it out. However, in children, the blood–brain barrier is not yet fully developed, so lead exposure in children can affect their mental development.

Acute lead poisoning is rare; however, chronic lead poisoning is not uncommon. In adults, chronic lead poisoning sometimes results in the painful gastrointestinal symptoms known as lead colic. Because of the pain, lead colic often compels exposed persons to seek medical help, whereupon an accurate diagnosis of lead exposure can prevent the development of more serious problems. Lead can also affect the neuromuscular system, decreasing muscle tone in the wrists and feet. Exposure to lead also affects the body's blood-forming system. Lead can interfere with the synthesis of heme (part of the oxygen-carrying compound hemoglobin), thereby causing anemia; it can also damage red blood cells in a condition known as basophilic stippling. The most serious effect of lead is the brain-degenerative condition called lead encephalopathy. This condition can occur in adults (Table 28.9). For example, historians now speculate that Caligula, the most insane of the Roman emperors, suffered from lead encephalopathy caused by eating food from lead-containing pewter dishes. Today, however, this condition is more common in children and can be quite serious; approximately 25% of children with lead encephalopathies die, and about 40% of the survivors experience neurologic after effects. Finally, lead has been shown to affect the kidneys of laboratory animals, causing impairment of function and cancer.

Cadmium, like lead, has many sources. It is a by-product of lead and zinc mining, it is used as a pigment, and it is found in corrosion-resistant coatings and nickel–cadmium batteries. It is also released when fossil fuels are burned. Cadmium can enter drinking water when corrosive water contacts certain types of water piping. Historically it

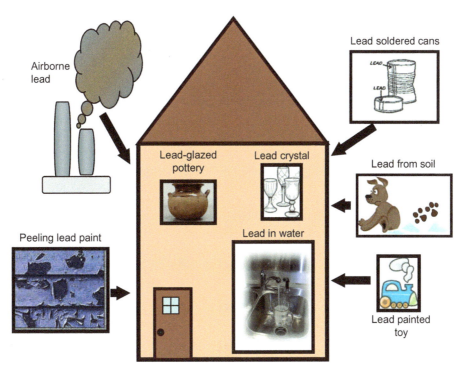

FIG. 28.18 Sources of lead in the environment.

TABLE 28.9 Toxic Responses to Increased Lead Concentrations in the Blood

Blood Lead Concentration (µg/L)		Health Effects	
		Adults	Children
100	↓	Hypertension may begin to occur	Crosses placenta, developmental toxicity
			Impairment of IQ
			Increased erythrocyte protoporphyrin
200	↓		Beginning impairment of nerve conduction velocity
300	↓	Systolic hypertension, decreased hearing	Impaired vitamin D metabolism
400	↓	Infertility in males, renal effects, neuropathy	Hemoglobin synthesis impaired
		Fatigue, headache, abdominal pain	
500	↓	Decreased heme synthesis, decreased hemoglobin, anemia, intestinal symptoms, headache, tremor	Abdominal pain, neuropathy Encephalopathy, anemia, nephropathy, seizures
		Lethargy, seizures	
1000		Encephalopathy	

Modified from Sullivan, J.B., Krieger, G.R., 1992. Hazardous materials toxicology: clinical principles of environmental health. Williams and Wilkins, Baltimore, MD.

entered the food chain through application of sewage sludge to the land, where it was taken up by plants and stored in leaves and seeds. However, point source controls have reduced this problem (see also Chapter 23). Cadmium's adverse effect on health was originally made public by an incident in Japan, where rice paddies were contaminated with cadmium-rich drainage from zinc mines. Rice grown on the paddies concentrated the cadmium, and those eating it suffered such characteristic symptoms as easily broken bones and extreme joint pain; thus cadmium poisoning is known as *itai-itai* ("ouch-ouch") disease.

Since cadmium is water soluble, only 1%–5% of a given dose is absorbed in the GI tract (although 10%–40% can be absorbed through the lungs), and cadmium distributes to the kidney and liver (Fig. 28.19). Acute cadmium poisoning causes GI disturbances. Chronic cadmium poisoning most severely affects the kidney. Animal studies have shown cadmium to be carcinogenic, and some researchers have suggested that it may increase the incidence of prostate cancer in elderly men.

Arsenic occurs naturally in bedrock and soil (Fig. 28.20) and is a waste product from smelting operations as well as the manufacture of products such as pesticides and herbicides. Airborne particles of arsenic may travel considerable distances from the point of origin. Arsenic is also readily taken up by food plants, with the degree of uptake being dependent on soil pH. Arsenic is considered an essential dietary element, although in very small amounts. There are four main types of arsenic: organoarsenics, arsenate, arsenite, and arsine gas.

Arsenic can be excreted relatively quickly, and it has a half-life of 2 days. Although the effects of acute arsenic poisoning are seen primarily in mystery fiction, chronic arsenic poisoning manifests in a wide variety of chronic toxic effects. Many of these effects stem from the capacity of arsenic to increase the permeability of capillaries in various locations in the body, and the structural similarity of arsenate to phosphate, for which it can substitute. Increased permeability is, however, harmful; it allows plasma to leak into the tissues, sometimes leading to severe diarrhea and kidney injury. Arsenic also has the potential to damage the central nervous system, inflaming peripheral nerves and causing brain injuries, and damage the liver, infiltrating fatty deposits and causing tissue necrosis. Moreover, the EPA has classified arsenic as a Type A human carcinogen (i.e., a human carcinogen based on epidemiological studies), with skin and lung cancer as the two principal types of cancer arising from arsenic exposure.

Mercury is a naturally occurring metal dispersed throughout the ecosystem. Mercury contamination of the environment is caused by both natural and anthropogenic sources. Natural sources include volcanic action, erosion of Hg-containing sediments, and gaseous emissions from the earth's crust. The majority of Hg comes from anthropogenic sources or activities including mining, combustion of fossil fuels (Hg content of coal is about 1 ppm), transporting Hg ores, processing pulp and paper, incineration, use of Hg compounds as seed dressings in agriculture, and exhaust from metal smelters (Yu, 2001). In addition, Hg waste is found as a by-product of chlorine manufacturing plants, used batteries, light bulbs, and gold recovery processes.

Mercury compounds are added to paints as preservatives. In addition, Hg is used in jewelry making, pesticides, and other manufacturing processes. The light emitted by electrical discharge through Hg vapor is rich in ultraviolet rays, and lamps of this kind in fused quartz envelopes are widely used as sources of UV light. High-pressure Hg-vapor lamps are now widely used for lighting streets and highways.

In the United States, the largest user of Hg is the chlor-alkali industry, in which chlorine and caustic soda are produced by

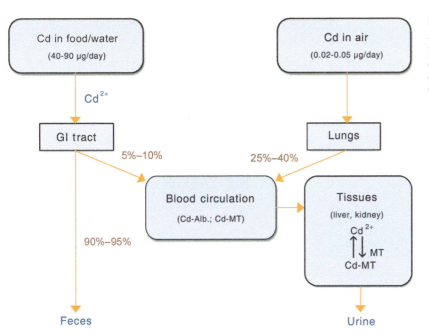

FIG. 28.19 Metabolism of Cd in humans. Cd-Alb: Cd attached to albumin; Cd-MT: Cd attached to metallothionein. *(From Yu, M.H., 2001. Environmental Toxicology. CRC Press, Boca Raton, FL. Reproduced by permission of Taylor and Francis Group, LLC, a division of Informa plc.)*

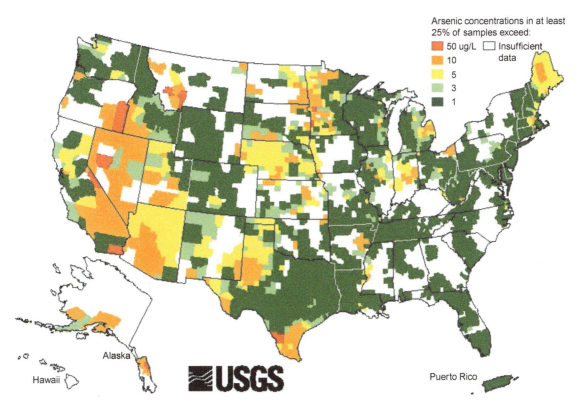

FIG. 28.20 Arsenic concentration in wells and springs (1973–2001). *(Available from: http://co.water.usgs.gov/trace/pubs/geo_v46n11/fig2.jpeg.)*

electrolysis of salt (NaCl) solution (Yu, 2001). In some methods of producing chlorine, an Hg cathode is used. The Na^+ ions discharge at the Hg surface, forming sodium amalgam. The resultant amalgam is continuously drained away and treated with water to produce NaOH solution and Hg:

$$Hg - Na \xrightarrow{H_2O} NaOH(solution) + Hg \qquad (28.1)$$

Various forms of Hg are present in the environment. Conversion of one form to another occurs in sediment, water, and air and is catalyzed by various biological

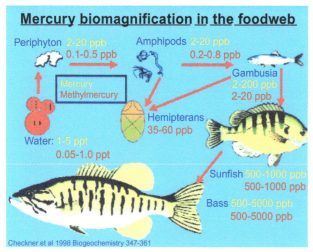

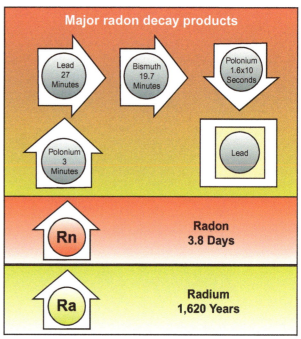

FIG. 28.21 Mercury magnification in the food web. *(From Cleckner, L. B., et al., Trophic transfer of methylmercury in the northern Everglades. Biogeochemistry, vol. 40, pp. 347–361. Reprinted with kind permission of Springer Science and Business Media.) (Available from: http:// energy.cr.usgs.gov/radon/georadon/page3.gif.)*

systems. In addition, Hg frequently finds it way to lakes and seas. Microorganisms then convert the elemental Hg into methylmercury (MeHg) through a process called methylation, allowing it to enter the food chain and be biomagnified (Fig. 28.21).

Mercury is a nerve toxin, and the main health concern is its effect on the brain, particularly in the growing fetus and the young. The phrase "mad as a hatter" stems from the mercury poisoning of hat makers, who used the metal for curing felt. Mercury can damage reproduction in mammals by interfering with the formation of sperm. Neurological and reproductive effects have also been seen in birds. In fish, its effects include a decreased sense of smell, damage to the gills, blindness, and changes in the ability to absorb nutrients in the intestines. Plants can also be sensitive to mercury, and high concentrations can lead to reduced growth.

28.10.2 Inorganic Radionuclides

Certain unstable elements spontaneously decay into different atomic configurations, in the process releasing radiation consisting of alpha particles, beta particles, or gamma rays. These particles and rays can damage living tissue and/ or cause cancer to develop, with the degree of damage depending on the type of radiation and means of exposure (i.e., inhalation, ingestion, or external radiation). As an element undergoes radioactive decay, it progresses through a series of atomic configurations, known as isotopes. Isotopes are simply atoms identified by atomic weight, a number indicating the number of neutral and charged particles in the nucleus (Fig. 28.22). The uranium in pitchblende, for example, is a mixture of three isotopes that have atomic weights of 234, 235, and 238.

FIG. 28.22 Natural decay of uranium. *(Available from: http://energy.cr. usgs.gov/radon/georadon/page3.gif.)*

Radionuclides are atoms (nuclei) that are undergoing spontaneous decay, emitting radiation as they disintegrate to form isotopes of lower atomic weight. The radioactivity of an isotope is expressed in units called picocuries (pCi),[2] which represents the isotope's number of disintegrations per second; 1 pCi is equal to 3.7×10^{-2} disintegrations per second. (*Note*: Radionuclides continue to decay until they achieve a stable configuration, which is often the configuration of another element altogether.)

Radioactive decay is a natural process. In fact, everyone is exposed to some background radiation both from cosmic rays and from radioactive soil and rock (Fig. 28.23). Three naturally occurring series of isotopes arise from the decay of the isotopes uranium-238, uranium-235, and thorium-232. These naturally occurring radionuclides can be found in drinking water, with the uranium-238 series (whose decay isotopes include uranium-234, radium-226, and radon-222) and the thorium series (whose decay products include radium-228) being of greatest concern. Synthetic radioactive isotopes, such as strontium-90, also pose health risks, but such isotopes generally occur in lower concentrations in the environment than

---

2. Another unit commonly used to express radioactivity is the becquerel (Bq), which is defined as disintegrations per second (s^{-1}). 1 Bq = 3.7×10^2 pCi.

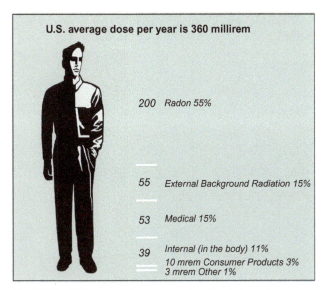

FIG. 28.23   U.S. average dose per year is 360 millirem. *(From www. ocrwm.doe.gov.)*

the naturally occurring radionuclides. However, site-specific contamination, such as nuclear waste disposal sites and nuclear power plant accidents (e.g., Chernobyl), may release concentrations of anthropogenic radionuclides that threaten human and environmental health.

The concern over radionuclides focuses largely on their potential to cause cancer. Radium-226, which has a half-life of 1622 years, is perhaps the single most important radioactive isotope found in drinking water. It is deposited in bone and can cause bone cancer. One of the decay products of radium-226 is radon gas, that is, the isotope radon-222, which has a half-life of 3.85 days. Inhalation of the short-lived decay products of radon-222 can cause lung cancer; however, less is known about the risks of ingested radon. This isotope has recently become the subject of great public concern because it has been discovered in homes and other buildings. Most of the total amount of radon-222 that enters homes comes through the soil, but it can also enter by degassing from a dissolved state, in drinking and washing water. This degassing occurs when water is heated and/or aerated, as for example, during showering, bathing, and clothes and dish washing.

28.10.3   Insecticides

Insecticides can be divided into organochlorine, organophosphorus, and carbamate compounds, as well as botanical insecticides (see also Chapters 12 and 14). Within each group, the pesticides have similar characteristics.

Organochlorine Insecticides: This category, which was developed in the 1930s and 1940s, includes the chlorinated ethanes, chlorinated cyclodienes, and other chlorinated compounds. Dichlorodiphenyltrichlorethane (DDT) is the most famous of the chlorinated insecticides. First synthesized in 1943, it was used extensively (worldwide) in agriculture from the end of World War II until 1972, when it was banned in the United States. It was first used to control disease-carrying insects such as body lice and mosquitoes that spread malaria. DDT also provided effective against a variety of agricultural pests and was extensively used on crops. This highly lipid-soluble compound is stored in fat—in fact, the fat of most U.S. residents contains DDT concentrations of $5–7 \, mg \, kg^{-1}$. DDT is very persistent in the environment and is *biomagnified* (Fig. 28.24) in the food chain. That is, smaller organisms absorb the compound, then they are eaten by larger organisms, and the progression continues until DDT attains a relatively high concentration in macrovertebrates such as fish, which are then eaten by humans and other large animals.

In general, DDT is toxic to humans and most other higher animal life only in extremely high doses. However, because of its low toxicity, it was applied in much greater quantities than were necessary. Then, in the 1960s, the effects of these massive applications became noticeable. For example, certain birds, such as the peregrine falcon, began to produce overly fragile egg shells that broke before hatching, thereby threatening their survival as a species. Fish, too, are extremely vulnerable to DDT, and die-offs occurred following heavy rains, when the pesticide was washed into streams and rivers.

DDT and other chlorinated hydrocarbons are very resistant to metabolic breakdown. In animals and humans, DDT is degraded to DDE (1,1-dichloro-2,2-bis (*p*-chlorophenyl ethylene)) or dichlorodiphenyl dichloroethylene) or DDD (1,1-dichloro-2, 2-bis(*p*-chlorophenyl ethane)) (Fig. 28.25).

The other chlorinated hydrocarbons, such as lindane, toxaphene, mirex, and kepone, are similar to DDT. In general, then, we can say that all organochlorine insecticides cause some central nervous system (CNS) stimulation, increase cancer incidence in laboratory animals, and persist in the environment to some degree.

Organophosphorus Insecticides: Organophosphorus insecticides are the most toxic of the insecticides; they are dangerous not only to insects but also to mammals. Many of the compounds, such as parathion, paraoxon and tetram are in the "supertoxic" category for humans. Human fatal doses for those toxicants are $<5 \, mg/kg$. As little as 2 mg of parathion has been known to kill children (Yu, 2001).

Organophosphorus compounds do not persist in the environment and have an extremely low potential to produce cancer; thus insecticides based on these compounds have largely replaced the chlorinated hydrocarbon insecticides. However, these phosphorous-containing compounds have a much higher acute toxicity in humans than

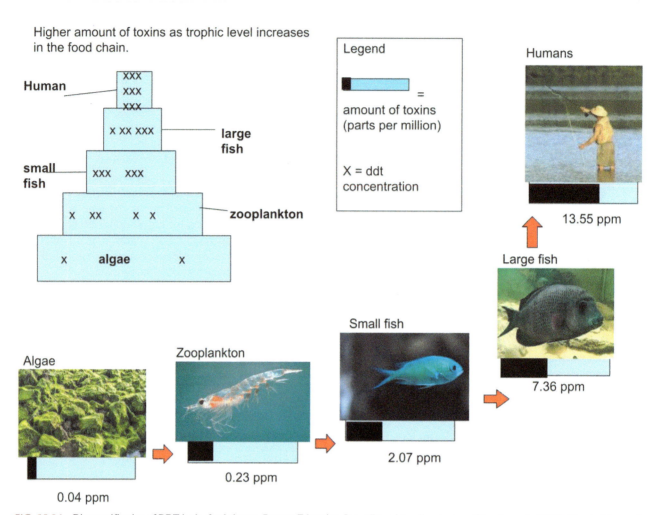

FIG. 28.24   Biomagnification of DDT in the food change. Pearson Education, Inc., *(From https://commons.wikimedia.org/wiki/File:The_build_up_of_ toxins_in_a_food_chain.svg.)*

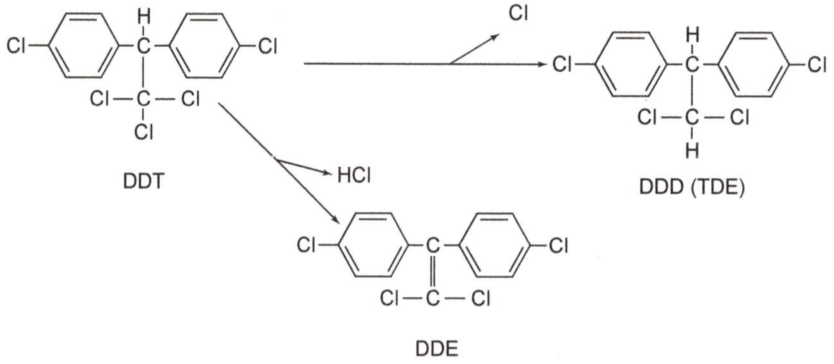

FIG. 28.25   Metabolism of DDT.

the organochlorines, and are the most frequent cause of human insecticide poisoning. Fortunately, both laboratory tests and antidotes are available for acute organophosphorus poisoning.

A typical organophosphorus insecticide is parathion, which must be metabolized to the compound paraoxon to exert its toxic effect. This toxic effect stems from the compound's ability to inhibit the enzyme cholinesterase, a crucial chemical for the regulation of the nerve transmitter acetylcholine. Thus acute effects of poisoning with organophosphorus insecticides include fibrillation of muscles, low heart rate, paralysis of respiratory muscles, confusion,

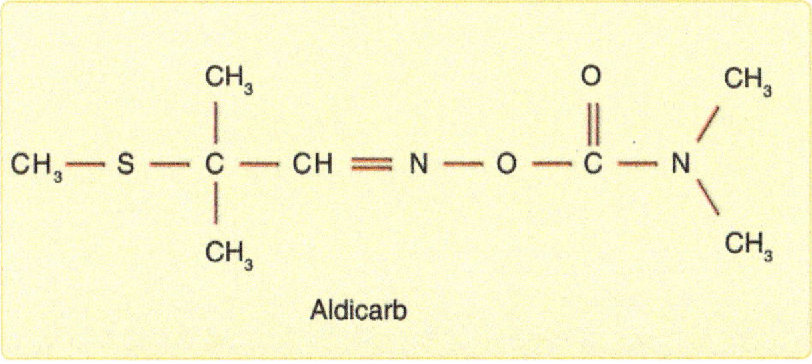

FIG. 28.26 Chemical structure of aldicarb.

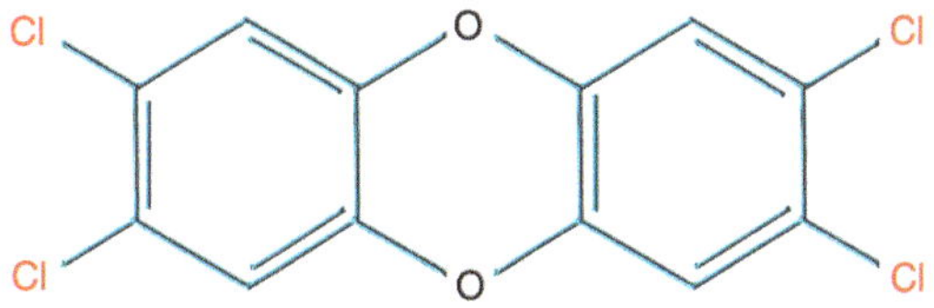

FIG. 28.27 Chemical structure of dioxin (TCDD).

convulsions, and eventually death. Other organophosphorus pesticides, such as malathion, are less toxic in acute doses than parathion.

Carbamate Insecticides: The carbamate insecticides, which include carbaryl and aldicarb (Fig. 28.26), have toxicities very similar to those of the organophosphorus insecticides. Like the organophosphorus pesticides, these widely used chemicals also act by inhibiting cholinesterase, but the toxic effects of carbamates may be more easily reversed than those of the organophosphorus compounds. In addition, current evidence does not seem to suggest carcinogenicity as a toxic effect of the carbamates. Although most of these chemicals are not persistent in the environment, aldicarb may be the exception. Used on potato crops in Long Island, New York, aldicarb has contaminated groundwater there. It has been estimated that levels of $6\,\mu g\,L^{-1}$ may persist up to 20 years.

28.10.4 Herbicides

In the United States, herbicides—chemicals that kill plants—are used in greater quantities than insecticides. Herbicides are added directly to the soil; thus they can, in some cases, readily enter the groundwater. Many herbicides originally on the U.S. market have subsequently been banned. Chlorophenoxy compounds include 2,4-dichlorophenoxy-acetic acid (2,4-D); 2,4,5-trichlorophenoxyacetic acid (2,4,5-T); and 2,4,5-trichlorophenoxypropionic acid (2,4,5-TP or silvex). These herbicides act as growth hormones, forcing plant growth to outstrip the ability to provide nutrients. Although these compounds can have toxic effects on the liver, kidney, and central nervous system, clinical reports of poisoning are rare. The compounds, which have a half-life of about 24h, are rapidly excreted in humans via urine.

One of the best known, and most controversial, of the chlorophenoxy compounds is 2,4,5-T, which was combined with 2,4-D to create the defoliant Agent Orange, used in the Vietnam War. However, during the industrial synthesis of 2,4,5-T and 2,4-D, a hazardous by-product can also be inadvertently formed: tetrachlorodioxin (TCDD) (Fig. 28.27) or *dioxin* for short. Dioxin, which can also be produced during the combustion of certain substances, is the most toxic manufactured chemical known. In sufficient doses, dioxin is a potent teratogen and carcinogen in laboratory animals, and causes liver injury and general tissue wasting. At lower doses, it causes a form of acne called chloracne, which concentrates between the eyes and hairline.

Clinical reports of acute dioxin poisoning are rare. Dramatic interspecies differences exist for the effects of dioxin; the LD_{50} for guinea pigs is about 1/10,000 of the LD_{50} for hamsters. Fortunately, current evidence indicates that the human reaction to dioxin tends to resemble that of the hamster rather than that of the guinea pig. However, dioxin contamination must be considered a serious environmental problem. In Times Beach, Missouri, for example, dioxin contamination has forced the abandonment of 800 homes since 1984.

28.10.5 Halogenated Hydrocarbons

Halogenated hydrocarbons are common because they are widely used as effective, yet relatively nonflammable solvents, unlike kerosene or gasoline. Halogenated hydrocarbons are also formed during the chlorination of drinking water when chlorine combines with organic material in the water. One of the oldest and simplest of these compounds is carbon tetrachloride (CCl_4), which was extensively used as a solvent and dry cleaning agent; it is also so nonflammable that it was used in fire extinguishers. (The use of carbon tetrachloride as a dry cleaning agent was banned after it was shown to cause liver damage.)

The halogenated hydrocarbons (including trichloroethene, halothane, and chloroform) tend to be similar to carbon tetrachloride in their health effects. In high doses, carbon tetrachloride causes CNS depression—so much so that it was once used as an anesthetic. It can also sensitize the heart muscle to catecholamines (hormones such as epinephrine) and thus can cause heart attacks. It can also cause kidney injury, liver injury, and cancer in laboratory animals. High blood alcohol levels can act as a potentiator for carbon tetrachloride's damaging effects on internal organics. It owes its toxicity to the fact that it is converted in the liver to carbon trichloride ($\cdot CCl_3$), which is a free radical capable of inducing peroxidation of lipid double bonds and poisoning protein-synthesizing enzymes.

Table 28.10 summarizes the health effects of various halogenated hydrocarbons. Plus signs (+) indicate a harmful effect, minus signs (–) indicate a lack of effect, and both a plus and a minus indicates a less significant effect. Among the methanes is chloroform, a trihalomethane formed during the chlorination of drinking water. Chloroethene (vinyl chloride) receives three plus signs under cancer because it has been established as a probable human carcinogen. Also noteworthy are trichloroethene (TCE) and tetrachloroethene (PCE, perchlorethylene). These chemicals are very common contaminants of groundwater (see Chapter 15). TCE is listed as a known human carcinogen whereas PCE is listed as likely to be carcinogenic to humans. Also, it has been observed that when PCE and TCE degrade naturally in groundwater under anaerobic conditions, vinyl chloride is formed as a degradation product. The toxic effects of other common hazardous compounds are shown in Table 28.11.

QUESTIONS AND PROBLEMS

1. What event in the year 1962 had an impact in creating an interest in the environment? Why?
2. Why are small animals used in laboratory tests involving toxic materials?
3. How is safety of chemical substances defined with regard to exposure?
4. What is exposure concentration? Exposure dose? How are the two related?
5. Adults are assumed to consume 2 L of water daily. If a substance is present at $10 \, mg \, L^{-1}$, give the average individual intake of the substance.
6. Explain the similarities and differences between carcinogens, mutagens, and teratogens.
7. What is LD_{50}? How is it used?
8. What are the advantages of short-term toxicity testing?
9. What role do initiators, promoters, and suppressor genes play in cancer formation?

TABLE 28.10 Health Effects of Some Chlorinated Halogenated Hydrocarbons, (+) = Harmful Effect, (–) = No Effect, (±) = A Less Significant Effect

	CNS Depression	Sensitization of Heart	Liver Injury	Kidney Injury	Cancer
Methanes					
Carbon tetrachloride	+	+	++++	++	+++
Chloroform	+	+	+++	+++	+++
Dichloromethane (methylene chloride)	+	–	±	–	+++
Ethanes					
1,1-Dichloroethane	+	+	+	+	++
1,2-Dichloroethane	+	NDA[a]	+	+	+++
1,1,1-Trichloroethane	+	+	±	–	–
1,1,2,2-Tetrachloroethane	+	NDA	++	++	NDA
Hexachloroethane	+	NDA	NDA	+	+
Ethylenes					
Chloroethene (vinyl chloride)	+	NDA	++	–	++++
1,1-Dichloroethene (vinylidine chloride)	+	NDA	+++	–	+
1,2-trans-Dichloroethene	+	NDA	++	NDA	NDA
Trichloroethene	+	+	+	±	++++
Tetrachloroethene (perchloroethylene)	0	–	±	±	+++

[a]NDA = No data available.

Adapted from U.S. EPA, 1989. Risk assessment, management and communication of drinking water contamination. EPA/625/4-89/024, Washington, DC, U.S. Environmental Protection Agency.

TABLE 28.11 Toxic Effects of Common Hazardous Compounds

Chemical	Acute Effects	Chronic Effects
Aliphatic hydrocarbons		
Alkanes	Central nervous system impairment	No known carcinogenicity or other chronic effects
Monocyclic aromatic hydrocarbons		
Toluene	Central nervous system depression, including agitation, delirium, coma	Central nervous system impairment; no proven carcinogenic activity
Xylenes	Liver toxicity, including steatosis, hepatic cell necrosis, and partial tract enlargement	
Nonhalogenated solvents		
Acetone	Low toxicity, some CNS effects	Minimal chronic effects
Pesticides		
Aldrin and Dieldrin	Tremors, seizures, and coma	Carcinogenicity
DDT	Affects sodium–potassium pump of neural membrane pesticides produce headaches, nausea, vomiting, and tremors	Minimal chronic effects in humans; no proven human carcinogenicity
Pentachlorophenol	Uncoupling of oxidative phosphorylation	Liver toxicity, including fatty tissue infiltration and elevated enzymes
Industrial intermediates		
Phenol	A range of health effects including cardiac dysrhythmia, dermal necrosis, and elevated liver enzymes	Not a human carcinogen; some evidence that phenol is a complete carcinogen in mice
Chlorinated benzenes	Dizziness, headaches	Porphyria cutanea, aplastic anemia, leukemia
Polychlorinated biphenyls		
PCBs	Minimal acute toxicity (0.5 g/kg to 11.3 g/kg)	Chloracne; increased liver enzymes; possible reproduction effects; act as cancer promoters
Dioxins and Furans		
PCDDs/PCDFs	Chloracne, headaches, peripheral neuropathy	Induction of microsomal enzymes; altered liver metabolism; altered T-cell subsets; immunotoxicity; strongly implicated in carcinogenicity (may be a promoter)
Inorganic compounds		
Arsenic	Loss of blood, intestinal injuries, acute respiratory failure	Myelogeneous leukemia, cancer of skin, lungs, lymph glands, bladder, kidney, prostate, and liver
Cadmium	Vomiting, cramping, weakness, and diarrhea	Oral ingestion results in renal necrosis and dysfunction; induces lung, prostate, kidney, and stomach cancer in animals; no documented human cancer
Hexavalent chromium	Readily absorbed by the skin, where it acts as an irritant and immune-system sensitizer; oral absorption results in acute renal failure	
Mercury	Central nervous system impairment including injury to motor neurons; renal dysfunction	Central nervous system dysfunction, memory deficits, decrease in psychomotor skill, tremors. Immune system effects resulting in allergic contact dermatitis
Nickel	Not highly toxic; headache, shortness of breath	

Modified from Watts, R.J., 1998. Hazardous Wastes: Sources, Pathways, Receptors. Wiley, New York. From U.S. EPA, 1989. Risk assessment, management and communication of drinking water contamination. EPA/625/4-89/024, Washington, DC, U.S. Environmental Protection Agency.

10. Can one predict if a new chemical is a carcinogen by its chemical structure? How would you do this?

11. Give an example of a chemical that is made more toxic through biotransformation in the body.

12. What are some of the eliminations in using epidemiological studies to assess the toxicity or carcinogenicity of a chemical? Do negative findings indicate that the substance is not a carcinogen? Why or why not?

13. What are some bioassays that can be used to assess the toxicity of effluents from wastewater treatment plants?

14. A section of DNA has the following sequence of nitrogenous bases.

    If a frameshift mutation occurs at $\otimes$, list the triplet codes before and after the mutation.

    A person is exposed to a highly water-soluble chemical with a constant ingestion rate of 0.12 mg/h. Its transformation in the liver is described by a first-order rate constant of $0.1\,h^{-1}$. The compound and its metabolites are cleared through the kidneys at a rate of 0.03 mg/h. If the volume of body fluids is 15 L, determine the steady-state concentration of the chemical in the person's vascular system.

15. List two lipophilic substances that cannot be biotransformed by Phase I and II reactions.

16. List the following contaminants in increasing order of storage in the body:

    Acetone
    DDT
    Phenol
    Lead

17. How does chemical structure affect the toxicity of a chemical?

18. How do halogenated compounds get into drinking water?

REFERENCES

Bitton, G., 1999. Wastewater Microbiology. Wiley-Liss, New York.

Hughes, W.W., 1996. Essentials of Environmental Toxicology. Taylor and Francis, Washington, DC..

Rodricks, J.V., 2007. Calculated Risks: The Toxicity and Human Health Risks of Chemicals in Our Environment. Cambridge University Press, Cambridge, England.

U.S. EPA. (1989) Risk Assessment, Management and Communication of Drinking Water Contamination. EPA/625/4–89/024. U.S. Environmental Protection Agency, Washington, DC.

Watts, R.J., 1998. Hazardous Wastes: Sources, Pathways, Receptors. Wiley, New York.

Yu, M.H., 2001. Environmental Toxicology. CRC Press, Boca Raton, FL.

FURTHER READING

Chiras, D.D., 2016. Environmental Science, 10th ed. Benjamin/Cummings, Menlo Park, CA.

Cleckner, L.B., Garrison, P.J., Hurley, J.P., Olson, M.L., Krabbenhoft, D.P., 1998. Trophic transfer of methylmercury in the northern Everglades. Biogeochemistry 40, 347–361.

Hodgson, E., Levi, P.E., 2010. A Textbook of Modern Toxicology, 4th ed. 2010.

Landis, W.G., Yu, M.H., 1995. Introduction to Environmental Toxicology. Lewis Publishers, Boca Raton, FL.

Maraham, S.E., 2003. Toxicological Chemistry and Biochemistry, 3rd ed. CRC Press, Boca Raton, FL.

Stine, K.E., Brown, T.M., 2015. Principles of Toxicology. John Wiley and Sons, New York.

Tortora, G.J., Funke, B.R., Case, C.L., 2018. Microbiology, An Introduction, 13th ed. Pearson Education Inc., Benjamin Cummings, San Francisco, CA.

Chapter 29

Risk Assessment

C.P. Gerba

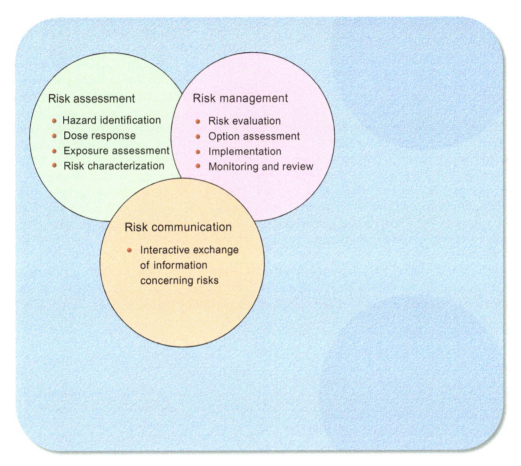

Structure of risk analysis.

29.1 THE CONCEPT OF RISK ASSESSMENT

Risk, which is common to all life, is an inherent property of everyday human existence. It is therefore a key factor in all decision making. Risk assessment or analysis, however, means different things to different people: Wall Street analysts assess financial risks and insurance companies calculate actuarial risks, while regulatory agencies estimate the risks of fatalities from nuclear plant accidents, the incidence of cancer from industrial emissions, and habitat loss associated with increases in human populations. What all

these seemingly disparate activities have in common is the concept of a measurable phenomenon called risk that can be expressed in terms of probability. Thus we can define *risk assessment* as the process of estimating both the probability that an event will occur, and the probable magnitude of its adverse effects—economic, health/safety-related, or ecological—over a specified time period. For example, we might determine the probability that a chemical reactor will fail and the probable effect of its sudden release of contents on the immediate area in terms of injuries and property loss over a period of days. In

Environmental and Pollution Science. https://doi.org/10.1016/B978-0-12-814719-1.00029-X

addition, we might estimate the probable incidence of cancer in the community where the chemical was spilled. Or, in yet another type of risk assessment, we might calculate the health risks associated with the presence of pathogens in drinking water or pesticide in food.

There are, of course, several varieties of risk assessment. Risk assessment as a formal discipline emerged in the 1940s and 1950s, paralleling the rise of the nuclear industry. Safety-hazard analyses have been used since at least the 1950s in the nuclear, petroleum-refining, and chemical-processing industries, as well as in aerospace. Health risk assessments, however, had their beginnings in 1976 with the EPA's publication of the *Carcinogenic Risk Assessment Guidelines*.

In this chapter, we are concerned with two types of risk assessment:

- *Health-based risks*: For these risks, the focus is on general human health, mainly outside the workplace. Health-based risks typically involve high-probability, low-consequence, chronic exposures whose long latency periods and delayed effects make cause-and-effect relationships difficult to establish. This category also includes microbial risks, which usually have acute short-term effects. However, the consequences of microbial infection can persist throughout an individual's lifetime.

- *Ecological risks*: For these risks, the focus is on the myriad interactions among populations, communities, and ecosystems (including food chains) at both the micro and the macro level. Ecological risks typically involve both short-term catastrophes, such as oil spills, and long-term exposures to hazardous substances.

- Whatever its focus, the *risk assessment process* consists of four basic steps:
  - *Hazard identification*: Defining the hazard and nature of the harm; for example, identifying a chemical contaminant, say, lead or carbon tetrachloride, and documenting its toxic effects on humans.
  - *Exposure assessment*: Determining the sources, pathways, and routes of exposure of a contaminating agent or stressor in the environment and determining the magnitude (amount), frequency, and duration of exposure to estimate its rate of intake in target organisms (receptors). An example would be determining the concentration of aflatoxin in peanut butter and calculating the dose an "average" person would receive.
  - *Dose-response assessment*: Quantifying the adverse effects arising from exposure to a hazardous agent or stressor based on the degree of exposure. This assessment is usually expressed mathematically as a plot showing the response in living organisms to increasing doses of the agent.

  - *Risk characterization*: Estimating the potential impact of a hazard based on the severity of its effects and the amount of exposure.

Once the risks are characterized, various regulatory options are evaluated in a process called *risk management*, which includes consideration of social, political, and economic issues, as well as the engineering problems inherent in a proposed solution. One important component of risk management is *risk communication*, which is the interactive process of information and opinion exchange among individuals, groups, and institutions. Risk communication includes the transfer of risk information from expert to non-expert audiences. In order to be effective, risk communication must provide a forum for balanced discussions of the nature of the risk, lending a perspective that allows the benefits of reducing the risk to be weighed against the costs (see Chapter 31 for more details).

In the United States, the passage of federal and state laws to protect public health and the environment has expanded the application of risk assessment. Major federal agencies that routinely use risk analysis include the Food and Drug Administration (FDA), the Environmental Protection Agency (EPA), and the Occupational Safety and Health Administration (OSHA). Together with state agencies, these regulatory agencies use risk assessment in a variety of situations (Information Box 29.1).

Risk assessment provides an effective framework for determining the relative urgency of problems and the

---

INFORMATION BOX 29.1 Applications of Risk Assessment

- Setting standards for concentrations of toxic chemicals or pathogenic microorganisms in water or food.
- Conducting baseline analyses of contaminated sites or facilities to determine the need for remedial action and the extent of cleanup required.
- Performing cost/benefit analyses of contaminated-site cleanup or treatment options (including treatment processes to reduce exposure to pathogens).
- Developing cleanup goals for contaminants for which no federal or state authorities have promulgated numerical standards; evaluating acceptable variance from promulgated standards and guidelines (e.g., approving alternative concentration limits).
- Constructing "what-if" scenarios to compare the potential impact of remedial or treatment alternatives and to set priorities for corrective action.
- Evaluating existing and new technologies for effective prevention, control, or mitigation of hazards and risks.
- Articulating community public health concerns and developing consistent public health expectations among different localities.

allocation of resources to reduce risks. Using the results of risk analyses, we can target prevention, remediation, and control efforts toward areas, sources, or situations in which the greatest risk reductions can be achieved with the resources available. However, risk assessment is not an absolute procedure carried out in a vacuum; rather, it is an evaluative, multifaceted, comparative process. Thus, to evaluate risk, we must inevitably compare one risk to a host of others. In fact, the comparison of potential risks associated with several problems or issues has developed into a subset of risk assessment called *comparative risk assessment*. Some commonplace risks are presented in Table 29.1. Here we see, for example, that risks from chemical exposure are fairly small relative to those associated with driving a car or smoking cigarettes.

Comparing different risks allows us to comprehend the uncommon magnitudes involved and to understand the level, or magnitude, of risk associated with a particular hazard. But comparison with other risks cannot itself establish the *acceptability* of a risk. Thus the fact that the chance of death from a previously unknown risk is about the same as that from a known risk does not necessarily imply that the two risks are equally acceptable. Generally, comparing risks along a single dimension is not helpful when the risks are widely perceived as qualitatively different. Rather, we must take into account certain qualitative factors that affect risk perception and evaluation when selecting risks to be compared. Some of these qualifying factors are listed in Table 29.2. We must also understand

the underlying premise that *voluntary risk is always more acceptable than involuntary risk.* For example, the same people who cheerfully drive their cars every day—thus incurring a 2:100 lifetime risk of death by automobile—are quite capable of refusing to accept the 6:10,000

TABLE 29.1 Examples of Some Commonplace Risks in the United States[a]

Risk	Lifetime Risk of Mortality
Cancer from cigarette smoking (one pack per day)	1:4
Death in a motor vehicle accident	2:100
Homicide	1:100
Home accident deaths	1:100
Cancer from exposure to radon in homes	3:1000
Exposure to the pesticide aflatoxin in peanut butter	6:10,000
Diarrhea from rotavirus	1:10,000
Exposure to typical EPA maximum chemical contaminant levels	1:10,000– 1:10,000,000

[a]Based on data in Wilson, R., Crouch, E.A.C., 1987. Risk assessment and comparisons: an introduction. Science 236, 267–270 and Gerba, C.P., Rose, J.B., 1992. Comparative Environmental Risks. Lewis Publishers, Boca Raton, FL. From Pollution Science, Academic Press, San Diego, CA, 1996.

TABLE 29.2 Factors Affecting Risk Perception and Risk Analysis

Factor	Conditions Associated With Increased Public Concern	Conditions Associated With Decreased Public Concern
Catastrophic potential	Fatalities and injuries grouped in time and space	Fatalities and injuries scattered and random
Familiarity	Unfamiliar	Familiar
Understanding	Mechanisms or process not understood	Mechanisms or process understood
Controllability (personal)	Uncontrollable	Controllable
Voluntariness of exposure	Involuntary	Voluntary
Effects on children	Children specifically at risk	Children not specifically at risk
Effects manifestation	Delayed effects	Immediate effects
Effects on future generations	Risk to future generations	No risk to future generations
Victim identity	Identifiable victims	Statistical victims
Dread	Effects dreaded	Effects not dreaded
Trust in institutions	Lack of trust in responsible institutions	Trust in responsible institutions
Media attention	Much media attention	Little media attention
Accident history	Major and sometimes minor accidents	No major or minor accidents
Equity	Inequitable distribution of risks and benefits	Equitable distribution of risks and benefits
Benefits	Unclear benefits	Clear benefits
Reversibility	Effects irreversible	Effects reversible
Origin	Caused by human actions or failures	Caused by acts of nature

Source: From Pollution Science, Academic Press, San Diego, CA, 1996.

involuntary risk of eating peanut butter contaminated with aflatoxin.

In considering risk, then, we must also understand another principle—the *de minimis* principle, which means that there are some levels of risk so trivial that they are not worth further consideration. However attractive, this concept is difficult to define, especially if we are trying to find a *de minimis* level acceptable to an entire society. Understandably, regulatory authorities are reluctant to be explicit about an "acceptable" risk. (How much aflatoxin would you consider acceptable in *your* peanut butter and jelly sandwich? How many dead insect parts?) But it is generally agreed that a lifetime risk on the order of one in a million (in the range of 10^{-6}) is sufficiently small to be acceptable for the general public. Although the origins and precise meaning of a one-in-a-million acceptable risk remain obscure, its impact on product choices, operations, and costs is very real—running, for example, into hundreds of billions of dollars in hazardous waste site cleanup decisions alone. While one in a million is the typical value, the levels of acceptable risk can vary over a range of 1 in 10,000 (10^{-4}) to 10^{-6}, or even greater (more stringent). Levels of risk at the less stringent end of the range (10^{-4} rather than 10^{-6}) may be acceptable if just a few people are exposed rather than the entire populace. For example, workers dealing with food additives can often tolerate higher levels of risk than can the public at large. These higher levels are justified because workers tend to be a relatively homogeneous, healthy group and because employment is voluntary; however, the same level of risks would generally not be acceptable for those same food additives for the general public.

29.2 THE PROCESS OF RISK ASSESSMENT

29.2.1 Hazard Identification

The first step in risk assessment is to determine the nature of the hazard. A *stressor* (or agent) is a substance, circumstance, or energy field that has the inherent ability to impose adverse effects upon an organism. The environment is subject to many different stressors, including chemicals, microorganisms, ionizing radiation, and rapid changes in temperatures. For pollution-related problems, the hazard in question is usually a specific chemical, a physical agent (such as irradiation), or a microorganism identified with a specific illness or disease. Thus the hazard identification component of an environmental risk assessment consists of a review of all relevant information bearing on whether or not an agent or stressor poses a specific threat. For example, in the *Guidelines for Carcinogen Risk Assessment* (U.S. EPA, 1986), the following information is evaluated for a potential carcinogen:

- Physical/chemical properties, routes, and patterns of exposure
- Structure/activity relationships of the substance

TABLE 29.3 EPA Categories for Carcinogenic Groups

Class	Description
A	Human carcinogen
B	Probable carcinogen
B_1	Linked human data
B_2	No evidence in humans
C	Possible carcinogen
D	No classification
E	No evidence

From United States Environmental Protection Agency. (U.S. EPA): Guidelines for carcinogen risk assessment, Federal Register, 24 September 1986.

- Absorption, distribution, metabolism, and excretion characteristics of the substance in the body
- The influence of other toxicological effects
- Data from short-term tests in living organisms
- Data from long-term animal studies
- Data from human studies

Once these data are reviewed, the animal and human data are both separated into groups characterized by degree of evidence:

- Sufficient evidence of carcinogenicity
- Limited evidence of carcinogenicity
- Inadequate evidence
- No data available
- No evidence of carcinogenicity

The available information on animal and human studies is then combined into a weight-of-evidence classification scheme to assess the likelihood of carcinogenicity. This analysis gives more weight to human than to animal evidence (when it is available) and includes several groupings (Table 29.3).

Clinical studies of disease can be used to identify very large risks (between 1/10 and 1/100), most epidemiological studies can detect risks down to 1/1000, and very large epidemiological studies can examine risks in the 1/10,000 range. However, risks lower than 1/10,000 cannot be studied with much certainty using epidemiological approaches. Since regulatory policy objectives generally strive to limit risks below 1/100,000 for life-threatening diseases like cancer, these lower risks are often estimated by extrapolating from the effects of high doses given to animals.

29.2.2 Exposure Assessment

Exposure assessment is the process of measuring or estimating the magnitude, frequency, and duration of exposure to an environmental agent or stressor. Contaminant sources, release mechanisms, transport, and transformation characteristics are all important aspects of exposure assessment. The nature of the source and release mechanisms will determine the type and amount of contaminant that is present, and the location of the contaminant within the various environmental compartments (atmosphere, soil, surface water, groundwater). Once released, many contaminants can transfer to different compartments, and be transformed or degraded during transport. As a result, the concentration of a contaminant that a receptor is exposed to (i.e., the dose) at a particular point of contact (e.g., at one's home) is influenced by many factors. This explains why it is critical to understand the factors and processes influencing the transport and fate of a contaminant in the environment (see Chapters 7–9).

An *exposure pathway* is the course that a hazardous agent takes from a source to a receptor (e.g., human or animal) located at a specified point of contact. The transport from source to receptor occurs via environmental carriers or media—generally, air (volatile compounds, particulates), water (soluble compounds), or solids (soil, food). Exposure pathways for human and ecological risk are illustrated in Fig. 29.1. An exception is electromagnetic radiation, which needs no medium.

Exposure to contaminants can occur via inhalation, ingestion of water or food, or absorption through the skin upon dermal contact. These are the routes of exposure, or *exposure routes*. The characteristics of the receptor play a critical role in exposure. For example, the location and activity patterns of the exposed population influence the chances that a receptor will come into contact with a contaminant, and partly determine the frequency and duration of contact. Therefore it is important to understand the behavior of the receptors in addition to contaminant transport behavior. The quantification of exposure, intake, or potential dose can involve equations with three sets of variables:

- Concentrations of chemicals or microbes in the media
- Exposure rates (magnitude, frequency, duration)
- Quantified biological characteristics of receptors (e.g., body weight, absorption capacity for chemicals; level of immunity to microbial pathogens)

Exposure concentrations are derived from measured and/or modeled data. Ideally, exposure concentrations should be measured at the points of contact between the environmental media and current or potential receptors. It is usually possible to identify potential receptors and exposure points from field observations and other information. However, it is seldom possible to anticipate all potential exposure points and measure all environmental concentrations under all conditions. In practice, a combination of monitoring and modeling data, together with a great deal of professional judgment, is required to estimate exposure concentrations.

In order to assess exposure rates via different exposure pathways, we have to consider and weigh many factors. For example, in estimating exposure to a substance via drinking water, we first have to determine the average daily consumption of that water. But this is not as easy as it sounds. Studies have shown that daily fluid intake varies greatly from individual to individual. Moreover, water intake depends on how much fluid is consumed as drinking water, and how much is ingested in the form of soft drinks and other non-drinking-water sources. Drinking water intake also changes significantly with age (Fig. 29.2), body weight, diet, and climate. Because these factors are so variable, the EPA has suggested a number of very conservative "default" exposure values that can be used when assessing contaminants in drinking water, vegetables, soil, and the like (Table 29.4).

One important route of exposure is the food supply. Toxic substances are often bioaccumulated, or concentrated, in plant and animal tissues, thereby exposing humans who ingest those tissues as food. Moreover, many toxic substances tend to be biomagnified in the food chain, so that animal tissues contain relatively high concentrations of toxins. Take fish, for example. It is relatively straightforward to estimate concentrations of contaminants in water. Thus we can use a *bioconcentration factor* (BCF) to estimate the tendency for a substance in water to accumulate in fish tissue. The concentration of a chemical in fish can be estimated by multiplying its concentration in water by the BCF. The greater the value of the BCF, the more the chemical accumulates in the fish and the higher the risk of exposure to humans.

The units of BCF—liters per kilogram ($L\,kg^{-1}$)—are chosen to allow the concentration of a chemical to be expressed as milligrams per liter ($mg\,L^{-1}$) of water and the concentration in fish to be in milligrams per kilogram ($mg\,kg^{-1}$) of fish body weight. In Table 29.5, we see the BCFs of several common organic and inorganic chemicals. Note the high values of BCF for the chlorinated hydrocarbon pesticides like dichlorodiphenyltrichloroethane (DDT) and the polychlorinated biphenyls (PCBs). This exemplifies the concern we have with such compounds, as discussed in Chapter 12.

29.2.3 Dose-Response Assessment

Chemicals and other contaminants are not equal in their capacity to cause adverse effects. To determine the capacity of agents to cause harm, we need quantitative toxicity data. Some toxicity data are derived from occupational, clinical, and epidemiological studies. Most toxicity data, however, come from animal experiments in which researchers expose laboratory animals, mostly mice and rats, to increasingly higher concentrations or doses and observe their corresponding effects. The result of these experiments is the *dose-response relationship*—a quantitative relationship that indicates the agent's degree of toxicity to exposed species. Dose is normalized as milligrams of substance or

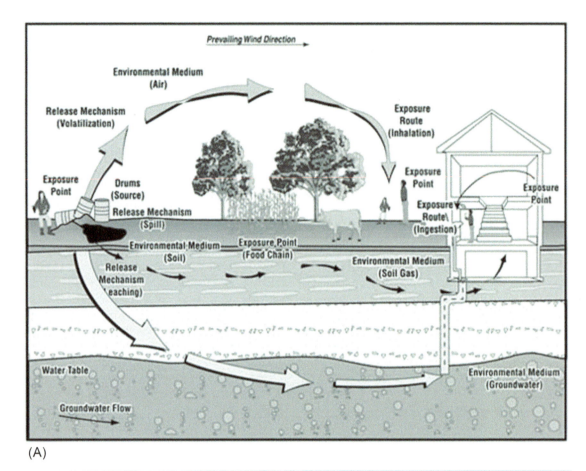

(A)

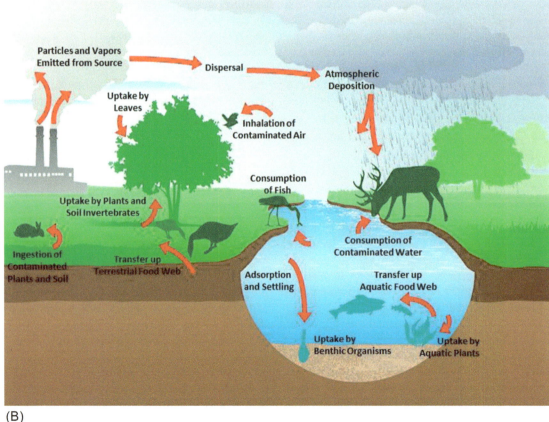

(B)

FIG. 29.1 Exposure pathways for (A) human risk [Available from: https://www.atsdr.cdc.gov/hac/phamanual/ch6.html) and (B) ecological risk [Available from: https://www.epa.gov/ecobox/epa-ecobox-tools-exposure-pathways-exposure-pathways-era].

impairment (e.g., bronchitis or emphysema arising from smoke damage), to death.

The goal of a dose-response assessment is to obtain a mathematical relationship between the amount (concentration) of a toxicant or microorganism to which a human is exposed and the risk of an adverse outcome from that dose. The data resulting from experimental studies is presented as a dose-response curve, as shown in Fig. 29.3. The abscissa describes the dose, while the ordinate measures the risk that some adverse health effect will occur. In the case of a pathogen, for instance, the ordinate may represent the risk of infection, and not necessarily illness.

Dose-response curves derived from animal studies must be interpreted with care. The data for these curves are necessarily obtained by examining the effects of large doses on test animals. Because of the costs involved, researchers are limited in the numbers of test animals they can use—it is both impractical and cost prohibitive to use thousands (even millions) of animals to observe just a few individuals that show adverse effects at low doses (e.g., risks of 1:1000 or 1:10,000). Researchers must therefore extrapolate low-dose responses from their high-dose data. And therein lies the problem: dose-response curves are subject to controversy because their results change depending on the method chosen to extrapolate from the high doses actually administered to laboratory test subjects to the low doses humans are likely to receive in the course of everyday living.

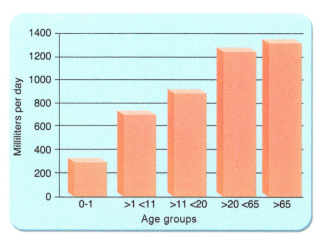

FIG. 29.2 Average drinking water ingestion rates in the United States by age. *(From Roseberry, A.M., Burmaster, D.E., 1992. Lognormal distributions for water intake by children and adults. Risk Anal. 12, 99–104.)*

pathogen ingested, inhaled, or absorbed (in the case of chemicals) through the skin per kilogram of body weight per day ($mg\,kg^{-1}\,day^{-1}$). Responses or effects can vary widely— from no observable effect, to temporary and reversible effects (e.g., enzyme depression caused by some pesticides or diarrhea caused by viruses), to permanent organ injury (e.g., liver and kidney damage caused by chlorinated solvents, heavy metals, or viruses), to chronic functional

TABLE 29.4 EPA Standard Default Exposure Factors

Land Use	Exposure Pathway	Daily Intake	Exposure Frequency (days/year)	Exposure Duration (years)
Residential	Ingestion of potable water	$2\,L\,day^{-1}$	350	30
	Ingestion of soil and dust	200 mg (child)	350	6
		100 mg (adult)		24
	Inhalation of contaminants	$20\,m^3$ (total)	350	30
		$15\,m^3$ (indoor)		
Industrial and commercial	Ingestion of potable water	1 L	250	25
	Ingestion of soil and dust	50 mg	250	25
	Inhalation of contaminants	$20\,m^3$ (workday)	250	25
Agricultural	Consumption of homegrown produce	42 g (fruit)	350	30
		80 g (vegetable)		
Recreational	Consumption of locally caught fish	54 g	350	30

Modified from Kolluru RV: Environmental strategies handbook, New York, 1993, McGraw-Hill. From Pollution Science, Academic Press, San Diego, CA, 1996.

TABLE 29.5 Bioconcentration Factors (BCFs) for Various Organic and Inorganic Compounds

Chemical	BCF (L kg^{-1})
Aldrin	28
Benzene	44
Cadmium	81
Chlordane	14,000
Chloroform	3.75
Copper	200
DDT	54,000
Formaldehyde	0
Nickel	47
PCBs	100,000
Trichloroethylene	10.6
Vinyl chloride	1.17

From United States Environmental Protection Agency (U.S. EPA): Risk assessment, management and communication of drinking water contamination. EPA/625/4-89/024, Washington, DC, 1990.

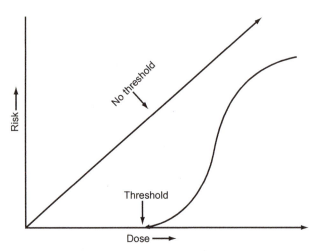

FIG. 29.3   Relationship between a threshold and nonthreshold response.

This controversy revolves around the choice of several mathematical models that have been proposed for extrapolation to low doses. Unfortunately, no model can be proved or disproved from the data, so there is no way to know which model is the most accurate. The choice of models is therefore strictly a policy decision, which is usually based on conservative assumptions. Thus, for noncarcinogenic chemical responses, the assumption is that some *threshold* exists below which there is no toxic response; that is, no adverse effects will occur below some very low dose (say, one in a million) (Fig. 29.3). Carcinogens, however, are considered *nonthreshold*—that is, the conservative assumption is that exposure to any amount of carcinogen creates some likelihood of cancer. This means that the only "safe" amount of carcinogen is zero, so the dose-response plot is required to go through the origin (0), as shown in Fig. 29.3.

There are many mathematical models to choose from, including the one-hit model, the multistage model, the multihit model, and the probit model. The characteristics of these models for nonthreshold effects are listed in Table 29.6.

The *one-hit model* is the simplest model of carcinogenesis in which it is assumed:

1. That a single chemical "hit," or exposure, is capable of inducing malignant change (i.e., a single hit causes irreversible damage of DNA, leading to tumor development). Once the biological target is hit, the process leading to tumor formation continues independently of dose.

2. That this change occurs in a single stage.

The *multistage model* assumes that tumors are the result of a sequence of biological events, or stages. In simplistic terms, the biological rationale for the multistage model is that there are a series of biological stages that a chemical must pass through (e.g., metabolism, covalent bonding, DNA repair, and so on) without being deactivated, before the expression of a tumor is possible.

The rate at which the cell passes through one or more of these stages is a function of the dose rate. The multistage model also has the desirable feature of producing a linear relationship between risk and dose.

The *multihit model* assumes that a number of dose-related hits are needed before a cell becomes malignant. The most important difference between the multistage and multihit model is that in the multihit model, all hits must result from the dose, whereas in the multistage model, passage through some of the stages can occur spontaneously. The practical implication of this is that the multihit models are generally much flatter at low doses and consequently predict a lower risk than the multistage model.

The *probit model* is not derived from mechanistic assumptions about the cancer process. It may be thought of as representing distributions of tolerances to carcinogens in a large population. The model assumes that the probability of the response (cancer) is a linear function of the log of the dose (log normal). While these models may be appropriate for acute toxicity they are considered questionable for carcinogens. These models would predict the lowest level of risk of all the models.

TABLE 29.6 Primary Models Used for Assessment of Nonthreshold Effects

Model[a]	Comments
One-hit	Assumes (1) a single stage for cancer and (2) malignant change induced by one molecular or radiation interaction *Very conservative*
Linear multistage	Assumes multiple stages for cancer *Fits curve to the experimental data*
Multihit	Assumes several interactions needed before cell becomes transformed *Least conservative model*
Probit	Assumes probit (lognormal) distribution for tolerances of exposed population *Appropriate for acute toxicity; questionable for cancer*

[a]*All these models assume that exposure to the pollutant will always produce an effect, regardless of dose.*

Modified from Cockerham LG, Shane BS: Basic environmental toxicology, Boca Raton, FL, 1994, CRC Press. From Pollution Science Academic Press, San Diego, CA, 1996.

The effect of each model on estimating risk for a given chemical is presented in Table 29.7 and Fig. 29.4. As we can see, the choice of models results in order-of-magnitude differences in estimating the risk at low levels of exposure.

The *linear multistage model*, a modified version of the multistage model, is the EPA's model of choice, because this agency chooses to err on the side of safety and overemphasize risk. This model assumes that there are multiple stages for cancer (i.e., a series of mutations or biotransformations) involving many carcinogens, cocarcinogens, and promoters (see Chapter 28) that can best be modeled by a

TABLE 29.7 Lifetime Risks of Cancer Derived From Different Extrapolation Models

Model Applied	Lifetime Risk (1.0 mg kg^{-1} day^{-1}) of Toxic Chemical[a]	
One-hit	6.0×10^{-5}	(1 in 17,000)
Multistage	6.0×10^{-6}	(1 in 167,000)
Multihit	4.4×10^{-7}	(1 in 2.3 million)
Probit	1.9×10^{-10}	(1 in 5.3 billion)

[a]*All risks for a full lifetime of daily exposure. The lifetime is used as the unit of risk measurement, because the experimental data reflect the risk experienced by animals over their full lifetimes. The values shown are upper confidence limits on risk.*

Source: United States Environmental Protection Agency (U.S. EPA): Risk assessment, management and communication of drinking water contamination. EPA/625/4-89/024, Washington, DC, 1990. From Pollution Science, Academic Press, San Diego, CA, 1996.

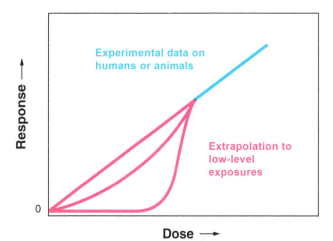

FIG. 29.4 Extrapolation of dose-response curves. *(Adapted from United States Environmental Protection Agency (U.S. EPA): Risk assessment, management and communication of drinking water contamination. EPA/625/4-89/024, Washington, DC, 1990. From Pollution Science, Academic Press, San Diego, CA, 1996.)*

series of mathematical functions. At low doses, the slope of the dose-response curve produced by the linear multistage model is called the *potency factor* (PF) or *slope factor* (SF) (Fig. 29.5), which is the reciprocal of the concentration of chemical measured in milligrams per kilogram of animal body weight per day, that is, $1/(\text{mg kg}^{-1} \text{ day}^{-1})$, or the risk produced by a lifetime *average dose* (AD) of $1 \text{ mg kg}^{-1} \text{ day}^{-1}$. Thus the dose-response equation for a carcinogen is

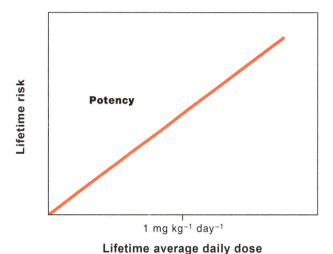

FIG. 29.5 Potency factor is the slope of the dose-response curve at low doses. At low doses, the slope of the dose-response curve produced by the multistage model is called the potency factor. It is the risk produced by a lifetime average dose of $1 \text{ mg kg}^{-1} \text{ day}^{-1}$. *(Adapted from United States Environmental Protection Agency (U.S. EPA): Risk assessment, management and communication of drinking water contamination. EPA/625/4-89/024, Washington, DC, 1990. From Pollution Science Academic Press, San Diego, CA, 1996.)*

$$\text{Lifetime Risk} = \text{AD} \times \text{PF} \qquad (29.1)$$

The probability of *getting* cancer (not the probability of *dying* of cancer) and the associated dose, consist of an average taken over an assumed 70-year human lifetime. This dose is called the lifetime average daily dose or *chronic daily intake*.

The dose-response effects for noncarcinogens allow for the existence of thresholds, that is, a certain quantity of a substance or dose below which there is *no observable toxic effect* (NOAEL; see Chapter 28) by virtue of the body's natural repair and detoxifying capacity. If a NOAEL is not available, a LOAEL (*lowest observed adverse effect level*) may be used, which is the lowest observed dose or concentration of a substance at which there is a detectable adverse health effect. When a LOAEL is used instead of a NOAEL, an additional uncertainty factor is normally applied. Examples of toxic substances that have thresholds are heavy metals and polychlorinated biphenyls (PCBs). These thresholds are represented by the *reference dose*, or RfD, of a substance, which is the intake or dose of the substance per unit body weight per day ($\text{mg}\,\text{kg}^{-1}\,\text{day}^{-1}$) that is likely to pose no appreciable risk to human populations, including such sensitive groups as children (Table 29.8). A dose-response plot for carcinogens therefore goes through this reference point (Fig. 29.6).

In general, substances with relatively high slope factors and low reference doses tend to be associated with higher toxicities. The RfD is obtained by dividing the NOAEL (see Chapter 28) by an appropriate uncertainty factor, sometimes called a *safety factor* or *uncertainty factor*. A 10-fold uncertainty factor is used to account for differences in sensitivity between the most sensitive individuals in an exposed human population. These include pregnant women, young children, and the elderly, who are more sensitive than "average" people. Another factor of 10 is added when the NOAEL is

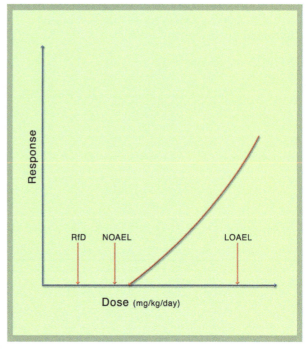

FIG. 29.6 Relationships between RfD, NOAEL, and LOAEL for noncarcinogens.

based on animal data that are extrapolated to humans. In addition, another factor of 10 is sometimes applied when questionable or limited human and animal data are available. The general formula for deriving an RfD is

$$\text{RfD} = \frac{\text{NOAEL}}{\text{VF}_1 \times \text{VF}_2 \times \cdots \times \text{VF}_n} \qquad (29.2)$$

where VF_i are the uncertainty factors. As the data become more uncertain, higher safety factors are applied. For example, if data are available from a high-quality epidemiological study, a simple uncertainty factor of 10 may be used by simply dividing the original value for RfD by 10 to arrive at a new value of RfD, which reflects the concern for safety. The RfDs of several noncarcinogenic chemicals are presented in Table 29.8.

The RfD (Fig. 29.6) can be used in quantitative risk assessments by using the following relationship:

$$\text{Risk} = \text{PF}\,(\text{CDI} - \text{RfD}) \qquad (29.3)$$

where CDI is the chronic daily intake, and the *potency factor* (PF) is the slope of the dose-response curve. Table 29.9 contains potency factors for some potential carcinogens:

$$\begin{aligned} &\text{CDI}\,\left(\text{mg}\,\text{kg}^{-1}\,\text{day}^{-1}\right) \\ &\quad = \frac{\text{Average daily dose}\,\left(\text{mg}\,\text{day}^{-1}\right)}{\text{Body weight}\,(\text{kg})} \end{aligned} \qquad (29.4)$$

This type of risk calculation is rarely performed. In most cases, the RfD is used as a simple indicator of potential risk in practice. That is, the chronic daily intake is simply

TABLE 29.8 Chemical RfDs for Chronic Noncarcinogenic Effects of Selected Chemicals

Chemical	RfD ($\text{mg}\,\text{kg}^{-1}\,\text{day}^{-1}$)
Acetone	0.1
Cadmium	0.0005
Chloroform	0.01
Methylene chloride	0.06
Phenol	0.04
Polychlorinated biphenyl	0.0001
Toluene	0.3
Xylene	2.0

From United States Environmental Protection Agency (U.S. EPA): Risk assessment, management and communication of drinking water contamination. EPA/625/4-89/024, Washington, DC, 1990.

TABLE 29.9 Toxicity Data for Selected Potential Carcinogens

Chemical	Potency Factor Oral Route (mg kg day^{-1})
Arsenic	1.75
Benzene	2.9×10^{-2}
Carbon tetrachloride	0.13
Chloroform	6.1×10^{-3}
DDT	0.34
Dieldrin	30
Heptachlor	3.4
Methylene chloride	7.5×10^{-3}
Polychlorinated biphenyls (PCBs)	7.7
2,3,7,8-TCDD (dioxin)	1.56×10^{5}
Tetrachloroethylene	5.1×10^{-2}
Trichloroethylene (TCE)	1.1×10^{-2}
Vinyl chloride	2.3

From U.S. EPA. Available from: www.epa.gov/iris.

compared with the RfD, then, if the CDI is below the RfD, it is assumed that the risk is negligible for almost all members of an exposed population.

29.2.4 Risk Characterization

The final phase of risk assessment process is risk characterization. In this phase, exposure and dose-response assessments are integrated to yield probabilities of effects occurring in humans under specific exposure conditions. Quantitative risks are calculated for appropriate media and pathways. For example, the risks of lead in water are estimated over a lifetime assuming: (1) that the exposure of 21 of water per day is ingested over a 70-year lifetime; and (2) that different concentrations of lead occur in the drinking water. This information can then be used by risk managers to develop standards or guidelines for specific toxic chemicals or infectious microorganisms in different media, such as the drinking water or food supply.

29.2.4.1 Cancer Risks

If the dose-response curve is assumed to be linear at low doses for a carcinogen, then:

Incremental lifetime risk of cancer $= (CDI)(PF)$ (29.5)

The linearized multistage model assumptions (see Table 29.6) estimates the risk of getting cancer, which is not necessarily the same as the risk of dying of cancer, so it should be even more conservative as an upper-bound estimate of cancer deaths. Potency factors can be found in the EPA database on toxic substances called the Integrated Risk Information System (IRIS) (see Information Box 29.2). Table 29.9 contains the potency factor for some of these chemicals.

The mean exposure concentration of contaminants is used with exposed population variables and the assessment determined variables to estimate contaminant intake. The general equation for chemical intake is

INFORMATION BOX 29.2 Integrated Risk Information System (IRIS)

The Integrated Risk Information System (IRIS), prepared and maintained by the U.S. Environmental Protection Agency (U.S. EPA), is an electronic database containing information on human health effects that may result from exposure to various chemicals in the environment (www.epa.gov/iris). IRIS was initially developed for EPA staff in response to a growing demand for consistent information on chemical substances for use in risk assessments, decision making, and regulatory activities. The information in IRIS is intended for those without extensive training in toxicology, but with some knowledge of health sciences. The heart of the IRIS system is its collection of computer files covering individual chemicals. These chemical files contain descriptive and quantitative information in the following categories:

- Oral reference doses and inhalation reference concentrations (RfDs) for chronic noncarcinogenic health effects.
- Hazard identification, oral slope factors, and oral and inhalation unit risks for carcinogenic effects.

Oral RfD Summary for Arsenic:

Critical Effect	Experimental Doses[a]	UF	RFD
Hyperpigmentation, keratosis and possible vascular complications Human chronic oral exposure (Tseng, 1977; Tseng et al., 1968)	NOAEL: 0.009 mg/L converted to 0.0008 mg/kg-day LOAEL: 0.17 mg/L converted to 0.014 mg/kg-day	3	3E-4 mg/kg-day

[a]*Conversion Factors—NOAEL was based on an arithmetic mean of 0.009 mg/L in a range of arsenic concentration of 0.001–0.017 mg/L. This NOAEL also included estimation of arsenic from food. Since experimental data were missing, arsenic concentrations in sweet potatoes and rice were estimated as 0.002 mg/day. Other assumptions included consumption of 4.5 L water/day and 55 kg body weight. NOAEL = [(0.009 mg/L × 4.5 L/day) + 0.002 mg/day]/55 kg = 0.0008 mg/kg-day. The LOAEL dose was estimated using the same assumptions as the NOAEL starting with an arithmetic mean water concentration from Tseng (1977) of 0.17 mg/L. LOAEL = [(0.17 mg/L × 4.5 L/day) + 0.002 mg/day]/55 kg = 0.014 mg/kg-day.*

UF = Uncertainty Factor or Safety Factor.

$$CDI = \frac{C \times CR \times EFD}{BW} \times \frac{1}{AT} \qquad (29.6)$$

where

CDI = chronic daily intake; the amount of chemical at the exchange boundary (mg/kg-day)

C = average exposure concentration over the period (e.g., $mg\,L^{-1}$ for water or $mg\,m^{-3}$ for air)

CR = contact rate, the amount of contaminated medium contacted per unit time ($L\,day^{-1}$ or $m^3\,day^{-1}$)

EFD = exposure frequency and duration, a variable that describes how long and how often exposure occurs. The EFD is usually divided into two terms:

EF = exposure frequency (days/year) and

ED = exposure duration (years)

BW = average body mass over the exposure period (kg)

AT = averaging time; the period over which the exposure is averaged (days).

Determination of accurate intake data is sometimes difficult; for example, exposure frequency and duration vary among individuals and must often be estimated; site-specific information may be available; and professional judgment may be necessary. Equations for estimating daily contamination intake rates from drinking water, the air, and contaminated food, and for dermal exposure while swimming, have been reported by the EPA. Two of the most common routes of exposure are through drinking contaminated water and breathing contaminated air. The intake for ingestion of waterborne chemicals is (see Example Calculations 29.1 and 29.2)

$$CDI = \frac{CW \times IR \times EF \times ED}{BW \times AT} \qquad (29.7)$$

where

CDI = chronic daily intake by ingestion (mg/kg-day)

CW = chemical concentration in water ($mg\,L^{-1}$)

IR = ingestion rate ($L\,day^{-1}$)

EF = exposure frequency (days/year)

ED = exposure duration (years)

BW = body weight (kg)

AT = averaging time (period over which the exposure is averaged—days)

Some of the values used in Eq. (29.5) are

CW: site-specific measured or modeled value

IR: $2\,L\,day^{-1}$ (adult, 90th percentile); $1.4\,L\,day^{-1}$ (adult, average)

EF: pathway-specific value (dependent on the frequency of exposure-related activities)

ED: 70 years (lifetime; by convention); 30 years [national upper-bound time (90th percentile) at one residence]; 9 years [national median time (50th percentile) at one residence]

BW: 70 kg (adult, average); Age-specific values

Example Calculation 29.1 Estimation of an Oral Chronic Daily Intake

The mean concentration of 1,2-dichlorobenzene in a water supply is $1.7\,\mu g\,L^{-1}$. Determine the chronic daily intake for a 70-kg adult. Assume that 2 L of water are consumed per day.

Solution

The chronic daily intake (CDI) may be calculated using Eq. (29.7).

$$CDI = \frac{C \times CR \times EF \times ED}{BW \times AT}$$

where

CDI = chronic daily intake (mg/kg-day)

$C = 1.7\,\mu g\,L^{-1} = 0.0017\,mg\,L^{-1}$

$CR = 2\,L\,day^{-1}$

EF = 365 days/year

ED = 30 years (standard exposure duration for an adult exposed to a noncarcinogenic)

BW = 70 kg

AT = 365 days/year × 30 years = 10,950 days

Substituting values into the equation yields the chronic daily intake

$$CDI = \frac{0.0017 \times 2 \times 365 \times 30}{70 \times 10,950}$$
$$= 4.86 \times 10^{-5}\ mg/10^{-5}\ mg/kg\ day$$

Example Calculation 29.2 Application of Hazard Index and Incremental Carcinogenic Risk Associated With Chemical Exposure

A drinking water supply is found to contain $0.1\,mg\,L^{-1}$ of acetone and $0.1\,mg\,L^{-1}$ of chloroform. A 70-kg adult drinks 2 L per day of this water for 5 years. What would be the hazard index and the carcinogenic risk from drinking this water?

First, we need to determine the average daily doses (ADDs) for each of the chemicals and then their individual hazard quotients.

For Acetone

$$ADD = \frac{\left(0.1\,mgL^{-1}\right)\left(2L\,day^{-1}\right)}{70\,kg}$$
$$= 2.9 \times 10^{-3}\ mg\ kg^{-1}\ day^{-1}$$

From Table 29.5, the RfD for acetone is $0.1\,mg\,kg^{-1}\,day^{-1}$

$$\text{Hazard quotient (HQ)} = \frac{2.9 \times 10^{-3}\,mg\,kg^{-1}\,day^{-1}}{0.1}$$
$$= 0.029$$

For Chloroform

$$ADD = \frac{\left(0.1\,mgL^{-1}\right)\left(2L\,day^{-1}\right)}{70\,kg}$$
$$= 2.9 \times 10^{-3}\ mg\ kg^{-1}\ day^{-1}$$

From Table 29.5, the RfD value for chloroform is $0.01\,\text{mg kg}^{-1}\,\text{day}^{-1}$

$$HQ = \frac{2.9 \times 10^{-3}\,\text{mg kg}^{-1}\,\text{day}^{-1}}{0.01}$$

$$= 0.029$$

Thus

Hazard Index $= 0.029 + 0.29 = 0.319$

Since the hazard index is less than 1.0, the water is considered safe. Notice that we did not need to take into consideration that the person drank the water for 5 years.

The incremental carcinogenic risk associated with chloroform is determined as follows:

Risk = (CDI) (Potency factor)

$$CDI = \frac{(0.1\,\text{mg L}^{-1})(2\,\text{L day}^{-1}\,(365\,\text{days yr}^{-1})\,(5\,\text{yrs}))}{(70\,\text{kg})\,(365\,\text{days yr}^{-1})(70\,\text{yrs})}$$

$$= 4.19 \times 10^{-5}\,\text{mg kg}^{-1}\,\text{day}^{-1}$$

From Table 29.6, the potency factor for chloroform is 6.1×10^{-3}

$$\text{Risk} = (\text{CDI}) (\text{Potency factor})$$

$$\text{Risk} = \left(4.19 \times 10^{-5}\,\text{mg kg}^{-1}\,\text{day}^{-1}\right)$$

$$\left(6.1 \times 10^{-3}\,\text{mg kg}^{-1}\,\text{day}^{-1}\right) = 2.55 \times 2.55 \times 10^{-7}$$

From a cancer risk standpoint, the risk over this period of exposure is less than the 10^{-6} goal.

AT: pathway-specific period of exposure for noncarcinogenic effects (i.e., ED × 365 days/year), and 70-year lifetime for carcinogenic effects (i.e., 70 years × 365 days/year), averaging time.

29.2.4.2 Noncancer Risks

Noncancer risks are expressed in terms of a hazard quotient (HQ) for a single substance, or hazard index (HI) for multiple substances and/or exposure pathways (see Example Calculation 29.2)

Hazard quotient (HQ)

$$= \frac{\text{Average daily dose during exposure period}\,\left(\text{mg kg}^{-1}\,\text{day}^{-1}\right)}{\text{RfD}\,\left(\text{mg kg}^{-1}\,\text{day}^{-1}\right)}$$

$$(29.8)$$

Unlike a carcinogen, the toxicity is important only during the time of exposure, which may be one day, a few days, or years. The HQ has been defined so that if it is less than 1.0, there should be no significant risk or systemic toxicity. Ratios above 1.0 could represent a potential risk, but there is no way to establish that risk with any certainty.

When exposure involves more than one chemical, the sum of the individual hazard quotients for each chemical is generally used as a measure of the potential for harm. This sum is called the hazard index (HI):

$$\text{HI} = \text{Sum of hazard quotients} \quad (29.9)$$

29.2.4.3 Uncertainty Analysis

Uncertainty is inherent in every step of the risk assessment process. Thus, before we can begin to characterize any risk, we need some idea of the nature and magnitude of uncertainty in the risk estimate. Sources of uncertainty include:

- Extrapolation from high to low doses
- Extrapolation from animal to human responses
- Extrapolation from one route of exposure to another
- Limitations of analytical methods
- Estimates of exposure

Although the uncertainties are generally much larger in estimates of exposure and the relationships between dose and response (e.g., the percent mortality), it is important to include the uncertainties originating from all steps in a risk assessment in risk characterization.

Two approaches commonly used to characterize uncertainty are sensitivity analyses and Monte Carlo simulations. In *sensitivity analyses*, we simply vary the uncertain quantities of each parameter (e.g., average values, high and low estimates), usually one at a time, to find out how changes in these quantities affect the final risk estimate. This procedure gives us a range of possible values for the overall risk and tells us which parameters are most crucial in determining the size of the risk. In a *Monte Carlo simulation*, however, we assume that all parameters are random or uncertain.

Thus, instead of varying one parameter at a time, we use a computer program to select parameter distributions randomly every time the model equations are solved, the procedure being repeated many times. The resulting output can be used to identify values of exposure or risk corresponding to a specified probability, say, the 50th percentile or 95th percentile.

29.2.4.4 Risk Projections and Management

In the risk characterization phase, exposure and dose-response assessments are integrated to yield probabilities of effects occurring in humans under specific exposure conditions. Quantitative risks are calculated for appropriate media and pathways. For example, the risks of lead in water are estimated over a lifetime, assuming (1) that the exposure is 2 L of water ingested per day over a 70-year lifetime and (2) that different concentrations of lead occur in the drinking water. This information can then be used by risk managers to develop standards or guidelines for specific toxic chemicals or infectious microorganisms in different media, such as the drinking water or food supply.

29.2.4.5 Hazardous Waste Risk Assessment

Hazardous waste risk assessments are a key part of the Comprehensive Environmental Response, Compensation, and Liability Act (CERCLA). Risk assessments are performed to assess health and ecological risks at Superfund sites and to evaluate the effectiveness of remedial alternatives for attaining site cleanup objectives. Since specific cleanup requirements have not been established for most contaminants under CERCLA, each site is assessed on an individual basis and cleaned up to a predetermined level of risk, such as 1 cancer case per 1,000,000 people. Risks may be different from one site to the next, depending on characteristics of the site and the potential for exposure.

For example, at one site, a high level of contaminants may be present (10,000 mg kg^{-1} of soil), but there is no nearby population, there is a large distance to groundwater, and the soils are of low permeability. Based on a risk assessment, the best remedial action for the site may be to leave the contaminated soil in place, where natural attenuation processes will eventually result in its degradation. Removing the contaminated soil with disposal in a landfill or in situ treatment may result in a higher exposure risk due to release of wind-blown dusts that may expose workers at the site. In contrast, a soil contaminated with 10 mg of hazardous material per kg of soil may be considered a greater risk if the site has sandy soil, shallow groundwater, and nearby drinking water wells, and is located near a school. Cleanup to low levels would be necessary in this case to protect human health (Fig. 29.7).

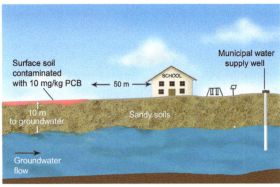

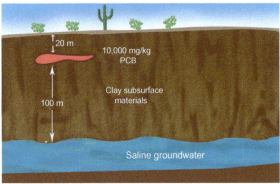

FIG. 29.7 Two extremes of potential risk from contaminated sites. Site A is a high-risk site with potential for contaminant migration from the source to nearby receptors. Site B, although characterized by a higher source concentration, has minimal potential for contaminant migration and risk. *(Modified from Watts RJ: Hazardous wastes: sources, pathways, receptors, New York, 1998, John Wiley & Sons.)*

29.3 ECOLOGICAL RISK ASSESSMENT

Ecological risk assessment is a process that evaluates the probability that adverse ecological effects will occur as the result of exposure to one or more stressors (Fig. 29.1B). Ecological risk assessment may evaluate one or more stressors and ecological components (e.g., specific organisms, populations, communities, or ecosystems). Ecological risks may be expressed as true probabilistic estimates of adverse effects (as is done with carcinogens in human health risk assessment), or they may be expressed in a more qualitative manner.

In the United States, the Comprehensive Environmental Response, Compensation, and Liability Act (CERCLA) (otherwise known as the Superfund), the Resource Conservation and Recovery Act (RCRA), and other regulations require an ecological assessment as part of all remedial investigation and feasibility studies. Pesticide registration, which is required under the Federal Insecticide, Fungicide, and Rodenticide Act (FIFRA), must also include an ecological assessment. In the CERCLA/RCRA context, a typical objective is to determine and document actual or potential effects of contaminants on ecological receptors and habitats as a basis for evaluating remedial alternatives in a scientifically defensible manner.

The four major phases or steps in ecological assessment (Fig. 29.8) are as follows:

- Problem formulation and hazard identification
- Exposure assessment
- Ecological effects/toxicity assessment
- Risk characterization

An ecological risk assessment may be initiated under many circumstances—the manufacturing of a new chemical, evaluation of cleanup options for a contaminated site, or the planned filling of a marsh, among others. The problem-formulation process begins with an evaluation of the stressor characteristics, the ecosystem at risk, and the likely ecological effects. An endpoint is then selected. An *endpoint* (Fig. 29.9) is a characteristic of an ecological component, *e.g.*, the mortality of fish) that may be affected by a stressor. Two types of endpoints are generally used: assessment endpoints and measurement endpoints. *Assessment endpoints* are particular environmental values to be protected. Such endpoints, which are recognized and valued by the public, drive the decisions made by official risk managers. *Measurement endpoints* are qualitatively or quantitatively measurable factors. Suppose, for example, a community that values the quality of sports fishing in the area is worried about the effluent from a nearby paper mill. In this case, a decline in the trout population might serve as the assessment endpoint, while the increased mortality of minnows, as evaluated by laboratory studies, might be the measurement endpoint. Thus risk managers would use the quantitative data gathered on the surrogate minnow population to develop management

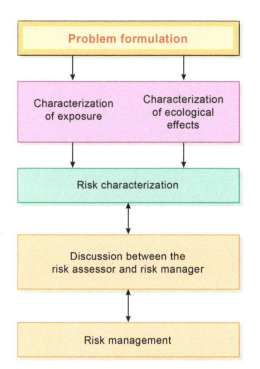

FIG. 29.8 Framework for ecological risk assessment. *(Adapted from United States Environmental Protection Agency (U.S. EPA): Framework for ecological risk assessment. EPA 1630/R-92/001, Washington, DC, 1992a.)*

INFORMATION BOX 29.3 Examples of a Management Goal, Assessment Endpoint, and Measures

Goal: Viable, self-sustaining coho salmon population that supports a subsistence and sport fishery.

Assessment Endpoint: Coho salmon breeding success, fry survival, and adult return rates.

Measures of Effects

- Egg and fry response to low dissolved oxygen
- Adult behavior in response to obstacles
- Spawning behavior and egg survival with changes in sedimentation

Measures of Ecosystem and Receptor Characteristics

- Water temperature, water velocity, and physical obstructions
- Abundance and distribution of suitable breeding substrate
- Abundance and distribution of suitable food sources for fry
- Feeding, resting, and breeding behavior
- Natural reproduction, growth, and mortality rates

Measures of Exposure

- Number of hydroelectric dams and associated ease of fish passage
- Toxic chemical concentrations in water, sediment, and fish tissue
- Nutrient and dissolve oxygen levels in ambient waters
- Riparian cover, sediment loading, and water temperature

FIG. 29.9 Ecological risk assessment.

strategies designed to protect the trout population (see Information Box 29.3).

Exposure assessment is a determination of the environmental concentration range of a particular stressor and the actual dose received by the *biota* (all the plants and animals) in a given area. The most common approach to exposure analysis is to measure actual concentrations of a stressor and combine these measurements with assumptions about contact and uptake by the biota. For example, the exposure of simple aquatic organisms to chemicals can often be measured simply as the concentration of that chemical in the water because the physiologic systems of these organisms are assumed to be in equilibrium with the surrounding water. Stressor measurements can also be combined with quantitative parameters describing the frequency and magnitude of contact. For example, concentrations of chemicals or microorganisms in food items can be combined with ingestion rates to estimate dietary exposure. Exposure assignment is, however, rarely straightforward. Biotransformations may occur, especially for heavy metals such as mercury. Such transformations may result in the formation of even more toxic forms of the stressor. Researchers must therefore use mathematical models to predict the fate and resultant exposure to a stressor and to determine the outcome of a variety of scenarios.

The purpose of evaluating ecological effects is to identify and quantify the adverse effects elicited by a stressor and, to the extent possible, to determine cause-and-effect relationships. During this phase, toxicity data are usually compiled and compared.

Generally, there are acute and chronic data for the stressor acting on one or several species. Field observations can provide additional data, and so can controlled-microcosm and large-scale tests.

The process of developing a stressor-response profile is complex because it inevitably requires models, assumptions, and extrapolations. For example, the relationship between measurement and assessment endpoint is an assumption. It is often expressly stated in the model used, but when it is not specifically stated, it is left to professional judgment. In addition, the stressor-response profile is analogous to a dose-response curve in the sense that it involves extrapolations; in this case, though, a single-species toxicity test is extrapolated to the community and ecosystem level. One of the difficulties in the quantification of the stressor-response profile is that many of the quantitative extrapolations are drawn from information that is qualitative in nature. For example, when we use *phylogenic extrapolation* to transfer toxicity data from one species to another species—or even to a whole class of organisms—we are assuming a degree of similarity based on qualitative characteristics. Thus, when we use green algal toxicity test data to represent all photosynthetic eukaryotes (which we often do), we must remember that all photosynthetic eukaryotes are not, in fact, green algae. Because many of the responses are extrapolations based on models ranging from the molecular to the ecosystem level, it is critically important that uncertainties and assumptions be clearly delineated.

Risk assessment consists of comparing the exposure and stressor-response profiles to estimate the probability of effects, given the distribution of the stressor within the system. As you might expect, this process is extraordinarily difficult to accomplish. In fact, our efforts at predicting adverse effects have been likened to the weather forecaster's prediction of rain (Landis and Ho-Yu, 1995). Thus the predictive process in ecological risk assessment is still very much an art form, largely dependent on professional judgment.

Conceptual model diagrams can be used to better visualize potential impacts (Fig. 29.10). They may be based on theory and logic, empirical data, mathematical models, or probability models. These diagrams are useful tools for communicating important pathways in a clear and concise way. They can be used to ask new questions about relationships that help generate plausible risk hypotheses.

29.4 MICROBIAL RISK ASSESSMENT

Outbreaks of waterborne disease caused by microorganisms usually occur when the water supply has been obviously and significantly contaminated. In such high-level cases, the exposure is manifest, and cause and effect are relatively easy to determine. However, exposure to low-level microbial contamination is difficult to determine epidemiologically. We know, for example, that long-term exposure to

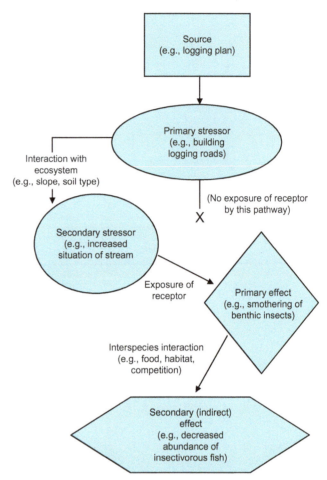

FIG. 29.10 Conceptual model for logging. *(Source: Available from: www.epa.gov.)*

microbes can have a significant impact on the health of individuals within a community, but we need a way to measure that impact.

For some time, methods have been available to detect the presence of low levels (1 organism per 1000 L) of pathogenic organisms in water, including enteric viruses, bacteria, and protozoan parasites. The trouble is that the risks posed to the community by these low levels of pathogens in a water supply over time are not like those posed by low levels of chemical toxins or carcinogens. For example, it takes just one amoeba in the wrong place at the wrong time to infect one individual, whereas that same individual would have to consume some quantity of a toxic chemical to be comparably harmed. Microbial risk assessment is therefore a process that allows us to estimate responses in terms of the *risk of infection* in a quantitative fashion. Microbial risk generally follows the steps used in other health-based risk assessments—hazard identification, exposure assessment, dose-response, and risk characterization. The differences are in the specific assumptions, models, and extrapolation methods used.

Hazard identification in the case of pathogens is complicated because several outcomes—from asymptomatic

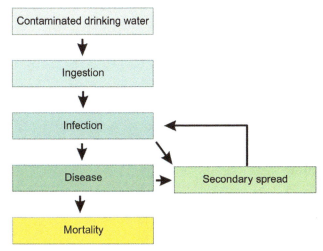

FIG. 29.11 Outcomes of enteric viral exposure. *(From Pollution Science Academic Press, San Diego, CA, 1996.)*

infection to death (see Fig. 29.11)—are possible, and these outcomes depend upon the complex interaction between the pathogenic agent (the "infector") and the host (the "infectee"). This interaction, in turn, depends on the -characteristics of the host as well as the nature of the pathogen. Host factors, for example, include preexisting immunity, age, nutrition, ability to mount an immune response, and other nonspecific host factors. Agent factors include type and strain of the organism as well as its capacity to elicit an immune response.

Among the various outcomes of infection is the possibility of *subclinical infection.* Subclinical (asymptomatic) infections are those in which the infection (growth of the microorganism within the human body) results in no obvious illness such as fever, headache, or diarrhea. That is, individuals can host a pathogen microorganism—and transmit it to others—without ever getting sick themselves. The ratio of clinical to subclinical infection varies from pathogen to pathogen, especially in viruses, as presented in Table 29.10. Poliovirus infections, for instance, seldom result in obvious clinical symptoms; in fact, the proportion of individuals developing clinical illness may be less than 1%. However, other enteroviruses, such as the coxsackie viruses, may exhibit a greater proportion. In many cases, as in that of rotaviruses, the probability of developing clinical illness appears to be completely unrelated to the dose an individual receives via ingestion. Rather, the likelihood of developing clinical illness depends upon the type and strain of the virus as well as host age, nonspecific host factors, and possibly preexisting immunity. The incidence of clinical infection can also vary from year to year for the same virus, depending on the emergence of new strains.

Another outcome of infection is the development of clinical illness. Several host factors play a major role in this outcome. The age of the host is often a determining factor.

In the case of hepatitis A, for example, clinical illness can vary from about 5% in children less than 5 years of age to 75% in adults. Similarly, children are more likely to develop rotaviral gastroenteritis than are adults. Immunity is also an important factor, albeit a variable one. That is, immunity may or may not provide long-term protection from reinfection, depending on the enteric pathogen. It does not, for example, provide long-term protection against the development of clinical illness in the case of the norovirus or *Giardia.* However, for most enteroviruses and for the hepatitis A virus, immunity from reinfection is believed to be lifelong. Other undefined host factors may also control the odds of developing illness. For example, in experiments with norovirus, human volunteers who did not become infected upon an initial exposure to the virus also did not respond to a second exposure. In contrast, those volunteers who developed gastroenteritis upon the first exposure also developed illness after the second exposure.

The ultimate outcome of infection—mortality—can be caused by nearly all enteric organisms. The factors that control the prospect of mortality are largely the same factors that control the development of clinical illness. Host age, for example, is significant. Thus mortality for hepatitis A and poliovirus is greater in adults than in children. In general, however, one can say that the very young, the elderly, and the immunocompromised are at the greatest risk of a fatal outcome of most illnesses (Gerba et al., 1996). For example, the case-fatality rate (%) for *Salmonella* in the general population is 0.1%, but it has been observed to be as high as 3.8% in nursing homes (Table 29.11). In North America and Europe, the reported case-fatality rates (i.e., the ratio of cases to fatalities reported as a percentage of persons who die) for enterovirus infections range from less than 0.1% to 0.94%, as presented in Table 29.12. The case-fatality rate for common enteric bacteria ranges from 0.1% to 0.2% in the general population. Enteric bacterial diseases can be treated with antibiotics, but no treatment is available for enteric viruses.

Recognizing that microbial risk involves a myriad of pathogenic organisms capable of producing a variety of outcomes that depend on a number of factors—many of which are undefined—one must now face the problem of exposure assessment, which has complications of its own. Unlike chemical-contaminated water, microorganism-contaminated water does not have to be consumed to cause harm. That is, individuals who do not actually drink, or even touch, contaminated water also risk infection because pathogens—particularly viruses—may be spread by person-to-person contact or subsequent contact with contaminated inanimate objects (such as toys). This phenomenon is described as the *secondary attack rate,* which is reported as a percentage. For example, one person infected with poliovirus can transmit it to 90% of the persons with whom he or she associates. This secondary spread of viruses has been well documented

TABLE 29.10 Ratio of Clinical to Subclinical Infections With Enteric Viruses

Virus	Frequency of Clinical Illness[a] (%)
Poliovirus 1	0.1–1
Coxsackie	
A16	50
B2	11–50
B3	29–96
B4	30–70
B5	5–40
Echovirus	
Overall	50
9	15–60
18	Rare–20
20	33
25	30
30	50
Hepatitis A (adults)	75
Rotavirus	
(Adults)	56–60
(Children)	28
Astrovirus (adults)	12–50

[a]The percentage of the individuals infected who develop clinical illness.

From Gerba, C.P., Rose, J.B., 1992. Comparative Environmental Risks. Lewis Publishers, Boca Raton, FL. From Pollution Science Academic Press, San Diego, CA, 1996.

TABLE 29.12 Case-Fatality Rates for Enteric Viruses and Bacteria

Organism	Case-Fatality Rate (%)
Viruses	
Poliovirus 1	0.90
Coxsackie	
A2	0.50
A4	0.50
A9	0.26
A15	0.12
Coxsackie B	0.59–0.94
Echovirus	
6	0.29
9	0.27
Hepatitis A	0.30
Rotavirus	
(Total)	0.01
(Hospitalized)	0.12
Norwalk	0.0001
Astrovirus	0.01
Bacteria	
Shigella	0.2
Salmonella	0.1
Escherichia coli 0157:H7	0.2
Campylobacter jejuni	0.1

From Gerba and Rose (1993) and Gerba, C.P., Rose, J.B., Haas, C.N., 1996. Sensitive populations: who is at the greatest risk? Int. J. Food Microbiol. 301, 113–123. From Pollution Science Academic Press, San Diego, CA, 1996.

TABLE 29.11 Case Fatality Observed for Enteric Pathogens in Nursing Homes Versus General Population

Organism	Case Fatality (%) in General Population	Case Fatality (%) in Nursing Homes
Campylobacter jejuni	0.1	1.1
Escherichia coli 0157:H7	0.2	11.8
Salmonella	0.1	3.8
Rotavirus	0.01	1.0

Modified from Gerba, C.P., Rose, J.B., Haas, C.N., 1996. Sensitive populations: who is at the greatest risk? Int. J. Food Microbiol. 301, 113–123.

for waterborne outbreaks of several diseases, including that caused by norovirus, whose secondary attack rate is about 30%.

The question of dose is another problem in exposure assessment. How does one define "dose" in this context? To answer this question, researchers have conducted a number of studies to determine the infectious dose of enteric microorganisms in human volunteers. Such human experimentation is necessary because determination of the infectious dose in animals and extrapolation to humans is often impossible. In some cases, for example, humans are the primary or only known host. In other cases, such as that of *Shigella* or norovirus, infection can be induced in laboratory-held primates, but it is not known whether the infectious dose data can be extrapolated to humans. Much of the existing data on infectious doses of viruses has been obtained with attenuated vaccine viruses or with avirulent laboratory-grown strains, so that the likelihood of serious illness is minimized. An example of a dose-response curve for a human feeding study with rotavirus is shown in Fig. 29.12.

In the microbiological literature, the term *minimum infectious dose* is used frequently, implying that a threshold dose exists for microorganisms. In reality, the term used usually refers to the ID_{50} dose at which 50% of the animals or humans exposed became infected or exhibit any symptoms of an illness. Existing infectious dose data are compatible with nonthreshold responses, and the term "infectivity" is probably more appropriate when referring to differences in the likelihood of an organism causing an infection. For example, the probability of a given number of ingested rotaviruses causing diarrhea is greater than that for *Salmonella*. Thus the infectivity of rotavirus is greater than that of *Salmonella*.

Next, one must choose a dose-response model, whose abscissa is the dose and whose ordinate is the risk of infection (see Fig. 29.12). The choice of model is critical so that risks are not greatly overestimated or underestimated. A modified exponential (beta-Poisson distribution) or a logprobit (simple lognormal, or exponential, distribution) model may be used to describe the probability of infection in human subjects for many enteric microorganisms (Haas, 1983). These models have been found to best fit experimental data. For the beta-Poisson model, the probability of infection from a single exposure, P, can be described as follows:

$$P = 1 - (1 + N/\beta)^{-\alpha} \quad (29.10)$$

where N is the number of organisms ingested per exposure and α and β represent parameters characterizing the host-virus interaction (dose-response curve). Some values for α and β for several enteric waterborne pathogens are presented in Table 29.13; these values were determined from human studies. For some microorganisms, an *exponential model* may better represent the probability of infection.

$$P = 1 - \exp(-rN) \quad (29.11)$$

In this equation, r is the fraction of the ingested microorganisms that survive to initiate infections (host-microorganism interaction probability). Table 29.13 shows examples of results of both models for several organisms.

These models define the probability of the microorganisms overcoming the host defenses (including stomach pH, finding a susceptible cell, nonspecific immunity, and so on) to establish an infection in the host. When one uses these models, one estimates the probability of becoming infected after ingestion of various concentrations of pathogens. For example, Example Calculation 29.3 shows how to calculate the risk of acquiring a viral infection from consumption of contaminated oysters containing echovirus 12 using Eq. (29.10).

Annual and lifetime risks can also be determined, again assuming a Poisson distribution of the virus in the material consumed (assuming daily exposure to a constant concentration of viral contamination), as follows:

$$P_A = 1 - (1 - P)^{365} \quad (29.12)$$

where P_A is the annual risk (365 days) of contracting one or more infections, and

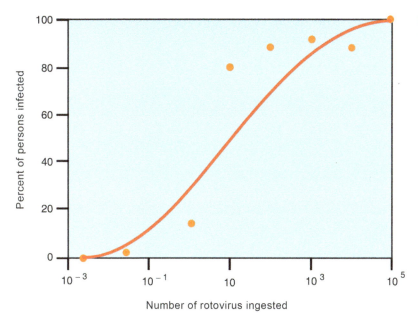

FIG. 29.12 Dose-response for human rotavirus by oral ingestion.

TABLE 29.13 Best-Fit Dose-Response Parameters for Enteric Pathogen Ingestion Studies

Microorganism	Best Model	Model Parameters
Echovirus 12	Beta-Poisson	$\alpha = 0.374$
		$\beta = 186,69$
Rotavirus	Beta-Poisson	$\alpha = 0.26$
		$\beta = 0.42$
Poliovirus 1	Exponential	$r = 0.009102$
Poliovirus 1	Beta-Poisson	$\alpha = 0.1097$
		$\beta = 1524$
Poliovirus 3	Beta-Poisson	$\alpha = 0.409$
		$\beta = 0.788$
Cryptosporidium	Exponential	$r = 0.004191$
Giardia lamblia	Exponential	$r = 0.02$
Salmonella	Exponential	$r = 0.00752$
Escherichia coli	Beta-Poisson	$\alpha = 0.1705$
		$\beta = 1.61 \times 10^6$

Modified from Regli, S., Rose, J.B., Haas, C.N., Gerba, C.P., 1991. Modeling the risk from Giardia and viruses in drinking water. J. Am. Water Works Assoc. 83, 76–84. From Pollution Science Academic Press, San Diego, CA, 1996.

Example Calculation 29.3 Application of a Virus Risk Model to Characterize Risks from Consuming Shellfish

It is well known that infectious hepatitis and viral gastroenteritis are caused by consumption of raw or, in some cases, cooked clams and oysters. The concentration of echovirus 12 was found to be 8 plaque-forming units per 100 g in oysters collected from coastal New England waters. What are the risks of becoming infected and ill from echovirus 12 if the oysters are consumed? Assume that a person usually consumes 60 g of oyster meat in a single serving:

$$\frac{8\,PFU}{100\ g} = \frac{N}{60\ g} \quad N = 4.8\ PFU\ consumed$$

From Table 29.13, $\alpha = 0.374$, $\beta = 186.64$. The probability of infection from Eq. (29.10) is then

$$P = 1 - \left(1 + \frac{4.8}{186.69}\right)^{-0.374} = 9.4 \times 10^{-3}$$

If the percent of infections that result in risk of clinical illness is 50%, then from Eq. (29.14) one can calculate the risk of clinical illness:

Risk of clinical illness $= \left(9.4 \times 10^{-3}\right)(0.50) = 4.7 \times 10^{-3}$.

If the case-fatality rate is 0.001%, then from Eq. (29.15)

Risk of mortality $= \left(9.4 \times 10^{-3}\right)(0.50)(0.001) = 4.7 \times 10^{-6}$

If a person consumes oysters 10 times a year with 4.8 PFU per serving, then one can calculate the risk of infection in one year from Eq. (29.12)

Annual risk $= P_A = 1 - \left(1 - 9.4 \times 10^{-3}\right)^{365} = 9.7 \times 10^{-1}$

$$P_L = 1 - (1-P)^{25,550} \tag{29.13}$$

where P_L is the lifetime risk (assuming a lifetime of 70 years $= 25,550$ days) of contracting one or more infections.

Risks of clinical illness and mortality can then be determined by incorporating terms for the percentage of clinical illness and mortality associated with each particular virus:

Risk of clinical illness $= PI \tag{29.14}$

Risk of mortality $= PIM \tag{29.15}$

where I is the percentage of infections that result in clinical illness and M is the percentage of clinical cases that result in mortality.

Application of this model allows estimation of the risks of infection, development of clinical illness, and mortality for different levels of exposure. As presented in Table 29.14,

TABLE 29.14 Risk of Infection, Disease, and Mortality for Rotavirus

Virus Concentration	Risk	
Per 100 L	Daily	Annual
Infection		
100	9.6×10^{-2}	1.0
1	1.2×10^{-3}	3.6×10^{-1}
0.1	1.2×10^{-4}	4.4×10^{-2}
Disease		
100	5.3×10^{-2}	5.3×10^{-1}
1	6.6×10^{-4}	2.0×10^{-1}
0.1	6.6×10^{-5}	2.5×10^{-2}
Mortality		
100	5.3×10^{-6}	5.3×10^{-5}
1	6.6×10^{-8}	2.0×10^{-5}
0.1	6.6×10^{-9}	2.5×10^{-6}

Modified from Gerba, C.P., Rose, J.B., 1992. Comparative environmental risks. Lewis Publishers, Boca Raton, FL. From Pollution Science © 1996, Academic Press, San Diego, CA.

for example, the estimated risk of infection from 1 rotavirus in 100 L of drinking water (assuming ingestion of 2 L per day) is 1.2×10^{-3}, or almost 1 in 1000 for a single-day exposure. This risk would increase to 3.6×10^{-1}, or approximately one in three, on an annual basis. As can be seen from this table, the risk of developing a clinical illness also appears to be significant for exposure to low levels of rotavirus in drinking water. See Example Calculation 29.4.

The EPA recommends that any drinking water treatment process should be designed to ensure than human populations are not subjected to risk of infection greater than 1:10,000 for a yearly exposure. To achieve this goal, it would appear from the data presented in Table 29.10 that the virus concentration in drinking water would have to be less than 1 per 1000 L. Thus if the average concentration of enteric viruses in untreated water is 1400/1000 L, treatment plants should be designed to remove at least 99.99% of the virus present in the raw water. A further application of this approach is to define the required treatment of a water source in terms of the concentration of a disease-causing organism in that supply. Thus the more contaminated the raw water source, the more treatment is required to reduce the risk to an acceptable level. An example of this application is shown in Fig. 29.13. The plausibility of validation of microbial risk assessment models has been examined by using data from foodborne outbreaks in which information has been available on exposure and outcomes (Rose et al., 1995; Crockett et al., 1996). These studies suggest that microbial risk assessment can give reasonable estimates of illness from exposure to contaminated foods (Table 29.15).

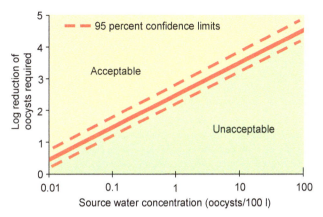

FIG. 29.13 Relationship of influent *Cryptosporidium* concentration and log reduction by treatment necessary to produce acceptable water. *(From Haas et al., 1996. Reprinted from J. AWWA 88(9) (September 1996), by permission. Copyright © 1996, American Water Works Association.)*

In summary, risk assessment is a critical tool for decision making in the regulatory arena. This approach is used to explain chemical and microbial risks, as well as ecosystem impacts. The results of such assessments can be used to inform risk managers of the probability and extent of environmental impacts resulting from exposure to different levels of stress (contaminants). Moreover, this process, which allows the quantification and comparison of diverse risks, lets risk managers use the maximum amount of complex information in the decision-making process. This information can also be used to weigh the costs and benefits of control options and to develop standards or treatment options (see Information Box 29.4).

Example Calculation 29.4 Risk Assessment for Rotavirus in Drinking Water

Pathogen identified	Rotavirus
↓	↓
Dose-Response Model (based on human ingestion studies)	Best fit for data is the Beta Poisson Model $P = (1 + N/\beta)^{-\alpha}$ $\alpha = 0.2631$ $\beta = 0.42$
↓	↓
Exposure (field studies on concentration in drinking water)	4 Rotavirus/1000 L
↓	↓
Risk Characterization	Risk of Infection Assumes: 2 L/day of drinking water ingested. Thus $N = 0.008$/day Risk of Infection/day = 1:200 Risk of Infection/year $P_A = 1 - (1 - P)^{365}$ $P_A = 1 : 2$

TABLE 29.15 Comparison of Outbreak Data to Model Predictions for Assessment of Risks Associated With Exposure to *Salmonella*

Food	Dose CFU	Amount Consumed	Attack Rate (%)	Predicted P (%)
Water	17	1 L	12	12
Pancretin	200	7 doses	100	77
Ice cream	102	1 portion	52	54
Cheese	100–500	28 g	28–36	53–98
Cheese	10^5	100 g	100	>99.99
Ham	10^6	50–100 g	100	>99.99

Source: Rose, J.B., Haas, C.N., Gerba, C.P., 1995. Linking microbiological criteria for foods with quantitative risk assessment. J. Food Safety 15, 111–132. From Environmental Microbiology, Academic Press, San Diego, CA, 2000.

INFORMATION BOX 29.4 How Do We Set Standards for Pathogens in Drinking Water?

In 1974, the U.S. Congress passed the Safe Drinking Water Act, giving the U.S. Environmental Protection Agency (EPA) the authority to establish standards for contaminants in drinking water. Through a risk analysis approach, standards have been set for many chemical contaminants in drinking water. Setting standards for microbial contaminants proved more difficult because (1) methods for the detection of many pathogens are not available, (2) days to weeks are sometimes required to obtain results, and (3) costly and time-consuming methods are required. To overcome these difficulties, coliform bacteria had been used historically to assess the microbial quality of drinking water. However, by the 1980s it had become quite clear that coliform bacteria did not indicate the presence of pathogenic waterborne *Giardia* or enteric viruses. Numerous outbreaks had occurred in which coliform standards were met, because of the greater resistance of viruses and *Giardia* to disinfection. A new approach was needed to ensure the microbial safety of drinking water.

To achieve this goal a new treatment approach was developed called the Surface Treatment Rule (STR). As part of the STR, all water utilities that use surface waters as their source of potable water would be required to provide filtration to remove *Giardia* and enough disinfection to kill viruses. The problem facing the EPA was how much removal should be required. To deal with this issue, the EPA for the first time used a microbial risk assessment approach. The STR established that the goal of treatment was to ensure that microbial illness from *Giardia lamblia* infection should not be any greater than 1 per 10,000 exposed persons annually (10^{-4} per year). This value is close to the annual risk of infection from waterborne disease outbreaks in the United States. Based on the estimated concentration of *Giardia* and enteric viruses in surface waters in the United States from the data available at the time, it was required that all drinking water treatment plants be capable of removing 99.9% of the *Giardia* and 99.99% of the viruses. In this manner it was hoped that the risk of infection of 10^{-4} per year would be achieved. The STR went into effect in 1991.

To better assess whether the degree of treatment required is adequate, the EPA developed the Information Collection Rule, which required major drinking water utilities that use surface waters to analyze these surface water for the presence of *Giardia, Cryptosporidium,* and enteric viruses for a period of almost 2 years. From this information, the EPA set treatment control requirements to ensure that the 10^{-4} yearly risk is met. Utilities that have heavily contaminated source water are required to achieve greater levels of treatment (see Fig. 29.13).

QUESTIONS AND PROBLEMS

1. List the four steps in a formal risk assessment.
2. Why do we use safety factors in risk assessment?
3. What is the most conservative dose-response curve? What does it mean?
4. What is the difference between risk assessment and risk management?
5. What are some of the differences between the risks posed by chemicals and those posed by microorganisms?
6. Suppose a 50-kg individual drinks $2\,L\,day^{-1}$ of chloroform and $0.1\,mg\,L^{-1}$ phenol. What is the hazard index? Is there cause for concern?
7. Estimate the cancer risk for a 70-kg individual consuming 1.5 L of water containing trichloroethylene (TCE) per day for 70 days.
8. Calculate the risk of infection from rotavirus during swimming in polluted water. Assume 30 mL of water is ingested during swimming and the concentration of rotavirus was 1 per 100 L. What would the risk be in a year if a person went swimming 5 times and 10 times in the same water with the same concentration of virus?
9. What is a NOAEL and how does it differ from a LOAEL?
10. If 10 oocysts of *Cryptosporidium* are detected in 100 L of surface water, how much reduction (in $\log_{10}$) by a water treatment plant is required to achieve a 1:10,000 annual risk of infection?
11. Give an example of a nonthreshold response for a chemical toxin.
12. What is the difference between a stressor and a receptor? Give an example of a chemical stressor and a receptor in an aquatic system. What endpoint would you use?
13. Draw an exposure pathway for pathogens for the disposal of raw sewage into the ocean. Consider likely routes of ingestions and inhalation. As a risk manager, what options may you have to reduce the risks of exposure?
14. Using the U.S. Environmental Protection Agency IRIS database (www.epa.gov/iris), find the critical effect, uncertainty factor, and NOAEL, LOAEL, and RfD for mercury, chromium, and chloroform. In drinking water, which one would be the most toxic?

REFERENCES

Crockett, C.S., Haas, C.N., Fazil, A., Rose, J.B., Gerba, C.P., 1996. Prevalence of shigellosis: consistency with dose-response information. Int. J. Food Prot. 30, 87–99.

Gerba, C.P., Rose, J.B., Haas, C.N., 1996. Sensitive populations: who is at the greatest risk? Int. J. Food Microbiol. 301, 113–123.

Haas, C.N., 1983. Estimation of risk due to low levels of microorganisms: a comparison of alternative methodologies. Am. J. Epidemiol. 118, 573–582.

Landis, W.G., Ho-Yu, M.H., 1995. Introduction to Environmental Toxicology. Lewis Publishers, Boca Raton, FL.

Rose, J.B., Haas, C.N., Gerba, C.P., 1995. Linking microbiological criteria for foods with quantitative risk assessment. J Food Safety 15, 111–132.

Tseng, W.P., 1977. Effects and dose response relationships of skin cancer and blackfoot disease with arsenic. Environ. Health Perspect. 19, 109–119.

Tseng, W.P., Chu, H.M., How, S.W., Fong, J.M., Lin, C.S., Yeh, S., 1968. Prevalence of skin cancer in an endemic area of chronic arsenic in Taiwan. J. Natl. Cancer Inst. 40, 453–463.

United States Environmental Protection Agency. (U.S. EPA), (Septermber 1986) Guidelines for carcinogen risk assessment. Federal Reg. 24.

FURTHER READING

Cockerham, L.G., Shane, B.S., 1994. Basic Environmental Toxicology. CRC Press, Boca Raton, FL.

Covello, V., von Winterfieldt, D., Slovic, P., 1986. Risk communication: a review of the literature. Risk Anal. 3, 171–182.

Gerba, C.P., Rose, J.B., 1992. Estimating viral disease risk from drinking water. In: Cothern, C.R. (Ed.), Comparative Environmental Risks. Lewis Publishers, Boca Raton, FL.

Gerba, C.P., Rose, J.B., Haas, C.N., Crabtree, K.D., 1997. Waterborne rotavirus: a risk assessment. Water Res. 12, 2929–2940.

Haas, C.N., Crockett, C.S., Rose, J.B., Gerba, C.P., Fazil, A.M., 1996. Assessing the risk posed by oocysts in drinking water. J. Am. Water Works Assoc. 88 (9), 131–136.

Haas, C.N., Rose, J.B., Gerba, C.P., 1999. Quantitative Microbial Risk Assessment. John Wiley & Sons, New York.

ILSI Risk Science Institute Pathogen Risk Assessment Working Group, 1996. A conceptual framework to assess the risks of human disease following exposure to pathogens. Risk Anal. 16, 841–848.

Kammen D.M. and Hassenzahl D.M. Should We Risk It? Princeton University Press, Princeton, NJ.

Kolluru, R.V., 1993. Environmental Strategies Handbook. McGraw-Hill, New York.

National Research Council, 1983. Risk Assessment in the Federal Government: Managing the Process. National Academy Press, Washington, DC.

National Research Council, 1989. Improving Risk Communication. National Academy Press, Washington, DC.

National Research Council, 1991. Frontiers in Assessing Human Exposure. National Academy Press, Washington, DC.

Regli, S., Rose, J.B., Haas, C.N., Gerba, C.P., 1991. Modeling the risk from Giardia and viruses in drinking water. J. Am. Water Works Assoc. 83, 76–84.

Rodricks, J.V., 1992. Calculated Risks. Understanding the Toxicity and Human Health Risks of Chemicals in Our Environment. Cambridge University Press, Cambridge, England.

Roseberry, A.M., Burmaster, D.E., 1992. Lognormal distributions for water intake by children and adults. Risk Anal. 12, 99–104.

Straub, T.M., Pepper, I.L., Gerba, C.P., 1993. Hazards from pathogenic microorganisms in land-disposed sewage sludge. Rev. Environ. Contam. Toxicol. 132, 55–91.

United States Environmental Protection Agency (U.S. EPA), (1990) Risk assessment, management and communication of drinking water contamination. EPA/625/4–89/024, Washington, DC.

United States Environmental Protection Agency (U.S. EPA), (1992a) Framework for ecological risk assessment. EPA 1630/R–92/001, Washington, DC.

United States Environmental Protection Agency (U.S. EPA), (1992b) Dermal exposure assessment: principles and applications. EPA 600/8-91/011B, Washington, DC.

United States Environmental Protection Agency (U.S. EPA), (2016) Guidelines for human exposure assessment. Draft. Available from: https://www.epa.gov/sites/production/files/2016-02/documents/guidelines_for_human_exposure_assessment_peer_review_draftv2.pdf.

Ward, R.L., Berstein, D.I., Young, E.C., 1986. Human rotavirus studies in volunteers of infectious dose and serological response to infection. J. Infect. Dis. 154, 871–877.

Watts, R.J., 1998. Hazardous Wastes: Sources, Pathways, Receptors. John Wiley & Sons, New York.

Wilson, R., Crouch, E.A.C., 1987. Risk assessment and comparisons: an introduction. Science 236, 267–270.

Chapter 30

Environmental Laws and Regulations

C.P. Gerba and M.L. Brusseau

The U.S. Congress frequently imposes federal laws and regulations that affect the environment.

30.1 REGULATORY OVERVIEW

In the United States, most environmental legislation consists of federal legislation that has been enacted within the last 40 years (Table 30.1). In addition, there are many state environmental laws and regulations that have been patterned after—and work together with—the federal programs. Federal legislation is, however, the basis for the development of regulations designed to protect our air, water, and food supply, and to control pollutant discharge. Enactment of new legislation empowers the executive branch to develop environment-specific regulations and to implement their enforcement. Among other things, regulations may involve the development of standards for waste discharge, requirements for the cleanup of polluted sites, or guidelines for waste disposal. The Environmental Protection Agency (EPA) is the federal agency responsible for development of federal environmental regulations and enforcement.

When a new environmental law is approved by Congress and the president, it is called an *Act* (Fig. 30.1), and the text of the Act is known as a public statute. Some of the better known laws related to the environment are the

Clean Air Act, the Clean Water Act, and the Safe Drinking Water Act. Once the Act is passed, the House of Representatives standardizes the text of the law and publishes it in the United States Code. The U.S. Code is the official record of all federal laws. After the law becomes official, how is it put into practice? Laws often do not include all the details. The U.S. Code would not tell you, for example, what the speed limit is in front of your house. In order to make the laws work on a day-to-day level, Congress authorizes certain government agencies—including the Environmental Protection Agency (EPA)—to create regulations. Regulations set specific rules about what is legal and what is not. For example, a regulation issued by EPA to implement the Clean Air Act might state what levels of a pollutant—such as sulfur dioxide—are safe. It would tell industries how much sulfur dioxide they can legally emit into the air, and what the penalty would be if they emit too much. Once the law is in effect, it is the responsibility of the EPA to enforce it.

During the development of regulations, the agency conducts research and then publishes a proposed regulation in the Federal Register so that members of the public can consider it and send their comments to the agency. The agency considers all the comments, revises the regulation, and issues a final *Rule*. At each stage in the process, the agency publishes a notice in the Federal Register. Once a regulation is completed and has been printed in the Federal Register as a final rule, it is "codified" by being published in the *Code of Federal Regulations* (CFR) (Fig. 30.1). The CFR is the official record of all regulations created by the federal government. It is divided into 50 volumes, called titles, each of which focuses on a particular area. Almost all environmental regulations appear in Title 40.

30.2 NATIONAL ENVIRONMENTAL POLICY ACT

This Act was passed in 1970 and requires federal agencies to assess the environmental effects and related social and economic effects of their proposed actions prior to making decisions. The range of actions covered by NEPA is broad and includes: (1) activities of Federal agencies, and (2) any

Environmental and Pollution Science. https://doi.org/10.1016/B978-0-12-814719-1.00030-6

TABLE 30.1 Scope of Federal Regulations Governing Environmental Pollution

Federal Regulations	Purpose/Scope
Policy	
National Environmental Policy Act (NEPA). Enacted 1970	This act declares a national policy to promote efforts to prevent or eliminate damage to the environment. It requires federal agencies to assess environmental impacts of implementing major programs and actions early in the planning stage
Pollution Prevention Act of 1990	The basic objective is to prevent or reduce pollution at the source instead of an end-of-pipe control approach
Water	
Clean Water Act. Enacted 1948	Eliminates discharge of pollutants into navigable waters. It is the prime authority for water pollution control programs
1977 Amendment of the Clean Water Act	Covers regulation of sludge application by federal and state government
Safe Drinking Water Act (SDWA). Enacted 1974	Protects sources of drinking water and regulates using proper water treatment techniques using drinking water standards based on maximum contaminant levels (MCLs)
Clean air	
Clean Air Act. Enacted 1970	This act, which amended the Air Quality Act of 1967, is intended to protect and enhance the quality of air sources. Sets a goal for compliance with ambient air quality standards
Clean Air Amendments of 1977	Purpose is to define issues to prevent industries from benefiting economically from noncompliance
Clear Air Amendments of 1990	Basic objectives of Clean Air Act of 1990 are to address acid precipitation and power plant emissions
Hazardous substances	
Comprehensive Environmental Response, Compensation, and Liability Act (CERCLA). Enacted 1980	The act, known as Superfund, provides an enforcement agency the authority to respond to releases of hazardous wastes. The act amends the Solid Waste Disposal Act. Note: also affects water resources
Superfund Amendments and Reauthorization Act (SARA). Enacted 1986	This act revises and extends CERCLA by the addition of new authorities known as the Emergency Planning and Community Right-to-Know Act of 1986. Involves toxic chemical recall reporting. Note: also affects water resources
Toxic Substances Control Act. Enacted in 1976	This act sets up the toxic substances program administered by the EPA. The act also regulates labeling and disposal of polychlorinated biphenyls (PCBs)
Amendment of 1986	Addresses issues of inspection and removal of asbestos
Resource Conservation and Recovery Act (RCRA). Enacted in 1976. Amended in 1984	This completely revised the Solid Waste Disposal Act. As it exists now, it is a culmination of legislation dating back to the passage of the Solid Waste Disposal Act of 1965. Defines hazardous wastes. It requires tracking of hazardous waste. Regulates facilities that burn wastes and oils in boilers and industrial furnaces. Requires inventory of hazardous waste sites
Federal Insecticide, Fungicide, and Rodenticide Act (FIFRA). Enacted 1947	Regulates the use and safety of pesticide products
FIFRA Amendments of 1972	Intended to ensure that the environmental harm does not outweigh the benefits

major project, whether on a federal, state, or local level, that involves federal funding, or work performed by the federal government, or permits issued by a federal agency. This means that the Act covers not only government-initiated projects, but also any action that is entirely funded and managed by private sector entities where a federal permit is required. Because each and every project may have an impact on land, air, water, wildlife, or humans, the Act covers essentially any major project irrespective of the origin.

Bill Proposed

↓

Passed by Congress

↓

Approved by the President

↓

U.S. House of
Representatives standardizes
the text of the law (Act)

↓

Published in the United
States Code, the official
record of all federal laws

↓

Environmental Protection
Agency develops a regulation

↓

Proposed Regulation published
in the Federal Register
CRF Title 40: Protection
of the Environment

↓

Public comments
reviewed

↓

Final Rule published in the
Federal Register

FIG. 30.1 The process for the development of federal regulations.

This Act originated the Environmental Assessment (EA), whose purpose is to determine the significance of the proposed project's environmental outcomes and to look at alternative approaches. An objective of the EA is to provide sufficient evidence and analysis for determining whether it is necessary to prepare an EIS (Environmental Impact Statement) and to aid an agency's compliance with NEPA when no EIS is necessary. The purpose of an EIS is to help public officials make informed decisions based on the relevant environmental consequences and the alternatives available for a proposed project. Through the requirements for EAs and EISs, this Act has a critical influence on the development and redevelopment of land.

30.3 THE SAFE DRINKING WATER ACT

In 1974, the passage of the *Safe Drinking Water Act* (SDWA) gave the federal government overall authority for the protection of public drinking water supplies. Previous to that time, the individual states had primary authority for development and enforcement of standards. Under this authority, specific standards were promulgated for contaminant concentrations and minimum water treatment. *Maximum contaminant levels* (MCL) or *maximum contaminant level goals* (MCLG) were set for specific contaminants in drinking water (see Chapter 12). Whereas an MCL is an achievable, required level, an MCLG is a desired goal, which may or may not be achievable. For example, the MCLG for enteric viruses in drinking water is zero, since even one ingested virus may cause illness. We cannot always reach zero, but we are obliged to try. The standards must be met for those constituents on the Primary standards list. In contrast, the standards are not mandatory for those constituents listed as Secondary contaminants.

This Act covers all public water supply systems that serve more than 24 people or have more than 14 individual user connections. Thus the Act does not cover very small systems. It also does not cover private wells. It is the responsibility of the owners of small systems and private wells to have the water tested.

Within the provisions of the SDWA, there are a number of specific rules such as the *Surface Treatment Rule*, which requires all water utilities in the United States to provide filtration and disinfection to control waterborne disease from *Giardia* and enteric viruses. These rules are developed through a process that allows input from the regulated community, special interest groups, the general public, and the scientific community. This process is outlined in Fig. 30.2.

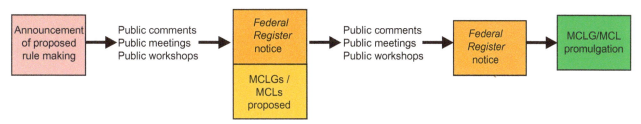

FIG. 30.2 Regulatory development processes under the Safe Drinking Water Act. A series of steps are followed in the development of new regulations for contaminants in drinking water or processes for their control. These steps usually involve a series of public notices in the publication the *Federal Register* and meetings to allow for comment from the public, environmental groups, and the regulated industry before the final regulation is developed.

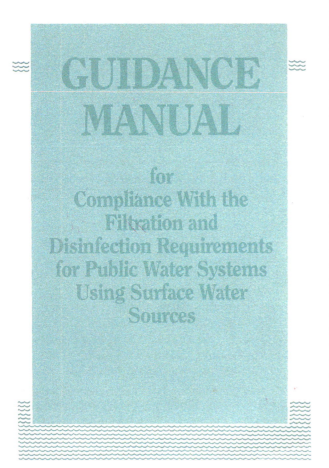

FIG. 30.3 Example of a guidance manual designed to aid water utilities to meet treatment rules under the Safe Drinking Water Act. *(From EPA.)*

To aid utilities, the EPA has published guidance manuals on the types of treatment that have been developed (Fig. 30.3).

Another key provision of the SDWA is the Unregulated Contaminant Monitoring Rule (UCMR) focused on emerging contaminants (ECs) in drinking water. The issue of emerging contaminants was discussed in Chapter 12. The EPA is required to routinely identify and analyze emerging contaminants and provide guidance to states, local officials, and the public about the potential public health risks and acceptable contamination levels for these materials. As a part of this responsibility, the EPA must develop the Contaminant Candidate List—a list of contaminants that:

- Are not regulated by the National Primary Drinking Water Regulations
- Are known or anticipated to occur in public water systems

- May warrant regulation under the SDWA due to potential toxicity to humans

Other key sections of the SDWA provide for the establishment of state programs to enforce the regulations. Individual states are responsible for developing their own regulatory programs, which are then submitted to the EPA for approval. The state programs must set drinking water standards equal to, or more stringent than, the federal standards. Such programs must also issue permits to facilities that treat drinking water supplies and develop wellhead protection areas for groundwater sources of potable water.

30.4 THE CLEAN WATER ACT

The *Clean Water Act* began with the Water Pollution Control Act of 1948 and was the first law to deal with comprehensive water pollution control. This Act was designed to protect the navigable surface waters of the United States (Fig. 30.4). The United States Supreme Court in 2006 concluded that waters covered by this Act "includes only those relatively permanent, standing or continuously flowing bodies of water forming geographic features that are described in ordinary parlance as streams[,] ... oceans, rivers, [and] lakes." Two important components comprise the Act—a permit system to control discharge of pollutants into water bodies, and a water quality standards system to preserve water quality.

The method used by the Clean Water Act to limit pollutant discharge is the permit. Anyone—public or

FIG. 30.4 The Clean Water Act was designed to help protect the recreational uses of surface waters from pollutants. *(Courtesy Wisconsin Department of Natural Resources. http://infotrek.er.usgs.gov/doc/beach/closed_04.gif.)*

private—engaged in construction or operation of a facility that may discharge anything—animal, vegetable, or mineral (or even energetic, as in the case of heat)—into navigable fresh waters must first obtain a permit. Permit applications must include a certification that the discharge meets applicable provisions of the act under the *National Pollution Discharge and Elimination Standards* (NPDES). (Permits for a discharge into ocean water are issued under separate guidelines from the EPA.) The permit specifies the total maximum daily discharge allowed for each contaminant. Note that the NPDES system does not eliminate pollutant discharge; rather, it controls discharge to levels that are set to protect human health and the environment. The *priority pollutant list* was established for the pollutants under current regulation.

Under the terms of this act, effluent standards are set and permits are issued on the basis of these standards for existing and new sources of water pollution. These are source-specific limitations. Also, the act lists categories of such point sources as sewage treatment plants, for which the EPA must issue standards of performance. The EPA may provide a list of toxic pollutants and set effluent limitations based on the best available technology economically achievable for designated point sources. In addition, the EPA may issue pretreatment standards (i.e., treatment preceding discharge of wastes into sewers) for toxic pollutants.

In addition to direct discharge (e.g., sewage treatment plants), EPA has established regulations and a permit program to regulate storm water discharges. Industries and municipalities are required to obtain storm water permits that incorporate storm water management plans and *Best Management Practices* (BMPs).

This Act established the Water Quality Standards system to protect public health and welfare, enhance the quality of water, and serve the purposes of the Clean Water Act (the Act). A water quality standard defines the water quality goals of a water body, or portion thereof, by designating the use or uses to be made of the water and by setting criteria that protect the designated uses. States are responsible for the establishment and enforcement of the water quality standards, that is, they must develop and submit to the EPA a procedure for applying and enforcing these standards.

We see that the standards serve the dual purposes of establishing the water quality goals for a specific water body and also serve as the regulatory basis for the establishment of water-quality-based treatment controls. They preserve water quality for the protection and propagation of fish, shellfish, and wildlife, for recreation in and on the water, and take into consideration their use and value as public water supplies, and for use in agricultural, industrial, and other purposes including navigation. The standards can vary as a function of the designated use. Many water bodies will have more than one designated use; in these cases, the most stringent standard will apply.

30.5 COMPREHENSIVE ENVIRONMENTAL RESPONSE, COMPENSATION, AND LIABILITY ACT

In the late 1970s, when the now-infamous Love Canal landfill in upstate New York was revealed as a major environmental catastrophe, the attendant publicity spurred Congress to pass the *Comprehensive Environmental Response, Compensation, and Liability Act* (CERCLA) of 1980 (see Fig. 30.5). This act makes owners and operators of hazardous waste disposal sites liable for cleanup costs and property damage. Transporters and producers must also bear some of the financial burden. This legislation also established a multibillion dollar cleanup fund (known as Superfund). Of that sum, 90% was to come from taxes levied on the production of oil and chemicals by U.S. industries and the rest from taxpayers. This tax has since expired and has not yet been renewed by congress. Until it is, all funds will originate from taxpayers.

CERCLA establishes *strict liability* (liability without proof of fault) for the cleanup of facilities by "responsible parties," which are those entities deemed responsible for generating the contamination. The courts have found under CERCLA that these parties are "jointly and severally" liable; that is, each and every ascertainable party is liable for the full cost of remediation, regardless of the level of "guilt" a party may have had in creating a particular polluted site.

The Superfund is exclusively dedicated to clean up sites that pose substantial threats to human health and well-being, that is, *imminent* hazards. Moreover, the Superfund provides only for the cleanup of contaminated areas and compensation for damage to property. It cannot be used to reimburse or compensate victims of illegal dumping of

FIG. 30.5 CERCLA provides authority to identify and control release of toxic substances. *(From the Office of Response and Restoration, National Ocean Service, National Oceanic and Atmospheric Administration. http://response.restoration.noaa.gov/cpr/phototour/strandley.)*

hazardous wastes for personal injury or death. Victims must take their complaints to the courts.

Details on how CERCLA is implemented to clean up hazardous waste sites are presented in Chapter 19.

30.6 FEDERAL INSECTICIDE AND RODENTICIDE ACT

Because of the potentially harmful effects of pesticides on wildlife and humans, Congress enacted the *Federal Insecticide, Fungicide, and Rodenticide Act* (FIFRA) in 1947. This act, which has been periodically amended, requires that all commercial pesticides (including disinfectants against microorganisms) be approved and registered by the EPA (Fig. 30.6).

FIFRA include several key provisions: (1) studies by the manufacturer of the risks posed by pesticides (requiring registration); (2) classification and certification of pesticides by specific use (as a way to control exposure); (3) restriction (or suspension) of the use of pesticides that are harmful to health or the environment; and (4) enforcement of the previous requirements through inspections, labeling, notices, and state regulation. For example, every pesticide container must have a label indicating that its contents have been registered with the EPA.

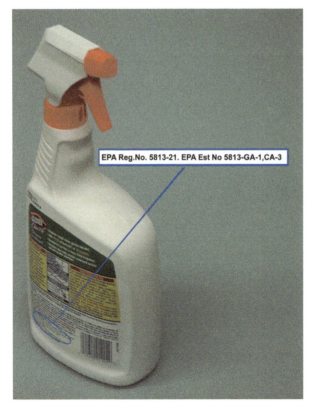

EPA Reg.No. 5813-21. EPA Est No 5813-GA-1,CA-3

FIG. 30.6 All pesticides, including disinfectants must be registered with the EPA under FIFRA. *(Photo courtesy K.L. Josephson.)*

30.7 CLEAN AIR ACT

The Clean Air Act, originally passed in 1970, required the EPA to establish *National Ambient Air Quality Standards* (NAAQS) for several outdoor pollutants, such as suspended particulate matter, sulfur dioxide, ozone, carbon monoxide, nitrogen oxides, hydrocarbons, and lead. (Of these, ozone is the most pervasive and least likely to meet the standard; see also Chapter 17.) Each standard specifies the maximum allowable level, averaged over a specific period of time. These primary ambient air quality standards are designed to protect human health. Secondary ambient air quality standards are designed to maintain visibility and to protect crops, buildings, and water supplies.

Each state must monitor the quality of its ambient air for purposes of determining attainment with the NAAQS. Each state is then divided into Air Quality Regions based on compliance with NAAQS. These regions are designated to be *Attainment Areas, Unclassifiable Areas*, and *Nonattainment Areas* for each of the six criteria pollutants. For example, an Air Quality Region can be designated as an Attainment Area for Ozone, but Nonattainment for PM_{10}. Nonattainment Areas can be further classified from Moderate to Extreme.

This act, which sets air pollution control requirements for various geographic areas of the United States, also deals with the control of tailpipe emissions for motor vehicles. Requirements compel automobile manufacturers to improve design standards to limit carbon monoxide, hydrocarbon, and nitrogen oxide emissions. For cities or areas where the ozone and carbon monoxide concentrations are high, reformulated and oxygenated gasolines are required. This act also addresses power plant emissions of sulfur dioxide and nitrogen oxide, which can generate acid rain (see Chapter 4).

30.8 RESOURCE CONSERVATION AND RECOVERY ACT (RCRA)

The Federal Solid Waste Disposal Act, as amended by the *Resource Conservation and Recovery Act* (RCRA), was adopted in 1976 to combat the problems associated with the unregulated land disposal of hazardous wastes. Its primary goals are to protect human health and the environment from hazards posed by waste disposal, conserve energy and natural resources through waste recycling and recovery, reduce or eliminate as expeditiously as possible the generation of hazardous waste, and ensure that wastes are managed in a manner that is protective of human health and the environment.

The most important feature of RCRA provides "cradle-to-grave" management of "hazardous waste" from the point of generation through transportation and its treatment, storage, and disposal. RCRA requires generators to characterize their wastes, properly manage their wastes on site, and then to manifest all wastes sent off-site for disposal. Transporters of hazardous waste are required to fill in their portion of the manifest, comply with RCRA and federal Department of Transportation placarding requirements, and must deliver the hazardous waste to an approved "Treatment, storage and disposal facility." Those facilities must meet strict performance standards and must control their air emission. They must also monitor and prevent discharges to groundwater. RCRA also prohibits land disposal of certain untreated hazardous wastes.

30.9 THE POLLUTION PREVENTION ACT

The *Pollution Prevention Act of 1990* established as national policy the following waste management hierarchy designed to prevent pollution and encourage recycling:

1. Prevention—to eliminate or reduce pollution at the source whenever feasible.
2. Recycling—to recycle unpreventable wastes in an environmentally safe manner whenever feasible.
3. Treatment—to treat unpreventable, unrecyclable wastes to applicable standards prior to release or transfer.
4. Disposal—to safely dispose of wastes that cannot be prevented, recycled, or treated.

The EPA, which is integrating pollution prevention into all its programs and activities, has developed unique voluntary reduction programs with public and private sectors.

30.10 BROWNFIELDS ACT

This Act, officially known as the Small Business Liability Relief and Brownfields Revitalization Act, was passed in 2002. A brownfield is a property whose expansion, redevelopment, or reuse may be complicated by the presence or potential presence of a hazardous substance. Examples of brownfield sites include abandoned industrial or commercial properties (manufacturing plants, service stations, dry cleaners) and closed military bases. The EPA estimates that there are approximately 450,000 brownfields in the United States. Many of these sites contain hazardous contaminants due to prior activities.

The presence of an abandoned property is often a blight for the local community. Thus local politicians and the public are often interested in redeveloping these properties for a number of reasons. These include increasing the local tax base; facilitating job growth; using existing infrastructure; taking development pressures off of undeveloped, open land; and improving environmental quality.

The act clarified CERCLA liability protections to promote redevelopment of these sites. This entailed:

- Bona Fide Prospective Purchasers—A release from liability of prospective purchasers even if they knew about the existence of contamination, but all contamination took place prior to purchase.
- Innocent Landowner Defense—If, for instance, a landowner happened to lease their property to a polluter and can prove by a preponderance of fact of not being not aware of any release of hazardous material and did not consent to its release, the landowner can qualify for an exemption.

30.11 OTHER REGULATORY AGENCIES AND ACCORDS

Many nations and international organizations have developed environmental laws and guidelines. For example, the *European Economic Community* (EEC) has developed a series of recommendations for standards and guidelines for control of air and waterborne pollutants. The World Health Organization has also published guidelines for the levels of contaminants in air and water, and provisions for their control (World Health Organization, 2017) (Fig. 30.7). The United Nations through the development of

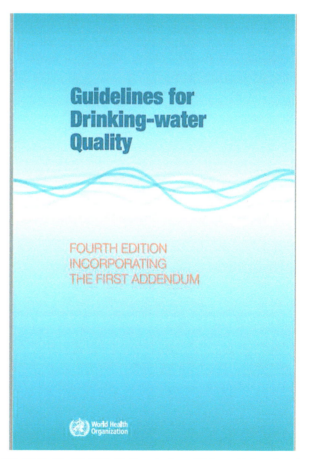

FIG. 30.7 The World Health Organization publishes guidelines for safe levels of contamination in water and air.

international agreements has supported work on global-scale environmental issues (see Chapter 32).

QUESTIONS AND PROBLEMS

1. What are the fundamental differences between an act, a regulation, and a rule?
2. Which regulations potentially affect the safety of drinking water?
3. Which regulations potentially affect contaminants in air?
4. Which acts deal with hazardous wastes, and what is the responsibility of each?

REFERENCE

World Health Organization, 2017. Guidelines for Drinking Water Quality, fourth ed. Switzerland, World Health Organization, Geneva.

FURTHER READING

Pontius, F.W., 2003. Drinking Water Regulation and Health. John Wiley & Sons, New York.

Chapter 31

Environmental Justice

Mónica Ramirez-Andreotta

Widespread environmental pollution became an issue in the United States due to a number of factors. One primary factor was that economic growth was the sole focus of development, with minimal concerns given to the potential impacts on the environment or human health. In other words, development was not carried out in a sustainable manner, but rather based on economics. As discussed in this chapter, a major factor in this was that no economic value was attributed to environmental resources such as soil, air, and water, which means no costs were apportioned for their impairment or degradation. These and other factors coalesced and led to the situation where soil, air, and water pollution was endemic across the United States in the mid-20th century. As we will discuss in this Chapter, the distribution of pollution, and the resultant impacts on human health and well-being, are generally spread nonuniformly among the populace.

31.1 START OF THE ENVIRONMENTAL MOVEMENT

There are pivotal periods in U.S. history in which individuals and new social movements contested the dominant attitude toward the environment (Cox and Pezzullo, 2016). With regard to pollution, there was the public health and the environment movement (1960–70s), and the environmental justice movement (1980–90s). The U.S. environmental movement began with Rachel Carson (1962) and others and focused on wilderness and wildlife preservation, resource conservation, and pollution abatement (Bullard, 1990). The first Earth Day began on April 22, 1970, where ~20 million people advocated for environmental controls on industrial pollution. The U.S. Environmental Protection Agency was born in 1970, along with a suite of environmental legislative landmarks (see Chapter 30). These activities led to major improvements in water and air quality, and consequently the health and well-being of all. Despite these milestones in environmental history, it was continually challenging to hold polluting entities accountable for their waste and resultant impacts on individuals across the United States and globally (Cox and Pezzullo, 2016). Communities were increasingly worried about the contamination of water and soil. Prominent historical examples include the small New York Community of Love Canal (1978), which was built atop a hazardous waste site, and Times Beach Missouri (1982) where residents were forced to leave their town because of elevated levels of dioxins in soils. These and other examples helped usher in the CERCLA program to deal with uncontrolled hazardous waste sites (Chapters 19 and 30). Another program, the Resource Conservation and Recovery Act was developed to manage the production, storage, and disposal of hazardous materials.

31.2 THE ROLE OF SOCIOECONOMIC STATUS AND RACE IN THE EXPOSURE TO POLLUTION

Despite the major advances made in the 1970s to improve environmental quality, U.S. environmental organizations failed to recognize the full spectrum of problems faced by urban residents and those in certain rural communities. As a result, environmental quality challenges in urban and rural settings continued. This led to the development of disparities in the distribution of environmental health effects. Health disparities are differences in health outcomes that are closely linked with social, economic, and environmental disadvantage. Environmental factors such as air, water, soil, and food are fundamental determinants of our health and well-being. Just like access to green space and high quality foods can improve human health, environmental factors can also lead to disease and health disparities when the environments where people live, work, learn, and play are toxic. These health inequalities are considered unnecessary, avoidable, and unfair/unjust (Commission on Social Determinants of Health, World Health Organization, 2008).

In 1968, Dr. Martin Luther King went to Memphis, Tennessee to work with African American sanitation workers who were on strike for better sanitation conditions and wages. In 1971, there was an Urban Environment Conference, which grew out of an effort to provide a forum to discuss issues of joint concern to urban reform groups, environmentalists, and organized labor. Residents of these

Environmental and Pollution Science. https://doi.org/10.1016/B978-0-12-814719-1.00031-8

communities felt dumped upon and that certain communities were targeted as "sacrifice zones" (Bullard and Johnson, 2000). With roots in the civil rights movement, the Environmental Justice (EJ) movement emerged from local communities' struggles with toxic contamination in the United States. This was later termed environmental racism, which is the disproportionate impact of environmental hazards on people of color. Cox and Pezzullo (2016, p. 42) further define this term to mean: "…not only threats to their health from hazardous waste landfills, incinerators, agricultural pesticides, sweatshops, polluting factories, but also the disproportionate burden that these practices place on the people of color and the workers and residents of low-income communities."

In 1990, the father of environmental justice Robert Bullard articulated the nuances of the time in his renowned book, Dumping in Dixie, Class and Environmental Quality. Bullard describes that there are three competing advocacy groups, as presented in Information Box 31.1.

The motivations of these groups were typically not aligned, which led to competition and lack of collaboration. For example, environmentalists did not realize the social implications of the not-in-my-backyard (NIMBY) phenomenon (Morrison, 1986), meaning when environmentalists advocated for the removal of, or petitioned against the placement of hazardous wastes, garbage dumps, and polluting industries in their local region, these unwanted land uses typically ended up in poor, powerless communities rather than in affluent suburbs.

Dumping in Dixie, Class and Environmental Quality focuses on the southern United States since it was undergoing significant development at the time. More than 17 million new jobs were added in the South between 1960 and 1985. The south had its own ecology, later termed "southern ecology" by Historian David R. Goldfield. The "Southern ecology has been shaped largely by excessive economic boosterism, a pro-business climate, lax enforcement of environmental regulations, and industrial strategies that had little regard for environmental cost" (Bullard, 1990). Bullard further describes how economic boosters convinced minority leaders in the south that environmental regulations were bad for business, even when the decisions had adverse

impacts on the less advantaged. When environmental proposals were made, employees were threatened by plant closures, layoffs, and economic dislocation. This was later termed "job blackmail, meaning businesses were" threatening their employees with a "choice" between their jobs and their health, and employers worked to make the public believe there were no alternatives to "business as usual" (KazIis and Grossman, Fear at Work, p. 37). In general, toxic dumping and the location of undesirable land uses have followed the "path of least resistance," where minority and poor communities have been disproportionately burdened with these types of externalities. Government and private industry have continued to follow the path of least resistance when addressing externalities such as pollution discharges, waste disposal, and nonresidential activities that may pose a health threat to nearby communities.

As noted by Robert Bullard "… all of the issues of environmental racism and environmental justice don't just deal with people of color. We are just as much concerned with inequities in Appalachia, for example, where white people are basically dumped on because of lack of economic and political clout and lack of having a voice to say no and that's environmental injustice." Appalachian residents in southeastern Ohio who live along the Ohio River are disproportionately subject to industrial pollution, such as perfluorooctanoic acid, or C8, a chemical used in numerous consumer products (Information Box 31.2).

This exploitative pattern can also be seen in communities neighboring resource extraction sites. These actions by industry have created economically impaired regions that, in turn, limit political debate and activism around environmental quality in order to have or maintain economic development and employment (Kozlowski and Perkins, 2015). Community members living adjacent to active resource extraction sites may feel as though they have to choose between economic growth and environmental quality (Ramirez-Andreotta et al., 2016). This challenge is exacerbated when information and power imbalances exist between the affected community and government and industry stakeholders, also known as information disparities.

INFORMATION BOX 31.1 The Three Major Competing Advocacy Groups

In 1990, the father of environmental justice Robert Bullard articulated the nuances of the time in his renowned book, *Dumping in Dixie, Class and Environmental Quality*. Bullard describes that there are three competing advocacy groups:

- Environmentalists: Concerned about leisure and recreation, wildlife and wilderness preservation, resource conservation, pollution abatement, industrial regulation

- Social justice advocates: Major concerns are about basic civil rights, social equity, expanded opportunity, economic mobility, and institutional discrimination
- Economic boosters: Chief concerns are regarding profit maximization, industrial expansion, economic stability, and deregulation

INFORMATION BOX 31.2 Appalachian Residents and Perfluorooctanoic Acid (C8)

Appalachian residents in southeastern Ohio who live along the Ohio River are disproportionately subject to industrial pollution, such as perfluorooctanoic acid (see Chapter 12), or C8, a chemical used in numerous consumer products (Kozlowski and Perkins, 2015). The Appalachian region faces a long history of exploitation by industries that establish and abandon hazardous facilities with minimal regard for the health of nearby residents (Glasmeier and Farrigan, 2003). Kozlowski and Perkins (2015) posit, "…white, working-class communities

must challenge the notion of privilege that defends the contaminated status quo." Scholars working in Appalachia have documented the severe environmental health risks people are willing to accept in order to maintain employment, specifically in the context of immobility and economic uncertainty. Interestingly these findings are similar to what Bullard documented in some of the predominately black communities he studied in the 1980s.

31.3 THE ENVIRONMENTAL JUSTICE MOVEMENT AND ITS EVOLUTION INTO A FIELD OF STUDY

It is important to note that environmental justice is both a social movement and a field of study (Bullard and Johnson, 2000). As a field of study, scholars aim to address inequitable distributions of environmental health risks from exposure to pollution and environmental hazards (Bullard and Johnson, 2000). The environmental justice movement began in 1982 when the state of North Carolina had to identify a location to dispose of PCB-laden soils. They settled on Shocco, Warren County, NC a small African-American community. Local residents and supporters tried to stop the state's plan by filing two lawsuits, but failed. When the trucks started rolling into Warren County, they were met with a nonviolent protest formed by the National Association for the Advancement of Colored People and others. More than 500 protesters were arrested for acts of nonviolent civil disobedience including Dr. Benjamin F. Chavis, Jr., from the United Church of Christ, and Delegate Walter Fauntroy, then a member of the United States House of Representatives from the District of Columbia. Regardless of the protest and activism, the hazardous waste landfill was still placed in Warren County. Shortly after, definitive studies emerged demonstrating the link between minority status, low socioeconomic status, and community proximity to toxic landfills (e.g., US General Accounting Office, 1983; United Church of Christ, 1987).

The environmental justice movement links social justice and environmental quality and was in response to environmental injustices that occurred in Warren County and elsewhere. Now, Environmental justice is defined by the US EPA as: "the fair treatment and meaningful involvement of all people regardless of race, color, national origin, or income with respect to the development, implementation, and enforcement of environmental laws, regulations, and policies" (US EPA, 2016). Major events in the history of the movement are presented in Information Box 31.3.

The current literature on environmental justice comprises a wide range of quantitative studies consistently

concluding that environmental risk burdens, known or potential, are distributed inequitably across racial/ethnic minorities and individuals with lower socioeconomic status (Chakraborty et al., 2014). Within the context of social justice, we also must consider other variables that impact justice issues in our society and perpetuate environmental injustices. These additional factors are listed and defined below.

31.3.1 Factors

Privilege: In sociology, this concept is a social theory that special rights or advantages are available only to a particular person or group of people. For example, in the United States, privilege is granted to people who have membership in one or more dominant social identity groups, such as those in the highest economic status.

Limited political influence: Lacking the social and economic resources to influence political decisions; this is likely to be the case in areas of greater concentrations of poverty (Schulz and Northridge, 2004).

Linguistic isolation: People who do not speak the dominant language may be left out of the dialog thus impairing their participation in the development, implementation, and enforcement of environmental laws, regulations, and policies.

Class differences: A social context that individuals inhabit over time and a fundamental lens through which we see ourselves and others (Kraus et al., 2012). Because people of lower socioeconomic status have fewer resources and opportunities than those of relatively higher status, they tend to believe that external, uncontrollable social forces and others' power have correspondingly greater influence over their lives (Kraus et al., 2012).

Rural health: A rural health determinant has been proposed as a determinant of health outcomes, suggesting that there may be cultural and environmental factors exclusive to towns, regions, or economic types (e.g., farming, mining, manufacturing, or federal/state government dependent) that may affect health behavior and health (Hartley, 2004). To effectively address rural health disparities one must

INFORMATION BOX 31.3 Major Environmental Justice History Milestones

1982—The U.S. Government Accounting Office completed the *Siting of hazardous waste landfills and their correlation with racial and economic status of surrounding communities* study confirming the pattern of disproportionate exposure to environmental hazards.

1987—The Commission for Racial Justice's 1987 study *Toxic Wastes and Race in the United States* revealed that race is a major factor related to the presence of hazardous wastes in residential communities throughout the United States.

1990—Robert Bullard wrote *Dumping in Dixie, Class and Environmental Quality*—describing more cases and incidences of environmental pollution correlated with low-income communities of color, predominately in the South.

1991—On October 27, 1991, 650 grassroots and national leaders from around the world came together for the First National People of Color Environmental Leadership Summit. They Adopted 17 Principles of Environmental Justice. These 17 principles were and continue to be used for EJ organizing (see Information Box 31.4 for details).

1993—US EPA creates the National Environmental Justice Advisory Committee.

1994—Environmental Justice Meeting for the first research grant program on environmental justice and health by the National Institute of Environmental Health Sciences.

1994—President Clinton's issuance of Executive Order 12898, Federal Action to Address Environmental Justice in Minority Populations and Low-Income Populations.

2002—Second National People of Color Environmental Leadership Summit.

2010—Lisa Jackson, US EPA Administrator, hosted a White House forum and initiated meetings across the country. Stated they would focus on green jobs in disadvantaged communities.

2014—A set of strategies proposed by the USEPA to reinvigorate environmental justice efforts called "Plan EJ 2014." This also acknowledged the 20th anniversary of President Clinton's issuance of Executive Order 12898, Federal Actions to Address Environmental Justice in Minority Populations and Low-Income Populations.

INFORMATION BOX 31.4 Principles of Environmental Justice

1. Environmental justice affirms the sacredness of Mother Earth, ecological unity and the interdependence of all species, and the right to be free from ecological destruction.

2. Environmental justice demands that public policy be based on mutual respect and justice for all peoples, free from any form of discrimination or bias.

3. Environmental justice mandates the right to ethical, balanced and responsible uses of land and renewable resources in the interest of a sustainable planet for humans and other living things.

4. Environmental justice calls for universal protection from nuclear testing, extraction, production, and disposal of toxic/hazardous wastes and poisons and nuclear testing that threaten the fundamental right to clean air, land, water, and food.

5. Environmental justice affirms the fundamental right to political, economic, cultural, and environmental self-determination of all peoples.

6. Environmental justice demands the cessation of the production of all toxins, hazardous wastes, and radioactive materials, and that all past and current producers be held strictly accountable to the people for detoxification and the containment at the point of production.

7. Environmental justice demands the right to participate as equal partners at every level of decision-making including needs assessment, planning, implementation, enforcement, and evaluation.

8. Environmental justice affirms the right of all workers to a safe and healthy work environment, without being forced to choose between an unsafe livelihood and unemployment. It also affirms the right of those who work at home to be free from environmental hazards.

9. Environmental justice protects the right of victims of environmental injustice to receive full compensation and reparations for damages as well as quality health care.

10. Environmental justice considers governmental acts of environmental injustice a violation of international law, the Universal Declaration On Human Rights, and the United Nations Convention on Genocide.

11. Environmental justice must recognize a special legal and natural relationship of Native Peoples to the U.S. government through treaties, agreements, compacts, and covenants affirming sovereignty and self-determination.

12. Environmental justice affirms the need for urban and rural ecological policies to clean up and rebuild our cities and rural areas in balance with nature, honoring the cultural integrity of all our communities, and providing fair access for all to the full range of resources.

13. Environmental justice calls for the strict enforcement of principles of informed consent, and a halt to the testing of experimental reproductive and medical procedures and vaccinations on people of color.

14. Environmental justice opposes the destructive operations of multinational corporations.

15. Environmental justice opposes military occupation, repression and exploitation of lands, peoples and cultures, and other life forms.

16. Environmental justice calls for the education of present and future generations which emphasizes social and environmental issues, based on our experience and an appreciation of our diverse cultural perspectives.

17. Environmental justice requires that we, as individuals, make personal and consumer choices to consume as little of Mother Earth's resources and to produce as little waste as possible; and make the conscious decision to challenge and reprioritize our lifestyles to ensure the health of the natural world for present and future generations.

(***Source:*** *Text adapted from Principles of Environmental Justice, adopted on October 27, 1991, in Washington, D.C., Energy Justice Network. https://www.ejnet.org/ej/principles.html*).

acknowledge the nexus between individuals, culture, and environment, which is also the case when working with EJ communities.

Information disparities: Knowledge differences between the potentially affected community and government and industry stakeholders (Emmett and Desai, 2010). Forms of information disparity are provided in Information Box 31.5.

Health issues: Grineski (2007) argues that health outcomes—instead of proxies for health is a next step in quantitative environmental justice research. Correlating exposure to pollution and health issues and disparities needs to be considered and documented to fully understand environmental injustices.

31.3.2 Visualizing Environmental Justice

In the United States, people of color, low-income communities, and tribal populations have been, and continue to be, disproportionately exposed to environmental conditions that can harm their health. To improve our understanding of the convergence of contaminant exposures and sociodemographic information, the United States Environmental Protection Agency developed EJSCREEN. The EJ SCREEN "is a single, nationally consistent tool that can be used by EPA, its governmental partners, and the public to understand environmental and demographic characteristics of locations throughout the United States" (U.S. EPA, 2016). The EJSCREEN was released in 2015 and is an accomplishment under the 1994 EO 12898. It includes 11 environmental indicators and six demographic indicators (see Information Box 31.6 and Table 31.1).

31.3.3 Climate Justice

Global climate change is one of the most pressing issues for society (see Chapter 25). Temperature increases will range from 0.03 to 4.8°C by 2100, depending on mitigation strategies implemented (Pachauri et al., 2014). Major emitters of carbon dioxide are coal-fired power plants, chemical producers, mining operations, and vehicles. Fossil fuel combustion in higher income countries and the burning of biomass in lower income countries account for 85 percent of airborne particulate pollution (Das and Norton, 2017). The effects of climate change on public health will be substantial as there is already a disproportionate distribution of risk in our society based on socioeconomic factors, such as education level, ethnicity, and poverty level. Thus we can anticipate that climate change will serve to perpetuate these disparities in health (Frumkin et al., 2008).

Termed a "cruel irony," climate justice stresses how those living in poverty contribute the least to climate change but suffer the most consequences (Farber, 2012). This is illustrated in Fig. 31.1. This is further exacerbated by their voices being left out of solution discussions. Building upon the principles of environmental justice (Information Box 31.1), the Bali Principles of Climate Justice were proposed in 2002. Climate justice seeks to remedy this by providing a platform for disadvantaged voices to be heard and to create community-based solutions. Just as seen with the environmental justice movement, these principles forced the discussion to not only be a scientific–technical debate, but also one about ethics that focused on human rights and justice (Agyeman et al., 2010).

INFORMATION BOX 31.5 Forms of Information Disparities

As outlined by (Emmett and Desai, 2010), potentially affected communities may lack the following when compared to other stakeholders:

- Technical expertise
- Representation on Advisory Committees
- Access to information/ability to conduct Toxicological Testing
- Ability to conduct/ability to conduct epidemiologic studies
- Capacity to develop detailed submissions (e.g., public comments to regulatory processes)

- Lack of detailed day-to-day information about presence/levels of chemical in environmental media including water.
- Previous knowledge of chemicals in environmental media
- Authority or ability to influence level of discharges or control measures at industrial facilities
- Access to scientific publications

INFORMATION BOX 31.6 A Tool to Understand Environmental and Demographic Characteristics.

EJSCREEN, "an environmental justice mapping and screening tool that provides EPA with a nationally consistent dataset and approach for combining environmental and demographic indicators" (US EPA, 2016). The EJ Screening Tool also provides the EJ indexes, which summarize how environmental indicators and demographics come together in the same location. For example, the EJ Index for traffic would combine Traffic proximity and volume with the reported low-income and minority population size residing in the selected Census block group.

TABLE 31.1 Components of the U.S. EPA's EJ Screening Tool

Environmental Indicators	Demographic Indicators
National-scale air toxics assessment (NATA) air toxics cancer risk	Percent low-income
NATA respiratory hazard index	Percent minority
NATA diesel PM	Less than high school education
Particulate matter	Linguistic isolation
Ozone	Individuals under age 5
Traffic proximity and volume	Individuals over age 64
Lead paint indicator	
Proximity to risk management plan sites (places where is potential for a chemical accident)	
Proximity to treatment storage and disposal facilities	
Proximity to national priorities list (NPL) sites	
Wastewater dischargers indicator (stream proximity and toxic concentration)	

From https://www.epa.gov/ejscreen/what-ejscreen.

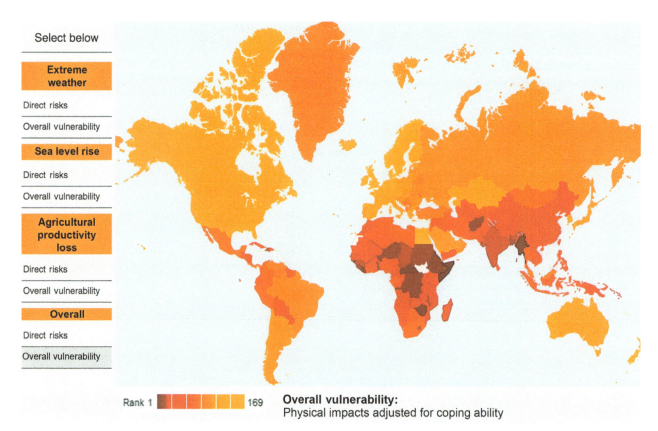

FIG. 31.1 Mapping the Impacts of Climate Change; Quantifying Vulnerability to Climate Change: Implications for Adaptation Assistance, and informs his project mapping the impacts of climate change. *(Prepared by David Wheeler, Center for Global Development, University of Colorado, Boulder).*

31.4 METHODS TO ADDRESS ENVIRONMENTAL INJUSTICES

As outlined before and throughout this book, challenges at hazardous waste and contaminated sites are persistent, complex, and multifactorial. They encompass cleanup and environmental remediation challenges as well as issues related to human health and well-being. The persistence of contamination and environmental injustices can be attributed in part to the lack of collaboration, information transfer, and partnership building among government, the affected community, scientists, site owners, industry, and other interested parties. Issues pertaining to public participation in environmental decision-making, collaboration between all stakeholders and the affected communities, and communicating risk to the communities add an additional layer of complexity for which most environmental scientists have not been trained to manage. Traditionally, environmental scientists and engineers are not introduced to the nuances of environmental communication and the social sphere associated with contaminated sites. Hence, they may not be aware of the voices and practices that various stakeholders and community groups use when discussing environmental issues, and they are not instructed in how to raise public awareness, or work with the communities neighboring contamination (Cox and Pezzullo, 2016). This set of deficiencies is likely to hinder the effectiveness of environmental scientists in their efforts to address environmental issues. Understanding how to work with all stakeholders, build partnerships, elicit local knowledge, and increase community capacity can enhance their success.

Communication about environmental issues and solutions is too often restricted to the technical sphere, and thus excludes those who are most affected, such as the communities neighboring contaminated sites (Cox and Pezzullo, 2016). Issues beyond the technical aspects of site remediation can hinder the cleanup of a site and as a result, a legacy of mistrust can permeate the relationship between regulatory officials, scientists, and the affected communities. Efforts to address and resolve local environmental issues are most effective when scientists from various disciplines, regulatory officials, industry, and the affected community are fully engaged working toward a unified solution. To effect meaningful changes in the environments and health of communities, community-based organizations and leaders must engage the larger public and work in coalition with government agencies, academic institutions, public and private foundations, policymakers, legal experts, and local businesses (Shepard et al., 2002). Freudenberg (2004) argues that it is at the community level where environmental health issues should be confronted because it is frequently the site for health promotion interventions and because it is the place where the individual is confronted

with socioeconomic factors that then influence specific patterns of health and disease.

How do regulatory agencies engage the communities neighboring contaminated sites, and are these community involvement mechanisms effective? Chess and Purchell (1999) reviewed the usefulness of traditional public participation methods (public meetings, workshops, and Community Advisory Committees) and developed a set of "public participation rules of thumb" to improve public involvement efforts. These rules are: clarify goals, begin participation early and invest in advanced planning, modify participation format to community style and needs, provide multiple forms of public participation, and collect feedback on public participation efforts to determine whether they worked (Chess and Purchell, 1999). In 2008 the National Research Council, 2008 published a report entitled: "Public Participation in Environmental Assessment and Decision Making," emphasizing the importance of public participation in environmental issues. When done well, public participation improves the quality and legitimacy of a decision, builds the capacity of all involved to engage in the public process, and leads to better results in terms of environmental quality and other social objectives such as environmental health.

Research translation and community engagement activities may engage different constituencies and have different goals, but it is important to recognize that stakeholders often share common goals and objectives in the cleanup of contaminated sites. The challenges associated with research translation and community engagement are as follows:

- What methods need to be employed to advance the communication of complex problems found at contaminated sites?
- How can we increase public participation in environmental decision-making?
- How can we increase collaboration between other stakeholders and the affected communities?
- How do we effectively communicate risk to the affected communities?
- How do we build the ability of communities to take action at the individual, community, and policy level to improve environmental health (community capacity)?

As discussed before, the importance of developing and supporting full participation by communities and community members affected by a given site has been well established. However, the method by which to most effectively accomplish such participation has yet to be determined. Many disciplines have been grappling with the challenges associated with environmental and risk communication, public participation in environmental data generation and decision-making, and increasing community capacity. Ramirez-Andreotta et al. (2015) posit that it is

INFORMATION BOX 31.7 The Environmental Research Translation Framework

Proposed ERT framework:

1. Provides a team with scientists and practitioners from various disciplines to help tackle the life-world problem (a *transdisciplinary team*)
2. Ensures that a true collaboration is taking place (*effective collaboration*)
3. Calls for a place-specific strategy to ensure effective bidirectional communication efforts (*information transfer*)
4. Values the observations, interests, and knowledge generated by the affected community, encourages their

participation in the investigatory process to increase the colearning, coproduction of data, and the level of informed decision-making by all involved (*public participation in environmental projects*)
5. Involves the affected groups in risk communication and places the risk in context (*cultural model of risk communication*).

---

advantageous to synthesize and integrate the observations and approaches of these disciplines to develop an effective toolkit for environmental scientists. After reviewing selected public participation approaches to research and education, Ramirez-Andreotta et al. developed an integrated approach for environmental research translation (ERT) to facilitate participation of and communication with communities affected by contamination. This approach can be employed at any contaminated site (regardless of contaminant or demographics) to address the injustices described before, and to improve collaboration, bidirectional information transfer, and partnership building among scientists, the affected community, and other stakeholders.

Incorporating participatory methods into an environmental scientist's toolkit will improve the effectiveness and the authenticity of their community engagement activities. As a means to promote interaction and communication among involved parties at contaminated sites, Ramirez-Andreotta et al. (2014) developed the Environmental Research Translation (ERT) concept and framework, presented in Information Box 31.7 and in Fig. 31.2. To implement ERT, we must draw upon and bring together existing participatory approaches to improve interactions between the community, stakeholders, and academic institutions for effective communication of, and action toward reducing, environmental human health risks. The concepts and approaches described below were intentionally selected to properly equip an environmental scientist to interact effectively with communities and other stakeholders.

Knowledge generation and the application of knowledge to contribute to solving societal problems can be achieved if scientists and practitioners from a wide range of disciplines work together with stakeholders (Rosenfield, 1992). When combining these disciplines, innovative ideas and solutions may emerge—and this is the reason for having a transdisciplinary team. Transdisciplinary investigatory efforts transcend disciplinary boundaries and is driven by the need to solve problems of the life-world and improve the human condition (Rosenfield, 1992).

Maintaining effective collaboration with government agencies, community members, and other members of the

public sphere is paramount to translation efforts. There are many stakeholders involved at contaminated sites, and it is crucial to establish transparency and equity between all parties. To do this, a collaborative approach to communication is essential. Collaboration can be defined as constructive, open, civil communication, with emphasis on learning, some degree of power sharing, and leveling of the playing field (Walker, 2004). Collaboration invites all stakeholders to engage in problem-solving discussions rather than advocacy and debate, and community-based collaboration can arise in order to address a specific issue in the local community (Cox and Pezzullo, 2016).

It is vital to recognize one's audience and select the appropriate information transfer mechanism based upon the audience type (e.g., stakeholder, directly affected public) (National Research Council, 2008). In general, a regulatory agency might appreciate an executive summary providing a concise introduction to an environmental issue currently confronting the state, or nation, and possible remediation options. Conversely, a member of the affected

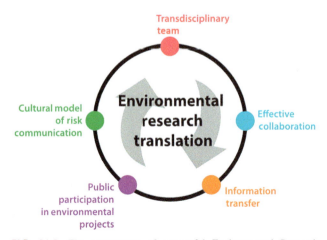

FIG. 31.2 The components of successful Environmental Research Translation programs at contaminated sites. *(From Ramirez-Andreotta, et al., 2014. Environmental research translation: enhancing interactions with communities at contaminated sites. Sci. Total Environ. 497–498, 651–664.)*

community may prefer information specific to the chemical of concern, the risks they pose, and possible methods to reduce exposure. To select which information-transfer mechanism to use for affected communities specifically, that is, delineating the place-specific strategy, it is crucial to learn the community's ecology and the social context in which the environmental contamination and human health risks are embedded (Caron and Serrell, 2009). By understanding the community's ecology, one is able to establish and ensure a two-way dialog with affected communities.

Community members living in contaminated communities often are the first to identify adverse ecological and health outcomes associated with toxic exposures (Brown and Mikkelsen, 1990). Further, *citizen-driven data collection* pertaining to environmental contamination and disasters is often initiated due to unknown and unassessed risks that lay people see in their daily lives (McCormick, 2012), and this step involves the integration of community members in the project and the coproduction of knowledge (Corburn, 2005). Thus it is important to incorporate participatory data collection processes and the community's experiences, which allows for new information regarding environmental contaminants or exposure routes to be introduced that may be not be collected or addressed in a typical expert-only-led environmental science or health investigation and/or risk assessment. Furthermore, having an expert-driven only project may, unfortunately, ignore the role of lay knowledge in research, overlook the applicability of the study findings to improve regulation, and may not prioritize the need to transparently communicate the results of the study to the community (Cohen et al., 2012).

How to successfully communicate results and risk is another key factor for successful engagement. The method in which the results and risk are communicated should be tailored to the community's need so they may make more informed decisions. Particularly, it is important to present specific steps the community can take to assert some level of control in their lives and reduce their exposure to potential environmental hazards.

To successfully communicate risk, it is important to put the risk in context, make comparisons with other risks, and encourage a dialog. If all or most of the other ERT steps have been met, encouraging a dialog should not be difficult. This step is about communicating and discussing risk assessment and management. For example, community members should be able to witness and perform the calculations used to complete the risk assessment (i.e., daily dose of the contaminant of concern and the excess cancer risk with the community) (Ramirez-Andreotta et al., 2015) and have a discussion regarding the uncertainties associated with the current risk analysis paradigm.

31.5 CONCLUSION

In a recent report prepared by the Lancet Commission on pollution and health, they state: "For decades, pollution and its harmful effects on people's health, the environment, and the planet have been neglected both by Governments and the international development agenda." Yet, pollution is a major cause of disease and death in the world today, responsible for an estimated 13 million premature deaths annually (see Chapter 26). The report further explains that 92% of pollution-related deaths occur in low- and middle-income countries and that children face the highest risks because small exposures to chemicals in utero and in early childhood can result in lifelong disease, disability, premature death, as well as reduced learning and earning potential. Unfortunately, health disparities as a result of environmental pollution exist in our society. Environmental injustices are linked to health disparities. These health inequalities are considered unnecessary, avoidable, and unfair/unjust. The factors that perpetuate injustice are resolvable and require understanding, collaboration, and a paradigm shift in the way we approach and engage in the solution-generating dialog. By combining and adopting the principles of environmental justice (Information Box 31.5) along with ERT (Information Box 31.7), one can ethically, collaboratively, and holistically generate novel solutions to environmental health challenges.

QUESTIONS AND PROBLEMS

1. What are the major differences between environmental racism and environmental justice?
2. What is the definition of health disparities? How does environmental quality impact health outcomes?
3. In this chapter, three types of advocacy groups are mentioned? Define each and provide an example of each type. Explain why these three groups might have competing agendas.
4. Within the context of social justice, we much consider other variables besides race and socioeconomic status that impact and perpetuate environmental injustices. Please list and describe these other factors.
5. As described before, EJSCREEN is an environmental justice mapping and screening tool that provides EPA with a nationally consistent dataset and approach for combining environmental and demographic indicators. For this question:
   a. Go to the http://www.epa.gov/ejscreen. Learn how to use the mapping tool.
   b. Once you understand how to use the mapping and screening tool, select a specific neighborhood of your choice. Explore and report the neighborhood's

status. Is it above 50th percentile for any of the environmental indicators? For demographic indicators? If so, list which ones?

c. Now select the EJ index, which is a combination of environmental and demographic information. Would you call that neighborhood or specific area an EJ community? Why? Thoroughly explain you rationale.

REFERENCES

Agyeman, J., Bullard, R.D., Evans, B., 2010. Exploring the nexus: bringing together sustainability, environmental justice and equity. Space Polity 6, 77–90.

Brown, P., Mikkelsen, E.J., 1990. No Safe Place: Toxic Waste, Leukemia, and Community Action. University of California Press, Berkeley, CA.

Bullard, R., Johnson, S., 2000. Environmental justice: grassroots activism and its impact on public policy decision making. J. Soc. Issues 56, 555–578.

Bullard, R.D., 1990. Dumping in Dixie: Race, Class, and Environmental Quality. Westview, Boulder, CO.

Caron, R.M., Serrell, N., 2009. Community ecology and capacity: keys to progressing the environmental communication of wicked problems. Appl. Environ. Educ. Commun. 8, 195–203.

Chakraborty, J., Collins, T.W., Grineski, S.E., Montgomery, M.C., Hernandez, M., 2014. Comparing disproportionate exposure to acute and chronic pollution risks: a case study in Houston, Texas. Risk Anal. 34, 2005–2020.

Chess, C., Purchell, K., 1999. Public Participation and the environment: do we know what works? Environ. Sci. Technol. 30 (16), 2685–2692.

Cohen, A., Lopez, A., Malloy, N., Morello-Frosch, R., 2012. Our environment, our health: a community-based participatory environmental health survey in Richmond. California Health Educ. Behav. 39 (2), 198–209.

Commission on Social Determinants of Health, World Health Organization, 2008. Available from http://www.who.int/social_determinants/thecommission/finalreport/en/. (Accessed 13 October 2017).

Corburn, J., 2005. Street Science: Community Knowledge and Environmental Health Justice. MIT Press, MA.

Cox, R., Pezzullo, P.C., 2016. Environmental Communication and the Public Sphere, fourth ed. Sage Publications, Inc., CA.

Das, P., Norton, R., 2017. Pollution, health, and the planet: time for decisive action. Lancet. Available from: http://www.thelancet.com/journals/lancet/article/PIIS0140-6736(17)32588-6/fulltext. (Accessed 29 January 2018).

Emmett, E., Desai, C., 2010. Community first communication: reversing information disparities to achieve environmental justice. Environ. Justice 3, 79–84.

Farber, D.A., 2012. Climate justice. Mich. L. Rev. 110, 985. Available from https://repository.law.umich.edu/mlr/vol110/iss6/5. (Accessed 29 January 2018).

Freudenberg, N., 2004. Community capacity for environmental health promotion: determinants and implications for practice. Health Educ. Behav. 31 (4), 472–490.

Frumkin, H., Hess, J., Luber, G., Malilay, J., McGeehin, M., 2008. Climate change: the public health response. Am. J. Public Health 98, 435–445.

Glasmeier, A., Farrigan, T., 2003. Poverty, sustainability, and the culture of despair: can sustainable development strategies support poverty alleviation in America's most environmentally challenged communities? Ann. Am. Acad. Pol. Soc. Sic. 590, 131–149.

Grineski, S.E., 2007. Incorporating health outcomes into environmental justice research: the case of children's asthma and air pollution in Phoenix, Arizona. Environ. Hazard. 7, 360–371.

Hartley, D., 2004. Rural health disparities, population health, and rural culture. Am. J. Public Health 94, 1675–1678.

Kozlowski, M., Perkins, H.A., 2015. Environmental justice in Appalachia Ohio? An expanded consideration of privilege and the role it plays in defending the contaminated status quo in a white, working-class community. Local Environ. 1, 1–17.

Kraus, M.W., Piff, P.K., Mendoza-Denton, R., Rheinschmidt, M.L., Keltner, D., 2012. Social class, solipsism, and contextualism: how the rich are different from the poor. Psychol. Rev. 119 (3), 546–572.

McCormick, S., 2012. After the cap: risk assessment, citizen science and disaster recovery. Ecol. Soc. 17 (4), 31–40.

Morrison, D.E., 1986. How and why environmental consciousness has trickled down. In: Schnaiberg, A., Watts, N., Zimmermann, K. (Eds.), Distributional Conflict in Environmental-Resource Policy. St. Martin's Press, New York, pp. 187–220.

National Research Council, 2008. Public Participation in Environmental Assessment and Decision Making. National Academy Press, Washington, DC.

Pachauri, R.K., Allen, M.R., Barros, V.R., Broome, J., Cramer, W., Christ, R., … Dubash, N.K., 2014. Climate Change 2014: synthesis Report (Contribution of Working Groups I, II and III to the Fifth Assessment Report of the Intergovernmental Panel on Climate Change, p. 151). Intergovernmental Panel on Climate Change, Geneva, Switzerland.

Ramirez-Andreotta, M.D., Brody, J.G., Lothrop, N., Loh, M., Beamer, P.I., Brown, P., 2016. Improving environmental health literacy and justice through environmental exposure results communication. Int. J. Environ. Res. Public Health 13 (7), 690–717. 27399755.

Ramirez-Andreotta, M.D., Brusseau, M.L., Artiola, J.F., Maier, R.M., Gandolfi, A.J., 2014. Environmental research translation: enhancing interactions with communities at contaminated sites. Sci. Total Environ. 497-498, 651–664. 25173762.

Ramirez-Andreotta, M.D., Brusseau, M.L., Artiola, J.F., Maier, R.M., Gandolfi, A.J., 2015. Building a co-created citizen science program with gardeners neighboring a superfund site: the Gardenroots case study. Int. Pub. Health J. 7 (1), 139–153. 25954473.

Rosenfield, P.L., 1992. The potential of transdisciplinary research for sustaining and extending linkages between the health and social sciences. Soc. Sci. Med. 35, 1343–1357.

Schulz, A., Northridge, M.E., 2004. Social determinants of health: implications for environmental health promotion. Health Educ. Behav. 31 (4), 455–471.

Shepard, P., Northridge, M.E., Prakash, S., Stover, G., 2002. Preface: advancing environmental justice through community-based participatory research. Environ. Health Perspect. 110 (2), 139–140.

U.S. General Accounting Office, 1983. Siting of Hazardous Waste Landfills and Their Correlation with Racial and Economic Status of Surrounding Communities. Washington, DC, United States General Accounting Office.

United Church of Christ, 1987. Toxic Wastes and Race in the United States: A National Report on the Racial and Socio-Economic

Characteristics of Communities Surrounding Hazardous Waste Sites. United Church of Christ, Commission for Racial Justice, New York, NY.

U.S. EPA, 2016. Environmental Justice. Available from: https://www.epa.gov/environmentaljustice. (Accessed 29 January 2017).

Walker, G.B., 2004. The roadless areas initiative as national policy: is public participation an oxymoron? In: Depoe, S.P., Delicath, J.W., Aepli Elsenbeer, M.-F. (Eds.), Communication and Public Participation in Environmental Decision Making. State University of New York Press, NY.

FURTHER READING

Bullard, R., Wright, B.H., 1987. Environmentalism and the politics of equity: emergent trends in the Black community. Mid-Am. Rev. Soc. 7 (2), 21–38.

Graham, G.N., 2016. Why your zip code matters more than your genetic code: promoting healthy outcomes from mother to child. Breastfeed. Med. 11, 396–397. https://doi.org/10.1089/bfm.2016.0113.

U.S. Environmental Protection Agency, (2017). EJSCREEN: environmental justice screening and mapping tool. Available from: https://www.epa.gov/EJSCREEN. (Accessed 29 January 2017).

Chapter 32

Sustainable Development and Other Solutions to Pollution and Global Change

M.L. Brusseau

Sustainable Development Goals from the UN Sustainable Development Summit in 2015. *(From The Sustainable Development Agenda © United Nations. Reprinted with the permission of the United Nations.)*

The impacts of human activity on our environment have led to myriad issues and concerns, as highlighted in several chapters of this text. Many past activities have been so damaging that they have led to irrevocable changes, such as large-scale land disturbances (deforestation and desertification), habitat loss, and species extinction (Chapter 25). As the global human population continues to grow, the demands placed on the environment will continue to increase. It is imperative that we minimize current and future adverse environmental impacts of human activities so that the environment can continue to support human and other life. This imperative is captured by the concept of sustainability and sustainable development. In addition, it is critical that we improve the ability of individuals and communities, as well as their supporting infrastructure and environment, to resist and recover from environmental impacts. These concepts are the focus of this chapter.

32.1 LAWS, REGULATIONS, AND TREATIES

Most countries have laws and regulations governing human activity to reduce environmental pollution and other impacts. Such laws and regulations for the United States are covered in Chapter 30. As discussed in several prior chapters, implementation of these laws and regulations has greatly reduced the impact of human activities on the environment and the associated impacts to human health.

Environmental and Pollution Science. https://doi.org/10.1016/B978-0-12-814719-1.00032-X

These laws and regulations are typically developed and implemented at graduated scales. For the United States, the U.S. EPA promulgates nationwide regulations based on federal laws enacted by Congress. Individual states often implement their own regulations to supplement the federal ones. Counties and cities may also implement their own as well. This multilevel approach can lead to a "patchwork" of regulations that can limit effectiveness. Similarly, each country has their own set of laws and regulations that vary in levels of strictness and effectiveness. This type of approach is a particular hindrance to addressing regional or global scale issues.

Bilateral and multination treaties are a primary means used to address larger scale issues. These constitute agreements arranged between two or more countries and may have names such as a protocol or accord in addition to treaty. Treaties are used for many different purposes. There are several well-known treaties for the environment.

One environmental treaty considered to be the most successful to date is the *Montreal Protocol on Substances that Deplete the Ozone Layer*, or "Montreal Protocol," for short. Growing worldwide use of chlorofluorocarbons (CFCs), together with evidence of CFC-induced decline of stratospheric O_3 concentrations over Antarctica, convinced most nations to sign the Montreal Protocol (see Information Box 32.1), which called for a complete phase out of the production of CFCs by the year 2000.

Another major environmental treaty is the *Stockholm Convention on Persistent Organic Pollutants*. This is an international treaty, signed in 2001 and effective from May 2004, developed to eliminate or restrict the production and use of persistent organic pollutants (POPs). These are chemical substances that have great persistence in the environment, that bioaccumulate through the food web, and pose a risk of causing adverse effects to human health and the environment. Examples of POPs were mentioned in Chapter 12.

The *Minamata Convention on Mercury* is an international treaty designed to protect human health and the environment from anthropogenic emissions and releases of mercury. The critical issue of mercury pollution and the impact on human health was discussed in Chapter 26, with specific mention of the Minamata pollution event after which this treaty is named. Delegates representing close to 140 countries approved this Convention in Geneva on January 19, 2013. The objective of this agreement is to reduce mercury pollution and thereby reduce human health impacts.

A series of treaties have been developed to address global climate change. *The United Nations Framework Convention on Climate Change* (UNFCCC) was adopted in 1992 and entered into force in 1994. The objective of the UNFCCC is to achieve "stabilization of greenhouse gas concentrations in the atmosphere at a level that would prevent dangerous anthropogenic interference with the climate system. Such a level should be achieved within a time-frame sufficient to allow ecosystems to adapt naturally to climate change, to ensure that food production is not threatened and to enable economic development to proceed in a sustainable manner." (http://unfccc.int/files/essential_background/background_ publications_htmlpdf/application/pdf/conveng.pdf)

The framework sets nonbinding limits on greenhouse gas emissions for individual countries and contains no enforcement mechanisms. The framework outlines how additional international treaties may be negotiated to specify further action toward accomplishing the objective of the UNFCCC. The Kyoto Protocol, adopted in Kyoto, Japan in 1997, and the Paris Climate Accord, adopted in Paris in 2015, are two major follow-ups to the UNFCCC. This mitigation approach of reducing the emission of greenhouse gases, particularly CO_2, is generally viewed as the most effective alternative available to reducing the future impacts of global climate change.

INFORMATION BOX 32.1 Montreal Protocol: International Cooperation to Reduce an Environmental Risk From Ozone Depletion

Research in 1974 at the University of California, Irvine, indicated that synthetic chemicals known as chlorofluorocarbons (CFCs) used as refrigerants and propellants in aerosol cans were slowly diffusing into the stratosphere, breaking down, and releasing chlorine atoms that catalytically destroyed stratospheric ozone. In 1985, observational evidence from the British Antarctic Survey of ozone layer depletion (the ozone hole) over Antarctica spurred developed (including the United States) and developing nations to continue negotiations to significantly reduce the production and use of CFCs. In 1987, the Montreal Protocol on Substances that Deplete the Ozone Layer was adopted. It initially required a 50% reduction of use of CFCs by the year 2000 and a freeze on the use of halons

(fire-extinguishing compounds consisting of bromine, fluorine, and carbon) with a 10-year grace period for developing nations. In 1989, the Montreal Protocol entered into force with 13 developed nations announcing a phaseout of CFCs by 1997. Because of mounting scientific evidence of ozone depletion by CFCs, adherents to the Protocol in 1990 agreed to a total ban on CFCs and halons by 2000. The treaty has been ratified by 197 parties, including 196 countries and the European Union, making it the first universally ratified treaty in United Nations history. Because of the banning of CFCs and halons, the ozone hole has stopped growing and appears to be shrinking. Continued adherence to the protocol is likely to lead to continued improvements during the coming years.

The development and adoption of international treaties is a very complex process. It is typically fraught with many differences of opinion and competing interests among (and sometimes within) individual countries. Such differences can lead to agreements that are not as effective as they could be. Once adopted, their continued implementation can be subject to changes of political climate within countries. However, the Montreal Protocol illustrates that multinational treaties can be an effective approach to address large-scale environmental issues.

32.2 SUSTAINABILITY AND SUSTAINABILITY GOALS

The United Nations has been integral to the widespread development of sustainability. The *United Nations Conference on the Human Environment* was held in Stockholm, Sweden in 1972. A declaration was published as part of this conference that recognized the interdependence of humans and the environment:

"Man is both creature and moulder of his environment, which gives him physical sustenance and affords him the opportunity for intellectual, moral, social and spiritual growth. In the long and tortuous evolution of the human race on this planet a stage has been reached when, through the rapid acceleration of science and technology, man has acquired the power to transform his environment in countless ways and on an unprecedented scale. Both aspects of man's environment, the natural and the man-made, are essential to his well-being and to the enjoyment of basic human rights the right to life itself." (http://webarchive.loc.gov/all/20150314024203/http%3A//www.unep.org/Documents.Multilingual/Default.asp?documentid%3D97%26articleid%3D1503)

The Conference called upon "Governments and peoples to exert common efforts for the preservation and improvement of the human environment, for the benefit of all the people and for their posterity." A set of 26 principles for supporting this effort was released.

A seminal event in advancing the concept of sustainability occurred in 1987 with the release of *Our Common Future*, also known as the Brundtland Report, from the United Nations World Commission on Environment and Development (WCED). The report formalized the concept of sustainable development and gave this often-cited definition: "Sustainable development is development that meets the needs of the present without compromising the ability of future generations to meet their own needs."

This work led to the *United Nations Conference on Environment and Development* (UNCED), also known as the Earth Summit, held in Rio de Janeiro in June 1992. The *Rio Declaration on Environment and Development* was produced as part of this conference. It consists of 27 principles developed to help guide countries in developing and implementing sustainable development efforts. Examples of the principles are presented in Information Box 32.2. Of note, the Rio Declaration contains many concepts central to sustainable development, such as the critical role of humans (Principle 1), the idea of considering the needs of not only ourselves but also of future generations (Principle 3), the need for protecting the environment (Principles 4 and 11), the importance of involving the public in planning and decision-making (Principle 10), concepts of the "precautionary principle" (Principle 15), the "polluter pays principle" (Principle 16), and the need for impact assessments (Principle 17).

The *Millennium Summit* was a meeting of world leaders held at the United Nations headquarters in New York City

---

INFORMATION BOX 32.2 Selected Principles from the Rio Declaration

Principle 1—Human beings are at the centre of concerns for sustainable development. They are entitled to a healthy and productive life in harmony with nature.

Principle 3—The right to development must be fulfilled so as to equitably meet developmental and environmental needs of present and future generations.

Principle 4—In order to achieve sustainable development, environmental protection shall constitute an integral part of the development process and cannot be considered in isolation from it.

Principle 10—Environmental issues are best handled with the participation of all concerned citizens, at the relevant level....

Principle 11—States shall enact effective environmental legislation....

Principle 15—In order to protect the environment, the precautionary approach shall be widely applied by States

according to their capabilities. Where there are threats of serious or irreversible damage, lack of full scientific certainty shall not be used as a reason for postponing cost-effective measures to prevent environmental degradation.

Principle 16—National authorities should endeavour to promote the internalization of environmental costs and the use of economic instruments, taking into account the approach that the polluter should, in principle, bear the cost of pollution, with due regard to the public interest and without distorting international trade and investment.

Principle 17—Environmental impact assessment, as a national instrument, shall be undertaken for proposed activities that are likely to have a significant adverse impact on the environment and are subject to a decision of a competent national authority.

(**Source:** *Available from: https://en.wikisource.org/wiki/Rio_Declaration_on_Environment_and_Development*).

in 2000. The *Millennium Declaration* was adopted at this meeting. It has eight chapters and key objectives. Objective 4 deals with the environment. A set of goals were developed to promote sustainability efforts by the nations. These goals are presented in Fig. 32.1. Note goal seven, "ensure environmental sustainability."

The *United Nations Conference on Sustainable Development* (UNCSD), also known as Rio 2012 or Earth Summit 2012, was the third international conference on sustainable development aimed at reconciling the economic and environmental goals of the global community (United Nations, 2012). The conference had three objectives: (1) Securing renewed political commitment for sustainable development, (2) assessing the progress and implementation gaps in meeting previous commitments, and (3) addressing new and emerging challenges. The conference led to publication of "The Future We Want." This document includes discussion of the need to develop a new set of sustainable development goals. These goals were formally adopted at

the *UN Sustainable Development Summit* in 2015 in New York City. The 17 *Sustainable Development Goals* (SDGs) are presented in Fig. 32.2. Note the four goals directly linked to the environment: 6. Clean Water and Sanitation, 13. Climate Action, 14. Life Below Water, and 15. Life on Land. However, it is important to recognize that other goals have indirect influence from and on the environment.

32.3 SUSTAINABILITY AND RELATED CONCEPTS

The concept of sustainability reconciles the need for and associated impacts of human development with the imperative to sustain and preserve the environment and the ecosystem services provided. In other words, it is a recognition that human activities can adversely impact the environment, and that these adverse impacts can reduce or eliminate the services that the environment provides to support human well-being, and as such it is in our best interests to minimize the adverse impacts of human activities on the environment. Sustainability also recognizes the complexity and interconnectedness of human-environment interactions. A shorthand definition of sustainability could be given as "a comprehensive consideration of the consequences of our actions."

There are related concepts that have developed that acknowledge and attempt to address the essential ideas of sustainability. A few primary examples are presented in this section.

32.3.1 The "Rs"

We encountered three famous "Rs" in Chapters 19 and 20—Restoration, Reclamation, and Remediation—that focus on repairing environmental damage. There is another set of "Rs" focused on sustainability—preventing or reducing damage. The well-known original three for waste management are—Reduce, Reuse, and Recycle. More recently, a fourth has been added—Recover. These "Rs" capture the concept of minimizing the demands humans place on the environment and its resources.

The "R" concept has been extended to broaden the types of activities included to provide more comprehensive contributions to sustainability. One example is the 6 "Rs" = Refuse, Reduce, Reuse, Recycle, Repair, and Rethink from https://practicalaction.org/. Refuse is added to bring in the idea of eliminating a certain activity altogether, while Rethink was added to capture the idea of finding alternatives to the action in question.

Other "Rs," such as Responsibility can be added to capture the idea that human behavior is a critical component of sustainable actions. A comprehensive but not complete set of "Rs" grouped into five categories is presented in Information Box 32.3.

FIG. 32.1   Millennium development goals from the 2000 UN Millennium Summit. *(Reproduced with permission from UNDP Brazil.)*

FIG. 32.2 Sustainable development goals from the UN sustainable development summit in 2015. *From The Sustainable Development Agenda © United Nations. (Reprinted with the permission of the United Nations.)*

INFORMATION BOX 32.3 The "Rs" of Sustainability

1. Rethink-Reinvent-Replace
2. Refuse-Reduce
3. Repair-Remediate-Reclaim-Rehabilitate-Restore-Renew
4. Recycle-Recover-Reuse-Repurpose
5. Responsibility-Respect-Relationships

32.3.2 The Nexus Concept

One meaning of *Nexus* is "a means of connection." The Nexus concept is being employed in environmental science to recognize the interconnectedness of human-environment interactions. The Soil Health-Human Health Nexus was discussed in Chapter 27. The *Water-Energy Nexus* is another example. This concept was developed to formally recognize the interconnections between energy production and use and water production and use. For example, energy, typically in the form of electricity, is required to extract, treat, and deliver potable water. Concomitantly, water is used in multiple phases of energy resource extraction (e.g., oil and gas production) and energy production (e.g., electricity generation). An illustration of the interconnection between energy and water is presented in Fig. 32.3 (U.S. Department of Energy, 2006).

An international conference titled "Water, Energy and Food Security Nexus—Solutions for the Green Economy," or the Bonn2011 Nexus conference for short, was held in Bonn, Germany in 2011. This was held as a contribution to the Rio 2012 UN Conference on Sustainable Development. Food was added to the Water-Energy Nexus concept in recognition that (a) food security is one of the critical issues of the 21st century, and (b) water security, energy security, and food security are interconnected. For example:

– Agriculture accounts for 70% of global water withdrawal
– The food production and supply chain accounts for about 30% of total global energy consumption
– Roughly 75% of all industrial water withdrawals are used for energy production
– 90% of global power generation is water intensive

This concept provides consideration of the effect of changes to one component on the other two. For example, what are the impacts of long-term droughts on energy consumption and food production? One such impact would be that extended periods of drought lead to lower water levels in

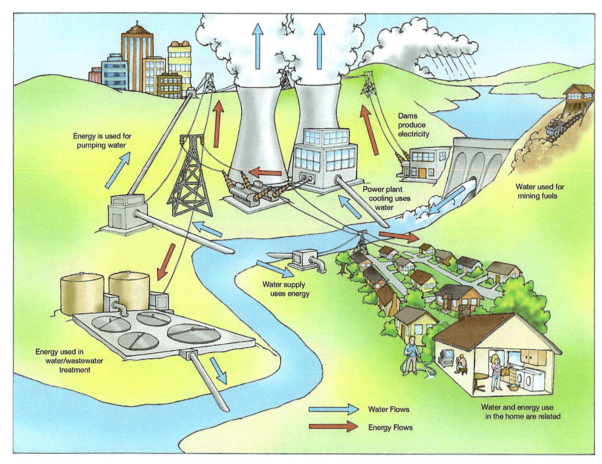

FIG. 32.3   Water-Energy Nexus—examples of interconnections between energy and water. *(From DOE: Energy demands on water resources. Report to Congress on the Interdependency of Energy and Water. U.S. Department of Energy, 2006, p. 13. Available from: https://www.circleofblue.org/wp-content/uploads/2010/09/121-RptToCongress-EWwEIAcomments-FINAL2.pdf.)*

lakes and reservoirs, which can greatly affect the generation of hydroelectric power. This is illustrated by what has happened at Lake Mead and Hoover dam, one of the largest electricity generating plants in the southwest United States. Over the past 15 years, the water level in the lake has dropped by ~130 ft. This has resulted in a reduction of more than 20% in power production. Such reductions at this and other dams will have a significant impact on the power supply in the southwest United States. In addition, the extended drought would likely lead to reduced volumes of water available for various uses, and agricultural use is often one of the first to be reduced in such periods. This could in turn lead to changes in the types or amounts of agricultural production in the region.

In the Water-Energy-Food Nexus, "Water" represents the environment and its associated ecosystem services. One could expand the Nexus concept to include the other components of the environment and their ecosystem services. For example, we could add Air to the Nexus. We know that energy production can impact air quality (e.g., smog), as can food production (e.g., feedlot methane, dust). In turn, air quality can affect food production. We could

also add Soil to the Nexus, to capture the interconnections highlighted in Chapter 27. Ultimately, we should perhaps rephrase the Water-Energy-Food Nexus to the *Environment-Energy-Food Nexus*. This revised term fully captures the concept that energy production/use and food production/consumption are both intimately connected to the environment and vice versa.

32.3.3   Conservation and Resource Management

As discussed in Chapter 6, the environment provides numerous ecosystem services that are critical to human life. Also as noted therein, the issue of resource overuse and the Tragedy of the Commons is a common issue across the world. Two concepts related to sustainability have been developed to address this issue of resource overuse: conservation and natural resource management. An entire section (Section 2) of the aforementioned 1992 Rio Declaration is focused on conservation and management of resources.

Conservation is the concept of the ethical use of natural resources in such a manner that will maintain the long-term health and functioning of the ecosystem. The "R's" concept, such as Refuse, Reduce, Reuse, Recycle, Repair, and Rethink, embodies the essence of resource conservation. Conservation efforts can be implemented in different ways. One approach is to develop preserves, such as national parks and nature reserves, that limit or disallow particular types of resource use. One of the largest scale examples of this approach is the establishment of protected area in the Amazon Basin to protect the Amazon Rainforest.

Another major way in which to implement conservation is to regulate resource use. This is accomplished through natural resource management, also referred to as environmental resource management. Natural resource management is inherently complex due to the complexity of ecosystems themselves and to that of human interactions with ecosystems. As discussed in Chapter 6, ecosystems are composed of many components and are very dynamic, with a multitude of interacting parts. A change to one component may have far-reaching and/or long-term impacts to other components and to the system as a whole. In addition, the human dimension is present, involving various stakeholders and their often-competing interests, and the social, cultural, political, and economic ramifications of management decisions.

32.4 SUSTAINABLE DEVELOPMENT

The application of sustainability concepts and goals to a particular human activity or action is *sustainable development*. Recall the definition from Our Common Future: "Sustainable development is development that meets the needs of the present without compromising the ability of future generations to meet their own needs." This simple principle is in reality a very complex endeavor to implement. Some of the important aspects of sustainable development are covered in this section.

32.4.1 General Concepts

Sustainable development is often discussed in terms of three key components or pillars, Economic, Environment, and Social. A Venn diagram of sustainability has been developed by practitioners to represent the three pillars and their interconnectedness (Fig. 32.4). In essence, the diagram presents the idea that sustainable development can be achieved when economic development is conducted in a manner that preserves and protects the environment and its resources while supporting individual and community well-being.

This concept has been presented in related form in business through the concept of the "triple bottom line." The term is a reference to the traditional focus of companies

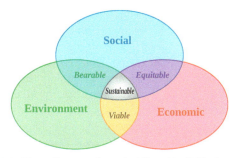

FIG. 32.4 Venn diagram of sustainability. *(Available from: https:// commons.wikimedia.org/wiki/File:Sustainable_development.svg)*

on the "bottom line," meaning profit or loss. It is now recognized that business entities need to consider environmental and social performance along with financial. As described by the International Institute for Sustainable Development, the broader concept of *corporate social responsibility* (CSR) has evolved in response to the recognition that corporations can no longer act as isolated economic entities operating in detachment from broader society. In essence, CSR promotes adopting business strategies and activities that meet the needs of the enterprise and its stakeholders today while protecting, sustaining, and enhancing the human and natural resources that will be needed in the future.

A related concept is that of "*social license to operate*." This refers to the degree of acceptance that local communities and stakeholders have for a particular organization and their operations. The concept has evolved recently from the broader concept of CSR. It promotes the idea that institutions and companies need "social permission" to conduct their business, as well as regulatory permission. Social license to operate has become a critical issue, for example, in the mining industry.

The *Precautionary Principle* is a guiding concept whose implementation could reduce the adverse impacts of humans on the environment and on their own well-being. This concept originated in the 1980s, and for example was embodied in Principle 15 of the Rio Declaration as noted before. The principle captures the idea that decision-makers need to anticipate and consider adverse effects of an action before it occurs. Additionally, the responsibility lies with the proponent of the action to establish that the proposed action will not or is unlikely to cause adverse impacts. This is in contrast to the standard approach wherein an action is implemented without prior full consideration of potential adverse impacts, and the public bear the burden of proof (and the adverse impacts). For example, as noted in Chapter 12, thousands of chemicals are routinely produced and used in consumer products. The impacts of many of these chemicals on the environment and human health have not been tested. The precautionary principle would deem that the chemical manufacturers would be required to test all chemicals for potential adverse impacts prior to their widespread use. While the formal concept of the precautionary principle has been around for just a few decades, the idea that caution should be exercised in making decisions has been present for a long time. This is embodied in common sayings such as "a stich in time saves nine," "look before you leap," and "better safe than sorry."

Green Development is a related term for sustainable development and captures the idea that environmental and social impacts need to be considered in development. Hand in hand with green development is the concept of *Green Technology*. This refers to the design and production of products that consider sustainability, cradle-to-grave management, reduced resource use, and other environmental impacts. The development and application of green technologies is needed to support green development. Some examples are as follows.

Green or Sustainable Architecture is focused on developing buildings that are built and operate efficiently to conserve resources. Primary considerations include (a) the use of building materials that are produced sustainably or are recycled, (b) energy efficiency and the use of renewable energy sources (e.g., solar panels, wind turbines, heat pumps), (c) on-site waste management (e.g., composting toilets, food waste composting gardens, gray water management), and (d) designed in harmony with its surroundings.

A subset of green architecture is *Green Infrastructure*, which is defined by the U.S. Environmental Protection Agency as a cost-effective, resilient approach to managing wet weather impacts that provides many community benefits. In short, it is a decentralized, dwelling-based approach to deal with stormwater runoff while also reducing water use and demand. Rainwater harvesting is a prime example of green infrastructure. Rainwater harvesting systems collect and store rainfall for later use. When designed appropriately, they reduce stormwater runoff and provide a source of water for the dwelling inhabitants. This practice could be particularly valuable in arid regions, where it could reduce demands on increasingly limited water supplies. See Fig. 32.5 for an example of a rainwater harvesting system.

Green Chemistry, or sustainable chemistry, is the design of chemical products and processes that reduce or eliminate the use or generation of hazardous substances. Green chemistry applies across the life cycle of a chemical product, including its design, manufacture, use, and ultimate disposal. Green chemistry attempts to reduce or prevent pollution by minimizing or eliminating hazardous chemical feedstocks, reagents, solvents, and products. The 12 principles of green chemistry are presented in Information Box 32.4.

A well-known example of green development and technology is green or *Renewable Energy*. This constitutes energy from sources that are naturally replenished over a human life scale. These include solar energy, wind energy, geothermal energy, hydropower (rivers, waves, and tides), and biofuels. The concept of renewable energy in terms of sustainability is discussed in the following subsection.

32.4.2   Issues in Sustainable Development

It is important to recognize that sustainable development is not just an outcome. It is also a process, with particular modes of thinking and operation. Human interactions with the environment are complex, comprising a dense web of interactions among individual components of a system. Effective consideration of human-environment interactions, and successful sustainable development, therefore, requires a shift in our approach to studying and solving environmental issues. Previously, the environment, ecosystems, and human-environment interactions were treated with a reductionist approach, wherein a single individual component was studied in isolation. This is also termed piecewise thinking. It is now recognized that a *systems thinking* approach is needed for effective sustainable development. This approach considers the connections among the individual components of a system, and that systems are more than the sum of their parts.

Many adverse environmental impacts originating from human activities are unintended consequences of the action or activity. These unintended consequences in many cases could have been anticipated with a systems thinking approach. In short, sustainable development requires comprehensive consideration of the consequences of our actions. An example of unintended consequences is the issue of MTBE in groundwater.

As noted in Information Box 15.5, methyl tertiary-butyl ether (MTBE) is a prevalent groundwater contaminant. The primary source of MTBE groundwater contamination was its use as a gasoline additive. The 1990 Clean Air Act amendments required the use of reformulated gasoline through the addition of oxygenates, which are organic compounds that contain oxygen in their molecular structure. The purpose of this effort was to enhance fuel combustion and reduce air pollution such as ground-level ozone and smog generated by automobiles. MTBE was an initial primary oxygenate used for this purpose. The physico-chemical properties of MTBE that made it an ideal gasoline additive unfortunately also made it an "ideal" groundwater contaminant. It has essentially infinite solubility in water, and it is relatively persistent. This resulted in relatively rapid migration to groundwater from leaking storage tanks or spills, and resultant widespread groundwater contamination. As noted in Chapter 12, MTBE was added to the Contaminant Candidate List in 1998 due to its widespread occurrence and potential human health effects. Full consideration of its potential environmental transport and fate behavior and recognition of its potential health risks, that

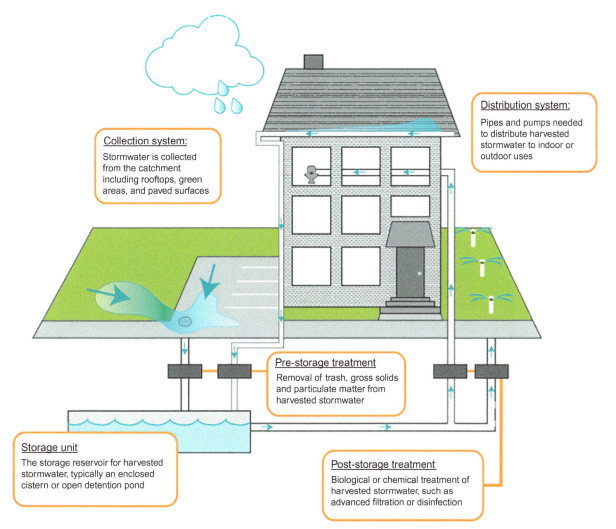

FIG. 32.5 Diagram of a rainwater harvesting system for a home. *(Source: https://stormwater.pca.state.mn.us/index.php?title=Overview_for_ stormwater_and_rainwater_harvest_and_use/reuse. Reprinted with permission from Minnesota Pollution Control Agency.)*

The labels in the diagram read:

Collection system: Stormwater is collected from the catchment including rooftops, green areas, and paved surfaces

Distribution system: Pipes and pumps needed to distribute harvested stormwater to indoor or outdoor uses

Pre-storage treatment Removal of trash, gross solids and particulate matter from harvested stormwater

Storage unit The storage reservoir for harvested stormwater, typically an enclosed cistern or open detention pond

Post-storage treatment Biological or chemical treatment of harvested stormwater, such as advanced filtration or disinfection

INFORMATION BOX 32.4 Green Chemistry's 12 Principles

1. *Prevent waste*: Design chemical syntheses to prevent waste. Leave no waste to treat or clean up.
2. *Maximize atom economy*: Design syntheses so that the final product contains the maximum proportion of the starting materials. Waste few or no atoms.
3. *Design less hazardous chemical syntheses*: Design syntheses to use and generate substances with little or no toxicity to either humans or the environment.
4. *Design safer chemicals and products*: Design chemical products that are fully effective yet have little or no toxicity.
5. *Use safer solvents and reaction conditions*: Avoid using solvents, separation agents, or other auxiliary chemicals. If you must use these chemicals, use safer ones.
6. Increase energy efficiency: Run chemical reactions at room temperature and pressure whenever possible.
7. *Use renewable feedstocks*: Use starting materials (also known as feedstocks) that are renewable rather than depletable. The source of renewable feedstocks is often agricultural products or the wastes of other processes; the source of depletable feedstocks is often fossil fuels (petroleum, natural gas, or coal) or mining operations.

8. *Avoid chemical derivatives*: Avoid using blocking or protecting groups or any temporary modifications if possible. Derivatives use additional reagents and generate waste.
9. *Use catalysts, not stoichiometric reagents*: Minimize waste by using catalytic reactions. Catalysts are effective in small amounts and can carry out a single reaction many times. They are preferable to stoichiometric reagents, which are used in excess and carry out a reaction only once.
10. *Design chemicals and products to degrade after use*: Design chemical products to break down to innocuous substances after use so that they do not accumulate in the environment.
11. *Analyze in real time to prevent pollution*: Include in-process, real-time monitoring and control during syntheses to minimize or eliminate the formation of byproducts.
12. *Minimize the potential for accidents*: Design chemicals and their physical forms (solid, liquid, or gas) to minimize the potential for chemical accidents including explosions, fires, and releases to the environment.

(**Source:** *Available from: The U.S. Environmental Protection Agency,* https:// www.epa.gov/greenchemistry/basics-green-chemistry#definition).

is, a systems thinking approach, would have identified that it was likely to contaminate groundwater, and therefore should not be used in the oxygenate program. The use of MTBE has since been replaced with ethanol.

Another consideration for sustainable development is if an action can truly be considered as sustainable if it uses nonrenewable resources. For example, can there truly be such things as "sustainable mining" and "clean coal," or are they oxymorons? By its nature, mining typically involves the extraction of resources that are not replaced within human life scales—the definition of a nonrenewable resource. However, sustainable mining incorporates the concepts of minimizing the adverse environmental and social impacts of mining while also maximizing resource recovery and reuse. Thus while some human activates may not be sustainable in the strictest sense, efforts can be made to reduce the impacts of these actions.

On the other end of the resource-use spectrum is renewable energy. These energy sources are labeled as renewable and are also typically presented as "clean" energies. However, is any renewable energy truly renewable or clean? For example, hydroelectric power requires the placement of dams and the flooding of large expanses of habitat. How sustainable is this in situations of severe, long-term drought? Also, what about the severe impacts on the local ecosystem of flooding an entire valley?

Another example is wind energy. According to a DOE report, a typical wind turbine contains 89% steel, 5.8% fiberglass, 1.6% copper, 1.3% concrete (primarily cement, water, aggregates, and steel reinforcement), 0.8% aluminum, and 0.4% core materials (primarily foam, plastic, and wood) by weight (U.S. Department of Energy, 2008). One standard wind turbine contains more than 3000 kg of copper and ~300 tons of steel. All of these materials need to be mined, processed, and manufactured. This means extraction of nonrenewable resources is required, as well as energy and water use. For example, the USGS (2011a) estimated that between 15,000 and 40,000 tons of copper would be required annually, or <2% of the U.S. apparent consumption of refined copper in 2008, assuming the goal of supplying 20% of US electricity demand from wind power by 2030. This amount is the equivalent to a portion of annual production of a typical single copper mine in the United States. In addition, with an operational lifetime of 20–30 years, disposal of nonreusable windmill components needs to be considered. Other primary environmental impacts of wind energy are animal deaths (birds, bats) and impact on scenic views. Considering all of these environmental impacts, is wind energy truly clean? And, given that wind turbine construction requires mining of nonrenewable resources, is wind energy truly renewable?

Clearly, neither wind energy nor any other energy source can be considered to be completely clean or renewable. In reality, the concept of renewable energy is relative, with different energy sources existing along a spectrum of sustainability. For example, it has been estimated that a wind farm needs to operate for a year to several years to "break even" with coal power in terms of greenhouse gas emissions generated for turbine and windfarm construction (Scientific American, 2013). However, in the long term, wind energy clearly produces significantly less carbon emissions compared to coal—~800–1000 g of CO_2-equivalent per kilowatt hour of electricity produced for coal versus ~12 for wind energy.

One other factor critical to sustainable development is human behavior dynamics, for example, how humans respond to change. Humans often respond in unanticipated and unexpected ways. This can complicate planning and decision-making for sustainable development, as well as lead to unexpected results. A recent example is that of solid-state lighting and light pollution.

For the past decade and a half there has been a major effort to shift from traditional forms of lighting (incandescent and fluorescent) to solid-state lighting (light-emitting diodes, LED). This shift was heralded to reduce overall energy consumption. For example, the U.S. Department of Energy estimates that switching to solid-state lighting could reduce national lighting energy use by 75% by 2035. One unanticipated and unwanted outcome of this switch is an apparent increase in light pollution. A recent study indicates that there has been a significant increase in lighting use over the past 5 years due to the reduced costs associated with solid-state lighting. This increase in lighting use has led to an increase in light pollution, as illustrated in Fig. 32.6 (Kyba et al., 2017). Planning and implementation of the sustainable development of solid-sate lighting did not account for human behavior dynamics—a typical human response that if something is cheaper to buy or use, we often will buy/use more of that item.

The preceding examples highlight that sustainable development is not a binary—yes or no—proposition. Instead, it consists of a spectrum from unsustainable to poorly sustainable to moderately sustainable to highly sustainable actions. Effective sustainable development requires systems thinking, full consideration of all interconnections and potential impacts, and consideration of human behavior dynamics. In this manner, alternatives can be effectively screened and the most sustainable solution selected.

32.5 IMPLEMENTING AND ASSESSING SUSTAINABLE DEVELOPMENT

Implementing sustainable development requires information about the planned activity, the system, and interconnections. Effective evaluation of the potential outcomes and

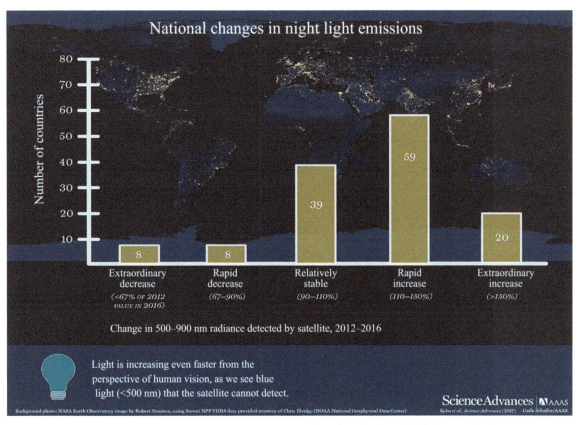

FIG. 32.6   Changes in National Nighttime light emissions. *(From Kyba et al., 2017. Sci. Adv. 3, e1701528. Reprinted with permission from AAAS)*

consequences of an action using this information requires decision-making tools. Once an action has been implemented, the system needs to be monitored to evaluate the actual outcomes and assess sustainability performance. Assessment methods are required for this purpose. These topics are discussed in this section.

32.5.1   Life-Cycle Analysis

Life-cycle analysis (LCA), also known as life-cycle assessment, is a primary tool used to support decision-making for sustainable development. According to the U. S. Environmental Protection Agency, LCA is a tool to evaluate the potential environmental impacts of a product, material, process, or activity. Crucially, an LCA is a comprehensive method for assessing all direct and indirect environmental impacts across the full life cycle of a product system, from materials acquisition, to manufacturing, to use, and to final disposition (disposal or reuse). The "cradle-to-grave" accounting concept of LCA is depicted in Fig. 32.7.

The application of LCA helps to promote the sustainable design and redesign of products and processes, leading to reduced overall environmental impacts and the reduced use and release of nonrenewable or toxic materials. LCA

FIG. 32.7   Components of life-cycle analysis. *(Available from: https://www.nist.gov/systems-integration-division/lifecycle-graphic)*

studies identify key materials and processes within the products' life cycles that are likely to pose the greatest impacts, including resource demand and human health impacts. These assessments delineate the full benefits and

INFORMATION BOX 32.5 Four Components of Life-cycle Analysis and Example Outcomes

The LCA process is a systematic, phased approach and consists of four components: goal definition and scoping, inventory analysis, impact assessment, and interpretation.

1. Goal definition and scoping

   - Define and describe the product, process or activity. Establish the context in which the assessment is to be made and identify the boundaries and environmental effects to be reviewed for the assessment.

2. Inventory analysis

   - Identify and quantify energy, water and materials usage and environmental releases (e.g., air emissions, solid waste disposal, waste water discharges).

3. Impact assessment

   - Assess the potential human and ecological effects of energy, water, and material usage and the environmental releases identified in the inventory analysis.

4. Interpretation

   - Evaluate the results of the inventory analysis and impact assessment to select the preferred product, process or service with a clear understanding of the uncertainty and the assumptions used to generate the results.

By performing an LCA, decision-makers can for example:

- Develop a systematic evaluation of the environmental consequences associated with a given product.
- Analyze the environmental trade-offs associated with one or more specific products/processes to help gain stakeholder (state, community, etc.) acceptance for a planned action.
- Quantify environmental releases to air, water, and land in relation to each life cycle stage and/or major contributing process.
- Assess the human and ecological effects of material consumption and environmental releases to the local community, region, and world.
- Compare the health and ecological impacts between two or more rival products/processes or identify the impacts of a specific product or process.

(**Source:** *Modified from U.S. Environmental Protection p0355 Agency, Life cycle assessment: principles and practice. EPA/600/R-06/060, May 2006, Cincinnati, OH, 2006).*

costs of a product or process, which allows decision-makers to select the most effective solution.

The LCA process is a systematic, phased approach and consists of four components: goal definition and scoping, inventory analysis, impact assessment, and interpretation, shown in Information Box 32.5. Illustrative outcomes of conducting an LCA are also presented in Information Box 32.5. A brief history of LCA is presented in Information Box 32.6 (U.S. Environmental Protection Agency, 2006).

32.5.2 Sustainability Performance Assessment

Assessing how well an activity meets sustainability goals requires an articulated set of goals and objectives, metrics to measure performance, and a framework and methodology for conducting the assessment. Development of frameworks and metrics has progressed over the past two decades. A key development was the publication of the BellagioSTAMP principles for development of effective sustainability assessment and reporting efforts.

The 10 Bellagio Principles were originally developed in 1996 by a group of international sustainability measurement experts meeting in Bellagio, Italy. The International Institute for Sustainable Development (IISD) and the Organization for Economic Co-operation and Development's (OECD) Measuring the Progress of Societies

initiative revised them in 2009 into eight more concise principles (Table 32.1). The Bellagio Principles are not a framework in itself; instead, they are designed and used to evaluate existing and proposed sustainability assessment frameworks.

The standard approach for conducting a performance assessment is to (a) specify the time period for assessment—typically an annual basis, (b) collect data and information, (c) analyze the data, and (d) report the results. The data are gathered and analyzed based on a set of metrics or "indicators." These indicators are measures, for example, of resource use, waste generation, and other quantifiable impacts for the environment. There are also indicators for economic and social impacts. An example of a set of performance indicators for the mining sector is presented in Fig. 32.8.

The Global Reporting Initiative (GRI[1]), started in 1997, is the most widely adopted framework for sustainability reporting. The GRI sustainability reporting framework

---

1. **GRI Boiler plate**. GRI is an independent international organization that has pioneered sustainability reporting since 1997. GRI helps businesses and governments worldwide understand and communicate their impact on critical sustainability issues such as climate change, human rights, governance and social wellbeing. This enables real action to create social, environmental and economic benefits for everyone. The GRI Sustainability Reporting Standards are developed with true multistakeholder contributions and rooted in the public interest.

INFORMATION BOX 32.6 A Brief History of Life-Cycle Assessment

Life Cycle Analysis (LCA) had its beginnings in the 1960s. Concerns over the limitations of raw materials and energy resources sparked interest in finding ways to cumulatively account for energy use and to project future resource supplies and use. In one of the first publications of its kind, Harold Smith reported his calculation of cumulative energy requirements for the production of chemical intermediates and products at the World Energy Conference in 1963.

Later in the 1960s, global modeling studies published in *The Limits to Growth* (Meadows et al., 1972) and *A Blueprint for Survival* (Goldsmith et al., 1972) resulted in predictions of the effects of the world's changing populations on the demand for finite raw materials and energy resources. The predictions for rapid depletion of fossil fuels and climatological changes resulting from excess waste heat stimulated more detailed calculations of energy use and output in industrial processes. During this period, about a dozen studies were performed to estimate costs and environmental implications of alternative sources of energy.

In 1969, researchers initiated an internal study for The Coca-Cola Company that laid the foundation for the current methods of life cycle inventory analysis in the United States. In a comparison of different beverage containers to determine which container had the lowest releases to the environment and least affected the supply of natural resources, this study quantified the raw materials and fuels used and the environmental loadings from the manufacturing processes for each container. Other companies in both the United States and Europe performed similar comparative life cycle inventory analyses in the early 1970s.

At that time, many of the available sources were derived from publicly available sources such as government documents or technical papers, as specific industrial data were not available.

The process of quantifying the resource use and environmental releases of products became known as a Resource and Environmental Profile Analysis (REPA), as practiced in the United States. In Europe, it was called an Ecobalance. In 1991, concerns over the inappropriate use of LCAs to make broad marketing claims made by product manufacturers resulted in a statement issued by eleven State Attorneys General in the USA denouncing the use of LCA results to promote products until uniform methods for conducting such assessments are developed and a consensus reached on how this type of environmental comparison can be advertised nondeceptively. This action, along with pressure from other environmental organizations to standardize LCA methodology, led to the development of the LCA standards in the International Standards Organization (ISO) 14,000 series (1997–2002).

In 2002, the United Nations Environment Programme (UNEP) joined forces with the Society of Environmental Toxicology and Chemistry (SETAC) to launch the Life Cycle Initiative, an international partnership. The three programs of the Initiative aim at putting life cycle thinking into practice and at improving the supporting tools through better data and indicators.

(**Source:** *From U.S. Environmental Protection Agency, Life cycle assessment: principles and practice. EPA/600/R-06/060, May 2006, Cincinnati, OH, 2006*).

TABLE 32.1 Bellagio Stamp Principles

Principles	Description
1. Guiding vision	Assessing progress toward sustainable development is guided by the goal to deliver well-being within the capacity of the biosphere to sustain it for future generations
2. Essential considerations	Sustainability assessments consider: The underlying social, economic, and environmental system as a whole and the interactions among its components; The adequacy of governance mechanisms; Dynamics of current trends and drivers of change and their interactions; Risks, uncertainties, and activities that can have an impact across boundaries; Implications for decision-making, including trade-offs and synergies
3. Adequate scope	Sustainability assessments adopt: Appropriate time horizon to capture both short and long-term effects of current policy decisions and human activities; Appropriate geographical scope ranging from local to global
4. Framework and indicators	Sustainability assessments are based on: A conceptual framework that identifies the domains that core indicators have to cover; The most recent and reliable data, projections and models to infer trends and build scenarios; Standardized measurement methods, wherever possible, in the interest of comparability; Comparison of indicator values with targets and benchmarks, where possible
5. Transparency	The assessment of progress toward sustainable development: Ensures the data, indicators and results of the assessment are accessible to the public; Explains the choices, assumptions and uncertainties determining the results of the assessment; Discloses data sources and methods; Discloses all sources of funding and potential conflicts of interest
6. Effective communication	In the interest of effective communication, to attract the broadest possible audience and to minimize the risk of misuse, Sustainability Assessments: Use clear and plain language; Present information in a fair and objective way, that helps to build trust; Use innovative visual tools and graphics to aid interpretation and tell a story; Make data available in as much detail as reliable and practical
7. Broad participation	To strengthen their legitimacy and relevance, sustainability assessments should: Find appropriate ways to reflect the views of the public, while providing active leadership; Engage early on with users of the assessment so that it best fits their needs
8. Continuity and capacity	Assessments of progress toward sustainable development require: Repeated measurement; Responsiveness to change; Investment to develop and maintain adequate capacity; Continuous learning and improvement

Available from: https://www.icmm.com/en-gb/about-us/member-commitments/icmm-10-principles.

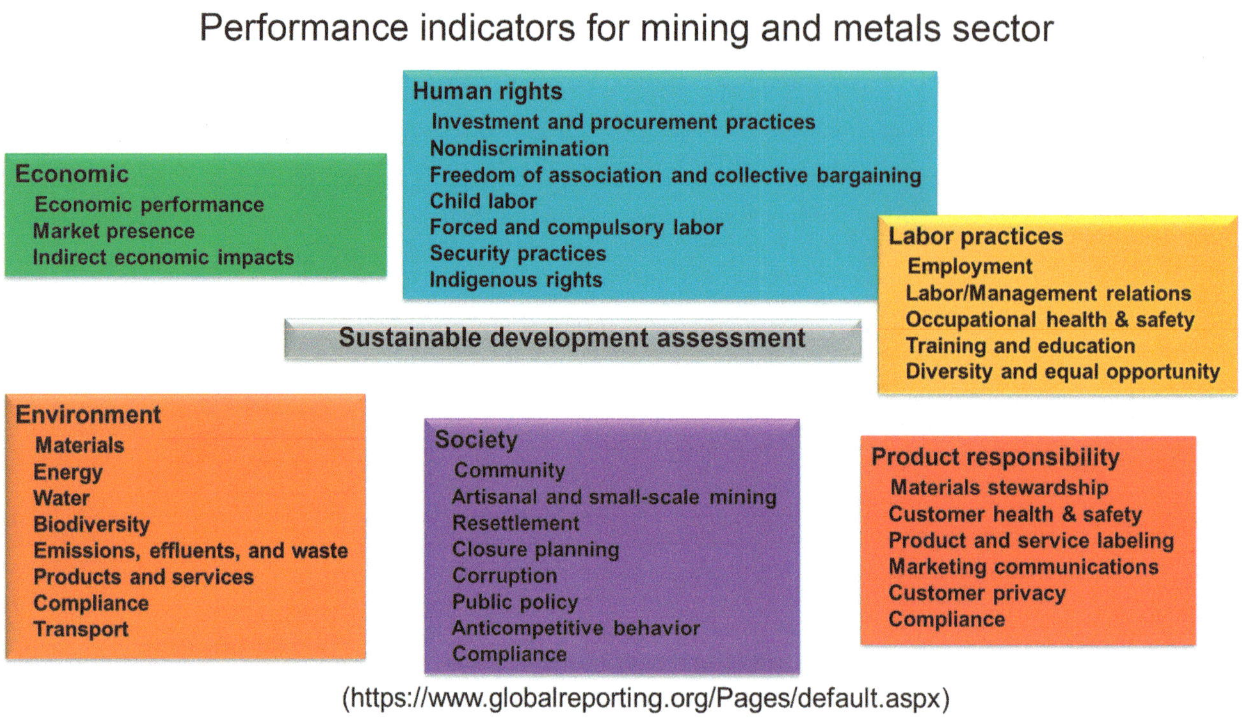

Performance indicators for mining and metals sector

Economic
- Economic performance
- Market presence
- Indirect economic impacts

Human rights
- Investment and procurement practices
- Nondiscrimination
- Freedom of association and collective bargaining
- Child labor
- Forced and compulsory labor
- Security practices
- Indigenous rights

Labor practices
- Employment
- Labor/Management relations
- Occupational health & safety
- Training and education
- Diversity and equal opportunity

Sustainable development assessment

Environment
- Materials
- Energy
- Water
- Biodiversity
- Emissions, effluents, and waste
- Products and services
- Compliance
- Transport

Society
- Community
- Artisanal and small-scale mining
- Resettlement
- Closure planning
- Corruption
- Public policy
- Anticompetitive behavior
- Compliance

Product responsibility
- Materials stewardship
- Customer health & safety
- Product and service labeling
- Marketing communications
- Customer privacy
- Compliance

(https://www.globalreporting.org/Pages/default.aspx)

FIG. 32.8 Performance indicators for the mining and metals sector. *Compiled from the Global Reporting Initiative, https://www.globalreporting.org.*

has three main elements that provide guidance on how and what to report:

1. *The Sustainability Reporting Guidelines:* The guidelines consist of quality and quantity principles (materiality, stakeholder inclusiveness, sustainability context, and completeness) as well as over 130 managerial and performance indicators in several thematic categories (organizational, managerial, economic, environmental, social, human rights, society, and product responsibility issues).
2. *Sector Supplements:* The supplements complement provide additional guidance for a given sector, with sector-specific performance indicators. For example, there is a Mining and Metals Sector Supplement.
3. *Indicator Protocols:* The protocols exist for each of the performance indicators contained in the guidelines. They provide definitions and technical and methodological guidance to assist those preparing reports and to ensure consistency in the interpretation of the performance indicators.

Sustainability assessment and reporting has become a common tool among industries. For example, the International Council on Mining and Metals (ICMM), established in 2001, is a global industry organization that represents 21 mining and metals companies as well as 35 national and regional mining associations and global commodity associations in sustainability related issues. The ICMM has developed a Sustainable Development Framework

(SDF) for sustainability assessment (ICMM, 2015). Participating members are expected to implement the SDF and publish independently verified sustainability reports regarding their performance. The SDF consists of 10 principles (Table 32.2), which are based on issues identified in the Mining, Minerals, and Sustainable Development project—a 2-year project that included consultation with stakeholders to identify key issues relating to mining and sustainable development.

The "Public Reporting" component of the SDF states that member companies are committed to publicly report on their sustainable development performance on an annual basis, in line with the guidelines and protocols set by the Global Reporting Initiative. ICMM members are to report specifically with the GRI's Mining and Metals Sector Supplement. "Independent assurance" was added to the SDF in 2008, requiring members to obtain independent third-party assurance to review and assess the quality of their sustainability reports.

The GRI and other frameworks provide methods to conduct sustainability performance assessments. There are ongoing issues with conducting successful assessments. For example, the ideal set of performance indictors has yet to be defined. In addition, the ideal set may change from location to location, and some indicators may be difficult or impossible to measure. Another notable limitation is that indicators are typically evaluated in isolation and the synergies among sustainability components are largely overlooked.

TABLE 32.2 The ICMM's Sustainable Development Principles

Principle	
1	Apply ethical business practices and sound systems of corporate governance and transparency to support sustainable development
2	Integrate sustainable development in corporate strategy and decision-making processes
3	Respect human rights and the interests, cultures, customs and values of employees and communities affected by our activities
4	Implement effective risk-management strategies and systems based on sound science and which account for stakeholder perceptions of risks
5	Pursue continual improvement in health and safety performance with the ultimate goal of zero harm
6	Pursue continual improvement in environmental performance issues, such as water stewardship, energy use and climate change
7	Contribute to the conservation of biodiversity and integrated approaches to land-use planning
8	Facilitate and support the knowledge-base and systems for responsible design, use, reuse, recycling and disposal of products containing metals and minerals
9	Pursue continual improvement in social performance and contribute to the social, economic and institutional development of host countries and communities
10	Proactively engage key stakeholders on sustainable development challenges and opportunities in an open and transparent manner. Effectively report and independently verify progress and performance

Most assessment frameworks have another significant limitation—a lack of geographical or spatial focus. Because performance is typically reported at the level of the "organization" (e.g., the entire mining company) instead of by facility (e.g., specific mine), it provides a one-size-fits-all approach, with no consideration of site-specific conditions or properties. The organizations aggregate data from all of their facilities/sites into a single report in an effort to promote standardization. However, when indicators are aggregated across geographical sites, the sustainability performance evaluation can become too broad to identify local issues. Brusseau and colleagues recently developed a set of components for an enhanced sustainability performance assessment framework (Table 32.3, Virgone et al., 2018).

32.6 ENGINEERED INTERVENTIONS

In many cases, laws, regulations, and treaties are not enacted or are incomplete to address a given issue. In addition, sustainable development and management actions may be insufficient or ineffective for resolving that same issue. In such cases, engineered interventions may be an option to reduce environmental and human health impacts of disturbances. The use of engineered approaches for remediation and restoration of contaminated or disturbed sites was discussed in Chapters 19 and 20. This section deals with two entirely different types of engineered intervention—Infrastructure Enhancement and Geoengineering.

Public infrastructure such as water and wastewater conveyance and treatment systems, roads and bridges,

dams, and electrical power distribution systems are typically in poor condition in most countries due to inadequate maintenance. In addition, the existing infrastructure is often inadequate for current population levels. Even newer infrastructure systems such as telecommunications have been shown to be susceptible to the impacts of natural disasters such as hurricanes.

Every 4 years, the American Society of Civil Engineers produces a Report Card for America's Infrastructure (ASCE, 2017). The report card grades the condition and performance of 16 categories of U.S. infrastructure—assigning letter grades for each one based on the physical conditions for different systems, and needed investments for improvement. The most recent report, released in 2017, assigned an aggregate grade of D+. The conditions associated with this grade are "The infrastructure is in fair to poor condition and mostly below standard, with many elements approaching the end of their service life. A large portion of the system exhibits significant deterioration. Condition and capacity are of serious concern with strong risk of failure." They estimated that approximately $4.6 trillion is required to restore all infrastructure needing improvement, and that there is a $2 trillion gap between available funds and the total needed. They also estimate that failing to close this infrastructure investment gap brings serious economic consequences:

- $3.9 trillion in losses to the U.S. GDP by 2025;
- $7 trillion in lost business sales by 2025; and
- 2.5 million lost American jobs in 2025.

TABLE 32.3 Components of an Enhanced Sustainability Assessment Framework

Ideal Components	Definition of Ideal Components
Component 1: *Foundation* • Integrations between sustainability categories • Economic valuation of resources • Stakeholder/community involvement	The *Foundation* of the framework provides a comprehensive model of sustainability through the integration of the three pillars (social, environmental, and economic) by: (1) Defining and requiring integrated indicators in sustainability reports (2) Requiring stakeholder/community involvement through outreach and active participation (3) Requiring that the economic value of ecosystem services and health impacts be calculated and reported.
Component 2: *Focus* • Temporal orientation • Geographic scope	The *Focus* of the framework includes: (1) A comprehensive assessment that requires collecting annual data, compiling it with previous years data, and analyzing trends and forecasting future trends. (2) An assessment that is specific to the mining site, which requires that data collection and reporting be completed at each site separately.
Component 3: *Breadth* • Types and quantity of indicators • Scale effects	The *Breadth* of the framework encompasses: (1) Performance indicators will be specific and explicitly defined. (2) The quantity of indicators should remain concise so that operations of all scales can afford the necessary time and resources to carry out the assessment.
Component 4: *Quality Assurance* • External verification • Public access to information • Disclosure of assumptions, uncertainties, and indicators not evaluated • Research translation/common language	The framework incorporates *Quality Assurance* by: (1) Requiring third party verification and routine audits of reports. (2) Requiring that reports will have full disclosure of information regarding assumptions, uncertainties, and indicators not evaluated. (3) Requiring that reports are publically available and written in language that makes the information relevant and accessible to all parties.
Component 5: *Relevance* • Decision making context	The *Relevance* of the framework is the basis of why sustainability reports are important. It explains how the information included in reports can be applied through: (1) Providing a well defined and "clear from the start" method for using reported information to inform decisions.

From Virgone, K., Ramirez-Andreotta, M., Mainhagu, J., Brusseau, M.L., 2018. Effective integrated frameworks for assessing mining sustainability. Environ. Geochem. Health, 2018, https://doi.org/10.1007/s10653-018-0128-6.

The conditions described before increase the vulnerability of infrastructure systems to disturbances induced by natural disaster events and climate change. This in turn increases the susceptibility of the population to suffer adverse effects of such events. Unfortunately, these issues are rarely a focus of political or societal attention until after a natural disaster.

Infrastructure Enhancement is focused on making improvements to existing public, and in some cases private, infrastructure so that they can better withstand the impacts of natural disasters, climate change, population growth, and other factors. In other words, infrastructure enhancement projects are designed to improve the resiliency of infrastructure systems. Such projects can take many forms, but typically involve repair and renovation of existing systems to improve capacity and stability. In some cases, replacement or new systems are needed.

A prime example of a common infrastructure enhancement effort is public works projects to update and increase the capacity of flood control systems. Unfortunately, these projects are often only implemented by a city after a major natural disaster. Clearly, it would be much better to implement these projects before a natural disaster

occurs. However, such projects are expensive, and there is generally a lack of funds and political will to implement the projects.

The issue of aging and inadequate infrastructure is a major concern with respect to the projected impacts of global climate change. Impacts such as increased severity of natural disasters, greater intensity of heat and cold waves, and sea-level rise will stress many infrastructure systems. Improvements will be needed to increase the resiliency of the systems.

Infrastructure enhancement is focused on local-scale interventions to address the impacts of events that are often caused by or at least influenced by global-scale processes. In contrast, *geoengineering* is focused on developing large-scale engineered solutions to regional and global-scale issues. The term geoengineering is generally used in conjunction with global climate change. For example, the Royal Society of London, the United Kingdom's national academy of sciences, has defined geoengineering as "the deliberate large-scale intervention in the Earth's climate system, in order to moderate global warming" (Royal Society, 2009). These interventions generally fall into

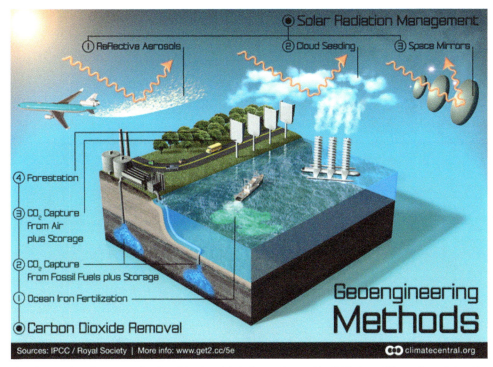

FIG. 32.9 Examples of geoengineering approaches for mitigating global climate change. *(Available from: Climate Central, http://www.climatecentral. org/news/geoengineering-could-cut-global-rainfall-study-finds-16699)*

two categories: (a) Carbon dioxide removal (CDR) techniques that address the root cause of climate change by permanently removing CO_2 from the atmosphere, and (b) solar radiation management (SRM) techniques that attempt to offset the effects of increased greenhouse gas concentrations by reducing the amount of solar radiation that reaches or is absorbed by the earth (Fig. 32.9).

As the name implies, carbon dioxide removal methods seek to remove carbon dioxide from the atmosphere and store it in some manner. Proposed methods include those that directly remove carbon dioxide and other gases from the atmosphere, as well as indirect methods that seek to promote natural processes that sequester CO_2. Example CDR methods are presented in Information Box 32.7.

Related to this approach is the concept of carbon capture and storage. This technique is being tested as a means by which to capture CO_2 produced, for example, during

---

INFORMATION BOX 32.7 Examples of Climate Geoengineering Methods

1. Carbon dioxide removal methods:
   - Large-scale physical or chemical treatment methods to convert atmospheric CO_2 into solid or liquid forms.
   - Converting agricultural and other green waste into biochar, which can be mixed into soil, thereby "locking-in" carbon that would otherwise have been released if the waste was burned or decayed.
   - Modifying land ecosystems to improve carbon capture and storage capacity, such as through developing or increasing forested areas.
   - Ocean fertilization via addition of nutrients such as iron to enhance the growth of phytoplankton that remove CO_2.

2. Solar Radiation Management Methods:
   - Placing larger mirrors between the Earth and Sun, either in the atmosphere or in space, to reflect radiation before it reaches Earth.

   - Injecting reflective particles such as sulfate aerosols into the stratosphere to reflect incoming radiation. This process is designed to mimic the natural process that occurs after large volcanic eruptions, when aerosols ejected into the atmosphere leads to global surface cooling.
   - Increasing the reflectivity of clouds (i.e., brightening) over oceans by injecting aerosols such as sea salt.
   - Increasing the reflectivity of the Earth's surface through various means, such as modifying the surfaces of sea ice and glaciers, using pale-colored roofing materials for high-population centers, and using surface coverings over large expanses of desert.

(**Source:** *Additional information is available in Royal Society, 2009*).

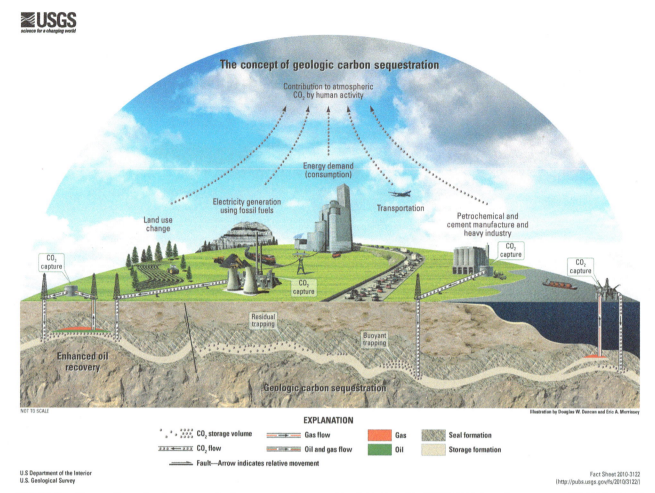

FIG. 32.10   Diagram illustrating the concept of carbon capture and storage. *(Available from: The USGS, 2011, https://pubs.usgs.gov/fs/2010/3122/pdf/ FS2010-3122.pdf)*

electricity generation at fossil-fuel power plants and storing it in some manner. The typical storage method being investigated involves injecting the CO_2 deep underground into geologic formations (see Fig. 32.10, USGS, 2011b).

Solar radiation management (SRM) techniques reduce solar radiation by deflecting sunlight away from the Earth or by increasing the reflectivity (albedo) of the atmosphere or the Earth's surface. Methods to accomplish this are presented in Information Box 32.7.

Geoengineering for climate modification is currently a very controversial topic. There is great uncertainty regarding the effectiveness of the methods, and many scientific challenges would need to be overcome before their implementation. There are also major concerns over potential unintended consequences and the possibility of inequitable distribution of benefits. Some are also concerned that a focus on geoengineering, which is addressing the symptom rather than the cause, would reduce the urgency or will to focus directly on reducing greenhouse gas emissions (i.e., the cause). These issues will continue to be hotly debated as we move further into the 21st century.

32.7   EDUCATION, COMMUNICATION, AND RESILIENCE

As discussed throughout this text, human actions have many deleterious impacts on the environment. And, in many cases, these impacts in turn negatively affect human health and well-being. As the global human population continues to grow, demands placed on the environment and associated ecosystem services will continue to increase. This chapter has focused on the several tools available to reduce and mitigate these impacts.

Taking the example of global climate change, we observe the difficulty in developing unanimously supported and implemented multinational treaties to mitigate greenhouse gas emissions. Sustainable development methods are being implemented at several levels to lessen human impacts and reduce the effects of global climate change. Scientific and technological innovations continue to enhance our ability to do so. However, these efforts are being implemented in a nonuniform and piecemeal basis, limiting their overall effectiveness.

The prior section presented the concept of improving the resiliency of our infrastructure to better cope with the effects of global climate change. The concept of social resilience, the ability of individual, communities, and societies to cope with negative impacts of events was discussed in Chapter 26. Finally, the concept of ecological resilience, the ability of an ecosystem to cope with a disturbance, was introduced in Chapter 6. These interrelated concepts of resiliency provide an additional means by which to prepare for and lessen the impacts of future global climate change.

As we have seen throughout history, significant change often occurs only after the general populace demands it. Environmental issues often receive minimal attention until they are at a critical level. This illustrates the great importance of education and communication. The issues related to environmental education and communication, forms of community and public engagement, were presented in Chapter 31. It is essential that we continue to strive to improve environmental awareness across the broad expanse of the public, and to enhance communication amongst all stakeholders and interested parties, which comprise all of us.

QUESTIONS AND PROBLEMS

1. What does sustainability mean to you?
2. What are the advantages of life-cycle analysis?
3. Conduct a life-cycle assessment of photovoltaic solar energy and discuss in terms of its overall sustainability.
4. Develop a more comprehensive set of examples than provided in the chapter for the Water-Energy-Food Nexus.
5. Select a specific large-scale geoengineering project and discuss the pros and cons of its implementation. What are some examples of potential unintended consequences?
6. Research and discuss a recent example of a multinational environmental agreement.
7. Discuss an example of green chemistry application for a product or item you use in your daily life.

REFERENCES

ASCE, 2017. Infrastructure Report Card: A Comprehensive Assessment of America's Infrastructure. American Society of Civil Engineers. Available from: https://www.infrastructurereportcard.org/.

ICMM, 2015. ICMM 10 principles for sustainable development in the mining and metals industry. https://www.icmm.com/en-gb/about-us/member-commitments/icmm-10-principles.

Kyba, C.C.M., et al., 2017. Artificially lit surface of Earth at night increasing in radiance and extent. Sci. Adv. 3 (11), e1701528. https://doi.org/10.1126/sciadv.1701528.

Royal Society, 2009. Geoengineering the Climate: Science, Governance and Uncertainty. The Royal Society, London.

Scientific American, 2013. Renewable energy's hidden costs. Available from: https://www.scientificamerican.com/article/renewable-energys-hidden-costs/.

United Nations, 2012. The future we want. Available from: http://www.un.org/disabilities/documents/rio20_outcome_document_complete.pdf.

U.S. Department of Energy (DOE), 2006. Energy demands on water resources. Report to Congress on the Interdependency of Energy and Water. U.S. Department of Energy. Available from: https://www.circleofblue.org/wp-content/uploads/2010/09/121-RptToCongress-EWwEIAcomments-FINAL2.pdf.

U.S. Department of Energy (DOE), 2008. 0% Wind energy by 2030 increasing wind energy's contribution to U.S. electricity supply. DOE/GO-102008-2567. https://www.nrel.gov/docs/fy08osti/41869.pdf.

U.S. Environmental Protection Agency, 2006. Life cycle assessment: principles and practice. EPA/600/R-06/060, May 2006, OH, Cincinnati.

United States Geological Survey (USGS), 2011a. Wind energy in the United States and materials required for the land-based Wind Turbine Industry from 2010 through 2030. Scientific investigations report 2011-503. USGS, Reston, VA. https://pubs.usgs.gov/sir/2011/5036/sir2011-5036.pdf.

United States Geological Survey (USGS), 2011b. The concept of geologic carbon sequestration. Fact Sheet 2010–3122. https://pubs.usgs.gov/fs/2010/3122/pdf/FS2010-3122.pdf.

Virgone, K., Ramirez-Andreotta, M., Mainhagu, J., Brusseau, M.L., 2018. Effective integrated frameworks for assessing mining sustainability. Environ. Geochem. Health. https://doi.org/10.1007/s10653-018-0128-6.

Chapter 33

Epilogue: Is the Future of Pollution History?

M.L. Brusseau

A pollution free sunrise. *Photo courtesy K.L. Josephson.*

33.1 HUMAN IMPACTS TO EARTH AND THE ENVIRONMENT

Throughout the history of life on Earth, the nature, number, and very existence of living organisms has been shaped by Earth's environment. We are clearly living in a time when a species, humans, is now causing significant change to Earth's environment and systems. Many scientists are calling this current epoch the *Anthropocene*, to denote this fact.

The Earth is often divided into four major spheres—atmosphere, hydrosphere, lithosphere, and biosphere. Environmental systems have been defined by interactions within and among these four spheres. A fifth sphere, the *technosphere*, may now be added. The technosphere is defined by the Oxford dictionary as "The sphere or realm of human technological activity; the technologically modified environment." The technosphere represents all of the buildings, structures, devices, and modifications that humans have created to support their existence on Earth—from houses, factories, and farms, to planes, trains, ships, and automobiles, to roads and airports, to reservoirs, aqueducts, and dams, to communication and power generation/

transmission systems, to soccer and other sports stadia, to mines and oil and gas fields, to all of the devices and products humans use, to waste treatment and storage facilities. It also includes the human systems that exist within society—cultural, social, politcal, and economic interactions among human populations. A recent study estimated that the current aggregate mass of the technosphere is approximately 30 trillion tons. This represents a mass of more than 50 kilograms for every square meter of the Earth's surface (Zalasiewicz et al., 2016). With a current world population of 7.6 billion, the mass of the technosphere represents almost 4000 tons per person. Perhaps it is time for the technosphere to be put on a diet?

As we have seen throughout many chapters of this text, most human activities pollute or otherwise disturb the environment. While these impacts have been occurring since the dawn of humankind, the impacts are starting to have significant effects on the Earth in this period of the Anthropocene. In other words, the technosphere has a major impact on the health of the other four Earth spheres. Furthermore, we have learned that pollution and disruption of the environment can have severe direct and indirect consequences for human health and well-being. It should now be clear that it is in our own best interests to protect the environment.

33.2 DEMONSTRATING THE CONSEQUENCES OF POLLUTION

Several chapters in the text highlighted the myriad impacts of pollution on the environment, in terms of degraded air, water, soil, and habitats. These impacts have significant, often severe, consequences for ecosystems and the plants and animals that live within them. Unfortunately, such impacts to "nature and wildlife" are not typically a priority concern for some individuals or nations. This has resulted in alternative approaches for characterizing and demonstrating the significance of environmental pollution.

The first approach is the concept of ecosystem services, discussed in Chapter 6. This concept provides a means to directly illustrate the many ways in which the environment supports human life and well-being. The second approach is

Environmental and Pollution Science. https://doi.org/10.1016/B978-0-12-814719-1.00033-1

the determination of the impacts of pollution in terms of human health and mortality (Chapter 26). In the third approach, efforts are now being made to determine the monetary costs associated with the disruption of ecosystem service benefits and the impairment and loss of human life. Enhancing the visibility of environmental pollution and its effects through these and other efforts will hopefully strengthen our interest in protecting the environment.

Air pollution can be used as an illustrative example of these approaches for characterizing the human-related impacts of pollution. Air pollution is pervasive throughout the world. Approximately 9 out of every 10 people breathe outdoor air polluted above World Health Organization guideline levels (WHO, 2016). Ambient (outdoor) air pollution originates from many human activities, as well as some natural sources (Chapter 17), and has numerous impacts on ecosystems, ecosystem services, and human health (Fig. 33.1). In addition, indoor air pollution, particularly from cooking and heating, is a major concern for many developing countries.

As discussed in Chapter 26, studies have estimated that environmental pollution causes approximately 12 million premature human deaths globally each year. This represents more than 20% of all deaths. Environmental factors in aggregate are the single greatest risk factor for human mortality. More than 7 million deaths are attributed to indoor and outdoor air pollution alone, by far the single largest source of environment-related deaths (GBD, 2016; Prüss-Ustün et al., 2016). This is greater than the number of deaths attributed to tobacco use, being overweight, alcohol use, and unsafe sex. In fact, it is greater than every other single risk factor except for high blood pressure. The number of premature deaths attributable to air pollution is projected to increase to between 9 and 12 million annually by 2060 in the absence of more stringent policies (OECD, 2016; Landrigan et al., 2017).

The factors that contribute to the human welfare costs of pollution were presented in Chapter 26. It is estimated that the global human welfare costs due to air pollution total more than $5 trillion annually (WBG, 2016; UNEP, 2017). These costs represent significant fractions of gross domestic product worldwide (Fig. 33.2). The costs are projected to rise to more than $25 trillion annually by 2060 in the absence of more stringent control measures. Note that these costs do not include the economic impacts of damage

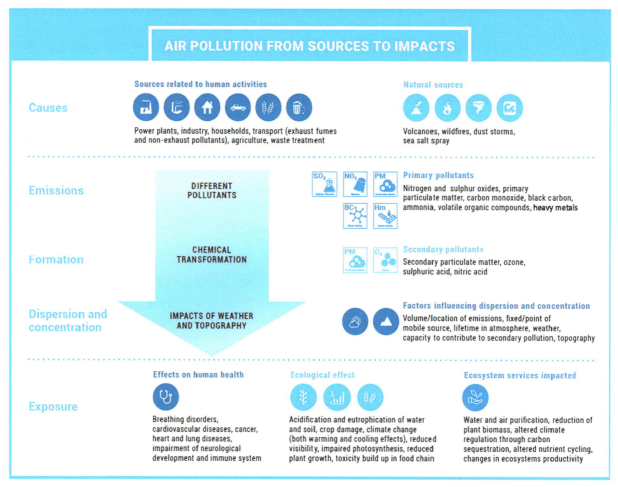

FIG. 33.1 Air pollution from sources to impacts. *(Available from: https://environmentlive.unep.org/airpollution.)*

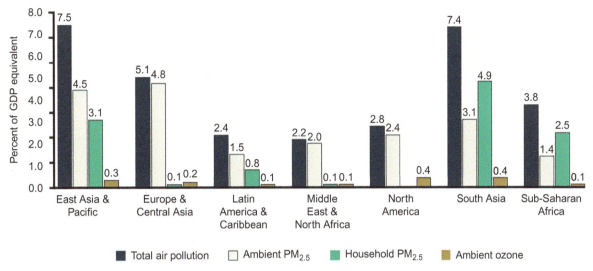

FIG. 33.2   Human welfare losses due to air pollution in 2013 by region. *(From WBG, World Bank Group and Institute for Health Metrics and Evaluation: The cost of air pollution: strengthening the economic case for action. International Bank for Reconstruction and Development/The World Bank, Washington, DC, 2016.)*

to ecosystems, wildlife, or crop production, or the impacts associated with interactions between air pollution and global climate change.

By these measures of human mortality and economic costs accrued, air pollution may be considered to be the most critical current environmental human health issue. Hence, a major effort should be focused on improving air quality. The status across nations in implementing various approaches for improving air quality is presented in Fig. 33.3. We see that while progress is being made, we remain far from optimal action. The central question is, what is impeding our advancement in controlling and preventing air pollution, and more generally, all forms of pollution and environmental disturbance?

33.3   CONSTRAINTS TO MAKING POLLUTION HISTORY

Several constraints exist that limit our ability to solve environmental issues. Major factors include the following.

Failure to Enact and Enforce Regulations: Enactment of environmental laws and associated regulations is a primary means by which to control and reduce pollution. However, promulgation and enforcement of environmental regulations is often constrained by a number of factors. Select primary factors are briefly discussed.

One primary factor is the opposition of vested interests, operating under flawed, outdated paradigms. These paradigms include: (1) the assumption that pollution and disease are the inevitable and unavoidable consequences of economic development, (2) that costs associated with environmental regulations are a drain on corporate profitability,

(3) that enactment of environmental regulations leads to unemployment, and (4) that costs accrued from pollution are negligible and cannot be quantified.

Research has clearly demonstrated that the application of effective environmental laws and regulations produces significant economic benefits, in addition to reducing deleterious environmental and human health impacts. For example, the U.S. Environmental Protection Agency analyzed the costs and monetized benefits of the Clean Air Act between 1970 and 1990 (EPA, 1997). They determined that the monetized benefits totaled more than $1 trillion annually. Essentially all costs savings accrued from human health-related benefits. In comparison, the costs of compliance averaged approximately $25 billion annually. Subtracting costs from benefits yields a net monetized benefit of approximately $1 trillion annually. The EPA estimated in a more recent prospective study covering the period between 1990 and 2020 that the net monetized benefits associated with the Clean Air Act average greater than $1 trillion annually, compared to estimated annual costs of compliance of approximately $50 billion (EPA, 2011). These results are consistent with those of the first study. Thus every dollar invested in air pollution control results in approximately $30 in benefits.

The benefits also typically outweigh the costs for other environmental regulations as well. For example, an analysis of the costs and benefits of 39 major environmental regulations enforced by the U.S. EPA over the period of 2006–2016 showed a benefit–cost ratio of approximately 9 to 1 (OMB, 2017), meaning that every dollar invested resulted in approximately $9 in benefits. These estimated monetized benefits do not include economic gains associated with the pollution control industry—employment and profit

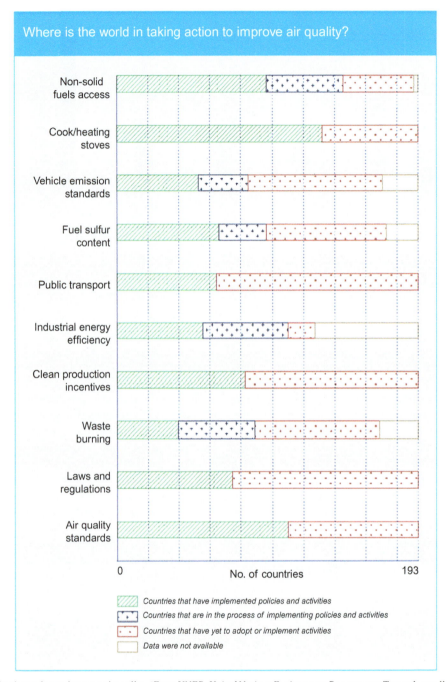

FIG. 33.3 Status of actions taken to improve air quality. *(From UNEP, United Nations Environment Programme: Towards a pollution-free planet, 2017. Available from: http://web.unep.org/environmentassembly/report-executive-director-towards-pollution-free-planet.)*

accruing from the design, manufacture, sales, and servicing of pollution control equipment. Research has also demonstrated that environmental regulation costs are a small percentage of industry revenues. According to 2005 data from U.S. manufacturers, their total pollution abatement spending represented less than 1% of the $4.74 trillion value of the goods they shipped (EPA, 2018). In addition, studies have shown that environmental regulations have not resulted in loss of employment (EPA, 2018). Finally, as

illustrated in the preceding section for air pollution, the monetary costs associated with environment pollution have now been determined and they are enormous.

Another major factor limiting the implementation of environmental regulations is the different stages of development across nations, which leads to differences in critical issues of concern and the availability of supportive infrastructure and institutional capacity. Thus the degree of regulatory action varies from country to country. For example, some nations face severe issues of widespread food or water scarcity, poverty, or sociopolitical conflict. Such issues can limit the resources and capacity available to address environmental issues. Building environmental capacity in these cases would require development of critical institutional capacity, supportive infrastructure, and conflict resolution.

Many current critical environmental issues are global in scale (Chapter 25). As discussed in Chapter 32, global-scale environmental issues require cooperation across nations. Ineffective multinational cooperation impedes implementation of effective policies and limits progress in resolving environmental issues.

Fragmented Approaches: The fragmented approaches that are often used is another limitation to resolving pollution issues. As discussed in Chapter 32, sustainable solutions require a systems thinking as opposed to piecewise (fragmented) approach. When examining the environmental impacts caused by a particular activity, we tend to focus on a single medium or portion of the environment. This compartmentalization of systems and components is especially pervasive in the regulatory and policy arena. Numerous environmental laws exist for controlling pollution and other environmental impacts of human activity (Chapter 30). These laws and regulations have often been developed in isolation, which has resulted in sets of fractionated, and sometimes overlapping and/or contradictory rules that are less effective as a whole for protecting the environment.

Alternatively, systems thinking is an integrated approach that considers each medium and component of the environment as part of a whole. This approach allows us to evaluate the synergistic and antagonistic interactions that can and do occur in real systems. In addition, it is critical to incorporate the social science aspects of environmental issues into policy considerations. Such an approach is fundamentally inter-, multi-, and transdisciplinary in nature, requiring the active collaboration of many individuals trained in such disparate disciplines as physical sciences, engineering, medicine, economics, sociology, and public policy. This type of approach is captured in the concept of *convergence*. As defined in a recent National Academy of Sciences report (NAS, 2014), "*convergence is an approach to problem solving that cuts across disciplinary boundaries. It integrates knowledge, tools, and ways of thinking from life and health sciences, physical, mathematical, and computational sciences, engineering disciplines, and beyond to form a comprehensive synthetic framework for tackling scientific and societal challenges that exist at the interfaces of multiple fields.*" This concept focuses not only on the convergence of expertise and knowledge, but also on the formation of the partnerships required to support the convergence. Although difficult to achieve, this approach will be important for solving complex pollution issues.

Legacy Contamination, ECs, Analytical Methods, and Toxicology: The issues of legacy contamination, emerging contaminants (ECs), continued improvements in analytical methods, and advances in toxicity testing combine for another constraint to solving pollution issues. Some contaminants that entered the environment years ago prior to current environmental regulations are very persistent in the environment, meaning they remain present and toxic for decades (Chapter 12). The same contaminants are typically also difficult and costly to remove from the environment. In combination, these legacy contaminants continue to pose problems for long time frames. In addition, new contaminants continue to be discovered in the environment (Chapter 12). Thus this issue of emerging contaminants continues to pose new environmental problems to address.

The continued development of new and improved analytical methods provides the means to detect ever lower concentrations of contaminants in environmental samples. Thus contaminants are now detected in samples that would have yielded no detection using the older, less-sensitive methods. Therefore samples that would previously have been deemed to be uncontaminated would now be found in fact to be contaminated. Examples of this phenomenon include the chlorinated solvents in the 1970s and 80s and more recently 1,4-dioxane (Chapter 15). This issue raises the question of what is an acceptable level of pollution and to what degree should we clean up pollution? These questions, in turn, are mediated in part by toxicological assessments.

Advances in methods of toxicological assessment and in the number of contaminants assessed can lead to changes in hazard characterization profiles of polluted environmental media. An example of this was the reduction in the MCL for arsenic in U.S. drinking water from 50 to $10\,\mu g/L$. This has led to large volumes of surface and groundwater being reclassified as contaminated and needing treatment.

Data Collection and Dissemination: Gathering and disseminating environmental data is another issue constraining investigation and resolution of pollution issues. Environmental data are the cornerstone that supports actions toward solving environmental issues. Such data are required to uncover the existence of an issue, to define the magnitude and importance of the issue, and to monitor changes as a function of natural perturbations or anthropogenic interventions.

We have the scientific and technological means to characterize environmental systems for pollution and disturbance (see Chapter 10). However, there are several factors that constrain full-scale implementation. One is the desire and funding to implement effective monitoring programs across all environments in all regions of the world. Another is development of standardized procedures for collection, processing, and storing enormous volumes of data. A third is providing uniform, transparent access to all parties, including the public.

Mixed Messaging: As we have learned, there are a plethora of environmental issues, most of which affect human health and well-being and environmental quality to some degree. In this digital age of communication, we typically have a multitude of news stories, reports, and postings for each and every issue, each one being touted as a dire, critical issue for humankind. This has resulted in a cacophony of mixed messaging. For example, climate change was stated to be the most pressing public health problem in a recent National Academy of Sciences report (NAS, 2018). Conversely, a recent report by the United Nations Environment Programme stated that air pollution is the world's single greatest environmental risk to human health (UNEP, 2017). So, which one is correct? The recent advancements made in quantifying the risks and costs associated with environmental issues provide a tool that can help to clarify and delineate the most significant factors.

Public Participation and Education: In the past, regulators, industry, and policy makers operated without much involvement of the communities affected by a particular environmental issue. It is now recognized that it is critical to engage the affected communities when working toward solutions to environmental issues (Chapter 31). More broadly, it is critical to involve the general public in decision-making on environmental issues. This includes supporting education programs to enhance environmental and environmental-health literacy, and promoting effective communication efforts among all parties. As recommended by the National Academy of Sciences (NAS, 2008), "*Public participation should be fully incorporated into environmental assessment and decision-making processes, and it should be recognized by government agencies and other organizers of the processes as a requisite of effective action,*

not merely a formal procedural requirement." Ultimately, it falls to the government—and thus to the general public—to determine the level of effort expended on addressing environmental issues.

33.4 SOLUTIONS TO MAKING POLLUTION HISTORY

We have the scientific knowledge to address many of the current environmental issues, as presented in many chapters of this book. This knowledge has led to the development of numerous approaches available to resolve pollution issues. These include methods for reclamation, remediation, and restoration of polluted environments, covered in Chapters 19 and 20, and methods for waste management and reuse, covered in Chapters 21–24. Different types of engineered interventions to manage or reduce pollution effects are available or under development, as discussed in Chapter 32.

Despite the availability of these various tools, environmental pollution problems continue to exist. Some of the primary reasons for this were discussed in the preceding section. The net result is that there is typically a significant time lag between the generation of scientific knowledge about a specific environmental issue and implementation of actions to address the issue, as illustrated in Fig. 33.4.

Given the intractability of many environmental pollution problems, and the great costs required to solve them, it should be obvious that the most beneficial and cost-effective approach to pollution management is to stop it before it starts. Thus *pollution prevention* is a widely accepted goal—a goal that is being implemented to an increasing degree.

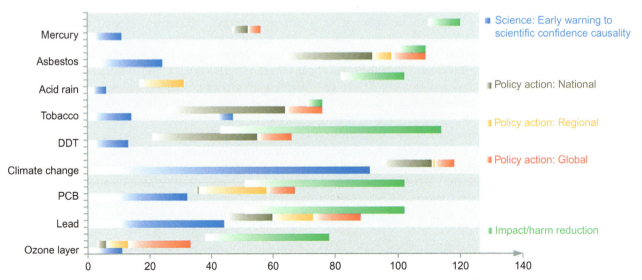

FIG. 33.4 Time lag in years between the generation of scientific knowledge about select environmental issues and the implementation of actions to address that issue. (*From Global Sustainable Development Report 2015 © 2015 United Nations. Reprinted with the permission of the United Nations.*)

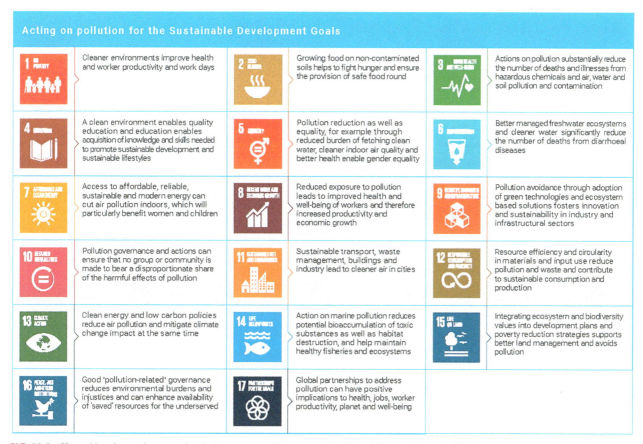

FIG. 33.5 How addressing environmental pollution supports achievement of the United Nations 17 sustainable development goals. *(From UNEP, United Nations Environment Programme: Towards a pollution-free planet, 2017. Available from: http://web.unep.org/environmentassembly/report-executive-director-towards-pollution-free-planet.)*

Pollution prevention can take many forms. For example, best management practices have been developed to optimize the amount of water, fertilizers, and pesticides employed for crop production, that is, to use only the amounts necessary to support plant growth. Manufacturing industries are developing safer chemicals, optimizing chemical use, and minimizing waste generation—so-called green chemistry and green technology. Life cycle analysis is increasingly being used to optimize and reduce the use of nonrenewable resources. These are all manifestations of sustainable development, as discussed in Chapter 32. As noted in that chapter, the United Nations developed 17 sustainable development goals designed to promote environmental protection and human health and well-being. Fig. 33.5 illustrates how resolving environmental pollution issues can support achievement of each one of the 17 goals.

33.5   PRIORITIES AND MOVING FORWARD

We have covered a myriad of environmental issues in this text. Considering that human and economic resources are finite, it is necessary to prioritize the most critical issues upon which to focus our efforts over the next few decades. In this regard, it is instructive to examine the stated environmental priorities of the primary national and international environmental agencies. The priorities for the U.S. Environmental Protection Agency, the African Union, the European Union, the Association of Southeast Asian Nations, the Union of South American Nations, and the United Nations Environment Programme (UNEP) are presented in Information Boxes 33.1–33.6.

The first five information boxes represent priorities for individual major regions of the world, whereas the sixth, presented by UNEP, represents what they deem to be the priority issues globally. Several themes are common throughout the sets of priorities. Preservation of water resources is mentioned in some form in all five of the regional lists. Climate change is also mentioned in all five lists. Water and climate are the only two issues present in each of the five regional lists. These two are also present in the UNEP global list.

Water pollution is estimated to be responsible for approximately 2 million human deaths annually (Chapter 26), the second largest source after air pollution

INFORMATION BOX 33.1 Seven Priorities for EPA's Future

- Taking Action on Climate Change
- Improving Air Quality
- Assuring the Safety of Chemicals
- Cleaning Up Our Communities
- Protecting America's Waters
- Expanding the Conversation on Environmentalism and Working for Environmental Justice
- Building Strong State and Tribal Partnerships
  Available from: https://blog.epa.gov/blog/2010/01/seven-priorities-for-epas-future-2/. Accessed July 2018.

INFORMATION BOX 33.2 Environmental Priorities of the African Union

- Biodiversity, conservation and, sustainable natural resource management
- Water security
- Climate resilience and natural disasters preparedness and prevention
- Renewable energy
  Available from: http://www.un.org/en/africa/osaa/pdf/au/agenda2063-first10yearimplementation.pdf. Accessed July 2018.

INFORMATION BOX 33.3 Environmental Priorities of the European Union

- Protect, conserve and enhance the Union's natural capital: Natural capital refers to the biodiversity that provides goods and services we rely on, from fertile soil and productive land and seas to fresh water and clean air. It includes vital services such as pollination of plants, natural protection against flooding, and the regulation of our climate.
- Turn the Union into a resource-efficient, green, and competitive low-carbon economy: This entails improving resource efficiency across the economy, including energy production and use, improvements to the environmental performance of products over their life cycle, and reductions in the environmental impact of consumption, including issues such as cutting food waste and using biomass in a sustainable way.
- Safeguard the Union's citizens from environment-related pressures and risks to health and wellbeing: This includes challenges such as air and water pollution, excessive noise, chemicals, and the impacts of climate change.
- Improved implementation, integration, and investment: This comprises improved implementation of existing environmental legislation, enhanced integration of environmental concerns into other policy areas, adequate accounting of the costs of environmental impacts, and increased investment in products, services, and policies.
- Increased information: Scientific research, monitoring and reporting environmental developments mean that our understanding of the environment is constantly increasing. This knowledge base should be made more accessible to citizens and policymakers to ensure policy continues to draw on a sound understanding of the state of the environment.
- Sustainable cities: Europe is densely populated and 80% of its citizens are likely to live in or near a city by 2020. Cities often share a common set of problems such as poor air quality, high levels of noise, greenhouse gas emissions, water scarcity, and waste.
  Available from: http://ec.europa.eu/environment/action-programme/. Accessed July 2018.

INFORMATION BOX 33.4 Environmental Priorities of the Association of Southeast Asian Nations

- Nature conservation and biodiversity
- Coastal and marine environment
- Water resources management
- Environmentally sustainable cities
- Climate change
- Chemicals and waste
- Environmental education and sustainable consumption and production
  Available from: http://asean.org/asean-socio-cultural/asean-ministerial-meeting-on-environment-amme/overview/. Accessed July 2018.

of environment-related deaths. In addition, water scarcity and the attendant consequences is a primary critical issue for many regions of the world. Clearly, the inclusion of water on all of the priority lists is well justified. Global climate change is anticipated to have a myriad of effects on the environment that are projected to cascade through a widespread range of impacts to human well-being. In particular, global climate change will likely exacerbate the severity of other issues present in the lists. In addition, some of the potential consequences may span multiple human generations.

Biodiversity and conservation of natural resources is present in four of the five regional lists and is present in the UNEP list. Clearly, sustainable resource use is central for reducing and preventing scarcity issues and minimizing disturbances of ecosystems and the services they provide. Chemical safety is present in three of the five regional lists and also in the UNEP list. The three lists for which this issue is present are associated with regions that are the most economically developed and for which the legacy of poor chemical waste management practices is more established. It would be of great benefit to the countries whose economies are in the midst of expansion to employ pollution prevention strategies to reduce the future occurrence of chemical contamination issues suffered previously by other nations.

As discussed in Section 33.5

INFORMATION BOX 33.5 Environment-Related Goals of the Union of South American Nations

- Energy integration for the sustainable and fair use of the resources of our Region
- Infrastructure development to guarantee the interconnection of the region and our peoples according to criteria of sustainable social and economic development
- Protection of our biodiversity, water resources and ecosystems as well as cooperation among Member States in matters of disaster prevention and the fight against the causes and effects of climate change

Available from: http://www.unasursg.org/en/node/180. Accessed July 2018.

Interestingly, air pollution is mentioned in only two of the regional lists. As discussed in Section 33.5 and in Chapter 26, air pollution is responsible for by far more environmental-related human deaths than any other factor. In addition, the global human welfare costs due to air pollution total more than $5 trillion annually and are a significant fraction of global GDP. Given the enormous costs of air pollution, it is apparent that it should be a critical priority for all regions. This is particularly true for developing nations, where the increase in industrial and energy production associated with continued economic development and population growth will increase air pollution and its impacts.

INFORMATION BOX 33.6 Priority Areas of the United Nations Environment Programme

1. Chemicals and Waste

    Chemicals are integral to our life, but they also can have major impacts on the environment and human health. So too do the various types of wastes produced by human activities. It is critical to develop the science, policies, and legal, and institutional frameworks for sound chemical and waste management.

2. Climate Change

    Climate change is expected to have unprecedented implications on where people can settle, grow food, build cities, and rely on functioning ecosystems for the services they provide. Efforts to address these issues include pursuit of low carbon emission development and boosting the capacity of communities and nations to adapt and be resilient to climate change.

3. Disasters and Conflicts

    Since the beginning of the 21st century, the world has witnessed more than 2500 environmental disasters and 40 major conflicts. These events—which have affected more than two billion people—destroy infrastructure, displace populations, and fundamentally undermine human security. They also compound poverty and limit sustainable development. Work is needed to minimize the threats to human well-being from the environmental causes and consequences of disasters and conflicts while building resilience to future crises.

4. Ecosystem Management

    Humans depend on healthy and productive ecosystems to meet their basic needs, but many people's needs are not being met sustainably—if at all. An estimated 795 million people suffer from hunger and 1.2 billion live in water-stressed areas. At the same time, biodiversity loss and ecosystem degradation are expected to continue, or even accelerate. By 2030, the world will require 40% more water, 50% more food, 40% more energy and 40% more timber and fibre. The only way we can meet these demands is by managing our ecosystems smartly and sustainably. Integrated ecosystem management aims to sustain ecosystems to meet both ecological and human needs.

5. Environmental Governance

    "We may sometimes think of the law as something divorced from our day-to-day lives, but for many of us it is the reason we can still breathe fresh air, drink clean water, and sleep safely at night. Sadly, it is also the reason why so many cannot." [Erik Solheim, UN Environment Executive Director]. In our globalized world, environmental threats require effective responses that promote peace, justice, development, and the fullfilment of environmental and human rights.

6. Environment Under Review

    With the latest advances in Big Data, Earth observation technologies, and citizen science, anyone with an internet connection can now access data on the environment in near real time. This is powerful information that can inform the development of robust environmental policies and help us achieve the goals of the 2030 Agenda on Sustainable Development.

7. Resource Efficiency

    The unsustainable use of resources has triggered critical scarcities and caused climate change and widespread environmental degradation—all of which have negative impacts on the well-being of the planet and its people. At the same time, more than 10% of the world population continues to live in extreme poverty, unable to meet even their most basic needs. Responding to this dual challenge will require innovative policies, redirected investment, environmentally sound technologies, international cooperation, and capacity development to support countries to transition to inclusive green economies. Producers will need to change how they design, source, manufacture, and market their products. Consumers will need to incorporate environmental and social concerns into their consumption decisions and adopt sustainable lifestyles.

Source: Available from: http://web.unep.org/about/how-we-operate/priority-areas#. Accessed July 2018.

Different metrics can be used to assess the impacts of environmental pollution and disturbances. Examples include economic costs due to infrastructure damage, economic loss due to reduced availability of natural resources and services, numbers of affected or displaced people, numbers of deaths, and the costs accrued from disability and death. Based on the measures of loss of human life and the associated human welfare costs that have been estimated for the various environmental factors (see Chapter 26), one might conclude that air pollution is the most critical environmental issue to address over the next few decades. Would we arrive at the same conclusion if a different set of metrics were used? In addition, this assessment is based on global-scale impacts. We need to also keep in mind the local-scale aspects of environmental health—for example, addressing air pollution does not solve the concerns of community members affected by chemical pollution of their drinking water. This illustrates the need for policies and interventions at both the community and global scales of environmental health.

Determining a priority focus is just the first step in the process. Developing solutions requires determination of the causes of pollution and the various factors that affect those causes. For example, particulate matter is a primary contributor to air pollution. A major source of particulate matter is fossil fuel use for energy production and transportation. Thus solutions for reducing particulate matter in air would need to involve changes to the methods of energy production and transportation. We know that fossil fuel consumption is also a major cause of elevated CO_2 levels in the atmosphere, which is contributing to global climate change. Another major source of particulate matter is dust from wind erosion of denuded land, which often occurs in arid environments experiencing drought. And, global climate change is likely going to increase the magnitudes and rates of drought. These examples illustrate the interconnectedness of environmental issues. This brings us back to the systems thinking approach necessary for the development of sustainable solutions. We need this convergence approach to identify the keystone factors at the heart of critical environmental issues, ones for which effective interventions can produce the greatest benefits for the health and welfare of humans and the environment.

Scientific and technological advances over the past decades have provided us with the knowledge and means to address and solve many if not most of our critical environmental issues. However, the environment continues to be polluted and resources continue to be used unsustainably. What is it, then, that prevents us from making the future of pollution history? The answer is complex, of course, and depends on a number of different factors, including financial, political, and societal—in other words the human dimensions of pollution discussed before and in other chapters of this book. To be effective environmental

scientists in the 21st century, we need to not only understand the natural science of pollution, but also the human dimensions to pollution. We hope that the information, concepts, and ideas presented in this textbook will help support the development of effective environmental scientists who will contribute to turning the page on environmental pollution, thereby making it history.

QUESTIONS AND PROBLEMS

1. Discuss why air pollution is or is not the most critical environmental issue.
2. Discuss why global climate change is or is not the most critical environmental issue.
3. What factors should be considered when assessing the importance of an environmental issue?

REFERENCES

Environmental Protection Agency, EPA, 1997. The benefits and costs of the clean air act, 1970 to 1990. U.S. Environmental Protection Agency.
Environmental Protection Agency, EPA, 2011. The benefits and costs of the clean air act from 1990 to 2020. U.S. Environmental Protection Agency, Office of Air and Radiation.
Environmental Protection Agency, EPA (2018). The clean air act and the economy. Available from: https://www.epa.gov/clean-air-act-overview/clean-air-act-and-economy#invest. Last accessed July 2018.
GBD 2015 Risk Factors Collaborators, 2016. Global, regional, and national comparative risk assessment of 79 behavioural, environmental and occupational, and metabolic risks or clusters of risks, 1990–2015: a systematic analysis for the global burden of disease. Lancet 388, 1659–1724.
Landrigan, P.J., et al., 2017. The Lancet commission on pollution and health. Lancet. https://doi.org/10.1016/S0140-6736(17)32345-0.
National Academies of Sciences, NAS, 2008. Public Participation in Environmental Assessment and Decision Making. The National Academies Press, Washington, DC.
National Academies of Sciences, NAS, 2014. Convergence: Facilitating Transdisciplinary Integration of Life Sciences, Physical Sciences, Engineering, and Beyond. The National Academies Press, Washington, DC.
National Academies of Sciences, NAS, 2018. Protecting the health and well-being of communities in a changing climate. In: Proceedings of a Workshop. The National Academies Press, Washington, DC. https://doi.org/10.17226/24846.
OECD, 2016. The Economic Consequences of Outdoor Air Pollution. OECD Publishing, Paris, FR.
Office of Management and Budget, OMB, 2017. Draft Report to Congress on the Benefits and Costs of Federal Regulations and Agency Compliance with the Unfunded Mandates Reform Act. Office of Management and Budget, Office of Information and Regulatory Affairs.
Prüss-Üstün, A., Wolf, J., Corvalan, C., Bos, R., Neira, M., 2016. Preventing Disease Through Healthy Environments. World Health Organization, Geneva.
UNEP, United Nations Environment Programme, 2017. Towards a pollution-free planet. Available from: http://web.unep.org/environmentassembly/report-executive-director-towards-pollution-free-planet.

WBG, World Bank Group and Institute for Health Metrics and Evaluation, 2016. The Cost of Air Pollution: Strengthening the Economic Case for Action. International Bank for Reconstruction and Development/The World Bank, Washington, DC.

WHO, World Health Organization (2016). Ambient air pollution: a global assessment of exposure and burden of disease. Available from: http://apps.who.int/iris/bitstream/10665/250141/1/9789241511353-eng.pdf.

Zalasiewicz, J., et al., 2016. Scale and diversity of the physical technosphere: a geological perspective. Anthropocene Rev.. https://doi.org/10.1177/2053019616677743.

FURTHER READING

United Nations, 2015. Global Sustainable Development Report 2015. Available from: https://sustainabledevelopment.un.org/content/documents/1758GSDR%202015%20Advance%20Unedited%20Version.pdf.

Index

Note: Page numbers followed by *f* indicate figures, *t* indicate tables, and *b* indicate boxes.